The Ricci Flow: Techniques and Applications

Part I: Geometric Aspects

Mathematical
Surveys
and
Monographs

Volume 135

The Ricci Flow: Techniques and Applications

Part I: Geometric Aspects

Bennett Chow
Sun-Chin Chu
David Glickenstein
Christine Guenther
James Isenberg
Tom Ivey
Dan Knopf
Peng Lu
Feng Luo
Lei Ni

American Mathematical Society

EDITORIAL COMMITTEE

Jerry L. Bona Peter S. Landweber
Michael G. Eastwood Michael P. Loss

J. T. Stafford, Chair

2000 *Mathematics Subject Classification.* Primary 53C44, 53C25, 58J35, 35K55, 35K05.

For additional information and updates on this book, visit
www.ams.org/bookpages/surv-135

ISBN 978-0-8218-3946-1

Copying and reprinting. Individual readers of this publication, and nonprofit libraries acting for them, are permitted to make fair use of the material, such as to copy a chapter for use in teaching or research. Permission is granted to quote brief passages from this publication in reviews, provided the customary acknowledgment of the source is given.

Republication, systematic copying, or multiple reproduction of any material in this publication is permitted only under license from the American Mathematical Society. Requests for such permission should be addressed to the Acquisitions Department, American Mathematical Society, 201 Charles Street, Providence, Rhode Island 02904-2294, USA. Requests can also be made by e-mail to reprint-permission@ams.org.

© 2007 by Bennett Chow. All rights reserved.
Printed in the United States of America.

∞ The paper used in this book is acid-free and falls within the guidelines
established to ensure permanence and durability.
Visit the AMS home page at http://www.ams.org/

10 9 8 7 6 5 4 3 2 1   12 11 10 09 08 07

Contents

Preface	ix
What this book is about	ix
Highlights of Part I	xi
Acknowledgments	xiii
Contents of Part I of Volume Two	xvii

Chapter 1. Ricci Solitons — 1
1. General solitons and their canonical forms — 2
2. Differentiating the soliton equation — local and global analysis — 6
3. Warped products and 2-dimensional solitons — 11
4. Constructing the Bryant steady soliton — 17
5. Rotationally symmetric expanding solitons — 26
6. Homogeneous expanding solitons — 32
7. When breathers and solitons are Einstein — 41
8. Perelman's energy and entropy in relation to Ricci solitons — 44
9. Buscher duality transformation of warped product solitons — 46
10. Summary of results and open problems on Ricci solitons — 50
11. Notes and commentary — 52

Chapter 2. Kähler–Ricci Flow and Kähler–Ricci Solitons — 55
1. Introduction to Kähler manifolds — 55
2. Connection, curvature, and covariant differentiation — 62
3. Existence of Kähler–Einstein metrics — 70
4. Introduction to the Kähler–Ricci flow — 74
5. Existence and convergence of the Kähler–Ricci flow — 80
6. Survey of some results for the Kähler–Ricci flow — 95
7. Examples of Kähler–Ricci solitons — 97
8. Kähler–Ricci flow with nonnegative bisectional curvature — 103
9. Matrix differential Harnack estimate for the Kähler–Ricci flow — 109
10. Linear and interpolated differential Harnack estimates — 118
11. Notes and commentary — 124

Chapter 3. The Compactness Theorem for Ricci Flow — 127
1. Introduction and statements of the compactness theorems — 127
2. Convergence at all times from convergence at one time — 132
3. Extensions of Hamilton's compactness theorem — 138

4.	Applications of Hamilton's compactness theorem	142
5.	Notes and commentary	148

Chapter 4. Proof of the Compactness Theorem — 149
1. Outline of the proof — 149
2. Approximate isometries, compactness of maps, and direct limits — 150
3. Construction of good coverings by balls — 158
4. The limit manifold $(\mathcal{M}^n_\infty, g_\infty)$ — 165
5. Center of mass and nonlinear averages — 175
6. Notes and commentary — 187

Chapter 5. Energy, Monotonicity, and Breathers — 189
1. Energy, its first variation, and the gradient flow — 190
2. Monotonicity of energy for the Ricci flow — 197
3. Steady and expanding breather solutions revisited — 203
4. Classical entropy and Perelman's energy — 214
5. Notes and commentary — 219

Chapter 6. Entropy and No Local Collapsing — 221
1. The entropy functional $\mathcal{W}$ and its monotonicity — 221
2. The functionals μ and ν — 235
3. Shrinking breathers are shrinking gradient Ricci solitons — 242
4. Logarithmic Sobolev inequality — 246
5. No finite time local collapsing: A proof of Hamilton's little loop conjecture — 251
6. Improved version of no local collapsing and diameter control — 264
7. Some further calculations related to $\mathcal{F}$ and $\mathcal{W}$ — 273
8. Notes and commentary — 284

Chapter 7. The Reduced Distance — 285
1. The $\mathcal{L}$-length and distance for a static metric — 286
2. The $\mathcal{L}$-length and the L-distance — 288
3. The first variation of $\mathcal{L}$-length and existence of $\mathcal{L}$-geodesics — 296
4. The gradient and time-derivative of the L-distance function — 306
5. The second variation formula for $\mathcal{L}$ and the Hessian of L — 312
6. Equations and inequalities satisfied by L and ℓ — 322
7. The ℓ-function on Einstein solutions and Ricci solitons — 335
8. $\mathcal{L}$-Jacobi fields and the $\mathcal{L}$-exponential map — 345
9. Weak solution formulation — 363
10. Notes and commentary — 379

Chapter 8. Applications of the Reduced Distance — 381
1. Reduced volume of a static metric — 381
2. Reduced volume for Ricci flow — 386
3. A weakened no local collapsing theorem via the monotonicity of the reduced volume — 399
4. Backward limit of ancient κ-solution is a shrinker — 406

5.	Perelman's Riemannian formalism in potentially infinite dimensions	417
6.	Notes and commentary	432

Chapter 9. Basic Topology of 3-Manifolds 433
1. Essential 2-spheres and irreducible 3-manifolds 433
2. Incompressible surfaces and the geometrization conjecture 435
3. Decomposition theorems and the Ricci flow 439
4. Notes and commentary 442

Appendix A. Basic Ricci Flow Theory 445
1. Riemannian geometry 445
2. Basic Ricci flow 456
3. Basic singularity theory for Ricci flow 465
4. More Ricci flow theory and ancient solutions 470
5. Classical singularity theory 474

Appendix B. Other Aspects of Ricci Flow and Related Flows 477
1. Convergence to Ricci solitons 477
2. The mean curvature flow 482
3. The cross curvature flow 490
4. Notes and commentary 500

Appendix C. Glossary 501

Bibliography 513

Index 531

Preface

Trinity: It's the question that drives us, Neo. It's the question that brought you here. You know the question just as I did.
Neo: What is the Matrix? ...
Morpheus: Do you want to know what it is? The Matrix is everywhere.... Unfortunately, no one can be told what the Matrix is. You have to see it for yourself.
− From the movie "The Matrix".

What this book is about

This is the sequel to the book "The Ricci Flow: An Introduction" by two of the authors [**108**]. In the previous volume (henceforth referred to as Volume One) we laid some of the foundations for the study of Richard Hamilton's Ricci flow. The Ricci flow is an evolution equation which deforms Riemannian metrics by evolving them in the direction of minus the Ricci tensor. It is like a heat equation and tries to smooth out initial metrics. In some cases one can exhibit global existence and convergence of the Ricci flow. A striking example of this is the main result presented in Chapter 6 of Volume One: Hamilton's topological classification of closed 3-manifolds with positive Ricci curvature as spherical space forms. The idea of the proof is to show, for any initial metric with positive Ricci curvature, the normalized Ricci flow exists for all time and converges to a constant curvature metric as time approaches infinity. Note that on any closed 2-dimensional manifold, the normalized Ricci flow exists for all time and converges to a constant curvature metric. Many of the techniques used in Hamilton's original work in dimension 2 have influenced the study of the Ricci flow in higher dimensions. In this respect, of special note is Hamilton's 'meta-principle' of considering *geometric quantities which either vanish or are constant on gradient Ricci solitons.*

It is perhaps generally believed that the Ricci flow tries to make metrics more homogeneous and isotropic. However, for general initial metrics on closed manifolds, singularities may develop under the Ricci flow in dimensions as low as 3.[1] In Volume One we began to set up the study of singularities by discussing curvature and derivative of curvature estimates, looking at how generally dilations are done in all dimensions, and studying

---

[1] For noncompact manifolds, finite time singularities may even occur in dimension 2.

aspects of singularity formation in dimension 3. In this volume, we continue the study of the fundamental properties of the Ricci flow with particular emphasis on their application to the study of singularities. We pay particular attention to dimension 3, where we describe some aspects of Hamilton's and Perelman's nearly complete classification of the possible singularities.[2]

As we saw in Volume One, Ricci solitons (i.e., self-similar solutions), differential Harnack inequalities, derivative estimates, compactness theorems, maximum principles, and injectivity radius estimates play an important role in the study of the Ricci flow. The maximum principle was used extensively in the 3-dimensional results we presented. Some of the other techniques were presented only in the context of the Ricci flow on surfaces. In this volume we take a more detailed look at these general topics and also describe some of the fundamental new tools of Perelman which almost complete Hamilton's partial classification of singularities in dimension 3. In particular, we discuss Perelman's energy, entropy, reduced distance, and some applications. Much of Perelman's work is independent of dimension and leads to a new understanding of singularities. It is difficult to overemphasize the importance of the reduced distance function, which is a space-time distance-like function (not necessarily nonnegative!) which is intimately tied to the geometry of solutions of the Ricci flow and the understanding of forming singularities. We also discuss stability and the linearized Ricci flow. Here the emphasis is not just on one solution to the Ricci flow, but on the dependence of the solutions on their initial conditions. We hope that this direction of study may have applications to showing that certain singularity types are not generic.

This volume is divided into two parts plus appendices. For the most part, the division is along the lines of whether the techniques are geometric or analytic. However, this distinction is rather arbitrary since the techniques in Ricci flow are often a *synthesis of geometry and analysis*. The first part is intended as an introduction to some basic geometric techniques used in the study of the singularity formation in general dimensions. Particular attention is paid to finite time singularities on closed manifolds, where the spatial maximum of the curvature tends to infinity in finite time. We also discuss some basic 3-manifold topology and reconcile this with some classification results for 3-dimensional finite time singularities. The partial classification of such singularities is used in defining Ricci flow with surgery. In particular, given a good enough understanding of the singularities which can occur in dimension 3, one can perform topological-geometric surgeries on solutions to the Ricci flow either right before or at the singularity time. One would then like to continue the solution to the Ricci flow until the next singularity and iterate this process. In the end one hopes to infer the existence of a geometric decomposition on the underlying 3-manifold. This is what Hamilton's program aims to accomplish and this is the same framework on which

---

[2]Not all singularity models have been classified, even for finite time solutions of the Ricci flow on closed 3-manifolds. Apparently this is independent of Hamilton's program for Thurston geometrization.

Perelman's work is based. In view of the desired topological applications of Ricci flow, in Chapter 9 we give a more detailed review of 3-manifold topology than was presented in Volume One. We hope to discuss the topics of nonsingular solutions (and their variants), where one can infer the existence of a geometric decomposition, surgery techniques, and more advanced topics in the understanding of singularities elsewhere.[3]

The second part of this volume emphasizes analytic and geometric techniques which are useful in the study of Ricci flow, again especially in regards to singularity analysis. We hope that the second part of this volume will not only be helpful for those wishing to understand analytic and geometric aspects of Ricci flow but that it will also provide tools for understanding certain technical aspects of Ricci flow. The appendices form an eclectic collection of topics which either further develop or support directions in this volume.

We have endeavored to make each of the chapters as self-contained as possible. In this way it is hoped that this volume may be used not only as a text for self-study, but also as a reference for those who would like to learn any of the particular topics in Ricci flow. To aid the reader, we have included a detailed guide to the chapters and appendices of this volume and in the first appendix we have also collected the most relevant results from Volume One for handy reference.

For the reader who would like to learn more about details of Perelman's work on Hamilton's program, we suggest the following excellent sources: Kleiner and Lott [**231**], Sesum, Tian, and Wang [**326**], Morgan and Tian [**273**], Chen and Zhu [**81**], Cao and Zhu [**56**], and Topping [**356**]. For further expository accounts, please see Anderson [**5**], Ding [**126**], Milnor [**267**], and Morgan [**272**]. Part of the discussion of Perelman's work in this volume was derived from notes of four of the authors [**102**].

Finally a word about notation; if an unnumbered formula appears on p. $\heartsuit\spadesuit$ of Volume One, we refer to it as (V1-p. $\heartsuit\spadesuit$); if the equation is numbered $\diamondsuit.\clubsuit$, then we refer to it as (V1-$\diamondsuit.\clubsuit$).

Highlights of Part I

In Part I of this volume we continue to lay the foundations of Ricci flow and give more geometric applications. We also discuss some aspects of Perelman's work on the Ricci flow.[4] Some highlights of Part I of this volume are the following:

(1) Proof of the existence of the Bryant steady soliton and rotationally symmetric expanding gradient Ricci solitons. Examples of homogeneous Ricci solitons. Triviality of breather solutions (no nontrivial steady or expanding breathers result). The Buscher duality transformation of gradient Ricci solitons of warped product type. An

---

[3]Some of these topics will appear in Part II of this volume.
[4]Further treatment of Perelman's work will appear in Part II and elsewhere.

open problem list on the geometry and classification of Ricci solitons.

(2) Introduction to the Kähler–Ricci flow. Long-time existence of the Kähler–Ricci flow on Kähler manifolds with first Chern class having a sign. Convergence of the Kähler–Ricci flow on Kähler manifolds with negative first Chern class. Construction of the Koiso solitons and other U(n)-invariant solitons. Differential Harnack estimates and their applications under the assumption of nonnegative bisectional curvature. A survey of uniformization-type results for complete noncompact Kähler manifolds with positive curvature.

(3) Proof of the global version of Hamilton's Cheeger–Gromov-type compactness theorem for the Ricci flow. We take care to follow Hamilton and prove the compactness theorem for the Ricci flow in the category of pointed solutions with the convergence in C^∞ on compact sets. Outline of the proof of the local version of the aforementioned result. Application to the existence of singularity models.

(4) A unified approach to Perelman's monotonicity formulas for energy and entropy and the expander entropy monotonicity formula. Perelman's λ-invariant and application to the second proof of the no nontrivial steady or expanding breathers result. Other entropy results due to Hamilton and Bakry-Emery.

(5) Proof of the no local collapsing theorem assuming only an upper bound on the scalar curvature. Relation of no local collapsing and Hamilton's little loop conjecture. Perelman's μ- and ν-invariants and application to the proof of the no shrinking breathers result. Discussion of Topping's diameter control result. Relation between the variation of the modified scalar curvature and the linear trace Harnack quadratic. Second variation of energy and entropy.

(6) Theory of the reduced length. Comparison between the reduced length for static metrics and solutions of the Ricci flow. The $\mathcal{L}$-length, L-, $\bar{L}$-, and ℓ-distances and the first and second variation formulas for the $\mathcal{L}$-length. Existence of $\mathcal{L}$-geodesics and estimates for their speeds. Formulas for the gradient and time-derivative of the L-distance function and its local Lipschitz property. Formulas for the Laplacian and Hessian of L and differential inequalities for L, $\bar{L}$, and ℓ including a space-time Laplacian comparison theorem. Upper bound for the spatial minimum of ℓ. Formulas for ℓ on Einstein and gradient Ricci soliton solutions. $\mathcal{L}$-Jacobi fields, the $\mathcal{L}$-Jacobian, and the $\mathcal{L}$-exponential map, and their properties. Estimate for the time-derivative of the $\mathcal{L}$-Jacobian. Bounds for ℓ, its space-derivative, and its time-derivative. Properties of Lipschitz functions applied to ℓ and equivalence of notions of supersolutions in view of differential inequalities for ℓ.

(7) Applications of the reduced distance. Reduced volume of a static metric and its monotonicity. Monotonicity formula for the reduced volume and application to weakened no local collapsing for complete (possibly noncompact) solutions of the Ricci flow with bounded curvature. Certain backward limits of ancient κ-solutions are shrinking gradient Ricci solitons.

(8) A survey of basic 3-manifold topology and a brief description of the role of Ricci flow as an approach to the geometrization conjecture.

(9) Concise summary of the contents of Volume One including some main formulas and results. Formulas for the change in geometric quantities given a variation of the metric, evolution of geometric quantities under Ricci flow, maximum principles, curvature estimates, classical singularity theory including applications of classical monotonicity formulas, ancient 2-dimensional solutions, Hamilton's partial classification of 3-dimensional finite time singularities.

(10) List of some results in the basic theory of Ricci flow and the background Riemannian geometry. Bishop–Gromov volume comparison theorem, Laplacian and Hessian comparison theorems, Calabi's trick, geometry at infinity of gradient Ricci solitons, dimension reduction, properties of ancient solutions, existence of necks using the combination of classical singularity theory in dimension 3 and no local collapsing.

(11) Discussion of some results on the asymptotic behavior of complete solutions of the Ricci flow on noncompact manifolds diffeomorphic to Euclidean space. A brief discussion of the mean curvature flow (MCF) of hypersurfaces in Riemannian manifolds. Huisken's monotonicity formula for MCF of hypersurfaces in Euclidean space, including a generalization by Hamilton to MCF of hypersurfaces in Riemannian manifolds. Short-time existence (Buckland) and monotonicity formulas (Hamilton) for the cross curvature flow of closed 3-manifolds with negative sectional curvature.

Acknowledgments

We would like to thank the following mathematicians for helpful discussions and/or encouragement: Sigurd Angenent, Robert Bryant, Esther Cabezas-Rivas, Huai-Dong Cao, Xiaodong Cao, Jim Carlson, Albert Chau, Bing-Long Chen, Xiuxiong Chen, Li-Tien Cheng, Shiu-Yuen Cheng, Yuxin Dong, Klaus Ecker, David Ellwood, Bob Gulliver, Hongxin Guo, Emmanuel Hebey, Gerhard Huisken, Bruce Kleiner, Brett Kotschwar, Junfang Li, Peter Li, John Lott, Robert McCann, John Morgan, Andre Neves, Hugo Rossi, Rick Schoen, Natasa Sesum, Weimin Sheng, Luen-Fai Tam, Gang Tian, Peter Topping, Yuan-Long Xin, Nolan Wallach, Jiaping Wang, Guofang Wei, Neshan Wickramasekera, Jon Wolfson, Deane Yang, Rugang Ye, Yu Yuan, Qi Zhang, Yu Zheng, and Xi-Ping Zhu. We would like to thank Jiaping

Wang for advice on how to structure the book. We are especially grateful to Esther Cabezas-Rivas for a plethora of helpful suggestions; we have incorporated a number of her suggestions essentially verbatim in the first part of this volume.

The authors would like to express special thanks to the following people. Helena Noronha, formerly a geometric analysis program director at NSF, for her encouragement and support during the early stages of this project. Ed Dunne and Sergei Gelfand for their help, encouragement and support to make the publication of these books possible with the AMS. We are especially grateful and indebted to Ed Dunne for his constant help and encouragement throughout the long process of writing this volume. We thank Arlene O'Sean for her expert and wonderful help as the copy and production editor.

During the preparation of this volume, Bennett Chow was partially supported by NSF grants DMS-9971891, DMS-020392, DMS-0354540 and DMS-0505507. Christine Guenther was partially supported by the Thomas J. and Joyce Holce Professorship in Science. Jim Isenberg was partially supported by NSF grant PHY-0354659. Dan Knopf was partially supported by NSF grants DMS-0511184, DMS-0505920, and DMS-0545984. Peng Lu was partially supported by NSF grant DMS-0405255. Feng Luo was partially supported by NSF grant DMS-0103843. Lei Ni was partially supported by NSF grants DMS-0354540 and DMS-0504792. Bennett Chow and Lei Ni were partially supported by NSF FRG grant DMS-0354540 (jointly with Gang Tian).

Bennett Chow would like to thank Richard Hamilton and Shing-Tung Yau for making the study of the Ricci flow possible for him and for their help and encouragement over a period of many years. He would like to thank East China Normal University, Mathematical Sciences Research Institute in Berkeley, National Center for Theoretical Sciences in Hsinchu, Taiwan, and Université de Cergy-Pontoise.

Ben expresses extra special thanks to Classic Dimension for continued support, faith, guidance, encouragement, and inspiration. He would like to thank his parents, Yutze and Wanlin, for their encouragement and love throughout his life. Ben lovingly dedicates this book to his daughters Michelle and Isabelle.

Sun-Chin Chu would like to thank Nai-Chung Leung and Wei-Ming Ni for their encouragement and help over the years. Sun-Chin would like to thank his parents for their love and support throughout his life and dedicates this book to his family.

David Glickenstein would like to thank his wife, Tricia, and his parents, Helen and Harvey, for their love and support. Dave dedicates this book to his family.

Christine Guenther would like to thank Jim Isenberg as a friend and colleague for his guidance and encouragement. She thanks her family, in particular Manuel, for their constant support and dedicates this book to them.

Jim Isenberg would like to thank Mauro Carfora for introducing him to Ricci flow. He thanks Richard Hamilton for showing him how much fun it can be. He dedicates this book to Paul and Ruth Isenberg.

Tom Ivey would like to thank Robert Bryant and Andre Neves for helpful comments and suggestions.

Dan Knopf thanks his colleagues and friends in mathematics, with whom he is privileged to work and study. He is especially grateful to Kevin McLeod, whose mentorship and guidance has been invaluable. On a personal level, he thanks his family and friends for their love, especially Dan and Penny, his parents, Frank and Mary Ann, and his wife, Stephanie.

Peng Lu benefits from the notes on Perelman's work by Bruce Kleiner and John Lott. Peng thanks Gang Tian for encouragement and help over the years. Peng thanks his family for support. Peng dedicates this book to his grandparents and wishes them well in their world.

Feng Luo would like to thank the NSF for partial support.

Lei Ni would like to thank Jiaxing Hong and Yuanlong Xin for initiating his interests in geometry and PDE, Peter Li and Luen-Fai Tam for their teaching over the years and for collaborations. In particular, he would like to thank Richard Hamilton and Grisha Perelman, from whose papers he learned much of what he knows about Ricci flow.

Sun-Chin Chu, David Glickenstein, Christine Guenther, Jim Isenberg, Tom Ivey, Dan Knopf, Peng Lu, Feng Luo, and Lei Ni would like to collectively thank Bennett Chow for organizing this project.[5]

> Bennett Chow, UC San Diego and East China Normal University
> Sun-Chin Chu, National Chung Cheng University
> David Glickenstein, University of Arizona
> Christine Guenther, Pacific University
> Jim Isenberg, University of Oregon
> Tom Ivey, College of Charleston
> Dan Knopf, University of Texas, Austin
> Peng Lu, University of Oregon
> Feng Luo, Rutgers University
> Lei Ni, UC San Diego
>
> rfv2@math.ucsd.edu
> August 23, 2006

---

[5]Bennett Chow would like to thank his coauthors for letting him ride on their coattails.

Contents of Part I of Volume Two

"There must be some way out of here, ..."
– From "All Along the Watchtower" by Bob Dylan

We describe the main topics considered in each of the chapters of Part I of this volume.

Chapter 1. We consider the self-similar solutions to the Ricci flow, called Ricci solitons. These special solutions evolve purely by homotheties and diffeomorphisms. When the diffeomorphisms are generated by gradient vector fields, the Ricci solitons are called gradient. We begin by presenting a canonical form for gradient Ricci solitons and by systematically differentiating the gradient Ricci soliton equations to obtain higher-order equations. These equations play an important role in the qualitative study of gradient Ricci solitons.

Warped products provide elegant and important examples of solitons, such as the cigar metric, which is an explicit steady Ricci soliton defined on $\mathbb{R}^2$ and which is conformal to the Euclidean metric, positively curved and asymptotic to a cylinder. We also discuss the construction of the higher-dimensional generalization of the cigar, the rotationally symmetric Bryant soliton defined on $\mathbb{R}^n$. Interestingly, the qualitative behavior of the Bryant soliton is quite different from the cigar. We also construct rotationally symmetric expanding gradient Ricci solitons.

An interesting class of Ricci solitons is the homogenous Ricci solitons. It is notable that in dimensions as low as 3, there exist expanding homogeneous Ricci solitons which are not gradient. We especially discuss the 3-dimensional case of expanding Ricci solitons.

The scarcity of Ricci solitons on closed manifolds is exhibited by the fact that the only steady or expanding Ricci solitons on such manifolds are Einstein metrics. Moreover, in dimensions 2 and 3, the only shrinking Ricci solitons on closed manifolds are Einstein. In dimension 2, this follows from the Kazdan-Warner identity and in dimension 3 this relies on a pinching estimate for the curvature due independently to Hamilton and Ivey.

The consideration of gradient Ricci solitons, in particular those geometric quantities which either vanish or are constant on gradient Ricci solitons, has played an important role in the discovery of monotonicity formulas. We briefly introduce Perelman's energy and entropy functionals from this point of view. These functionals will be considered in more detail in Chapters 5 and 6.

Ricci solitons were actually introduced first in the physics literature, where nontrivial ones were called quasi-Einstein metrics. So it is perhaps not surprising that some aspect of duality theory is related to Ricci solitons. We discuss gradient Ricci solitons in the form of warped products with tori and the Buscher duality transformation of these special solitons.

We conclude the chapter with a summary of results and open problems on Ricci solitons.

A fundamental aspect of Ricci solitons is that they occur as singularity models, i.e., limits of dilations of a singular solution. In particular, for metrics with nonnegative curvature operators and whose scalar curvature attains its maximum in space and time, the limits of Type II singularities are steady Ricci solitons. One proof of this relies on the matrix differential Harnack inequality and the strong maximum principle for tensors, whose proofs are given in Part II. In the chapter on differential Harnack inequalities in Part II we shall motivate the consideration of the Harnack quadratic by differentiating the expanding gradient Ricci soliton equation to obtain the matrix Harnack quantity, which vanishes in certain directions on expanding gradient solitons.

Chapter 2. We discuss the Kähler–Ricci flow, which is simply the Ricci flow on Kähler manifolds. In the compact case, the Ricci flow preserves the Kähler structure of the metric. Because of the interaction of the complex structure with the evolving metric, a rich field has developed in the study of the Kähler–Ricci flow.

We begin by giving a basic introduction to Kähler geometry. This introduction is not meant as a replacement of the standard texts on Kähler geometry, but rather an attempt to make the book more self-contained. We encourage the novice to read other texts (some of these are cited in the notes and commentary) either before or in conjunction with this chapter. We emphasize local coordinate calculations in holomorphic coordinates in the style of the book by Morrow and Kodaira [**275**].

To put the study of the Kähler–Ricci flow in a broader context, we give a brief summary of some results on the existence and uniqueness of Kähler–Einstein metrics. Many of the results in this field are deep and we encourage the interested reader to consult the original papers or other sources.

Our study of the Kähler–Ricci flow begins with the fundamental result of H.-D. Cao on the long-time existence and convergence on closed manifolds. Long-time existence holds independently of whether the first Chern class is negative, zero, or positive. Convergence to a Kähler–Einstein metric holds in the cases where the Chern class is either negative or zero.

There has been substantial progress on the Kähler–Ricci flow in both the compact and the complete noncompact cases. We briefly survey some results in this area.

There are a number of interesting Kähler–Ricci solitons. We present the construction of some of these solitons, including the Koiso soliton. All of the known examples have some sort of symmetry. It is hoped that the study of Kähler–Ricci solitons may shed some light on the problem of formulating weak solutions to the Kähler–Ricci flow as a way of canonically flowing past a singularity.

Finally we discuss nonnegativity of curvature and the Kähler–Ricci flow. A particularly natural condition is nonnegative bisectional curvature. This condition is preserved under the Kähler–Ricci flow and H.-D. Cao has proved a differential matrix Harnack estimate under this curvature condition (assuming bounded curvature). The trace form of the matrix estimate may be generalized to a differential Harnack estimate which ties more closely to the heat equation. We present a family of such inequalities and discuss some applications.

Chapter 3. The study of the limiting behavior of solutions begins with Hamilton's Cheeger–Gromov-type compactness theorem for the Ricci flow, which is the subject of this chapter. One considers sequences of pointed solutions to the Ricci flow and one attempts to extract a limit of a subsequence. Such sequences arise when studying singular solutions by taking sequences of points and times approaching the singularity time and dilating the solutions by the curvatures at these basepoints. In order to extract a limit, we assume that the injectivity radii at the basepoints and the curvatures everywhere are bounded. By Shi's local derivative bounds, we get pointwise bounds on all the derivatives of the curvatures. This enables us to prove convergence in C^∞ on compact subsets for a subsequence.

We prove the compactness theorem for solutions (time-dependent) from the compactness theorem for metrics (time-independent), which will be proved in the next chapter. We also consider a local version of the compactness theorem and discuss the application of the compactness theorem to the existence of singularity models for solutions of the Ricci flow assuming an injectivity radius estimate (such an estimate holds for finite time solutions on closed manifolds by Perelman's no local collapsing theorem).

The outline of the proof of the compactness theorem is as follows. One first proves a compactness theorem for pointed Riemannian manifolds with bounded injectivity radii, curvatures and derivatives of curvature. To prove the compactness theorem for pointed solutions $\{\mathcal{M}, g_k(t), O_k\}_{k \in \mathbb{N}}$ to the Ricci flow from this, one observes the following. By Shi's estimate and the compactness theorem for pointed Riemannian manifolds there exists a subsequence $g_k(t_0)$ which converges for a fixed time t_0. The bounds on the curvatures and their derivatives also imply that the metrics $g_k(t)$ are uniformly equivalent on compact time intervals and the covariant derivatives of the metrics $g_k(t)$ with respect to a fixed metric g are bounded. The compactness theorem for solutions then follows from the Arzela–Ascoli theorem.

We also briefly discuss the Cheeger–Gromov-type compactness theorems for both Kähler metrics and solutions of the Kähler–Ricci flow. The only issue in applying the Riemannian compactness theorem is showing that the limiting metric/solution is Kähler; fortunately this is easily handled.

Chapter 4. In this chapter we first give an outline of the proof of the Cheeger–Gromov compactness theorem for pointed Riemannian manifolds. The proof of the compactness theorem for pointed manifolds is rather technical and involves a few steps. The main step is to define, after passing to a subsequence, approximate isometries Ψ_k from balls $B(O_k, k)$ in $\mathcal{M}_k$ to balls $B(O_{k+1}, k+1)$ in $\mathcal{M}_{k+1}$. The manifold $\mathcal{M}_\infty$ is defined as the direct limit of the directed system Ψ_k. Convergence to $\mathcal{M}_\infty$ and the completeness of this limit follows from Ψ_k being approximate isometries.

The ideas in the proof of the main step are as follows. In each of the manifolds $\mathcal{M}_k$ in an appropriate subsequence, starting with the origins $O_k = x_k^0$, one constructs a net (sequence) of points $\{x_k^\alpha\}_{\alpha=0}^{N(k)}$ which will be the centers of balls B_k^α of the appropriate radii (technically one considers balls of different radii for what follows). By passing to a subsequence and appealing to the Arzela–Ascoli theorem repeatedly, we may assume these Riemannian balls have a limit as $k \to \infty$ for each α. Furthermore, we may also assume that the balls cover larger and larger balls centered at O_k and that the intersections of balls B_k^α and B_k^β are independent of k in the limit. Choosing frames at the centers of these balls yields local coordinate charts H_k^α (this depends on a decay estimate for the injectivity radius and our choice of the radii of the balls) and we can define overlap maps $J_k^{\alpha\beta} = \left(H_k^\beta\right)^{-1} \circ H_k^\alpha$. By passing to a subsequence, we may assume the $J_k^{\alpha\beta}$ converge as $k \to \infty$ for each α and β. The local coordinate charts also define maps between manifolds by $F_{k\ell}^\alpha = H_\ell^\alpha \circ (H_k^\alpha)^{-1}$. Now we can define approximate isometries $F_{k\ell} : B(O_k, k) \to \mathcal{M}_\ell$ by taking a partition of unity and averaging the maps $F_{k\ell}^\alpha$. Technically, this is accomplished using the so-called center of mass and nonlinear averages technique. This brings us to the remaining step, which is to show that these maps are indeed approximate isometries.

Chapter 5. In Volume One we saw the integral monotonicity formula for Hamilton's entropy for solutions to the Ricci flow on surfaces with positive curvature. There we also saw various curvature pinching estimates, the gradient of the scalar curvature estimate, and higher derivative of curvature estimates. Other monotonicity-type formulas, for the evolution of the lengths and areas of stable minimal geodesics and surfaces, yielded injectivity radius estimates in various special cases in low dimensions. In a generalized sense, all of these estimates may be thought of as monotonicity formulas.

In this chapter we address Perelman's energy formula. One of the main ideas here is the introduction of an auxiliary function, which serves several purposes. It fixes the volume form, it satisfies a backward heat-type equation, it is used to understand the action of the diffeomorphism group, and it relates to gradient Ricci solitons. We discuss the first variation formula for the energy functional and the modified Ricci flow as the gradient flow for this energy. The nonexistence of steady and expanding breathers on closed manifolds, originally proved by one of the authors, may be proved using the energy functional and an associated invariant. We also discuss the classical entropy and its relation to Perelman's energy.

Chapter 6. In this chapter we discuss Perelman's remarkable entropy functional. This functional is actually an energy-entropy quantity which combines Perelman's energy with the classical entropy using a positive parameter τ which, in the context of Ricci flow, plays the dual roles of the scale and minus time. We compute the first variation of the entropy and derive its monotonicity under the Ricci flow coupled to the adjoint heat equation. The entropy formula, via the consideration of test functions concentrated at points, leads to a volume noncollapsing result for *all* solutions to the Ricci flow on closed manifolds, called no local collapsing. This result also yields a strong injectivity radius estimate and rules out the formation of the cigar soliton as a finite time singularity model on closed manifolds.

By minimizing the entropy functional over all functions satisfying a constraint and then minimizing over all scales τ, we obtain two geometric invariants, one depending on the metric and scale and one depending only on the metric. The consideration of these invariants is useful in proving the nonexistence of nontrivial *shrinking* breather solutions on closed manifolds.

We also provide an improved version of the no local collapsing theorem, also due to Perelman. Our presentation is based on the diameter bound result of Topping, whose proof we also sketch.

Finally we discuss variational formulas for the modified scalar curvature, which is an integrand for Perelman's entropy functional, the second variation of energy and entropy functionals, and also a matrix Harnack-type calculation for the adjoint heat equation coupled to the Ricci flow.

Chapter 7. In this chapter we give a detailed introduction to Perelman's reduced distance, also called the ℓ-function. To reduce technicalities, we first consider the analogous function corresponding to fixed Riemannian metrics, which is simply the function $d(p,q)^2/4\tau$. In the Ricci flow case we first consider the $\mathcal{L}$-length.

After presenting some basic properties of the $\mathcal{L}$-length, we compute its first variation formula and discuss the existence of $\mathcal{L}$-geodesics, which are the critical points of the $\mathcal{L}$-length. Associated to the $\mathcal{L}$-length is the L-distance, which is obtained by taking the infimum of the $\mathcal{L}$-length over paths with given endpoints. We compute the first space- and time-derivatives of the L-distance. This is partially analogous to the Gauss lemma in Riemannian geometry. Next we compute the second variation formula for the $\mathcal{L}$-length, and motivated by space-time considerations, we express the formula in terms of Hamilton's matrix Harnack quadratic. This second variation formula yields an estimate for the Hessian of the L-distance (and hence for the reduced distance). Next we derive a number of differential equalities and inequalities for the reduced distance. These inequalities are the basis for the use of the reduced distance in the study of singularity formation under the Ricci flow. We also consider the reduced distance in the special cases of Einstein solutions, and more generally, gradient Ricci solitons.

There is a whole space-time geometry associated to the $\mathcal{L}$-length and reduced distance. We consider the notions of $\mathcal{L}$-Jacobi fields and $\mathcal{L}$-exponential map. We derive properties of these objects including the $\mathcal{L}$-Jacobi equation, bounds for $\mathcal{L}$-Jacobi fields, the $\mathcal{L}$-cut locus, and $\mathcal{L}$-Jacobian. We derive bounds for the reduced distance, its spatial gradient, and its time-derivative. Since the reduced distance is a Lipschitz function, we recall the basic properties of Lipschitz functions and formulate the precise sense in which differential inequalities for the reduced distance hold.

Chapter 8. We discuss applications of the study of the reduced distance to the study of finite time singularities for the Ricci flow. First we consider the reduced volume associated to a static metric. This is simply the integral of the transplanted Euclidean heat kernel using the exponential map based at some point. In the case of nonnegative Ricci curvature, the static metric reduced volume is monotonically nonincreasing. This corresponds to the fact that the reduced volume integrand is a weak subsolution to the heat equation. With analogies to no local collapsing in mind, we relate the static metric reduced volume to volume ratios of balls.

Next we consider Perelman's reduced volume for the Ricci flow. For *all* solutions of the Ricci flow on closed manifolds, the reduced volume is monotonically nondecreasing. We present various heuristic proofs and then justify these proofs using the basic properties of the reduced distance as a Lipschitz function and the $\mathcal{L}$-Jacobian developed in the previous chapter.

We prove a weakened version of the no local collapsing theorem using the reduced volume monotonicity. This proof is somewhat technical since one needs some estimates for the $\mathcal{L}$-exponential map. Its advantage over the entropy proof given in Chapter 6 is that it holds for complete solutions of the Ricci flow on *noncompact* manifolds with bounded curvature. Perelman's no local collapsing theorem tells us that singularity models in dimension 3 are ancient κ-solutions. To obtain more canonical limits, one often needs to take backward limits in time and rescale to obtain new ancient κ-solutions. The reduced distance function may be used to show that certain backward limits of ancient κ-solutions are shrinking gradient Ricci solitons. In dimension 3 such a soliton must either be a spherical space form, the cylinder $S^2 \times \mathbb{R}$, or its $\mathbb{Z}_2$-quotient. This has important consequences for singularity formation in dimension 3.

Chapter 9. In this chapter we give a survey of some of the basic 3-manifold topology which is related to the Ricci flow.

Appendix A. We review the contents of Volume One and other aspects of basic Riemannian geometry and Ricci flow.

Appendix B. In many cases, solutions to the Ricci flow limit to Ricci solitons. We present some low-dimensional results of this type for certain classes of complete solutions with bounded curvature on noncompact manifolds diffeomorphic to Euclidean space. We also discuss the mean curvature flow and the cross curvature flow.

Appendix C. This is a glossary of terms related to the study of the Ricci flow.

CHAPTER 1

Ricci Solitons

> The art of doing mathematics consists in finding that special case which contains all the germs of generality. – David Hilbert

> The last thing one knows when writing a book is what to put first.
> – Blaise Pascal

The notion of soliton solutions to evolution equations first appeared in connection with the modelling of shallow water waves and the Korteweg-de Vries equation [**131**]. In this context, a soliton is, in the memorable phrase of Scott-Russell, a 'wave of translation', i.e., a solitary water wave moving by translation, without losing its shape. More generally, we now think of solitons as self-similar solutions, i.e., *solutions which evolve along symmetries of the flow*. In the case of the Ricci flow, these symmetries are scalings and diffeomorphisms.

In this chapter we study Ricci solitons and in particular the special case of gradient Ricci solitons. In later chapters we shall further see how these special solutions of the Ricci flow motivate the general analysis of the Ricci flow through monotonicity formulas and their subsequent applications. In particular, Ricci solitons have inspired the entropy and Harnack estimates, the space-time formulation of Ricci flow, and the reduced distance and reduced volume. Furthermore, the entropy and reduced volume monotonicity formulas have the geometric application of no local collapsing, which is fundamental in the study of singularities (see the diagram below). Gradient Ricci solitons also model the high curvature regions of singular solutions. This motivates trying to classify gradient Ricci solitons, especially in low dimensions.[1]

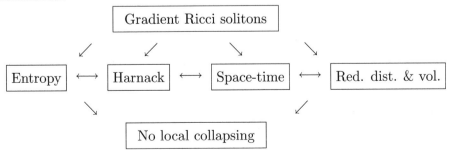

---

[1] "Red. dist. & vol." is an abbreviation for reduced distance and volume, to be discussed in Chapter 7.

Some highlights of this chapter are a systematic description of the equations obtained from differentiating the gradient Ricci soliton equation, the constructions of the Bryant steady Ricci soliton and the rotationally symmetric expanding Ricci soliton with positive curvature operator, examples of homogeneous Ricci solitons, introduction to Perelman's energy and entropy functionals via gradient Ricci solitons, and the Buscher duality transformation.

1. General solitons and their canonical forms

We begin by recalling the following.

DEFINITION 1.1 (General Ricci soliton). A solution $g(t)$ of the Ricci flow on $\mathcal{M}^n$ is a **Ricci soliton** (or **self-similar solution**) if there exist a positive function $\sigma(t)$ and a 1-parameter family of diffeomorphisms $\varphi(t) : \mathcal{M} \to \mathcal{M}$ such that

$$g(t) = \sigma(t)\varphi(t)^* g(0). \tag{1.1}$$

Let $\mathfrak{Met}$ denote the space of Riemannian metrics on a differentiable manifold $\mathcal{M}$, and let $\mathfrak{Diff}$ denote the group of diffeomorphisms of $\mathcal{M}$. Consider the quotient map $\pi : \mathfrak{Met} \to \mathfrak{Met}/\mathfrak{Diff} \times \mathbb{R}_+$, where $\mathbb{R}_+$ acts by scalings. One verifies that a Ricci soliton is a solution $g(t)$ of the Ricci flow for which $\pi(g(t))$ is independent of t, i.e., stationary.

We start by looking at what initial conditions give rise to Ricci solitons. Differentiating (1.1) yields

$$-2\operatorname{Rc}(g(t)) = \dot{\sigma}(t)\varphi(t)^* g_0 + \sigma(t)\varphi(t)^* (\mathcal{L}_X g_0), \tag{1.2}$$

where $g_0 = g(0)$, $\mathcal{L}$ denotes the Lie derivative, X is the time-dependent vector field such that $X(\varphi(t)(p)) = \frac{d}{dt}(\varphi(t)(p))$ for any $p \in \mathcal{M}$, and $\dot{\sigma} \doteqdot \frac{d\sigma}{dt}$.

DEFINITION 1.2. For obvious reasons, we say that $g(t)$ is **expanding**, **steady**, or **shrinking** at a time t_0 if $\dot{\sigma}(t_0)$ is >0, $=0$, or <0, respectively.

Since $\operatorname{Rc}(g(t)) = \varphi(t)^* \operatorname{Rc}(g_0)$, we can drop the pullbacks in (1.2) and get

$$-2\operatorname{Rc}(g_0) = \dot{\sigma}(t)g_0 + \mathcal{L}_{\widetilde{X}(t)} g_0, \tag{1.3}$$

where $\widetilde{X}(t) = \sigma(t)X(t)$. Although g_0 is independent of time, both $\dot{\sigma}(t)$ and $\widetilde{X}(t)$ may depend on time. For example, static Euclidean space $(\mathcal{M}^n, g(t)) \equiv (\mathbb{R}^n, g_{\text{can}})$ is a stationary solution to the Ricci flow and, as such, is a steady Ricci soliton; however, this solution may also be considered as a Ricci soliton which *expands* or *shrinks* modulo diffeomorphisms. In particular, given any function $\sigma(t) > 0$ with $\sigma(0) = 1$, consider the diffeomorphisms $\varphi(t) : \mathbb{R}^n \to \mathbb{R}^n$ defined by $\varphi(t)(x) = \sigma(t)^{-1/2} x$ for $x \in \mathbb{R}^n$. Then $\varphi(t)^* g_{\text{can}} = \sigma(t)^{-1} g_{\text{can}}$. Since $g(t) \equiv g_{\text{can}}$, we may rewrite this as

$$g(t) = g_{\text{can}} = \sigma(t)\varphi(t)^* g(0). \tag{1.4}$$

By choosing $\sigma(t)$ so that $\dot{\sigma}(t)$ changes sign, this soliton may both expand and shrink at different times.

On the other hand, in general, a Ricci soliton solution is evolving purely by scaling (modulo diffeomorphisms). Since the time-derivative of the metric is equal to the negative of twice the Ricci tensor, which is scale-invariant, it is thus natural to ask whether one can put the general soliton equation in a canonical form where $\sigma(t)$ is equal to a linear function. The following gives us a condition under which we may assume that a Ricci soliton defined as in (1.1) has such a form.

PROPOSITION 1.3 (Canonical form for a general soliton). *Let $(\mathcal{M}^n, g(t))$ be a Ricci soliton, and assume that the solution of the Ricci flow with initial metric $g_0 = g(0)$ is unique among soliton solutions. Then there exist diffeomorphisms $\psi(t): \mathcal{M} \to \mathcal{M}$ and a constant $\varepsilon \in \mathbb{R}$ such that*

$$(1.5) \qquad g(t) = (1 + \varepsilon t)\,\psi(t)^* g_0.$$

PROOF. Differentiating (1.3) with respect to time gives

$$(1.6) \qquad \ddot{\sigma}(t) g_0 + \mathcal{L}_{\dot{\widetilde{X}}(t)} g_0 = 0.$$

Case 1. If $\ddot{\sigma}(t) \equiv 0$, then $\sigma(t) = 1 + \varepsilon t$ for some constant ε. Hence, by (1.1), we may simply take $\psi(t) = \varphi(t)$ to obtain (1.5).

Case 2. If $\ddot{\sigma}(t)$ is not identically zero, then let $Y_0 = -\dot{\widetilde{X}}(t_0) / \ddot{\sigma}(t_0)$ at some t_0 where $\ddot{\sigma}(t_0) \neq 0$. We then have

$$(1.7) \qquad \mathcal{L}_{Y_0} g_0 = g_0.$$

Substituting (1.7) into (1.3), we have

$$-2\operatorname{Rc}(g_0) = \mathcal{L}_{\dot{\sigma}(t) Y_0 + \widetilde{X}(t)} g_0$$

for all t. Consider the vector field

$$X_0 \doteqdot \dot{\sigma}(0) Y_0 + \widetilde{X}(0).$$

Then

$$-2\operatorname{Rc}(g_0) = \mathcal{L}_{X_0} g_0.$$

Let $\psi(t)$ be the 1-parameter group of diffeomorphisms generated by X_0. Then it is easy to check that $\tilde{g}(t) = \psi(t)^* g_0$ satisfies the Ricci flow with the same initial conditions g_0 and is a steady soliton. Thus, by our uniqueness assumption for soliton solutions to the Ricci flow with initial metric g_0, by replacing $\varphi(t)$ by $\psi(t)$, we have $\sigma(t) \equiv 1$ in (1.1). $\square$

REMARK 1.4. The proof shows that under the uniqueness assumption, a Ricci soliton not in canonical form can be made a steady soliton.

If $(\mathcal{M}^n, g(t))$ is a Ricci soliton where (1.5) holds, then we say that the soliton is in **canonical form**. By rescaling, we may assume that $\varepsilon = -1, 0,$ or 1; these cases correspond to solitons of *shrinking*, *steady*, or *expanding* type, respectively.

A way to circumvent the undesirableness of the uniqueness assumption in Proposition 1.3 is as follows. Since the geometry of any Ricci soliton $g(t)$ is the same as that of g_0, we will start with g_0 and then construct another Ricci soliton in canonical form and with the same initial metric g_0. Choose any time t_0 and let $\varepsilon \doteqdot \dot{\sigma}(t_0)$ and $X_0 \doteqdot \widetilde{X}(t_0)$, so that equation (1.3) becomes at $t = t_0$

$$-2\operatorname{Rc}(g_0) = \varepsilon g_0 + \mathcal{L}_{X_0} g_0.$$

We will now drop the subscripts on X and g. Using indices, the equation above now reads

(1.8) $$-2R_{ij} = \nabla_i X_j + \nabla_j X_i + \varepsilon g_{ij},$$

where $X_i = g_{ij} X^j$ are the components of $X^\flat$, the covariant tensor (1-form) obtained from X by lowering indices using g. A triple (g, X, ε) (or pair (g, X) if we suppress the dependence on ε) consisting of a metric and a vector field that satisfies (1.8) for some constant ε is called a **Ricci soliton structure**.[2] We say that X is the vector field the soliton is flowing along.

REMARK 1.5 (Solitons and normalized Ricci flow). If (g, X) is a Ricci soliton structure on a compact manifold $\mathcal{M}$, then g evolves purely by diffeomorphisms under the normalized (constant volume) Ricci flow.

DEFINITION 1.6 (Gradient Ricci soliton). A Ricci soliton structure (g, X) is a **gradient soliton structure** if there exists a function f (called the **potential function**) such that $X^\flat = df$. In this case, (1.8) becomes

(1.9) $$R_{ij} + \nabla_i \nabla_j f + \frac{\varepsilon}{2} g_{ij} = 0.$$

The following, whose proof is elementary, shows that given a complete gradient Ricci soliton structure, we can construct a gradient Ricci soliton in canonical form (see Theorem 4.1 on p. 154 of [**111**] or Kleiner and Lott [**231**]). In particular, the result below illustrates the sense in which a Ricci soliton structure may be regarded as initial data for a Ricci solution, i.e., for a self-similar solution to Ricci flow.

PROPOSITION 1.7 (Gradient soliton structures and canonical forms). *Suppose $(g_0, \nabla f_0, \varepsilon)$ is a complete gradient Ricci soliton structure on $\mathcal{M}^n$. Then there exists a solution $g(t)$ of the Ricci flow with $g(0) = g_0$, diffeomorphisms $\varphi(t)$ with $\varphi(0) = \operatorname{id}_\mathcal{M}$, and functions $f(t)$ with $f(0) = f_0$ defined for all t with*

(1.10) $$\tau(t) \doteqdot 1 + \varepsilon t > 0,$$

such that

---

[2]Below, we will sometimes denote a Ricci soliton structure by $(\mathcal{M}^n, g, X)$, in order to emphasize the underlying manifold, e.g. when $\mathcal{M}$ is a Lie group.

(1) $\varphi(t) : \mathcal{M} \to \mathcal{M}$ is the 1-parameter family of diffeomorphisms generated by $X(t) \doteqdot \frac{1}{\tau(t)} \operatorname{grad}_{g_0} f_0$; that is,

$$\frac{\partial}{\partial t} \varphi(t)(x) = \frac{1}{\tau(t)} (\operatorname{grad}_{g_0} f_0)(\varphi(t)(x)),$$

(2) $g(t)$ is the pull-back by $\varphi(t)$ of g_0 up to the scale factor $\tau(t)$,

(1.11) $$g(t) = \tau(t) \varphi(t)^* g_0,$$

(3) $f(t)$ is the pull-back by $\varphi(t)$ of f_0:

(1.12) $$f(t) = f_0 \circ \varphi(t) = \varphi(t)^* f_0.$$

Moreover,

(1.13) $$\operatorname{Rc}(g(t)) + \nabla^{g(t)} \nabla^{g(t)} f(t) + \frac{\varepsilon}{2\tau} g(t) = 0,$$

where $\nabla^{g(t)}$ denotes the covariant derivative with respect to $g(t)$, and

(1.14) $$\frac{\partial f}{\partial t}(t) = \left| \operatorname{grad}_{g(t)} f(t) \right|^2_{g(t)}.$$

EXERCISE 1.8. Prove Proposition 1.7. HINT: Verify the equations in the order in which they are presented.

Because of the proposition above, we shall at some times consider gradient Ricci soliton structures and at other times consider gradient Ricci solitons in canonical form.

Now we give a couple of examples of Ricci solitons in canonical form. We are already acquainted with solutions which evolve purely by homothety, e.g., these solutions correspond to Einstein metrics. If g_0 is an Einstein metric with Einstein constant λ (i.e., $\operatorname{Rc}(g_0) = \lambda g_0$), then

$$g(t) = (1 - 2\lambda t) g_0$$

satisfies the Ricci flow

$$\frac{\partial}{\partial t} g = -2 \operatorname{Rc}(g),$$

since $\operatorname{Rc}(g) = \operatorname{Rc}(g_0) = \lambda g_0$. An Einstein manifold with $\lambda > 0$ is necessarily compact (thanks to Myers's theorem) and, as the metric approaches zero, the manifold shrinks to a point in finite time. Einstein metrics are stationary points for the normalized Ricci flow on a closed manifold. Following up on our previous discussion of static Euclidean space, we have the following.

EXAMPLE 1.9 (Gaussian soliton). Regard Euclidean space as a Ricci soliton in canonical form, so that (1.4) holds with $\sigma(t) = 1 + \varepsilon t$ for some $\varepsilon \in \mathbb{R}$. For $\varepsilon \neq 0$, the Euclidean solution is called the **Gaussian soliton**. By differentiating (1.4) and multiplying the result by $\sigma(t)$, we obtain

(1.15) $$0 = -2 \operatorname{Rc}(g_{\operatorname{can}}) = \mathcal{L}_{\nabla f} g_{\operatorname{can}} + \varepsilon g_{\operatorname{can}},$$

where

(1.16) $$f(x) = -\frac{\varepsilon |x|^2}{4}$$

is the exponent in the Gaussian function. Here we used the standard identity $\mathcal{L}_{\nabla f} g_{\text{can}} = 2\nabla\nabla f$.

2. Differentiating the soliton equation — local and global analysis

2.1. Differentiating general solitons. Let (g, X) be a Ricci soliton structure (gradient or not) on $\mathcal{M}^n$:

$$2R_{ij} + \nabla_i X_j + \nabla_j X_i + \varepsilon g_{ij} = 0. \tag{1.17}$$

Condition (1.17) places a strong condition on g and X. For example, contracting (1.17) with g (tracing) gives

$$R + \operatorname{div} X + \frac{n\varepsilon}{2} = 0, \tag{1.18}$$

where $\operatorname{div} X \doteq g^{ij}\nabla_i X_j$ and R is the scalar curvature. If $\mathcal{M}$ is closed, then this implies

$$\varepsilon = -\frac{2r}{n}, \tag{1.19}$$

where $r \doteq \int_{\mathcal{M}} R\, d\mu/\operatorname{vol}(\mathcal{M})$ denotes the **average scalar curvature**.

Furthermore, we also have the following.

LEMMA 1.10. *If (g, X) is a Ricci soliton structure on $\mathcal{M}^n$, then*

$$\Delta X_i + R_{i\ell}\, g^{\ell m} X_m = 0, \tag{1.20}$$

or more invariantly,

$$\Delta X^\flat + \operatorname{Rc}\left(X^\flat\right) = 0,$$

where $\operatorname{Rc} : T^*\mathcal{M} \to T^*\mathcal{M}$ is defined by $\operatorname{Rc}(\alpha)_i \doteq R_{i\ell}\, g^{\ell m}\alpha_m$.

PROOF. Taking the divergence of (1.17) and applying the second contracted Bianchi identity (V1-3.13) and the Ricci identity (V1-p. 286b), we obtain

$$-\nabla_i R = -2g^{jk}\nabla_j R_{ik} = g^{jk}\nabla_j (\nabla_i X_k + \nabla_k X_i)$$
$$= g^{jk}\left(\nabla_i \nabla_j X_k - R_{jik\ell} g^{\ell m} X_m\right) + \Delta X_i$$
$$= -\nabla_i R + R_{i\ell}\, g^{\ell m} X_m + \Delta X_i,$$

where we have used (1.18). The lemma follows from cancelling the $-\nabla_i R$ terms. □

Lastly, computing the scalar curvature of the evolving metric (1.11) and comparing its time-derivative (at $t = 0$) with its evolution under the Ricci flow gives

LEMMA 1.11. *If (g, X) is a Ricci soliton structure on $\mathcal{M}^n$, then*

$$\Delta R + 2|\operatorname{Rc}|^2 = \mathcal{L}_X R - \varepsilon R. \tag{1.21}$$

PROOF. The left-hand side (LHS) is just the usual expression for $\frac{\partial R}{\partial t}$ under the Ricci flow. To obtain the equality, we compute $\frac{\partial R}{\partial t}$ by way of the equality $R(\tau \varphi^* g) = \tau^{-1} \varphi^* (R(g))$. Hence $\frac{\partial R}{\partial t} = \mathcal{L}_X R - \varepsilon R$. □

REMARK 1.12. Equation (1.21) can also be derived by taking the divergence of (1.20), commuting derivatives and using the contracted second Bianchi identity; however this is more tedious.

To classify solitons, we must use global techniques like the maximum principle. The following is in Proposition 5.20 on p. 117 and Lemma 9.15 on p. 271 of Volume One (see Hamilton [**186**] and one of the authors [**218**]).

PROPOSITION 1.13 (Expanding and steady solitons on closed manifolds are Einstein). *Any expanding or steady Ricci solution (g, X) on a closed n-dimensional manifold $\mathcal{M}^n$ is Einstein. Any shrinking Ricci solution on a closed n-dimensional manifold has positive scalar curvature.*

PROOF. By (1.21),

$$(1.22) \qquad \Delta R - \langle \nabla R, X \rangle + 2|\operatorname{Rc}|^2 + \varepsilon R = 0.$$

We rewrite this as

$$(1.23) \quad \Delta \left(R + \frac{n\varepsilon}{2} \right) - \left\langle \nabla \left(R + \frac{n\varepsilon}{2} \right), X \right\rangle + 2 \left| \operatorname{Rc} + \frac{\varepsilon}{2} g \right|^2 - \varepsilon \left(R + \frac{n\varepsilon}{2} \right) = 0.$$

At any point $x_0 \in \mathcal{M}$ such that $R(x_0) = R_{\min}$, we have

$$2 \left| \operatorname{Rc} + \frac{\varepsilon}{2} g \right|^2 - \varepsilon \left(R + \frac{n\varepsilon}{2} \right) \leq 0.$$

If (g, X) is expanding or steady, then $\varepsilon \geq 0$. Since $R(x_0) + \frac{n\varepsilon}{2} = R(x_0) - r$ is nonpositive, this implies $\left| \operatorname{Rc} + \frac{\varepsilon}{2} g \right|^2 (x_0) = 0$. Tracing then implies

$$R_{\min} = R(x_0) = -\frac{n\varepsilon}{2} = r,$$

and hence $R \equiv r = -\frac{n\varepsilon}{2}$. Substituting back into (1.23), we conclude that $\operatorname{Rc} \equiv -\frac{\varepsilon}{2} g$.

If (g, X) is shrinking, then $\varepsilon < 0$. Applying the weak maximum principle to (1.22), we have $R \geq 0$. By the strong maximum principle, either $R \equiv 0$ or $R > 0$. If $R \equiv 0$, then (1.22) would imply $\operatorname{Rc} \equiv 0$, contradicting the assumption that (g, X) is shrinking. Hence $R > 0$. □

We will return to the consideration of shrinking solitons on closed manifolds in Section 7 of this chapter.

2.2. Differentiating gradient solitons. The gradient Ricci soliton structure equation

$$(1.24) \qquad R_{ij} + \nabla_i \nabla_j f + \frac{\varepsilon}{2} g_{ij} = 0$$

relates the Ricci tensor to the Hessian of f. First note that tracing this gives

$$(1.25) \qquad R + \Delta f + \frac{n\varepsilon}{2} = 0.$$

Now we proceed to systematically differentiate the function f up to fourth order, commute pairs of covariant derivatives, trace, and apply (1.24) and the second Bianchi identity to relate derivatives of the curvature and f.

We begin with considering three derivatives of f since for one derivative there is nothing to do, and a pair of covariant derivatives acting on a function commute. Since $\nabla_k [\nabla_i, \nabla_j] f = 0$, the only nontrivial commutator is

$$\nabla_i \nabla_j \nabla_k f - \nabla_j \nabla_i \nabla_k f = [\nabla_i, \nabla_j] \nabla_k f.$$

By (1.24), the commutator formula, and $\nabla g = 0$, we have

$$(1.26) \qquad -\nabla_i R_{jk} + \nabla_j R_{ik} = -R_{ijk\ell} \nabla_\ell f.$$

Taking the trace by multiplying by g^{jk} and using the contracted second Bianchi identity, we get

$$(1.27) \qquad \frac{1}{2} \nabla_i R = R_{i\ell} \nabla_\ell f.$$

Note that (1.26) is antisymmetric in i and j, and in particular, (1.27) is the only equation obtained by tracing (1.26).

Next we consider equations obtained by commuting four derivatives of f. The only essentially new equation is obtained by considering $\nabla_i [\nabla_j, \nabla_k] \nabla_\ell f$. The quantity $\nabla_i \nabla_j [\nabla_k, \nabla_\ell] f$ is zero and $[\nabla_i, \nabla_j] \nabla_k \nabla_\ell f$ yields only a standard commutator formula. We have

$$\nabla_i [\nabla_j, \nabla_k] \nabla_\ell f = \nabla_i \nabla_j \nabla_k \nabla_\ell f - \nabla_i \nabla_k \nabla_j \nabla_\ell f,$$

which implies

$$\nabla_i (-R_{jk\ell m} \nabla_m f) = -\nabla_i \nabla_j R_{k\ell} + \nabla_i \nabla_k R_{j\ell}$$

and hence

$$(1.28) \quad \nabla_i \nabla_j R_{k\ell} - \nabla_i \nabla_k R_{j\ell} = \nabla_i R_{jk\ell m} \nabla_m f - R_{jk\ell m} \left(R_{im} + \frac{\varepsilon}{2} g_{im} \right).$$

First we trace by g^{ij}. Commuting derivatives and applying the contracted second Bianchi identity yield

$$\Delta R_{k\ell} - \frac{1}{2} \nabla_k \nabla_\ell R + 2 R_{ik\ell m} R_{im} - R_{km} R_{m\ell} + \frac{\varepsilon}{2} R_{k\ell}$$
$$= \nabla_i R_{ik\ell m} \nabla_m f$$
$$= (-\nabla_\ell R_{km} + \nabla_m R_{k\ell}) \nabla_m f.$$

Since (1.28) is antisymmetric in j and k, we have that tracing (1.28) by g^{jk} is zero, whereas tracing by g^{ik} is equivalent to tracing by g^{ij}. The remaining trace is by $g^{i\ell}$. We leave it as an exercise for the reader to show that taking this trace yields $(-\nabla_j R_{km} + \nabla_k R_{jm}) \nabla_m f = 0$, which is nothing new since it follows directly from (1.26).

In conclusion, the new identities we obtain by differentiating f up to fourth order are

$$-\nabla_i R_{jk} + \nabla_j R_{ik} = -R_{ijk\ell}\nabla_\ell f, \tag{1.29}$$

$$\nabla_i\nabla_j R_{k\ell} - \nabla_i\nabla_k R_{j\ell} = \nabla_i R_{jk\ell m}\nabla_m f - R_{jk\ell m}\left(R_{im} + \frac{\varepsilon}{2}g_{im}\right) \tag{1.30}$$

and the traces are

$$\frac{1}{2}\nabla_i R = R_{i\ell}\nabla_\ell f, \tag{1.31}$$

$$M_{k\ell} = (-\nabla_\ell R_{km} + \nabla_m R_{k\ell})\nabla_m f, \tag{1.32}$$

$$\Delta R + 2|\mathrm{Rc}|^2 + \varepsilon R = \nabla f \cdot \nabla R, \tag{1.33}$$

where

$$M_{k\ell} \doteq \Delta R_{k\ell} - \frac{1}{2}\nabla_k\nabla_\ell R + 2R_{ik\ell m}R_{im} - R_{km}R_{m\ell} + \frac{\varepsilon}{2}R_{k\ell}.$$

The last equation, which is the trace of the equation above it, is a special case of (1.21).

REMARK 1.14. The quantities in (1.29) and (1.32) appear in the matrix Harnack quadratic discussed in Part II of this volume (see also subsection 4.2 of Appendix A).

We will finish this subsection with an application of Lemma 1.10 to gradient solitons. Substituting $X = \nabla f$ in (1.20) yields

$$\Delta(\nabla_i f) + R_{i\ell}g^{\ell m}\nabla_m f = 0.$$

On the other hand, commuting covariant derivatives gives

$$\Delta(\nabla_i f) = \nabla_i(\Delta f) + R_{ik}g^{k\ell}\nabla_\ell f.$$

Combining these equations with (1.25) and (1.9) yields

$$0 = \nabla_i R + 2g^{k\ell}\nabla_i\nabla_k f\nabla_\ell f + \varepsilon\nabla_i f = \nabla_i(R + |\nabla f|^2 + \varepsilon f),$$

proving the following.

PROPOSITION 1.15 (Constant gradient quantity on solitons). *If $(g, \nabla f, \varepsilon)$ is a gradient Ricci soliton structure on a manifold $\mathcal{M}^n$, then*

$$R + |\nabla f|^2 + \varepsilon f \equiv \mathrm{const} \tag{1.34}$$

is constant in space. Consequently, by (1.25),

$$R + 2\Delta f - |\nabla f|^2 - \varepsilon f \equiv \mathrm{const}. \tag{1.35}$$

Formula (1.34) is used in the study of the geometric properties of gradient Ricci solitons; see Chapter 9 of [**111**] for an exposition. A significance of the quantity on the LHS of (1.35) will be exhibited in the use of (5.42) and (5.43) to prove energy monotonicity in Chapter 5.

With additional hypotheses, we can determine the constant in the proposition above (see also Theorem 20.1 in [**186**]).

COROLLARY 1.16. *If $(g, \nabla f)$ is a steady gradient Ricci soliton structure on $\mathcal{M}^n$ with positive Ricci curvature and if the scalar curvature attains its maximum at a point O, then*

$$R + |\nabla f|^2 \equiv R(O).$$

PROOF. We have $R_{ij} = -\nabla_i \nabla_j f > 0$ and by Proposition 1.15, $R + |\nabla f|^2$ is constant. On the other hand, at O we have $0 = \nabla_i |\nabla f|^2 = -2R_{ij}g^{jk}\nabla_k f$, which implies that $\nabla f(O) = 0$. □

REMARK 1.17. If we only assume the Ricci curvature is nonnegative, then there is a counterexample by simply taking the flat metric on the xy-plane and letting f be the x-coordinate, in which case $R_{ij} = -\nabla_i \nabla_j f = 0$, so that $R + |\nabla f|^2 \equiv \text{const} > R(O) = 0$.

Quantities which are constant on gradient Ricci solitons are useful in the study of their geometries; see Chapter 9 of [**111**] for an exposition of some examples of this.

2.3. Ricci soliton structures and exterior differential systems. We end this section with a local analysis of Ricci soliton structures. From (1.17), which defines a Ricci soliton structure (g, X), one might hope to obtain more stringent tensorial local conditions on g and/or X by a combination of differentiation, contracting, and equating mixed partials. However, no further lower-order (i.e., first-order in X and second-order in g) identities arise this way. This follows from applying the machinery of **exterior differential systems** to the soliton equation (1.17), written in local coordinates as a system of PDE that is second-order in the entries of g and first-order in the components of X. The theory of exterior differential systems — in particular, the **Cartan-Kähler Theorem**[3] — is able to predict when a system of PDE has solutions, and how large the solution space is, provided the system passes a test indicating that it is **involutive**.[4] In addition, if the system is involutive, then any k-jet of a solution can be extended to a $(k+1)$-jet of a solution, and so on to a convergent power series solution.

In the case of (1.17), in order to obtain an involutive system, one has to reduce the enormous size of the solution space (which is due to the diffeomorphism invariance of the soliton condition) by adding the requirement that the local coordinates are harmonic functions with respect to the metric g. Since $\Delta = g^{ij}\left(\partial_i \partial_j - \Gamma_{ij}^k \partial_k\right)$, this is a first-order condition on the entries of g and takes the form

(1.36) $$g^{ij}\Gamma_{ij}^k = 0,$$

where Γ_{ij}^k are the Christoffel symbols. This is just another variation on **DeTurck's trick** for proving short-time existence for the Ricci flow (see

---

[3]See Theorem III.2.2 and Corollary III.2.3 on pp. 80–88 in [**36**] or Theorem 7.3.3 on pp. 254–256 in [**223**].

[4]See Chapter III, pp. 58–65 of [**36**] or Chapter 7, p. 256 in [**223**].

Sections 4.3 and 5 of Chapter 3 of [**108**]). Once this condition is added to (1.17), the system becomes involutive (see [**221**] for the proof). Furthermore, involutivity implies that Cauchy problems for the augmented system (1.17) and (1.36) are locally solvable; for example, the 1-jet of g may be arbitrarily prescribed along a hypersurface transverse to a given vector field X. Thus, Ricci soliton structures (g, X) locally depend, modulo diffeomorphisms, on $n(n+1)$ functions of $n-1$ variables (i.e., the components of g, and their transverse derivatives, along the hypersurface).

3. Warped products and 2-dimensional solitons

In this section, we review some of the examples of 2-dimensional Ricci solitons, such as the cigar, which have been constructed to date; the Bryant soliton is discussed in Section 4 of this chapter and examples of Kähler–Ricci solitons will be discussed in the next chapter. In Section 1 of Appendix B we will give some conditions under which the Ricci flow converges to some of the solitons discussed here.

Recall from Lemma 5.96 on p. 168 of Volume One that any ancient solution of the Ricci flow on a surface of positive curvature that attains its maximum curvature in space and time is the cigar. Similarly, any ancient solution of the Ricci flow on a manifold of positive curvature operator that attains its maximum curvature in space and time is a steady gradient Ricci soliton. In dimension 3, it is conjectured that such a soliton is the Bryant soliton. This is one reason for focusing our attention on the cigar and Bryant solitons.

3.1. Solitons and Killing vector fields on surfaces. If (g, X) is a soliton structure on a surface $\mathcal{M}^2$, then X is a conformal vector field. This is simply because $\nabla_i X_j + \nabla_j X_i = -(R + \varepsilon) g_{ij}$. If $(g, \nabla f)$ is a gradient soliton, then $J(\nabla f)$ is a Killing vector field (where $J : T\mathcal{M} \to T\mathcal{M}$ is the complex structure, defined as counterclockwise rotation of tangent vectors by $90°$). To see this, we observe that since ∇f is a conformal vector field on a surface,

$$\left(\mathcal{L}_{J(\nabla f)} g\right)_{ij} = \nabla_i \left(J_j^k \nabla_k f\right) + \nabla_j \left(J_i^k \nabla_k f\right) = J_j^k \nabla_i \nabla_k f + J_i^k \nabla_j \nabla_k f$$
$$= -\left(\frac{R}{2} + \frac{\varepsilon}{2}\right) \left(J_j^k g_{ik} + J_i^k g_{jk}\right) = 0,$$

where the components J_i^j are defined by $J\left(\frac{\partial}{\partial x^i}\right) \doteqdot J_i^j \frac{\partial}{\partial x^j}$. That is, $J(\nabla f)$ is a Killing vector field.

In dimension 2 we have the following.

LEMMA 1.18. *A surface with a Killing vector field is locally a warped product. In particular, a gradient soliton on a surface is locally a warped product.*

PROOF. Let $(\mathcal{M}^2, g)$ be a Riemannian surface (not necessarily complete) with a Killing vector field K. Let $x \in \mathcal{M}$ and define a smooth unit speed path $\gamma : (r_0 - \varepsilon, r_0 + \varepsilon) \to \mathcal{M}$ by $\dot{\gamma}(r) = \frac{J(K)}{|J(K)|}(\gamma(r))$, $\gamma(r_0) = x$. Define a 1-parameter family of smooth paths $\beta_r : (\theta_0 - \varepsilon, \theta_0 + \varepsilon) \to \mathcal{M}$ by $\dot{\beta}_r(\theta) = K(\beta_r(\theta))$, $\beta_r(\theta_0) = \gamma(r)$, and a 1-parameter family of smooth unit speed paths $\gamma_\theta : (r_0 - \varepsilon, r_0 + \varepsilon) \to \mathcal{M}$ by $\dot{\gamma}_\theta(r) = \frac{J(K)}{|J(K)|}(\gamma_\theta(r))$, $\gamma_\theta(r_0) = \beta_{r_0}(\theta)$. Note that $\gamma_{\theta_0} = \gamma$.

CLAIM.
$$\beta_r(\theta) = \gamma_\theta(r).$$

PROOF OF CLAIM. This follows from
$$\left[K, \frac{J(K)}{|J(K)|}\right] = \frac{1}{|J(K)|}[K, J(K)] - \frac{1}{2}\frac{1}{|J(K)|^3} K|J(K)|^2 J(K) = 0$$
since $[K, J(K)] = 0$ and $K|J(K)|^2 = 0$.

Hence (r, θ) defines local coordinates on a neighborhood of x. Define $f(r) \doteq |K|(\beta_r)$, where we are using the fact that $|K|$ is constant on β_r. The metric is given by
$$g = \left\langle \frac{J(K)}{|J(K)|}, \frac{J(K)}{|J(K)|} \right\rangle dr^2 + 2\left\langle K, \frac{J(K)}{|J(K)|} \right\rangle dr d\theta + \langle K, K \rangle d\theta^2$$
$$= dr^2 + f(r)^2 d\theta^2.$$
□

It can be shown on a surface that the integral curves of a Killing vector field $J(\nabla f)$ are closed loops. In particular one can prove the following *without* using the **Uniformization Theorem** (see [86]).

THEOREM 1.19 (Shrinking surface soliton is spherical). *If $(\mathcal{M}^2, g)$ is a shrinking gradient soliton on a closed surface, then g has constant positive curvature.*

More generally, if $(\mathcal{M}^2, g)$ is a shrinking gradient soliton on a closed, orientable 2-dimensional orbifold with isolated singularities, then g is rotationally symmetric with positive curvature. When the 2-orbifold is bad, i.e., not covered by a smooth surface, a unique Ricci soliton exists, which is a nontrivial shrinking gradient Ricci soliton (see [180] and [372]). Topologically, a closed, bad 2-orbifold has either one or two singular points [343].

REMARK 1.20. There do not exist complete shrinking Ricci solitons on noncompact surfaces.

3.2. Warped products. Many known examples of gradient Ricci solitons have been constructed in the form of *warped products*. In particular, we consider a metric on the product $\mathcal{M}^{n+1} = \mathcal{I} \times \mathcal{N}^n$ of the form

(1.37) $$g = dr^2 + w(r)^2 \tilde{g},$$

where r is the standard coordinate on an interval $\mathcal{I} \subset \mathbb{R}$, $\tilde{g}$ is a given metric on an n-dimensional manifold $\mathcal{N}$, and $w(r) > 0$ is the warping function, which scales distances along the $\mathcal{N}$-factors in the product.

Several well-known metrics may be expressed as warped products, at least on an open subset. When $\mathcal{N}$ is the unit circle $\mathcal{S}^1$, with $\tilde{g} = d\theta^2$ for the standard coordinate θ, then $w(r) = r$, $w(r) = \sin r$, $w(r) = \sinh r$ give, respectively, the Euclidean plane, the round sphere (with Gauss curvature $+1$), and the hyperbolic plane (with Gauss curvature -1) with the origin omitted. When $\mathcal{N} = \mathcal{S}^n$ with the standard metric $\tilde{g} = g_{\text{can}}$ of constant curvature $+1$, then $w(r) = r$ gives the standard (flat) metric on $\mathbb{R}^{n+1}$ with the origin omitted. In each of these cases, the metric extends smoothly across the origin. When $\mathcal{N} = \mathcal{S}^n$, Lemma 2.10 on p. 29 in Volume One (i.e., Lemma A.2 of this volume) gives sufficient conditions for smoothly closing off the Riemannian manifold $\mathcal{M}$, by a point at one (or both) ends of the interval $\mathcal{I}$. Namely, assuming we wish to close off as $r \searrow 0$, we need $\lim_{r \to 0} w(r) = 0$ and $\lim_{r \to 0} w'(r) = 1$ (the prime denotes the derivative with respect to r).

For constructing Ricci solitons, it is convenient that $\tilde{g}$ be a sufficiently 'nice' metric on $\mathcal{N}$; for the rest of this section, we will assume that $\tilde{g}$ is an Einstein metric, with $\text{Rc}(\tilde{g}) = \rho \tilde{g}$. An easy calculation in moving frames gives the following (see Proposition 9.106 on p. 266 of Besse [27]).

LEMMA 1.21 (Ricci tensor and Hessian of warped product). *If $\tilde{g}$ is an Einstein metric on a manifold $\mathcal{N}^n$, with Einstein constant ρ, then the Ricci tensor of the warped product metric (1.37) is given by*[5]

$$(1.38) \qquad \text{Rc}(g) = -n\frac{w''}{w}dr^2 + \left(\rho - ww'' - (n-1)(w')^2\right)\tilde{g}.$$

Furthermore, if f is any function of the radial coordinate r, then the Hessian of f with respect to g is given by

$$(1.39) \qquad \nabla\nabla f = f''(r)dr^2 + ww'f'\tilde{g}.$$

For example, if $f(r) = \int_a^r w(t)dt$, then $\nabla\nabla f = w'(r)g$. Conversely, by an argument of Cheeger and Colding, warped products may essentially be characterized by the existence of a function whose Hessian is some function times the metric; see §1 in [71] for details.

3.3. Constructing the cigar and other 2-dimensional solitons.

For the sake of starting with something familiar, before we study the Bryant soliton, we will briefly repeat the construction of the cigar metric, which is discussed at greater length in Chapter 2 of Volume One.

Suppose we wish to construct a complete, steady, rotationally symmetric gradient soliton metric on $\mathbb{R}^2$. Such a metric will be a warped product (1.37),

---

[5]In the case where $(\mathcal{N}, \tilde{g})$ is the unit n-sphere, this follows from (1.58).

where $\mathcal{N} = \mathcal{S}^1$, and it is natural to assume that $\text{Rc}(g) + \nabla^2 f = 0$ for a radial function $f(r)$. Using (1.38), (1.39) with $\rho = 0$ and $n = 1$ gives

(1.40) $$wf'' - w'' = 0 = w(w'f' - w'').$$

Integrating $wf'' - w'f' = 0$ gives

(1.41) $$f' = 2aw$$

for some constant a, whereupon $w'f' - w''$ integrates to

(1.42) $$w' - aw^2 = b.$$

Using the closure conditions $w(0) = 0$, $w'(0) = 1$, we get $b = 1$. Integrating, we obtain a smooth odd function $w(r)$ whose type depends on the sign of a:

- for $a = 0$, $w(r) = r$, giving the flat metric;
- for $a = \alpha^2$, $w(r) = \dfrac{1}{\alpha}\tan(\alpha r)$;
- for $a = -\alpha^2$, $w(r) = \dfrac{1}{\alpha}\tanh(\alpha r)$.

The third case is the **cigar soliton** (see [**180**], [**373**])

(1.43) $$g_{\text{cig}} = dr^2 + \frac{1}{\alpha^2}\tanh^2(\alpha r)d\theta^2,$$

where by (1.41) we see that the potential function may be taken to be $f(r) = -2\log\cosh(\alpha r)$. The Gauss curvature is $K = 2\alpha^2\text{sech}^2(\alpha r) > 0$.

In the second case, the metric (see [**125**])

(1.44) $$g_{\text{xpd}} = dr^2 + \frac{1}{\alpha^2}\tan^2(\alpha r)d\theta^2,$$

$0 < r < \pi/(2\alpha)$, which we call the **exploding soliton**, is not complete, since $\tan(\alpha r) \to \infty$ at a finite distance away from the origin. The potential function is $f(r) = -2\log\cos(\alpha r)$ and the Gauss curvature is $K = -2\alpha^2\sec^2(\alpha r) < 0$.

The above steady gradient solitons are defined on topological disks. By taking $b = -1$, we have the following steady solitons on the punctured disk. For $a = -\alpha^2$, taking $w(r) = \dfrac{1}{\alpha}\coth(\alpha r)$ yields

$$g = dr^2 + \frac{1}{\alpha^2}\coth^2(\alpha r)d\theta^2,$$

which has potential $f(r) = -2\log\sinh(\alpha r)$ and $K = -2\alpha^2\text{csch}^2(\alpha r) < 0$. For $a = \alpha^2$, taking $w(r) = \dfrac{1}{\alpha}\cot(\alpha r)$ yields

$$g = dr^2 + \frac{1}{\alpha^2}\cot^2(\alpha r)d\theta^2,$$

which has potential $f(r) = 2\log\sin(\alpha r)$ and curvature $K = 2\alpha^2\cot(\alpha r) > 0$. Both of the above metrics are incomplete because of their behavior near $r = 0$.

Likewise, taking $b = 0$, we have the ODE $w' = aw^2$. If $a = -\alpha^2$, then $w(r) = \frac{1}{\alpha^2 r + C}$, $f(r) = -2\log(\alpha^2 r + C)$ and $K = -\frac{2}{(r + C\alpha^{-2})^2}$.

EXERCISE 1.22. Show that if w is a solution to (1.42), i.e., $w' - aw^2 = b$, then $v \doteqdot 1/w$ satisfies
$$v' + bv^2 = -a.$$

Related to the above exercise, in Section 9 of this chapter we shall see Buscher duality exhibited in the above solitons.

EXERCISE 1.23. Determine all of the solutions of the rotationally symmetric steady gradient Ricci soliton equation on a surface.

3.4. A metric transformation on surfaces related to circle actions. It is interesting to search for duality transformations, besides Buscher duality, for Ricci solitons. One transformation for rotationally symmetric metrics on surfaces is the following, which is related to the study of collapsing sequences of solutions of the Ricci flow on 3-manifolds. Given a rotationally symmetric metric $g = dr^2 + w(r)^2 d\theta^2$ on a surface $\mathcal{M}^2$, we may define another metric on $\mathcal{M}^2$ by

$$\hat{g} \doteqdot dr^2 + \frac{w(r)^2}{\alpha^2 w(r)^2 + 1} d\theta^2.$$

A more general transformation is given by Cheeger [**69**]. The inverse transformation is

$$\breve{g} \doteqdot dr^2 + \frac{w(r)^2}{1 - \alpha^2 w(r)^2} d\theta^2,$$

which is defined as long as $0 < w(r) < 1/\alpha$.

EXERCISE 1.24. Consider the $\mathcal{S}^1$ action on $(\mathcal{M}^2, g) \times \mathcal{S}^1(1/\alpha)$, where $\mathcal{S}^1(\rho) \doteqdot \mathbb{R}/2\pi\rho\mathbb{Z}$, defined by
$$\phi(r, \theta, \eta) \doteqdot (r, \theta + \phi, \eta + \phi), \quad \phi \in \mathcal{S}^1.$$
Show that $\hat{g}$ is the quotient metric on $[(\mathcal{M}^2, g) \times \mathcal{S}^1(1/\alpha)]/\mathcal{S}^1$.

SOLUTION. See Proposition 2 of [**103**].

We now consider some examples. The inverse transformation of the cigar,
$$\breve{g}_{\text{cig}} = dr^2 + \frac{1}{\alpha^2} \sinh^2(\alpha r) d\theta^2,$$
is the hyperbolic metric of constant curvature $K \equiv -\alpha^2$. The transformation of the exploding soliton,
$$\hat{g}_{\text{xpd}} = dr^2 + \frac{1}{\alpha^2} \sin^2(\alpha r) d\theta^2,$$
is the spherical metric of constant curvature $K \equiv \alpha^2$. Since g_{xpd} is only defined for $0 < r < \pi/(2\alpha)$, we see that $\hat{g}_{\text{xpd}}$ is a hemisphere. In summary, we have the following types of examples:

(1) a constant negative curvature metric on a surface transforming to a steady soliton,
(2) an incomplete steady soliton on a surface transforming to an incomplete metric with constant positive curvature.

3.5. Classifying 2-dimensional solitons.

PROPOSITION 1.25 (Surface solitons conformal to $\mathbb{R}^2$). *The only complete steady gradient solitons conformal to the standard metric on $\mathbb{R}^2$ are the cigar and the flat metric.*

REMARK 1.26. We have not assumed the curvature is bounded or has a sign. Note that since a steady gradient soliton is an ancient solution, by applying the maximum principle to the evolution equation for the scalar curvature, one sees that a complete steady Ricci soliton on a surface with curvature bounded from below is either flat or has positive curvature (see [**111**]). For a similar result to Proposition 1.25, see Corollary 1.28 below.

PROOF. Consider the pullback of the steady gradient soliton metric under the map from the cylinder $\mathcal{S}^1 \times \mathbb{R}$ to $\mathbb{R}^2$ defined by

$$x = e^u \cos v, \qquad y = e^u \sin v,$$

where (u, v) are coordinates on $\mathcal{S}^1 \times \mathbb{R}$ and $\mathcal{S}^1 = \mathbb{R}/2\pi\mathbb{Z}$. The pullback is a steady gradient soliton on the cylinder, and the vector field X that it is flowing along is conformal. Hence, the complexification[6] of X is a holomorphic vector field and is of the form $h(w)\partial/\partial w$, where $w = u + iv$ and $h(w)$ is a 2π-periodic entire (analytic) function. Since X is the gradient of a function f, then X can have no closed orbits, and hence any zeros of h must be simple (by virtue of appealing to a local power series expansion). Consequently, critical points of f can only be local maxima and minima, so that $\text{index}_p (\nabla f) = 1$ at any critical point p. Since $\chi \left(\mathcal{S}^1 \times \mathbb{R} \right) = 0$, the Poincaré–Hopf Theorem, which says that the sum of the indices of ∇f is the Euler characteristic, implies that f has no critical points, hence h has no zeros. The periodicity then implies that h is a constant. Moreover, X must be a constant times $\partial/\partial u$, since otherwise the corresponding vector field $\hat{X}$ on $\mathbb{R}^2$ has orbits which spiral into the origin.

Hence we know that $\hat{X} = r\partial/\partial r$ up to multiple, and then by the argument of subsection 3.1 of this chapter, $J(\hat{X}) = \partial/\partial\theta$ is a Killing field, and so by Lemma 1.18 the metric on $\mathbb{R}^2$ is rotationally symmetric (and a warped product). Then we may appeal to the above calculation of solutions to (1.40). □

---

[6]The complexification of a vector field X is defined to be $X^{(1,0)} \doteq \frac{1}{2} (X + iJ(X))$. For example, the complexification of $r\frac{\partial}{\partial r} = x\frac{\partial}{\partial x} + y\frac{\partial}{\partial y}$ is $z\frac{\partial}{\partial z} = \frac{1}{2}\left(x\frac{\partial}{\partial x} + y\frac{\partial}{\partial y} + i\left(y\frac{\partial}{\partial x} - x\frac{\partial}{\partial y}\right)\right)$. See the next chapter for a more detailed discussion of Kähler manifolds (real surfaces are 1-complex dimensional Kähler manifolds). Since we are on a surface, the complexification of a conformal vector field is a holomorphic vector field.

We have the following from Theorem 13 on p. 61 and Theorem 10 on p. 55 of Huber [**210**] (see also §10 of Li [**251**]).

THEOREM 1.27 (Conformal structure of surface with finite total curvature). *If $(\mathcal{M}^2, g)$ is a complete Riemannian surface with $\int_\mathcal{M} K_- d\mu < \infty$, where $K_- \doteqdot \max\{-K, 0\}$, then $(\mathcal{M}, g)$ is conformal to a closed Riemannian surface with a finite number of points removed. Furthermore, by the Cohn–Vossen inequality*

$$\int_\mathcal{M} K_+ d\mu \leq \int_\mathcal{M} K_- d\mu + 2\pi \chi(\mathcal{M}) < \infty,$$

where $K_+ \doteqdot \max\{K, 0\}$.

By the first part of the above theorem, a complete noncompact surface with positive curvature, which we know is diffeomorphic to the plane, must be conformal to the plane. From this and Proposition 1.25 we conclude the following.

COROLLARY 1.28 (Steady surface soliton with $R > 0$ is cigar). *If $(\mathcal{M}^2, g)$ is a complete steady gradient Ricci soliton with positive curvature, then $(\mathcal{M}, g)$ is the cigar.*

This result gives us a classification of complete steady Ricci solitons on surfaces with curvature bounded from below since such solutions are either flat or have positive curvature.

4. Constructing the Bryant steady soliton

We may generalize the cigar metric to a rotationally symmetric steady gradient Ricci soliton in higher dimensions on $\mathbb{R}^{n+1}$ by setting $\mathcal{N} = \mathcal{S}^n$, the unit sphere with constant sectional curvature $+1$.[7] As the following calculations parallel unpublished work of Robert Bryant for $n = 2$, we will refer to the complete metrics obtained as **Bryant solitons**. The Bryant soliton is a **singularity model** for the degenerate neckpinch, a finite time singularity which is expected to form for some (nongeneric) initial data on closed manifolds. (See Section 6 in Chapter 2 of Volume One.)

REMARK 1.29. A singularity model is a **long existing solution** (i.e., the time interval of existence has infinite duration) of the Ricci flow obtained from a **singular solution** (i.e., a solution defined on a maximal time interval $[0, T)$) of the Ricci flow as the limit of dilations about a sequence of points and times approaching the singularity time T.

---

[7]For background on Einstein metrics which are warped products over 1-dimensional bases, see the notes and commentary at the end of this chapter.

4.1. Setting up the ODE for Bryant solitons.

With $\rho = n - 1$, substituting (1.38) and (1.39) into the steady gradient soliton condition $\mathrm{Rc}(g) + \nabla\nabla f = 0$ gives the following system of ODE for w and f:

$$(1.45) \qquad f'' = n\frac{w''}{w}, \qquad ww'f' = ww'' - (n-1)(1-(w')^2).$$

Since $R = -\Delta f$ for a steady gradient soliton, Proposition 1.15 implies that $\Delta f - |\nabla f|^2 = C$, a constant. On the other hand, tracing (1.39) gives

$$\Delta f = f'' + n\frac{w'f'}{w}.$$

Using this expression for Δf, we obtain

$$w^2 f'' + nww'f' - w^2\left(f'\right)^2 = Cw^2.$$

Eliminating f'' and w'' using (1.45) gives a first integral of our ODE system:

$$(1.46) \qquad 2nww'f' + n(n-1)(1-(w')^2) - w^2(f')^2 = Cw^2.$$

The analysis of solutions to (1.45) is simplified by using variables that are invariant under the symmetries of the system (i.e., translating r, translating f, and simultaneously scaling r and w). We choose new dependent variables x and y and independent variable t (not to be confused with time), such that

$$x \doteqdot w', \qquad y \doteqdot nw' - wf', \qquad dt \doteqdot \frac{dr}{w}.$$

From (1.46) we have

$$(1.47) \qquad nx^2 - y^2 + n(n-1) = Cw^2.$$

Then by (1.45)

$$\frac{dx}{dt} = ww'' = ww'f' + (n-1)(1-(w')^2),$$

$$\frac{dy}{dt} = wy' = -ww'f'.$$

Thus the ODE system (1.45) becomes

$$(1.48) \qquad \frac{dx}{dt} = x^2 - xy + n - 1, \qquad \frac{dy}{dt} = x(y - nx).$$

Substituting $C = 0$ in the first integral (1.47) gives an invariant hyperbola $y^2 - nx^2 = n(n-1)$ for the system (1.48); see Figure 1 on the next page.

REMARK 1.30. When $n = 1$, equation (1.48) becomes

$$(1.49) \qquad \frac{dx}{dt} = x(x-y), \qquad \frac{dy}{dt} = x(y-x),$$

which implies $x + y = \mathrm{const}$. Assuming, $x, y \to 1$ as $t \to -\infty$, we have $\frac{dx}{dt} = 2x(x-1)$. The solution of (1.49) with $x(t)$ decreasing is given by

$$x = \frac{1}{1+e^{2t}}, \qquad y = 2 - \frac{1}{1+e^{2t}}.$$

4. CONSTRUCTING THE BRYANT STEADY SOLITON

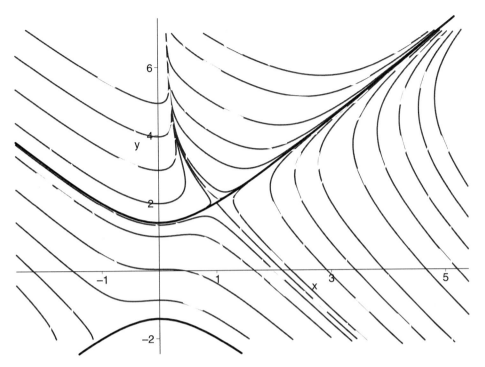

FIGURE 1. Phase portrait for system (1.48), $n = 2$, drawn using Maple.

Since $\frac{d}{dt} \log w = x$, we have $w = \left(e^{-2t} + 1\right)^{-1/2}$. Moreover

$$r = \int w \, dt = \ln\left(e^t + \sqrt{1 + e^{2t}}\right)$$
$$= \operatorname{arcsinh}\left(e^t\right).$$

Hence $e^t = \sinh r$, so that $w(r) = \left(\operatorname{csch}^2 r + 1\right)^{-1/2} = \tanh r$, and we have the cigar soliton.

EXERCISE 1.31. For $n = 1$, what is the solution with $x, y \to 1$ as $t \to -\infty$, and $x(t)$ increasing?

We will establish the existence of a complete steady gradient Ricci soliton on $\mathbb{R}^{n+1}$ by doing phase plane analysis on the system (1.48). Note that the stationary solutions of (1.48) satisfy $y = nx$ and $x^2 - 1 = 0$, so that the stationary points are $(x, y) = (1, n)$ and $(x, y) = (-1, -n)$. These are both saddle points since the linearization of (1.48) at $(1, n)$ is[8]

$$\frac{du}{dt} = -(n-2)u - v,$$
$$\frac{dv}{dt} = -nu + v,$$

and the negative of this at $(-1, -n)$.[9] Since we want the metric to close up smoothly as $r \to 0$, by Lemma A.2, we will need

$$x \to 1 \quad \text{and} \quad y \to n.$$

Therefore, we will limit our attention to the two trajectories that emerge from the saddle point $(1, n)$ as t increases. (These, plus the point itself, comprise the unstable manifold of the saddle point; the stable manifold is the hyperbola.) Given either of these trajectories, we can reconstruct functions w, r and f of t by successively integrating

(1.50) $$\frac{dw}{w} = x\, dt, \qquad dr = w\, dt, \qquad df = (nx - y)dt.$$

Furthermore, we can choose constants of integration such that $w \searrow 0, r \searrow 0$, and $df/dr \to 0$ as $t \to -\infty$. It then follows that $f(r)$ and $w(r)$ smoothly extend to even and odd functions of r, respectively (cf. [217], Proposition 5.2).

4.2. Phase plane analysis of the right-hand trajectory. The linearization of (1.48) at the saddle point shows that the right-hand trajectory emerges from the saddle point into the region where $dx/dt > 0$ and $dy/dt < 0$. Because trajectories along the boundary of this region only point into the region, x continues to increase and y continues to decrease. Linearization also shows that $y - nx$ is initially negative along the trajectory, and the equation

$$\frac{d}{dt}(y - nx) = x(y - nx) - n\frac{dx}{dt}$$

shows that $y - nx$ is negative and monotone decreasing. Because $x \geq 1$ and $y - nx < -C$ for some positive constant C when t is sufficiently large,

---

[8]The linearization of the general system

$$\frac{dx}{dt} = f(x, y), \qquad \frac{dy}{dt} = g(x, y)$$

at (x, y) is

$$\frac{d\bar{x}}{dt} = \frac{\partial f}{\partial x}(x, y)\bar{x} + \frac{\partial f}{\partial y}(x, y)\bar{y},$$
$$\frac{d\bar{y}}{dt} = \frac{\partial g}{\partial x}(x, y)\bar{x} + \frac{\partial g}{\partial y}(x, y)\bar{y}.$$

[9]The eigenvalues of the matrix

$$\begin{pmatrix} 2-n & -1 \\ -n & 1 \end{pmatrix}$$

are 2 and $1 - n$, with corresponding eigenvectors $\langle -1, n \rangle$ and $\langle 1, 1 \rangle$, respectively.

the differential equation for y in (1.48) shows that there exists t_0 such that $y \leq 0$ for $t \geq t_0$. Therefore,

$$\frac{dx}{dt} \geq x^2 + n - 1$$

and $\lim_{t \nearrow T} x = +\infty$ for some finite $T > t_0$.

PROPOSITION 1.32. *The metric associated to this trajectory is incomplete.*

PROOF. We will show that there exists a constant $a \in (0, 1)$ such that

(1.51) $$x \leq \frac{a}{T - t}$$

for t sufficiently close to T. From this inequality, it follows by integration that $w \leq C(T-t)^{-a}$ for some constant C and hence that $r = \int w \, dt$ is finite as $t \nearrow T$.

To establish the upper bound on x, suppose we knew that there is a positive constant δ and a $t_1 \in (t, T)$ such that

(1.52) $$y + \delta x \leq 0 \text{ for all } t \geq t_1.$$

Then $dx/dt \geq (1+\delta)x^2 + n - 1$, and this implies that

$$\frac{d}{dt} \arctan\left(\frac{x}{\sqrt{a(n-1)}}\right) \geq \sqrt{\frac{n-1}{a}}, \qquad a \doteq 1/(1+\delta).$$

Integrating from t to T on both sides and solving for x give the inequality (1.51).

To establish (1.52), note that

$$\frac{d}{dt}(y + \delta x) = x(y - nx) + \delta(x^2 - xy + n - 1).$$

When $y + \delta x = 0$, the right-hand side equals $(\delta^2 - n)x^2 + (n-1)\delta$. Let x_0 be the x-intercept of the trajectory, and let $\delta_0 > 0$ be small enough so that

(1.53) $$(\delta_0^2 - n)x^2 + (n-1)\delta_0 < 0 \text{ for all } x > x_0.$$

Let (x_1, y_1) be any point on the trajectory such that $x_1 > x_0$ and $y_1 < 0$, and choose $\delta \leq \delta_0$ sufficiently small so that $y_1 + \delta x_1 \leq 0$. Then (1.53) shows that this inequality persists for all later values of t. □

4.3. Phase plane analysis of the left-hand trajectory. We now turn to the analysis of the other trajectory emerging from saddle point $(1, n)$. By the same reasoning as before, we see that x is decreasing, y is increasing, and $y - nx$ is positive along this trajectory. Moreover,

$$x \to 0_+ \quad \text{and} \quad y \to +\infty$$

as t increases. In order to prove that the corresponding metric is complete, we need to know the limiting behavior of w and r. To do this, we introduce new phase plane coordinates

$$X \doteq \sqrt{n}\frac{x}{y}, \qquad Y \doteq \frac{\sqrt{n(n-1)}}{y}.$$

As $x \to 0_+$ and $y \to +\infty$,

$$X \to 0_+ \quad \text{and} \quad Y \to 0_+.$$

(The constants are chosen so that the invariant hyperbola becomes the unit circle in the XY-plane.) With a new independent variable s such that $ds \doteq y\, dt$, we have

(1.54) $$\frac{dX}{ds} = X^3 - X + \alpha Y^2, \qquad \frac{dY}{ds} = Y\left(X^2 - \alpha X\right),$$

where $\alpha \doteq 1/\sqrt{n}$. In these coordinates, the first integral corresponds to a Lyapunov function $L \doteq X^2 + Y^2$ which satisfies $dL/ds = 2X^2(L-1)$. Since $L < 1$ for this trajectory, L is strictly decreasing but $dL/ds \geq 2L(L-1)$. So, the trajectory approaches the origin exponentially as $s \to +\infty$. Since

(1.55) $$d(\log w) = x\, dt = \frac{1}{\sqrt{n}X - 1}\frac{dY}{Y},$$

it suffices to find the limiting behavior of X in terms of Y. Note that for X and Y small, (1.54) implies

$$\frac{dX}{ds} \approx -X + \alpha Y^2, \qquad \frac{dY}{ds} \approx -\alpha XY.$$

In particular, we expect $-X + \alpha Y^2 \to 0$ as $s \to +\infty$. Indeed, we have the following.

LEMMA 1.33.
$$\lim_{s \to +\infty} \frac{X}{Y^2} = \alpha$$

and

$$\lim_{s \to +\infty} \frac{X - \alpha Y^2}{Y^4} = 2\alpha^3,$$

where $\alpha = 1/\sqrt{n}$.

PROOF. One might try to apply l'Hôpital's Rule to find the limit of X/Y^2. However,

(1.56) $$2\frac{\frac{d}{ds}X}{\frac{d}{ds}Y^2} = \frac{X}{Y^2}\left(\frac{X^2 - 1}{X^2 - \alpha X}\right) + \frac{1}{\sqrt{n}X^2 - X}$$

shows that this is not straightforward, since the limit of X/Y^2 is also involved in the right-hand side. Let $g(s) = X$ and $h(s) = Y^2$ and apply the Cauchy mean value formula (see, e.g., [**10**]):

$$g'(c)(h(a) - h(b)) = h'(c)(g(a) - g(b))$$

for some c between a and b. Letting $b \to +\infty$ gives $g'(c)/h'(c) = g(a)/h(a)$. Substituting in from (1.56) gives

$$\frac{X}{Y^2}\left(\frac{X^2-1}{X^2-\alpha X}\right) + \frac{1}{\sqrt{n}X^2 - X}\bigg|_{s=c} = 2\frac{X}{Y^2}\bigg|_{s=a}.$$

Multiplying through by $X^2 - \alpha X$ (evaluated at c) and taking $a \to +\infty$ (which forces $c \to +\infty$) give

$$-\lim_{s \to +\infty} \frac{X}{Y^2} + \alpha = 0.$$

However, to get zero on the right-hand side, we need to first know that X/Y^2 is bounded as $s \to +\infty$.

We know that $X/Y^2 \geq 0$. To obtain an upper bound, we use the differential equation

$$\frac{d}{ds}\left(\frac{X}{Y^2}\right) = \alpha - \frac{X}{Y^2}\left(X^2 - 2\alpha X + 1\right).$$

Since the trajectory approaches the origin in the XY-plane exponentially, there exists a $k > 0$ such that

$$\frac{d}{ds}\left(\frac{X}{Y^2}\right) \leq \left(\frac{X}{Y^2} - \alpha\right)(e^{-ks} - 1) + \alpha e^{-ks}.$$

Comparing with the solution of the ODE

$$du/ds = u(e^{-ks} - 1) + \alpha e^{-ks}$$

shows that X/Y^2 is bounded above. In particular, for $s \geq s_0$

$$\frac{X}{Y^2}(s) \leq u(s) = e^{-\left(\frac{1}{k}e^{-ks}+s\right)}\left(\int_{s_0}^{s} e^{\frac{1}{k}e^{-k\bar{s}}+\bar{s}}\alpha e^{-k\bar{s}}d\bar{s} + e^{-\frac{1}{k}}\frac{X}{Y^2}(s_0)\right).$$

The second limit, which gives the next term in the power series expansion of X in terms of Y near the origin, follows by applying the same arguments to the equations

$$\frac{d}{ds}\left(\frac{X-\alpha Y^2}{Y^4}\right) = (-1 - 3X^2 + 4\alpha X)\left(\frac{X-\alpha Y^2}{Y^4}\right) - \frac{\alpha}{Y^2}(X^2 - 2\alpha X)$$

and

$$\frac{\frac{d}{ds}(X-\alpha Y^2)}{\frac{d}{ds}Y^4} = \left(\frac{X-\alpha Y^2}{Y^4}\right)\left(\frac{-1}{4(X^2-\alpha X)}\right) - \frac{\alpha}{2Y^2} + \frac{X^3}{4Y^4(X^2-\alpha X)}.$$

$\square$

EXERCISE 1.34. Verify the second limit.

Now we derive the asymptotic behavior of w and r. We have

$$\frac{d}{dY}(\log w + \log Y) = \frac{X}{Y(X-\alpha)},$$

which by Lemma 1.33 approaches zero as $Y \to 0$. Hence, wY approaches some positive constant C as $Y \to 0$. Next, note that

$$\frac{dr}{w} = \frac{dY}{\sqrt{n(n-1)}(X^2 - \alpha X)}$$

shows that r is monotone increasing as Y decreases to zero, because $X - \alpha = \alpha(nx - y)/y$ is negative along this trajectory. Furthermore, the limiting behavior of X and w implies that there is a constant ε such that

$$-\frac{dr}{dY} = \frac{w}{\sqrt{n(n-1)}X(\alpha - X)} > \frac{C}{\sqrt{n-1}(1+\varepsilon)Y^3}$$

for Y sufficiently close to zero. We can similarly find a constant ε' such that

$$-\frac{dr}{dY} < \frac{2(1+\varepsilon')C}{\sqrt{n-1}Y^3}$$

for Y sufficiently close to zero. Thus, we see by integration that r is $O(Y^{-2})$ as Y approaches zero. In particular, $r \to +\infty$. Therefore, the metric is complete. This proves the following.

THEOREM 1.35 (Existence of Bryant soliton). *There exists a complete, rotationally symmetric steady gradient soliton on $\mathbb{R}^{n+1}$ which is unique up to homothety.*

REMARK 1.36. Since $w = O(Y^{-1})$ and $r = O(Y^{-2})$, we have $w(r) = O(r^{1/2})$. This tells us the Bryant soliton is like a paraboloid. Throughout this section, by $x = O(y^p)$ we mean there exists a positive constant C such that $C^{-1}y^p \leq x \leq Cy^p$. This is an abuse of notation since usually $x = O(y^p)$ means $|x| \leq Cy^p$ for some C.

4.4. Geometric properties of Bryant solitons. It is interesting to contrast the asymptotic behavior of the curvature for these metrics with that of the cigar metric. Recall from Chapter 2 of Volume One that the Gauss curvature of the cigar falls off exponentially as a function of the distance to the origin; in fact, since $w(r)$ is asymptotically constant as $r \to \infty$, the cigar is asymptotic to a cylinder. On the other hand, in a higher-dimensional warped product, the n-dimensional volume of the sphere at distance r from the origin is $w(r)^n \operatorname{vol}(\mathcal{S}^n)$, where $\operatorname{vol}(\mathcal{S}^n)$ is the volume of the standard n-sphere. Since $w(r) = O(r^{1/2})$, we see that the sphere volume is unbounded.

From the limits in Lemma 1.33, we obtain the asymptotics of the derivatives of w as functions of r:

$$w' = x = \sqrt{n-1}\, X/Y = O(Y) = O(r^{-1/2})$$

and
$$w'' = \frac{x^2 - xy + n - 1}{w} = \frac{(n-1)\left(X^2 - \sqrt{n}\left(X - \alpha Y^2\right)\right)}{wY^2}$$
(1.57)
$$= O(-Y^3) = O(-r^{-3/2})$$

as $r \to \infty$.

A moving frames calculation like that mentioned above[10] shows that the sectional curvatures of the Bryant solitons are

(1.58) $$\nu_1 = \frac{1 - (w')^2}{w^2}, \qquad \nu_2 = -\frac{w''}{w},$$

where ν_1 is the curvature for planes tangent to the spheres, and ν_2 for planes tangent to the radial direction. In particular, $2\nu_1$ is the eigenvalue of the curvature operator corresponding to the eigenspace

$$E_1 \doteqdot \{\alpha \wedge \beta : \alpha, \beta \in \Lambda^1\left(\mathcal{S}^n \times \{r\}\right)\}$$

and $2\nu_2$ is the eigenvalue corresponding to the eigenspace

$$E_2 \doteqdot \{\alpha \wedge dr : \alpha \in \Lambda^1\left(\mathcal{S}^n \times \{r\}\right)\}.$$

From the asymptotic behavior of w and its first two derivatives, we obtain $\nu_1 = O(r^{-1})$ and $\nu_2 = O(r^{-2})$. So, the sectional curvatures decay inverse linearly in terms of the distance from the origin. Like the cigar, the curvature is positive:

LEMMA 1.37 (Curvature of Bryant soliton). *The curvature operator of the Bryant soliton is strictly positive (i.e., $\nu_1 > 0$ and $\nu_2 > 0$) away from the origin, and these curvatures have a positive limit at the origin. Moreover,*

$$\nu_1 = O(r^{-1}) \quad \text{and} \quad \nu_2 = O(r^{-2}).$$

PROOF. The positivity of ν_1 for $r > 0$ follows from the fact that $w' = x$ is strictly less than 1. To show that ν_2 is positive for $r > 0$, we will show that w'' is negative. From (1.57), it suffices to show that $x^2 - xy + n - 1$ is negative, or equivalently that $X^2 - \sqrt{n}X + Y^2$ is negative. Linearizing the system (1.48) about the saddle point shows that $x^2 - xy + n - 1$ is initially negative when the trajectory emerges. Moreover,

$$\frac{d}{ds}(X^2 - \sqrt{n} + Y^2) = X^2(X^2 + Y^2 - 1) + (X^2 - 1)(X^2 - \sqrt{n}X + Y^2)$$

shows that if $X^2 - \sqrt{n} + Y^2$ were ever equal to zero in the middle of the trajectory, it would have a negative derivative. So, w'' is negative for all $r > 0$.

At the origin, all sectional curvatures are equal. The scalar curvature R satisfies $R + |\nabla f|^2 = C$, and we know C must be positive because

$$R = n(n-1)\nu_1 + 2n\nu_2$$

---

[10]See, for example, the solution to Exercise 1.188 in [**111**].

is positive everywhere else. The curvatures approach zero as $r \to +\infty$, so R has a positive maximum value at some point. Because $f' = (y - nx)/w$ is positive for $r > 0$, this must occur at the origin. $\square$

We end this section by noting that the preceding construction can be generalized to the case where the fiber is a product of an Einstein manifold and a sphere (see [**219**]).

THEOREM 1.38 (Steady solitons on doubly-warped products). *Given an Einstein manifold $(\mathcal{N}^m, g_\mathcal{N})$ with positive Ricci curvature, there exists a 1-parameter family of complete steady gradient soliton metrics on $\mathbb{R}^{n+1} \times \mathcal{N}$ in the form of **doubly-warped products**[11]:*

$$g = dr^2 + v(r)^2 g_{\text{can}} + w(r)^2 g_\mathcal{N},$$

where g_{can} is the standard metric on $\mathcal{S}^n$, $n \geq 1$.

These solitons have positive Ricci curvature when $r > 0$. However, the sectional curvature along the copies of $\mathcal{N}$ can be negative.

5. Rotationally symmetric expanding solitons

In Section 4 in Chapter 2 of Volume One, we described the construction of a 1-parameter family of rotationally symmetric expanding gradient Ricci solitons on $\mathbb{R}^2$ (see also Gutperle, Headrick, Minwalla, and Schomerus [**174**]). These solitons are asymptotic to a cone (with any cone angle in the interval $(0, 2\pi)$ possible) and have curvature exponentially decaying as a function of the distance to the origin (see Exercise 4.15 and Corollary 9.60 of [**111**]). In this section we consider the problem of constructing rotationally symmetric expanding gradient Ricci solitons on $\mathbb{R}^{n+1}$ for $n \geq 2$. In particular, we generalize the construction from the previous section on the Bryant soliton to show that there exists a 1-parameter family of rotationally symmetric complete expanding gradient solitons on $\mathbb{R}^{n+1}$.

5.1. Setting up the ODE for expanding gradient solitons.
To start, we begin with the expanding gradient soliton condition

$$\operatorname{Rc}(g) + \nabla\nabla f + \lambda g = 0, \qquad \lambda = \frac{\varepsilon}{2} > 0.$$

Substituting (1.38) and (1.39) into this equation gives

$$0 = -n\frac{w''}{w}dr^2 + \left(\rho - ww'' - (n-1)(w')^2\right)\widetilde{g} \\ + \left(f''dr^2 + ww'f'\widetilde{g}\right) + \lambda\left(dr^2 + w^2\widetilde{g}\right),$$

where $w(r)$ is the warping function and $\widetilde{g}$ is the metric on S^n with Einstein constant ρ. Taking $\rho = n - 1$ and collecting the components of the metric

---

[11]This construction was inspired by a similar construction for Einstein metrics by Bérard-Bergery [**24**].

5. ROTATIONALLY SYMMETRIC EXPANDING SOLITONS

on $\mathbb{R}^{n+1}$, we get a system of two second-order ODEs for f and w as functions of r:

$$f'' + \lambda = nw''/w, \qquad ww'f' + \lambda w^2 = ww'' + (n-1)((w')^2 - 1).$$

Again, Proposition 1.15, together with equation (1.25), provides a first integral for this system:

$$f'' + n(w'f'/w) - (f')^2 - 2\lambda f = C.$$

We can reduce the order of the system by again introducing variables that are invariant under the symmetries of translating r and translating f:

$$x \doteq w', \qquad y \doteq nw' - wf'.$$

Because simultaneous scaling of r and w is no longer a symmetry of the system, we must also retain w as a variable, yielding the following first-order system:

(1.59)
$$\begin{aligned} dw/dt &= xw, \\ dx/dt &= x^2 - xy + \lambda w^2 + n - 1, \\ dy/dt &= x(y - nx) + \lambda w^2, \end{aligned}$$

where the parameter t is related to r by $dt = w^{-1} dr$. (Notice that the first integral is not invariant under the translation symmetries, so it does not give an invariant manifold for this system.)

Clearly, solutions of the steady soliton system (1.48) are also solutions of (1.59) for which $w = 0$. (Of course, in the steady case, the warping function is not identically zero, but it is recovered by the quadrature $d(\log w) = x\, dt$.) To get a metric that closes up smoothly at the origin, we need a trajectory that emerges from the singular point $P = (0, 1, n)$ as r increases. (We take w, x, y, in that order, as coordinates on the phase space $\mathbb{R}^3$.)

Linearization at P shows that there is a 2-dimensional unstable manifold passing through P, whose tangent space at P is spanned by the vectors $\langle 1, 0, 0\rangle$ and $\langle 0, -1, n\rangle$.[12] The trajectory corresponding to the steady soliton lies in this surface and is tangent to $\langle 0, -1, n\rangle$ at P. Of course, we are only

---

[12]The linearization of (1.59) at P is

$$\begin{aligned} \frac{d\bar{w}}{dt} &= \bar{w}, \\ \frac{d\bar{x}}{dt} &= (2-n)\bar{x} - \bar{y}, \\ \frac{d\bar{y}}{dt} &= -n\bar{x} + \bar{y}. \end{aligned}$$

Note that the last two equations appeared in the linearization for the steady soliton equation and the eigenvalues of the matrix

$$\begin{pmatrix} 2-n & -1 \\ -n & 1 \end{pmatrix}$$

are 2 and $1-n$, with corresponding eigenvectors $\langle -1, n\rangle$ and $\langle 1, 1\rangle$, respectively.

interested in solutions in the half-space where w is positive. There is a 1-parameter family of these trajectories in the unstable manifold, and they are tangent to the vector $\langle 1, 0, 0 \rangle$ as they exit from P. (This vector is associated to the positive eigenvalue closest to zero.) Among these is the hyperbolic metric, for which $f' = 0$, $y = nx$, and the warping function is

$$w = \sqrt{n/\lambda} \sinh\left(\sqrt{\lambda/n}\, r\right), \qquad y = n \cosh\left(\sqrt{\lambda/n}\, r\right).$$

5.2. Analysis of a 1-parameter family of trajectories. For the rest of this section, we will consider only the 1-parameter family of trajectories that lie in the quadrant of the unstable manifold of P, between the hyperbolic metric and the steady soliton, i.e., those for which w and $y - nx$ are positive for small values of r. Among these is the flat **Gaussian soliton** on $\mathbb{R}^{n+1}$ described in Example 1.9; the corresponding solution of (1.59) is

$$w = r, \quad x = 1, \quad y = n + \lambda r^2.$$

In order to show that the metrics corresponding to our family of trajectories are complete and to study their curvatures, we again introduce rescaled coordinates

$$W \doteqdot \frac{w}{y}, \qquad X \doteqdot \sqrt{n}\,\frac{x}{y}, \qquad Y \doteqdot \frac{\sqrt{n(n-1)}}{y}.$$

We introduce a new independent variable s such that $ds = y\, dt$, whereupon system (1.59) becomes

(1.60)
$$\frac{dW}{ds} = W(X^2 - \lambda W^2),$$
$$\frac{dX}{ds} = X^3 - X + \alpha Y^2 + \lambda(\sqrt{n} - X)W^2, \qquad \alpha \doteqdot 1/\sqrt{n},$$
$$\frac{dY}{ds} = Y(X^2 - \alpha X - \lambda W^2).$$

In these coordinates, $P = (0, \alpha, \sqrt{1 - \alpha^2})$, and the hyperbolic trajectory has X identically equal to α, approaching the critical point $H = (1/\sqrt{n\lambda}, \alpha, 0)$ as $r \to \infty$. The trajectories we are considering lie in the unstable manifold of P, and for small r, in the region where $X < \alpha$ and $W > 0$. The flat trajectory satisfies $Y = \sqrt{n-1}\, X$ and $X^2 - \alpha X + \lambda W^2 = 0$.

LEMMA 1.39. *These trajectories remain in the region defined by*

$$0 < X < \alpha, \qquad 0 < Y \le \sqrt{1 - X^2}, \qquad 0 < W \le 1/\sqrt{n\lambda}.$$

PROOF. Along the plane where $X = \alpha$,

$$dX/ds = \alpha^3 - \alpha + \alpha Y^2 + \lambda(\sqrt{n} - \alpha)W^2.$$

Let Q stand for the quantity on the right, which must be negative for r small. The following equation shows that Q remains negative if $X < \alpha$:

$$dQ/ds = \alpha XY^2(X - \alpha) + \lambda(\sqrt{n} - \alpha)(X^2 - \alpha^2)W^2 - \lambda Q W^2.$$

This establishes that $X < \alpha$ always along these trajectories, and the equation
$$dX/ds = \alpha Y^2 + \lambda\sqrt{n}\,W^2 \text{ when } X = 0$$
shows that X remains positive.

Let $L = X^2 + Y^2$. Then
$$dL/ds = 2(L-1)(X^2 - \lambda W^2) + 2\lambda\sqrt{n}\,W^2(X - \alpha)$$
shows that $L \leq 1$ is preserved. Finally, when $X < \alpha$,
$$dW/ds \leq W(\alpha^2 - \lambda W^2),$$
showing that W cannot exceed $\alpha/\sqrt{\lambda} = 1/\sqrt{n\lambda}$. $\square$

LEMMA 1.40. *As $s \to +\infty$, these trajectories approach the origin in WXY coordinates, and the corresponding metrics are complete.*

PROOF. Because the right-hand sides of the system (1.60) are polynomial, parameter s is unbounded along these trajectories. Because the trajectories are bounded within the region given by Lemma 1.39, each trajectory must limit to either the critical point H or to the origin O as $s \to +\infty$. However, linearization shows that the tangent space of the stable manifold of H intersects the closure of the region given in Lemma 1.39 only in the line through H tangent to the hyperbolic trajectory. Thus, all the trajectories under consideration limit to the origin.

To show that the metrics corresponding to these trajectories are complete, note that

$$(1.61) \qquad \int dr = \int W\,ds = \int \frac{dW}{X^2 - \lambda W^2}.$$

Along these trajectories, W must eventually become a decreasing function of s. Then it suffices to show that there is a point (W_0, X_0, Y_0) along one of our trajectories such that the limit

$$\lim_{\varepsilon \to 0} \int_\varepsilon^{W_0} \frac{dW}{\lambda W^2 - X^2}$$

is infinity; this follows from $\lambda W^2 - X^2 < \lambda W^2$. $\square$

The flat trajectory divides our chosen quadrant of the unstable manifold near S into regions of negative sectional curvature (bordering the hyperbolic trajectory) and regions of positive sectional curvature (bordering the steady soliton trajectory). The following lemma shows that these signs persist; thus, there exist 1-parameter families of rotationally symmetric expanding solitons of strictly positive and of strictly negative sectional curvature.

PROPOSITION 1.41. *Excluding the flat metric, the sectional curvatures of these metrics are either strictly positive or strictly negative for all r.*

PROOF. In these coordinates, the sectional curvatures are

$$\nu_1 = \frac{1-(w')^2}{w^2} = \frac{Y^2-(n-1)X^2}{n(n-1)W^2}, \quad \nu_2 = -\frac{(X^2+Y^2+n\lambda W^2-\sqrt{n}\,X)}{nW^2},$$

for 2-planes tangent to and perpendicular to, respectively, the orbits of the rotational symmetry. The evolution equations for the numerators are

$$\frac{1}{2}\frac{d}{ds}(Y^2-(n-1)X^2) = (X^2-\lambda W^2-\alpha X)(Y^2-(n-1)X^2)$$
$$+ \alpha(n-1)X(-(X^2+Y^2+n\lambda W^2-\sqrt{n}\,X))$$

and

$$\frac{1}{2}\frac{d}{ds}(-(X^2+Y^2+n\lambda W^2-\sqrt{n}\,X))$$
$$= -(X^2-\lambda W^2+\tfrac{1}{2}(\alpha X-1))(X^2+Y^2+n\lambda W^2-\sqrt{n}\,X)$$
$$+ \frac{\alpha}{2}X(Y^2-(n-1)X^2).$$

Observing that the sign of the first nonconstant factor in the second line of each equation is positive, we conclude that the signs $\nu_1 > 0, \nu_2 > 0$ and $\nu_1 < 0, \nu_2 < 0$ are both preserved. $\square$

We will now obtain the asymptotic behavior of the curvature relative to distance r.

PROPOSITION 1.42. *Assuming that the sectional curvatures are negative, then they both decay at least quadratically in r, and for large r, the warping function $w(r)$ is bounded between two linear functions of r.*

PROOF. First, we establish that, as the trajectories approach the origin, the distance r is asymptotic to a constant times $1/W$. We will do this by establishing a positive lower bound for W^2/X, by examining the equations

$$\frac{d}{ds}\left(\frac{W^2}{X}\right) = \frac{W^2}{X}\left(X^2-\lambda W^2+1-\frac{(\alpha Y^2+\lambda\sqrt{n}\,W^2)}{X}\right),$$
$$\frac{d}{ds}\left(\frac{Y^2}{X}\right) = \frac{Y^2}{X}\left(X^2-\lambda W^2+1-2\alpha X-\frac{(\alpha Y^2+\lambda\sqrt{n}\,W^2)}{X}\right).$$

For r sufficiently large, we can bound the nonconstant terms in the parentheses (those that do not involve dividing by X) by a small number ε. Then, if we set $\widetilde{Y} \doteq \alpha Y^2/X$ and $\widetilde{W} \doteq \lambda\sqrt{n}\,W^2/X$, we have

$$\frac{d}{ds}(\widetilde{W}+\widetilde{Y}) \geq (\widetilde{W}+\widetilde{Y})\left(1-\varepsilon-\widetilde{Y}-\widetilde{W}\right).$$

Hence, $\widetilde{W}+\widetilde{Y} \geq c$ for some positive constant c. On the other hand, because sectional curvature ν_1 is negative, $Y^2/X \leq (n-1)X$, and so $\lim_{s\to+\infty}\widetilde{Y} = 0$.

Hence $W^2/X \geq c_1$ for some positive c_1, for r sufficiently large. Thus, the denominator in the integral (1.61) is bounded below,

$$\lambda W^2 - X^2 \geq \lambda W^2 - \frac{W^4}{(c_1)^2},$$

and it follows that $r = O(1/W)$.

The orbital sectional curvature satisfies

$$\nu_1 = \frac{Y^2 - (n-1)X^2}{n(n-1)W^2} > -\frac{X^2}{nW^2} \geq -\frac{W^2}{n(c_1)^2}$$

and so decays at least as fast as $1/r^2$. The equation

(1.62) $$\nu_2 + (n-1)\nu_1 = -\frac{(X^2 - \alpha X + \lambda W^2)}{W^2}$$

shows that in order for ν_2 to also decay to zero, it is necessary that

(1.63) $$\lim_{s \to +\infty} X/W^2 = \lambda\sqrt{n}.$$

We already know that X/W^2 is bounded above, and we can then apply the Cauchy mean value theorem (as in Lemma 1.33) to the equation

$$2\frac{\frac{d}{ds}X}{\frac{d}{ds}W^2} = \frac{\frac{X}{W^2}\left(X^2 - \lambda W^2 + \alpha\frac{Y^2}{X} - 1\right) + \lambda\sqrt{n}}{X^2 - \lambda W^2}$$

to obtain the desired limit. Next, dividing (1.62) by W^2 gives

$$\frac{-\nu_2}{W^2} = (n-1)\frac{\nu_1}{W^2} + \frac{X^2}{W^4} + \frac{\lambda - \alpha\frac{X}{W^2}}{W^2}.$$

In order to show that ν_2 decays at least as fast as $1/r^2$, we need only show that the last term is bounded. This follows from the differential equation

$$\frac{d}{ds}\left(\frac{\lambda - \alpha\frac{X}{W^2}}{W^2}\right) = (-1 + 2\lambda W^2 - 3X^2)\left(\frac{\lambda - \alpha\frac{X}{W^2}}{W^2}\right)$$
$$+ \lambda(X - \alpha)\frac{X}{W^2} - \alpha^2\frac{Y^2}{W^4},$$

where the boundedness of the terms on the second line follows from the limit (1.63) and the negative curvature condition $Y^2 \leq (n-1)X^2$.

To establish the last assertion, it suffices to show that $dw/dr = x = \sqrt{n-1}\,X/Y$ is bounded above and below for r sufficiently large. First, because of (1.62) we can say that

$$\frac{d}{ds}\left(\frac{W^2}{Y}\right) = (X^2 + \alpha X - \lambda W^2)\frac{W^2}{Y} \leq 2X^2\frac{W^2}{Y}.$$

Because of the limit (1.63),

$$\frac{d}{ds}\log\frac{W^2}{Y} \leq CW^4$$

for some constant $C > 0$. Because $dr = W\,ds$ and $W = O(1/r)$, then

$$\frac{d}{dr}\log\frac{W^2}{Y} \leq Cr^{-3},$$

from which it follows that W^2/Y is bounded above for large r. Thus, there are positive constants c_1, c_2 such that

$$\frac{W^2}{c_1} < X < c_1 W^2, \qquad \frac{W^2}{c_2} < Y < c_2 W^2$$

for sufficiently large r, and the bounds on X/Y follow. □

Much more detailed information about the solutions in the positive and negative curvature cases (as well as the steady case) has been obtained by Robert Bryant (in unpublished notes, using different coordinates). In particular, Bryant proved

PROPOSITION 1.43 (Bryant). *Each positive curvature solution defines a complete, rotationally symmetric expanding soliton, whose sectional curvature decays proportionally to $1/r^2$.*

6. Homogeneous expanding solitons

6.1. Existence. Homogeneous spaces are among the nicest examples of Riemannian manifolds.[13] Einstein metrics are among the nicest examples of Riemannian metrics. It is thus natural to ask if a given homogeneous space $\mathcal{M}^n = G/K$ admits a G-invariant Einstein metric. If $\mathcal{M}$ is closed, the answer is frequently 'yes'. For example, consider the following result of Böhm and Kerr [29].

THEOREM 1.44. *Let $\mathcal{M}^n$ be a closed, simply-connected homogeneous space. If $n < 12$, then $\mathcal{M}$ admits a homogeneous Einstein metric.*

On the other hand, Wang and Ziller's example $SU(4)/SU(2)$ of a 12-dimensional homogeneous space that admits no homogenous Einstein metric shows that the answer is 'no' in general and that the dimension restriction above is sharp [365].

Every known example of a noncompact, nonflat homogeneous space that admits an Einstein metric is isomorphic to a solvable Lie group S. (See [198] and [317].) Moreover, for all known examples, the Einstein metric is of standard type.

DEFINITION 1.45. A left-invariant metric g on a solvable Lie group S, regarded as an inner product on the Lie algebra $\mathfrak{s}$, is said to be of **standard type** if the orthogonal complement with respect to g of the derived algebra $[\mathfrak{s},\mathfrak{s}]$ forms an abelian subalgebra $\mathfrak{a}$ of $\mathfrak{s}$.

---

[13]See Chapter 7 of [27], for example.

Many noncompact homogeneous spaces admit no Einstein metrics whatsoever. For instance, there is the following result of Milnor [**266**, Theorem 2.4].

THEOREM 1.46. *If the Lie algebra of a Lie group N is nilpotent but not commutative, then the Ricci curvature of any left-invariant metric on N has mixed sign.*

The Ricci soliton structure equation

(1.64) $$\mathrm{Rc} = -\lambda g - \mathcal{L}_X g$$

illustrates the sense in which a Ricci soliton may be regarded as a generalization of an Einstein metric. Therefore, it is natural to look for Ricci solitons on homogeneous spaces, such as nonabelian nilpotent Lie groups, that do not admit any Einstein metric. Existence and uniqueness of Ricci solitons on such spaces has been investigated by Lauret [**244**].

To describe his results, we must recall some notation from the cohomology of a Lie algebra $\mathfrak{g}$, relative to its adjoint representation. (See [**232**] and [**369**].) A k-cochain is a skew-symmetric k-linear map $\mathfrak{g} \times \cdots \times \mathfrak{g} \to \mathfrak{g}$. Denote the vector space of all k-cochains by C^k, noting the natural identifications $C^0 = \mathfrak{g}$, $C^1 = \mathrm{End}(\mathfrak{g}) = \mathfrak{g}^* \otimes \mathfrak{g}$, and $C^2 = \Lambda^2(\mathfrak{g}^*) \otimes \mathfrak{g}$. The **Lie algebra cohomology** of $\mathfrak{g}$ relative to its **Chevalley complex**,

$$H^k(\mathfrak{g}) = \ker(\delta^k)/\mathrm{im}(\delta^{k-1}),$$

is derived from the coboundary operators $\delta_k : C^k \to C^{k+1}$ defined by

$$\delta_k(A)(X_1, \ldots, X_{k+1}) = \sum_{1 \leq j \leq k+1} (-1)^{j+k} [X_j, A(X_1, \ldots, \hat{X}_j, \ldots, X_{k+1})]$$
$$+ \sum_{1 \leq i < j \leq k+1} (-1)^{i+j} A(X_1, \ldots, \hat{X}_i, \ldots, \hat{X}_j, \ldots, [X_i, X_j]).$$

In particular, one has $\delta_0(X)(Y) = -[Y, X] = \mathrm{ad}_X(Y)$ for all $X, Y \in \mathfrak{g}$, and

$$\delta_1(A)(X, Y) = [X, A(Y)] - [Y, A(X)] - A([X, Y])$$

for all $A \in \mathrm{End}(\mathfrak{g})$ and $X, Y \in \mathfrak{g}$. A **derivation** of $\mathfrak{g}$ is an element of $\ker(\delta_1)$.

Let g be a left-invariant metric on a simply-connected Lie group G. Regarding the Ricci curvature Rc of g as an endomorphism, Lauret considers the condition

(1.65) $$\delta_1(\mathrm{Rc})(\cdot, \cdot) = -\lambda[\cdot, \cdot] \quad (\lambda \in \mathbb{R}),$$

which is easily seen to be equivalent to

(1.66) $$\mathrm{Rc} = -\lambda I + D \quad (\lambda \in \mathbb{R},\ D \in \ker(\delta_1)).$$

Notice that equations (1.65) and (1.66) relate the geometry of (G, g) to algebraic data of $\mathfrak{g}$. This turns out to be a productive point of view. We now survey some of the results it generates, referring the reader to [**244**] for the detailed proofs.

One begins with the observation that conditions (1.64)–(1.66) are equivalent for a nilpotent group.

LEMMA 1.47. *Let N be a simply-connected nilpotent Lie group with Lie algebra $\mathfrak{n}$ and a left-invariant metric g. Then there exist $\lambda \in \mathbb{R}$ and $X \in \mathfrak{n}$ solving the Ricci soliton structure equation (1.64) if and only if there exist $\lambda \in \mathbb{R}$ and a derivation $D \in \ker(\delta_1)$ solving (1.66).*

DEFINITION 1.48. A **standard metric solvable extension** of $(\mathfrak{n}, g)$ is a pair $(\mathfrak{s}, \tilde{g})$, where $\mathfrak{s} = \mathfrak{a} \oplus \mathfrak{n}$ is a solvable Lie algebra such that $[\cdot, \cdot]_{\mathfrak{s}}|_{\mathfrak{n} \times \mathfrak{n}} = [\cdot, \cdot]_{\mathfrak{n}}$ and $\tilde{g}$ is an inner product of standard type such that $[\mathfrak{s}, \mathfrak{s}]_{\mathfrak{s}} = \mathfrak{n} = \mathfrak{a}^{\perp}$ and $\tilde{g}|_{\mathfrak{n} \times \mathfrak{n}} = g$.

Recalling Definition 1.45, note that for $\tilde{g}$ to be of standard type is equivalent to $\mathfrak{a}$ being abelian. The following result relates Ricci soliton structures on nilpotent Lie groups with Einstein metrics of standard type on solvable Lie groups.

THEOREM 1.49. *Let N be a simply-connected nilpotent Lie group with Lie algebra $\mathfrak{n}$ and a left-invariant metric g. Then (N, g) admits a Ricci soliton structure if and only if $(\mathfrak{n}, g)$ admits a standard metric solvable extension $(\mathfrak{s} = \mathfrak{a} \oplus \mathfrak{n}, \tilde{g})$ such that the simply-connected solvable Lie group $(S, \tilde{g})$ is Einstein.*

Although its proof is nonconstructive, Theorem 1.49 is a very useful criterion. For example, any generalized Heisenberg group [25] and many other two-step nilpotent Lie groups admit a Ricci soliton structure. On the other hand, if $\mathfrak{n}$ is characteristically nilpotent, then N admits no such structure. (See [244] for more examples.)

If a Ricci soliton structure does exist on N, it is essentially unique.

THEOREM 1.50. *Let N be a simply-connected nilpotent Lie group with Lie algebra $\mathfrak{n}$. If g and g' are both Ricci soliton metrics, then there exist $a > 0$ and $\eta \in \mathrm{Aut}(\mathfrak{n})$ such that $g' = a\eta(g)$.*

REMARK 1.51. If one regards a Ricci soliton structure as a 3-tuple (N, g, X) satisfying (1.64), then uniqueness of X must be understood modulo addition of a Killing vector field. (See Example 1.57 below.)

Some partial answers are known regarding the existence of Ricci soliton structures on broader classes of Lie groups. For example, if G is semisimple, then condition (1.66) implies condition (1.64). However, (1.66) has no solutions other than Einstein metrics:

THEOREM 1.52. *A left-invariant metric g on a simply-connected semisimple Lie group G satisfies (1.66) if and only if $D = 0$, hence if and only if g is Einstein.*

It is not known in general whether or not a noncompact, semisimple Lie group admits an Einstein metric.

Lauret also gives a variational characterization of Ricci solitons [**244**]. Roughly speaking, this says that any simply-connected nilpotent Lie group (N, g) admitting a Ricci soliton structure is a critical point of a functional that measures, in a certain sense, how far (N, g) is from being Einstein.

We now make this precise. It will be more convenient to fix an inner product and allow the Lie algebra brackets to vary. (Compare with Examples 1.63 and 1.64 below.) Specifically, fix an n-dimensional inner product space $(\mathfrak{n}, \langle \cdot, \cdot \rangle)$ and let $\mathcal{A} = \Lambda^2(\mathfrak{n}^*) \otimes \mathfrak{n}$ denote the space of all skew-symmetric bilinear forms on $\mathfrak{n}$. (Compare to $C_2 \supseteq \mathrm{im}(\delta_1)$ defined above.) Let $\mathcal{N}$ denote the subspace of all nilpotent elements of $\mathcal{A}$ that satisfy the Jacobi identity. Then $\mathcal{N}$ may be regarded as the space of all nilpotent Lie brackets on $\mathfrak{n}$. To each $\mu \in \mathcal{N}$, associate the simply-connected Lie group N_μ and the left-invariant metric g_μ on N_μ determined by $\langle \cdot, \cdot \rangle$. Notice that $\mathcal{N}$ is invariant under the natural action of the group $\mathrm{GL}(\mathfrak{n})$. Thus the $\mathrm{GL}(\mathfrak{n})$ orbit of $\mu \in \mathcal{N}$ corresponds to all simply-connected homogeneous nilpotent Lie groups N isomorphic to N_μ, and the $\mathrm{O}(\mathfrak{n})$ orbit of μ corresponds to all (N, g) isometric to (N_μ, g_μ).

The fixed metric $\langle \cdot, \cdot \rangle$ on $\mathfrak{n}$ induces a metric $(\cdot, \cdot)$ on $\mathcal{A}$ defined by

$$(\mu, \nu) = \sum_{i,j,k} \langle \mu(X_i, X_j), X_k \rangle \langle \nu(X_i, X_j), X_k \rangle$$

when $\{X_1, \ldots, X_n\}$ is any orthonormal basis of $\mathfrak{n}$. Let

$$\mathcal{N}_1 = \{\mu \in \mathcal{N} : (\mu, \mu) = 1\}.$$

Using the Ricci endomorphism, Lauret defines a functional $F : \mathcal{N}_1 \to \mathbb{R}$ by

$$F(\mu) = \mathrm{tr}(\mathrm{Rc}(g_\mu)^2)$$

and proves

THEOREM 1.53. *Let $\mu \in \mathcal{N}_1$. Then (N_μ, g_μ) satisfies (1.64)–(1.66) (i.e. admits a Ricci soliton structure) if and only if μ is a critical point of F.*

To interpret this result, observe that $E(\mu) = \left| \mathrm{Rc}(g_\mu) - \frac{1}{n} R(g_\mu) I \right|^2$ measures the trace-free part of the Ricci endomorphism, i.e., how far g_μ is from being Einstein. But for $\mu \in \mathcal{N}_1$, one has

$$E(\mu) = \mathrm{tr}(\mathrm{Rc}(g_\mu)^2) - \frac{1}{n} R(g_\mu)^2$$
$$= F(\mu) - \frac{1}{16n}.$$

Hence a local minimum μ of F in $\mathcal{N}_1$ should correspond to a homogeneous space (N_μ, g_μ) that is closest to Einstein among all nearby (N_ν, g_ν).

6.2. Construction. The first explicit examples of Ricci soliton structures on Lie groups were constructed by Baird and Danielo [15] and independently by Lott [256]. We discuss [15] in this section and [256] in subsection 6.3, below.

Baird and Danielo discovered soliton structures on 3-dimensional manifolds by studying semiconformal maps to Riemannian surfaces. Significantly, their constructions give the first known examples of nongradient soliton structures. We will only describe two of their conclusions, referring the reader to [15] for details of the method.

EXAMPLE 1.54. Let N denote the 3-dimensional Heisenberg group nil^3. Recall that N may be represented as the group of upper-triangular matrices

$$\left\{ \begin{pmatrix} 1 & u & v \\ 0 & 1 & w \\ 0 & 0 & 1 \end{pmatrix} : u, v, w \in \mathbb{R} \right\}$$

under matrix multiplication.

It is a general fact [14] that any simply-connected nilpotent Lie group is diffeomorphic to $\mathbb{R}^n$. So give $\mathbb{R}^3$ its standard coordinates (x_1, x_2, x_3) and define the frame field

$$F_1 = 2\frac{\partial}{\partial x_1}, \qquad F_2 = 2\left(\frac{\partial}{\partial x_2} - x_1 \frac{\partial}{\partial x_3}\right), \qquad F_3 = 2\frac{\partial}{\partial x_3}.$$

It is easy to check that all brackets $[F_i, F_j]$ vanish except $[F_1, F_2] = -2F_3$, hence that (F_1, F_2, F_3) is a nil^3-geometry frame.[15] The connection 1-forms may be displayed as

$$\begin{pmatrix} \nabla_{F_1} F_1 & \nabla_{F_1} F_2 & \nabla_{F_1} F_3 \\ \nabla_{F_2} F_1 & \nabla_{F_2} F_2 & \nabla_{F_2} F_3 \\ \nabla_{F_3} F_1 & \nabla_{F_3} F_2 & \nabla_{F_3} F_3 \end{pmatrix} = \begin{pmatrix} 0 & -F_3 & F_2 \\ F_3 & 0 & -F_1 \\ F_2 & -F_1 & 0 \end{pmatrix}.$$

Using the dual field

$$\omega^1 = \frac{1}{2} dx_1, \qquad \omega^2 = \frac{1}{2} dx_2, \qquad \omega^3 = \frac{1}{2}(x_1\, dx_2 + dx_3),$$

define a left-invariant metric on N by

$$g = 4\left(\omega^1 \otimes \omega^1 + \omega^2 \otimes \omega^2 + \omega^3 \otimes \omega^3\right).$$

Recalling the standard formula

$$\langle R(X, Y)Y, X \rangle = \frac{1}{4}|(\operatorname{ad} X)^* Y + (\operatorname{ad} Y)^* X|^2 - \langle (\operatorname{ad} X)^* X, (\operatorname{ad} Y)^* Y \rangle$$
$$- \frac{3}{4}|[X, Y]|^2 - \frac{1}{2}\langle [[X, Y], Y], X \rangle - \frac{1}{2}\langle [[Y, X], X], Y \rangle,$$

it is straightforward to compute that

$$\operatorname{Rc}(g) = -2(\omega^1 \otimes \omega^1) - 2(\omega^2 \otimes \omega^2) + 2(\omega^3 \otimes \omega^3).$$

---

[14]The exponential map of a connected, simply-connected, nilpotent Lie group is a diffeomorphism. For instance, see [200].

[15]See Volume One, Chapter 1, Sections 3–4.

Define a vector field
$$X = -\frac{1}{2}x_1 F_1 - \frac{1}{2}x_2 F_2 - (\frac{1}{2}x_1 x_2 + x_3)F_3.$$

A calculation shows that the coordinates $(\nabla_i X^j)$ of $\nabla X = \nabla_i X^j \, \omega^i \otimes F_j$ correspond to the matrix

$$(\nabla_i X^j) = \begin{pmatrix} -1 & -(\frac{1}{2}x_1 x_2 + x_3) & -\frac{1}{2}x_2 \\ \frac{1}{2}x_1 x_2 + x_3 & -1 & \frac{1}{2}x_1 \\ \frac{1}{2}x_2 & -\frac{1}{2}x_1 & -2 \end{pmatrix}.$$

It is then easy to see that
$$-2\operatorname{Rc}(g) = \mathcal{L}_X g + 3g,$$
hence that (N, g, X) is a Ricci soliton structure.

REMARK 1.55. Because the 1-form metrically dual to X is not closed, it follows that $X \neq \operatorname{grad} f$ for any soliton potential function f.

REMARK 1.56. Compact locally homogeneous manifolds with nil^3 geometry occur as mapping tori of $\Upsilon_A : T^2 \to T^2$ induced by $A = \begin{pmatrix} 1 & k \\ 0 & 1 \end{pmatrix} \in \operatorname{SL}(2, \mathbb{Z})$ with $k \neq 0$. The left-invariant metric g is compatible with any compact quotient, but the soliton structure is never compatible with compactification. Indeed, the scalar curvature of g is $R = -1/2$, while every compact Ricci soliton of nonpositive scalar curvature is Einstein.

EXAMPLE 1.57. Let S denote the simply-connected 3-dimensional solvable Lie group $\operatorname{sol}^3 = \mathbb{R} \ltimes \mathbb{R}^2$, where the action of $u \in \mathbb{R}$ sends $(v, w) \in \mathbb{R}^2$ to $(e^u v, e^{-u} w)$. One may also regard S as the group of rigid motions of Minkowski 2-space.

S is diffeomorphic to $\mathbb{R}^3$. In standard coordinates (x_1, x_2, x_3) on $\mathbb{R}^3$, define the frame field

$$F_1 = 2\frac{\partial}{\partial x_1}, \qquad F_2 = 2(e^{-x_1}\frac{\partial}{\partial x_2} + e^{x_1}\frac{\partial}{\partial x_3}), \qquad F_3 = 2(e^{-x_1}\frac{\partial}{\partial x_2} - e^{x_1}\frac{\partial}{\partial x_3}).$$

Its bracket relations are $[F_1, F_2] = -2F_3$, $[F_2, F_3] = 0$, and $[F_3, F_1] = 2F_2$. So (F_1, F_2, F_3) is a sol^3-geometry frame. The connection 1-forms may be displayed as

$$(\nabla_{F_i} F_j) = \begin{pmatrix} 0 & 0 & 0 \\ 2F_3 & 0 & -4F_1 \\ 2F_2 & -4F_1 & 0 \end{pmatrix}.$$

Using the dual field
$$\omega^1 = \frac{1}{2} dx_1, \qquad \omega^2 = \frac{1}{4}(e^{x_1} dx_2 + e^{-x_1} dx_3), \qquad \omega^3 = \frac{1}{4}(e^{x_1} dx_2 - e^{-x_1} dx_3),$$
define a left-invariant metric on S by
$$g = 4(\omega^1 \otimes \omega^1) + 8(\omega^2 \otimes \omega^2) + 8(\omega^3 \otimes \omega^3).$$

The Ricci tensor of this metric is simply
$$\mathrm{Rc}(g) = -8(\omega^1 \otimes \omega^1).$$

Given any $\mu \in \mathbb{R}$, define a vector field
$$X = \mu\left[-F_1 - e^{-x_1}x_3 F_2 + e^{-x_1}x_3 F_3\right] + (1-\mu)\left[F_1 - e^{x_1}x_2 F_2 - e^{x_1}x_2 F_3\right].$$

The computation
$$\mathcal{L}_X g = -32(\omega^2 \otimes \omega^2) - 32(\omega^3 \otimes \omega^3)$$
implies
$$-2\,\mathrm{Rc}(g) = \mathcal{L}_X g + 4g,$$
and hence (S, g, X) is a Ricci soliton structure.

REMARK 1.58. As in the previous example, $X \neq \mathrm{grad}\, f$ for any soliton potential function f.

REMARK 1.59. Compact locally homogeneous manifolds with sol^3 geometry are mapping tori of $\Upsilon_A : T^2 \to T^2$ induced by $A \in \mathrm{SL}(2, \mathbb{Z})$ with eigenvalues $\lambda_- < 1 < \lambda_+$. As in the previous example, the soliton structure cannot descend to any compact quotient, because the scalar curvature of g is $R = -2$.

6.3. Type III singularity models. An important reason for studying shrinking or steady solitons is that they can provide valuable information about finite time singularities of Ricci flow. For example, the (Type I) neckpinch singularity is modeled in all dimensions $n \geq 3$ by the shrinking gradient cylinder soliton
$$(\mathbb{R} \times \mathcal{S}^{n-1},\ g = ds^2 + 2(n-1)g_{\mathrm{can}},\ X = \mathrm{grad}(s^2/4)).$$
(See Section 5 in Chapter 2 of Volume One and [7, 8].) The conjectured (Type II) degenerate neckpinch (Section 6 in Chapter 2 of Volume One) is expected to be modeled on the Bryant soliton, discussed above. We shall now see that homogeneous expanding solitons can model infinite time behavior of Ricci flow.

DEFINITION 1.60. A Type III solution of Ricci flow $(\mathcal{M}^n, g(t))$ exists for $t \in [0, \infty)$ (i.e., is immortal) and satisfies
$$\sup_{\mathcal{M} \times [0, \infty)} t|\mathrm{Rm}| < \infty.$$

There are many examples of Type III solutions (e.g., manifolds with nil^3 or sol^3 geometry) that collapse with bounded curvature. As $t \to \infty$, these examples exhibit pointed **Gromov–Hausdorff convergence** to lower-dimensional manifolds. (See Sections 6 and 7 in Chapter 1 of Volume One.) For such solutions, it is not possible to form a limit solution $(\mathcal{M}_\infty^n, g_\infty(t))$ in a naive way. However, Lott has shown that such solutions may have limits, properly understood, which turn out to be expanding homogeneous solitons [256]. We will describe only two of his results, omitting many details. To discuss convergence of Ricci flow solutions, it is necessary to anticipate some material from Chapter 3. In particular, we refer the reader to Definition 3.6 for the notion of Cheeger–Gromov convergence in the C^∞-topology of a sequence of pointed solutions to the Ricci flow.

Now suppose that $(\mathcal{M}^n, g(t))$ is a Type III solution of Ricci flow. Fix an origin $x \in \mathcal{M}$. For each $s > 0$, there is a rescaled pointed solution of Ricci flow $(\mathcal{M}, g_s(t), x)$ defined for $t \in [0, \infty)$ by

$$g_s(t) = \frac{1}{s} g(st).$$

Lott proves the following:

THEOREM 1.61. *Let $(\mathcal{M}^n, g(t))$ be a Type III solution of Ricci flow. If the limit*

$$(\mathcal{M}^n_\infty, g_\infty(t), x_\infty) = \lim_{s \to \infty} (\mathcal{M}, g_s(t), x)$$

exists, then $(\mathcal{M}_\infty, g_\infty(t), x_\infty)$ is an expanding Ricci soliton.

In dimension $n = 3$, one does not have to assume existence of a limit.

THEOREM 1.62. *Let $(\mathcal{M}^3, g(t))$ solve Ricci flow on a simply-connected homogeneous space $\mathcal{M}^3 = G/K$. Here, G is a unimodular Lie group and K is a compact isotropy subgroup. Then there exists a limit*

$$(\mathcal{M}^3_\infty, g_\infty(t), x_\infty) = \lim_{s \to \infty} (\mathcal{M}^3, g_s(t), x),$$

which is an expanding homogeneous soliton on a (possibly different) Lie group.

Note that the diffeomorphisms with respect to which the convergence of Definition 3.6 occurs become singular as $s \to \infty$.

We will illustrate the content of Theorem 1.62 by relating it to Examples 1.54 and 1.57 (isometric versions of which were also discovered independently by Lott).

EXAMPLE 1.63. Let (N, g, X) denote the nil^3 Ricci soliton structure of Example 1.54. With respect to the frame $\beta = (F_1, F_2, F_3)$ constructed there, one may regard the fixed metric g as the matrix

$$g_\beta = \begin{pmatrix} 4 & 0 & 0 \\ 0 & 4 & 0 \\ 0 & 0 & 4 \end{pmatrix}.$$

Now consider the time-dependent frame $\alpha(t) = \beta A(t)$ given by

$$A(t) = \begin{pmatrix} 0 & 0 & a(t) \\ 0 & a(t) & 0 \\ -2a^2(t) & 0 & 0 \end{pmatrix}, \qquad a(t) = \sqrt{\frac{1}{12} t^{-2/3}}.$$

With respect to the frame $\alpha(t)$, one obtains the identification
$$g_{\alpha(t)} = \begin{pmatrix} \frac{1}{9}t^{-4/3} & 0 & 0 \\ 0 & \frac{1}{3}t^{-2/3} & 0 \\ 0 & 0 & \frac{1}{3}t^{-2/3} \end{pmatrix}.$$

The limit soliton metric is
$$g_\infty(t) = 3t g_{\alpha(t)}.$$

EXAMPLE 1.64. Let (S, g, X) denote the sol^3 Ricci soliton structure of Example 1.57. With respect to the frame $\beta = (F_1, F_2, F_3)$ constructed there, regard the fixed metric g as the matrix
$$g_\beta = \begin{pmatrix} 4 & 0 & 0 \\ 0 & 8 & 0 \\ 0 & 0 & 8 \end{pmatrix}.$$

Now consider the time-dependent frame $\alpha(t) = \beta A(t)$ given by
$$A(t) = \begin{pmatrix} 0 & -\frac{1}{2} & 0 \\ a(t) & 0 & 0 \\ 0 & 0 & a(t) \end{pmatrix}, \qquad a(t) = \sqrt{\frac{1}{32}t}.$$

With respect to the frame $\alpha(t)$, one gets the identification
$$g_{\alpha(t)} = \begin{pmatrix} \frac{1}{4}t^{-1} & 0 & 0 \\ 0 & 1 & 0 \\ 0 & 0 & \frac{1}{4}t^{-1} \end{pmatrix}.$$

The limit soliton metric is
$$g_\infty(t) = 4t g_{\alpha(t)}.$$

Lott also discovered a 4-dimensional example [**256**]. We will describe its (time-independent) Ricci soliton structure, leaving construction of the corresponding Ricci flow solution as an interesting exercise for the reader.

EXAMPLE 1.65. Let N denote the simply-connected 4-dimensional nilpotent Lie group nil^4 with bracket relations
$$[F_1, F_4] = F_2, \qquad [F_2, F_4] = F_3,$$
and all other $[F_i, F_j] = 0$.

The frame field defined in standard coordinates (x_1, x_2, x_3, x_4) on $\mathbb{R}^4$ by
$$F_1 = \frac{\partial}{\partial x_1}, \quad F_2 = \frac{\partial}{\partial x_2}, \quad F_3 = \frac{\partial}{\partial x_3}, \quad F_4 = x_1 \frac{\partial}{\partial x_2} + x_2 \frac{\partial}{\partial x_3} + \frac{\partial}{\partial x_4}$$
realizes these relations. The connection 1-forms are
$$(\nabla_{F_i} F_j) = \frac{1}{2}\begin{pmatrix} 0 & -F_4 & 0 & F_2 \\ -F_4 & 0 & -F_4 & F_1+F_3 \\ 0 & -F_4 & 0 & F_2 \\ -F_2 & F_1-F_3 & F_2 & 0 \end{pmatrix}.$$

Using the dual field
$$\omega^1 = dx_1, \qquad \omega^2 = dx_2 - x_1 dx^4, \qquad \omega^3 = dx_3 - x_2 dx^4, \qquad \omega^4 = dx^4,$$
define a left-invariant metric on S by
$$g = \omega^1 \otimes \omega^1 + \omega^2 \otimes \omega^2 + \omega^3 \otimes \omega^3 + \omega^4 \otimes \omega^4.$$
Its Ricci tensor is
$$\mathrm{Rc}(g) = -\frac{1}{2}(\omega^1 \otimes \omega^1) + \frac{1}{2}(\omega^3 \otimes \omega^3) - (\omega^4 \otimes \omega^4).$$
Define a vector field
$$X = -2x_1 F_1 + (-3x_2 + x_1 x_4) F_2 + (-4x_3 + x_2 x_4) F_3 - x_4 F_4.$$
Then one has
$$\mathcal{L}_X g = -2(\omega^1 \otimes \omega^1) - 3(\omega^2 \otimes \omega^2) - 4(\omega^3 \otimes \omega^3) - (\omega^4 \otimes \omega^4)$$
and thus
$$-2\,\mathrm{Rc}(g) = \mathcal{L}_X g + 3g.$$
Therefore, (N, g, X) is a Ricci soliton structure.

7. When breathers and solitons are Einstein

In this section, we review some of the significant results to date on classifying the breathers and Ricci soliton metrics that exist on a given type of manifold. As noted in the introduction, a shrinking or expanding soliton on a closed manifold will evolve purely by diffeomorphisms under the normalized Ricci flow. More generally, we define a **breather** for the (normalized or unnormalized) Ricci flow to be a solution $g(t)$ for which there exists a period T and a diffeomorphism ϕ such that
$$g(t+T) = \phi^* g(t).$$
So a breather solution is a *periodic orbit* in the space of metrics modulo diffeomorphisms, as compared to a Ricci soliton, which is a fixed point.

The following result is the analogue of Proposition 1.13 for breather solutions to the normalized flow (see [**218**]).

PROPOSITION 1.66 (Steady and expanding breathers are Einstein). *Any breather or soliton for the normalized Ricci flow on a closed manifold $\mathcal{M}^n$ is either Einstein with constant scalar curvature $R \leq 0$ or it has positive scalar curvature.*

PROOF. Under the normalized flow, the scalar curvature satisfies
$$(1.67) \qquad \frac{\partial R}{\partial t} = \Delta R + 2|\mathring{\mathrm{Rc}}|^2 + \frac{2}{n} R(R - r),$$
where $\mathring{\mathrm{Rc}}$ denotes the traceless part of the Ricci tensor and r is the average scalar curvature. By compactness in space and periodicity in time, there exists a point $p \in \mathcal{M}$ where R attains its global minimum $R_{\min}$. Since $\Delta R \geq 0$ and $\partial R/\partial t = 0$ at p, then (1.67) implies that $R_{\min}(R_{\min} - r) \leq 0$,

with equality only if $\overset{\circ}{\operatorname{Rc}} = 0$. Thus, if $R_{\min} < 0$, then R is constant at this time. By applying the same argument at every point, we get $\overset{\circ}{\operatorname{Rc}} \equiv 0$ and the metric is Einstein.

Otherwise, assume $R_{\min} = 0$; by (1.67), we know that $\Delta R \leq 0$ at p. Let Ω be the open set where $\Delta R < 0$ at this time. If Ω is nonempty, then R can only attain its infimum on $\overline{\Omega}$ (which is zero) at a point on $\partial \Omega$. Applying the Hopf maximum principle shows that the outward normal derivative of R must be negative there; but this is impossible because $\nabla R = 0$ there. Thus, Ω is empty and $\Delta R \geq 0$ everywhere. Since R achieves its maximum somewhere, by the maximum principle R is constant (in fact, identically zero), and the metric is Einstein by applying the same argument as above. $\square$

REMARK 1.67. Note that a breather with positive scalar curvature on a closed manifold is a shrinking breather. In Chapter 6 we will describe Perelman's result that shrinking breathers on closed manifolds are Ricci solitons. The proof of this uses his entropy formula. In Chapter 5 another proof that steady or expanding breathers are Einstein will be given.

It was shown in Proposition 5.10 of Volume One that the only solitons for the normalized flow on compact surfaces were constant curvature metrics. We will now prove the generalization of this to compact 3-manifolds, by first obtaining a curvature pinching estimate for the sectional curvatures. As a corollary to Theorem A.31, we have the following (see Hamilton [**186**] and one of the authors [**218**]).

COROLLARY 1.68 (Hamilton–Ivey estimate). *Assume the normalization* $\inf_{x \in \mathcal{M}^3} \nu(x, 0) \geq -1$ *on the initial metric, where* $\nu(x, t)$ *denotes the smallest eigenvalue of the curvature operator. There exists a continuous positive nondecreasing function* $\psi : \mathbb{R} \to \mathbb{R}$ *with* $\psi(u)/u$ *decreasing for* $u > 0$ *and* $\psi(u)/u \to 0$ *as* $u \to \infty$, *such that for any solution* $(\mathcal{M}^3, g(t))$ *of the Ricci flow on a closed 3-manifold, we have* $\nu \geq -\psi(R)$. *That is,*

$$\operatorname{Rm} \geq -\psi(R) \operatorname{id},$$

where $\operatorname{id} : \Lambda^2 \to \Lambda^2$ *is the identity.*

PROOF. By Theorem A.31, wherever $\nu < 0$, we have $R \geq |\nu|(\log|\nu| - 3)$. In particular, if $\nu \leq -e^6$, then $R \geq \frac{1}{2}|\nu| \log |\nu|$. The function $f(u) = u \log u$ is increasing for $u \geq 1/e$ and hence has an inverse $f^{-1} : [-1/e, \infty) \to [1/e, \infty)$. If $\nu \leq -e^6$, then $R \geq 3e^6$ and $|\nu| \leq f^{-1}(2R)$. If we let

$$\psi(u) = \begin{cases} f^{-1}(2u) & \text{if } u \geq 3e^6, \\ e^6 & \text{if } u < 3e^6, \end{cases}$$

then $\nu \geq -\psi(R)$ at all points in space and time. It is easy to see that the function ψ has all of the properties claimed in the statement of the corollary. $\square$

REMARK 1.69. A version of Corollary 1.68 for the Ricci tensor may be proved directly from the evolution of the curvature operator eigenvalues, using the maximum principle for systems. See [**218**] for details.

COROLLARY 1.70 (Ricci flow on 3-manifolds: R bounds Rm). *For any solution $(\mathcal{M}^3, g(t))$ of the Ricci flow on a closed 3-manifold, if R is uniformly bounded, then $|\text{Rm}|$ is also uniformly bounded.*

PROOF. If $R \leq C$, then $\nu \geq -C'$ for some constant C'. If $\lambda \geq \mu \geq \nu$ are the eigenvalues of Rm, then $R = \lambda + \mu + \nu$ and $\lambda \leq R - 2\nu \leq C + 2C'$. □

A nice application of the Hamilton–Ivey estimate is the following (see [**218**]).

THEOREM 1.71 (Shrinking breathers on closed 3-manifolds are Einstein). *The only solitons (or breathers) for the normalized Ricci flow on a closed connected 3-manifold $\mathcal{M}$ are constant sectional curvature metrics.*

The example of Koiso's shrinking soliton (see Section 7 of Chapter 2) shows that this result cannot be extended to dimension 4.

PROOF. By Proposition 1.66, either the metric is Einstein (which is equivalent to constant curvature in dimension 3) or $R_{\min} > 0$; in the latter case, the breather is a shrinking breather. By Corollary 1.68, for the unnormalized Ricci flow,

$$\frac{\text{Rm}}{R} \geq -\frac{\psi(R)}{R} \text{id} \to 0 \text{ as } R \to \infty,$$

and $R_{\min}(t) \to \infty$ as $t \to T$, where $[0, T)$ is the maximal time interval of existence. So the sectional curvature becomes asymptotically nonnegative under the unnormalized flow. Since we assume our solution to be either a soliton or a breather, the curvature must have been nonnegative to begin with.

In [**179**], Hamilton has shown that either the sectional curvature becomes strictly positive immediately or $\mathcal{M}$ splits locally as a product of a 1-dimensional flat factor and a surface with positive curvature, and this splitting is preserved by the flow. In the former case, we know g converges to a metric of constant positive sectional curvature under the normalized flow. To rule out the latter case, consider the evolution equation for r under the normalized flow:

$$\frac{d}{dt} \int R d\mu = -\int \left\langle \frac{\partial g}{\partial t}, \text{Rc} - \frac{1}{2} Rg \right\rangle d\mu,$$

where the volume is fixed at one and the pointwise inner product is given by contraction using the metric. We can calculate the integrand on the right as

$$\left\langle \frac{2}{3} rg - 2\text{Rc}, \text{Rc} - \frac{1}{2} Rg \right\rangle = -\frac{1}{3} rR + R^2 - 2|\text{Rc}|^2.$$

Because of the flat factor, $R^2 = 2|\operatorname{Rc}|^2$, and therefore $dr/dt = r^2/3$. This implies that r increases without bound, which is impossible for a soliton or a breather. □

A similar (but more elaborate) argument based on a pinching set can be used to prove that a nontrivial soliton for the normalized Ricci flow on a compact Kähler surface must have curvature at least as negative as Koiso's example; see [**222**].

8. Perelman's energy and entropy in relation to Ricci solitons

The notion of gradient Ricci soliton has motivated the discovery of monotonicity formulas for the Ricci flow, which in turn have useful geometric applications. Here we consider some monotone integral quantities. In particular, in Chapter 5 we shall further study **Perelman's energy functional**:

$$(1.68) \qquad \mathcal{F}(g, f) = \int_{\mathcal{M}} \left(R + |\nabla f|^2 \right) e^{-f} d\mu,$$

where $(\mathcal{M}^n, g)$ is a closed Riemannian manifold and $f : \mathcal{M} \to \mathbb{R}$. As we will see in (5.31) and (5.41), this functional is nondecreasing under the following set of evolution equations (see below for a motivation for considering (1.70)):

$$(1.69) \qquad \frac{\partial g_{ij}}{\partial t} = -2R_{ij},$$

$$(1.70) \qquad \frac{\partial f}{\partial t} = -\Delta f + |\nabla f|^2 - R.$$

In particular, we have the following.

THEOREM 1.72 (Energy monotonicity). *For any solution $(g(t), f(t))$ of (1.69)–(1.70) on a closed manifold $\mathcal{M}^n$, we have*

$$\frac{d}{dt} \mathcal{F}(g(t), f(t)) = 2 \int_{\mathcal{M}} |R_{ij} + \nabla_i \nabla_j f|^2 e^{-f} d\mu \geq 0.$$

Hence $\frac{d}{dt}\mathcal{F}(g(t), f(t)) = 0$ at some time t_0 if and only if

$$(1.71) \qquad R_{ij} + \nabla_i \nabla_j f \equiv 0$$

at time t_0. That is, $g(t_0)$ is a steady gradient Ricci soliton flowing along $\nabla f(t_0)$.

We can motivate the consideration of equations (1.69)–(1.70) by seeing how it relates to a steady soliton g flowing along a gradient vector field ∇f and in canonical form. By (1.14) and (1.25),

$$\frac{\partial f}{\partial t} = |\nabla f|^2 = |\nabla f|^2 - R - \Delta f,$$

which is (1.70).

A similar consideration for *shrinking* gradient solitons can be used to motivate the study of Perelman's entropy functional $\mathcal{W}(g, f, \tau)$ discussed in

8. PERELMAN'S ENERGY AND ENTROPY IN RELATION TO RICCI SOLITONS

Chapter 6 and the associated equations for g, f and $\tau > 0$. In particular, analogous to $\mathcal{F}$ we define in (6.1) the **entropy**:

$$\mathcal{W}(g, f, \tau) \doteqdot \int_{\mathcal{M}} \left[\tau \left(R + |\nabla f|^2 \right) + f - n \right] (4\pi\tau)^{-n/2} e^{-f} d\mu.$$

Under the system of equations

(1.72) $$\frac{\partial}{\partial t} g_{ij} = -2R_{ij},$$

(1.73) $$\frac{\partial f}{\partial t} = -\Delta f + |\nabla f|^2 - R + \frac{n}{2\tau},$$

(1.74) $$\frac{d\tau}{dt} = -1,$$

we shall show in (6.17) the following.

THEOREM 1.73 (Entropy monotonicity). *If $(g(t), f(t), \tau(t))$, $\tau(t) > 0$, is a solution of (1.72)–(1.74) on a closed manifold $\mathcal{M}^n$, then*

$$\frac{d}{dt} \mathcal{W}(g(t), f(t), \tau(t))$$
$$= \int_{\mathcal{M}} 2\tau \left| R_{ij} + \nabla_i \nabla_j f - \frac{g_{ij}}{2\tau} \right|^2 (4\pi\tau)^{-n/2} e^{-f} d\mu \geq 0.$$

Note that the right-hand side (RHS) vanishes, i.e., $\frac{d}{dt}\mathcal{W}(g(t), f(t), \tau(t)) = 0$, if and only if $g(t)$ is a shrinking gradient Ricci soliton. The above monotonicity formulas beautifully display the utility of considering Ricci solitons.

REMARK 1.74. Assuming that $(\mathcal{M}^n, g(t))$ is a shrinking gradient Ricci soliton in canonical form (1.13) with $\lambda = -1$, we have

$$\operatorname{Rc}(g(t)) + \nabla^{g(t)} \nabla^{g(t)} f(t) - \frac{1}{2\tau} g(t) = 0,$$

where $\frac{d\tau}{dt} = -1$, and hence (1.14) implies f satisfies (1.73).

Finally we consider the *expander* entropy of Feldman, Ilmanen, and one of the authors [**143**]. Define the functional

$$\mathcal{W}_+(g, f_+, \tau) \doteqdot \int_{\mathcal{M}} \left[\tau \left(R + |\nabla f_+|^2 \right) - f_+ + n \right] (4\pi\tau)^{-n/2} e^{-f_+} d\mu.$$

Under the system

(1.75) $$\frac{\partial}{\partial t} g_{ij} = -2R_{ij},$$

(1.76) $$\frac{\partial f_+}{\partial t} = -\Delta f_+ + |\nabla f_+|^2 - R - \frac{n}{2\tau},$$

(1.77) $$\frac{d\tau}{dt} = 1,$$

we have the following.

THEOREM 1.75 (Expander entropy monotonicity formula). *For a solution $(g(t), f_+(t), \tau(t))$, $\tau(t) > 0$, of (1.75)–(1.77) on a closed manifold $\mathcal{M}^n$,*

$$\frac{d}{dt}\mathcal{W}_+(g(t), f_+(t), \tau(t))$$
$$= \int_{\mathcal{M}} 2\tau \left| R_{ij} + \nabla_i \nabla_j f_+ + \frac{g_{ij}}{2\tau} \right|^2 (4\pi\tau)^{-n/2} e^{-f_+} d\mu \geq 0.$$

Here $\frac{d}{dt}\mathcal{W}_+(g(t), f_+(t), \tau(t)) = 0$ if and only if $g(t)$ is an expanding gradient Ricci soliton.

Recall that from Proposition 1.7

(1.78) $$S^\varepsilon_{ij} \doteq R_{ij} + \nabla_i \nabla_j f + \frac{\varepsilon}{2\tau} g_{ij} \stackrel{\mathrm{G}}{=} 0,$$

where $\tau(t) \doteq \varepsilon t + 1 > 0$ and $\stackrel{\mathrm{G}}{=}$ denotes an equality which holds for a gradient Ricci soliton $(\mathcal{M}^n, g(t), f(t), \varepsilon)$ in canonical form.

EXERCISE 1.76. Show that

(1)

(1.79) $$\frac{\partial f}{\partial t} \stackrel{\mathrm{G}}{=} -\Delta f - R + |\nabla f|^2 - \frac{n\varepsilon}{2\tau},$$

(2)

(1.80) $$V_\varepsilon \doteq \tau \left(R + 2\Delta f - |\nabla f|^2 \right) - \varepsilon(f - n) \stackrel{\mathrm{G}}{=} C(t).$$

Moreover, if f has a critical point in space, then $C(t)$ is independent of t.

9. Buscher duality transformation of warped product solitons

In this section we describe an interesting duality transformation for solutions of the modified Ricci flow on certain warped products which in particular take gradient Ricci solitons to gradient Ricci solitons. The consideration of these warped products with tori of potentially infinite dimensions leads to Perelman's energy functional.

9.1. A metric duality transformation. Let $(\mathcal{M}^n, g)$ be a Riemannian manifold and let $(\mathcal{P}^q, h)$ be a flat manifold such as a torus or Euclidean space. Given a function $A : \mathcal{M} \to (0, \infty)$, consider the **warped product manifold** $(\mathcal{M}, g) \times_A (\mathcal{P}, h)$ which is the Riemannian manifold $(\mathcal{M} \times \mathcal{P}, g + Ah)$. The **Buscher duality transformation** (see [38], [39]) takes the metric

$$^\flat g \doteq g + Ah$$

to the metric

$$^\sharp g \doteq g + A^{-1} h.$$

(This is a special case of *T*-**duality** in string theory.) Although its definition is simple, the transformation has some surprising properties. Let $\{x^i\}_{i=1}^n$

and $\{y^\alpha\}_{\alpha=1}^q$ be local coordinates on $\mathcal{M}$ and $\mathcal{P}$, respectively, such that $h_{\alpha\beta} \doteq h\left(\frac{\partial}{\partial y^\alpha}, \frac{\partial}{\partial y^\beta}\right) = \delta_{\alpha\beta}$. We then have

$$^\flat g_{ij} = g_{ij}, \qquad ^\sharp g_{ij} = g_{ij},$$
$$^\flat g_{\alpha\beta} = A\delta_{\alpha\beta}, \qquad ^\sharp g_{\alpha\beta} = A^{-1}\delta_{\alpha\beta},$$

and the rest of the components are zero, where

$$^\flat g_{ij} \doteq {}^\flat g\left(\frac{\partial}{\partial x^i}, \frac{\partial}{\partial x^j}\right), \qquad ^\flat g_{\alpha\beta} \doteq {}^\flat g\left(\frac{\partial}{\partial y^\alpha}, \frac{\partial}{\partial y^\beta}\right),$$

and similarly for $^\sharp g$.

The explicit formulas below are from Haagensen [175] (see also (1.38) or §J in Chapter 9 of Besse [27] for curvature formulas for warped products).

LEMMA 1.77. *The Christoffel symbols of $^\flat g$ are*

$$^\flat\Gamma_{ij}^k = \Gamma_{ij}^k,$$
$$^\flat\Gamma_{i\alpha}^\beta = \delta_\alpha^\beta \frac{1}{2}\nabla_i \log A,$$
$$^\flat\Gamma_{\alpha\beta}^i = -\delta_{\alpha\beta}\frac{A}{2}\nabla^i \log A,$$
$$^\flat\Gamma_{\alpha\beta}^\gamma = {}^\flat\Gamma_{\alpha j}^i = {}^\flat\Gamma_{ij}^\alpha = 0,$$

and likewise, the Christoffel symbols of the dual metric $^\sharp g$ are

$$^\sharp\Gamma_{ij}^k = \Gamma_{ij}^k,$$
$$^\sharp\Gamma_{i\alpha}^\beta = -\delta_\alpha^\beta \frac{1}{2}\nabla_i \log A,$$
$$^\sharp\Gamma_{\alpha\beta}^i = \delta_{\alpha\beta}\frac{1}{2A}\nabla^i \log A,$$
$$^\sharp\Gamma_{\alpha\beta}^\gamma = {}^\sharp\Gamma_{\alpha j}^i = {}^\sharp\Gamma_{ij}^\alpha = 0.$$

LEMMA 1.78. *The Ricci tensor of $^\flat g$ is given by*

$$^\flat R_{\alpha\beta} = -\frac{A}{2}\left[\Delta \log A + \frac{q}{2}|\nabla \log A|^2\right]\delta_{\alpha\beta},$$
$$^\flat R_{\alpha i} = 0,$$
$$^\flat R_{ij} = R_{ij} - \frac{q}{2}\nabla_i\nabla_j \log A - \frac{q}{4}\nabla_i \log A \nabla_j \log A,$$

and the Ricci tensor of $^\sharp g$ is

$$^\sharp R_{\alpha\beta} = -\frac{1}{2A}\left(-\Delta \log A + \frac{q}{2}|\nabla \log A|^2\right)\delta_{\alpha\beta},$$
$$^\sharp R_{\alpha i} = 0,$$
$$^\sharp R_{ij} = R_{ij} + \frac{q}{2}\nabla_i\nabla_j \log A - \frac{q}{4}\nabla_i \log A \nabla_j \log A.$$

LEMMA 1.79. *The scalar curvature of $^\flat g$ and $^\sharp g$ are given by*

$$^\flat R = R - q\Delta \log A - \frac{q(q+1)}{4}|\nabla \log A|^2,$$

$$^\sharp R = R + q\Delta \log A - \frac{q(q+1)}{4}|\nabla \log A|^2.$$

EXERCISE 1.80. Show that if $A \doteqdot \exp\left(-\frac{2}{q}f\right)$ (i.e., $\log A = -\frac{2}{q}f$), then

$$^\flat R_{ij} = R_{ij} + \nabla_i \nabla_j f - \frac{1}{q}\nabla_i f \nabla_j f,$$

$$^\flat R = R + 2\Delta f - \frac{q+1}{q}|\nabla f|^2,$$

$$^\sharp R = R - 2\Delta f - \frac{q+1}{q}|\nabla f|^2.$$

Hence

$$\lim_{q \to \infty} R^\flat = R + 2\Delta f - |\nabla f|^2,$$

$$\lim_{q \to \infty} {}^\flat R_{ij} = R_{ij} + \nabla_i \nabla_j f$$

are Perelman's modified Ricci tensor and scalar curvature (see also (5.18) and (5.19)).

As a consequence of the above exercise, if h has unit volume, we then have

$$\int_{\mathcal{M} \times \mathcal{P}} R\left({}^\flat g\right) d\mu_{^\flat g} = \int_\mathcal{M} \left(R + 2\Delta f - \frac{q+1}{q}|\nabla f|^2\right) e^{-f} d\mu_g$$

$$= \int_\mathcal{M} \left(R + \frac{q-1}{q}|\nabla f|^2\right) e^{-f} d\mu_g.$$

Taking $q \to \infty$, this limits to $\mathcal{F}(g, f)$ defined in (1.68).

9.2. Buscher duality. For a warped product solution of the modified Ricci flow of the above type we have the following.

THEOREM 1.81 (Buscher duality). *If*

$$^\flat g(t) = g(t) + \exp\left(-\frac{2}{q}f(t)\right) \sum_{\alpha=1}^q dy^\alpha \otimes dy^\alpha,$$

$t \in \mathcal{I}$, *satisfy the modified Ricci flow*

(1.81) $$\frac{\partial}{\partial t} {}^\flat g_{ab} = -2\left({}^\flat R_{ab} + 2 {}^\flat \nabla_a {}^\flat \nabla_b {}^\flat \phi\right),$$

where $^\flat \phi$ is a function on $\mathcal{M} \times \mathcal{I}$, then the dual metrics

$$^\sharp g(t) = g(t) + \exp\left(\frac{2}{q}f(t)\right) \sum_{\alpha=1}^q dy^\alpha \otimes dy^\alpha$$

satisfy

(1.82) $$\frac{\partial}{\partial t} {}^\sharp g_{ab} = -2\left({}^\sharp R_{ab} + 2\,{}^\sharp\nabla_a\,{}^\sharp\nabla_b\,{}^\sharp\phi\right),$$

where
$${}^\sharp\phi \doteq {}^\flat\phi + f.$$

REMARK 1.82. Note that ${}^\flat\nabla_\alpha f = {}^\sharp\nabla_\alpha f = 0$.

The functions ${}^\flat\phi$ and ${}^\sharp\phi$ are called **dilatons** and the transformation from ${}^\flat\phi$ to ${}^\sharp\phi$ is called the **dilaton shift**. The Buscher duality transformation takes solutions of the modified Ricci flow with dilaton ${}^\flat\phi$ to solutions of the modified Ricci flow with dilaton ${}^\sharp\phi$.

LEMMA 1.83 (Buscher duality preserves modified Ricci tensor after dilaton shift).

$${}^\sharp R_{\alpha\beta} + 2\,{}^\sharp\nabla_\alpha\,{}^\sharp\nabla_\beta\,{}^\sharp\phi = -\frac{1}{A^2}\left({}^\flat R_{\alpha\beta} + 2\,{}^\flat\nabla_\alpha\,{}^\flat\nabla_\beta\,{}^\flat\phi\right),$$

$${}^\sharp R_{\alpha i} + 2\,{}^\sharp\nabla_\alpha\,{}^\sharp\nabla_i\,{}^\sharp\phi = {}^\flat R_{\alpha i} + 2\,{}^\flat\nabla_\alpha\,{}^\flat\nabla_i\,{}^\flat\phi = 0,$$

$${}^\sharp R_{ij} + 2\,{}^\sharp\nabla_i\,{}^\sharp\nabla_j\,{}^\sharp\phi = {}^\flat R_{ij} + 2\,{}^\flat\nabla_i\,{}^\flat\nabla_j\,{}^\flat\phi.$$

REMARK 1.84. If $\log A = -\frac{2}{q}f$, then

$${}^\flat R_{ij} + 2\,{}^\flat\nabla_i\,{}^\flat\nabla_j\,{}^\flat\phi = R_{ij} + \nabla_i\nabla_j f - \frac{1}{q}\nabla_i f \nabla_j f + 2\nabla_i\nabla_j\,{}^\flat\phi,$$

$${}^\sharp R_{ij} + 2\,{}^\sharp\nabla_i\,{}^\sharp\nabla_j\,{}^\sharp\phi = R_{ij} - \nabla_i\nabla_j f - \frac{1}{q}\nabla_i f \nabla_j f + 2\nabla_i\nabla_j\,{}^\sharp\phi.$$

Suppose $\frac{\partial}{\partial t}{}^\flat g_{ab} = -2\,{}^\flat R_{ab}$ so that ${}^\flat\phi = 0$. By Exercise 1.80, taking the limit as $q \to \infty$, the equation for the metric $g_{ij}(t)$ on the base manifold is

$$\frac{\partial}{\partial t} g_{ij} = -2\left(R_{ij} + \nabla_i\nabla_j f\right).$$

Note that $d\mu\left({}^\flat g\right) = e^{-f}d\mu \wedge dy^1 \wedge \cdots \wedge dy^q$. The **effective action** of $\left({}^\flat g, {}^\flat\phi\right)$ is (e.g., see Alvarez and Kubyshin [**2**], equation (26))

$$S\left({}^\flat g, {}^\flat\phi\right) = \int_{\mathcal{M}\times\mathcal{P}} \left(R\left({}^\flat g\right) + 4\left|\nabla\,{}^\flat\phi\right|^2\right) e^{-2\,{}^\flat\phi} d\mu\left({}^\flat g\right)$$

$$= \int_{\mathcal{M}} \left(R + 2\Delta f - \frac{q+1}{q}|\nabla f|^2 + 4\left|\nabla\,{}^\flat\phi\right|^2\right) e^{-2\,{}^\flat\phi} e^{-f} d\mu,$$

and similarly for the dual pair $\left({}^\sharp g, {}^\sharp\phi\right)$. (Note that by taking $\phi^\flat = 0$, we have $S\left({}^\flat g, 0\right) = \int_{\mathcal{M}\times\mathcal{P}} R\left({}^\flat g\right) d\mu_{{}^\flat g}$.) Buscher duality preserves the effective action:

$$S\left({}^\sharp g, {}^\sharp\phi\right) = S\left(g^\flat, \phi^\flat\right).$$

9.3. Examples. The Buscher dual of a rotationally symmetric soliton solution on a surface

$$^{\flat}g = dr^2 + w(r)^2 d\theta^2$$

is the soliton

$$^{\sharp}g = dr^2 + w(r)^{-2} d\theta^2.$$

For example,

$^{\flat}g$	$^{\sharp}g$
$dr^2 + \tanh^2 r \, d\theta^2$	$dr^2 + \coth^2 r \, d\theta^2$
$dr^2 + \tan^2 r \, d\theta^2$	$dr^2 + \cot^2 r \, d\theta^2$
$dr^2 + r^2 d\theta^2$	$dr^2 + r^{-2} d\theta^2$
$dr^2 + d\theta^2$	$dr^2 + d\theta^2$

Let $(\mathcal{N}^m, h)$ be an Einstein metric with $\mathrm{Rc} = \varepsilon g$, $\varepsilon \in \mathbb{R}$. Consider the doubly-warped product metric on $\mathbb{R}^2 \times \mathcal{N}^m$:

$$^{\flat}g \doteqdot ds^2 + F(s)^2 d\theta^2 + G(s)^2 h,$$

where s is the radial coordinate, θ is the coordinate on the circle, and F and G are positive functions. The gradient Ricci soliton equation $^{\flat}R_{ab} + 2 \, ^{\flat}\nabla_a \, ^{\flat}\nabla_b f = 0$ becomes (see equation (1) in [**219**]) the following system for the triple (F, G, f):

$$(1.83) \quad -2f'' = -\frac{G''}{G} - \frac{F''}{F},$$

$$(1.84) \quad F'' = -n\frac{F'G'}{G} - f'F',$$

$$(1.85) \quad \frac{G''}{G} = \frac{\varepsilon}{G^2} - (n-1)\left(\frac{G'}{G}\right)^2 - \frac{F'G'}{FG} - f'\frac{G'}{G}.$$

The Buscher dual metric is

$$^{\sharp}g = ds^2 + \frac{1}{F(s)^2} d\theta^2 + G(s)^2 h.$$

We know that $^{\sharp}g$ is a steady gradient Ricci soliton if $^{\flat}g$ is, and so we leave it to the reader to verify that indeed the triple $\left(\frac{1}{F}, G, f - \log F\right)$ is a solution to (1.83)–(1.85) if (F, G, f) is.

10. Summary of results and open problems on Ricci solitons

In this section we collect some known results and open problems about the properties and classification of gradient Ricci solitons. Besides the discussion earlier in this chapter, one may consult [**111**] for some of the proofs.

10. SUMMARY OF RESULTS AND OPEN PROBLEMS ON RICCI SOLITONS

10.1. Gradient Ricci solitons on surfaces. The following is a compendium of known results about complete 2-dimensional gradient Ricci solitons with *bounded curvature* (shrinkers, steadies, and expanders). In proving them, one may use the fact that if $(\mathcal{M}^2, g)$ admits a nontrivial Killing vector field X which vanishes at some point $O \in \Sigma$, then $(\mathcal{M}, g)$ is rotationally symmetric.

(1) A shrinker has constant positive curvature. In particular, the underlying surface is compact.
(2) A steady is either flat or the cigar.
(3) A compact expander has constant negative curvature (if $\chi(\mathcal{M}) < 0$, then there are no nonzero conformal Killing vector fields).
(4) An expander with positive curvature is rotationally symmetric and unique up to homothety (see [**241**] and [**111**]).

In part (4), one can apply the arguments of subsection 3.1 of this chapter and the following result (see [**286**] and [**111**]).

THEOREM 1.85. *If $(\mathcal{M}^n, g)$ is a gradient Ricci soliton on a noncompact manifold with $R_{ij} \geq \varepsilon R g_{ij}$ for some $\varepsilon > 0$, where $R \geq 0$, then R decays exponentially in distance to a fixed origin. In particular, if $(\mathcal{M}^2, g)$ is an expanding gradient Ricci soliton on a surface, then R decays exponentially.*

A complete proof, using a more direct method, is given in [**241**].

PROBLEM 1.86. Are there any other expanders on a surface diffeomorphic to $\mathbb{R}^2$ besides the positively curved rotationally symmetric expander and the hyperbolic disk?

PROBLEM 1.87. Do all complete 2-dimensional gradient Ricci solitons have bounded curvature? Are all complete 2-dimensional Ricci solitons gradient?

Note that in dimension 3 there are complete homogeneous expanding solitons which are not gradient; see Section 5 of this chapter.

10.2. Gradient Ricci solitons on 3-manifolds. In dimension 3 we have the following results. This subsection is abbreviated since we pose some more problems in the next subsection for dimensions at least 3.

(1) (Perelman) Any nonflat shrinker with bounded nonnegative sectional curvature is isometric to either a quotient of the 3-sphere or a quotient of $\mathcal{S}^2 \times \mathbb{R}$. In particular, any shrinker with bounded positive sectional curvature is isometric to a shrinking solution with constant positive sectional curvature.
(2) There exists a rotationally symmetric steady with positive sectional curvature, namely the Bryant soliton.
(3) There exists a rotationally symmetric expander with positive sectional curvature.

PROBLEM 1.88. Are there any 3-dimensional steady gradient solitons besides a flat solution, the Bryant soliton, and a quotient of the product of the cigar and $\mathbb{R}$? This is equivalent to asking if a steady gradient Ricci soliton with $n = 3$ and $\operatorname{sect}(g(t)) > 0$ is isometric to a Bryant soliton.

10.3. Gradient Ricci solitons in higher dimensions. Here we assume $n \geq 3$. Note that any expanding or steady Ricci soliton on a closed n-dimensional manifold is Einstein.

PROBLEM 1.89. Is a shrinking gradient Ricci soliton with $n \geq 4$ and $\operatorname{Rm}(g(t)) > 0$ compact?

PROBLEM 1.90. Is a compact shrinking gradient Ricci soliton, $n \geq 4$, and $\operatorname{Rm}(g(t)) > 0$ isometric to a shrinking constant positive sectional curvature solution?[16]

PROBLEM 1.91. Are there n-dimensional steadies with positive curvature operator besides the Bryant soliton?

PROBLEM 1.92. Are there any n-dimensional expanders with positive curvature operator besides the rotationally symmetric one?

PROBLEM 1.93. Does there exist an expander with $n \geq 3$ and positively pinched Ricci curvature, i.e., $R_{ij} \geq \varepsilon R g_{ij}$ for some $\varepsilon > 0$, where $R > 0$? By Theorem 1.85, such an expander has R decaying exponentially.

11. Notes and commentary

Section 1. The first occurrence of the notion of a Ricci soliton in the literature is in Friedan [**145**], where non-Einstein Ricci solitons are called **quasi-Einstein metrics**. See Besse [**27**] for a comprehensive treatment of Einstein manifolds.

The Gaussian soliton first appeared in §2.1 of Perelman [**297**].

In the case of surfaces we have encountered Ricci solitons in Chapter 5 of Volume One. There, solitons motivated several aspects of Hamilton's original proof of convergence of the Ricci flow for surfaces with $R(g_0) > 0$, including the scalar curvature, Harnack, and entropy estimates, as well as the estimate which shows the metric approaches a soliton. See in Volume One, Corollary 5.17 on p. 115, Proposition 5.57 on p. 145, Proposition 5.39 on p. 134, and Corollary 5.35 on p. 130.

Expanding solitons are of interest because they constitute borderline cases for Hamilton's Harnack inequality. (Expanding solitons also model the formation of certain Type III singularities.) In fact, as we mentioned in Section 2 of this chapter and will discuss in more detail in Part II, the consideration of quantities which vanish on solitons led to the discovery of the Harnack quadratic. Steady solitons, where the isometry class of the metric is independent of $t \in (-\infty, \infty)$, occur as limits of Type II singularities.

---

[16]The answer to this problem is 'yes' and follows from the recent work of Böhm and Wilking [**30**] (see Part II of this volume).

Section 4. Einstein metrics are special cases of Ricci solitons. In general, if a warped product metric g is Einstein, then the fiber $(\mathcal{P}, \tilde{g})$ is Einstein. If the base is 1-dimensional, then provided $w(r)$ is not constant, the metric
$$g = dr^2 + w(r)^2 \tilde{g}$$
satisfies $\operatorname{Rc}(g) = \lambda g$ if and only if
$$w'(r)^2 + \frac{\lambda}{n} w(r)^2 = \frac{\rho}{n-1},$$
where $\operatorname{Rc}(\tilde{g}) = \rho \tilde{g}$ and $n = \dim \mathcal{P}$. Without loss of generality (e.g., up to scaling and diffeomorphism), we have the following cases:

$\rho = \frac{R(\tilde{g})}{n+1}$	$\lambda = \frac{R(g)}{n}$	$w(r)$	g
$-(n-1)$	$-n$	$\cosh r$	including hyperbolic with two ends
0	$-n$	e^r	including hyperbolic cusp
$n-1$	$-n$	$\sinh r$	including hyperbolic space
$n-1$	0	r	Ricci flat cone
$n-1$	n	$\sin r$	including sphere
0	0	1	Ricci flat product

As a consequence, we have the following (see Theorem 9.110 on p. 268 of [**27**]).

THEOREM 1.94 (Einstein warped products over 1-dimensional base). *If a warped product $(\mathcal{M}, g)$ over a 1-dimensional base, with the dimension of the fiber at least 2, is a complete Einstein manifold, then either*

(1) *g is a Ricci flat product,*
(2) *$\mathcal{M}$ is topologically a cone on $\mathcal{P}$ and the fiber is Einstein with positive scalar curvature, or*
(3) *the base is $\mathbb{R}$, the fiber is Einstein with nonpositive scalar curvature, and g has negative scalar curvature.*

CHAPTER 2

Kähler–Ricci Flow and Kähler–Ricci Solitons

> Symmetry, as wide or narrow as you may define its meaning, is one idea by which man through the ages has tried to comprehend and create order, beauty, and perfection. – Hermann Weyl

The Kähler–Ricci flow is simply an abbreviation for the Ricci flow on Kähler manifolds. In this chapter we first review some basic definitions and properties for Kähler manifolds and survey some of the fundamental results on the existence of Kähler–Einstein metrics. Then we discuss elementary properties of the Kähler–Ricci flow and state some of the fundamental long-time existence and convergence results. We also give a survey of Kähler–Ricci solitons.

Some other highlights of this chapter are an exposition of tensor calculations in holomorphic coordinates, the proof of the long-time existence and convergence of the Kähler–Ricci flow on Kähler manifolds with $c_1 < 0$, construction of the Koiso solitons and other $U(n)$-invariant solitons, proofs of differential Harnack estimates under the assumption of nonnegative holomorphic bisectional curvature, and a survey of uniformization-type results for complete noncompact Kähler manifolds with positive bisectional curvature.

We assume the reader either has some knowledge of Kähler geometry or will read other references on Kähler geometry along with this chapter, so we make no attempt to be completely self-contained.[1] The latter chapters do not depend on this chapter and the reader interested only in *Riemannian* Ricci flow may skip this chapter.

1. Introduction to Kähler manifolds

In this section we introduce the basic concepts of complex and Kähler manifolds including the point of view of having an almost complex structure on an even-dimensional Riemannian manifold satisfying natural properties.

Let $\mathcal{M}$ be a real $2n$-dimensional differentiable manifold. A **system of holomorphic coordinates** on $\mathcal{M}$ is a collection $\{z_i : U_i \to z_i(U_i) \subset \mathbb{C}^n\}$, where $\{U_i\}$ is a cover of $\mathcal{M}$ and z_i are homeomorphisms such that the maps

$$z_i \circ z_j^{-1} : z_j(U_i \cap U_j) \to z_i(U_i \cap U_j)$$

---

[1] See the notes and commentary at the end of this chapter for some references on complex manifolds and Kähler geometry.

are holomorphic (complex analytic), and hence biholomorphic, whenever $U_i \cap U_j \neq \varnothing$. Two systems of holomorphic coordinates $\{U_i, z_i\}$ and $\{V_j, w_j\}$ on $\mathcal{M}$ are **equivalent** if whenever $U_i \cap V_j \neq \varnothing$,

$$w_j \circ z_i^{-1} : z_i\left(U_i \cap V_j\right) \to w_j\left(U_i \cap V_j\right)$$

is a biholomorphism. A **complex structure** on a differentiable manifold $\mathcal{M}$ is an equivalence class of systems of holomorphic coordinates. A **complex manifold** is simply a differentiable manifold with a complex structure. A complex manifold has a natural real analytic structure and is orientable (using holomorphic coordinates $\left\{z^\alpha \doteqdot x^\alpha + \sqrt{-1}y^\alpha\right\}_{\alpha=1}^n$, an orientation is determined by requiring that the frame $\left\{\frac{\partial}{\partial x^1}, \frac{\partial}{\partial y^1}, \ldots, \frac{\partial}{\partial x^n}, \frac{\partial}{\partial y^n}\right\}$ be positively oriented). In this chapter and only in this chapter we shall use $\mathcal{M}^n$ to denote a complex manifold $\mathcal{M}$ of *complex* dimension n (half of the real dimension), i.e., $n \doteqdot \dim_{\mathbb{C}} \mathcal{M} = \frac{1}{2} \dim_{\mathbb{R}} \mathcal{M}$.

A real submanifold $\mathcal{N}$ of a complex manifold $\mathcal{M}^n$ is a **complex submanifold** if for every $p \in \mathcal{N}$ there exist holomorphic coordinates $\{z^\alpha\}_{\alpha=1}^n$ in a neighborhood $\mathcal{U}$ of p such that

$$\mathcal{N} \cap \mathcal{U} = \left\{ q \in \mathcal{U} : z^{k+1}(q) = \cdots = z^n(q) = 0 \right\}.$$

Clearly a complex submanifold is itself a complex manifold. In the above definition, k is the complex dimension of $\mathcal{N}$.

Some elementary examples of complex manifolds are

(1) complex Euclidean space $\mathbb{C}^n$,
(2) complex projective space $\mathbb{C}P^n$,
(3) complex submanifold of $\mathbb{C}P^m$ = algebraic manifold = zero set of a finite number of homogeneous polynomials,
(4) complex torus $\mathbb{C}^n / \Gamma$, where Γ is a lattice.

A complex structure defines at each point $p \in \mathcal{M}$ a map $J : T\mathcal{M}_p \to T\mathcal{M}_p$ by $J = d\left(z^{-1} \circ \sqrt{-1} \circ z\right)$, where z are holomorphic coordinates defined in a neighborhood of p (this definition is independent of the choice of z). Since $\sqrt{-1} \circ \sqrt{-1} = -\operatorname{id}_{\mathbb{C}^n}$, it is easy to see $J^2 = -\operatorname{id}_{T\mathcal{M}}$. In general, given an even-dimensional differentiable manifold $\mathcal{M}$, an automorphism $J : T\mathcal{M} \to T\mathcal{M}$ is called an **almost complex structure** if $J^2 = -\operatorname{id}_{T\mathcal{M}}$. (A manifold with an almost complex structure is called an **almost complex manifold**.) Hence a complex structure induces an almost complex structure. We say that an almost complex structure is **integrable** if there exists a complex structure which induces the almost complex structure.

Given an almost complex manifold $(\mathcal{M}, J)$, the **Nijenhuis tensor** is defined by

$$N_J(X, Y) \doteqdot [JX, JY] - J[JX, Y] - J[X, JY] - [X, Y]$$

for $X, Y \in T\mathcal{M}$. By the **Newlander-Nirenberg Theorem**, a necessary and sufficient condition that an almost complex structure be integrable is that the Nijenhuis tensor vanish.

1. INTRODUCTION TO KÄHLER MANIFOLDS

It is interesting that $\mathcal{S}^6$ admits an almost complex structure. In fact $\mathcal{S}^2$ and $\mathcal{S}^6$ are the only even-dimensional spheres which admit almost complex structures (see §8 of Chern [95] for example). The following is a long-standing unsolved question.

PROBLEM 2.1 (Existence of complex structure on $\mathcal{S}^6$). Does $\mathcal{S}^6$ admit a complex structure? More generally, one may ask which closed even-dimensional manifolds admit almost complex structures but not complex structures.

Let $(\mathcal{M}, J)$ be an almost complex manifold. A vector field V is an **infinitesimal automorphism of the almost complex structure** if the Lie derivative of J with respect to V is zero, i.e.,

$$\mathcal{L}_V J = 0. \tag{2.1}$$

Note that

$$(\mathcal{L}_V J)(W) = \mathcal{L}_V (JW) - J(\mathcal{L}_V W)$$
$$= [V, JW] - J([V, W]),$$

so (2.1) is equivalent to $J([V, W]) = [V, JW]$ for any vector field W.

There are various equivalent ways to define a Kähler manifold. We say that a Riemannian manifold $(\mathcal{M}, g)$ with an almost complex structure $J : T\mathcal{M} \to T\mathcal{M}$ is a **Kähler manifold** if the metric g is J-**invariant** (or **Hermitian**):

$$g(JX, JY) = g(X, Y)$$

and J is parallel:

$$\nabla J = 0 \quad \text{or equivalently,} \quad \nabla_X (JY) = J(\nabla_X Y)$$

for all X, Y, where ∇ is the Riemannian covariant derivative.[2] The metric g is called a **Kähler metric**.

Note that almost complex structures which yield Kähler manifolds are necessarily integrable. Indeed, by the Newlander-Nirenberg Theorem, we only need to check that the Nijenhuis tensor vanishes for a Kähler manifold:

$$N_J(X, Y) = \nabla_{JX} JY - \nabla_{JY} JX - J(\nabla_{JX} Y - \nabla_Y JX)$$
$$- J(\nabla_X JY - \nabla_{JY} X) - \nabla_X Y + \nabla_Y X$$
$$= -[J(\nabla_{JX} Y) - \nabla_{JX} JY] + [J(\nabla_{JY} X) - \nabla_{JY} JX]$$
$$+ [J(\nabla_Y JX) - \nabla_Y J(JX)] - [J(\nabla_X JY) - \nabla_X J(JY)]$$
$$= 0.$$

A complex manifold $(\mathcal{M}, J)$ is called a Kähler manifold if it admits a Kähler metric.

---

[2]Since $0 = (\nabla_X J)(Y) = \nabla_X (JY) - J(\nabla_X Y)$.

Given an almost complex manifold $(\mathcal{M}, J)$, the **complexified tangent bundle** is $T_{\mathbb{C}}\mathcal{M} \doteqdot T\mathcal{M} \otimes_{\mathbb{R}} \mathbb{C}$. For each $p \in \mathcal{M}$ we may extend the almost complex structure J to a complex linear map

$$J_{\mathbb{C}} : T_{\mathbb{C}}\mathcal{M}_p \to T_{\mathbb{C}}\mathcal{M}_p.$$

Since $(J_{\mathbb{C}})^2 = -\operatorname{id}_{T_{\mathbb{C}}\mathcal{M}_p}$, the eigenvalues of $J_{\mathbb{C}}$ are $\sqrt{-1}$ and $-\sqrt{-1}$. We define the **holomorphic tangent bundle** by

$$T^{1,0}\mathcal{M} \doteqdot \left\{ V \in T_{\mathbb{C}}\mathcal{M} : J_{\mathbb{C}}(V) = \sqrt{-1} V \right\}$$

and the **anti-holomorphic tangent bundle** by

$$T^{0,1}\mathcal{M} \doteqdot \left\{ V \in T_{\mathbb{C}}\mathcal{M} : J_{\mathbb{C}}(V) = -\sqrt{-1} V \right\}.$$

This gives us a decomposition

$$T_{\mathbb{C}}\mathcal{M} = T^{1,0}\mathcal{M} \oplus T^{0,1}\mathcal{M}.$$

A vector in $T^{1,0}\mathcal{M}$ is called type $(1,0)$ and a vector in $T^{0,1}\mathcal{M}$ is called type $(0,1)$. Alternatively, we can obtain the above decomposition by decomposing a (real) tangent vector $V \in T\mathcal{M}$ as $V = V^{1,0} + V^{0,1}$, where

$$V^{1,0} \doteqdot \frac{1}{2}\left(V - \sqrt{-1} JV\right) \in T^{1,0}\mathcal{M},$$

$$V^{0,1} \doteqdot \frac{1}{2}\left(V + \sqrt{-1} JV\right) \in T^{0,1}\mathcal{M}.$$

I.e.,

$$T^{1,0}\mathcal{M} = \left\{ V^{1,0} : V \in T\mathcal{M} \right\},$$
$$T^{0,1}\mathcal{M} = \left\{ V^{0,1} : V \in T\mathcal{M} \right\}.$$

Since $T_{\mathbb{C}}\mathcal{M}$ is the complexification of a real vector space, the **complex conjugate** (bar operation), $\overline{X + \sqrt{-1} Y} \doteqdot X - \sqrt{-1} Y$, where $X, Y \in T\mathcal{M}$, is defined on $T_{\mathbb{C}}\mathcal{M}$. Given $Z \in T_{\mathbb{C}}\mathcal{M}$, the **real part** of Z is defined by

$$\operatorname{Re}(Z) \doteqdot \frac{1}{2}\left(Z + \bar{Z}\right).$$

Note that if $V \in T\mathcal{M}$, then

$$2\operatorname{Re}\left(V^{1,0}\right) = 2\operatorname{Re}\left(V^{0,1}\right) = V.$$

The almost complex structure satisfies

$$J_{\mathbb{C}}(\bar{V}) = \overline{J_{\mathbb{C}}(V)}, \quad V \in T_{\mathbb{C}}\mathcal{M},$$

and we have $\overline{T^{1,0}\mathcal{M}} = T^{0,1}\mathcal{M}$ and $\overline{T^{0,1}\mathcal{M}} = T^{1,0}\mathcal{M}$.

The **complexified cotangent bundle** is $T_{\mathbb{C}}^*\mathcal{M} = (T_{\mathbb{C}}\mathcal{M})^* = T^*\mathcal{M} \otimes_{\mathbb{R}} \mathbb{C}$, which decomposes into

$$T_{\mathbb{C}}^*\mathcal{M} = \Lambda^{1,0}\mathcal{M} \oplus \Lambda^{0,1}\mathcal{M},$$

where $\Lambda^{1,0}\mathcal{M} \doteqdot \left(T^{1,0}\mathcal{M}\right)^*$ and $\Lambda^{0,1}\mathcal{M} \doteqdot \left(T^{0,1}\mathcal{M}\right)^*$.

A covariant tensor of **type** (p,q) is a section of $\bigotimes^{p,q} \mathcal{M} \doteq \left(\bigotimes^p \Lambda^{1,0}\mathcal{M}\right) \otimes \left(\bigotimes^q \Lambda^{0,1}\mathcal{M}\right)$.[3] The complex conjugate extends to the tensor bundles and

$$\overline{\bigotimes^{p,q}\mathcal{M}} = \bigotimes^{q,p}\mathcal{M}.$$

We say that a differential $(p+q)$-form η is of **type** (p,q) if it is a section of the vector bundle $\Lambda^{p,q}\mathcal{M} \doteq \left(\Lambda^p T^{1,0}\mathcal{M}\right) \wedge \left(\Lambda^q T^{0,1}\mathcal{M}\right) \subset \Lambda^{p+q}T_{\mathbb{C}}\mathcal{M}$. We denote the space of (p,q)-forms by $\Omega^{p,q}(\mathcal{M})$. We say that a (p,p)-tensor (or form) η is **real** if $\bar{\eta} = \eta$.

When we are considering complex manifolds, it is most natural to carry out calculations in holomorphic coordinates. Let $\{z^\alpha\}$ be local holomorphic coordinates. We may write $z^\alpha \doteq x^\alpha + \sqrt{-1}y^\alpha$, where x^α and y^α are real-valued functions. Define

$$dz^\alpha \doteq dx^\alpha + \sqrt{-1}dy^\alpha,$$
$$d\bar{z}^\alpha \doteq dx^\alpha - \sqrt{-1}dy^\alpha$$

and

$$\frac{\partial}{\partial z^\alpha} \doteq \frac{1}{2}\left(\frac{\partial}{\partial x^\alpha} - \sqrt{-1}\frac{\partial}{\partial y^\alpha}\right),$$
$$\frac{\partial}{\partial \bar{z}^\alpha} \doteq \frac{1}{2}\left(\frac{\partial}{\partial x^\alpha} + \sqrt{-1}\frac{\partial}{\partial y^\alpha}\right),$$

so that

$$dz^\alpha\left(\frac{\partial}{\partial z^\beta}\right) = \delta^\alpha_\beta, \qquad d\bar{z}^\alpha\left(\frac{\partial}{\partial \bar{z}^\beta}\right) = \delta^\alpha_\beta,$$
$$dz^\alpha\left(\frac{\partial}{\partial \bar{z}^\beta}\right) = d\bar{z}^\alpha\left(\frac{\partial}{\partial z^\beta}\right) = 0.$$

Note that $\overline{\frac{\partial}{\partial z^\alpha}} = \frac{\partial}{\partial \bar{z}^\alpha}$ and $\overline{\frac{\partial}{\partial \bar{z}^\beta}} = \frac{\partial}{\partial z^\beta}$. The holomorphic tangent bundle $T^{1,0}\mathcal{M}$ is locally the span of the vectors $\left\{\frac{\partial}{\partial z^\alpha}\right\}_{\alpha=1}^n$. Given two overlapping local holomorphic coordinates $\{z^\alpha\}$ and $\{w^\beta\}$, the transition matrix relating $\frac{\partial}{\partial z^\alpha}$ and $\frac{\partial}{\partial w^\beta}$ is $\left\{\frac{\partial z^\alpha}{\partial w^\beta}\right\}_{\alpha,\beta=1}^n$, which satisfies

$$\frac{\partial}{\partial w^\beta} = \sum_{\alpha=1}^n \frac{\partial z^\alpha}{\partial w^\beta} \cdot \frac{\partial}{\partial z^\alpha}.$$

The anti-holomorphic tangent bundle $T^{0,1}\mathcal{M}$ is locally the span of $\left\{\frac{\partial}{\partial \bar{z}^\beta}\right\}_{\beta=1}^n$.

NOTATION 2.2. *Henceforth we shall use the **Einstein summation convention** where each pair of repeated indices consisting of an upper index and a lower index is summed from 1 to n (sometimes, as a reminder, we include the summation symbol, however we always sum over repeated indices unless otherwise indicated).*

---

[3]Not to be confused with a (p,q)-tensor in Riemannian geometry which is a section of $\left(\bigotimes^p T^*M\right) \otimes \left(\bigotimes^q TM\right)$.

EXERCISE 2.3. Given a (p,q)-form
$$\eta = \eta_{i_1\cdots i_p \bar{j}_1 \cdots \bar{j}_q} dz^{i_1} \wedge \cdots \wedge dz^{i_p} \wedge dz^{\bar{j}_1} \wedge \cdots \wedge dz^{\bar{j}_q},$$
show that its complex conjugate is given in local coordinates by
$$\bar{\eta} \doteqdot (-1)^{pq} \overline{\eta_{i_1\cdots i_p \bar{j}_1 \cdots \bar{j}_q}} dz^{j_1} \wedge \cdots \wedge dz^{j_q} \wedge dz^{\bar{i}_1} \wedge \cdots \wedge dz^{\bar{i}_p} \in \Omega^{q,p}(\mathcal{M}).$$

The exterior differentiation operator d maps $\Lambda^{p,q}\mathcal{M}$ into $\Lambda^{p+1,q}\mathcal{M} \oplus \Lambda^{p,q+1}\mathcal{M}$. Corresponding to this decomposition of the image of d, we have
$$d \doteqdot \partial + \bar{\partial},$$
where
$$\partial : \Lambda^{p,q}\mathcal{M} \to \Lambda^{p+1,q}\mathcal{M},$$
$$\bar{\partial} : \Lambda^{p,q}\mathcal{M} \to \Lambda^{p,q+1}\mathcal{M}.$$

We extend the Riemannian metric g complex linearly to define
$$g_{\mathbb{C}} : T_{\mathbb{C}}\mathcal{M}_p \times T_{\mathbb{C}}\mathcal{M}_p \to \mathbb{C}.$$
Similarly
$$\nabla_{\mathbb{C}} : T_{\mathbb{C}}\mathcal{M} \times C^{\infty}(T_{\mathbb{C}}\mathcal{M}) \to C^{\infty}(T_{\mathbb{C}}\mathcal{M})$$
is the complex linear extension of $\nabla : T\mathcal{M} \times C^{\infty}(T\mathcal{M}) \to C^{\infty}(T\mathcal{M})$ with the convention that $\nabla_X Y \doteqdot \nabla(X,Y)$ and $(\nabla_{\mathbb{C}})_X Y \doteqdot \nabla_{\mathbb{C}}(X,Y)$. The complex linear extension of Rm is denoted by $\mathrm{Rm}_{\mathbb{C}}$.

Let
$$g_{\alpha\bar{\beta}} \doteqdot g_{\mathbb{C}}\left(\frac{\partial}{\partial z^{\alpha}}, \frac{\partial}{\partial \bar{z}^{\beta}}\right), \qquad g_{\bar{\beta}\alpha} \doteqdot g_{\mathbb{C}}\left(\frac{\partial}{\partial \bar{z}^{\beta}}, \frac{\partial}{\partial z^{\alpha}}\right).$$

Since $\overline{g_{\mathbb{C}}(V,W)} = g_{\mathbb{C}}(\bar{V},\bar{W}) = g_{\mathbb{C}}(\bar{W},\bar{V})$, these coefficients satisfy the Hermitian condition:
$$g_{\alpha\bar{\beta}} = \overline{g_{\beta\bar{\alpha}}} = g_{\bar{\beta}\alpha}.$$

Similarly, we define
$$g_{\alpha\beta} \doteqdot g_{\mathbb{C}}\left(\frac{\partial}{\partial z^{\alpha}}, \frac{\partial}{\partial z^{\beta}}\right), \qquad g_{\bar{\alpha}\bar{\beta}} \doteqdot g_{\mathbb{C}}\left(\frac{\partial}{\partial \bar{z}^{\alpha}}, \frac{\partial}{\partial \bar{z}^{\beta}}\right).$$

We claim $g_{\alpha\beta} = g_{\bar{\alpha}\bar{\beta}} = 0$. Indeed, if $X, Y \in T^{1,0}\mathcal{M}$, then
$$g_{\mathbb{C}}(X,Y) = g_{\mathbb{C}}(JX, JY) = g_{\mathbb{C}}(\sqrt{-1}X, \sqrt{-1}Y) = -g_{\mathbb{C}}(X,Y),$$
which implies $g_{\mathbb{C}}(X,Y) = 0$. Similarly, if $X, Y \in T^{0,1}\mathcal{M}$, then we also have $g_{\mathbb{C}}(X,Y) = 0$. Thus, in local holomorphic coordinates, the Kähler metric takes the form
$$g_{\mathbb{C}} = g_{\alpha\bar{\beta}}\left(dz^{\alpha} \otimes d\bar{z}^{\beta} + d\bar{z}^{\beta} \otimes dz^{\alpha}\right).$$

EXERCISE 2.4. Show that if $a = \sqrt{-1} a_{\alpha\bar{\beta}} dz^{\alpha} \wedge d\bar{z}^{\beta}$ is a $(1,1)$-form, then a is real if and only if
$$\overline{a_{\alpha\bar{\beta}}} = a_{\beta\bar{\alpha}}.$$

Given a J-invariant symmetric 2-tensor b, we define a J-invariant 2-form β by
$$\beta(X,Y) \doteq b(JX,Y).$$
(We check that $\beta(Y,X) = b(JY,X) = b(-Y,JX) = -\beta(X,Y)$.) We call β the **associated 2-form** to the symmetric 2-tensor b. The **Kähler form** ω, on a Riemannian manifold $(\mathcal{M}, J, g)$ with an almost complex structure and whose metric is Hermitian, is defined to be the 2-form associated to g:
$$\omega(X,Y) \doteq g(JX,Y),$$
which is a real $(1,1)$-form.

EXERCISE 2.5. Show that if g is a Kähler metric, then ω is a closed 2-form. In fact, ω is parallel.

HINT: See [**27**], Proposition 2.29 on p. 70.

SOLUTION: We compute
$$\begin{aligned}(\nabla_X \omega)(Y,Z) &= X(\omega(Y,Z)) - \omega(\nabla_X Y, Z) - \omega(Y, \nabla_X Z) \\ &= X(g(JY,Z)) - g(J(\nabla_X Y), Z) - g(JY, \nabla_X Z) \\ &= (\nabla_X g)(JY,Z) = 0,\end{aligned}$$
where we used the definition of ω, $J(\nabla_X Y) = \nabla_X(JY)$, and g is parallel.

On a Kähler manifold $(\mathcal{M}, J, g)$, in local holomorphic coordinates, the Kähler form is
$$\omega = \sqrt{-1}\sum_{\alpha,\beta=1}^{n} g_{\alpha\bar{\beta}} dz^\alpha \wedge d\bar{z}^\beta.$$

REMARK 2.6. In our convention, $dz^\alpha \wedge d\bar{z}^\beta \left(\frac{\partial}{\partial z^\gamma}, \frac{\partial}{\partial \bar{z}^\delta}\right) = \frac{1}{2}\delta^\alpha_\gamma \delta^\beta_\delta$. Note that for the standard Euclidean $dx^2 + dy^2$ metric on $\mathbb{C}^n$, we have $g_{\alpha\bar{\beta}} = \frac{1}{2}\delta_{\alpha\beta}$ and $g_{\mathbb{C}} = \frac{1}{2}\sum_{\alpha=1}^{n}(dz^\alpha \otimes d\bar{z}^\alpha + d\bar{z}^\alpha \otimes dz^\alpha)$.

Since g is Kähler, we have the **Kähler identities**:
$$(2.2) \qquad \frac{\partial}{\partial z^\gamma} g_{\alpha\bar{\beta}} = \frac{\partial}{\partial z^\alpha} g_{\gamma\bar{\beta}},$$
which are equivalent to $d\omega = 0$. The real cohomology class
$$[\omega] \in H^{1,1}(\mathcal{M};\mathbb{R}) \subseteq H^2(\mathcal{M};\mathbb{R})$$
is called the **Kähler class** of ω.

EXERCISE 2.7 (Characterization of Kähler condition). If $(\mathcal{M}, J, g)$ is a triple consisting of a Riemannian manifold, an almost complex structure, and a Hermitian metric, then $d\omega = 0$ if and only if $\nabla J = 0$. That is, g is Kähler if and only if the real $(1,1)$-form ω is closed.

An oriented Riemannian surface $(\mathcal{M}, g)$ has a natural complex structure and g is a Kähler metric with respect to this complex structure. In particular, define the almost complex structure $J : T\mathcal{M} \to T\mathcal{M}$ as counterclockwise rotation by $90°$ with respect to the orientation and metric (clearly $J^2 = -\text{id}_{T\mathcal{M}}$). Since $\dim_{\mathbb{R}} \mathcal{M} = 2$, we have $d\omega = 0$, which implies $\nabla J = 0$.

2. Connection, curvature, and covariant differentiation

In this section we consider the connection, curvature, and covariant differentiation on a Kähler manifold $(\mathcal{M}^n, J, g)$. Our emphasis, in the style of the book by Morrow and Kodaira [**275**], is to calculate geometric quantities in local holomorphic coordinates, where the formulas are particularly elegant and simpler than their Riemannian counterparts. These calculations shall prove useful in our study of the Kähler–Ricci flow.

The **Christoffel symbols** of the **Levi-Civita connection**, defined by

$$(\nabla_{\mathbb{C}})_{\frac{\partial}{\partial z^\alpha}} \frac{\partial}{\partial z^\beta} \doteq \sum_{\gamma=1}^{n} \left(\Gamma^{\gamma}_{\alpha\beta} \frac{\partial}{\partial z^\gamma} + \Gamma^{\bar\gamma}_{\alpha\beta} \frac{\partial}{\partial \bar z^\gamma} \right),$$

$$(\nabla_{\mathbb{C}})_{\frac{\partial}{\partial z^\alpha}} \frac{\partial}{\partial \bar z^\beta} \doteq \sum_{\gamma=1}^{n} \left(\Gamma^{\gamma}_{\alpha\bar\beta} \frac{\partial}{\partial z^\gamma} + \Gamma^{\bar\gamma}_{\alpha\bar\beta} \frac{\partial}{\partial \bar z^\gamma} \right),$$

etc., are zero unless all the indices are unbarred or all the indices are barred. To see this, from the Riemannian formula for the Christoffel symbols and since we are complex linearly extending the covariant derivative, we have the following using (2.2).

LEMMA 2.8 (Kähler Christoffel symbols). *Let $g^{\alpha\bar\beta}$ be defined by $g_{\gamma\bar\beta} g^{\alpha\bar\beta} = \delta^{\alpha}_{\gamma}$. Then, in holomorphic coordinates, we have*

(2.3) $$\Gamma^{\gamma}_{\alpha\beta} = \frac{1}{2} g^{\gamma\bar\delta} \left(\frac{\partial}{\partial z^\alpha} g_{\beta\bar\delta} + \frac{\partial}{\partial z^\beta} g_{\alpha\bar\delta} - \frac{\partial}{\partial \bar z^\delta} g_{\alpha\beta} \right) = g^{\gamma\bar\delta} \frac{\partial}{\partial z^\alpha} g_{\beta\bar\delta}$$

and

$$\Gamma^{\gamma}_{\alpha\beta} = \Gamma^{\gamma}_{\beta\alpha}$$

(in fact, the last equality is equivalent to the Kähler condition).[4] *Similarly*

$$\Gamma^{\gamma}_{\alpha\bar\beta} = \frac{1}{2} g^{\gamma\bar\delta} \left(\frac{\partial}{\partial z^\alpha} g_{\bar\beta\bar\delta} + \frac{\partial}{\partial \bar z^\beta} g_{\alpha\bar\delta} - \frac{\partial}{\partial \bar z^\delta} g_{\alpha\bar\beta} \right) = 0,$$

$\Gamma^{\gamma}_{\bar\alpha\bar\beta} = 0$, *and so forth.*

EXERCISE 2.9. Show that

$$\Gamma^{\bar\gamma}_{\bar\alpha\bar\beta} = \overline{\Gamma^{\gamma}_{\alpha\beta}}.$$

---

[4]We leave this as an exercise or see for example Theorem 5.1 in [**275**].

2. CONNECTION, CURVATURE, AND COVARIANT DIFFERENTIATION

SOLUTION. We compute
$$\Gamma^{\bar{\gamma}}_{\bar{\alpha}\bar{\beta}} = \frac{1}{2}g^{\bar{\gamma}\delta}\left(\frac{\partial}{\partial \bar{z}^\alpha}g_{\bar{\beta}\delta} + \frac{\partial}{\partial \bar{z}^\beta}g_{\bar{\alpha}\delta} - \frac{\partial}{\partial z^\delta}g_{\bar{\alpha}\bar{\beta}}\right) = g^{\bar{\gamma}\delta}\frac{\partial}{\partial \bar{z}^\alpha}g_{\bar{\beta}\delta}$$
$$= \overline{g^{\gamma\bar{\delta}}\frac{\partial}{\partial z^\alpha}g_{\beta\bar{\delta}}} = \overline{\Gamma^{\gamma}_{\alpha\beta}}.$$

Similarly to $g_{\mathbb{C}}$ and $\nabla_{\mathbb{C}}$, we may define $\mathrm{Rm}_{\mathbb{C}}$ and $\mathrm{Rc}_{\mathbb{C}}$ as the complex multilinear extensions of Rm and Rc. The components of the curvature $(3,1)$-tensor $\mathrm{Rm}_{\mathbb{C}}$ are defined by

$$\mathrm{Rm}_{\mathbb{C}}\left(\frac{\partial}{\partial z^\alpha}, \frac{\partial}{\partial \bar{z}^\beta}\right)\frac{\partial}{\partial z^\gamma} \doteqdot R^{\delta}_{\alpha\bar{\beta}\gamma}\frac{\partial}{\partial z^\delta} + R^{\bar{\delta}}_{\alpha\bar{\beta}\gamma}\frac{\partial}{\partial \bar{z}^\delta},$$

$$\mathrm{Rm}_{\mathbb{C}}\left(\frac{\partial}{\partial z^\alpha}, \frac{\partial}{\partial \bar{z}^\beta}\right)\frac{\partial}{\partial \bar{z}^\gamma} \doteqdot R^{\delta}_{\alpha\bar{\beta}\bar{\gamma}}\frac{\partial}{\partial z^\delta} + R^{\bar{\delta}}_{\alpha\bar{\beta}\bar{\gamma}}\frac{\partial}{\partial \bar{z}^\delta},$$

etc. As a $(4,0)$-tensor the components of $\mathrm{Rm}_{\mathbb{C}}$, which are defined by

$$R_{\alpha\bar{\beta}\gamma\bar{\delta}} \doteqdot \mathrm{Rm}_{\mathbb{C}}\left(\frac{\partial}{\partial z^\alpha}, \frac{\partial}{\partial \bar{z}^\beta}, \frac{\partial}{\partial z^\gamma}, \frac{\partial}{\partial \bar{z}^\delta}\right),$$

satisfy

$$R_{\alpha\bar{\beta}\gamma\bar{\delta}} = g_{\eta\bar{\delta}}R^{\eta}_{\alpha\bar{\beta}\gamma}, \quad R_{\alpha\bar{\beta}\bar{\gamma}\delta} = g_{\delta\bar{\eta}}R^{\bar{\eta}}_{\alpha\bar{\beta}\bar{\gamma}},$$

etc.

The only nonvanishing components of the curvature $(3,1)$-tensor are

$$R^{\delta}_{\alpha\bar{\beta}\gamma}, \ R^{\bar{\delta}}_{\alpha\bar{\beta}\bar{\gamma}}, \ R^{\delta}_{\bar{\alpha}\beta\gamma}, \ R^{\bar{\delta}}_{\bar{\alpha}\beta\bar{\gamma}}.$$

In particular, $R^{\delta}_{\alpha\beta\gamma} = R^{\delta}_{\alpha\beta\bar{\gamma}} = 0$, etc. Hence, the only nonvanishing components of the curvature $(4,0)$-tensor are

$$R_{\alpha\bar{\beta}\gamma\bar{\delta}}, \ R_{\alpha\bar{\beta}\bar{\gamma}\delta}, \ R_{\bar{\alpha}\beta\gamma\bar{\delta}}, \ R_{\bar{\alpha}\beta\bar{\gamma}\delta}.$$

Since $\frac{\partial}{\partial z^\alpha}\Gamma^{\delta}_{\bar{\beta}\gamma} = 0$, we have

$$(2.4) \qquad R^{\delta}_{\alpha\bar{\beta}\gamma} = -\frac{\partial}{\partial \bar{z}^\beta}\Gamma^{\delta}_{\alpha\gamma} = -\frac{\partial}{\partial z^\alpha}g_{\gamma\bar{\eta}}\frac{\partial}{\partial \bar{z}^\beta}g^{\delta\bar{\eta}} - g^{\delta\bar{\eta}}\frac{\partial^2}{\partial z^\alpha \partial \bar{z}^\beta}g_{\gamma\bar{\eta}},$$

and thus we have the following lemma.

LEMMA 2.10 (Kähler Rm). *In holomorphic coordinates, the components of $\mathrm{Rm}_{\mathbb{C}}$ are given by*

$$(2.5) \qquad R_{\alpha\bar{\beta}\gamma\bar{\delta}} = -\frac{\partial^2}{\partial z^\alpha \partial \bar{z}^\beta}g_{\gamma\bar{\delta}} + g^{\lambda\bar{\mu}}\frac{\partial}{\partial z^\alpha}g_{\gamma\bar{\mu}}\frac{\partial}{\partial \bar{z}^\beta}g_{\lambda\bar{\delta}}.$$

We have the identities

$$R_{\alpha\bar{\beta}\gamma\bar{\delta}} = R_{\gamma\bar{\beta}\alpha\bar{\delta}} = R_{\alpha\bar{\delta}\gamma\bar{\beta}} = R_{\gamma\bar{\delta}\alpha\bar{\beta}}$$

and

$$\overline{R_{\alpha\bar{\beta}\gamma\bar{\delta}}} = R_{\beta\bar{\alpha}\delta\bar{\gamma}}.$$

The vanishing of some of the components of the curvature $(4,0)$-tensor is related to the following.

EXERCISE 2.11 (*J*-invariance of curvature). Show that for a Kähler manifold $\operatorname{Rm}(X, Y)$ is *J*-invariant, i.e., $\operatorname{Rm}(X, Y) JZ = J(\operatorname{Rm}(X, Y) Z)$.

SOLUTION. We compute
$$\begin{aligned}\operatorname{Rm}(X, Y) JZ &= \nabla_X \nabla_Y JZ - \nabla_Y \nabla_X JZ - \nabla_{[X,Y]} JZ \\ &= \nabla_X (J\nabla_Y Z) - \nabla_Y (J\nabla_X Z) - J(\nabla_{[X,Y]} Z) \\ &= J(\nabla_X \nabla_Y Z) - J(\nabla_Y \nabla_X Z) - J(\nabla_{[X,Y]} Z) \\ &= J(\operatorname{Rm}(X, Y) Z).\end{aligned}$$

The components of $\operatorname{Rc}_{\mathbb{C}}$ are defined similarly:
$$\operatorname{Rc}_{\mathbb{C}}\left(\frac{\partial}{\partial z^\alpha}, \frac{\partial}{\partial \bar{z}^\beta}\right) \doteqdot R_{\alpha\bar{\beta}},$$
$$\operatorname{Rc}_{\mathbb{C}}\left(\frac{\partial}{\partial z^\alpha}, \frac{\partial}{\partial z^\beta}\right) \doteqdot R_{\alpha\beta},$$

etc. Tracing (2.4), we see that components of the Ricci tensor $R_{\alpha\bar{\beta}} = R^\delta_{\delta\bar{\beta}\alpha}$ are given by

LEMMA 2.12 (Kähler Rc).

(2.6) $$R_{\alpha\bar{\beta}} = -\frac{\partial^2}{\partial z^\alpha \partial \bar{z}^\beta} \log \det\left(g_{\gamma\bar{\delta}}\right).$$

It is easy to see that $R_{\alpha\bar{\beta}} = \overline{R_{\beta\bar{\alpha}}}$ and $R_{\alpha\beta} = R_{\bar{\alpha}\bar{\beta}} = 0$.

EXERCISE 2.13. Show that for a Kähler manifold Rc is *J*-invariant, i.e., $\operatorname{Rc}(JX, JY) = \operatorname{Rc}(X, Y)$.

The **Ricci form** ρ is the 2-form associated to Rc,
$$\rho(X, Y) \doteqdot \frac{1}{2} \operatorname{Rc}(JX, Y),$$
which is also a real $(1, 1)$-form.

EXERCISE 2.14. Show that if g is a Kähler metric, then the Ricci form ρ is a closed 2-form.

HINT: See Proposition 2.47 on p. 74 of [**27**].

The real de Rham cohomology class $\left[\frac{1}{2\pi}\rho\right] \doteqdot c_1(\mathcal{M})$ is the **first Chern class** of $\mathcal{M}$. It is a beautiful fact that $\left[\frac{1}{2\pi}\rho\right]$ only depends on the complex structure of $\mathcal{M}$.

The Ricci form, which is a real $(1, 1)$-form, is in holomorphic coordinates:
$$\rho = \sqrt{-1} R_{\alpha\bar{\beta}} dz^\alpha \wedge d\bar{z}^\beta.$$

From (2.10) below and $\Gamma^\delta_{\alpha\beta} = \Gamma^\delta_{\beta\alpha}$ we have
$$\frac{\partial}{\partial z^\alpha} R_{\beta\bar{\gamma}} = \frac{\partial}{\partial z^\beta} R_{\alpha\bar{\gamma}},$$

which is equivalent to the Ricci form being closed: $d\rho = 0$. We may express (2.6) as
$$\rho = -\sqrt{-1}\partial\bar{\partial}\log\det\left(g_{\gamma\bar{\delta}}\right).$$
The **complex scalar curvature** is defined to be
$$R \doteqdot g^{\alpha\bar{\beta}} R_{\alpha\bar{\beta}}.$$
The complex scalar curvature is one-half of the Riemannian scalar curvature. (Exercise: Prove this.)

NOTATION 2.15. *In this chapter, R always denotes the complex scalar curvature whereas in the rest of the book, R always denotes the Riemannian scalar curvature. In this chapter we shall usually refer to the complex scalar curvature simply as the scalar curvature.*

Holomorphic coordinates $\{z^\alpha\}$ are said to be **unitary at a point** p if $g\left(\frac{\partial}{\partial z^\alpha}, \frac{\partial}{\partial \bar{z}^\beta}\right)(p) = \delta_{\alpha\beta}$. That is, $\left\{\frac{\partial}{\partial z^\alpha}(p)\right\}_{\alpha=1}^n$ is a unitary frame for $T^{1,0}\mathcal{M}_p$. In such a frame we have at p,

$$(2.7) \qquad R_{\alpha\bar{\beta}} = \sum_{\delta=1}^n R_{\alpha\bar{\beta}\delta\bar{\delta}} \qquad \text{and} \qquad R = \sum_{\alpha=1}^n R_{\alpha\bar{\alpha}}.$$

Note that for the standard Euclidean metric
$$g_{\mathbb{C}} = \frac{1}{2}\sum_{\alpha=1}^n \left(dw^\alpha \otimes d\bar{w}^\alpha + d\bar{w}^\alpha \otimes dw^\alpha\right)$$
on $\mathbb{C}^n$ the new coordinates $\left\{z^\alpha \doteqdot \frac{1}{\sqrt{2}}w^\alpha\right\}_{\alpha=1}^n$ are unitary.

Given $Z \in T^{1,0}\mathcal{M} - \{\vec{0}\}$, let $X \doteqdot \operatorname{Re}(Z) = \frac{1}{2}(Z + \bar{Z}) \in T\mathcal{M}$. The **holomorphic sectional curvature** in the direction Z is defined to be

$$(2.8) \qquad K_{\mathbb{C}}(Z) \doteqdot \frac{\operatorname{Rm}(X, JX, JX, X)}{|X|^4}$$
$$= \frac{\operatorname{Rm}\left(Z+\bar{Z}, \sqrt{-1}(Z-\bar{Z}), \sqrt{-1}(Z-\bar{Z}), Z+\bar{Z}\right)}{4|Z|^4}$$
$$= \frac{\operatorname{Rm}_{\mathbb{C}}(Z, \bar{Z}, Z, \bar{Z})}{|Z|^4}.$$

Hence $K_{\mathbb{C}}(Z) = \operatorname{Rm}_{\mathbb{C}}(V, \bar{V}, V, \bar{V})$, where $V \doteqdot Z/|Z|$. In particular, if $\{z^\alpha\}$ is unitary at p, then $K_{\mathbb{C}}\left(\frac{\partial}{\partial z^\alpha}\right) = R_{\alpha\bar{\alpha}\alpha\bar{\alpha}}$. We say that the holomorphic sectional curvature is **positive** (respectively, **nonnegative**) if $K_{\mathbb{C}}(Z) > 0$ (respectively, $K_{\mathbb{C}}(Z) \geq 0$) for all $Z \in T^{1,0}\mathcal{M} - \{\vec{0}\}$. By (2.8), positive (respectively, nonnegative) Riemannian sectional curvature implies positive (respectively, nonnegative) holomorphic sectional curvature.

Given $Z, W \in T^{1,0}\mathcal{M} - \{\vec{0}\}$, let $X \doteqdot \operatorname{Re}(Z)$ and $U \doteqdot \operatorname{Re}(W)$. The **(holomorphic) bisectional curvature** in the directions (Z, W) is defined

to be
$$K_\mathbb{C}(Z,W) \doteqdot \frac{\operatorname{Rm}(X,JX,JU,U)}{|X|^2|U|^2} = \frac{\operatorname{Rm}_\mathbb{C}(Z,\bar{Z},W,\bar{W})}{|Z|^2|W|^2}.$$

Clearly we have $K_\mathbb{C}(Z,Z) = K_\mathbb{C}(Z)$. We say that the bisectional curvature is **positive** (respectively, **nonnegative**) if $K_\mathbb{C}(Z,W) > 0$ (respectively, $K_\mathbb{C}(Z,W) \geq 0$) for all $Z,W \in T^{1,0}\mathcal{M} - \{\vec{0}\}$. Clearly positive (respectively, nonnegative) bisectional sectional curvature implies positive (respectively, nonnegative) holomorphic sectional curvature. We also have the following (see Proposition 1 on p. 32 of [**270**] for example).

LEMMA 2.16. *Positive (respectively, nonnegative) Riemannian sectional curvature implies positive (respectively, nonnegative) bisectional sectional curvature.*

Similarly to the notation for the components of the complex Riemann curvature tensor, we define
$$\nabla_\alpha R_{\beta\bar{\gamma}\delta\bar{\eta}} \doteqdot \left((\nabla_\mathbb{C})_{\frac{\partial}{\partial z^\alpha}} \operatorname{Rm}_\mathbb{C}\right)\left(\frac{\partial}{\partial z^\beta}, \frac{\partial}{\partial \bar{z}^\gamma}, \frac{\partial}{\partial z^\delta}, \frac{\partial}{\partial \bar{z}^\eta}\right).$$

The **second Bianchi identity** says that

(2.9) $$\nabla_\alpha R_{\beta\bar{\gamma}\delta\bar{\eta}} = \nabla_\beta R_{\alpha\bar{\gamma}\delta\bar{\eta}},$$

since
$$\left(\nabla_{\frac{\partial}{\partial z^\alpha}} \operatorname{Rm}_\mathbb{C}\right)\left(\frac{\partial}{\partial z^\beta}, \frac{\partial}{\partial \bar{z}^\gamma}, \frac{\partial}{\partial z^\delta}, \frac{\partial}{\partial \bar{z}^\eta}\right) + \left(\nabla_{\frac{\partial}{\partial \bar{z}^\gamma}} \operatorname{Rm}_\mathbb{C}\right)\left(\frac{\partial}{\partial z^\alpha}, \frac{\partial}{\partial z^\beta}, \frac{\partial}{\partial z^\delta}, \frac{\partial}{\partial \bar{z}^\eta}\right)$$
$$+ \left(\nabla_{\frac{\partial}{\partial z^\beta}} \operatorname{Rm}_\mathbb{C}\right)\left(\frac{\partial}{\partial \bar{z}^\gamma}, \frac{\partial}{\partial z^\alpha}, \frac{\partial}{\partial z^\delta}, \frac{\partial}{\partial \bar{z}^\eta}\right) = 0$$

(i.e., $\nabla_\alpha R_{\beta\bar{\gamma}\delta\bar{\eta}} + \nabla_\beta R_{\bar{\gamma}\alpha\delta\bar{\eta}} + \nabla_{\bar{\gamma}} R_{\alpha\beta\delta\bar{\eta}} = 0$, where $\nabla_{\bar{\gamma}} R_{\alpha\beta\delta\bar{\eta}} = 0$). Taking the trace, we have

(2.10) $$\nabla_\alpha R_{\beta\bar{\gamma}} = \nabla_\beta R_{\alpha\bar{\gamma}}.$$

Note that $\nabla_\alpha R_{\beta\bar{\gamma}} = \frac{\partial}{\partial z^\alpha} R_{\beta\bar{\gamma}} - \Gamma^\delta_{\alpha\beta} R_{\delta\bar{\gamma}}$.

The volume form is $d\mu_g \doteqdot \frac{1}{n!}\omega^n$, where $\omega^n \doteqdot \omega \wedge \cdots \wedge \omega$ (n times) and $n = \dim_\mathbb{C} \mathcal{M}$, which in local coordinates is
$$\omega^n = n!\left(\sqrt{-1}\right)^n \det(g_{\alpha\bar{\beta}}) dz^1 \wedge d\bar{z}^1 \wedge \cdots \wedge dz^n \wedge d\bar{z}^n.$$

The volume is

(2.11) $$\operatorname{Vol}(\mathcal{M}) = \operatorname{Vol}_g(\mathcal{M}) = \frac{1}{n!}\int_\mathcal{M} \omega^n.$$

The total scalar curvature may be written as

(2.12) $$\int_\mathcal{M} R\, d\mu = \frac{1}{(n-1)!} \int_\mathcal{M} \rho \wedge \omega^{n-1}.$$

Let $\nabla_{\bar{\beta}} \doteqdot \nabla_{\partial/\partial \bar{z}^\beta}$. The **Laplacian** acting on tensors is given by

$$(2.13) \quad \Delta \doteqdot \frac{1}{2} g^{\alpha\bar{\beta}} \left(\nabla_\alpha \nabla_{\bar{\beta}} + \nabla_{\bar{\beta}} \nabla_\alpha \right) = \frac{1}{2} \left(\nabla_\alpha \nabla_{\bar{\alpha}} + \nabla_{\bar{\alpha}} \nabla_\alpha \right).$$

NOTATION 2.17. *At this point we have begun to use the **extended Einstein summation convention** where not only pairs of repeated indices consisting of an upper index and a lower index are summed but also pairs of repeated lower indices consisting of a barred index and an unbarred index are summed. For example, $a_{\alpha\bar{\alpha}} \doteqdot \sum_{\alpha,\beta=1}^n g^{\alpha\bar{\beta}} a_{\alpha\bar{\beta}}$. Formulas which are expressed using this convention hold in the literal sense in unitary holomorphic coordinates at a point p. Usually we use the extended Einstein summation convention in lieu of computing in unitary coordinates at a point. In this sense, the formulas in (2.7) and throughout this chapter hold in arbitrary holomorphic coordinates.*

EXERCISE 2.18. Show that the Laplacian defined above in (2.13) is one-half of the Riemannian Laplacian.

Acting on functions, the Laplacian is

$$\Delta = g^{\alpha\bar{\beta}} \nabla_\alpha \nabla_{\bar{\beta}} = g^{\alpha\bar{\beta}} \frac{\partial^2}{\partial z^\alpha \partial \bar{z}^\beta},$$

since, acting on functions, $\nabla_\alpha \nabla_{\bar{\beta}} = \frac{\partial^2}{\partial z^\alpha \partial \bar{z}^\beta} = \nabla_{\bar{\beta}} \nabla_\alpha$. We also note that if ϕ is a real-valued function on $\mathcal{M}$, then

$$(2.14) \quad \frac{1}{2} |\operatorname{grad} \phi|^2 = g^{\alpha\bar{\beta}} \nabla_\alpha \phi \nabla_{\bar{\beta}} \phi \doteqdot |\nabla \phi|^2,$$

where the LHS is the Riemannian gradient. Likewise,

$$(2.15) \quad \frac{1}{2} |\operatorname{Hess} \phi|^2 = \left| \nabla_\alpha \nabla_{\bar{\beta}} \phi \right|^2 + \left| \nabla_\alpha \nabla_\beta \phi \right|^2,$$

where the LHS is the Riemannian Hessian.

In our calculation of evolution equations under the Kähler–Ricci flow we shall often use the following.

LEMMA 2.19 (Kähler commutator formulas). *On a Kähler manifold we have the following commutator formulas for covariant differentiation acting on tensors of type $(1,0)$, $(0,1)$, and $(1,1)$, respectively:*

$$(2.16) \quad \nabla_\alpha \nabla_{\bar{\beta}} a_\gamma - \nabla_{\bar{\beta}} \nabla_\alpha a_\gamma = -R^\delta_{\alpha\bar{\beta}\gamma} a_\delta,$$

$$(2.17) \quad \nabla_\alpha \nabla_{\bar{\beta}} b_{\bar{\gamma}} - \nabla_{\bar{\beta}} \nabla_\alpha b_{\bar{\gamma}} = -R^{\bar{\delta}}_{\alpha\bar{\beta}\bar{\gamma}} b_{\bar{\delta}} = \overline{R^\delta_{\beta\bar{\alpha}\gamma}} b_{\bar{\delta}},$$

$$(2.18) \quad \nabla_\alpha \nabla_{\bar{\beta}} a_{\gamma\bar{\delta}} - \nabla_{\bar{\beta}} \nabla_\alpha a_{\gamma\bar{\delta}} = -R_{\alpha\bar{\beta}\gamma\bar{\eta}} a_{\eta\bar{\delta}} + R_{\alpha\bar{\beta}\eta\bar{\delta}} a_{\gamma\bar{\eta}}.$$

Analogous formulas hold for higher degree tensors and forms (see Exercise 2.22).

REMARK 2.20. Note that $R_{\alpha\bar{\beta}\eta\bar{\delta}} = -R_{\alpha\bar{\beta}\bar{\delta}\eta}$, which exhibits the consistency of the formula above with the analogous formula in Riemannian geometry. We have also used the extended Einstein summation convention.

PROOF. Using the fact that the Christoffel symbols are zero unless all of the indices are unbarred or all are barred, we compute

$$\nabla_{\bar\beta} a_\gamma = \partial_{\bar\beta} a_\gamma$$

and

$$\nabla_\alpha \nabla_{\bar\beta} a_\gamma = \partial_\alpha \partial_{\bar\beta} a_\gamma - \Gamma^\delta_{\alpha\gamma} \partial_{\bar\beta} a_\delta,$$
$$\nabla_{\bar\beta} \nabla_\alpha a_\gamma = \partial_{\bar\beta} \left(\partial_\alpha a_\gamma - \Gamma^\delta_{\alpha\gamma} a_\delta \right).$$

Hence, using (2.4), we have

$$\nabla_\alpha \nabla_{\bar\beta} a_\gamma - \nabla_{\bar\beta} \nabla_\alpha a_\gamma = \partial_{\bar\beta} \Gamma^\delta_{\alpha\gamma} a_\delta = -R^\delta_{\alpha\bar\beta\gamma} a_\delta,$$

which is (2.16). Equation (2.17) is the complex conjugate of (2.16).

If a is a $(1,1)$-tensor, then we compute

$$\nabla_{\bar\beta} a_{\gamma\bar\delta} = \partial_{\bar\beta} a_{\gamma\bar\delta} - \Gamma^{\bar\varepsilon}_{\bar\beta\bar\delta} a_{\gamma\bar\varepsilon},$$
$$\nabla_\alpha a_{\gamma\bar\delta} = \partial_\alpha a_{\gamma\bar\delta} - \Gamma^\eta_{\alpha\gamma} a_{\eta\bar\delta}$$

and

$$\nabla_\alpha \nabla_{\bar\beta} a_{\gamma\bar\delta} = \partial_\alpha \left(\partial_{\bar\beta} a_{\gamma\bar\delta} - \Gamma^{\bar\varepsilon}_{\bar\beta\bar\delta} a_{\gamma\bar\varepsilon} \right) - \Gamma^\eta_{\alpha\gamma} \left(\partial_{\bar\beta} a_{\eta\bar\delta} - \Gamma^{\bar\varepsilon}_{\bar\beta\bar\delta} a_{\eta\bar\varepsilon} \right),$$
$$\nabla_{\bar\beta} \nabla_\alpha a_{\gamma\bar\delta} = \partial_{\bar\beta} \left(\partial_\alpha a_{\gamma\bar\delta} - \Gamma^\eta_{\alpha\gamma} a_{\eta\bar\delta} \right) - \Gamma^{\bar\varepsilon}_{\bar\beta\bar\delta} \left(\partial_\alpha a_{\gamma\bar\varepsilon} - \Gamma^\eta_{\alpha\gamma} a_{\eta\bar\varepsilon} \right).$$

Hence, after some cancellations, we obtain in holomorphic coordinates $\{z^\alpha\}$ at a point p where it is unitary,

$$\nabla_\alpha \nabla_{\bar\beta} a_{\gamma\bar\delta} - \nabla_{\bar\beta} \nabla_\alpha a_{\gamma\bar\delta} = -\partial_\alpha \Gamma^{\bar\varepsilon}_{\bar\beta\bar\delta} a_{\gamma\bar\varepsilon} + \partial_{\bar\beta} \Gamma^\eta_{\alpha\gamma} a_{\eta\bar\delta}$$
$$= R_{\alpha\bar\beta\varepsilon\bar\delta} a_{\gamma\bar\varepsilon} - R_{\alpha\bar\beta\gamma\bar\eta} a_{\eta\bar\delta}.$$

$\square$

LEMMA 2.21 (∇_α and ∇_β commute). *We have $[\nabla_\alpha, \nabla_\beta] = 0$ acting on tensors of any type. That is, if a is a (p,q)-tensor, then*

$$\nabla_\alpha \nabla_\beta a_{\gamma_1 \cdots \gamma_p \bar\delta_1 \cdots \bar\delta_q} - \nabla_\beta \nabla_\alpha a_{\gamma_1 \cdots \gamma_p \bar\delta_1 \cdots \bar\delta_q} = 0.$$

PROOF. We have

$$\nabla_\beta a_{\gamma_1 \cdots \gamma_p \bar\delta_1 \cdots \bar\delta_q} = \partial_\beta a_{\gamma_1 \cdots \gamma_p \bar\delta_1 \cdots \bar\delta_q} - \Gamma^\eta_{\beta\gamma_j} a_{\gamma_1 \cdots \gamma_{j-1} \eta \gamma_{j+1} \cdots \gamma_p \bar\delta_1 \cdots \bar\delta_q}.$$

It is easiest to compute the second covariant derivative in **normal holomorphic coordinates** centered at any given point $p \in \mathcal{M}$, where $\Gamma^\gamma_{\alpha\beta}(p) = 0$. In these coordinates we have at p,

$$\nabla_\alpha \nabla_\beta a_{\gamma_1 \cdots \gamma_p \bar\delta_1 \cdots \bar\delta_q} = \partial_\alpha \left(\partial_\beta a_{\gamma_1 \cdots \gamma_p \bar\delta_1 \cdots \bar\delta_q} - \Gamma^\eta_{\beta\gamma_j} a_{\gamma_1 \cdots \gamma_{j-1} \eta \gamma_{j+1} \cdots \gamma_p \bar\delta_1 \cdots \bar\delta_q} \right)$$
$$= \partial_\alpha \partial_\beta a_{\gamma_1 \cdots \gamma_p \bar\delta_1 \cdots \bar\delta_q} - \left(\partial_\alpha \Gamma^\eta_{\beta\gamma_j} \right) a_{\gamma_1 \cdots \gamma_{j-1} \eta \gamma_{j+1} \cdots \gamma_p \bar\delta_1 \cdots \bar\delta_q}.$$

2. CONNECTION, CURVATURE, AND COVARIANT DIFFERENTIATION

Hence

$$\nabla_\alpha \nabla_\beta a_{\gamma_1 \cdots \gamma_p \bar\delta_1 \cdots \bar\delta_q} - \nabla_\beta \nabla_\alpha a_{\gamma_1 \cdots \gamma_p \bar\delta_1 \cdots \bar\delta_q}$$
$$= \left(\partial_\beta \Gamma^\eta_{\alpha\gamma_j} - \partial_\alpha \Gamma^\eta_{\beta\gamma_j} \right) a_{\gamma_1 \cdots \gamma_{j-1} \eta \gamma_{j+1} \cdots \gamma_p \bar\delta_1 \cdots \bar\delta_q}.$$

Now from (2.3) we have at p,

$$\partial_\beta \Gamma^\eta_{\alpha\gamma_j} - \partial_\alpha \Gamma^\eta_{\beta\gamma_j} = g^{\eta\bar\lambda} \left(\partial_\beta \partial_\alpha g_{\gamma_j \bar\lambda} - \partial_\alpha \partial_\beta g_{\gamma_j \bar\lambda} \right) = 0,$$

and the lemma follows. □

EXERCISE 2.22. Compute the commutator formula for the operator $[\nabla_\alpha, \nabla_{\bar\beta}] \doteq \nabla_\alpha \nabla_{\bar\beta} - \nabla_{\bar\beta} \nabla_\alpha$ acting on (p,q)-tensors and forms.

SOLUTION TO EXERCISE 2.22. If a is a (p,q)-tensor, then

$$\nabla_\alpha \nabla_{\bar\beta} a_{\gamma_1 \cdots \gamma_p \bar\delta_1 \cdots \bar\delta_q} - \nabla_{\bar\beta} \nabla_\alpha a_{\gamma_1 \cdots \gamma_p \bar\delta_1 \cdots \bar\delta_q}$$
$$= -\sum_{i=1}^p R_{\alpha \bar\beta \gamma_i \bar\eta} a_{\gamma_1 \cdots \gamma_{i-1} \eta \gamma_{i+1} \cdots \gamma_p \bar\delta_1 \cdots \bar\delta_q}$$
$$+ \sum_{j=1}^q R_{\alpha \bar\beta \eta \bar\delta_j} a_{\gamma_1 \cdots \gamma_p \bar\delta_1 \cdots \bar\delta_{j-1} \bar\eta \bar\delta_{j+1} \cdots \bar\delta_q}.$$

The commutator $[\Delta, \nabla_{\bar\beta}]$, acting on functions, is

$$\Delta \nabla_{\bar\beta} f = \frac{1}{2} \left(\nabla_\alpha \nabla_{\bar\alpha} + \nabla_{\bar\alpha} \nabla_\alpha \right) \nabla_{\bar\beta} f$$
$$= \frac{1}{2} \nabla_\alpha \nabla_{\bar\beta} \nabla_{\bar\alpha} f + \frac{1}{2} \nabla_{\bar\beta} \nabla_{\bar\alpha} \nabla_\alpha f$$
(2.19)
$$= \frac{1}{2} R_{\gamma \bar\beta} \nabla_{\bar\gamma} f + \nabla_{\bar\beta} \Delta f.$$

LEMMA 2.23 (Kähler ∇ and Δ commutator). *The commutator $[\nabla_\alpha, \Delta]$, acting on $(0,1)$-forms, is given by*

(2.20) $\quad \nabla_\alpha \Delta a_{\bar\beta} - \Delta \nabla_\alpha a_{\bar\beta} = \frac{1}{2} \nabla_\alpha R_{\delta \bar\beta} a_{\bar\delta} - \frac{1}{2} R_{\alpha \bar\delta} \nabla_\delta a_{\bar\beta} + R_{\alpha \bar\gamma \delta \bar\beta} \nabla_\gamma a_{\bar\delta}.$

PROOF. We compute

$$2\left(\nabla_\alpha \Delta a_{\bar\beta} - \Delta \nabla_\alpha a_{\bar\beta} \right) = \nabla_\alpha \left(\nabla_\gamma \nabla_{\bar\gamma} + \nabla_{\bar\gamma} \nabla_\gamma \right) a_{\bar\beta} - \left(\nabla_\gamma \nabla_{\bar\gamma} + \nabla_{\bar\gamma} \nabla_\gamma \right) \nabla_\alpha a_{\bar\beta}$$
$$= \nabla_\gamma \left(\nabla_\alpha \nabla_{\bar\gamma} - \nabla_{\bar\gamma} \nabla_\alpha \right) a_{\bar\beta} + \left(\nabla_\alpha \nabla_{\bar\gamma} - \nabla_{\bar\gamma} \nabla_\alpha \right) \nabla_\gamma a_{\bar\beta}$$
$$= \nabla_\gamma \left(R_{\alpha \bar\gamma \delta \bar\beta} a_{\bar\delta} \right) - R_{\alpha \bar\gamma \gamma \bar\delta} \nabla_\delta a_{\bar\beta} + R_{\alpha \bar\gamma \delta \bar\beta} \nabla_\gamma a_{\bar\delta}$$
$$= \nabla_\alpha R_{\delta \bar\beta} a_{\bar\delta} - R_{\alpha \bar\delta} \nabla_\delta a_{\bar\beta} + 2 R_{\alpha \bar\gamma \delta \bar\beta} \nabla_\gamma a_{\bar\delta},$$

where we used $[\nabla_\alpha, \nabla_\gamma] = 0$ and $\nabla_\gamma R_{\alpha \bar\gamma \delta \bar\beta} = \nabla_\alpha R_{\delta \bar\beta}$ from (2.9). □

REMARK 2.24. By taking the complex conjugate of the lemma above, we have

(2.21) $\quad \nabla_{\bar\beta} \Delta a_\alpha - \Delta \nabla_{\bar\beta} a_\alpha = \frac{1}{2} \nabla_{\bar\beta} R_{\bar\delta \alpha} a_\delta - \frac{1}{2} R_{\bar\beta \delta} \nabla_{\bar\delta} a_\alpha + R_{\bar\beta \gamma \bar\delta \alpha} \nabla_{\bar\gamma} a_\delta.$

Since the Kähler–Ricci flow is a heat-type equation for Kähler metrics, some evolution equations we shall derive later in this chapter use the following commutator formula. If a is a time-dependent $(0,1)$-form, then under the Kähler–Ricci flow,
$$\left[\frac{\partial}{\partial t} - \Delta, \nabla_\alpha\right] a_{\bar\beta} = \frac{1}{2}\nabla_\alpha R_{\delta\bar\beta} a_{\bar\delta} - \frac{1}{2}R_{\alpha\bar\delta}\nabla_\delta a_{\bar\beta} + R_{\alpha\bar\gamma\delta\bar\beta}\nabla_\gamma a_{\bar\delta}.$$
Another basic formula is the commutator of the heat operator and the Hessian. First, by (2.19) and (2.20), we have for any function f on $\mathcal{M}$,
$$\nabla_\alpha \nabla_{\bar\beta} \Delta f = \Delta_L \nabla_\alpha \nabla_{\bar\beta} f$$
(2.22)
$$\doteq \Delta \nabla_\alpha \nabla_{\bar\beta} f + R_{\alpha\bar\beta\delta\bar\gamma}\nabla_\gamma \nabla_{\bar\delta} f - \frac{1}{2}R_{\alpha\bar\delta}\nabla_\delta \nabla_{\bar\beta} f - \frac{1}{2}R_{\gamma\bar\beta}\nabla_\alpha \nabla_{\bar\gamma} f,$$
where Δ_L is called the **(complex) Lichnerowicz Laplacian**. If f is a function also of time, then (2.22) tells us

(2.23) $$\nabla_\alpha \nabla_{\bar\beta}\left(\frac{\partial}{\partial t} - \Delta\right)f = \left(\frac{\partial}{\partial t} - \Delta_L\right)\nabla_\alpha \nabla_{\bar\beta} f.$$

Note that since $\nabla_\alpha \nabla_{\bar\beta} = \partial_\alpha \bar\partial_\beta$ acting on functions, we have $\frac{\partial}{\partial t}\left(\nabla_\alpha \nabla_{\bar\beta}\right) = 0$ when acting on functions.

REMARK 2.25. Formula (2.23) has the following Riemannian analogue. For a time-dependent function f and under the Ricci flow, we have

(2.24) $$\nabla_i \nabla_j \left(\frac{\partial}{\partial t} - \Delta\right) f = \left(\frac{\partial}{\partial t} - \Delta_L\right)\nabla_i \nabla_j f,$$

where Δ_L, defined by
$$\Delta_L v_{ij} \doteq \Delta v_{ij} + 2R_{kij\ell} v_{k\ell} - R_{ik} v_{jk} - R_{jk} v_{ik},$$
is the (Riemannian) Lichnerowicz Laplacian acting on symmetric 2-tensors (see Lemma 2.33 on p. 110 of [**111**]). In fact, (2.24) is a special case of (2.23).

3. Existence of Kähler–Einstein metrics

In this section we discuss what is known about the existence and uniqueness of Kähler–Einstein metrics, which are canonical (constant Ricci curvature) metrics on Kähler manifolds. The Kähler–Einstein equation is elliptic whereas the Kähler–Ricci flow, discussed beginning in the next section, may be considered as its parabolic analogue.

First recall the following $\partial\bar\partial$**-Lemma**, which is a consequence of the Hodge decomposition theorem.

LEMMA 2.26 (d-exact real $(1,1)$-form is $\partial\bar\partial$ of real-valued function). *Let $\mathcal{M}^n$ be a closed Kähler manifold. If $\bar\omega$ is an exact real $(1,1)$-form, then there exists a real-valued function ψ such that $\sqrt{-1}\partial\bar\partial\psi = \bar\omega$. That is,*
$$\frac{\partial^2}{\partial z^\alpha \partial \bar z^\beta}\psi = \bar\omega_{\alpha\bar\beta},$$

where $\bar{\omega} = \sqrt{-1}\bar{\omega}_{\alpha\bar{\beta}}dz^\alpha \wedge d\bar{z}^\beta$ and $\overline{\bar{\omega}_{\alpha\bar{\beta}}} = \bar{\omega}_{\beta\bar{\alpha}}$.

PROOF. This is a standard result in the theory of Kähler manifolds; see the book by Zheng [**383**] for example. $\square$

REMARK 2.27. More generally, we have the following. Let b be a (p,q)-form, where $p, q > 0$. If b is d-closed and either d-, ∂-, or $\bar{\partial}$-exact, then there exists a $(p-1, q-1)$-form φ such that $\partial\bar{\partial}\varphi = b$. When $p = q$ and b is real, we may take $\sqrt{-1}\varphi$ to be real. See Lemma 9.1 on p. 221 of [**383**] for example.

A fundamental problem in Kähler geometry is the **Calabi conjecture**, which was solved by Yau [**378, 379**] and says the following. (Also see Theorem 2.30 below.)

THEOREM 2.28 (Calabi conjecture: prescribing the Ricci form in a Kähler class). *Let $(\mathcal{M}^n, g_0)$ be a closed Kähler manifold with Kähler class $[\omega_0]$. For any closed real $(1,1)$-form $\bar{\omega} \in c_1(\mathcal{M})$, there exists a Kähler metric g with $[\omega] = [\omega_0]$ such that its Ricci form is the prescribed form:*

$$(2.25) \qquad \rho = 2\pi\bar{\omega}.$$

Since $\rho_0 = \sqrt{-1}\operatorname{Rc}(g_0)_{\alpha\bar{\beta}} dz^\alpha \wedge d\bar{z}^\beta$ is a real $(1,1)$-form in the same cohomology class $2\pi c_1(\mathcal{M})$ as $2\pi\bar{\omega} = 2\pi\sqrt{-1}\bar{\omega}_{\alpha\bar{\beta}}dz^\alpha \wedge d\bar{z}^\beta$, by Lemma 2.26 there exists a real-valued function f such that

$$\operatorname{Rc}(g_0)_{\alpha\bar{\beta}} - 2\pi\bar{\omega}_{\alpha\bar{\beta}} = \partial_\alpha\bar{\partial}_\beta f.$$

Therefore equation (2.25), which in local coordinates is $R_{\alpha\bar{\beta}} = 2\pi\bar{\omega}_{\alpha\bar{\beta}}$, is equivalent to

$$\partial_\alpha\bar{\partial}_\beta f = \operatorname{Rc}(g_0)_{\alpha\bar{\beta}} - R_{\alpha\bar{\beta}}$$
$$= -\frac{\partial^2}{\partial z^\alpha \partial\bar{z}^\beta} \log\det\left((g_0)_{\gamma\bar{\delta}}\right) + \frac{\partial^2}{\partial z^\alpha \partial\bar{z}^\beta}\log\det\left(g_{\gamma\bar{\delta}}\right).$$

Since $\mathcal{M}$ is closed, this implies

$$\log\frac{\det\left(g_{\gamma\bar{\delta}}\right)}{\det\left((g_0)_{\gamma\bar{\delta}}\right)} = f + \log C$$

for some constant $C > 0$. Since the real $(1,1)$-forms ω_0 and ω are in the same cohomology class, using Lemma 2.26 again, we see that there exists a real-valued function φ such that

$$g_{\gamma\bar{\delta}} = (g_0)_{\gamma\bar{\delta}} + \partial_\gamma\bar{\partial}_\delta\varphi.$$

Thus we may rewrite (2.25) as a complex **Monge–Ampère equation**:

$$\frac{\det\left((g_0)_{\gamma\bar{\delta}} + \partial_\gamma\bar{\partial}_\delta\varphi\right)}{\det\left((g_0)_{\gamma\bar{\delta}}\right)} = Ce^f.$$

Yau's proof of Theorem 2.28 involves solving the fully nonlinear equation above by using the **continuity method**. The proof was a tour de force.

One says that a Kähler metric g is **Kähler–Einstein** if $\rho = \lambda \omega$ for some $\lambda \in \mathbb{R}$. If $\mathcal{M}^n$ admits a Kähler–Einstein metric g, then

$$c_1(\mathcal{M}) = \left[\frac{1}{2\pi}\rho\right] = \left[\frac{\lambda}{2\pi}\omega\right].$$

(We have $\lambda = \frac{R}{n}$, where the (complex) scalar curvature R is constant.) Therefore, a necessary condition for the existence of a Kähler–Einstein metric on $\mathcal{M}$ is that its first Chern class have a sign. By having a sign we mean that $c_1(\mathcal{M}) = 0$, < 0, or > 0 if there exists a real $(1,1)$-form in the first Chern class which is zero, negative definite, or positive definite, respectively.

COROLLARY 2.29 ($c_1 = 0$: existence of Kähler Ricci flat metrics). *If $(\mathcal{M}^n, g_0)$ is a closed Kähler manifold with $c_1(\mathcal{M}) = 0$, then there exists a Kähler metric g with $[\omega] = [\omega_0]$ such that $\operatorname{Rc}(g) \equiv 0$.*

Kähler Ricci flat metrics are called **Calabi–Yau metrics** and Kähler manifolds with $c_1(\mathcal{M}) = 0$ are called **Calabi–Yau manifolds**. Another consequence of Theorem 2.28 is that if $c_1(\mathcal{M}) < 0$ (respectively, $c_1(\mathcal{M}) > 0$), then in each Kähler class there exists a metric with negative (respectively, positive) Ricci curvature.

When $c_1(\mathcal{M}) < 0$, we have the following result about the existence of Kähler–Einstein metrics conjectured by Calabi and proved by Aubin [**11, 12**] and Yau [**378, 379**]. Calabi proved that such a Kähler–Einstein metric is unique if it exists [**43**].

THEOREM 2.30 ($c_1 < 0$ Calabi conjecture: $R < 0$ Kähler–Einstein metrics). *If $\mathcal{M}^n$ is a closed complex manifold with $c_1(\mathcal{M}) < 0$, then there exists a Kähler–Einstein metric g on $\mathcal{M}$, which is unique up to homothety (scaling), with negative scalar curvature.*

A consequence of Theorem 2.30 is the following Chern number inequality:

$$(-1)^n c_1(\mathcal{M})^n \leq (-1)^n \frac{2(n+1)}{n} c_1(\mathcal{M})^{n-2} c_2(\mathcal{M}).$$

(See Yau [**378**] and also Corollary 9.6 on p. 226 of [**383**] for an exposition.)

REMARK 2.31. Another consequence of Theorem 2.30 is that if $(\mathcal{M}^2, g)$ is a closed Kähler surface homotopically equivalent to $\mathbb{C}P^2$, then $\mathcal{M}^2$ is biholomorphic to $\mathbb{C}P^2$ (see Yau [**378**]; earlier related work in all dimensions was done by Hirzebruch and Kodaira [**204**]).

If $c_1(\mathcal{M}) > 0$, however, there are obstructions to the existence of Kähler–Einstein metrics. An example is the **Futaki invariant** (see [**147**]). On a closed manifold $\mathcal{M}^n$ with $c_1 > 0$ (i.e., a **Fano manifold**), fix a Kähler metric g such that $[\omega]$ is a positive real multiple of $c_1(\mathcal{M})$. (This is the so-called

canonical case.) By scaling the metric, we may assume $[\omega] = c_1(\mathcal{M})$. By Lemma 2.26 there exists a smooth function $f : \mathcal{M} \to \mathbb{R}$ such that

$$\rho - 2\pi\omega = \sqrt{-1}\partial\bar{\partial}f.$$

(One can make f unique by the normalization $\int_{\mathcal{M}} e^{-f} d\mu = 1$.) Let $\eta(\mathcal{M})$ denote the space of (real) holomorphic vector fields on $\mathcal{M}$. The **Futaki functional** $\mathcal{F}_{[\omega]} : \eta(\mathcal{M}) \to \mathbb{C}$ is defined by

$$\mathcal{F}_{[\omega]}(V) \doteqdot \int_{\mathcal{M}} V(f) \, d\mu = \int_{\mathcal{M}} \langle V, \nabla f \rangle \, d\mu.$$

Futaki [**147**] showed that $\mathcal{F}_{[\omega]}$ is well defined, i.e., that it depends only on the homology class $[\omega]$. It is then clear that if $\mathcal{M}$ admits a Kähler–Einstein metric, then $\mathcal{F}_{[\omega]}$ vanishes. However, Tian has shown that $\eta(\mathcal{M}) = 0$ (which implies $\mathcal{F}_{[\omega]} = 0$) does not imply there exists a Kähler–Einstein metric.

The following uniqueness result in the $c_1 > 0$ case was proved by Bando and Mabuchi [**21**].

THEOREM 2.32 (Uniqueness of Kähler–Einstein metrics). *Let $(\mathcal{M}^n, g)$ be a closed Kähler manifold with $c_1(\mathcal{M}) > 0$. The Kähler–Einstein metric (with positive scalar curvature), if it exists, is unique up to scaling and the pull back by a biholomorphism of $\mathcal{M}$.*

The following result was proved by Andreotti and Frankel [**144**] for $n = 2$, Mabuchi [**260**] for $n = 3$, and Mori [**271**] and Siu and Yau [**336**] in all dimensions; Mori proved a more general algebraic-geometric result. Work on characterizing $\mathbb{C}P^n$ was done by Kobayashi and Ochiai [**237**].

THEOREM 2.33 (Frankel Conjecture). *If $(\mathcal{M}^n, g)$ is a closed Kähler manifold with positive bisectional curvature, then $\mathcal{M}^n$ is biholomorphic to $\mathbb{C}P^n$.*

In fact, if the bisectional curvature is nonnegative everywhere and positive at some point, then $\mathcal{M}$ is biholomorphic to $\mathbb{C}P^n$.

When $n = 2$ and $c_1(\mathcal{M}) > 0$, we have the following (see Tian [**345**]).

THEOREM 2.34 ($c_1 > 0$ surfaces: $R > 0$ Kähler–Einstein metrics). *If $\mathcal{M}^2$ is a closed complex surface with $c_1(\mathcal{M}) > 0$ and the Lie algebra of the automorphism group is reductive, then there exists a Kähler–Einstein metric g on $\mathcal{M}$ with positive scalar curvature.*

REMARK 2.35. Note that such surfaces are biholomorphic to $\mathbb{C}P^2$ blown up at p points, where $3 \leq p \leq 8$. On the other hand, for $n \geq 2$, $\mathbb{C}P^n$ blown up at 1 or 2 points does not admit a Kähler–Einstein metric (see p. 156 of Lichnerowicz [**254**] and Yau [**376**]). See Section 7 of this chapter for the existence of Kähler–Ricci solitons on $\mathbb{C}P^2$ blown up at 1 or 2 points.

There are a number of additional works related to stability and the existence of Kähler–Einstein metrics with $c_1 > 0$. Notably Aubin [**14**], Siu [**335**], Nadel [**281**], Tian [**346**], Donaldson [**129**], [**130**], and Phong and

Sturm [**304**]. The existence of Kähler–Einstein metrics on complete noncompact manifolds with $c_1 < 0$ and $c_1 = 0$ has been well studied (see Cheng and Yau [**94**] and Tian and Yau [**349**], [**350**] for example). For work on the existence of singular Kähler–Einstein metrics on certain classes of closed Kähler manifolds where c_1 does not have a sign, see Tsuji [**360**] (for some further recent work see Cascini and La Nave [**60**] and Song and Tian [**337**]).

Kähler–Einstein metrics are closely related to Kähler–Ricci solitons and hence the Kähler-Ricci flow which will be discussed next in this chapter.

4. Introduction to the Kähler–Ricci flow

In this section we introduce the Kähler–Ricci flow system and its equivalent formulation as a single parabolic Monge–Ampère equation. We discuss some basic estimates which may be proved using the maximum principle.

4.1. The Kähler–Ricci flow equation. Let $(\mathcal{M}^n, J)$ be a closed manifold with a fixed almost complex structure. Given a Riemannian metric g, we may define a 2-tensor ω by $\omega(X, Y) \doteqdot g(JX, Y)$. Recall that when g is Hermitian, ω is antisymmetric (i.e., defines a 2-form) and ω is called the Kähler form. If a solution $g(t)$ to the Ricci flow $\frac{\partial}{\partial t} g_{ij} = -2R_{ij}$ is Hermitian at some time t, then ω satisfies the equation $\frac{\partial}{\partial t} \omega = -2\rho$ at that time, where $\rho = \rho(t)$ is the Ricci form of $g(t)$. Hence

$$\frac{\partial}{\partial t}(d\omega) = d\left(\frac{\partial}{\partial t}\omega\right) = -2d\rho = 0$$

whenever $g(t)$ is Kähler (we can define even when g is not Hermitian). This suggests that if $g(0)$ is Kähler, then under the Ricci flow $g(t)$ is Kähler for all $t \geq 0$.

Consider the **Kähler–Ricci flow** equation

$$\tag{2.26} \frac{\partial}{\partial t} g_{\alpha\bar{\beta}} = -R_{\alpha\bar{\beta}}$$

for a 1-parameter family of Kähler metrics with respect to J, which is obtained from the Ricci flow by dropping the factor of 2. Now we derive the parabolic complex Monge–Ampère equation to which the Kähler–Ricci flow is equivalent. For a complete initial Kähler metric with bounded curvature, we will use this scalar equation to prove the short-time existence of a solution to the initial-value problem for the Kähler–Ricci flow. On a closed manifold we will use the scalar equation to prove that the Kähler property of an initial metric is preserved under the Ricci flow and to prove the long-time existence of solutions to the Kähler–Ricci flow.

By (2.26), $\frac{\partial}{\partial t}[\omega] = -[\rho(t)] = -[\rho(0)]$, so that the Kähler class of the metric at time t evolves linearly,

$$[\omega(t)] = [\omega(0)] - t[\rho(0)],$$

and the real $(1, 1)$-forms $\omega(t) - \omega(0) + t\rho(0)$ are exact for $t \geq 0$. Let $g^0_{\alpha\bar{\beta}} \doteqdot g_{\alpha\bar{\beta}}(0)$. Using Lemma 2.26, for each t there exists a real-valued function

$\varphi(t)$ defined on all of $\mathcal{M}$ such that

$$(2.27) \qquad g_{\alpha\bar{\beta}}(t) = g^0_{\alpha\bar{\beta}} + t\partial_\alpha \bar{\partial}_\beta \log \det g^0_{\gamma\bar{\delta}} + \partial_\alpha \bar{\partial}_\beta \varphi(t).$$

By (2.6) we have

$$R_{\alpha\bar{\beta}}(t) - R_{\alpha\bar{\beta}}(0) = -\partial_\alpha \bar{\partial}_\beta \log \frac{\det\left(g^0_{\gamma\bar{\delta}} + t\partial_\gamma \bar{\partial}_\delta \log \det g^0_{\mu\bar{\nu}} + \partial_\gamma \bar{\partial}_\delta \varphi(t)\right)}{\det g^0_{\gamma\bar{\delta}}}.$$

Hence, by differentiating (2.27), we obtain

$$\partial_\alpha \bar{\partial}_\beta \left(\frac{\partial}{\partial t} \varphi\right) = -R_{\alpha\bar{\beta}} - \partial_\alpha \bar{\partial}_\beta \log \det g^0_{\gamma\bar{\delta}}$$

$$= \partial_\alpha \bar{\partial}_\beta \log \frac{\det\left(g^0_{\gamma\bar{\delta}} + t\partial_\gamma \bar{\partial}_\delta \log \det g^0_{\mu\bar{\nu}} + \partial_\gamma \bar{\partial}_\delta \varphi(t)\right)}{\det g^0_{\gamma\bar{\delta}}}.$$

Hence we conclude that the Kähler–Ricci flow equation on a closed manifold is equivalent to the following **parabolic (scalar) complex Monge–Ampère equation**:

$$(2.28) \qquad \frac{\partial \varphi}{\partial t} = \log \frac{\det\left(g^0_{\gamma\bar{\delta}} + t\partial_\gamma \bar{\partial}_\delta \log \det g^0_{\mu\bar{\nu}} + \partial_\gamma \bar{\partial}_\delta \varphi(t)\right)}{\det g^0_{\gamma\bar{\delta}}} + c_1(t)$$

for some function of time $c_1(t)$. By standard parabolic theory, given any C^∞ initial function φ_0 on a complete Kähler manifold with bounded bisectional curvature, there exists a unique solution $\varphi(t)$ to (2.28) with $\varphi(0) = \varphi_0$, defined on some positive time interval $0 \leq t \leq \varepsilon$. We also have the following.

LEMMA 2.36 (The Kähler property is preserved under the Ricci flow). *If $(\mathcal{M}^n, J, g_0)$ is a closed Kähler manifold, then there exists a solution to the Kähler–Ricci flow $g(t)$, $0 \leq t \leq \varepsilon$, for some $\varepsilon > 0$ with $g(0) = g_0$. Furthermore $g(2t)$ is a solution of the (Riemannian) Ricci flow. Also any solution $\tilde{g}(t)$ of the (Riemannian) Ricci flow with $\tilde{g}(t) = g_0$ must be Kähler (preserving the compatibility with the almost complex structure).*

PROOF. Given g_0, we can find a solution $\varphi(t)$, $0 \leq t \leq \varepsilon$, of (2.28) with $c_1(t) \equiv 0$. From the derivation of (2.28), we know that $g(t)$ defined by (2.27) is a solution of (2.26). Hence $g(2t)$ is a solution of the Ricci flow. The last statement follows from the uniqueness of the initial-value problem for the Ricci flow. $\square$

REMARK 2.37. From the derivation of (2.28) it is clear that if we have a bounded C^4-solution $\varphi(t)$ for some $c_1(t)$ on any complex manifold (regardless of completeness and compactness), then we get a C^2-solution $g(t)$ defined by (2.27) to the Kähler–Ricci flow.

4.2. The normalized Kähler–Ricci flow equation. Let $(\mathcal{M}^n, J, g_0)$ be a closed Kähler manifold. Now we make the basic assumption (corresponding to the **canonical case**), holding for the rest of this section and the next section, that the first Chern class is a real multiple of the Kähler class, i.e., that

$$[\rho_0] = c[\omega_0]$$

for some $c \in \mathbb{R}$. Note that this is possible only if the first Chern class has a sign, i.e., is negative definite, zero, or positive definite. Comparing (2.11) and (2.12), we find that $c = \frac{r}{n}$, where $r \doteq \int_{\mathcal{M}} R_0 d\mu_{g_0} / \mathrm{Vol}_{g_0}(\mathcal{M})$ is the average (complex) scalar curvature, so that

$$\frac{r}{2\pi n}[\omega_0] = \frac{1}{2\pi}[\rho_0] = c_1(\mathcal{M}).$$

So r depends only on the cohomology class $[\omega_0]$, n, and $c_1(\mathcal{M})$.

The **normalized Kähler–Ricci flow** is

(2.29) $$\frac{\partial}{\partial t} g_{\alpha\bar{\beta}} = -R_{\alpha\bar{\beta}} + \frac{r}{n} g_{\alpha\bar{\beta}}, \quad \text{for } t \in [0, T).$$

The solution of (2.29) can be converted to the solution (2.26) by scaling the metric and reparametrizing time, and vice versa (see Section 9.1 in Chapter 6 of Volume One or subsection 9.1 below). Hence, from Lemma 2.36, we know that the initial-value problem for (2.29) with $g(0) = g_0$ has a solution for a short time.

By a derivation similar to that of (2.28), we get the following parabolic (scalar) complex Monge–Ampère equation, corresponding to (2.29) with $g_{\alpha\bar{\beta}}(t) = g_{\alpha\bar{\beta}}^0 + \partial_\alpha \bar{\partial}_\beta \varphi(t)$,

(2.30) $$\frac{\partial \varphi}{\partial t} = \log \frac{\det\left(g_{\gamma\bar{\delta}}^0 + \partial_\gamma \bar{\partial}_\delta \varphi\right)}{\det g_{\gamma\bar{\delta}}^0} + \frac{r}{n}\varphi - f_0 + c_1(t)$$

for some function of time $c_1(t)$. Here f_0 is defined by $R_{\alpha\bar{\beta}}(g_0) - \frac{r}{n} g_{\alpha\bar{\beta}}^0 = \partial_\alpha \bar{\partial}_\beta f_0$; this is possible because $\left[-\rho_0 + \frac{r}{n}\omega_0\right] = 0$.

4.3. Basic evolution equations. Let $g(t)$ be a solution of either the Kähler–Ricci flow or the normalized Kähler–Ricci flow. We define the **potential function** $f = f(t)$ by

(2.31) $$R_{\alpha\bar{\beta}}(g) - \frac{r}{n} g_{\alpha\bar{\beta}} = \partial_\alpha \bar{\partial}_\beta f = \nabla_\alpha \nabla_{\bar{\beta}} f.$$

This equation is solvable since $\left[-\rho + \frac{r}{n}\omega\right] = 0$ and by Lemma 2.26. Note that f is determined up to an additive constant. Taking the trace of (2.31), we have

(2.32) $$R - r = \Delta f.$$

Differentiating (2.3), we find that for both the Kähler–Ricci flow and the normalized Kähler–Ricci flow, the Christoffel symbols evolve by

$$\frac{\partial}{\partial t}\Gamma^{\gamma}_{\alpha\beta} = -g^{\gamma\bar{\delta}}\nabla_{\alpha}R_{\beta\bar{\delta}}. \tag{2.33}$$

The volume form and scalar curvature evolve according to the following.

LEMMA 2.38 (Evolution of $d\mu$ and R for normalized flow). *Under the normalized Kähler–Ricci flow (2.29),*

$$\frac{\partial}{\partial t}d\mu = (r - R)\,d\mu$$

and

$$\frac{\partial R}{\partial t} = \Delta R + \left|R_{\alpha\bar{\beta}}\right|^2 - \frac{r}{n}R. \tag{2.34}$$

In particular, since $\int_{\mathcal{M}}(r - R)\,d\mu = 0$, the normalized Kähler–Ricci flow preserves the volume.

PROOF. We first compute, using (2.29), that

$$\frac{\partial}{\partial t}\log\det g_{\gamma\bar{\delta}} = g^{\gamma\bar{\delta}}\frac{\partial}{\partial t}g_{\gamma\bar{\delta}} = r - R. \tag{2.35}$$

Hence

$$\frac{\partial}{\partial t}d\mu = (r - R)\,d\mu.$$

The evolution of the Ricci tensor is

$$\frac{\partial}{\partial t}R_{\alpha\bar{\beta}} = -\partial_{\alpha}\bar{\partial}_{\beta}\left(\frac{\partial}{\partial t}\log\det g_{\gamma\bar{\delta}}\right) = \partial_{\alpha}\bar{\partial}_{\beta}R. \tag{2.36}$$

From this and

$$\frac{\partial R}{\partial t} = g^{\alpha\bar{\beta}}\frac{\partial}{\partial t}R_{\alpha\bar{\beta}} - \frac{\partial}{\partial t}g_{\alpha\bar{\beta}}\cdot R_{\alpha\bar{\beta}}$$

we easily derive (2.34). □

EXERCISE 2.39 (Evolution of R for unnormalized flow). Show that under the Kähler–Ricci flow $\frac{\partial}{\partial t}g_{\alpha\bar{\beta}} = -R_{\alpha\bar{\beta}}$, we have $\frac{\partial}{\partial t}d\mu = -R\,d\mu$, $\frac{\partial}{\partial t}R_{\alpha\bar{\beta}} = \partial_{\alpha}\bar{\partial}_{\beta}R$, and

$$\frac{\partial R}{\partial t} = g^{\alpha\bar{\beta}}\frac{\partial}{\partial t}R_{\alpha\bar{\beta}} + \left|R_{\alpha\bar{\beta}}\right|^2 = \Delta R + \left|R_{\alpha\bar{\beta}}\right|^2. \tag{2.37}$$

EXERCISE 2.40 (Total and average scalar curvature evolution). Show that if $\mathcal{M}^n$ is closed, then under the Kähler–Ricci flow $\frac{\partial}{\partial t}g_{\alpha\bar{\beta}} = -R_{\alpha\bar{\beta}}$,

$$\frac{d}{dt}\int_{\mathcal{M}}R\,d\mu = \int_{\mathcal{M}}\left(\left|R_{\alpha\bar{\beta}}\right|^2 - R^2\right)d\mu,$$

and hence we have

$$\frac{dr}{dt} = \left(\int_{\mathcal{M}}d\mu\right)^{-1}\int_{\mathcal{M}}\left(\left|R_{\alpha\bar{\beta}}\right|^2 - R^2\right)d\mu + r^2.$$

REMARK 2.41 (1-dimensional normalized Kähler-Ricci flow). Recall the following facts, due to Hamilton [**180**], about the normalized Ricci flow on Riemannian surfaces $\frac{\partial}{\partial t} g = (r - R) g$. Throughout this remark, R and r denote the Riemannian scalar curvature and its average, respectively. Note that $\tilde{g}(t) \doteqdot g\left(\frac{1}{2} t\right)$ is a solution of the complex 1-dimensional normalized Kähler–Ricci flow. The potential function f, defined by (2.32) and normalized suitably by an additive constant, satisfies (see Lemma 5.12 on p. 113 of Volume One)

$$\text{(2.38)} \qquad \frac{\partial f}{\partial t} = \Delta f + rf.$$

By the maximum principle, this implies $|f| \leq Ce^{rt}$. The gradient quantity $H \doteqdot R - r + |\nabla f|^2$ satisfies (see Proposition 5.16 on p. 114 of Volume One)

$$\text{(2.39)} \qquad \frac{\partial H}{\partial t} = \Delta H - 2|M|^2 + rH,$$

where $M \doteqdot \nabla \nabla f - \frac{1}{2} \Delta f \cdot g$. By the maximum principle, we have (see Corollary 5.17 on p. 115 of Volume One)

$$-Ce^{rt} \leq R - r \leq H \leq Ce^{rt}.$$

This gives the exponential decay of $|R - r|$ when $r < 0$. The norm squared of the tensor M evolves by (see Corollary 5.35 on p. 130 of Volume One)

$$\text{(2.40)} \qquad \frac{\partial}{\partial t} |M|^2 = \Delta |M|^2 - 2|\nabla M|^2 - 2R|M|^2.$$

Generalizing the 1-dimensional formula (2.38) to higher dimensions, the potential satisfies a linear-type equation. (Strictly speaking, the equation is not linear since the Laplacian is with respect to the *evolving* metric.)

LEMMA 2.42 (The potential f satisfies a linear-type equation). *Under the normalized Kähler–Ricci flow on a closed manifold $\mathcal{M}^n$, the potential function f, defined by (2.31) and normalized by an additive constant, satisfies*

$$\text{(2.41)} \qquad \frac{\partial f}{\partial t} = \Delta f + \frac{r}{n} f.$$

PROOF. From (2.31) we compute

$$\partial_\alpha \bar{\partial}_\beta \left(\frac{\partial f}{\partial t} \right) = \frac{\partial}{\partial t} \left(\partial_\alpha \bar{\partial}_\beta f \right) = \frac{\partial}{\partial t} \left(R_{\alpha \bar{\beta}} - \frac{r}{n} g_{\alpha \bar{\beta}} \right)$$

$$= \partial_\alpha \bar{\partial}_\beta R - \frac{r}{n} \frac{\partial}{\partial t} g_{\alpha \bar{\beta}} = \partial_\alpha \bar{\partial}_\beta \left(\Delta f + \frac{r}{n} f \right).$$

Since $\mathcal{M}$ is closed, it follows that

$$\text{(2.42)} \qquad \frac{\partial f}{\partial t} = \Delta f + \frac{r}{n} f + c(t)$$

for some function $c(t)$ and the lemma follows from the fact that we have the freedom of adding a time-dependent constant in our choice of $f(x, t)$. □

COROLLARY 2.43 (Estimate for f). *If $\mathcal{M}^n$ is closed, then, for the function f given by Lemma 2.42, we have*

$$|f| \leq C e^{\frac{r}{n}t}. \tag{2.43}$$

This is the first hint that the Kähler–Ricci flow in the case where $c_1(\mathcal{M}) < 0$ is the easiest and that the case where $c_1(\mathcal{M}) > 0$ is the hardest.

We now compute for f in Lemma 2.42, for the normalized Kähler–Ricci flow, that

$$\frac{\partial}{\partial t}|\nabla_\alpha f|^2 = \Delta|\nabla_\alpha f|^2 - |\nabla_\alpha \nabla_\beta f|^2 - |\nabla_\alpha \nabla_{\bar{\beta}} f|^2 + \frac{r}{n}|\nabla_\alpha f|^2. \tag{2.44}$$

Define

$$h \doteq \Delta f + |\nabla_\alpha f|^2 = R - r + |\nabla_\alpha f|^2.$$

Similarly to (2.39), we have

LEMMA 2.44 (Ricci soliton gradient quantity evolution). *For the normalized Kähler–Ricci flow on a closed manifold $\mathcal{M}^n$,*

$$\frac{\partial h}{\partial t} = \Delta h - |\nabla_\alpha \nabla_\beta f|^2 + \frac{r}{n} h. \tag{2.45}$$

PROOF. We compute

$$\frac{\partial}{\partial t}(R - r) = \Delta(R - r) + |R_{\alpha\bar{\beta}}|^2 - \frac{r}{n} R$$

$$= \Delta(R - r) + |\nabla_\alpha \nabla_{\bar{\beta}} f|^2 + \frac{2r}{n}\Delta f + \frac{r^2}{n} - \frac{r}{n} R$$

$$= \Delta(R - r) + |\nabla_\alpha \nabla_{\bar{\beta}} f|^2 + \frac{r}{n}(R - r), \tag{2.46}$$

where we used (2.31) and (2.32). Equation (2.45) follows from summing this equation with (2.44). □

COROLLARY 2.45 (Estimate for R).

$$-C e^{\frac{r}{n}t} \leq R - r \leq C e^{\frac{r}{n}t}, \tag{2.47}$$

$$|\nabla f|^2 \leq C e^{\frac{r}{n}t}. \tag{2.48}$$

PROOF. By (2.46), the lower bound for $R - r$ follows from

$$\left(\frac{\partial}{\partial t} - \Delta\right)(R - r) \geq \frac{r}{n}(R - r).$$

To get the upper bound for $R - r$, we observe that by the maximum principle, we have

$$R - r \leq h \leq C e^{\frac{r}{n}t}.$$

This also implies (2.48) since $|\nabla f|^2 = h - (R - r) \leq h + C e^{\frac{r}{n}t}$. □

REMARK 2.46 (Exponential decay when $c_1 < 0$). When $c_1(\mathcal{M}) < 0$, so that $r < 0$, (2.47) says that R approaches its average exponentially fast. This suggests that the Kähler–Ricci flow converges to a Kähler–Einstein metric. Indeed, this is Theorem 2.50 below.

Here is an interesting equation due to Hamilton.

LEMMA 2.47 (Ricci soliton vanisher evolution equation). *For the potential function f in Lemma 2.42, we have*

$$(2.49) \quad \frac{\partial}{\partial t}|\nabla_\alpha \nabla_\beta f|^2 = \Delta |\nabla_\alpha \nabla_\beta f|^2 - |\nabla_\gamma \nabla_\alpha \nabla_\beta f|^2 - |\nabla_{\bar\gamma} \nabla_\alpha \nabla_\beta f|^2 \\ - 2R_{\alpha\bar\beta\gamma\bar\delta}\nabla_{\bar\alpha}\nabla_{\bar\gamma}f\nabla_\beta\nabla_\delta f.$$

PROOF. By (2.42) and the commutator formulas, we have

$$\frac{\partial}{\partial t}\left(\nabla_\alpha \nabla_\beta f\right) = \nabla_\alpha \nabla_\beta \left(\frac{\partial f}{\partial t}\right) - \left(\frac{\partial}{\partial t}\Gamma^\gamma_{\alpha\beta}\right)\nabla_\gamma f$$
$$= \nabla_\alpha \nabla_\beta \left(\Delta f + \frac{r}{n}f\right) + \nabla_\alpha R_{\beta\bar\gamma}\nabla_\gamma f.$$

On the other hand, for any function f,

$$\nabla_\alpha \nabla_\beta \Delta f = \nabla_\alpha \nabla_\beta \nabla_\gamma \nabla_{\bar\gamma} f = \nabla_\gamma \nabla_\alpha \nabla_{\bar\gamma} \nabla_\beta f$$
$$= \nabla_\gamma \nabla_{\bar\gamma} \nabla_\alpha \nabla_\beta f - \nabla_\gamma \left(R_{\alpha\bar\gamma\beta\bar\delta}\nabla_\delta f\right)$$
$$= \frac{1}{2}\left(\nabla_\gamma \nabla_{\bar\gamma} + \nabla_{\bar\gamma}\nabla_\gamma\right)\nabla_\alpha \nabla_\beta f - \nabla_\alpha R_{\beta\bar\delta}\nabla_\delta f - R_{\alpha\bar\gamma\beta\bar\delta}\nabla_\gamma \nabla_\delta f$$
$$- \frac{1}{2}\left(R_{\alpha\bar\gamma}\nabla_\gamma \nabla_\beta f + R_{\beta\bar\gamma}\nabla_\alpha \nabla_\gamma f\right)$$

since

$$\left(\nabla_\gamma \nabla_{\bar\gamma} - \nabla_{\bar\gamma}\nabla_\gamma\right)\nabla_\alpha \nabla_\beta f = -R_{\alpha\bar\gamma}\nabla_\gamma \nabla_\beta f - R_{\beta\bar\gamma}\nabla_\alpha \nabla_\gamma f.$$

Hence

$$\frac{\partial}{\partial t}\left(\nabla_\alpha \nabla_\beta f\right) = \Delta \nabla_\alpha \nabla_\beta f + \frac{r}{n}\nabla_\alpha \nabla_\beta f - R_{\alpha\bar\gamma\beta\bar\delta}\nabla_\gamma \nabla_\delta f \\ - \frac{1}{2}\left(R_{\alpha\bar\gamma}\nabla_\gamma \nabla_\beta f + R_{\beta\bar\gamma}\nabla_\alpha \nabla_\gamma f\right)$$

and one easily derives (2.49) from this. $\square$

PROBLEM 2.48. Find geometric applications of (2.49) in the study of the Kähler–Ricci flow.

EXERCISE 2.49. Show that when $\dim_\mathbb{C} \mathcal{M} = 1$,

$$|\nabla_\alpha \nabla_\beta f|^2 = \frac{1}{2}\left|\nabla_i \nabla_j f - \frac{1}{2}\Delta f g_{ij}\right|^2 \quad \text{and} \quad |\nabla_\alpha \nabla_{\bar\beta} f|^2 = \frac{1}{4}(\Delta f)^2.$$

Show also that (2.49) generalizes (2.40).

5. Existence and convergence of the Kähler–Ricci flow

In this section we present some of the proofs of the basic global existence and convergence results for the Kähler–Ricci flow due to H.-D. Cao [**46**].

5.1. Cao's existence and convergence theorem. When the first Chern class has a definite sign, either negative, zero, or positive, Cao proved that the normalized Kähler–Ricci flow exists for all time. Let **KRF** and **NKRF** denote the Kähler–Ricci flow and the normalized Kähler–Ricci flow, respectively.

THEOREM 2.50 (NKRF: $c_1 <, =, > 0$ global existence). *Let $(\mathcal{M}^n, g_0)$ be a closed Kähler manifold with*

(1) *either $c_1(\mathcal{M}) < 0$, $c_1(\mathcal{M}) = 0$, or $c_1(\mathcal{M}) > 0$, and*
(2) *$\frac{r_0}{n}[\omega_0] = 2\pi c_1(\mathcal{M})$.*

Then there exists a unique solution $g(t)$ of the normalized Kähler–Ricci flow defined for all $t \in [0, \infty)$ with $g(0) = g_0$.

When the first Chern class is nonpositive, Cao proved that the normalized Kähler–Ricci flow converges to a Kähler–Einstein metric in the same Kähler class as the initial metric.

THEOREM 2.51 (KRF: $c_1 \leq 0$ convergence). *Let $g(t)$ be a solution of the normalized Kähler–Ricci flow, as in Theorem 2.50, with $c_1(\mathcal{M}) \leq 0$. Then $g(t)$ converges exponentially fast in every C^k-norm to the unique Kähler–Einstein metric g_∞ in the Kähler class $[\omega_0]$.*

Note that the initial-value problem for the normalized Kähler–Ricci flow equation

$$\frac{\partial}{\partial t} g_{\alpha\bar{\beta}} = -R_{\alpha\bar{\beta}} + \frac{r}{n} g_{\alpha\bar{\beta}},$$

$$g_{\alpha\bar{\beta}}(0) = g^0_{\alpha\bar{\beta}},$$

is equivalent to the following parabolic Monge–Ampère equation for the **metric potential** function $\varphi(x,t)$:

(2.50) $$\frac{\partial \varphi}{\partial t}(x,t) = \log \frac{\det g_{\alpha\bar{\beta}}(x,t)}{\det g_{\alpha\bar{\beta}}(x,0)} + \frac{r}{n}\varphi(x,t) - f(x,0),$$

(2.51) $$\varphi(x,0) = 0,$$

where

(2.52) $$g_{\alpha\bar{\beta}}(x,t) \doteqdot g_{\alpha\bar{\beta}}(x,0) + \frac{\partial^2 \varphi}{\partial z^\alpha \partial \bar{z}^\beta}(x,t)$$

and $f(x,0)$ is the potential function of $R_{\alpha\bar{\beta}}(x,0) - \frac{r}{n} g_{\alpha\bar{\beta}}(x,0)$ defined by (2.31). To prove Theorem 2.50 and Theorem 2.51, it suffices to prove the long-time existence of $\varphi(x,t)$ and its convergence, respectively.

Presently we shall see that the reduction of the Kähler–Ricci flow to a single parabolic Monge–Ampère equation simplifies matters considerably. The main advantage is that it allows one to invoke the techniques of parabolic (and elliptic) PDE for *single* equations, such as maximum principles,

Moser iteration, and differential Harnack estimates (see the discussion below). The corresponding theory for *systems* is considerably harder, sometimes tractable only under more restrictive conditions such as the nonnegativity of the bisectional curvature (see Sections 6–8 of this chapter).

Since

$$\frac{\partial^2}{\partial z^\alpha \partial \bar{z}^\beta}\left(\frac{\partial \varphi}{\partial t}\right) = \frac{\partial}{\partial t}\left(\frac{\partial^2 \varphi}{\partial z^\alpha \partial \bar{z}^\beta}\right) = \frac{\partial}{\partial t} g_{\alpha\bar{\beta}} = -R_{\alpha\bar{\beta}} + \frac{r}{n} g_{\alpha\bar{\beta}}, \tag{2.53}$$

we have that

$$v(x,t) \doteqdot -\frac{\partial \varphi}{\partial t}(x,t)$$

is also a potential function of $R_{\alpha\bar{\beta}}(x,t) - \frac{r}{n}g_{\alpha\bar{\beta}}(x,t)$. From (2.50), (2.53), and (2.51), we compute

$$\frac{\partial v}{\partial t} = -\frac{\partial}{\partial t}\left(\frac{\partial \varphi}{\partial t}\right) = -g^{\alpha\bar{\beta}}\frac{\partial}{\partial t}g_{\alpha\bar{\beta}} - \frac{r}{n}\frac{\partial \varphi}{\partial t}$$

$$= \Delta v + \frac{r}{n}v$$

with the initial condition $v(x,0) = f(x,0)$. Therefore, if we insist, as in Lemma 2.42, that the potential function $f(x,t)$ satisfies the heat equation $\frac{\partial}{\partial t} f = \Delta f + \frac{r}{n} f$, we must have

$$f = -\frac{\partial \varphi}{\partial t}. \tag{2.54}$$

Recall from (2.43) that $|f| = \left|\frac{\partial \varphi}{\partial t}\right| \leq Ce^{\frac{r}{n}t}$. More precisely, we have the following.

LEMMA 2.52 (Time-derivative estimate for φ).

$$-C_1 e^{\frac{r}{n}t} \leq f(x,t) = -\frac{\partial \varphi}{\partial t} \leq C_2 e^{\frac{r}{n}t}, \tag{2.55}$$

where $C_1 \doteqdot -\min_{x \in \mathcal{M}^n} f(x,0)$ and $C_2 \doteqdot \max_{x \in \mathcal{M}} f(x,0)$.

5.2. Proof of Theorem 2.50. Theorem 2.50 is proved via a progression of estimates which culminates with a $C^{2,\alpha}$-estimate for $\varphi(t)$ on bounded time intervals. The C^0-estimate is the following.

LEMMA 2.53 (C^0-estimate: bound for φ—uniform when $c_1 \leq 0$). If $r \neq 0$, then

$$-\frac{C_2 n}{r}\left(e^{\frac{r}{n}t} - 1\right) \leq \varphi(x,t) \leq \frac{C_1 n}{r}\left(e^{\frac{r}{n}t} - 1\right). \tag{2.56}$$

If $r = 0$, then

$$-C_2 t \leq \varphi(x,t) \leq C_1 t.$$

PROOF. For the upper bound we compute

$$\varphi(x,t) = \varphi(x,0) + \int_0^t \frac{\partial \varphi}{\partial t}(\tau)\, d\tau \leq \int_0^t C_1 e^{\frac{r}{n}\tau} d\tau.$$

If $r \neq 0$, the integral on the RHS is equal to $\frac{C_1 n}{r}\left(e^{\frac{r}{n}t}-1\right)$. When $r=0$, we obtain $C_1 t$.

Similarly for the lower bound. □

REMARK 2.54. The qualitative dependence of the estimate (2.56) on the sign of r should be compared to Corollaries 2.43 and 2.45.

The next estimate, a bound for the determinant of the complex Hessian of φ, is also a straightforward application of the previous result and the estimates (2.55) and (2.56) to equation (2.50).

LEMMA 2.55 (Estimates for the volume form—uniform when $r \leq 0$). *If $r \leq 0$, then there exists a constant $C \geq 1$ such that*

$$\text{(2.57)} \qquad \frac{1}{C} \leq \frac{\det\left(g_{\alpha\bar{\beta}}(x,t)\right)}{\det\left(g_{\alpha\bar{\beta}}(x,0)\right)} \leq C$$

for all $x \in \mathcal{M}$ and $t \geq 0$. If $r > 0$, then

$$e^{-\frac{Cn}{r}\left(e^{\frac{r}{n}t}-1\right)} \leq \frac{\det\left(g_{\alpha\bar{\beta}}(x,t)\right)}{\det\left(g_{\alpha\bar{\beta}}(x,0)\right)} \leq e^{\frac{Cn}{r}\left(e^{\frac{r}{n}t}-1\right)}$$

for all $x \in \mathcal{M}$ and $t \geq 0$.

PROOF. Applying (2.47) to (2.35), we have

$$\text{(2.58)} \qquad \left|\frac{\partial}{\partial t} \log \frac{\det\left(g_{\alpha\bar{\beta}}(x,t)\right)}{\det\left(g_{\alpha\bar{\beta}}(x,0)\right)}\right| = |r - R| \leq C e^{\frac{r}{n}t}.$$

Hence, if $r \neq 0$, then

$$\left|\log \frac{\det\left(g_{\alpha\bar{\beta}}(x,t)\right)}{\det\left(g_{\alpha\bar{\beta}}(x,0)\right)}\right| \leq \frac{Cn}{r}\left(e^{\frac{r}{n}t}-1\right),$$

so that

$$e^{-\frac{Cn}{r}\left(e^{\frac{r}{n}t}-1\right)} \leq \frac{\det\left(g_{\alpha\bar{\beta}}(x,t)\right)}{\det\left(g_{\alpha\bar{\beta}}(x,0)\right)} \leq e^{\frac{Cn}{r}\left(e^{\frac{r}{n}t}-1\right)}.$$

In particular, if $r < 0$, then

$$e^{\frac{Cn}{r}} \leq \frac{\det\left(g_{\alpha\bar{\beta}}(x,t)\right)}{\det\left(g_{\alpha\bar{\beta}}(x,0)\right)} \leq e^{-\frac{Cn}{r}}.$$

When $r=0$, equation (2.58) is not strong enough to uniformly estimate $\frac{\det\left(g_{\alpha\bar{\beta}}(x,t)\right)}{\det\left(g_{\alpha\bar{\beta}}(x,0)\right)}$. In this case we use

$$\frac{\partial}{\partial t} \log \frac{\det\left(g_{\alpha\bar{\beta}}(x,t)\right)}{\det\left(g_{\alpha\bar{\beta}}(x,0)\right)} = r - R = -\Delta f = -\frac{\partial f}{\partial t},$$

which implies

$$\log \frac{\det\left(g_{\alpha\bar{\beta}}(x,t)\right)}{\det\left(g_{\alpha\bar{\beta}}(x,0)\right)} = -f(x,t) + f(x,0).$$

By the uniform bound (2.43) on f, we conclude
$$\frac{1}{\tilde{C}} \leq \frac{\det\left(g_{\alpha\bar{\beta}}(x,t)\right)}{\det\left(g_{\alpha\bar{\beta}}(x,0)\right)} \leq \tilde{C}$$
for some $\tilde{C} \geq 1$. □

REMARK 2.56. Alternatively, when $r \neq 0$, applying the estimates (2.55) and (2.56) to equation (2.50), we have
$$\left|\log \frac{\det g_{\alpha\bar{\beta}}(x,t)}{\det g_{\alpha\bar{\beta}}(x,0)}\right| = \left|\frac{\partial \varphi}{\partial t}(x,t) - \frac{r}{n}\varphi(x,t) + f(x,0)\right|$$
$$\leq \|f(\cdot,0)\|_\infty \left(e^{\frac{r}{n}t} + \left|e^{\frac{r}{n}t} - 1\right| + 1\right),$$
since $\max\{C_1, C_2\} \leq \|f(\cdot,0)\|_\infty$. In particular, if $r < 0$, then
$$e^{-2\|f(\cdot,0)\|_\infty} \leq \frac{\det\left(g_{\alpha\bar{\beta}}(x,t)\right)}{\det\left(g_{\alpha\bar{\beta}}(x,0)\right)} \leq e^{2\|f(\cdot,0)\|_\infty}.$$

Lemma 2.55 is the first step towards proving that $\nabla_\alpha \nabla_{\bar{\beta}} \varphi$ is bounded and that $g_{\alpha\bar{\beta}}(x,t)$ is equivalent to $g_{\alpha\bar{\beta}}(x,0)$, which in particular implies that $g_{\alpha\bar{\beta}}(x,t)$ is always positive definite.

Next we estimate the trace of $g_{\alpha\bar{\beta}}(x,t)$ with respect to $g_{\alpha\bar{\beta}}(x,0)$. Let

(2.59) $$Y(x,t) \doteqdot g^{\alpha\bar{\beta}}(x,0) g_{\alpha\bar{\beta}}(x,t)$$

be the trace-type quantity we want to estimate. As we shall see below, a bound for $Y(t)$ will imply a C^2-estimate for $\varphi(t)$. From (2.52) we have
$$Y = n + \Delta_{g(0)}\varphi,$$
(2.60) $$n = g^{\alpha\bar{\beta}}(t) g_{\alpha\bar{\beta}}(0) + \Delta_{g(t)}\varphi.$$

Hence an estimate for Y implies an estimate for $\Delta_{g(0)}\varphi$. Let λ_α denote the eigenvalues of $g_{\alpha\bar{\beta}}(t)$ with respect to $g_{\alpha\bar{\beta}}(0)$. Then

(2.61) $$Y = \sum_{\alpha=1}^n \lambda_\alpha$$

and the eigenvalues of $\left(\frac{\partial^2 \varphi}{\partial z^\alpha \partial \bar{z}^\beta}\right)$ with respect to $g_{\alpha\bar{\beta}}(0)$ are $\lambda_\alpha - 1$. If $Y \leq C$, then as long as $g_{\alpha\bar{\beta}}(t)$ is positive-definite, $\lambda_\alpha \leq C$ for each α. On the other hand, by Lemma 2.55, we have
$$\frac{1}{C} \leq \prod_{\alpha=1}^n \lambda_\alpha \leq C,$$
where for $r \leq 0$, the constant C is independent of time, whereas for $r > 0$, C depends on time but remains bounded as long as the solution exists (though the bound may tend to ∞ as time approaches ∞). Hence there exists a constant $c > 0$ such that $\lambda_\alpha \geq c$, where c is independent of time for $r \leq 0$ and may depend on time for $r > 0$. So indeed, $g_{\alpha\bar{\beta}}(t)$ remains positive-definite as long as the solution exists and we have $c \leq \lambda_\alpha \leq C'$, for some

constant $C' < \infty$. Thus *a bound on Y shall imply an estimate of the complex Hessian of φ*, i.e.,

$$\left|\nabla_\alpha\nabla_{\bar\beta}\varphi(t)\right|_{g(0)} \leq C' \tag{2.62}$$

for some $C' < \infty$. By abuse of notation, we shall call this the C^2-**estimate**.

REMARK 2.57. A quantity similar to Y also played an important role in the later work of Donaldson [**128**] on the Hermitian–Einstein flow.

We now turn to estimating Y. First we apply the heat operator to $\log Y$.

LEMMA 2.58. *We have*

$$\left(\frac{\partial}{\partial t} - \Delta\right)\log Y \leq -\frac{g^{\gamma\bar\delta}(t)g_{\alpha\bar\beta}(t)R^{\alpha\ \bar\beta}_{\ \gamma\bar\delta}(0)}{Y} + \frac{r}{n}, \tag{2.63}$$

where $\Delta \doteqdot \Delta_{g(t)}$ *and* $R^{\alpha\ \bar\beta}_{\ \gamma\bar\delta}(0) \doteqdot g^{\bar\eta\beta}(0)R^{\alpha}_{\ \gamma\bar\delta\eta}(0)$.

PROOF. From (2.59) and (2.29), we compute

$$\frac{\partial}{\partial t}\log Y = \frac{1}{Y}g^{\alpha\bar\beta}(0)\frac{\partial}{\partial t}g_{\alpha\bar\beta}(t) = \frac{-g^{\alpha\bar\beta}(0)R_{\alpha\bar\beta}(t) + \frac{r}{n}Y}{Y}. \tag{2.64}$$

Thus, to prove the lemma, it suffices to show that

$$Y\Delta\log Y \geq g^{\gamma\bar\delta}(t)g_{\alpha\bar\beta}(t)R^{\alpha\ \bar\beta}_{\ \gamma\bar\delta}(0) - g^{\alpha\bar\beta}(0)R_{\alpha\bar\beta}(t). \tag{2.65}$$

Given any point $x \in \mathcal{M}$, we will calculate in a local holomorphic coordinate system which is normal with respect to the metric $g(0)$ at x, so that $\frac{\partial}{\partial z^\alpha}g_{\beta\bar\gamma}(x,0) = 0$ and $g_{\alpha\bar\beta}(x,0) = \delta_{\alpha\beta}$. To simplify notation, we adopt the convention that the quantities below are at time t, unless there is a (0) after them, in which case they are at time 0. Since from (2.4), at x,

$$R^{\alpha}_{\ \gamma\bar\delta\mu}(0) = -g^{\alpha\bar\beta}(0)\frac{\partial^2}{\partial z^\gamma \partial z^{\bar\delta}}g_{\mu\bar\beta}(0) = -\frac{\partial^2}{\partial z^\gamma \partial z^{\bar\delta}}g_{\mu\bar\alpha}(0),$$

we compute that the Laplacian of Y at x is given by

$$\Delta Y = g^{\gamma\bar\delta}\frac{\partial^2}{\partial z^\gamma \partial z^{\bar\delta}}\left(g^{\alpha\bar\beta}(0)g_{\alpha\bar\beta}\right)$$

$$= g^{\gamma\bar\delta}g_{\alpha\bar\beta}\frac{\partial^2}{\partial z^\gamma \partial z^{\bar\delta}}g_{\beta\bar\alpha}(0) + g^{\gamma\bar\delta}g^{\alpha\bar\beta}(0)\frac{\partial^2}{\partial z^\gamma \partial z^{\bar\delta}}g_{\alpha\bar\beta}$$

$$= g^{\gamma\bar\delta}g_{\alpha\bar\beta}R^{\alpha\ \bar\beta}_{\ \gamma\bar\delta}(0) + g^{\alpha\bar\beta}(0)g^{\gamma\bar\delta}\frac{\partial^2}{\partial z^\gamma \partial z^{\bar\delta}}g_{\alpha\bar\beta}$$

$$= g^{\gamma\bar\delta}R^{\alpha\ \bar\beta}_{\ \gamma\bar\delta}(0)g_{\alpha\bar\beta} - g^{\alpha\bar\beta}(0)R_{\alpha\bar\beta} \tag{2.66}$$

$$\quad + g^{\alpha\bar\beta}(0)g^{\delta\bar\gamma}g^{\lambda\bar\eta}\frac{\partial}{\partial z^{\bar\delta}}g_{\alpha\bar\eta}\frac{\partial}{\partial z^\gamma}g_{\lambda\bar\beta},$$

where we used (2.5) and

$$R_{\alpha\bar\beta} = R^{\delta}_{\ \delta\bar\beta\alpha} = g^{\delta\bar\gamma}g^{\lambda\bar\eta}\frac{\partial}{\partial z^{\bar\delta}}g_{\alpha\bar\eta}\frac{\partial}{\partial z^\gamma}g_{\lambda\bar\beta} - g^{\delta\bar\eta}\frac{\partial^2}{\partial z^{\bar\delta}\partial z^{\bar\eta}}g_{\alpha\bar\beta} \tag{2.67}$$

(using the Kähler identities (2.2) to get the second equality in (2.67)). Note that
$$\nabla_\gamma Y = \frac{\partial}{\partial z^\gamma}\left(g^{\alpha\bar\beta}(0)g_{\alpha\bar\beta}\right) = g^{\alpha\bar\beta}(0)\frac{\partial}{\partial z^\gamma}g_{\alpha\bar\beta}$$
at x; this yields at x,
$$|\nabla Y|^2 = g^{\gamma\bar\delta}g^{\alpha\bar\beta}(0)g^{\nu\bar\mu}(0)\frac{\partial g_{\alpha\bar\beta}}{\partial z^\gamma}\frac{\partial g_{\nu\bar\mu}}{\partial z^{\bar\delta}}.$$
We claim that the Cauchy–Schwarz inequality gives
$$(2.68) \qquad |\nabla Y|^2 \leq Y \left(g(0)^{\nu\bar\beta}\, g^{\gamma\bar\delta}g^{\alpha\bar\mu}\frac{\partial g_{\alpha\bar\beta}}{\partial z^\gamma}\frac{\partial g_{\nu\bar\mu}}{\partial z^{\bar\delta}}\right).$$
By applying (2.68) to (2.66), we then obtain (2.65), as desired.

To prove (2.68), we first observe that we may further assume
$$g_{\alpha\bar\beta}(x,t) = \lambda_\alpha \delta_{\alpha\beta}$$
is diagonal, where λ_α is defined above. The LHS of (2.68) can be written at x as
$$|\nabla Y|^2 = \sum_{\alpha,\beta,\gamma} \frac{1}{\lambda_\gamma}\frac{\partial g_{\alpha\bar\alpha}}{\partial z^\gamma}\frac{\partial g_{\beta\bar\beta}}{\partial z^{\bar\gamma}}$$
$$\leq \sum_{\alpha,\beta}\left(\sum_\gamma \frac{1}{\lambda_\gamma}\left|\frac{\partial g_{\alpha\bar\alpha}}{\partial z^\gamma}\right|^2\right)^{\frac{1}{2}}\left(\sum_\eta \frac{1}{\lambda_\eta}\left|\frac{\partial g_{\beta\bar\beta}}{\partial z^\eta}\right|^2\right)^{\frac{1}{2}}$$
$$= \left(\sum_\alpha \left(\sum_\gamma \frac{1}{\lambda_\gamma}\left|\frac{\partial g_{\alpha\bar\alpha}}{\partial z^\gamma}\right|^2\right)^{\frac{1}{2}}\right)^2$$
$$= \left(\sum_\alpha (\lambda_\alpha)^{\frac{1}{2}}\left(\sum_\gamma \frac{1}{\lambda_\alpha \lambda_\gamma}\left|\frac{\partial g_{\alpha\bar\alpha}}{\partial z^\gamma}\right|^2\right)^{\frac{1}{2}}\right)^2$$
$$\leq \sum_\sigma \lambda_\sigma \cdot \sum_{\alpha,\gamma}\frac{1}{\lambda_\alpha \lambda_\gamma}\left|\frac{\partial g_{\alpha\bar\alpha}}{\partial z^\gamma}\right|^2$$
$$\leq \sum_\sigma \lambda_\sigma \cdot \sum_{\alpha,\beta,\gamma}\frac{1}{\lambda_\alpha \lambda_\gamma}\left|\frac{\partial g_{\alpha\bar\beta}}{\partial z^\gamma}\right|^2$$
$$= Y \left(g(0)^{\nu\bar\beta}\, g^{\gamma\bar\delta}g^{\alpha\bar\mu}\frac{\partial g_{\alpha\bar\beta}}{\partial z^\gamma}\frac{\partial g_{\nu\bar\mu}}{\partial z^{\bar\delta}}\right).$$
Thus the claimed inequality (2.68) and the lemma are both proved. $\square$

Next we prove the key C^2-estimate (2.62) via an application of the maximum principle to (2.63).

PROPOSITION 2.59 (C^2-estimate for φ). *Let $\varphi(t)$, $t \in [0,T)$, where $T \in (0,\infty]$, be a solution of the NKRF (2.50) on a closed Kähler manifold $(\mathcal{M}^n, g_0)$.*

(1) (Uniform estimate when $c_1 \leq 0$) If $c_1(\mathcal{M}) \leq 0$, then there exists a constant $C < \infty$ depending only on the initial metric such that

(2.69) $$Y(x,t) \leq C$$

for all $(x,t) \in \mathcal{M} \times [0,T)$. Hence there exists $C' < \infty$, independent of T, such that the complex Hessian of φ satisfies

(2.70) $$\left|\nabla_\alpha \nabla_{\bar\beta} \varphi(t)\right|_{g(0)} \leq C'$$

on $\mathcal{M} \times [0,T)$.

(2) (Time-dependent estimate when $c_1 > 0$) If $c_1(\mathcal{M}) > 0$, then there exist constants C and C', both depending on $g(0)$ and $T < \infty$, such that the above estimates (2.69) and (2.70) hold on $\mathcal{M} \times [0,T)$.

REMARK 2.60. Estimate (2.70) implies $\left|\Delta_{g(0)}\varphi\right| \leq C'$. On the other hand, by Lemma 2.53, we have $|\varphi| \leq C_0$ for some $C_0 < \infty$. Hence, by standard elliptic theory, we have for any $\alpha \in (0,1)$, $\|\varphi\|_{C^{1,\alpha}} \leq C_1$ for some $C_1 < \infty$ depending on α. However at this stage of the proof, it is not clear whether $|\text{Hess}\,\varphi|$ is bounded, where Hess denotes the real Hessian, but we do not need this.

PROOF. By the discussion above, regarding inequality (2.62), we only need to bound Y from above. Again, we calculate in local holomorphic coordinates around x where $g_{\alpha\bar\beta}(x,0) = \delta_{\alpha\beta}$, $\frac{\partial}{\partial z^\gamma} g_{\alpha\bar\beta}(x,0) = 0$, and $g_{\alpha\bar\beta}(x,t) = \lambda_\alpha \delta_{\alpha\beta}$. By (2.63) and $\lambda_\alpha \leq Y$, we have

(2.71) $$\left(\frac{\partial}{\partial t} - \Delta\right)\log Y \leq -\frac{R_{\alpha\bar\alpha\gamma\bar\gamma}(0)\lambda_\alpha}{Y\lambda_\gamma} + \frac{r}{n} \leq C_1 \sum_{\gamma=1}^n \frac{1}{\lambda_\gamma} + \frac{r}{n},$$

where C_1 is a constant depending only on a lower bound of the bisectional curvatures of $g(0)$. To control the bad terms on the RHS above, we consider equation (2.54)

(2.72) $$\left(\frac{\partial}{\partial t} - \Delta\right)\varphi = -f - \Delta\varphi = -f - n + \sum_{\gamma=1}^n \frac{1}{\lambda_\gamma},$$

where the second equality follows from (2.60). Consider the modified quantity
$$w = \log Y - (C_1 + 1)\varphi.$$
Combining (2.71) and (2.72), we have

(2.73) $$\left(\frac{\partial}{\partial t} - \Delta\right)w \leq -\sum_{\gamma=1}^n \frac{1}{\lambda_\gamma} + C_2,$$

where C_2 depends on C_1 and $\|f\|_\infty$ (if $r \leq 0$, then C_2 is independent of time, and if $r > 0$, then C_2 depends on time). By the maximum principle, (2.73) implies that w can be bounded above on $\mathcal{M} \times [0,T)$ by a constant C depending on T.

To get a bound for w independent of T when $r \leq 0$, we need to work harder. When $r \leq 0$, we shall use the term $-\sum_\gamma \frac{1}{\lambda_\gamma}$ to dominate (from below) a function of w which approaches $-\infty$ as $w \to \infty$. Using equation (2.50), $f = -\frac{\partial}{\partial t}\varphi$, and (2.61), we have

$$f(x,t) - f(x,0) + \frac{r}{n}\varphi(x,t) = -\log\frac{\det g_{\alpha\bar{\beta}}(x,t)}{\det g_{\alpha\bar{\beta}}(x,0)},$$

and hence

$$Y e^{f(x,t) - f(x,0) + \frac{r}{n}\varphi(x,t)} = \sum_\alpha \lambda_\alpha \cdot \prod_\gamma \frac{1}{\lambda_\gamma}$$

$$= \sum_\alpha \left(\prod_{\gamma \neq \alpha} \frac{1}{\lambda_\gamma}\right) \leq \left(\sum_\alpha \frac{1}{\lambda_\alpha}\right)^{n-1}$$

using a standard inequality (we dropped a factor of $\frac{1}{n^{n-2}}$ since it makes the inequality easier to see). Since f and φ are uniformly bounded, this implies that there exists a constant C_3, which only depends on the initial data, so that

$$(2.74) \qquad Y \leq C_3 \left(\sum_\alpha \frac{1}{\lambda_\alpha}\right)^{n-1}.$$

Notice that $e^w = e^{-(C_1+1)\varphi}Y$. We then have

$$(2.75) \qquad e^w \leq \left(C_4 \sum_\alpha \frac{1}{\lambda_\alpha}\right)^{n-1},$$

where $C_4 > 0$ only depends on the initial data. Combining (2.73) and (2.75), we have

$$(2.76) \qquad \left(\frac{\partial}{\partial t} - \Delta\right) w \leq -\frac{1}{C_4} e^{\frac{w}{n-1}} + C_2.$$

This implies that, at any time t where $w_{\max}(t) \geq (n-1)\log(C_2 C_4)$, we have $\frac{d}{dt} w_{\max}(t) \leq 0$. Hence, by the maximum principle,

$$\sup_{x \in \mathcal{M}} w(x,t) \leq \max\left\{\sup_{x \in \mathcal{M}} w(x,0), (n-1)\log(C_2 C_4)\right\}$$

for all $t \in [0,T)$, and hence Y is also bounded from above. When $r \leq 0$, both the constant C_2 and the function φ are uniformly bounded independent of T; hence Y is uniformly bounded independent of T. $\square$

Now we can complete the proof of Theorem 2.50.

PROOF OF THEOREM 2.50. Let $\varphi(t)$, $t \in [0,T)$, be a solution of the NKRF (2.50) on a closed Kähler manifold $(\mathcal{M}^n, g_0)$, where T is the maximal time of existence. If $T < \infty$, then by the C^2-estimate (i.e., the estimate for the complex Hessian of φ), the metrics $g(t)$ are uniformly equivalent to $g(0)$

on the time interval $[0, T)$. Once we have a uniform $C^{2,\alpha}$-estimate of φ on $[0, T)$ (we shall prove the $C^{2,\alpha}$-estimate in the next subsection), by choosing a time $t_0 < T$ close enough to T, the NKRF (2.50) with initial condition $\varphi(t_0)$ can be solved on $[t_0, t_0 + \varepsilon]$, where ε depends on the $C^{2,\alpha}$-estimate of φ but not t_0. If we choose $t_0 = T - \frac{\varepsilon}{2}$, this implies that we have a solution $\varphi(t)$ for $t \in [0, T + \frac{\varepsilon}{2}]$. This contradicts T being the maximal time of existence; hence the theorem is proved. $\square$

We shall supply the details for the $C^{2,\alpha}$-estimate in the next subsection.

5.3. The $C^{2,\alpha}$-estimate of φ. We now proceed to derive the $C^{2,\alpha}$-estimate for φ, which by (2.50) and (2.54) is a solution to the complex Monge–Ampère equation:

$$\log \det \left(g^0_{\alpha\bar{\beta}} + \varphi_{\alpha\bar{\beta}} \right) = \tilde{h} + \log \det g^0_{\alpha\bar{\beta}} \doteqdot h,$$

and

(2.77) $$\tilde{h}(x, t) = -f(x, t) + f(x, 0) - \frac{r}{n}\varphi(x, t).$$

By (2.70), there exists a constant $\Lambda > 0$ such that the Kähler metric $g_{\alpha\bar{\beta}} \doteqdot g^0_{\alpha\bar{\beta}} + \varphi_{\alpha\bar{\beta}}$ satisfies

(2.78) $$\frac{1}{\Lambda} \left(g^0_{\alpha\bar{\beta}} \right) \leq \left(g_{\alpha\bar{\beta}} \right) \leq \Lambda \left(g^0_{\alpha\bar{\beta}} \right).$$

By the fact that the bounded function f satisfies $\Delta f = R - r$ and that we have the scalar curvature bound (2.47), standard L^q theory (which only requires the uniform boundedness of the coefficient matrix $\left(g^{\alpha\bar{\beta}} \right)$ from above and below) implies that $\|f\|_{C^0(\mathcal{M})} + \|\nabla\bar{\nabla}f\|_{L^q(\mathcal{M})}$ is bounded and hence by (2.77), $\|\tilde{h}\|_{C^0(\mathcal{M})} + \|\nabla\bar{\nabla}\tilde{h}\|_{L^q(\mathcal{M})}$ is bounded (uniformly in t when $c_1(\mathcal{M}) \leq 0$) for any $q < \infty$ (independent of t).

Note that $\tilde{h}$ is globally defined whereas h is only locally defined. However, for compactly contained open subsets U of a holomorphic coordinate chart of g_0, $\|\tilde{h}\|_{C^0(U)} + \|\nabla\bar{\nabla}\tilde{h}\|_{L^q(U)}$ and $\|h\|_{C^0(U)} + \|\nabla\bar{\nabla}h\|_{L^q(U)}$ are equivalent. Let $B(R)$ denote the Euclidean ball of radius R centered at the origin in $\mathbb{C}^n$. Since $\mathcal{M}$ is compact, there exists a finite collection of open sets $\{U_k\}_{k=1}^{N_0}$ and normal holomorphic coordinates $z_k = \{z_k^\alpha\}_{\alpha=1}^n$ defined on U_k (independent of t) such that

$$B(3R_{k0}) \subset z_k(U_k) \quad \text{and} \quad \bigcup_{k=1}^{N_0} z_k^{-1}(B(R_{k0})) = \mathcal{M},$$

where $R_{k0} > 0$. Hence it suffices to prove the $C^{2,\alpha}$-estimate for φ in each open set $z_k^{-1}(B(R_{k0}))$ assuming that

(2.79) $$\|h\|_{C^0(U_k)} + \|\nabla\bar{\nabla}h\|_{L^q(U_k)} \leq C < \infty.$$

From now on we work in a fixed coordinate chart. More precisely, we use z_k to push forward our discussion to the Euclidean ball $B(3R_{k0})$. For simplicity we drop the indices k in our notation below. Since g is Kähler, we may write $g_{\alpha\bar{\beta}}$ locally as the complex (Hermitian) Hessian $u_{\alpha\bar{\beta}}$ of a function u. We shall show that the second derivative of u has bounded Hölder norm. Now the equation reads locally as

(2.80) $$\log\det\left(u_{\alpha\bar{\beta}}\right) = h.$$

It is convenient to write log det as a function $F(p)$, where p lies in the domain of positive definite Hermitian symmetric matrices. An important property that we shall make full use of is that $F(p) = \log\det(p)$ is a *concave function* of p, a fact which can be easily checked. Taking the derivative of (2.80), we have

$$\frac{\partial F}{\partial p_{\alpha\bar{\beta}}} u_{\alpha\bar{\beta}\gamma} = h_\gamma,$$

$$\frac{\partial^2 F}{\partial p_{\alpha\bar{\beta}}\partial p_{\mu\bar{\nu}}} u_{\alpha\bar{\beta}\gamma} u_{\mu\bar{\nu}\bar{\gamma}} + \frac{\partial F}{\partial p_{\alpha\bar{\beta}}} u_{\alpha\bar{\beta}\gamma\bar{\gamma}} = h_{\gamma\bar{\gamma}}$$

for each $\gamma = 1, \ldots, n$. By the concavity of F we have

(2.81) $$\frac{\partial F}{\partial p_{\alpha\bar{\beta}}} u_{\alpha\bar{\beta}\gamma\bar{\gamma}} \geq h_{\gamma\bar{\gamma}}.$$

On the other hand recall that

$$\frac{\partial F}{\partial p_{\alpha\bar{\beta}}} = u^{\alpha\bar{\beta}} = g^{\alpha\bar{\beta}},$$

where $\left(u^{\alpha\bar{\beta}}\right)$ is the inverse of $\left(u_{\alpha\bar{\beta}}\right) = \left(g_{\alpha\bar{\beta}}\right)$.[5] Therefore we can rewrite (2.81) as

(2.82) $$\Delta_u w \geq h_{\gamma\bar{\gamma}},$$

where

$$w \doteqdot u_{\gamma\bar{\gamma}}$$

and Δ_u denotes the Laplacian with respect to the metric $g_{\alpha\bar{\beta}}$.

For any $R \leq R_0$, let

(2.83) $$M(s) = \sup_{B(sR)} w \quad \text{and} \quad m(s) = \inf_{B(sR)} w.$$

Also define the **oscillation function**:

$$\omega(sR) \doteqdot M(s) - m(s).$$

The following weak form of the Harnack inequality, which holds in general for linear elliptic operators of divergence form, plays a crucial role in our estimate. One can find the proof of this result in various papers and

---

[5]This is not different from the real version of this formula used in Volume One, which is the variation formula

$$\frac{\partial}{\partial s}\log\det A_{ij} = \left(A^{-1}\right)^{ij}\frac{\partial}{\partial s}A_{ij}$$

for any invertible matrix A_{ij}.

books on PDE, such as Moser [**276**], Morrey [**274**], Gilbarg and Trudinger [**155**], Han and Lin [**195**] (e.g., see Theorem 4.15 on p. 83 of [**195**]).

THEOREM 2.61 (Harnack inequality). *Let $u : B(3R_0) \to \mathbb{R}$ be a C^2 function such that*

$$\frac{1}{\Lambda_0} \left(\delta_{\alpha\bar{\beta}}\right) \leq \left(u_{\alpha\bar{\beta}}\right) \leq \Lambda_0 \left(\delta_{\alpha\bar{\beta}}\right)$$

for some $\Lambda_0 \in [1, \infty)$. Suppose that a nonnegative function $v \in W^{2,2}(B(3R_0))$ and $\check{g} \in L^q(B(3R_0))$, for some $q > m/2$, satisfy

$$\Delta_u v \leq \check{g}$$

in the weak sense in $B(3R_0)$, where Δ_u denotes the Laplacian with respect to the metric $u_{\alpha\bar{\beta}}$. Then for any $0 < \theta \leq \tau < 1$ and $0 < p < \frac{m}{m-2}$, there exists a constant $C = C(p, q, 2n, \Lambda_0, \theta, \tau) < \infty$ such that for any $\rho \leq 2R_0$,

$$(2.84) \qquad \left(\frac{1}{\rho^m} \int_{B(\tau\rho)} v(y)^p \, dy\right)^{\frac{1}{p}} \leq C \left(\inf_{B(\theta\rho)} v + \rho^{2-\frac{m}{q}} \|\check{g}\|_{L^q(B(\rho))}\right).$$

REMARK 2.62. The reason for why we can apply this theorem to

$$\left(u_{\alpha\bar{\beta}}\right) = \left(g_{\alpha\bar{\beta}}\right) = \left(g^0_{\alpha\bar{\beta}} + \varphi_{\alpha\bar{\beta}}\right)$$

is that the C^2-estimate for φ yields (2.78).

Note that $\Delta_u (M(2) - w) \leq -h_{\gamma\bar{\gamma}}$ and $-h_{\gamma\bar{\gamma}} \in L^q$ for all $q \leq \infty$ (e.g., $q < \infty$), so we may apply the above theorem to $M(2) - w$ (for example, with $m = 2n$, $\rho = 2R$,[6] and $\theta = \tau = \frac{1}{2}$, so that $\theta\rho = \tau\rho = R$) to obtain for any $q > n$ and $0 < p < \frac{n}{n-1}$ that there exists $C = C(p, q, n, \Lambda) < \infty$ such that

$$\left(\frac{1}{R^{2n}} \int_{B(R)} (M(2) - w(y))^p dy\right)^{\frac{1}{p}}$$
$$(2.85) \qquad \leq C \left(M(2) - M(1) + R^{\frac{2(q-n)}{q}} \|h_{\gamma\bar{\gamma}}\|_{L^q(B(2R))}\right)$$

since $\inf_{B(R)} (-w) = M(1)$.

On the other hand, the concavity of F implies

$$F(u_{i\bar{j}}(x)) \leq F(u_{i\bar{j}}(y)) + \frac{\partial F}{\partial p_{i\bar{j}}}(u_{i\bar{j}}(y))(u_{i\bar{j}}(x) - u_{i\bar{j}}(y)).$$

Namely we have

$$(2.86) \qquad h(y) - h(x) \geq g^{i\bar{j}}(y) \left(u_{i\bar{j}}(y) - u_{i\bar{j}}(x)\right).$$

The following linear algebra fact enables us to estimate $\omega(R)$.

---

[6]Note that $R \leq R_0$.

LEMMA 2.63 (Linear algebra). *There exist unitary vectors $r_1, \ldots, r_N \in \mathbb{C}^n$ with the property that, as Hermitian symmetric positive definite matrices,*

$$(2.87) \qquad (g^{i\bar{j}}(y)) = \sum_{\nu=1}^{N} a_\nu(y) r_\nu \otimes \overline{r_\nu}.$$

Here $a_\nu(y) \in \mathbb{R}$ and $\frac{1}{A\Lambda} \leq a_\nu(y) \leq A\Lambda$ for some constant $A > 0$. Moreover we may assume that the first n vectors $r_1, \ldots, r_n$ form a unitary basis of $\mathbb{C}^n$.

REMARK 2.64. Let $\{e_i\}_{i=1}^n$ denote the standard basis for $\mathbb{C}^n$ and write $r_\nu \doteqdot \sum_{i=1}^n (r_\nu)^i e_i$ for each ν. By (2.87) we mean that

$$(g^{i\bar{j}}(y)) = \sum_{\nu=1}^{N} a_\nu(y) (r_\nu)^i \overline{(r_\nu)^j}.$$

EXERCISE 2.65. Prove the above linear algebra lemma.

Define

$$w_\nu \doteqdot \mathrm{Hess}(u)(r_\nu, \overline{r_\nu}) = u_{i\bar{j}} (r_\nu)_i \overline{(r_\nu)_j}.$$

Now we let $M_\nu(s)$ and $m_\nu(s)$ denote the quantities $M(s)$ and $m(s)$ defined by (2.83) using w_ν instead of w. Then (2.86) implies

$$(2.88) \qquad \sum_{\nu=1}^{N} a_\nu(y)(w_\nu(y) - w_\nu(x)) \leq h(y) - h(x).$$

Choosing $x \in \overline{B(2R)}$ to be a point where $w_1(x) = m_1(2)$, this in particular implies

$$a_1(y)(w_1(y) - w_1(x)) \leq h(y) - h(x) + \sum_{\nu \geq 2} a_\nu(y)(w_\nu(x) - w_\nu(y))$$

and hence

$$w_1(y) - m_1(2) \leq C(\Lambda, A)\left(R\|\nabla h\|_{C^0} + \sum_{\nu \geq 2}(M_\nu(2) - w_\nu(y))\right),$$

where we used $a_1(y) \geq \frac{1}{A\Lambda}$, and $a_\nu(y) \leq A\Lambda$ and $w_\nu(x) \leq M_\nu(2)$ for $\nu \geq 2$. Thus

$$\left(\frac{1}{R^{2n}} \int_{B(R)} (w_1(y) - m_1(2))^p dy\right)^{\frac{1}{p}} \leq C(\Lambda, A) R \|\nabla h\|_{C^0}$$

$$(2.89) \qquad + C(\Lambda, A) \sum_{\nu \geq 2} \left(\frac{1}{R^{2n}} \int_{B(R)} (M_\nu(2) - w_\nu(y))^p dy\right)^{\frac{1}{p}}.$$

5. EXISTENCE AND CONVERGENCE

On the other hand, applying (2.85) to bound the L^p-norm of $M_\nu(2) - w_\nu$, we have for each $\nu \geq 2$,

$$\left(\frac{1}{R^{2n}} \int_{B(R)} (M_\nu(2) - w_\nu(y))^p dy\right)^{\frac{1}{p}}$$
(2.90) $$\leq C \left(M_\nu(2) - M_\nu(1) + R^{\frac{2(q-n)}{q}} \|\operatorname{Hess}(h)(r_\nu, \overline{r_\nu})\|_{L^q(B(2R))} \right).$$

Combining (2.89) and (2.90), we have

(2.91) $$\left(\frac{1}{R^{2n}} \int_{B(R)} (w_1(y) - m_1(2))^p dy\right)^{\frac{1}{p}}$$
$$\leq C \left(\max_{\nu \geq 2} (M_\nu(2) - M_\nu(1)) + R\|\nabla h\|_{C^0} + R^{\frac{2(q-n)}{q}} \|\nabla \bar{\nabla} h\|_{L^q} \right).$$

Now let

$$\bar{\omega}(sR) \doteq \sum_{\nu=1}^{N} \omega_\nu(sR) \doteq \sum_{\nu=1}^{N} \left(\sup_{B(sR)} w_\nu - \inf_{B(sR)} w_\nu \right)$$
$$= \sum_{\nu=1}^{N} (M_\nu(s) - m_\nu(s)).$$

We then have

$$\left(\frac{1}{R^{2n}} \int_{B(R)} (w_1(y) - m_1(2))^p \right)^{\frac{1}{p}}$$
(2.92) $$\leq C \left(\bar{\omega}(2R) - \bar{\omega}(R) + R\|\nabla h\|_{C^0} + R^{\frac{2(q-n)}{q}} \|\nabla \bar{\nabla} h\|_{L^q} \right).$$

On the other hand, from (2.85) we also have

$$\left(\frac{1}{R^{2n}} \int_{B(R)} (M_1(2) - w_1(y))^p \right)^{\frac{1}{p}}$$
(2.93) $$\leq C \left(\bar{\omega}(2R) - \bar{\omega}(R) + R^{\frac{2(q-n)}{q}} \|\nabla \bar{\nabla} h\|_{L^q} \right).$$

Putting these together, we obtain

$$\omega_1(2R) = \left(\frac{1}{\operatorname{Vol} B(R)} \int_{B(R)} (M_1(2) - m_1(2))^p\right)^{\frac{1}{p}}$$

$$\leq \left(\frac{1}{\operatorname{Vol} B(R)} \int_{B(R)} (w_1(y) - m_1(2))^p\right)^{\frac{1}{p}}$$

$$+ \left(\frac{1}{\operatorname{Vol} B(R)} \int_{B(R)} (M_1(2) - w_1(y))^p\right)^{\frac{1}{p}}$$

$$\leq C\left(\bar{\omega}(2R) - \bar{\omega}(R) + R\|\nabla h\|_{C^0} + R^{\frac{2(q-n)}{q}}\|\nabla\bar{\nabla} h\|_{L^q}\right).$$

Since there is nothing special about the index 1, summing the corresponding upper bounds for $\omega_\nu(2R)$ implies

$$\bar{\omega}(2R) \leq C\left(\bar{\omega}(2R) - \bar{\omega}(R) + R\|\nabla h\|_{C^0} + R^{\frac{2(q-n)}{q}}\|\nabla\bar{\nabla} h\|_{L^q}\right)$$

with a different constant $C < \infty$. We conclude the following.

LEMMA 2.66 (Oscillation estimate). *There exists $\delta < 1$ (i.e., $\delta = 1 - \frac{1}{C}$) such that for any $R \leq R_0$ we have on $B(3R_0)$,*

(2.94) $$\bar{\omega}(R) \leq \delta \cdot \bar{\omega}(2R) + R\|\nabla h\|_{C^0} + R^{\frac{2(q-n)}{q}}\|\nabla\bar{\nabla} h\|_{L^q}.$$

Now since $R^{\frac{2(q-n)}{q}}\|\nabla\bar{\nabla} h\|_{L^q(U)}$ and $\|\nabla h\|_{C^0(U)}$ are bounded by (2.79), the Hölder continuity of $\nabla\bar{\nabla} u$ on $B(R_0)$ can be derived from (2.94) by a standard argument; see Moser [**276**], or Corollary 4.18 on p. 91 and Lemma 4.19 on p. 92 of Han and Lin [**195**], for example. Finally, the Hölder continuity of $\nabla\bar{\nabla} u$ is equivalent to the Hölder continuity of $\nabla\bar{\nabla}\varphi$.

5.4. Proof of Theorem 2.51. Finally, we give the proof of Theorem 2.51, i.e., the proof of the convergence of the normalized Kähler–Ricci flow in the case where $c_1 < 0$. Assume, without loss of generality, that $r = -n$, which can be achieved by scaling the initial metric g_0. Notice that we have that $f = -\frac{\partial \varphi}{\partial t}$ satisfies

$$\frac{\partial}{\partial t} f = \Delta_{g(t)} f - f$$

and $|f(x,t)| \leq Ce^{-t}$ and $|\nabla f|(x,t) \leq Ce^{-t}$. (The last inequality is by (2.48).) That is,

$$\|f\|_{C^1(\mathcal{M})} \leq C_1 e^{-t}$$

for some $C_1 < \infty$.

Now $\Delta_{g(t)} = g^{\alpha\bar{\beta}} \frac{\partial^2}{\partial z^\alpha \partial z^{\bar{\beta}}}$ and $g_{\alpha\bar{\beta}} = g^0_{\alpha\bar{\beta}} + \varphi_{\alpha\bar{\beta}}$. So the $C^{2,\alpha}$-estimate for φ implies a C^α-estimate for the coefficients $g^{\alpha\bar{\beta}}$. Thus we may apply the

parabolic Schauder estimate (e.g., Theorem 5 on p. 64 of Friedman [**146**]) to obtain
$$\|f\|_{C^{2,\alpha}(\mathcal{M})} \leq C_2 e^{-t}$$
for some $C_2 < \infty$. Iterating the Schauder estimate, we have $\|f\|_{C^{2m,\alpha}(\mathcal{M})} \leq C_m e^{-t}$ for some constants $C_m < \infty$ and all $m \in \mathbb{N}$. This implies the estimate $\left\|\frac{\partial \varphi}{\partial t}\right\|_{C^{2m,\alpha}(\mathcal{M})} \leq C_2 e^{-t}$ and hence implies the exponential convergence of $\varphi(\cdot, t) \to \varphi_\infty(\cdot)$ in C^∞ as $t \to \infty$ for some smooth function φ_∞. This proves that the normalized Kähler–Ricci flow converges in C^∞ to a Kähler–Einstein metric with negative scalar curvature. Theorem 2.51 is proved.

6. Survey of some results for the Kähler–Ricci flow

6.1. Closed Kähler manifolds with nonnegative bisectional curvature.
Using the short-time existence of the Kähler–Ricci flow, the result of Mori, Siu and Yau (Theorem 2.33) was generalized by Bando [**19**] when $n = 3$ and Mok [**269**] for $n \geq 4$.[7] Mok also used techniques from algebraic geometry.

THEOREM 2.67 (Kähler manifolds with nonnegative bisectional curvature). *If $(\mathcal{M}^n, g)$ is a closed Kähler manifold with nonnegative bisectional curvature, then its universal cover $\left(\widetilde{\mathcal{M}}^n, \tilde{g}\right)$ is isometrically biholomorphic to the product of complex Euclidean space, compact irreducible Hermitian symmetric spaces of rank at least 2, and complex projective spaces with Kähler metrics of nonnegative bisectional curvature.*

REMARK 2.68. Note that the above classification is up to isometry. Any complex projective space admits a metric with constant holomorphic sectional curvature (i.e., the Fubini-Study metric).

The proof of the theorem above uses the following result, proved by Bando for $n = 3$ and Mok for $n \geq 4$. We discuss this result further in Section 8 below.

THEOREM 2.69 (KRF: nonnegative bisectional curvature is preserved). *If $(\mathcal{M}^n, g(0))$ is a closed Kähler manifold with nonnegative bisectional curvature, then the solution $g(t)$ to the Kähler–Ricci flow has nonnegative bisectional curvature for all $t \geq 0$. If in addition $g(0)$ has positive Ricci curvature at one point, then $g(t)$ has positive holomorphic sectional curvature and positive Ricci curvature for all $t > 0$.*

Using the existence of a Kähler–Einstein metric, the convergence in the case of positive bisectional curvature was settled by Chen and Tian [**87**], [**88**].

---

[7]The case of nonnegative curvature operator was considered by Cao and one of the authors [**50**].

THEOREM 2.70 (KRF: compact positive bisectional curvature). *Suppose $(\mathcal{M}^n, g(0))$ is a closed Kähler manifold with nonnegative bisectional curvature everywhere and positive bisectional curvature at a point. Then the solution $g(t)$ to the normalized Kähler–Ricci flow, which has positive bisectional curvature for all $t > 0$, converges exponentially fast to the Fubini-Study metric of constant holomorphic sectional curvature on $\mathbb{C}P^n$.*

REMARK 2.71. Without using the existence of a Kähler–Einstein metric, Cao, Chen, and Zhu [**49**] proved a uniform curvature estimate (see Theorem 2.92).

6.2. Uniformization of noncompact Kähler manifolds with nonnegative bisectional curvature. In this subsection we recall Yau's fundamental conjecture on the uniformization of complete noncompact Kähler manifolds with nonnegative bisectional curvature.

CONJECTURE 2.72 (Noncompact Kähler uniformization $K_\mathbb{C}(V, W) > 0$). *If $(\mathcal{M}^n, g(0))$ is a complete noncompact Kähler manifold with positive bisectional curvature, then $\mathcal{M}$ is biholomorphic to $\mathbb{C}^n$.*

Using the Kähler–Ricci flow on noncompact manifolds, Chau and Tam [**65**] proved the following result, which affirms Yau's conjecture in the case of bounded curvature and **maximum volume growth**.

THEOREM 2.73 (KRF: noncompact positive bisectional curvature). *If $(\mathcal{M}^n, g(0))$ is a complete noncompact Kähler manifold with bounded positive bisectional curvature and maximum volume growth, then $\mathcal{M}$ is biholomorphic to $\mathbb{C}^n$.*

There have been a number of works on the Kähler–Ricci flow on noncompact manifolds with positive bisectional curvature. For example, the reader may consult Shi [**331**], Tam and one of the authors [**290**], [**292**], and Chen and Zhu [**79**], [**80**].

6.3. Limiting behavior of the Kähler–Ricci flow on closed manifolds. There are also the following results about the limiting behavior of the Kähler–Ricci flow due to Sesum [**323**].

THEOREM 2.74. *If $(\mathcal{M}^n, g(t))$, $t \in [0, \infty)$, is a solution to the Kähler–Ricci flow on a closed manifold with uniformly bounded Ricci curvature, then for any sequence $t_i \to \infty$ there exists a subsequence such that $(\mathcal{M}, g(t + t_i))$ converges to $(\mathcal{M}^n_\infty, g_\infty(t))$, where $g_\infty(t)$ is a solution to the Kähler–Ricci flow. The convergence is outside a set of real codimension 4.*

When $n = 2$, Sesum improved the above result to the following.

THEOREM 2.75. *If $(\mathcal{M}^2, g(t))$, $t \in [0, \infty)$, is a solution to the Kähler–Ricci flow on a closed manifold with uniformly bounded Ricci curvature, then for any sequence $t_i \to \infty$ there exists a subsequence such that $(\mathcal{M}, g(t + t_i))$ converges to $(\mathcal{M}^2_\infty, g_\infty(t))$, where $g_\infty(t)$ is a Kähler–Ricci soliton. The convergence is outside a finite number of points.*

For some other recent work on the Kähler–Ricci flow the reader is referred to Phong-Sturm [**305**], [**306**], Chen [**84**], Chen and Li [**85**], Song and Tian [**337**], Cascini and La Nave [**60**], Tian and Zhang [**351**], and [**234**].

7. Examples of Kähler–Ricci solitons

In this section, we provide a brief and regrettably incomplete sampling of some results on Kähler–Ricci solitons. These special solutions model singularities of the Kähler–Ricci flow. A **Kähler–Ricci soliton** is a Kähler manifold $(\mathcal{M}^n, g, J)$ such that the soliton structure equation

$$(2.95) \qquad \operatorname{Rc} + \lambda g + \frac{1}{2}\mathcal{L}_X g = 0$$

holds for some constant $\lambda \in \mathbb{R}$ and some real vector field X which is an **infinitesimal automorphism** (2.1) of the complex structure J. Note that X is an infinitesimal automorphism if and only if its $(1,0)$-part is holomorphic: $0 = \bar{\nabla}_\alpha X^\beta = \frac{\partial}{\partial \bar{z}^\alpha} X^\beta$. One imposes this requirement for the following reason. As we saw in Lemma 2.36, a solution of Ricci flow that starts with a Kähler metric on a complex manifold remains Kähler with respect to the same complex structure. On the other hand, if φ_t is any family of diffeomorphisms of $\mathcal{M}$, then each pullback $\varphi_t^*(g)$ is Kähler with respect to the complex structure $\varphi_t^*(J)$. Now consider the evolving metric $h(t) := (1 + \lambda t)\varphi_t^* g$, where φ_t is the family of diffeomorphisms generated by $\frac{1}{2(1+\lambda t)} X$. If X is an infinitesimal automorphism of the complex structure, then $\varphi_t^*(J) \equiv J$, which implies that $h(t)$ remains Kähler with respect to the same complex structure. Furthermore, using (2.95), it is easy to see that h solves the Kähler–Ricci flow: $\frac{\partial}{\partial t} h = -\operatorname{Rc}(h)$.

One may also define a Kähler–Ricci soliton to be a Kähler manifold $(\mathcal{M}^n, g, J)$ together with a constant $\lambda \in \mathbb{R}$ and a real vector field X satisfying the complex soliton equation

$$(2.96) \qquad R_{\alpha\bar{\beta}} + \lambda g_{\alpha\bar{\beta}} + \frac{1}{2}(\mathcal{L}_X g)_{\alpha\bar{\beta}} = 0.$$

Equation (2.96) is equivalent to the conjunction of equation (2.95) and the statement that X is holomorphic. Notice that if we restrict our attention to gradient solitons (so that X is the gradient of a real-valued function), then (2.96) is equivalent to (2.95) without any extra hypotheses. (See §2.2 of [**142**] for the detailed argument.)

7.1. Existence and uniqueness.
Any Kähler metric satisfying (2.96) with $X = 0$ is Kähler–Einstein. In this sense, Kähler–Einstein metrics may be regarded as trivial Kähler–Ricci solitons.[8] So if no Kähler–Einstein metric exists, a natural replacement is a Kähler–Ricci soliton. In fact, existence of a Kähler–Einstein metric and a nontrivial gradient Kähler–Ricci soliton

---

[8]Of course, there is nothing 'trivial' about Kähler–Einstein metrics!

are mutually exclusive: applying the **Futaki functional** $\mathcal{F}_{[\omega]}$ to the holomorphic vector field $X = \operatorname{grad} f$, one gets

$$\mathcal{F}_{[\omega]}(X) = \int_{\mathcal{M}} \langle X, X \rangle \, d\mu = \|X\|^2 > 0.$$

Moreover, Kähler–Ricci solitons on a compact Kähler manifold $(\mathcal{M}^n, J)$ are unique up to holomorphic automorphisms. (See Tian and Zhu [**352, 353, 354**].) Specifically, we have the following theorem.

THEOREM 2.76 (Uniqueness of Kähler-Ricci solitons). *Let $(\mathcal{M}^n, J)$ be a compact Kähler manifold. If metrics g and g' on $\mathcal{M}$ satisfy (2.95) with respect to holomorphic vector fields X and X', respectively, then there is an element σ in the identity component of the holomorphic automorphism group such that $g = \sigma^* g'$ and $X = (\sigma^{-1})_* X'$.*

For recent results on uniqueness and other properties of noncompact Kähler–Ricci solitons, see [**63, 64**], [**35**] and [**78**].

One might ask, therefore, whether there exists either a Kähler–Einstein metric or else a Kähler–Ricci soliton on every compact Kähler manifold $\mathcal{M}^n$ with $c_1(\mathcal{M}) > 0$. The answer is yes if $n \leq 2$. A compact complex surface with $c_1 > 0$ is $\mathbb{P}^2 \#_k \overline{\mathbb{P}}^2$ for some $k \in \{0, 1, \ldots, 8\}$. (Here and below, $\mathbb{P}^n = \mathbb{CP}^n$ is **complex projective space**.) A Kähler–Einstein metric exists for $k = 0$ and $3 \leq k \leq 8$. (See Theorem 2.34.) In the remaining cases $k = 1, 2$, there is a (non-Einstein) Kähler–Ricci soliton. (See [**239**], [**366**], [**47**], and Section 7.2 below.) In higher dimensions, however, the answer is no. There exist 3-dimensional compact complex manifolds that admit no Kähler–Einstein metric and no holomorphic vector fields, hence no Kähler–Ricci soliton structure. (See [**346**, §7] as well as [**215, 216**] and [**278**].) More generally, Tian and Zhu have exhibited a holomorphic invariant that generalizes the Futaki invariant and acts as an obstruction to the existence of a Kähler–Ricci soliton metric [**354**] on a compact complex manifold $(\mathcal{M}^n, J)$.

7.2. The Koiso solitons. As was noted in Proposition 1.13 or Proposition A.32, all compact steady or expanding solitons are Einstein. This is not true for shrinking solitons. The first examples of nontrivial (i.e. non-Einstein) compact shrinking solitons were discovered by Koiso [**239**] and independently by Cao [**47**]. These are Kähler metrics on certain k-twisted projective-line bundles $\mathbb{P}^1 \hookrightarrow \mathcal{F}_k^n \twoheadrightarrow \mathbb{P}^{n-1}$ first described by Calabi [**44**]. We will discuss their construction in considerable detail, because it serves as a prototype for later examples.

We begin with Calabi's bundle construction. $\mathbb{P}^{n-1}$ is covered by n charts $(\varphi_\alpha : \mathcal{U}_\alpha \to \mathbb{C}^{n-1})$, where $\mathcal{U}_\alpha = \{[x_1, \ldots, x_n] \in \mathbb{P}^{n-1} : x_\alpha \neq 0\}$ and $\varphi_\alpha : [x_1, \ldots, x_n] \mapsto (\frac{x_1}{x_\alpha}, \ldots, \frac{x_{\alpha-1}}{x_\alpha}, \frac{x_{\alpha+1}}{x_\alpha}, \ldots, \frac{x_n}{x_\alpha})$. (We write x_α instead of x^α here and in the next paragraph in order to simplify some formulas below.) In particular, one may define complex projective space by

7. EXAMPLES OF KÄHLER–RICCI SOLITONS

$\mathbb{P}^{n-1} = (\amalg_{\alpha=1}^n \varphi_\alpha(\mathcal{U}_\alpha))/\simeq$, where, for example,

$$\varphi_1(\mathcal{U}_1) \ni (z_1, \ldots, z_{n-1}) \simeq (\frac{1}{z_1}, \frac{z_2}{z_1}, \ldots, \frac{z_{n-1}}{z_1}) \in \varphi_2(\mathcal{U}_2).$$

Given $k \in \mathbb{N}$, we formally identify $\mathbb{P}^1 = \mathbb{C} \cup \{\infty\}$ and define the k-twisted bundle

$$\mathcal{F}_k^n = (\amalg_{\alpha=1}^n (\mathcal{U}_\alpha \times \mathbb{P}^1)) / \sim,$$

where $\mathcal{U}_\alpha \times \mathbb{P}^1 \ni ([x_1, \ldots, x_n]; \xi) \sim ([y_1, \ldots, y_n]; \eta) \in \mathcal{U}_\beta \times \mathbb{P}^1$ if and only if $[x_1, \ldots, x_n] = [y_1, \ldots, y_n]$ and $\eta = (\frac{x_\alpha}{y_\alpha})^k \xi$ for each α. Equivalently, one may define

$$\mathcal{F}_k^n = (\amalg_{\alpha=1}^n (\varphi_\alpha(\mathcal{U}_\alpha) \times \mathbb{P}^1)) / \approx,$$

where, for example,

$$\varphi_1(\mathcal{U}_1) \times \mathbb{P}^1 \ni (z_1, \ldots, z_{n-1}; \zeta) \approx (\frac{1}{z_1}, \frac{z_2}{z_1}, \ldots, \frac{z_{n-1}}{z_1}; z_1^k \zeta) \in \varphi_2(\mathcal{U}_2) \times \mathbb{P}^1.$$

Notice that $S_0 = \{[x_1, \ldots, x_n]; 0\}$ and $S_\infty = \{[x_1, \ldots, x_n]; \infty\}$ are two global sections of $\mathcal{F}_k^n$.

The key to constructing Kähler–Ricci solitons on $\mathcal{F}_k^n$ (as well as examples on other topologies to be considered below) will be to find a Kähler potential on $\mathbb{C}^n \backslash \{0\}$ satisfying certain symmetries and boundary conditions. To see why this is so, let $\widehat{\mathcal{F}}_k^n = \mathcal{F}_k^n \backslash (S_0 \cup S_\infty)$ and define $\psi : \mathbb{C}^n \backslash \{0\} \to \widehat{\mathcal{F}}_k^n$ so that

$$\psi : (x_1, \ldots, x_n) \mapsto ([x_1, \ldots, x_n]; x_\alpha^k)$$

if $x_\alpha \neq 0$. It is easy to see that $([x_1, \ldots, x_n]; x_\alpha^k) \sim ([x_1, \ldots, x_n]; x_\beta^k)$ whenever $x_\alpha \neq 0$ and $x_\beta \neq 0$, hence that ψ is well defined. The map ψ is clearly surjective. If $\psi(x_1, \ldots, x_n) = \psi(y_1, \ldots, y_n)$, where, say $x_\alpha \neq 0$ and $y_\beta \neq 0$, then

$$(\frac{x_1}{x_\alpha}, \ldots, \frac{x_{\alpha-1}}{x_\alpha}, \frac{x_{\alpha+1}}{x_\alpha}, \ldots, \frac{x_n}{x_\alpha}; x_\alpha^k) \approx (\frac{y_1}{y_\beta}, \ldots, \frac{y_{\beta-1}}{y_\beta}, \frac{y_{\beta+1}}{y_\beta}, \ldots, \frac{y_n}{y_\beta}; y_\beta^k).$$

The equivalence relation $\approx$ then implies that $y_\beta^k = x_\beta^k$, hence that $y_\beta = \theta x_\beta$ for some k-th root of unity θ. Because $[x_1, \ldots, x_n] = [y_1, \ldots, y_n]$, it follows that $y_\gamma = \theta x_\gamma$ for all $\gamma = 1, \ldots, n$, hence that ψ is a k-to-one map. Therefore, a Kähler potential P on $\mathbb{C}^n \backslash \{0\}$ will induce a well-defined Kähler metric on $\widehat{\mathcal{F}}_k^n$ provided that $\partial \bar{\partial} P(\theta x_1, \ldots, \theta x_n) = \partial \bar{\partial} P(x_1, \ldots, x_n)$.

With these considerations in mind, our method will be to construct a suitable Kähler potential $P : \mathbb{C}^n \backslash \{0\} \to \mathbb{R}$ whose asymptotics as $|z| \to 0$ and $|z| \to \infty$ ensure that the induced metric extends smoothly to S_0 and S_∞. If we are interested in shrinking solitons, what properties should P possess? Let's suppose that P determines a Kähler metric g. As above, we take the Kähler and Ricci forms to be $\omega = \sqrt{-1} g_{\alpha\bar{\beta}} \, dz^\alpha \wedge d\bar{z}^\beta$ and $\rho = \sqrt{-1} R_{\alpha\bar{\beta}} \, dz^\alpha \wedge d\bar{z}^\beta$, respectively. Then (locally) we have

$$g_{\alpha\bar{\beta}} = \frac{\partial^2}{\partial z^\alpha \partial \bar{z}^\beta} P$$

and
$$R_{\alpha\bar{\beta}} = -\frac{\partial^2}{\partial z^\alpha \partial \bar{z}^\beta} \log \det g,$$
exactly as in (2.6). If Q denotes the soliton potential function with gradient vector field X, then equation (2.96) reduces to
$$\frac{\partial^2}{\partial z^\alpha \partial \bar{z}^\beta}(\log \det g - Q - \lambda P) = 0.$$
Because we are interested in shrinking solitons, we may assume that $\lambda < 0$. Then (modifying the Kähler potential P by an element in the kernel of $\nabla\bar{\nabla}$ if necessary) we may assume that $Q = \log \det g - \lambda P$. This will give us a soliton provided $X = \text{grad } Q$ is holomorphic, that is, provided that
$$0 = \frac{\partial}{\partial \bar{z}^\alpha} X^\beta = \frac{\partial}{\partial \bar{z}^\alpha}\left(g^{\beta\bar{\gamma}} \frac{\partial}{\partial \bar{z}^\gamma} Q\right).$$
Substituting $Q = \log \det g - \lambda P$, we obtain a single fourth-order equation for the scalar function P, namely

(2.97) $$\frac{\partial}{\partial \bar{z}^\alpha}\left[g^{\beta\bar{\gamma}} \frac{\partial}{\partial \bar{z}^\gamma}(\log \det g - \lambda P)\right] = 0.$$

To proceed, we adopt the *Ansatz* that the potential P is invariant under the natural U(n) action in the sense that it is a function of $r = \log \sum_{\alpha=1}^n |z^\alpha|^2$ alone. In this case, setting $\varphi = P_r$, we have

(2.98) $$g_{\alpha\bar{\beta}} = e^{-r}\varphi \delta_{\alpha\beta} + e^{-2r}(\varphi_r - \varphi)\bar{z}^\alpha z^\beta,$$

so our P will be a Kähler potential if and only if φ and φ_r are everywhere positive. Now (following [**47**] and [**142**]) we can write (2.97) as the fourth-order ODE

(2.99) $$P_{rrrr} - 2\frac{P_{rrr}^2}{P_{rr}} + nP_{rrr} - (n-1)\frac{P_{rr}^3}{P_r^2} + \lambda(P_{rrr}P_r - P_{rr}^2) = 0.$$

We shall see that only two of the four arbitrary constants in its solution are geometrically significant.

To simplify (2.99), notice that $X^\alpha = g^{\alpha\bar{\beta}}\frac{\partial}{\partial \bar{z}^\beta}Q = \frac{Q_r}{P_{rr}}z^\alpha$ will be holomorphic if and only if $Q_r = \mu P_{rr}$ for some $\mu \in \mathbb{R}$. Since g will be Kähler–Einstein if $X = 0$, we may assume that $\mu \neq 0$. Substituting $Q = \log \det g - \lambda P$, one then obtains
$$(\log \varphi_r)_r + (n-1)(\log \varphi)_r - \mu\varphi_r - \lambda\varphi - n = 0,$$
which is a second-order equation for φ, hence a third-order equation for P. (This integration can also be accomplished by standard ODE techniques.) Because $\varphi_r > 0$ everywhere, one may regard r as a function of φ and hence may write $\varphi_r = F(\varphi)$. One finds (remarkably) that F satisfies a linear equation
$$F' + \left(\frac{n-1}{\varphi} - \mu\right)F - (n + \lambda\varphi) = 0,$$

whose solution is
$$\varphi_r = \varphi^{1-n} e^{\mu\varphi}(\nu + \lambda I_n + n I_{n-1}),$$
where ν is another arbitrary constant and $I_n = \int \varphi^n e^{-\mu\varphi}\, d\varphi$. This leads to
$$\varphi_r = \nu \varphi^{1-n} e^{\mu\varphi} - \frac{\lambda}{\mu}\varphi - \frac{\lambda+\mu}{\mu^{1+n}} \sum_{j=0}^{n-1} \frac{n!}{j!} \mu^j \varphi^{j+1-n},$$
which is a separable first-order equation for φ. The third and fourth arbitrary constants arise when one finds the implicit solution $r(\varphi)$ and then integrates φ to obtain P. These are geometrically insignificant, because they disappear in (2.97); the two geometric degrees of freedom are the parameters $\mu \neq 0$ and ν.

In summary, what we have thus far accomplished is to construct a potential function for a U(n)-invariant Kähler–Ricci soliton metric (2.98) on $\mathbb{C}^n \setminus \{0\}$, hence a possibly incomplete Kähler–Ricci soliton metric on $\widehat{\mathcal{F}}_k^n$. What remains is to choose μ and ν in order to get a complete metric on $\mathcal{F}_k^n$. Our choices of the two parameters will be determined by the two boundary conditions as $r \to \pm\infty$.

Since $\varphi_r > 0$, we may define $a < b \in [0,\infty]$ by $a = \lim_{r\to -\infty} \varphi(r)$ and $b = \lim_{r\to \infty} \varphi(r)$. If $a > 0$, one may write $P(r) = ar + p(e^{\alpha r})$ in a neighborhood of $|z| = 0$, with p smooth at zero, $p(0) = 0$, $p'(0) > 0$. Similarly, if $b < \infty$, one may write $P(r) = br + q(e^{-\beta r})$ in a neighborhood of $|z| = \infty$, with q smooth at zero, $q(0) = 0$, $q'(0) > 0$. One then takes advantage of the following observation.

LEMMA 2.77 (Calabi). *Assume that $k > 0$ and $a, b \in (0, \infty)$.*

(1) *When $\alpha = k$, the potential $P(r) = ar + p(e^{\alpha r})$ induces a smooth Kähler metric on a neighborhood of S_0 in $\mathcal{F}_k^n$. Any $\mathbb{P}^1$ in S_0 has area $a\pi$.*

(2) *When $\beta = k$, the potential $P(r) = br + q(e^{-\beta r})$ induces a smooth Kähler metric on a neighborhood of S_∞ in $\mathcal{F}_k^n$. Any $\mathbb{P}^1$ in S_∞ has area $b\pi$.*

For the proof, see [44] or [142, Lemma 4.2].

With more work, one finds that it is possible to satisfy both boundary conditions by appropriate choices of μ and ν. (See [47] or adapt the arguments in [142, §4.1].) Normalizing by fixing $\lambda = -1$, these choices yield $a = n - k$ and $b = n + k$. Since one needs $a > 0$, one obtains a unique gradient shrinking Kähler–Ricci soliton on $\mathcal{F}_k^n$ for each $k = 1, \ldots, n-1$. These are the Koiso solitons.

7.3. Other U(n)-invariant solitons. The construction we have described in Section 7.2 above has natural generalizations allowing the discovery of other explicit Kähler–Ricci soliton examples. Namely, one searches the 2-dimensional (μ, ν) parameter space of U(n)-invariant Kähler potentials

$P(r)$ on $\mathbb{C}^n\backslash\{0\}$ for those whose behavior at the boundary $|z| = 0$ implies that

1$_-$: the metric is completed by adding a smooth point at $r = -\infty$;
2$_-$: the metric is completed by adding an orbifold point at $r = -\infty$;
3$_-$: the metric is completed by adding a $\mathbb{P}^{n-1}$ at $r = -\infty$; or
4$_-$: the metric is complete as $r \to -\infty$;

and whose behavior at the boundary $|z| = \infty$ implies that

1$_+$: the metric is completed by adding a smooth point at $r = +\infty$;
2$_+$: the metric is completed by adding an orbifold point at $r = +\infty$;
3$_+$: the metric is completed by adding a $\mathbb{P}^{n-1}$ at $r = +\infty$; or
4$_+$: the metric is complete as $r \to +\infty$.

Of course, not all combinations of these alternatives are globally compatible. For example, it is easy to see that the growth condition $\varphi_r > 0$ prohibits completing the metric by adding a $\mathbb{P}^{n-1}$ at $|z| = 0$ and a smooth point at $|z| = \infty$. Nonetheless, this has been a productive line of research. In the remainder of this section, we will survey some of its results.

It is possible to add a smooth point at $|z| = 0$ and to construct a unique steady Kähler–Ricci soliton on $\mathbb{C}^n$ that is complete as $|z| \to \infty$. In complex dimension $n = 1$, this is just the cigar soliton discovered by Hamilton and discussed in Chapter 2 of Volume One; the examples in higher dimensions are due to Cao [**47**]. These solitons have the following asymptotic behavior: in the sphere S^{2n-1} at metric distance $r \gg 0$ from $|z| = 0$, the Hopf fibers $U(1) \cdot z$ have diameter $\mathcal{O}(1)$, while the $\mathbb{P}^{n-1}$ direction has diameter $\mathcal{O}(\sqrt{r})$. Accordingly, one calls this **cigar-paraboloid behavior**.

It is also possible to add a smooth point at $|z| = 0$ and to construct expanding Kähler–Ricci solitons on $\mathbb{C}^n$ that are complete as $|z| \to \infty$. There is in fact a 1-parameter family $(\mathbb{C}^n, g_\theta)_{\theta>0}$ of such examples in each dimension, due to Cao [**48**]. (It is a heuristic principle that expanding solitons are easier to find than their shrinking cousins. For these examples, satisfying the boundary condition at $|z| = 0$ reduces the parameter space by one dimension, but completion as $|z| \to \infty$ comes for free.) Each soliton $(\mathbb{C}^n, g_\theta)$ is asymptotic as $|z| \to \infty$ to the Kähler cone $(\mathbb{C}^n\backslash\{0\}, \hat{g}_\theta)$, where the metric $\hat{g}_\theta$ is induced by the Kähler potential $\hat{P}(r) = e^{\theta r}/\theta$.

For each $k = 2, 3, \ldots$, the authors of [**142**] add an orbifold point at $|z| = 0$ and a $\mathbb{P}^{n-1}$ at $|z| = \infty$ to construct a unique shrinking Kähler–Ricci soliton on an orbifold, which is called $\mathcal{G}_k^n$. The compact orbifold $\mathcal{G}_k^n$ may be regarded as $\mathbb{P}^n/\mathbb{Z}_k$ branched over the origin and the $\mathbb{P}^{n-1}$ at infinity. The orbifold singularity at the origin is modeled on $\mathbb{C}^n/\mathbb{Z}_k$.

For each dimension $n \geq 2$ and $k = 1, \ldots, n-1$, the authors of [**142**] add a k-twisted $\mathbb{P}^{n-1}$ at $|z| = 0$ and construct a unique shrinking Kähler–Ricci soliton metric that is complete as $|z| \to \infty$. The resulting soliton has the topology of the complex line bundle $\mathbb{C} \hookrightarrow L_{-k}^n \twoheadrightarrow \mathbb{P}^{n-1}$ characterized by $\langle c_1, [\Sigma] \rangle = -k$, where c_1 is the first Chern class of the bundle and $\Sigma \approx \mathbb{P}^1$ is a positively-oriented generator of $H_2(\mathbb{P}^{n-1}; \mathbb{Z})$. (For example, the total space of L_{-1}^n is simply $\mathbb{C}^n$ blown up at the origin.) As $|z| \to \infty$, the soliton metric is asymptotic to a Kähler cone $(\mathbb{C}^n \backslash \{0\}, \hat{g}_\theta)/\mathbb{Z}_k$, where $\theta = \theta(n, k)$.

For each dimension $n \geq 2$, Cao [47] adds an n-twisted $\mathbb{P}^{n-1}$ at $|z| = 0$ and constructs a unique complete steady Kähler–Ricci soliton on the total space of the bundle $\mathbb{C} \hookrightarrow L_{-n}^n \twoheadrightarrow \mathbb{P}^{n-1}$. The metric exhibits cigar-paraboloid behavior at infinity.

For each dimension $n \geq 2$ and $k = n+1, \ldots$, the authors of [142] add a $\mathbb{P}^{n-1}$ at $|z| = 0$ and construct a 1-parameter family of complete expanding Kähler–Ricci solitons on $\mathbb{C} \hookrightarrow L_{-k}^n \twoheadrightarrow \mathbb{P}^{n-1}$. The solutions are parameterized by $\theta > 0$, where $(\mathbb{C}^n \backslash \{0\}, \hat{g}_\theta)/\mathbb{Z}_k$ is the asymptotic Kähler cone at infinity.

8. Kähler–Ricci flow with nonnegative bisectional curvature

The study of the Kähler–Ricci flow of Kähler metrics with nonnegative bisectional curvature is somewhat analogous to the study of the Riemannian Ricci flow of metrics with nonnegative curvature operator. One aim is to uniformize Kähler metrics with nonnegative bisectional curvature in both the compact and noncompact setting. In particular, one would like to flow such metrics to canonical metrics, or to infer the existence of canonical metrics from the long-time behavior of the flow. One would also like to deduce properties of the underlying complex structure of the Kähler manifold, and when possible, classify the manifold up to biholomorphism.

8.1. Nonnegative bisectional curvature is preserved.
Consider the Kähler–Ricci flow $\frac{\partial}{\partial t} g_{\alpha \bar{\beta}} = -R_{\alpha \bar{\beta}}$. We shall prove that the Kähler–Ricci flow preserves the nonnegativity of the bisectional curvature. As in the real case, the key is Hamilton's weak maximum principle for tensors. This result was proved first by Bando for $n \leq 3$ and by Mok in any dimension. (See Theorem 2.69 above.) The result was also extended to the complete noncompact case by W.-X. Shi under the additional assumption of the bisectional curvature being bounded. We say that a Kähler metric has **quasi-positive Ricci curvature** if the Ricci curvature is nonnegative everywhere and positive at some point.

THEOREM 2.78 (Nonnegative bisectional curvature preserved). *The nonnegativity of the bisectional curvature is preserved under the Kähler–Ricci flow on closed Kähler manifolds. Moreover, if the initial metric also has quasi-positive Ricci curvature, then both the Ricci curvature and the holomorphic sectional curvature are positive for metrics at positive time.*

The basic computation in the proof of the above result is the following.

PROPOSITION 2.79 (Evolution equation for the curvature). *Under the Kähler–Ricci flow,*

(2.100)
$$\left(\frac{\partial}{\partial t} - \Delta\right) R_{\alpha\bar{\beta}\gamma\bar{\delta}} = R_{\alpha\bar{\mu}\nu\bar{\delta}}R_{\mu\bar{\beta}\gamma\bar{\nu}} - R_{\alpha\bar{\mu}\gamma\bar{\nu}}R_{\mu\bar{\beta}\nu\bar{\delta}} + R_{\alpha\bar{\beta}\nu\bar{\mu}}R_{\mu\bar{\nu}\gamma\bar{\delta}}$$
$$- \frac{1}{2}\left(R_{\alpha\bar{\mu}}R_{\mu\bar{\beta}\gamma\bar{\delta}} + R_{\mu\bar{\beta}}R_{\alpha\bar{\mu}\gamma\bar{\delta}} + R_{\gamma\bar{\mu}}R_{\alpha\bar{\beta}\mu\bar{\delta}} + R_{\mu\bar{\delta}}R_{\alpha\bar{\beta}\gamma\bar{\mu}}\right).$$

REMARK 2.80. The Riemannian analogue of this formula is given by Lemma 6.15 on p. 179 of [**108**].

In the proof of the proposition we find it convenient to use a formula relating ordinary derivatives and covariant derivatives at the center of normal holomorphic coordinates.

LEMMA 2.81 (Relation between ordinary and covariant derivatives). *If η is a closed $(1,1)$-form, then, at the center of normal holomorphic coordinates, we have*

(2.101)
$$\nabla_{\bar{\beta}}\nabla_{\alpha}\eta_{\gamma\bar{\delta}} = \frac{\partial^2}{\partial z^\alpha \partial \bar{z}^\beta}\eta_{\gamma\bar{\delta}} + \eta_{\lambda\bar{\delta}}R_{\alpha\bar{\beta}\gamma\bar{\lambda}},$$

(2.102)
$$\nabla_{\alpha}\nabla_{\bar{\beta}}\eta_{\gamma\bar{\delta}} = \frac{\partial^2}{\partial z^\alpha \partial \bar{z}^\beta}\eta_{\gamma\bar{\delta}} + \eta_{\gamma\bar{\lambda}}R_{\alpha\bar{\beta}\lambda\bar{\delta}}.$$

PROOF. We compute that at the center of normal holomorphic coordinates,
$$\nabla_{\bar{\beta}}\nabla_{\alpha}\eta_{\gamma\bar{\delta}} = \partial_{\bar{\beta}}\nabla_{\alpha}\eta_{\gamma\bar{\delta}} - \overline{\Gamma^{\varepsilon}_{\beta\delta}}\nabla_{\alpha}\eta_{\gamma\bar{\varepsilon}}$$
$$= \partial_{\bar{\beta}}\left(\partial_{\alpha}\eta_{\gamma\bar{\delta}} - \Gamma^{\varepsilon}_{\alpha\gamma}\eta_{\varepsilon\bar{\delta}}\right)$$
$$= \partial_{\bar{\beta}}\partial_{\alpha}\eta_{\gamma\bar{\delta}} - \partial_{\bar{\beta}}\Gamma^{\varepsilon}_{\alpha\gamma}\eta_{\varepsilon\bar{\delta}}$$
$$= \frac{\partial^2}{\partial z^\alpha \partial \bar{z}^\beta}\eta_{\gamma\bar{\delta}} + R^{\varepsilon}_{\alpha\bar{\beta}\gamma}\eta_{\varepsilon\bar{\delta}},$$

where we used (2.4) in the last line; this proves (2.101). Note that (2.102) is just the conjugate of (2.101). □

Now we give the

PROOF OF PROPOSITION 2.79. We compute the evolution equation for $R_{\alpha\bar{\beta}\gamma\bar{\delta}}$ at any point x and time t using normal holomorphic coordinates $\{z^\alpha\}$ centered at x with respect to $g(t)$. In such coordinates, $\frac{\partial g_{\alpha\bar{\beta}}}{\partial z^\gamma}(x,t) = 0$. Recall from (2.5) that

$$R_{\alpha\bar{\beta}\gamma\bar{\delta}} = -\frac{\partial^2 g_{\alpha\bar{\beta}}}{\partial z^\gamma \partial \bar{z}^\delta} + g^{\rho\bar{\sigma}}\frac{\partial g_{\alpha\bar{\sigma}}}{\partial z^\gamma}\frac{\partial g_{\rho\bar{\beta}}}{\partial \bar{z}^\delta}.$$

This implies
$$\frac{\partial}{\partial t} R_{\alpha\bar{\beta}\gamma\bar{\delta}} = -\frac{\partial^2}{\partial z^\gamma \partial \bar{z}^\delta}\left(\frac{\partial}{\partial t} g_{\alpha\bar{\beta}}\right) = \frac{\partial^2}{\partial z^\gamma \partial \bar{z}^\delta} R_{\alpha\bar{\beta}}$$
$$(2.103) \qquad = \nabla_\gamma \nabla_{\bar{\delta}} R_{\alpha\bar{\beta}} - R_{\alpha\bar{\lambda}} R_{\gamma\bar{\delta}\lambda\bar{\beta}},$$

where we used (2.102). Since (2.103) is tensorial, it holds in any holomorphic coordinate system. We wish to compare the above formula with

$$\Delta R_{\alpha\bar{\beta}\gamma\bar{\delta}} \doteq \frac{1}{2}\left(\nabla_\mu \nabla_{\bar{\mu}} + \nabla_{\bar{\mu}} \nabla_\mu\right) R_{\alpha\bar{\beta}\gamma\bar{\delta}}.$$

To this end we compute (apply the second Bianchi identity (2.9) and commute covariant derivatives (Exercise 2.22))

$$\begin{aligned}\nabla_\gamma \nabla_{\bar{\delta}} R_{\alpha\bar{\beta}} &= \nabla_\gamma \nabla_{\bar{\delta}} R_{\alpha\bar{\beta}\mu\bar{\mu}} = \nabla_\gamma \nabla_{\bar{\mu}} R_{\alpha\bar{\beta}\mu\bar{\delta}} \\ &= \nabla_{\bar{\mu}} \nabla_\mu R_{\alpha\bar{\beta}\gamma\bar{\delta}} - R_{\gamma\bar{\mu}\alpha\bar{\nu}} R_{\nu\bar{\beta}\mu\bar{\delta}} + R_{\gamma\bar{\mu}\nu\bar{\beta}} R_{\alpha\bar{\nu}\mu\bar{\delta}} \\ &\quad - R_{\gamma\bar{\mu}\mu\bar{\nu}} R_{\alpha\bar{\beta}\nu\bar{\delta}} + R_{\gamma\bar{\mu}\nu\bar{\delta}} R_{\alpha\bar{\beta}\mu\bar{\nu}}\end{aligned}$$

and

$$\begin{aligned}\nabla_{\bar{\mu}} \nabla_\mu R_{\alpha\bar{\beta}\gamma\bar{\delta}} &= \nabla_\mu \nabla_{\bar{\mu}} R_{\alpha\bar{\beta}\gamma\bar{\delta}} + R_{\mu\bar{\mu}\alpha\bar{\nu}} R_{\nu\bar{\beta}\gamma\bar{\delta}} - R_{\mu\bar{\mu}\nu\bar{\beta}} R_{\alpha\bar{\nu}\gamma\bar{\delta}} \\ &\quad + R_{\mu\bar{\mu}\gamma\bar{\nu}} R_{\alpha\bar{\beta}\nu\bar{\delta}} - R_{\mu\bar{\mu}\nu\bar{\delta}} R_{\alpha\bar{\beta}\gamma\bar{\nu}} \\ &= \nabla_\mu \nabla_{\bar{\mu}} R_{\alpha\bar{\beta}\gamma\bar{\delta}} + R_{\alpha\bar{\nu}} R_{\nu\bar{\beta}\gamma\bar{\delta}} - R_{\nu\bar{\beta}} R_{\alpha\bar{\nu}\gamma\bar{\delta}} \\ &\quad + R_{\gamma\bar{\nu}} R_{\alpha\bar{\beta}\nu\bar{\delta}} - R_{\nu\bar{\delta}} R_{\alpha\bar{\beta}\gamma\bar{\nu}}.\end{aligned}$$

Combining the formulas above yields

$$\begin{aligned}\frac{\partial}{\partial t} R_{\alpha\bar{\beta}\gamma\bar{\delta}} &= \nabla_{\bar{\mu}} \nabla_\mu R_{\alpha\bar{\beta}\gamma\bar{\delta}} - R_{\gamma\bar{\mu}\alpha\bar{\nu}} R_{\nu\bar{\beta}\mu\bar{\delta}} + R_{\gamma\bar{\mu}\nu\bar{\beta}} R_{\alpha\bar{\nu}\mu\bar{\delta}} \\ &\quad - R_{\gamma\bar{\nu}} R_{\alpha\bar{\beta}\nu\bar{\delta}} + R_{\gamma\bar{\mu}\nu\bar{\delta}} R_{\alpha\bar{\beta}\mu\bar{\nu}} - R_{\alpha\bar{\lambda}} R_{\gamma\bar{\delta}\lambda\bar{\beta}} \\ &= \Delta R_{\alpha\bar{\beta}\gamma\bar{\delta}} - R_{\gamma\bar{\mu}\alpha\bar{\nu}} R_{\nu\bar{\beta}\mu\bar{\delta}} + R_{\gamma\bar{\mu}\nu\bar{\beta}} R_{\alpha\bar{\nu}\mu\bar{\delta}} + R_{\gamma\bar{\mu}\nu\bar{\delta}} R_{\alpha\bar{\beta}\mu\bar{\nu}} \\ &\quad - \frac{1}{2}\left(R_{\alpha\bar{\nu}} R_{\nu\bar{\beta}\gamma\bar{\delta}} + R_{\nu\bar{\beta}} R_{\alpha\bar{\nu}\gamma\bar{\delta}} + R_{\gamma\bar{\nu}} R_{\alpha\bar{\beta}\nu\bar{\delta}} + R_{\nu\bar{\delta}} R_{\alpha\bar{\beta}\gamma\bar{\nu}}\right),\end{aligned}$$

and the proposition follows. $\square$

As a consequence of the proposition we have the following evolution equations for the bisectional curvature and the Ricci tensor.

COROLLARY 2.82 (Bisectional curvature evolution).

$$\begin{aligned}\left(\frac{\partial}{\partial t} - \Delta\right) R_{\alpha\bar{\alpha}\gamma\bar{\gamma}} &= \sum_{\mu,\nu=1}^n \left(|R_{\alpha\bar{\mu}\nu\bar{\gamma}}|^2 - |R_{\alpha\bar{\mu}\gamma\bar{\nu}}|^2 + R_{\alpha\bar{\alpha}\nu\bar{\mu}} R_{\mu\bar{\nu}\gamma\bar{\gamma}}\right) \\ (2.104) &\quad - \sum_{\mu=1}^n \operatorname{Re}\left(R_{\alpha\bar{\mu}} R_{\mu\bar{\alpha}\gamma\bar{\gamma}} + R_{\gamma\bar{\mu}} R_{\alpha\bar{\alpha}\mu\bar{\gamma}}\right).\end{aligned}$$

Here $\operatorname{Re}(A) = \frac{1}{2}(A + \bar{A})$ denotes the real part of a complex number A.

PROOF. Indeed, substituting $\bar\beta = \bar\alpha$ and $\bar\delta = \bar\gamma$ in (2.100), we have

$$\left(\frac{\partial}{\partial t} - \Delta\right) R_{\alpha\bar\alpha\gamma\bar\gamma} = R_{\alpha\bar\mu\nu\bar\gamma}R_{\mu\bar\alpha\gamma\bar\nu} - R_{\alpha\bar\mu\gamma\bar\nu}R_{\mu\bar\alpha\nu\bar\gamma} + R_{\alpha\bar\alpha\nu\bar\mu}R_{\mu\bar\nu\gamma\bar\gamma}$$
$$- \frac{1}{2}\left(R_{\alpha\bar\mu}R_{\mu\bar\alpha\gamma\bar\gamma} + R_{\mu\bar\alpha}R_{\alpha\bar\mu\gamma\bar\gamma} + R_{\gamma\bar\mu}R_{\alpha\bar\alpha\mu\bar\gamma} + R_{\mu\bar\gamma}R_{\alpha\bar\alpha\gamma\bar\mu}\right).$$

$\square$

COROLLARY 2.83 (Ricci tensor evolution). *The Ricci tensor satisfies the Lichnerowicz heat equation:*

(2.105) $$\frac{\partial}{\partial t}R_{\alpha\bar\beta} = \Delta R_{\alpha\bar\beta} + R_{\alpha\bar\beta\gamma\bar\delta}R_{\delta\bar\gamma} - R_{\alpha\bar\gamma}R_{\gamma\bar\beta} = \Delta_L R_{\alpha\bar\beta}.$$

REMARK 2.84. More generally, we say that a real $(1,1)$-tensor $h_{\alpha\bar\beta}$ satisfies the Lichnerowicz heat equation if

$$\frac{\partial}{\partial t}h_{\alpha\bar\beta} = \Delta_L h_{\alpha\bar\beta} \doteq \Delta h_{\alpha\bar\beta} + R_{\alpha\bar\beta\gamma\bar\delta}h_{\delta\bar\gamma} - \frac{1}{2}R_{\alpha\bar\gamma}h_{\gamma\bar\beta} - \frac{1}{2}R_{\gamma\bar\beta}h_{\alpha\bar\gamma}.$$

See also (2.22).

PROOF. Summing (2.100) over $\gamma = \delta$ from 1 to n, we have

$$\left(\frac{\partial}{\partial t} - \Delta\right) R_{\alpha\bar\beta} = \sum_{k=1}^{n} \left(\frac{\partial}{\partial t} - \Delta\right) R_{\alpha\bar\beta\gamma\bar\gamma} + R_{\delta\bar\gamma}R_{\alpha\bar\beta\gamma\bar\delta}$$
$$= R_{\alpha\bar\mu\nu\bar\gamma}R_{\mu\bar\beta\gamma\bar\nu} - R_{\alpha\bar\mu\gamma\bar\nu}R_{\mu\bar\beta\nu\bar\gamma} + R_{\alpha\bar\beta\nu\bar\mu}R_{\mu\bar\nu} + R_{\alpha\bar\beta\gamma\bar\delta}R_{\delta\bar\gamma}$$
$$- \frac{1}{2}\left(R_{\alpha\bar\mu}R_{\mu\bar\beta} + R_{\mu\bar\beta}R_{\alpha\bar\mu} + R_{\gamma\bar\mu}R_{\alpha\bar\beta\mu\bar\gamma} + R_{\mu\bar\gamma}R_{\alpha\bar\beta\gamma\bar\mu}\right)$$
$$= R_{\alpha\bar\beta\gamma\bar\delta}R_{\delta\bar\gamma} - R_{\alpha\bar\mu}R_{\mu\bar\beta},$$

after cancelling terms to get the last equality. $\square$

REMARK 2.85. Equation (2.105) may also be derived from (2.36), (2.10) and commuting covariant derivatives. In particular,

$$\Delta R_{\alpha\bar\beta} = \frac{1}{2}\left(\nabla_\gamma \nabla_{\bar\gamma} + \nabla_{\bar\gamma}\nabla_\gamma\right) R_{\alpha\bar\beta}$$
$$= \frac{1}{2}\nabla_\gamma \nabla_{\bar\beta} R_{\alpha\bar\gamma} + \frac{1}{2}\nabla_{\bar\gamma}\nabla_\alpha R_{\gamma\bar\beta}$$
$$= \nabla_\alpha \nabla_{\bar\beta} R - \frac{1}{2}R_{\gamma\bar\beta\alpha\bar\delta}R_{\delta\bar\gamma} + \frac{1}{2}R_{\delta\bar\beta}R_{\alpha\bar\delta}$$
$$+ \frac{1}{2}R_{\alpha\bar\delta}R_{\delta\bar\beta} - \frac{1}{2}R_{\bar\gamma\alpha\bar\beta\delta}R_{\gamma\bar\delta}$$

(2.106) $$= \nabla_\alpha \nabla_{\bar\beta} R - R_{\alpha\bar\beta\gamma\bar\delta}R_{\delta\bar\gamma} + R_{\alpha\bar\delta}R_{\delta\bar\beta}.$$

That is,

$$\nabla_\alpha \nabla_{\bar\beta} R = \Delta_L R_{\alpha\bar\beta}.$$

Based on the evolution equation (2.104) and Hamilton's maximum principle for tensors (see Chapter 4 of Volume One or Part II of this volume) we present the following.

PROOF OF THEOREM 2.78. (1) We first prove that the nonnegativity of the bisectional curvature is preserved under the flow. Analogous to Theorem 4.6 on p. 97 of Volume One, by the Kähler version of the maximum principle for tensors (see Proposition 1 in §4 of [**19**]), we need to show that the quadratic on the RHS of (2.104), i.e.,

$$(2.107) \quad Q_{\alpha\bar{\alpha}\gamma\bar{\gamma}} \doteqdot \sum_{\mu,\nu=1}^{n} \left(|R_{\alpha\bar{\mu}\nu\bar{\gamma}}|^2 - |R_{\alpha\bar{\mu}\gamma\bar{\nu}}|^2 + R_{\alpha\bar{\alpha}\nu\bar{\mu}} R_{\mu\bar{\nu}\gamma\bar{\gamma}} \right)$$
$$- \sum_{\mu=1}^{n} \operatorname{Re} \left(R_{\alpha\bar{\mu}} R_{\mu\bar{\alpha}\gamma\bar{\gamma}} + R_{\gamma\bar{\mu}} R_{\alpha\bar{\alpha}\mu\bar{\gamma}} \right),$$

satisfies the null eigenvector assumption. That is, we assume $R_{\alpha\bar{\alpha}\gamma\bar{\gamma}} = 0$ for some α and γ at some point x, and we shall prove that

$$(2.108) \quad Q_{\alpha\bar{\alpha}\gamma\bar{\gamma}} \geq 0$$

at x. First observe that since $R_{\alpha\bar{\alpha}\gamma\bar{\gamma}} = 0$ at x and the bisectional curvatures are nonnegative, we have at x,

$$\sum_{\mu=1}^{n} R_{\alpha\bar{\mu}} R_{\mu\bar{\alpha}\gamma\bar{\gamma}} = \sum_{\mu=1}^{n} R_{\gamma\bar{\mu}} R_{\alpha\bar{\alpha}\mu\bar{\gamma}} = 0.$$

By (2.107), in order to prove (2.108) at x, it suffices to show that

$$(2.109) \quad \sum_{\mu,\nu=1}^{n} R_{\alpha\bar{\alpha}\nu\bar{\mu}} R_{\mu\bar{\nu}\gamma\bar{\gamma}} \geq \sum_{\mu,\nu=1}^{n} \left(|R_{\alpha\bar{\mu}\gamma\bar{\nu}}|^2 - |R_{\alpha\bar{\mu}\nu\bar{\gamma}}|^2 \right).$$

We shall prove (2.109) below, but first we show how the positivity of the Ricci tensor and holomorphic sectional curvatures follow from the quasi-positivity of the Ricci curvature at $t = 0$.

(2) Recall that the Ricci tensor $R_{\alpha\bar{\beta}}$ satisfies the Lichnerowicz heat equation, so that by taking $\alpha = \beta$ in (2.105), we have

$$(2.110) \quad \frac{\partial}{\partial t} R_{\alpha\bar{\alpha}} = \Delta R_{\alpha\bar{\alpha}} + R_{\alpha\bar{\alpha}\gamma\bar{\delta}} R_{\delta\bar{\gamma}} - R_{\alpha\bar{\gamma}} R_{\gamma\bar{\alpha}}.$$

Since the nonnegativity of the bisectional curvature is preserved, by applying Hamilton's **strong maximum principle for tensors** to (2.110) (see Theorem A.53 and also Part II of this volume), the Ricci tensor becomes positive for all positive time. Now suppose there exists a space-time point (x_0, t_0), with $t_0 > 0$, at which some holomorphic sectional curvature $R_{\alpha\bar{\alpha}\alpha\bar{\alpha}}$ is zero. Since the holomorphic sectional curvature is nonnegative everywhere, by (2.104), we have at (x_0, t_0),

$$0 \geq \left(\frac{\partial}{\partial t} - \Delta \right) R_{\alpha\bar{\alpha}\alpha\bar{\alpha}} = \sum_{\mu,\nu=1}^{n} \left(2 |R_{\alpha\bar{\alpha}\nu\bar{\mu}}|^2 - |R_{\alpha\bar{\mu}\alpha\bar{\nu}}|^2 \right),$$

where we used $\sum_{\mu=1}^{n} \operatorname{Re}(R_{\alpha\bar{\mu}}R_{\mu\bar{\alpha}\alpha\bar{\alpha}} + R_{\alpha\bar{\mu}}R_{\alpha\bar{\alpha}\mu\bar{\alpha}}) = 0$ at (x_0, t_0). On the other hand, by (2.109) with $\alpha = \gamma$, we have

$$\sum_{\mu,\nu=1}^{n} |R_{\alpha\bar{\alpha}\nu\bar{\mu}}|^2 \geq \sum_{\mu,\nu=1}^{n} |R_{\alpha\bar{\mu}\alpha\bar{\nu}}|^2.$$

Hence we conclude

$$R_{\alpha\bar{\alpha}\nu\bar{\mu}} = R_{\alpha\bar{\mu}\alpha\bar{\nu}} = 0$$

at (x_0, t_0) for all μ, ν. This in turn implies $R_{\alpha\bar{\alpha}} = 0$, which is a contradiction.

(1) continued. We now verify (2.109). Consider the following Hermitian symmetric form defined by the bisectional curvatures:

$$\tilde{Q}(X, Y, s) \doteq \operatorname{Rm}_{\mathbb{C}}\left(\frac{\partial}{\partial z^{\alpha}} + sX, \overline{\frac{\partial}{\partial z^{\alpha}} + sX}, \frac{\partial}{\partial z^{\gamma}} + sY, \overline{\frac{\partial}{\partial z^{\gamma}} + sY}\right) \geq 0$$

for $X, Y \in T^{1,0}\mathcal{M}$ and $s \in \mathbb{R}$. At a point where the bisectional curvatures are nonnegative and $R_{\alpha\bar{\alpha}\gamma\bar{\gamma}} = 0$, we have $\tilde{Q}(X, Y, s) \geq 0$ and $\tilde{Q}(X, Y, 0) = 0$ for all $X, Y \in T^{1,0}\mathcal{M}$ and $s \in \mathbb{R}$. Therefore the second variation at $s = 0$ is nonnegative:

$$0 \leq \frac{d^2}{ds^2}\bigg|_{s=0} \tilde{Q}(X, Y, s)$$
$$= \operatorname{Rm}_{\mathbb{C}}\left(X, \overline{X}, \frac{\partial}{\partial z^{\gamma}}, \overline{\frac{\partial}{\partial z^{\gamma}}}\right) + \operatorname{Rm}_{\mathbb{C}}\left(\frac{\partial}{\partial z^{\alpha}}, \overline{\frac{\partial}{\partial z^{\alpha}}}, Y, \overline{Y}\right)$$
$$+ 2\operatorname{Re}\left(\operatorname{Rm}_{\mathbb{C}}\left(X, \overline{\frac{\partial}{\partial z^{\alpha}}}, Y, \overline{\frac{\partial}{\partial z^{\gamma}}}\right) + \operatorname{Rm}_{\mathbb{C}}\left(X, \overline{\frac{\partial}{\partial z^{\alpha}}}, \frac{\partial}{\partial z^{\gamma}}, \overline{Y}\right)\right).$$

In terms of a unitary $(1,0)$-frame $\{e_i\}_{i=1}^{n}$, we may write this as

$$R_{i\bar{j}\gamma\bar{\gamma}}X^i\overline{X^j} + 2\operatorname{Re}\left(R_{i\bar{\alpha}j\bar{\gamma}}X^iY^j + R_{i\bar{\alpha}\gamma\bar{j}}X^i\overline{Y^j}\right) + R_{\alpha\bar{\alpha}i\bar{j}}Y^i\overline{Y^j} \geq 0,$$

where $X \doteq \sum_{i=1}^{n} X^i e_i$ and $Y \doteq \sum_{j=1}^{n} Y^j e_j$. By Lemma 2.86 below, we have

$$\sum_{i,j=1}^{n} R_{i\bar{j}\gamma\bar{\gamma}}R_{\alpha\bar{\alpha}j\bar{\imath}} \geq \sum_{i,j=1}^{n} \left(|R_{i\bar{\alpha}j\bar{\gamma}}|^2 - |R_{i\bar{\alpha}\gamma\bar{j}}|^2\right).$$

The claimed inequality (2.109) follows and this completes the proof of Theorem 2.78. □

LEMMA 2.86. *Let $Q(X, Y)$ be a Hermitian symmetric quadratic form defined by*

$$Q(X, Y) = A_{i\bar{j}}X^i\overline{X^j} + 2\operatorname{Re}\left(B_{ij}X^iY^j + D_{i\bar{j}}X^i\overline{Y^j}\right) + C_{i\bar{j}}Y^i\overline{Y^j}.$$

If Q is semi-positive definite, then

$$\sum_{i,j=1}^{n} A_{i\bar{j}}C_{j\bar{\imath}} \geq \sum_{i,j=1}^{n} \left(|B_{ij}|^2 - |D_{i\bar{j}}|^2\right).$$

For the proof of this lemma, which is elementary in nature, we refer the reader to Mok [**269**].

9. Matrix differential Harnack estimate for the Kähler–Ricci flow

In this and the next section we discuss various differential Harnack estimates for the Kähler–Ricci flow and their geometric applications. Differential Harnack estimates for the Riemannian Ricci flow will be discussed in Part II. For Kähler–Ricci flow, a fundamental result is H.-D. Cao's differential Harnack estimate for solutions with nonnegative bisectional curvature (see [**46**]). Define
(2.111)
$$Z(X)_{\alpha\bar{\beta}} \doteqdot \frac{\partial}{\partial t}R_{\alpha\bar{\beta}} + R_{\alpha\bar{\gamma}}R_{\gamma\bar{\beta}} + \nabla_{\gamma}R_{\alpha\bar{\beta}}X^{\gamma} + \nabla_{\bar{\gamma}}R_{\alpha\bar{\beta}}X^{\bar{\gamma}} + R_{\alpha\bar{\beta}\gamma\bar{\delta}}X^{\gamma}X^{\bar{\delta}} + \frac{R_{\alpha\bar{\beta}}}{t}$$
for any $(1,0)$-vector $X = X^{\gamma}\frac{\partial}{\partial z^{\gamma}}$ and where $X^{\bar{\gamma}} \doteqdot \overline{X^{\gamma}}$.

THEOREM 2.87 (Kähler matrix Harnack estimate). *If $(\mathcal{M}^n, g(t))$ is a complete solution to the Kähler–Ricci flow with bounded nonnegative bisectional curvature, then*

(2.112) $$\left(Z(X)_{\alpha\bar{\beta}}\right) \geq 0$$

for any $(1,0)$-vector X.

This result may be considered as the space-time analogue of Theorem 2.78. We shall also see a similar analogy for the Riemannian Ricci flow in Part II, where Hamilton's matrix differential Harnack estimate will appear as the space-time analogue of the result that nonnegative curvature operator is preserved under the Ricci flow.

9.1. Trace differential Harnack estimate for the Kähler–Ricci flow. Taking the trace of the estimate (2.112) leads to the so-called **trace differential Harnack estimate**, after applying the second Bianchi identity.

COROLLARY 2.88 (The Kähler trace differential Harnack estimate). *Let $(\mathcal{M}^n, g(t))$ be a complete solution to the Kähler–Ricci flow with bounded nonnegative bisectional curvature. Then*

(2.113) $$\frac{\partial R}{\partial t} + \nabla_{\gamma}RX^{\gamma} + \nabla_{\bar{\gamma}}RX^{\bar{\gamma}} + R_{\alpha\bar{\beta}}X^{\alpha}X^{\bar{\beta}} + \frac{R}{t} \geq 0.$$

PROOF. By (2.112) we have
$$0 \leq g^{\alpha\bar{\beta}}Z_{\alpha\bar{\beta}} = g^{\alpha\bar{\beta}}\frac{\partial}{\partial t}R_{\alpha\bar{\beta}} + \left|R_{\alpha\bar{\beta}}\right|^2 + \nabla_{\gamma}RX^{\gamma} + \nabla_{\bar{\gamma}}RX^{\bar{\gamma}} + R_{\gamma\bar{\delta}}X^{\gamma}X^{\bar{\delta}} + \frac{R}{t},$$
and (2.113) follows from (2.37). □

When Rc > 0, the $(1,0)$-vector minimizing the LHS of (2.113) is $X^\gamma = -\left(\text{Rc}^{-1}\right)^{\gamma\bar{\rho}} \nabla_{\bar{\rho}} R$, where $\left(\text{Rc}^{-1}\right)^{\alpha\bar{\beta}} R_{\gamma\bar{\beta}} = \delta^\alpha_\gamma$. Hence, if Rc > 0, then (2.113) is equivalent to

$$\frac{\partial R}{\partial t} + \frac{R}{t} - \left(\text{Rc}^{-1}\right)^{\gamma\bar{\delta}} \nabla_\gamma R \nabla_{\bar{\delta}} R \geq 0.$$

Since $R_{\gamma\bar{\delta}} \leq R g_{\gamma\bar{\delta}}$, we have $-\left(\text{Rc}^{-1}\right)^{\gamma\bar{\delta}} \leq -\frac{1}{R} g^{\gamma\bar{\delta}}$, and hence

(2.114) $$\frac{\partial}{\partial t} \log(tR) - |\nabla \log(tR)|^2 = \frac{\partial}{\partial t} \log R + \frac{1}{t} - |\nabla \log R|^2 \geq 0.$$

Without assuming Rc > 0, we still obtain (2.114) by taking $X^\gamma = -\frac{1}{R} \nabla^\gamma R$ in (2.113) and using $R_{\alpha\bar{\beta}} \leq R g_{\alpha\bar{\beta}}$.

COROLLARY 2.89 (Integrated form of Kähler trace Harnack estimate). *If $(\mathcal{M}^n, g(t))$ is a complete solution to the Kähler–Ricci flow with bounded nonnegative bisectional curvature, then for any $x_1, x_2 \in \mathcal{M}$ and $0 < t_1 < t_2$, we have*

$$\frac{R(x_2, t_2)}{R(x_1, t_1)} \geq \frac{t_1}{t_2} e^{-\frac{1}{4}\Delta},$$

where $\Delta = \Delta(x_1, t_1; x_2, t_2) \doteqdot \inf_\gamma \int_{t_1}^{t_2} |\dot{\gamma}(t)|^2_{g(t)} dt$, and the infimum is taken over all paths $\gamma : [t_1, t_2] \to \mathcal{M}$ with $\gamma(t_1) = x_1$ and $\gamma(t_2) = x_2$.

PROOF. By the fundamental theorem of calculus and (2.114), we have for any $\gamma : [t_1, t_2] \to \mathcal{M}$ with $\gamma(t_1) = x_1$ and $\gamma(t_2) = x_2$,

$$\log \frac{t_2 R(x_2, t_2)}{t_1 R(x_1, t_1)} = \int_{t_1}^{t_2} \frac{d}{dt} [\log tR(\gamma(t), t)] dt$$

$$= \int_{t_1}^{t_2} \left[\left(\frac{\partial}{\partial t} \log tR\right)(\gamma(t), t) + \langle \nabla \log tR, \dot{\gamma}(t) \rangle_{g(t)}\right] dt$$

$$\geq \int_{t_1}^{t_2} \left[|\nabla \log tR|^2_{g(t)} + \langle \nabla \log tR, \dot{\gamma}(t) \rangle_{g(t)}\right] dt$$

$$\geq -\frac{1}{4} \int_{t_1}^{t_2} |\dot{\gamma}(t)|^2_{g(t)} dt.$$

The corollary follows from taking the infimum over all γ. $\square$

For the normalized Kähler–Ricci flow, we have the following.

COROLLARY 2.90 (Integrated trace Harnack for normalized Kähler–Ricci flow). *If $(\mathcal{M}^n, g(t))$ is a solution to the normalized Kähler–Ricci flow on a closed manifold with nonnegative bisectional curvature, then for any $x_1, x_2 \in \mathcal{M}$ and $0 < t_1 < t_2$,*

(2.115) $$\frac{R(x_2, t_2)}{R(x_1, t_1)} \geq \frac{1 - e^{-\frac{r}{n} t_1}}{1 - e^{-\frac{r}{n} t_2}} e^{-\frac{1}{4}\Delta},$$

where Δ is as above and $r \geq 0$ is the average (complex) scalar curvature.

REMARK 2.91. Note that $\frac{1 - e^{-\frac{r}{n} t_1}}{1 - e^{-\frac{r}{n} t_2}} \geq \frac{e^{\frac{r}{n} t_1} - 1}{e^{\frac{r}{n} t_2} - 1}$.

Before we prove the corollary, we first recall how to go from the Kähler–Ricci flow to the normalized Kähler–Ricci flow on closed manifolds. Let $g(t)$ be a solution to $\frac{\partial}{\partial t} g_{\alpha\bar{\beta}} = -R_{\alpha\bar{\beta}}$ and let $\tilde{g}(t) \doteqdot \psi(t) g(t)$, where $\psi(t)$ is to be defined below. We compute

$$\frac{\partial}{\partial t} \tilde{g}_{\alpha\bar{\beta}} = -\psi \tilde{R}_{\alpha\bar{\beta}} + \dot{\psi} g_{\alpha\bar{\beta}}$$

since $R_{\alpha\bar{\beta}} = \tilde{R}_{\alpha\bar{\beta}}$. Hence if we define a new time parameter $\tilde{t}$ by $\frac{\partial}{\partial \tilde{t}} = \frac{1}{\psi} \frac{\partial}{\partial t}$, we have

$$\frac{\partial}{\partial \tilde{t}} \tilde{g}_{\alpha\bar{\beta}} = -\tilde{R}_{\alpha\bar{\beta}} + \frac{\dot{\psi}}{\psi^2} \tilde{g}_{\alpha\bar{\beta}}.$$

In particular, to obtain the normalized Kähler–Ricci flow, where $\tilde{g}(t)$ remains in the same Kähler class, we set

$$\frac{\dot{\psi}(t)}{\psi(t)^2} = \frac{\tilde{r}}{n},$$

where $\tilde{r}$ is the average scalar curvature of $\tilde{g}(t)$, which is independent of time since $\tilde{g}(t)$ stays in the same Kähler class. Thus we take $\psi(t) \doteqdot \left(1 - \frac{\tilde{r}}{n} t\right)^{-1}$. Since $d\tilde{t} = \psi dt$, we may take $\tilde{t} \doteqdot -\frac{n}{\tilde{r}} \log\left(1 - \frac{\tilde{r}}{n} t\right)$. That is, $t = \frac{n}{\tilde{r}} \left(1 - e^{-\frac{\tilde{r}}{n} \tilde{t}}\right)$.

PROOF OF COROLLARY 2.90. Let $\tilde{g}(\tilde{t})$ be a solution of the normalized Kähler–Ricci flow $\frac{\partial}{\partial \tilde{t}} \tilde{g}_{\alpha\bar{\beta}} = -\tilde{R}_{\alpha\bar{\beta}} + \frac{\tilde{r}}{n} \tilde{g}_{\alpha\bar{\beta}}$. Then

$$g(t) \doteqdot \left(1 - \frac{\tilde{r}}{n} t\right)^{-1} \tilde{g}\left(-\frac{n}{\tilde{r}} \log\left(1 - \frac{\tilde{r}}{n} t\right)\right)$$

is a solution of the Kähler–Ricci flow and we have the estimate $\frac{R(x_2, t_2)}{R(x_1, t_1)} \geq \frac{t_1}{t_2} e^{-\frac{1}{4}\Delta}$. This implies

$$\frac{\tilde{R}(x_2, \tilde{t}_2)}{\tilde{R}(x_1, \tilde{t}_1)} \geq \frac{1 - e^{-\frac{\tilde{r}}{n} \tilde{t}_1}}{1 - e^{-\frac{\tilde{r}}{n} \tilde{t}_2}} e^{-\frac{1}{4}\tilde{\Delta}},$$

where

$$\tilde{\Delta}(x_1, \tilde{t}_1; x_2, \tilde{t}_2) = \inf_\gamma \int_{t_1}^{t_2} \left|\frac{d\gamma}{dt}(t)\right|^2_{g(t)} dt = \inf_\gamma \int_{\tilde{t}_1}^{\tilde{t}_2} \left|\frac{d\gamma}{d\tilde{t}}(\tilde{t})\right|^2_{\tilde{g}(\tilde{t})} d\tilde{t}$$

since $\frac{d}{dt} = \psi \frac{d}{d\tilde{t}}$, $g(t) = \psi^{-1} \tilde{g}(\tilde{t})$, and $dt = \psi^{-1} d\tilde{t}$. $\square$

Since $g(t)$ has nonnegative bisectional curvature, and in particular it has nonnegative Ricci curvature, under the normalized Kähler–Ricci flow we have $\frac{\partial}{\partial t} g_{\alpha\bar{\beta}} \leq \frac{r}{n} g_{\alpha\bar{\beta}}$, which implies $g(t) \leq e^{\frac{r}{n}(t-t_1)} g(t_1)$ for $t \geq t_1$. Hence (2.115) implies

$$\Delta \leq \inf_\gamma \int_{t_1}^{t_2} e^{\frac{r}{n}(t-t_1)} |\dot{\gamma}(t)|^2_{g(t_1)} dt = \frac{r}{n} \left(1 - e^{-\frac{r}{n}(t_2-t_1)}\right)^{-1} d^2_{g(t_1)}(x_1, x_2),$$

by taking $\gamma(t)$ to be a minimal geodesic, with respect to $g(t_1)$, joining x_1 to x_2 with speed $|\dot{\gamma}(t)|_{g(t_1)} = ce^{-\frac{r}{n}(t-t_1)}$, where

$$c = \frac{r}{n}\left(1 - e^{-\frac{r}{n}(t_2-t_1)}\right)^{-1} d_{g(t_1)}(x_1, x_2).$$

Thus
(2.116)
$$\frac{R(x_2, t_2)}{R(x_1, t_1)} \geq \frac{1-e^{-\frac{r}{n}t_1}}{1-e^{-\frac{r}{n}t_2}} \exp\left(-\frac{r}{4n}\left(1-e^{-\frac{r}{n}(t_2-t_1)}\right)^{-1} d^2_{g(t_1)}(x_1,x_2)\right).$$

Note by the inequality $x \leq e^x - 1$ that we have

$$\frac{R(x_2, t_2)}{R(x_1, t_1)} \geq \frac{1-e^{-\frac{r}{n}t_1}}{1-e^{-\frac{r}{n}t_2}} \exp\left(-\frac{1}{4}e^{\frac{r}{n}(t_2-t_1)}\frac{d^2_{g(t_1)}(x_1,x_2)}{t_2-t_1}\right),$$

which is the estimate we would have obtained from (2.115) by using $e^{\frac{r}{n}(t-t_1)} \leq e^{\frac{r}{n}(t_2-t_1)}$ and taking $\gamma(t)$ to be a minimal geodesic joining x_1 to x_2 with speed $|\dot{\gamma}(t)|_{g(t_1)} = \frac{d_{g(t_1)}(x_1,x_2)}{t_2-t_1}$.

9.2. Application of the trace estimate. A beautiful application of the trace differential Harnack estimate and Perelman's no local collapsing theorem is the following uniform bound for the curvatures of a solution of the normalized Kähler–Ricci flow on a closed manifold with nonnegative bisectional curvature. This proof is due to Cao, Chen, and Zhu [49] and gives a simple proof of an estimate of Chen and Tian [87], [88], who proved convergence of the Ricci flow for the normalized Kähler–Ricci flow on closed manifolds with positive bisectional curvature (Theorem 2.70).

THEOREM 2.92 (NKRF: $K_{\mathbb{C}}(V,W) \geq 0$ curvature estimate). *If $(\mathcal{M}^n, g_0)$ is a closed Kähler manifold with $\frac{r_0}{n}[\omega_0] = 2\pi c_1(\mathcal{M})$ and nonnegative bisectional curvature, then the solution $g(t)$ to the normalized Kähler–Ricci flow, with $g(0) = g_0$, has uniformly bounded curvature for all time.*

PROOF. Without loss of generality we may assume the solution is nonflat and the average scalar curvature r is equal to n, independent of time. Given any time $t > 1$, there exists a point $y \in \mathcal{M}$ such that $R(y, t+1) = n$. By (2.116) we have for any $x \in \mathcal{M}$

$$\frac{n}{R(x,t)} = \frac{R(y,t+1)}{R(x,t)} \geq \frac{1-e^{-t}}{1-e^{-(t+1)}} \exp\left(-\frac{d^2_{g(t)}(x,y)}{4(1-e^{-1})}\right).$$

Since $\frac{1-e^{-(t+1)}}{1-e^{-t}} \leq \frac{1-e^{-2}}{1-e^{-1}} = 1 + e^{-1}$ for $t \geq 1$, we conclude for any $x \in \mathcal{M}$,

(2.117) $$R(x,t) \leq n(1+e^{-1}) \exp\left(\frac{d^2_{g(t)}(x,y)}{4(1-e^{-1})}\right).$$

In particular, if $x \in B_t(y,1)$, then
$$R(x,t) \leq n\left(1+e^{-1}\right)\exp\left(\frac{1}{4(1-e^{-1})}\right).$$

Hence, since $g(t)$ has nonnegative bisectional curvature, the curvature of $g(t)$ is bounded[9] in $B_{g(t)}(y,1)$. By Perelman's no local collapsing theorem, which holds for a solution to the normalized Kähler–Ricci flow since its statement is scale-invariant and since the corresponding solution to the Kähler–Ricci flow must blow up in finite time, we conclude that there exists a constant $\kappa > 0$ depending only on the initial metric such that
$$\operatorname{Vol}_{g(t)}\left(B_{g(t)}(y,1)\right) \geq \kappa.$$

In particular, $\kappa > 0$ is independent of $t > 1$ and the choice of $y \in \mathcal{M}$ such that $R(y, t+1) = n$.

We may obtain a uniform diameter bound for $g(t)$ by Yau's argument. In particular, let $x \in \mathcal{M}$ be a point with $d(x,y) = d \geq 2$. Since $\operatorname{Rc} \geq 0$, by the Bishop–Gromov relative volume comparison theorem, we have

(2.118)
$$\frac{\operatorname{Vol}_{g(t)}\left(B_{g(t)}(x,d+1)\right) - \operatorname{Vol}_{g(t)}\left(B_{g(t)}(x,d-1)\right)}{\operatorname{Vol}_{g(t)}\left(B_{g(t)}(x,d-1)\right)} \leq \frac{(d+1)^n - (d-1)^n}{(d-1)^n}$$
$$\leq \frac{C(n)}{d}.$$

Since $B_{g(t)}(y,1) \subset B_{g(t)}(x,d+1) \setminus B_{g(t)}(x,d-1)$ and $B_{g(t)}(x,d-1) \subset B_{g(t)}(y, 2d-1)$, by (2.118), we have
$$\operatorname{Vol}_{g(t)}(\mathcal{M}) \geq \operatorname{Vol}_{g(t)}\left(B_{g(t)}(y, 2d-1)\right)$$
$$\geq \operatorname{Vol}_{g(t)}\left(B_{g(t)}(x, d-1)\right)$$
$$\geq \frac{\operatorname{Vol}_{g(t)}\left(B_{g(t)}(y,1)\right)}{C(n)} d.$$

Taking $d = \operatorname{diam}_{g(t)}(\mathcal{M})$, we have
$$\operatorname{diam}_{g(t)}(\mathcal{M}) \leq \frac{C(n)\operatorname{Vol}_{g(t)}(\mathcal{M})}{\operatorname{Vol}_{g(t)}\left(B_{g(t)}(y,1)\right)} \leq \frac{C(n)\operatorname{Vol}_{g(t)}(\mathcal{M})}{\kappa}.$$

Since under the normalized flow, $\operatorname{Vol}_{g(t)}(\mathcal{M})$ is constant, we obtain a uniform upper bound C for the diameter of $g(t)$.

Hence (2.117) implies
$$R(x,t) \leq n\left(1+e^{-1}\right)\exp\left(\frac{C^2}{4(1-e^{-1})}\right),$$
which is our desired uniform estimate for R. □

---

[9] We actually only need an upper bound on the scalar curvature for Perelman's no local collapsing theorem (Theorem 6.74).

9.3. Proof of the matrix Harnack estimate.
In this subsection we prove Theorem 2.87. Let

$$P_{\alpha\bar\beta\gamma} \doteqdot \nabla_\gamma R_{\alpha\bar\beta} + R_{\alpha\bar\beta\gamma\bar\delta} X_\delta, \qquad P_{\alpha\bar\beta\bar\gamma} \doteqdot \nabla_{\bar\gamma} R_{\alpha\bar\beta} + R_{\alpha\bar\beta\bar\gamma\delta} X_{\bar\delta}.$$

The following computation is the one corresponding to Proposition 2.79.

PROPOSITION 2.93 (Evolution of the Harnack quantity $Z_{\alpha\bar\beta}$). *Suppose that a vector field X satisfies*

$$\nabla_{\bar\delta} X_\gamma = \overline{\nabla_\delta X_{\bar\gamma}} = R_{\gamma\bar\delta} + \frac{1}{t} g_{\gamma\bar\delta},$$
$$\nabla_\delta X_\gamma = \nabla_{\bar\delta} X_{\bar\gamma} = 0,$$

and

$$\left(\frac{\partial}{\partial t} - \Delta\right) X^\gamma = \frac{1}{2} R^\gamma_\delta X^\delta - \frac{1}{t} X^\gamma.$$

Then $Z_{\alpha\bar\beta} = Z(X)_{\alpha\bar\beta}$ defined by (2.111) satisfies the evolution equation:

$$\left(\frac{\partial}{\partial t} - \Delta\right) Z_{\alpha\bar\beta} = R_{\alpha\bar\beta\gamma\bar\delta} Z_{\bar\gamma\delta} - \frac{1}{2}\left(R_{\alpha\bar\gamma} Z_{\gamma\bar\beta} + R_{\gamma\bar\beta} Z_{\alpha\bar\gamma}\right)$$
(2.119)
$$+ P_{\alpha\bar\delta\gamma} P_{\delta\bar\beta\bar\gamma} - P_{\alpha\bar\gamma\bar\delta} P_{\bar\beta\gamma\delta} - \frac{2}{t} Z_{\alpha\bar\beta}.$$

The proposition follows from Lemmas 2.94 and 2.95 below.

Assuming the proposition, we now prove the differential Harnack estimate, i.e., Proposition 2.93. Applying $Z_{\alpha\bar\beta}$ to a $(1,0)$-vector field W, we have the following general formula:

$$\left(\frac{\partial}{\partial t} - \Delta\right)\left(Z_{\alpha\bar\beta} W^\alpha W^{\bar\beta}\right) = \left(\left(\frac{\partial}{\partial t} - \Delta\right) Z_{\alpha\bar\beta}\right) W^\alpha W^{\bar\beta}$$
$$+ Z_{\alpha\bar\beta}\left(\left(\frac{\partial}{\partial t} - \Delta\right)\left(W^\alpha W^{\bar\beta}\right)\right)$$
$$+ \nabla_\tau Z_{\alpha\bar\beta} \nabla_{\bar\tau}\left(W^\alpha W^{\bar\beta}\right) + \nabla_{\bar\tau} Z_{\alpha\bar\beta} \nabla_\tau\left(W^\alpha W^{\bar\beta}\right),$$

where $W^{\bar\beta} \doteqdot \overline{W^\beta}$. Thus, if we have a null vector W of $Z_{\alpha\bar\beta}$ at a point (x_0, t_0) and if we extend W locally in space and time so that at (x_0, t_0)

(2.120) $$W_{\alpha,\bar\beta} = W_{\bar\alpha,\beta} = 0,$$

(2.121) $$\left(\frac{\partial}{\partial t} - \Delta\right) W^\alpha = 0,$$

where $W_{\bar{\alpha}} = g_{\beta\bar{\alpha}} W^{\beta}$, then (2.119) implies

$$\left(\frac{\partial}{\partial t} - \Delta\right) \left(Z_{\alpha\bar{\beta}} W_{\bar{\alpha}} W_{\beta}\right)$$
$$= \left(R_{\alpha\bar{\beta}\gamma\bar{\delta}} Z_{\delta\bar{\gamma}}\right) W_{\bar{\alpha}} W_{\beta} + \left(P_{\alpha\bar{\delta}\gamma} P_{\ell\bar{\beta}\bar{\gamma}} - P_{\alpha\bar{\delta}\bar{\gamma}} P_{\delta\bar{\beta}\gamma}\right) W_{\bar{\alpha}} W_{\beta}$$
$$- \frac{1}{2} \left(R_{\alpha\bar{\gamma}} Z_{\gamma\bar{\beta}} + Z_{\alpha\bar{\gamma}} R_{\gamma\bar{\beta}}\right) W_{\bar{\alpha}} W_{\beta} - \frac{2}{t} Z_{\alpha\bar{\beta}} W_{\bar{\alpha}} W_{\beta}$$
$$= R_{\alpha\bar{\beta}\gamma\bar{\delta}} Z_{\delta\bar{\gamma}} W_{\bar{\alpha}} W_{\beta} + |M_{\gamma\bar{\delta}}|^2 - |M_{\gamma\delta}|^2.$$

Here

$$M_{\bar{\delta}\gamma} = P_{\alpha\bar{\delta}\gamma} W_{\bar{\alpha}} \quad \text{and} \quad M_{\delta\gamma} = P_{\ell\bar{\beta}\gamma} W_{\beta}.$$

Now we use the facts that $Z_{\alpha\bar{\beta}} \geq 0$ (at least on all of $\mathcal{M} \times [0, t_0]$) and W is a null vector of $Z_{\alpha\bar{\beta}}$. An algebraic fact, similar to Lemma 2.86 and using a second variation computation similar to that in the proof of Theorem 2.78, shows that

$$R_{\alpha\bar{\beta}\gamma\bar{\delta}} Z_{\delta\bar{\gamma}} W_{\bar{\alpha}} W_{\beta} \geq |M_{\gamma\bar{\delta}}|^2 - |M_{\gamma\delta}|^2.$$

Thus at a point where W satisfies (2.120)–(2.121), we have

$$\left(\frac{\partial}{\partial t} - \Delta\right) \left(Z_{\alpha\bar{\beta}} W_{\bar{\alpha}} W_{\beta}\right) \geq 0.$$

Hence, by the maximum principle, on a closed manifold we have $Z_{\alpha\bar{\beta}} \geq 0$ on all of space and time. In the complete noncompact case, one can adapt the proof in Part II of this volume of Hamilton's matrix Harnack estimate for complete solutions to the Ricci flow with nonnegative curvature operator to this Kähler setting without significant modifications.

Now we give the two lemmas which are needed to complete the proof of Proposition 2.93.

LEMMA 2.94.

$$\left(\frac{\partial}{\partial t} - \Delta\right) \left(\Delta R_{\alpha\bar{\beta}} + R_{\alpha\bar{\beta}\gamma\bar{\delta}} R_{\bar{\gamma}\delta}\right)$$
$$= \frac{1}{2} \Delta R_{\alpha\bar{\rho}} R_{\rho\bar{\beta}} + \frac{1}{2} R_{\alpha\bar{\rho}} \Delta R_{\rho\bar{\beta}} + 2 R_{\alpha\bar{\rho},\gamma} R_{\rho\bar{\beta},\bar{\gamma}}$$
$$+ \frac{1}{2} R_{\delta\bar{\gamma}} \left(\nabla_{\bar{\delta}} \nabla_{\gamma} R_{\alpha\bar{\beta}} + \nabla_{\gamma} \nabla_{\bar{\delta}} R_{\alpha\bar{\beta}}\right) - \Delta \left(R_{\alpha\bar{\gamma}} R_{\gamma\bar{\beta}}\right)$$
$$+ \frac{1}{2} \left(\nabla_{\bar{\delta}} \nabla_{\gamma} R_{\alpha\bar{\beta}} + \nabla_{\gamma} \nabla_{\bar{\delta}} R_{\alpha\bar{\beta}} - R_{\alpha\bar{\rho}\gamma\bar{\delta}} R_{\rho\bar{\beta}} - R_{\alpha\bar{\rho}} R_{\rho\bar{\beta}\gamma\bar{\delta}}\right) R_{\delta\bar{\gamma}}$$

(2.122) $\qquad + 2 R_{\alpha\bar{\beta}\gamma\bar{\delta}} R_{\delta\bar{\rho}} R_{\rho\bar{\gamma}} + R_{\alpha\bar{\beta}\gamma\bar{\delta}} \frac{\partial}{\partial t} R_{\delta\bar{\gamma}}.$

PROOF. Let $h_{\alpha\bar\beta}$ be a real $(1,1)$-tensor. Using (2.33), we compute

$$\frac{\partial}{\partial t}\left(\nabla_{\bar\delta}\nabla_\gamma h_{\alpha\bar\beta}\right) = \nabla_{\bar\delta}\nabla_\gamma\left(\frac{\partial}{\partial t}h_{\alpha\bar\beta}\right) - \nabla_{\bar\delta}\left(\left(\frac{\partial}{\partial t}\Gamma^\rho_{\gamma\alpha}\right)h_{\rho\bar\beta}\right) - \left(\frac{\partial}{\partial t}\overline{\Gamma^\rho_{\delta\beta}}\right)\nabla_\gamma h_{\alpha\bar\rho}$$

$$= \nabla_{\bar\delta}\nabla_\gamma\left(\frac{\partial}{\partial t}h_{\alpha\bar\beta}\right) + \nabla_{\bar\delta}\left(g^{\rho\bar\sigma}\nabla_\gamma R_{\alpha\bar\sigma}h_{\rho\bar\beta}\right) + g^{\sigma\bar\rho}\nabla_{\bar\delta}R_{\bar\beta\sigma}\nabla_\gamma h_{\alpha\bar\rho}$$

(2.123)
$$= \nabla_{\bar\delta}\nabla_\gamma\left(\frac{\partial}{\partial t}h_{\alpha\bar\beta}\right) + \nabla_{\bar\delta}\nabla_\gamma R_{\alpha\bar\rho}h_{\rho\bar\beta}$$
$$+ \nabla_\gamma R_{\alpha\bar\rho}\nabla_{\bar\delta}h_{\rho\bar\beta} + \nabla_{\bar\delta}R_{\rho\bar\beta}\nabla_\gamma h_{\alpha\bar\rho}.$$

Taking the complex conjugate of (2.123), we have

(2.124)
$$\frac{\partial}{\partial t}\left(\nabla_\gamma\nabla_{\bar\delta}h_{\alpha\bar\beta}\right) = \nabla_\gamma\nabla_{\bar\delta}\left(\frac{\partial}{\partial t}h_{\alpha\bar\beta}\right) + \nabla_\gamma\nabla_{\bar\delta}R_{\rho\bar\beta}h_{\alpha\bar\rho}$$
$$+ \nabla_{\bar\delta}R_{\rho\bar\beta}\nabla_\gamma h_{\alpha\bar\rho} + \nabla_\gamma R_{\alpha\bar\rho}\nabla_{\bar\delta}h_{\rho\bar\beta}.$$

Next we compute, using (2.123), (2.124) and tracing, that

$$\frac{\partial}{\partial t}\left(\Delta h_{\alpha\bar\beta}\right) = \frac{1}{2}\frac{\partial}{\partial t}\left[g^{\gamma\bar\delta}\left(\nabla_{\bar\delta}\nabla_\gamma h_{\alpha\bar\beta} + \nabla_\gamma\nabla_{\bar\delta}h_{\alpha\bar\beta}\right)\right]$$
$$= \frac{1}{2}g^{\gamma\bar\delta}\frac{\partial}{\partial t}\left(\nabla_{\bar\delta}\nabla_\gamma h_{\alpha\bar\beta} + \nabla_\gamma\nabla_{\bar\delta}h_{\alpha\bar\beta}\right)$$
$$+ \frac{1}{2}R_{\delta\bar\gamma}\left(\nabla_{\bar\delta}\nabla_\gamma h_{\alpha\bar\beta} + \nabla_\gamma\nabla_{\bar\delta}h_{\alpha\bar\beta}\right)$$
$$= \Delta\left(\frac{\partial}{\partial t}h_{\alpha\bar\beta}\right) + \frac{1}{2}R_{\delta\bar\gamma}\left(\nabla_{\bar\delta}\nabla_\gamma h_{\alpha\bar\beta} + \nabla_\gamma\nabla_{\bar\delta}h_{\alpha\bar\beta}\right)$$
$$+ \frac{1}{2}\left(\nabla_{\bar\gamma}\nabla_\gamma R_{\alpha\bar\rho}h_{\rho\bar\beta} + \nabla_\gamma R_{\alpha\bar\rho}\nabla_{\bar\gamma}h_{\rho\bar\beta} + \nabla_{\bar\gamma}R_{\rho\bar\beta}\nabla_\gamma h_{\alpha\bar\rho}\right)$$
$$+ \frac{1}{2}\left(\nabla_\gamma\nabla_{\bar\gamma}R_{\rho\bar\beta}h_{\alpha\bar\rho} + \nabla_{\bar\gamma}R_{\rho\bar\beta}\nabla_\gamma h_{\alpha\bar\rho} + \nabla_\gamma R_{\alpha\bar\rho}\nabla_{\bar\gamma}h_{\rho\bar\beta}\right).$$

In particular, simplifying and taking $h_{\alpha\bar\beta}$ to be the Ricci tensor, we have

$$\left(\frac{\partial}{\partial t} - \Delta\right)\left(\Delta R_{\alpha\bar\beta}\right) = \Delta\left(R_{\alpha\bar\beta\gamma\bar\delta}R_{\delta\bar\gamma}\right) - \Delta\left(R_{\alpha\bar\gamma}R_{\gamma\bar\beta}\right)$$
$$+ \frac{1}{2}R_{\delta\bar\gamma}\left(\nabla_{\bar\delta}\nabla_\gamma R_{\alpha\bar\beta} + \nabla_\gamma\nabla_{\bar\delta}R_{\alpha\bar\beta}\right)$$
$$+ \frac{1}{2}\left(\nabla_{\bar\gamma}\nabla_\gamma R_{\alpha\bar\rho}R_{\rho\bar\beta} + \nabla_\gamma\nabla_{\bar\gamma}R_{\rho\bar\beta}R_{\alpha\bar\rho}\right)$$
$$+ \nabla_\gamma R_{\alpha\bar\rho}\nabla_{\bar\gamma}R_{\rho\bar\beta} + \nabla_{\bar\gamma}R_{\rho\bar\beta}\nabla_\gamma R_{\alpha\bar\rho}.$$

On the other hand, using (2.103), we compute

$$\frac{\partial}{\partial t}\left(R_{\alpha\bar\beta\gamma\bar\delta}R_{\bar\gamma\delta}\right) = \left(\nabla_\gamma\nabla_{\bar\delta}R_{\alpha\bar\beta} - R_{\alpha\bar\rho}R_{\rho\bar\beta\gamma\bar\delta}\right)R_{\bar\gamma\delta} + R_{\alpha\bar\beta\gamma\bar\delta}\frac{\partial}{\partial t}R_{\bar\gamma\delta}$$
$$+ 2R_{\alpha\bar\beta\gamma\bar\delta}R_{\delta\bar\rho}R_{\rho\bar\gamma}.$$

Hence
$$\left(\frac{\partial}{\partial t} - \Delta\right)\left(\Delta R_{\alpha\bar{\beta}} + R_{\alpha\bar{\beta}\gamma\bar{\delta}}R_{\bar{\gamma}\delta}\right)$$
$$= \frac{1}{2}\left(\nabla_{\bar{\gamma}}\nabla_{\gamma}R_{\alpha\bar{\rho}}R_{\rho\bar{\beta}} + \nabla_{\gamma}\nabla_{\bar{\gamma}}R_{\rho\bar{\beta}}R_{\alpha\bar{\rho}}\right) + 2\nabla_{\gamma}R_{\alpha\bar{\rho}}\nabla_{\bar{\gamma}}R_{\rho\bar{\beta}}$$
$$+ \frac{1}{2}R_{\delta\bar{\gamma}}\left(\nabla_{\bar{\delta}}\nabla_{\gamma}R_{\alpha\bar{\beta}} + \nabla_{\gamma}\nabla_{\bar{\delta}}R_{\alpha\bar{\beta}}\right) - \Delta\left(R_{\alpha\bar{\gamma}}R_{\gamma\bar{\beta}}\right)$$
$$+ \left(\nabla_{\gamma}\nabla_{\bar{\delta}}R_{\alpha\bar{\beta}} - R_{\alpha\bar{\rho}}R_{\rho\bar{\beta}\gamma\bar{\delta}}\right)R_{\bar{\gamma}\delta} + R_{\alpha\bar{\beta}\gamma\bar{\delta}}\frac{\partial}{\partial t}R_{\bar{\gamma}\delta}$$
$$+ 2R_{\alpha\bar{\beta}\gamma\bar{\delta}}R_{\delta\bar{\rho}}R_{\rho\bar{\gamma}}.$$

Finally we obtain (2.122) from the commutator equations:
$$\frac{1}{2}\left(\nabla_{\gamma}\nabla_{\bar{\delta}}R_{\alpha\bar{\beta}} - \nabla_{\gamma}\nabla_{\bar{\delta}}R_{\alpha\bar{\beta}}\right) = -\frac{1}{2}\left(R_{\gamma\bar{\delta}\alpha\bar{\rho}}R_{\rho\bar{\beta}} - R_{\gamma\bar{\delta}\rho\bar{\beta}}R_{\alpha\bar{\rho}}\right)$$

and
$$\nabla_{\bar{\gamma}}\nabla_{\gamma}R_{\alpha\bar{\rho}} - \nabla_{\gamma}\nabla_{\bar{\gamma}}R_{\alpha\bar{\rho}} = -\left(R_{\bar{\gamma}\gamma\alpha\bar{\sigma}}R_{\sigma\bar{\rho}} + R_{\bar{\gamma}\gamma\bar{\rho}\sigma}R_{\alpha\bar{\sigma}}\right)$$
$$= R_{\alpha\bar{\sigma}}R_{\sigma\bar{\rho}} - R_{\sigma\bar{\rho}}R_{\alpha\bar{\sigma}} = 0,$$

which imply
$$\nabla_{\bar{\gamma}}\nabla_{\gamma}R_{\alpha\bar{\rho}} = \nabla_{\gamma}\nabla_{\bar{\gamma}}R_{\alpha\bar{\rho}} = \Delta R_{\alpha\bar{\rho}}.$$
$\square$

LEMMA 2.95.
$$\left(\frac{\partial}{\partial t} - \Delta\right)\left(\nabla_{\gamma}R_{\alpha\bar{\beta}}X^{\gamma}\right)$$
$$= -\frac{1}{2}\left(R_{\alpha\bar{\rho}}\nabla_{\gamma}R_{\rho\bar{\beta}} + \nabla_{\gamma}R_{\alpha\bar{\rho}}R_{\rho\bar{\beta}} + \nabla_{\rho}R_{\alpha\bar{\beta}}R_{\gamma\bar{\rho}}\right)X^{\gamma}$$
$$+ \left(\nabla_{\rho}R_{\alpha\bar{\sigma}}R_{\gamma\bar{\beta}\sigma\bar{\rho}} - R_{\alpha\bar{\sigma}\gamma\bar{\rho}}\nabla_{\rho}R_{\sigma\bar{\beta}}\right)X^{\gamma} + \nabla_{\gamma}\left(R_{\alpha\bar{\beta}\rho\bar{\sigma}}R_{\sigma\bar{\rho}}\right)X^{\gamma}$$
(2.125)
$$+ \nabla_{\gamma}R_{\alpha\bar{\beta}}\left(\frac{\partial}{\partial t} - \Delta\right)X^{\gamma} - \nabla_{\bar{\delta}}\nabla_{\gamma}R_{\alpha\bar{\beta}}\nabla_{\delta}X^{\gamma} - \nabla_{\delta}\nabla_{\gamma}R_{\alpha\bar{\beta}}\nabla_{\bar{\delta}}X^{\gamma}.$$

PROOF. We compute using (2.105)
$$\left(\frac{\partial}{\partial t} - \Delta\right)\nabla_{\gamma}R_{\alpha\bar{\beta}} = \nabla_{\gamma}\left(\frac{\partial}{\partial t} - \Delta\right)R_{\alpha\bar{\beta}} + \left(\nabla_{\gamma}\Delta - \Delta\nabla_{\gamma}\right)R_{\alpha\bar{\beta}}$$
$$- \left(\frac{\partial}{\partial t}\Gamma^{\rho}_{\gamma\alpha}\right)R_{\rho\bar{\beta}}$$
$$= \nabla_{\gamma}\left(R_{\alpha\bar{\beta}\rho\bar{\delta}}R_{\delta\bar{\rho}} - R_{\alpha\bar{\rho}}R_{\rho\bar{\beta}}\right) + g^{\rho\bar{\sigma}}\nabla_{\gamma}R_{\alpha\bar{\sigma}}R_{\rho\bar{\beta}}$$
(2.126)
$$+ \frac{1}{2}\left(-R_{\sigma\bar{\beta}}\nabla_{\gamma}R_{\alpha\bar{\sigma}} + R_{\alpha\bar{\sigma}}\nabla_{\gamma}R_{\sigma\bar{\beta}}\right)$$
$$- R_{\gamma\bar{\rho}\alpha\bar{\sigma}}\nabla_{\rho}R_{\sigma\bar{\beta}} + R_{\gamma\bar{\rho}\sigma\bar{\beta}}\nabla_{\rho}R_{\alpha\bar{\sigma}} - \frac{1}{2}R_{\gamma\bar{\sigma}}\nabla_{\sigma}R_{\alpha\bar{\beta}}.$$

Here we also used (2.33) and the following general identity:

$$(\nabla_\gamma \Delta - \Delta \nabla_\gamma) h_{\alpha\bar{\beta}} = \frac{1}{2} \nabla_\rho (\nabla_\gamma \nabla_{\bar{\rho}} - \nabla_{\bar{\rho}} \nabla_\gamma) h_{\alpha\bar{\beta}}$$

$$+ \frac{1}{2} (\nabla_\gamma \nabla_{\bar{\rho}} - \nabla_{\bar{\rho}} \nabla_\gamma) \nabla_\rho h_{\alpha\bar{\beta}}$$

$$= \frac{1}{2} \nabla_\rho \left(-R_{\gamma\bar{\rho}\alpha\bar{\sigma}} h_{\sigma\bar{\beta}} + R_{\gamma\bar{\rho}\sigma\bar{\beta}} h_{\alpha\bar{\sigma}} \right)$$

$$+ \frac{1}{2} \left(-R_{\gamma\bar{\sigma}} \nabla_\sigma h_{\alpha\bar{\beta}} - R_{\gamma\bar{\rho}\alpha\bar{\sigma}} \nabla_\rho h_{\sigma\bar{\beta}} + R_{\gamma\bar{\rho}\sigma\bar{\beta}} \nabla_\rho h_{\alpha\bar{\sigma}} \right)$$

(2.127)
$$= \frac{1}{2} \left(-\nabla_\gamma R_{\alpha\bar{\sigma}} h_{\sigma\bar{\beta}} + \nabla_\gamma R_{\sigma\bar{\beta}} h_{\alpha\bar{\sigma}} \right)$$

$$- R_{\gamma\bar{\rho}\alpha\bar{\sigma}} \nabla_\rho h_{\sigma\bar{\beta}} + R_{\gamma\bar{\rho}\sigma\bar{\beta}} \nabla_\rho h_{\alpha\bar{\sigma}} - \frac{1}{2} R_{\gamma\bar{\sigma}} \nabla_\sigma h_{\alpha\bar{\beta}},$$

where we used the second Bianchi identity (2.9). Simplifying (2.126), we have

$$\left(\frac{\partial}{\partial t} - \Delta \right) \nabla_\gamma R_{\alpha\bar{\beta}} = -\frac{1}{2} \left(R_{\sigma\bar{\beta}} \nabla_\gamma R_{\alpha\bar{\sigma}} + R_{\alpha\bar{\sigma}} \nabla_\gamma R_{\sigma\bar{\beta}} + R_{\gamma\bar{\sigma}} \nabla_\sigma R_{\alpha\bar{\beta}} \right)$$

$$+ R_{\gamma\bar{\rho}\sigma\bar{\beta}} \nabla_\rho R_{\alpha\bar{\sigma}} - R_{\gamma\bar{\rho}\alpha\bar{\sigma}} \nabla_\rho R_{\sigma\bar{\beta}} + \nabla_\gamma \left(R_{\alpha\bar{\beta}\rho\bar{\delta}} R_{\delta\bar{\rho}} \right).$$

Equation (2.125) now follows from this and the general formula

$$\left(\frac{\partial}{\partial t} - \Delta \right) (\nabla_\gamma R_{\alpha\bar{\beta}} X^\gamma) = \left[\left(\frac{\partial}{\partial t} - \Delta \right) (\nabla_\gamma R_{\alpha\bar{\beta}}) \right] X^\gamma + \nabla_\gamma R_{\alpha\bar{\beta}} \left(\frac{\partial}{\partial t} - \Delta \right) X^\gamma$$

$$- \nabla_{\bar{\delta}} \nabla_\gamma R_{\alpha\bar{\beta}} \nabla_\delta X^\gamma - \nabla_\delta \nabla_\gamma R_{\alpha\bar{\beta}} \nabla_{\bar{\delta}} X^\gamma.$$

□

EXERCISE 2.96. Prove Proposition 2.93 using Lemmas 2.94 and 2.95.

10. Linear and interpolated differential Harnack estimates

In this section we consider a differential Harnack estimate related to the estimate of H.-D. Cao considered in the previous section. This estimate has applications in the study of the geometry and function theory of noncompact Kähler manifolds with nonnegative bisectional curvature.

Let $(\mathcal{M}^n, g(t))$, $t \in [0, T)$, be a complete noncompact solution of the Kähler–Ricci flow with bounded nonnegative bisectional curvature. By Shi's theorem, given an initial metric which is complete with bounded nonnegative bisectional curvature, such a solution exists, at least for some short time $T > 0$, with

$$|\nabla \operatorname{Rm}(x, t)| \leq \frac{C}{t^{1/2}}$$

for some $C < \infty$. Analogous to the Riemannian case (see Remark 2.25 in this chapter or Theorem 10.46 on p. 415 of [**111**]), we consider a solution to the **linearized Kähler–Ricci flow**. That is, we let $h_{\alpha\bar{\beta}}$ be a Hermitian symmetric (1, 1)-tensor satisfying the Kähler-Lichnerowicz Laplacian heat equation:

$$\left(\frac{\partial}{\partial t} - \Delta\right) h_{\alpha\bar{\beta}} = R_{\alpha\bar{\beta}\gamma\bar{\delta}} h_{\delta\bar{\gamma}} - \frac{1}{2}\left(R_{\alpha\bar{\gamma}} h_{\gamma\bar{\beta}} + R_{\gamma\bar{\beta}} h_{\alpha\bar{\gamma}}\right). \tag{2.128}$$

A bound for reasonable solutions is given by the following result (see Lemma 1.2 and Proposition 1.1 in Ni and Tam [**290**]).

PROPOSITION 2.97 (Exponential bound for $h_{\alpha\bar{\beta}}$). *Suppose a solution $h_{\alpha\bar{\beta}}$ of* (2.128) *satisfies, for some constants A and B, the following inequalities:*

$$\left|h_{\alpha\bar{\beta}}(x,0)\right| \le e^{A(1+r_0(x))} \tag{2.129}$$

and

$$\int_0^T \int_\mathcal{M} e^{-Br_0^2(x)} \left|h_{\alpha\bar{\beta}}(x,t)\right|^2 d\mu_{g(t)} dt < \infty. \tag{2.130}$$

Then there exists a constant $C < \infty$ such that

$$\left|h_{\alpha\bar{\beta}}(x,t)\right| \le e^{C(1+r_0(x))}.$$

Furthermore, if $\left(h_{\alpha\bar{\beta}}(x,0)\right) \ge 0$, then $\left(h_{\alpha\bar{\beta}}(x,t)\right) \ge 0$ for all $t > 0$.

The analogue of the linear trace differential Harnack estimate for the Riemannian Ricci flow, Theorem A.57, is as follows (see Theorem 1.2 on p. 633 of Ni and Tam [**290**]). The **Kähler linear trace differential Harnack quadratic** is defined by (compare with (A.27))

$$Z(h,V) \doteqdot \frac{1}{2} g^{\alpha\bar{\beta}} \left(\nabla_{\bar{\beta}} \operatorname{div}(h)_\alpha + \nabla_\alpha \operatorname{div}(h)_{\bar{\beta}}\right) + R_{\alpha\bar{\beta}} h_{\beta\bar{\alpha}}$$
$$+ g^{\alpha\bar{\beta}} \left(\operatorname{div}(h)_\alpha V_{\bar{\beta}} + \operatorname{div}(h)_{\bar{\beta}} V_\alpha\right) + h_{\alpha\bar{\beta}} V_\beta V_{\bar{\alpha}} + \frac{H}{t},$$

where V is a vector field of type $(1,0)$, $H \doteqdot g^{\alpha\bar{\beta}} h_{\alpha\bar{\beta}}$, and

$$\operatorname{div}(h)_\alpha \doteqdot g^{\gamma\bar{\beta}} \nabla_\gamma h_{\alpha\bar{\beta}}. \tag{2.131}$$

THEOREM 2.98 (Kähler linear trace differential Harnack estimate). *Suppose that $(\mathcal{M}^n, g(t))$, $t \in [0,T]$, is a complete solution of the Kähler–Ricci flow with bounded nonnegative bisectional curvature and $\left(h_{\alpha\bar{\beta}}\right) \ge 0$ is a solution of the Kähler-Lichnerowicz Laplacian heat equation* (2.128) *satisfying* (2.129) *and* (2.130). *Then*

$$Z(h,V) \ge 0$$

on $\mathcal{M} \times [0,T]$ for any vector field V of type $(1,0)$.

The proof of this theorem requires a number of calculations which we state and prove. In these calculations the theme is to derive a heat-type equation for each of the quantities under consideration. We then need to combine terms in a good way so that we obtain a supersolution to the heat equation. The way this is accomplished, as in the Riemannian case, is to look for terms which vanish on gradient Kähler–Ricci solitons.

LEMMA 2.99. *We have the following equations and their complex conjugates:*

(1)
$$(2.132) \quad \left(\frac{\partial}{\partial t} - \Delta\right) \operatorname{div}(h)_\alpha = R_{\mu\bar{\nu}}\nabla_\nu h_{\alpha\bar{\mu}} + h_{\bar{\mu}\nu}\nabla_\alpha R_{\mu\bar{\nu}} - \frac{1}{2}R_{\alpha\bar{\nu}}\operatorname{div}(h)_\nu,$$

(2)
$$(2.133) \quad \left(\frac{\partial}{\partial t} - \Delta\right)\left(g^{\alpha\bar{\beta}}\nabla_{\bar{\beta}}\operatorname{div}(h)_\alpha\right) = R_{\mu\bar{\alpha}}\nabla_{\bar{\mu}}\operatorname{div}(h)_\alpha + \nabla_{\bar{\alpha}}R_{\mu\bar{\nu}}\nabla_\nu h_{\alpha\bar{\mu}}$$
$$+ \nabla_\alpha R_{\mu\bar{\nu}}\nabla_{\bar{\alpha}}h_{\bar{\mu}\nu} + R_{\mu\bar{j}}\nabla_{\bar{\alpha}}\nabla_\nu h_{\alpha\bar{\mu}}$$
$$+ h_{\bar{\mu}\nu}\nabla_{\bar{\alpha}}\nabla_\alpha R_{\mu\bar{\nu}}.$$

PROOF. Using (2.128) and (2.131), we compute
$$\left(\frac{\partial}{\partial t} - \Delta\right)\nabla_\gamma h_{\alpha\bar{\beta}} = \nabla_\gamma\left(\frac{\partial}{\partial t} - \Delta\right)h_{\alpha\bar{\beta}} + (\nabla_\gamma\Delta - \Delta\nabla_\gamma)h_{\alpha\bar{\beta}} - \left(\frac{\partial}{\partial t}\Gamma^\delta_{\gamma\alpha}\right)h_{\delta\bar{\beta}}$$
$$= \nabla_\gamma\left(R_{\alpha\bar{\beta}\varepsilon\bar{\delta}}h_{\delta\bar{\varepsilon}} - \frac{1}{2}\left(R_{\alpha\bar{\varepsilon}}h_{\varepsilon\bar{\beta}} + R_{\varepsilon\bar{\beta}}h_{\alpha\bar{\varepsilon}}\right)\right)$$
$$- \frac{1}{2}\nabla_\gamma R_{\alpha\bar{\delta}}h_{\delta\bar{\beta}} + \frac{1}{2}\nabla_\gamma R_{\delta\bar{\beta}}h_{\alpha\bar{\delta}} - R_{\gamma\bar{\eta}\alpha\bar{\delta}}\nabla_\eta h_{\delta\bar{\beta}}$$
$$+ R_{\gamma\bar{\eta}\delta\bar{\beta}}\nabla_\eta h_{\alpha\bar{\delta}} - \frac{1}{2}R_{\gamma\bar{\delta}}\nabla_\delta h_{\alpha\bar{\beta}} + \nabla_\gamma R_{\alpha\bar{\delta}}h_{\delta\bar{\beta}},$$

since
$$2\left(\nabla_\gamma\Delta - \Delta\nabla_\gamma\right)h_{\alpha\bar{\beta}} = \nabla_\gamma\left(\nabla_\eta\nabla_{\bar{\eta}} + \nabla_{\bar{\eta}}\nabla_\eta\right)h_{\alpha\bar{\beta}} - \left(\nabla_\eta\nabla_{\bar{\eta}} + \nabla_{\bar{\eta}}\nabla_\eta\right)\nabla_\gamma h_{\alpha\bar{\beta}}$$
$$= \nabla_\eta\left(\nabla_\gamma\nabla_{\bar{\eta}} - \nabla_{\bar{\eta}}\nabla_\gamma\right)h_{\alpha\bar{\beta}} + \left(\nabla_\gamma\nabla_{\bar{\eta}} - \nabla_{\bar{\eta}}\nabla_\gamma\right)\nabla_\eta h_{\alpha\bar{\beta}}$$
$$= \nabla_\eta\left(-R_{\gamma\bar{\eta}\alpha\bar{\delta}}h_{\delta\bar{\beta}} + R_{\gamma\bar{\eta}\delta\bar{\beta}}h_{\alpha\bar{\delta}}\right)$$
$$- R_{\gamma\bar{\eta}\eta\bar{\delta}}\nabla_\delta h_{\alpha\bar{\beta}} - R_{\gamma\bar{\eta}\alpha\bar{\delta}}\nabla_\eta h_{\delta\bar{\beta}} + R_{\gamma\bar{\eta}\delta\bar{\beta}}\nabla_\eta h_{\alpha\bar{\delta}}$$
$$= -\nabla_\gamma R_{\alpha\bar{\delta}}h_{\delta\bar{\beta}} + \nabla_\gamma R_{\delta\bar{\beta}}h_{\alpha\bar{\delta}}$$
$$- 2R_{\gamma\bar{\eta}\alpha\bar{\delta}}\nabla_\eta h_{\delta\bar{\beta}} + 2R_{\gamma\bar{\eta}\delta\bar{\beta}}\nabla_\eta h_{\alpha\bar{\delta}} - R_{\gamma\bar{\delta}}\nabla_\delta h_{\alpha\bar{\beta}}.$$

Simplifying, we have
$$\left(\frac{\partial}{\partial t} - \Delta\right)\nabla_\gamma h_{\alpha\bar{\beta}} = \nabla_\gamma R_{\alpha\bar{\beta}\varepsilon\bar{\delta}}h_{\delta\bar{\varepsilon}} - R_{\gamma\bar{\eta}\alpha\bar{\delta}}\nabla_\eta h_{\delta\bar{\beta}} + R_{\gamma\bar{\eta}\delta\bar{\beta}}\nabla_\eta h_{\alpha\bar{\delta}}$$
$$+ R_{\alpha\bar{\beta}\varepsilon\bar{\delta}}\nabla_\gamma h_{\delta\bar{\varepsilon}} - \frac{1}{2}\left(R_{\alpha\bar{\varepsilon}}\nabla_\gamma h_{\varepsilon\bar{\beta}} + R_{\varepsilon\bar{\beta}}\nabla_\gamma h_{\alpha\bar{\varepsilon}} + R_{\gamma\bar{\delta}}\nabla_\delta h_{\alpha\bar{\beta}}\right).$$

Since $\operatorname{div}(h)_\alpha = g^{\gamma\bar{\beta}}\nabla_\gamma h_{\alpha\bar{\beta}}$, taking the trace and cancelling terms, we have
$$\left(\frac{\partial}{\partial t} - \Delta\right)\operatorname{div}(h)_\alpha = g^{\gamma\bar{\beta}}\left(\frac{\partial}{\partial t} - \Delta\right)\nabla_\gamma h_{\alpha\bar{\beta}} + R_{\beta\bar{\gamma}}\nabla_\gamma h_{\alpha\bar{\beta}}$$
$$= \nabla_\alpha R_{\varepsilon\bar{\delta}}h_{\delta\bar{\varepsilon}} + R_{\delta\bar{\eta}}\nabla_\eta h_{\alpha\bar{\delta}} - \frac{1}{2}R_{\alpha\bar{\varepsilon}}\nabla_\beta h_{\varepsilon\bar{\beta}},$$

which is (2.132).

Next we verify (2.133). Using (2.132) and (2.21), we compute

$$\left(\frac{\partial}{\partial t} - \Delta\right) \left(\nabla_{\bar{\beta}} \operatorname{div}(h)_{\alpha}\right)$$

$$= \nabla_{\bar{\beta}} \left(\frac{\partial}{\partial t} - \Delta\right) \operatorname{div}(h)_{\alpha} + \left(\nabla_{\bar{\beta}} \Delta - \Delta \nabla_{\bar{\beta}}\right) \operatorname{div}(h)_{\alpha}$$

$$= \nabla_{\bar{\beta}} \left(R_{\mu\bar{\nu}} \nabla_{\nu} h_{\alpha\bar{\mu}} + h_{\bar{\mu}\nu} \nabla_{\alpha} R_{\mu\bar{\nu}} - \frac{1}{2} R_{\alpha\bar{\nu}} \operatorname{div}(h)_{\nu}\right)$$

$$+ \frac{1}{2} \nabla_{\bar{\beta}} R_{\bar{\delta}\alpha} \operatorname{div}(h)_{\delta} - \frac{1}{2} R_{\bar{\beta}\delta} \nabla_{\bar{\delta}} \operatorname{div}(h)_{\alpha} + R_{\bar{\beta}\gamma\bar{\delta}\alpha} \nabla_{\bar{\gamma}} \operatorname{div}(h)_{\delta}$$

$$= \nabla_{\bar{\beta}} R_{\mu\bar{\nu}} \nabla_{\nu} h_{\alpha\bar{\mu}} + \nabla_{\bar{\beta}} h_{\bar{\mu}\nu} \nabla_{\alpha} R_{\mu\bar{\nu}}$$

$$+ R_{\mu\bar{\nu}} \nabla_{\bar{\beta}} \nabla_{\nu} h_{\alpha\bar{\imath}} + h_{\bar{\mu}\nu} \nabla_{\bar{\beta}} \nabla_{\alpha} R_{\mu\bar{\nu}} - \frac{1}{2} R_{\alpha\bar{\nu}} \nabla_{\bar{\beta}} \operatorname{div}(h)_{\nu}$$

$$- \frac{1}{2} R_{\bar{\beta}\delta} \nabla_{\bar{\delta}} \operatorname{div}(h)_{\alpha} + R_{\bar{\beta}\gamma\bar{\delta}\alpha} \nabla_{\bar{\gamma}} \operatorname{div}(h)_{\delta}.$$

Tracing and cancelling terms, we have

$$\left(\frac{\partial}{\partial t} - \Delta\right) \left(g^{\alpha\bar{\beta}} \nabla_{\bar{\beta}} \operatorname{div}(h)_{\alpha}\right)$$

$$= g^{\alpha\bar{\beta}} \left(\frac{\partial}{\partial t} - \Delta\right) \left(\nabla_{\bar{\beta}} \operatorname{div}(h)_{\alpha}\right) + R_{\beta\bar{\alpha}} \nabla_{\bar{\beta}} \operatorname{div}(h)_{\alpha}$$

$$= \nabla_{\bar{\alpha}} R_{\mu\bar{\nu}} \nabla_{\nu} h_{\alpha\bar{\mu}} + \nabla_{\bar{\alpha}} h_{\bar{\mu}\nu} \nabla_{\alpha} R_{\mu\bar{\nu}} + R_{\beta\bar{\alpha}} \nabla_{\bar{\beta}} \operatorname{div}(h)_{\alpha}$$

$$+ R_{\mu\bar{\nu}} \nabla_{\bar{\alpha}} \nabla_{\nu} h_{\alpha\bar{\mu}} + h_{\bar{\mu}\nu} \nabla_{\bar{\alpha}} \nabla_{\alpha} R_{\mu\bar{\nu}},$$

which is (2.133). $\square$

EXERCISE 2.100. Write down the formulas for the complex conjugate equations to (2.132) and (2.133).

Now let for $\varepsilon > 0$

$$\hat{Z} \doteqdot Z + \varepsilon \left(R + \frac{n}{t} + |V_{\alpha}|^2\right)$$

and

$$\tilde{h}_{\alpha\bar{\beta}} \doteqdot h_{\alpha\bar{\beta}} + \varepsilon g_{\alpha\bar{\beta}}.$$

Since Z is of the form $Z = A + B_{\bar{\alpha}} V_{\alpha} + B_{\alpha} V_{\bar{\alpha}} + \tilde{h}_{\alpha\bar{\beta}} V_{\bar{\alpha}} V_{\beta}$ and $\tilde{h}_{\alpha\bar{\beta}} \geq \varepsilon g_{\alpha\bar{\beta}} > 0$, at each (x, t) we have that $\hat{Z}$ attains its minimum for some V. By taking the first variation, we immediately see that

(2.134) $$\operatorname{div}(h)_{\alpha} + \tilde{h}_{\alpha\bar{\gamma}} V_{\gamma} = 0.$$

Differentiating this, we have

(2.135) $$\nabla_{\beta} \operatorname{div}(h)_{\alpha} + \nabla_{\beta} \tilde{h}_{\alpha\bar{\gamma}} V_{\gamma} + \tilde{h}_{\alpha\bar{\gamma}} \nabla_{\beta} V_{\gamma} = 0,$$

(2.136) $$\nabla_{\bar{\beta}} \operatorname{div}(h)_{\alpha} + \nabla_{\bar{\beta}} \tilde{h}_{\alpha\bar{\gamma}} V_{\gamma} + \tilde{h}_{\alpha\bar{\gamma}} \nabla_{\bar{\beta}} V_{\gamma} = 0.$$

In each of the above instances, we also have the complex conjugate equations; we leave it to the reader as an exercise to write these down. Recall that Cao's Kähler matrix differential Harnack quadratic is (2.111):

$$Z_{\alpha\bar{\beta}} = \Delta R_{\alpha\bar{\beta}} + R_{\alpha\bar{\beta}\gamma\bar{\delta}} R_{\bar{\gamma}\delta} + \nabla_\gamma R_{\alpha\bar{\beta}} V_{\bar{\gamma}} + \nabla_{\bar{\gamma}} R_{\alpha\bar{\beta}} V_\gamma + R_{\alpha\bar{\beta}\gamma\bar{\delta}} V_{\bar{\gamma}} V_\delta + \frac{R_{\alpha\bar{\beta}}}{t}.$$

From (2.132) and (2.133) and their conjugate equations, while substituting in (2.134), (2.135), (2.136) and their conjugate equations, we obtain

$$\left(\frac{\partial}{\partial t} - \Delta\right) \hat{Z} = Z_{\alpha\bar{\beta}} h_{\bar{\alpha}\beta} + \tilde{h}_{\gamma\bar{\delta}} \left(\nabla_\alpha V_{\bar{\gamma}} - R_{\alpha\bar{\gamma}} - \frac{1}{t} g_{\alpha\bar{\gamma}}\right) \left(\nabla_{\bar{\alpha}} V_\delta - R_{\bar{\alpha}\delta} - \frac{1}{t} g_{\bar{\alpha}\delta}\right)$$

(2.137)
$$\quad - \frac{1}{t} \left(-\tilde{h}_{\gamma\bar{\delta}} (\nabla_\delta V_{\bar{\gamma}} + \nabla_{\bar{\gamma}} V_\delta) + 2 R_{\delta\bar{\gamma}} h_{\gamma\bar{\delta}} + \frac{2(H + \varepsilon n)}{t} + 2\varepsilon R\right)$$

$$\quad + \tilde{h}_{\gamma\bar{\delta}} \nabla_{\bar{\alpha}} V_{\bar{\gamma}} \nabla_\alpha V_\delta.$$

From (2.134) and the trace of (2.136), we have

$$\hat{Z} = R_{\alpha\bar{\beta}} h_{\bar{\alpha}\beta} - \frac{1}{2} \tilde{h}_{\alpha\bar{\beta}} \nabla_{\bar{\alpha}} V_\beta - \frac{1}{2} \tilde{h}_{\beta\bar{\alpha}} \nabla_\alpha V_{\bar{\beta}} + \frac{H + \varepsilon n}{t} + \varepsilon R.$$

Substituting this into the RHS of (2.137) yields

$$Z_{\alpha\bar{\beta}} h_{\bar{\alpha}\beta} + \tilde{h}_{\gamma\bar{\delta}} \left(\nabla_\alpha V_{\bar{\gamma}} - R_{\alpha\bar{\gamma}} - \frac{1}{t} g_{\alpha\bar{\gamma}}\right) \left(\nabla_{\bar{\alpha}} V_\delta - R_{\bar{\alpha}\delta} - \frac{1}{t} g_{\bar{\alpha}\delta}\right) \geq 0.$$

Hence we have for the minimizer V satisfying (2.134)

$$\left(\frac{\partial}{\partial t} - \Delta\right) \left(t^2 \hat{Z}\right) \geq 0.$$

By applying the maximum principle (see pp. 639–640 of [**290**] for details), we may conclude that $t^2 \hat{Z} \geq 0$ for all $t > 0$.

We have the following matrix differential Harnack estimate due to one of the authors [**287**].

THEOREM 2.101 (Matrix interpolated differential Harnack estimate). Let $(\mathcal{M}^n, g(t))$ be a complete solution of the **ε-speed Kähler–Ricci flow**

(2.138) $$\frac{\partial}{\partial t} g_{\alpha\bar{\beta}} = -\varepsilon R_{\alpha\bar{\beta}},$$

where $\varepsilon > 0$, with bounded nonnegative bisectional curvature, and let u be a positive solution of the **forward conjugate heat equation**

(2.139) $$\frac{\partial u}{\partial t} = \Delta u + \varepsilon R u.$$

Then for any $(1,0)$-form V we have

(2.140) $$u_{\alpha\bar{\beta}} + \varepsilon u R_{\alpha\bar{\beta}} + \frac{u}{t} g_{\alpha\bar{\beta}} + u_\alpha V_{\bar{\beta}} + u_{\bar{\beta}} V_\alpha + u V_\alpha V_{\bar{\beta}} \geq 0.$$

Equivalently, $f \doteq \log u$ satisfies

(2.141) $$f_{\alpha\bar{\beta}} + \varepsilon R_{\alpha\bar{\beta}} + \frac{1}{t} g_{\alpha\bar{\beta}} \geq 0.$$

PROOF. First we observe that the equivalence of (2.140) and (2.141) follows from the fact that the minimizing $(1,0)$-form V_α for the LHS of (2.140) is equal to $-\frac{u_\alpha}{u}$ and dividing (2.140) by u. We compute

$$\frac{\partial f}{\partial t} = \Delta f + |\nabla_\gamma f|^2 + \varepsilon R.$$

Using (2.22) and commuting a pair of derivatives, we have

(2.142)
$$\begin{aligned}\frac{\partial}{\partial t} f_{\alpha\bar\beta} &= \nabla_\alpha \nabla_{\bar\beta}\left(\Delta f + |\nabla_\gamma f|^2 + \varepsilon R\right)\\ &= \Delta_L f_{\alpha\bar\beta} + \varepsilon \nabla_\alpha \nabla_{\bar\beta} R + f_{\alpha\gamma} f_{\bar\beta\bar\gamma} + f_{\alpha\bar\gamma} f_{\bar\beta\gamma}\\ &\quad + \nabla_{\bar\gamma} f \nabla_\gamma f_{\alpha\bar\beta} + \nabla_\gamma f \nabla_{\bar\gamma} f_{\alpha\bar\beta} + R_{\alpha\bar\gamma\delta\bar\beta} \nabla_{\bar\delta} f \nabla_\gamma f.\end{aligned}$$

Using the analogue of (2.105) for the ε-speed Kähler–Ricci flow,

$$\frac{\partial}{\partial t} R_{\alpha\bar\beta} = \varepsilon \Delta_L R_{\alpha\bar\beta} = \varepsilon \nabla_\alpha \nabla_{\bar\beta} R,$$

we then compute

$$\begin{aligned}\frac{\partial}{\partial t}\left(f_{\alpha\bar\beta} + \varepsilon R_{\alpha\bar\beta} + \frac{1}{t} g_{\alpha\bar\beta}\right) &= \Delta_L \left(f_{\alpha\bar\beta} + \varepsilon R_{\alpha\bar\beta}\right) + f_{\alpha\gamma} f_{\bar\beta\bar\gamma} + f_{\alpha\bar\gamma} f_{\bar\beta\gamma}\\ &\quad + \nabla_{\bar\gamma} f \nabla_\gamma f_{\alpha\bar\beta} + \nabla_\gamma f \nabla_{\bar\gamma} f_{\alpha\bar\beta} + R_{\alpha\bar\gamma\delta\bar\beta} \nabla_{\bar\delta} f \nabla_\gamma f\\ &\quad + \varepsilon^2 \Delta R_{\alpha\bar\beta} + \varepsilon^2 R_{\alpha\bar\beta\gamma\bar\delta} R_{\delta\bar\gamma} - \varepsilon^2 R_{\alpha\bar\gamma} R_{\gamma\bar\beta}\\ &\quad - \frac{1}{t^2} g_{\alpha\bar\beta} - \frac{\varepsilon}{t} R_{\alpha\bar\beta}.\end{aligned}$$

Hence letting $S_{\alpha\bar\beta} \doteqdot f_{\alpha\bar\beta} + \varepsilon R_{\alpha\bar\beta} + \frac{1}{t} g_{\alpha\bar\beta}$ and using (2.106), we have

$$\begin{aligned}\left(\frac{\partial}{\partial t} - \Delta_L\right) S_{\alpha\bar\beta} &= f_{\alpha\gamma} f_{\bar\beta\bar\gamma} + \varepsilon^2 \left(\Delta R_{\alpha\bar\beta} + R_{\alpha\bar\beta\gamma\bar\delta} R_{\delta\bar\gamma} + \frac{1}{\varepsilon t} R_{\alpha\bar\beta}\right)\\ &\quad - \varepsilon \nabla_{\bar\gamma} f \nabla_\gamma R_{\alpha\bar\beta} - \varepsilon \nabla_\gamma f \nabla_{\bar\gamma} R_{\alpha\bar\beta} + R_{\alpha\bar\gamma\delta\bar\beta} \nabla_{\bar\delta} f \nabla_\gamma f\\ &\quad + \nabla_{\bar\gamma} f \nabla_\gamma S_{\alpha\bar\beta} + \nabla_\gamma f \nabla_{\bar\gamma} S_{\alpha\bar\beta}\\ &\quad + \frac{1}{2} S_{\alpha\bar\gamma}\left(f_{\bar\beta\gamma} - \varepsilon R_{\gamma\bar\beta} - \frac{1}{t} g_{\gamma\bar\beta}\right)\\ &\quad + \frac{1}{2}\left(f_{\alpha\bar\gamma} - \varepsilon R_{\alpha\bar\gamma} - \frac{1}{t} g_{\alpha\bar\gamma}\right) S_{\bar\beta\gamma}.\end{aligned}$$

Now for the ε-speed Kähler–Ricci flow, by Cao's Kähler matrix differential Harnack estimate (2.111) with $X_\alpha = -\frac{1}{\varepsilon} \nabla_\alpha f$, we have

$$\begin{aligned}0 &\leq \Delta R_{\alpha\bar\beta} + R_{\alpha\bar\beta\gamma\bar\delta} R_{\delta\bar\gamma} + \frac{1}{\varepsilon t} R_{\alpha\bar\beta}\\ &\quad - \frac{1}{\varepsilon} \nabla_{\bar\gamma} f \nabla_\gamma R_{\alpha\bar\beta} - \frac{1}{\varepsilon} \nabla_\gamma f \nabla_{\bar\gamma} R_{\alpha\bar\beta}\\ &\quad + \frac{1}{\varepsilon^2} R_{\alpha\bar\gamma\delta\bar\beta} \nabla_{\bar\delta} f \nabla_\gamma f.\end{aligned}$$

Hence

$$\left(\frac{\partial}{\partial t} - \Delta_L\right) S_{\alpha\bar\beta} \geq f_{\alpha\gamma} f_{\bar\beta\bar\gamma} + \nabla_{\bar\gamma} f \nabla_\gamma S_{\alpha\bar\beta} + \nabla_\gamma f \nabla_{\bar\gamma} S_{\alpha\bar\beta}$$
$$+ \frac{1}{2} S_{\alpha\bar\gamma} \left(f_{\bar\beta\gamma} - \varepsilon R_{\gamma\bar\beta} - \frac{1}{t} g_{\gamma\bar\beta}\right)$$
$$+ \frac{1}{2} \left(f_{\alpha\bar\gamma} - \varepsilon R_{\alpha\bar\gamma} - \frac{1}{t} g_{\alpha\bar\gamma}\right) S_{\bar\beta\gamma}.$$

The estimate (2.141) follows from an application of the maximum principle; see [**287**] for details. □

Tracing (2.140), i.e., multiplying by $g^{\alpha\bar\beta}$ and summing, we have

COROLLARY 2.102 (Trace interpolated differential Harnack estimate). *Under the hypotheses of Theorem 2.101,*

$$\Delta u + \varepsilon u R + \frac{nu}{t} + g^{\alpha\bar\beta}\left(u_\alpha V_{\bar\beta} + u_{\bar\beta} V_\alpha + u V_\alpha V_{\bar\beta}\right) \geq 0,$$

which, by taking $V_\alpha = -\frac{u_\alpha}{u}$ and then dividing the resulting expression by u, implies the equivalent inequality

$$\Delta \log u + \varepsilon R + \frac{n}{t} \geq 0.$$

Let $N \doteqdot \int_\mathcal{M} u \log u \, d\mu$ be the (**classical**) **entropy** of u. We have under (2.139), $\frac{\partial}{\partial t} d\mu = -\varepsilon R d\mu$ (since $\frac{\partial}{\partial t} \det g_{\gamma\bar\delta} = -\varepsilon R \det g_{\gamma\bar\delta}$), and

$$\frac{dN}{dt} = \int_\mathcal{M} \left(\Delta u + \varepsilon R u + (\log u) \Delta u\right) d\mu$$
$$= \int_\mathcal{M} \left(\Delta \log u + \varepsilon R\right) u \, d\mu$$
$$\geq -\frac{n}{t} \int_\mathcal{M} u \, d\mu.$$

In other words,

(2.143) $$\frac{d}{dt} \int_\mathcal{M} u \log(t^n u) \, d\mu \geq 0.$$

11. Notes and commentary

Some books containing material on or devoted to complex manifolds and Kähler geometry, in essentially chronological order, are Weil [**370**], Chern [**95**], Goldberg [**157**], Kobayashi and Nomizu [**236**], Morrow and Kodaira [**275**], Griffiths and Harris [**166**], Aubin [**13**], Kodaira [**238**], Besse [**27**], Siu [**334**], Mok [**270**], Tian [**347**], Wells [**371**], and Zheng [**383**]. We refer the reader to these books for the proper study of Kähler geometry.

For the Ricci flow on real 2-dimensional orbifolds, see L.-F. Wu [**372**] and [**112**]. For the Ricci flow on noncompact Riemannian surfaces, see Wu [**373**],

Daskalopoulos and del Pino [**119**], Hsu [**206**], [**207**], and Daskalopoulos and Hamilton [**120**].

CHAPTER 3

The Compactness Theorem for Ricci Flow

> Although this may seem a paradox, all exact science is dominated by the idea of approximation. – Bertrand Russell

The compactness of solutions to geometric and analytic equations, when it is true, is fundamental in the study of geometric analysis. In this chapter we state and prove Hamilton's compactness theorem for solutions of the Ricci flow assuming Cheeger and Gromov's compactness theorem for Riemannian manifolds with bounded geometry (proved in Chapter 4). In Section 3 of this chapter we also give various versions of the compactness theorem for solutions of the Ricci flow.

Throughout this chapter, quantities depending on the metric g_k (or $g_k(t)$) will have a subscript k; for instance, ∇_k and Rm_k denote the Riemannian connection and Riemannian curvature tensor of g_k. Quantities without a subscript depend on the background metric g. Often we suppress the t dependence in our notation where it is understood that the metrics depend on time while being defined on a space-time set. Given a sequence of quantities indexed by $\{k\}$, when we talk about a subsequence, most of the time we shall still use the indices $\{k\}$ although we should use the indices $\{j_k\}$.

1. Introduction and statements of the compactness theorems

Given a sequence of solutions $(\mathcal{M}_k^n, g_k(t))$ to the Ricci flow, Hamilton's Cheeger–Gromov-type compactness theorem states that in the presence of injectivity radii and curvature bounds we can take a C^∞ limit of a subsequence. The role of the compactness theorem in Ricci flow is primarily to understand singularity formation. This is most effective when the compactness theorem is combined with monotonicity formulas and other geometric and analytic techniques, in part because these formulas and techniques enable us to gain more information about the limit and sometimes enable us to classify singularity models. This has been particularly successful in low dimensions. In latter parts of this volume we shall see some examples of this:

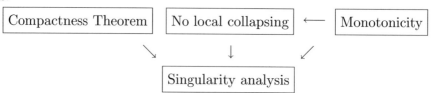

In general, there are three scenarios in which we shall apply the compactness theorem for the Ricci flow. The compactness result may be applied to study solutions $(\mathcal{M}^n, g(t))$ to the Ricci flow defined on time intervals (α, ω), where $\omega \leq \infty$ is maximal, i.e., the singularity time. To understand the limiting behavior of the solution $g(t)$ as t approaches ω, we shall take a sequence of times $t_k \to \omega$ and consider dilations of the solution $g(t)$ about the times t_k and a sequence of points $O_k \in \mathcal{M}$ by defining

$$(3.1) \qquad g_k(t) = K_k g\left(t_k + K_k^{-1} t\right),$$

where $K_k = |\mathrm{Rm}(O_k, t_k)|$ is the norm of $\mathrm{Rm}(g(t_k))$ at the point O_k. We are interested in determining when there exists a subsequence of pointed solutions to the Ricci flow $(\mathcal{M}, g_k(t), O_k)$ which limits to a complete solution $(\mathcal{M}_\infty^n, g_\infty(t), O_\infty)$. This limit solution reflects some aspects of what the singularity looks like near (O_k, t_k). Similarly, when $\alpha = -\infty$ for solution $(\mathcal{M}, g(t))$, which arises when we already have a (first) limit solution of a finite time singularity, we may consider sequences $t_k \to -\infty$ and take a second limit, now backward in time. Yet other limits that we shall consider arise from dimension reduction on a limit solution. Here t_k remains fixed whereas O_k tends to spatial infinity. Many of the topics in this volume are related to the study of the geometry (and topology) of the limits of these solutions when they exist.

1.1. Definition of convergence. Now we review the definition of C^∞-convergence on compact sets in a smooth manifold $\mathcal{M}^n$. By convergence on a compact set in C^p we mean the following.

DEFINITION 3.1 (C^p-convergence). Let $K \subset \mathcal{M}$ be a compact set and let $\{g_k\}_{k \in \mathbb{N}}$, g_∞, and g be Riemannian metrics on $\mathcal{M}$. For $p \in \{0\} \cup \mathbb{N}$ we say that g_k **converges in** C^p **to** g_∞ **uniformly on** K if for every $\varepsilon > 0$ there exists $k_0 = k_0(\varepsilon)$ such that for $k \geq k_0$,

$$\sup_{0 \leq \alpha \leq p} \sup_{x \in K} |\nabla^\alpha (g_k - g_\infty)|_g < \varepsilon,$$

where the covariant derivative ∇ is with respect to g.

Note that since we are on a compact set, the choice of metric g on K does not affect the convergence. For instance, we may choose $g = g_\infty$.

In regards to C^∞-convergence on *manifolds*, with the noncompact case in mind, we have the following. We say that a sequence of open sets $\{U_k\}_{k \in \mathbb{N}}$ in a manifold $\mathcal{M}^n$ is an **exhaustion of** $\mathcal{M}$ **by open sets** if for any compact set $K \subset \mathcal{M}$ there exists $k_0 \in \mathbb{N}$ such that $U_k \supset K$ for all $k \geq k_0$.

1. INTRODUCTION; STATEMENTS OF THE COMPACTNESS THEOREMS

DEFINITION 3.2 (C^∞-convergence uniformly on compact sets). Suppose $\{U_k\}_{k\in\mathbb{N}}$ is an exhaustion of a smooth manifold $\mathcal{M}^n$ by open sets and g_k are Riemannian metrics on U_k. We say that (U_k, g_k) **converges in C^∞ to** $(\mathcal{M}, g_\infty)$ **uniformly on compact sets in** $\mathcal{M}$ if for any compact set $K \subset \mathcal{M}$ and any $p > 0$ there exists $k_0 = k_0(K, p)$ such that $\{g_k\}_{k \geq k_0}$ converges in C^p to g_∞ uniformly on K.

In order to look at convergence of manifolds which come from dilations about a singularity, we must ensure that the form of convergence can handle diameters going to infinity. When this happens, a basepoint, or origin, is carried along with the manifold and the Riemannian metric to distinguish what parts of the manifolds in the sequence we are keeping in focus. This allows us to compare spaces that either have diameters going to infinity or are noncompact.

DEFINITION 3.3 (Pointed manifolds and solutions). A **pointed Riemannian manifold** is a 3-tuple $(\mathcal{M}^n, g, O)$, where $(\mathcal{M}, g)$ is a Riemannian manifold and $O \in \mathcal{M}$ is a choice of point (called the **origin**, or **basepoint**). If the metric g is complete, the 3-tuple is called a **complete pointed Riemannian manifold**. We say that $(\mathcal{M}^n, g(t), O)$, $t \in (\alpha, \omega)$, is a **pointed solution to the Ricci flow** if $(\mathcal{M}, g(t))$ is a solution to the Ricci flow.

REMARK 3.4. In [187] Hamilton considered marked Riemannian manifolds (and marked solutions to the Ricci flow), where one is also given a frame $F = \{e_a\}_{a=1}^n$ at O orthonormal with respect to the metric $g(0)$ with $0 \in (\alpha, \omega)$. Since for most applications, the choice of frame is not essential, we restrict ourselves to considering pointed Riemannian manifolds in this chapter.

Convergence of pointed Riemannian manifolds is defined in a way which takes into account the action of basepoint-preserving diffeomorphisms on the space of metrics.

DEFINITION 3.5 (C^∞-convergence of manifolds after diffeomorphisms). A sequence $\{(\mathcal{M}_k^n, g_k, O_k)\}_{k\in\mathbb{N}}$ of complete pointed Riemannian manifolds **converges** to a complete pointed Riemannian manifold $(\mathcal{M}_\infty^n, g_\infty, O_\infty)$ if there exist

(1) an exhaustion $\{U_k\}_{k\in\mathbb{N}}$ of $\mathcal{M}_\infty$ by open sets with $O_\infty \in U_k$ and
(2) a sequence of diffeomorphisms $\Phi_k : U_k \to V_k \doteqdot \Phi_k(U_k) \subset \mathcal{M}_k$ with $\Phi_k(O_\infty) = O_k$

such that $\left(U_k, \Phi_k^*\left[g_k|_{V_k}\right]\right)$ converges in C^∞ to $(\mathcal{M}_\infty, g_\infty)$ uniformly on compact sets in $\mathcal{M}_\infty$.

We shall also call the above convergence **Cheeger–Gromov convergence** in C^∞. The corresponding definition for sequences of pointed solutions of the Ricci flow is given by the following.

DEFINITION 3.6 (C^∞-convergence of solutions after diffeomorphisms). A sequence $\{(\mathcal{M}_k^n, g_k(t), O_k)\}_{k \in \mathbb{N}}$, $t \in (\alpha, \omega)$, of complete pointed solutions to the Ricci flow **converges** to a complete pointed solution to the Ricci flow $(\mathcal{M}_\infty^n, g_\infty(t), O_\infty)$, $t \in (\alpha, \omega)$, if there exist

(1) an exhaustion $\{U_k\}_{k \in \mathbb{N}}$ of $\mathcal{M}_\infty$ by open sets with $O_\infty \in U_k$, and
(2) a sequence of diffeomorphisms $\Phi_k : U_k \to V_k \doteqdot \Phi_k(U_k) \subset \mathcal{M}_k$ with $\Phi_k(O_\infty) = O_k$

such that $\left(U_k, \Phi_k^*\left[g_k(t)|_{V_k}\right]\right)$ converges in C^∞ to $(\mathcal{M}_\infty, g_\infty(t))$ uniformly on compact sets in $\mathcal{M}_\infty \times (\alpha, \omega)$.

REMARK 3.7. The last statement in the above definition is a slight abuse of notation; we really mean $\left(U_k \times (\alpha, \omega), \Phi_k^*\left[g_k(t)|_{V_k}\right] + dt^2\right)$ converges in C^∞ to $(\mathcal{M}_\infty \times (\alpha, \omega), g_\infty(t) + dt^2)$ uniformly on compact sets in $\mathcal{M}_\infty \times (\alpha, \omega)$ using Definition 3.2, where dt^2 is the standard metric on (α, ω).

When there is a bound on the curvatures (recall that when given a singular solution and a suitable sequence of space-time points, the choice of dilation factors is chosen to guarantee this for the associated sequence of solutions) and an injectivity radius estimate for a sequence of solutions to the Ricci flow, then the following compactness theorem provides a subsequence which will converge in the C^∞-Cheeger–Gromov sense. We end this subsection with a definition which is related to the assumption of bounded curvature.

DEFINITION 3.8 (Bounded geometry). We say that a sequence or family of Riemannian manifolds has **bounded geometry** if there exist positive constants C_p such that

$$|\nabla^p \operatorname{Rm}| \leq C_p$$

for all $p \in \mathbb{N} \cup \{0\}$ and for all metrics in this sequence or family. That is, the curvatures and their covariant derivatives of each order have uniform bounds.

1.2. Statements of the compactness theorems. Let $\operatorname{inj}_g(O)$ denote the injectivity radius of the metric g at the point O. For sequences of Riemannian manifolds we have the following convergence theorems (originally proven in [187]).

THEOREM 3.9 (Compactness for metrics). *Let $\{(\mathcal{M}_k^n, g_k, O_k)\}_{k \in \mathbb{N}}$ be a sequence of complete pointed Riemannian manifolds that satisfy*

(1) *(**uniformly bounded geometry**)*

$$\left|\nabla_k^p \operatorname{Rm}_k\right|_k \leq C_p \quad \text{on } \mathcal{M}_k$$

for all $p \geq 0$ and k where $C_p < \infty$ is a sequence of constants independent of k and

1. INTRODUCTION; STATEMENTS OF THE COMPACTNESS THEOREMS 131

(2) *(injectivity radius estimate)*

$$\operatorname{inj}_{g_k}(O_k) \geq \iota_0$$

for some constant $\iota_0 > 0$.

Then there exists a subsequence $\{j_k\}_{k\in\mathbb{N}}$ such that $\{(\mathcal{M}_{j_k}, g_{j_k}, O_{j_k})\}_{k\in\mathbb{N}}$ converges to a complete pointed Riemannian manifold $(\mathcal{M}_\infty^n, g_\infty, O_\infty)$ as $k \to \infty$.

For sequences of solutions to the Ricci flow the corresponding convergence theorem takes the following form.

THEOREM 3.10 (Compactness for solutions). *Let $\{(\mathcal{M}_k^n, g_k(t), O_k)\}_{k\in\mathbb{N}}$, $t \in (\alpha, \omega) \ni 0$, be a sequence of complete pointed solutions to the Ricci flow such that*

(1) *(uniformly bounded curvatures)*

$$|\operatorname{Rm}_k|_k \leq C_0 \quad \text{on } \mathcal{M}_k \times (\alpha, \omega)$$

for some constant $C_0 < \infty$ independent of k and
(2) *(injectivity radius estimate at $t = 0$)*

$$\operatorname{inj}_{g_k(0)}(O_k) \geq \iota_0$$

for some constant $\iota_0 > 0$.

Then there exists a subsequence $\{j_k\}_{k\in\mathbb{N}}$ such that $\{(\mathcal{M}_{j_k}, g_{j_k}(t), O_{j_k})\}_{k\in\mathbb{N}}$ converges to a complete pointed solution to the Ricci flow $(\mathcal{M}_\infty^n, g_\infty(t), O_\infty)$, $t \in (\alpha, \omega)$, as $k \to \infty$.

Note that the second theorem only supposes bounds on the curvature, not bounds on the derivatives of the curvature. This is because, for the Ricci flow, if the curvature is bounded on (α, ω), then all derivatives of the curvature are bounded at times $t > \alpha$ (see Chapter 7 of Volume One or Theorems A.29 and A.30 of this volume).[1] In particular all derivatives of the curvature are bounded at time $t = 0$ and we can apply Theorem 3.9 to $\{(\mathcal{M}_k^n, g_k(0), O_k)\}_{k\in\mathbb{N}}$.

In the next section we follow the proofs of Hamilton in [187]. We shall assume Theorem 3.9, which will be proven in Chapter 4. We will show that if there is a subsequence such that $(\mathcal{M}_k^n, g_k(0), O_k)$ converges to a complete limit $(\mathcal{M}_\infty^n, g_\infty(0), O_\infty)$, then there is a subsequence $(\mathcal{M}_k^n, g_k(t), O_k)$ which converges at all times.

---

[1] The bounds on the derivatives of Rm get worse as $t \to \alpha$.

2. Convergence at all times from convergence at one time

In this section we give the proof that the compactness theorem for Ricci flow (Theorem 3.10) follows from the compactness theorem at time $t = 0$ (Theorem 3.9). This is done by showing that bounds on the metric and covariant/time-derivatives of the metric at time $t = 0$ extend to bounds on the metric and covariant derivatives of the metric at subsequent times in the presence of bounds on the curvature and covariant derivatives of curvature (for all time). This is shown in subsection 2.1 below. The Arzela–Ascoli theorem is then used to show that these bounds on the covariant/time-derivatives of the metric imply that a subsequence converges to a solution of the Ricci flow for all times (in subsection 2.2.2 below).

2.1. Uniform derivative of metric bounds for all time.

In order to extend the convergence at one time to convergence at all times, the following derivative bounds need to be shown.

LEMMA 3.11 (Derivative of metric bounds at one time to all times). *Let $\mathcal{M}^n$ be a Riemannian manifold with a background metric g, let K be a compact subset of $\mathcal{M}$, and let g_k be a collection of solutions to the Ricci flow defined on neighborhoods of $K \times [\beta, \psi]$, where $t_0 \in [\beta, \psi]$. Suppose that*

(1) *the metrics $g_k(t_0)$ are all uniformly equivalent to g on K, i.e., for all $V \in T_x\mathcal{M}$, k, and $x \in K$,*

$$C^{-1} g(V, V) \leq g_k(t_0)(V, V) \leq C g(V, V),$$

where $C < \infty$ is a constant independent of V, k, and x; and

(2) *the covariant derivatives of the metrics $g_k(t_0)$ with respect to the metric g are all uniformly bounded on K, so that*

$$|\nabla^p g_k(t_0)| \leq C_p$$

for all k and $p \geq 1$, where $C_p < \infty$ is a sequence of constants independent of k; and

(3) *the covariant derivatives of the curvature tensors $\operatorname{Rm}_k(t)$ of the metrics $g_k(t)$ are uniformly bounded with respect to the metric $g_k(t)$ on $K \times [\beta, \psi]$:*

(3.2) $$|\nabla_k^p \operatorname{Rm}_k|_k \leq C_p'$$

for all k and $p \geq 0$, where C_p' is a sequence of constants independent of k.

Then the metrics $g_k(t)$ are uniformly equivalent to g on $K \times [\beta, \psi]$, e.g.,

(3.3) $$B(t, t_0)^{-1} g(V, V) \leq g_k(t)(V, V) \leq B(t, t_0) g(V, V),$$

where

$$B(t, t_0) = C e^{2\sqrt{n-1} C_0' |t - t_0|},$$

and the time-derivatives and covariant derivatives of the metrics $g_k(t)$ with respect to the metric g are uniformly bounded on $K \times [\beta, \psi]$, i.e., for each (p, q) there is a constant $\tilde{C}_{p,q}$ independent of k such that

$$\left| \frac{\partial^q}{\partial t^q} \nabla^p g_k(t) \right| \leq \tilde{C}_{p,q} \tag{3.4}$$

for all k.

REMARK 3.12. Since we often assume bounds on Rm whereas the metric evolves by Rc, we note

$$-\sqrt{n-1}\,|\mathrm{Rm}|\,g \leq \mathrm{Rc} \leq \sqrt{n-1}\,|\mathrm{Rm}|\,g.$$

Since we often interchange g and g_k norms, we recall the following elementary fact.

LEMMA 3.13 (Norms of tensors with respect to equivalent metrics). *Suppose that the metrics g and h are equivalent:*

$$C^{-1} g \leq h \leq C g.$$

Then for any (p, q)-tensor T, we have

$$|T|_h \leq C^{(p+q)/2} |T|_g. \tag{3.5}$$

PROOF. We can diagonalize g and h so that $g_{ij} = \delta_{ij}$ and $h_{ij} = \lambda_i \delta_{ij}$. The assumption implies $C^{-1} \leq \lambda_i \leq C$ for all i. Then

$$|T|_h^2 = \sum h_{k_1 \ell_1} \cdots h_{k_p \ell_p} h^{i_1 j_1} \cdots h^{i_q j_q} T^{k_1 \cdots k_p}_{i_1 \cdots i_q} T^{\ell_1 \cdots \ell_p}_{j_1 \cdots j_q}$$

$$\leq \sum_{k_1 \cdots k_p; i_1 \cdots i_q} \lambda_{k_1} \cdots \lambda_{k_p} (\lambda_{i_1})^{-1} \cdots (\lambda_{i_q})^{-1} T^{k_1 \cdots k_p}_{i_1 \cdots i_q} T^{k_1 \cdots k_p}_{i_1 \cdots i_q}$$

$$\leq C^{p+q} \sum_{k_1 \cdots k_p; i_1 \cdots i_q} T^{k_1 \cdots k_p}_{i_1 \cdots i_q} T^{k_1 \cdots k_p}_{i_1 \cdots i_q}.$$

□

PROOF OF LEMMA 3.11. For the first part, since

$$\frac{\partial}{\partial t} g_k(t)(V, V) = -2\,\mathrm{Rc}_k(t)(V, V)$$

and

$$|\mathrm{Rc}_k(t)(V, V)| \leq \sqrt{n-1}\,C'_0\,g_k(t)(V, V),$$

we can estimate the time-derivatives

$$\left| \frac{\partial}{\partial t} \log g_k(t)(V, V) \right| = \left| \frac{-2\,\mathrm{Rc}_k(t)(V, V)}{g_k(t)(V, V)} \right| \leq 2\sqrt{n-1}\,C'_0.$$

We have proved

$$\left| \frac{\partial}{\partial t} \log g_k(V, V) \right| \leq \bar{C}, \tag{3.6}$$

where $\bar{C} \doteqdot 2\sqrt{n-1}C_0'$.

Now we compute

$$\bar{C}|t_1 - t_0| \geq \int_{t_0}^{t_1} \left|\frac{\partial}{\partial t}\log g_k(t)(V,V)\right| dt$$

$$\geq \left|\int_{t_0}^{t_1} \frac{\partial}{\partial t}\log g_k(t)(V,V) dt\right|$$

$$= \left|\log \frac{g_k(t_1)(V,V)}{g_k(t_0)(V,V)}\right|,$$

or equivalently,

$$e^{-\bar{C}|t_1-t_0|}g_k(t_0)(V,V) \leq g_k(t_1)(V,V) \leq e^{\bar{C}|t_1-t_0|}g_k(t_0)(V,V).$$

Hence we have

$$C^{-1}e^{-\bar{C}|t_1-t_0|}g(V,V) \leq g_k(t_1)(V,V) \leq Ce^{\bar{C}|t_1-t_0|}g(V,V).$$

This completes the proof of (3.3).

For the second part we need to estimate the space- and time-derivatives of $g_k(t)$. We begin with estimating the first-order covariant derivatives of $g_k(t)$. Note that

$$\nabla_a(g_k)_{bc} = \frac{\partial}{\partial x^a}(g_k)_{bc} - \Gamma_{ab}^d(g_k)_{dc} - \Gamma_{ac}^d(g_k)_{bd},$$

so if we take the right combination, we see that

$$(g_k)^{ec}(\nabla_a(g_k)_{bc} + \nabla_b(g_k)_{ac} - \nabla_c(g_k)_{ab})$$
$$= 2(\Gamma_k)_{ab}^e - \Gamma_{ab}^e - (g_k)^{ec}\Gamma_{ac}^d(g_k)_{bd} - \Gamma_{ab}^e$$
$$- (g_k)^{ec}\Gamma_{bc}^d(g_k)_{ad} + (g_k)^{ec}\Gamma_{cb}^d(g_k)_{ad} + (g_k)^{ec}\Gamma_{ac}^d(g_k)_{bd}$$
(3.7) $\qquad = 2(\Gamma_k)_{ab}^e - 2\Gamma_{ab}^e.$

This implies that

(3.8) $$|\Gamma_k(t) - \Gamma|_k \leq \frac{3}{2}|\nabla g_k(t)|_k.$$

From

$$\nabla_a(g_k)_{bc} = (g_k)_{eb}[(\Gamma_k)_{ac}^e - \Gamma_{ac}^e] + (g_k)_{ec}[(\Gamma_k)_{ab}^e - \Gamma_{ab}^e],$$

we have

(3.9) $$|\nabla g_k(t)|_k \leq 2|\Gamma_k(t) - \Gamma|_k.$$

Hence the tensors $\nabla g_k(t)$ and $\Gamma_k(t) - \Gamma$ are equivalent.

We recall that the derivative of the Christoffel symbols (see, for instance, (6.1) on p. 175 of Volume One) is

$$\frac{\partial}{\partial t}(\Gamma_k)_{ab}^c = -(g_k)^{cd}[(\nabla_k)_a(\text{Rc}_k)_{bd} + (\nabla_k)_b(\text{Rc}_k)_{ad} - (\nabla_k)_d(\text{Rc}_k)_{ab}].$$

2. CONVERGENCE AT ALL TIMES FROM CONVERGENCE AT ONE TIME

So as tensors, we find that

$$\left|\frac{\partial}{\partial t}(\Gamma_k - \Gamma)\right|_k \leq 3\left|\nabla_k(\operatorname{Rc}_k)\right|_k \leq 3\sqrt{n-1}C_1'.$$

Thus

$$3\sqrt{n-1}C_1'|t_1 - t_0| \geq \int_{t_0}^{t_1}\left|\frac{\partial}{\partial t}(\Gamma_k(t) - \Gamma)\right|_k dt$$

$$\geq \left|\int_{t_0}^{t_1}\frac{\partial}{\partial t}(\Gamma_k(t) - \Gamma)\,dt\right|_k$$

$$\geq |\Gamma_k(t_1) - \Gamma|_k - |\Gamma_k(t_0) - \Gamma|_k.$$

Hence we have a bound

$$|\Gamma_k(t) - \Gamma|_k \leq 3\sqrt{n-1}C_1'|t - t_0| + |\Gamma_k(t_0) - \Gamma|_k$$

(3.10)
$$\leq 3\sqrt{n-1}C_1'|t - t_0| + \frac{3}{2}C^{3/2}C_1$$

using (3.8) and (3.5). Since $|t - t_0| \leq \psi - \beta$, we have in (3.3): $B(t, t_0) \leq B(\psi, \beta)$ for all $t \in [\beta, \psi]$. Thus by (3.9) and (3.10),

(3.11) $$|\nabla g_k(t)| \leq B(t, t_0)^{3/2}|\nabla g_k(t)|_k \leq \tilde{C}_{1,0},$$

where

$$\tilde{C}_{1,0} \doteqdot B^{3/2}(\psi, \beta)\left(6\sqrt{n-1}C_1'(\psi - \beta) + 3C^{3/2}C_1\right).$$

This proves (3.4) for $p = 1$ and $q = 0$.

Next we prove inductively that for $p \geq 1$,

$$|\nabla^p \operatorname{Rc}_k| \leq C_p''|\nabla^p g_k| + C_p''' \quad \text{and} \quad |\nabla^p g_k| \leq \tilde{C}_{p,0}$$

(where C_p'', C_p''', and $\tilde{C}_{p,0}$ are independent of k). If $p = 1$, then using (3.8) and (3.10),

$$|\nabla \operatorname{Rc}_k|(t) \leq B(t, t_0)^{3/2}|(\nabla - \nabla_k)\operatorname{Rc}_k + \nabla_k \operatorname{Rc}_k|_k$$

$$\leq B(t, t_0)^{3/2}(|\Gamma - \Gamma_k|_k|\operatorname{Rc}_k|_k + |\nabla_k \operatorname{Rc}_k|_k)$$

$$\leq B(t, t_0)^{3/2}\left(\left(3\sqrt{n-1}C_1'|\psi - \beta| + \frac{3}{2}C^{3/2}C_1\right)C_0' + C_1'\right).$$

If the estimates hold for $p < N$ with $N \geq 2$, then we will prove them for $p = N$. First we have

$$|\nabla^N \operatorname{Rc}_k| = \left|\sum_{i=1}^{N}\nabla^{N-i}(\nabla - \nabla_k)\nabla_k^{i-1}\operatorname{Rc}_k + \nabla_k^N \operatorname{Rc}_k\right|$$

$$\leq \sum_{i=1}^{N}\left|\nabla^{N-i}(\nabla - \nabla_k)\nabla_k^{i-1}\operatorname{Rc}_k\right| + \left|\nabla_k^N \operatorname{Rc}_k\right|.$$

Note that, using (3.7), we can rewrite $\nabla - \nabla_k = \Gamma - \Gamma_k$ as a sum of terms of the form ∇g_k. When $i = 1$, we can bound
$$\left|\nabla^{N-1}(\nabla - \nabla_k)\operatorname{Rc}_k\right|$$
by a sum of terms of the form $\left|\nabla^{N-j}g_k\right|\left|\nabla^j\operatorname{Rc}_k\right|$, $0 \le j \le N-1$. When $2 \le i \le N$, we can bound
$$\left|\nabla^{N-i}(\nabla - \nabla_k)\nabla_k^{i-1}\operatorname{Rc}_k\right|$$
by a sum of terms of the form $\left|\nabla^{N-i-j+1}g_k\right|\left|\nabla^j\nabla_k^{i-1}\operatorname{Rc}_k\right|$, $0 \le j \le N-i$. We can also bound
$$\left|\nabla^j\nabla_k^{i-1}\operatorname{Rc}_k\right| = \left|((\nabla - \nabla_k) + \nabla_k)^j \nabla_k^{i-1}\operatorname{Rc}_k\right|$$
by a sum of terms which are products of $\left|\nabla_k^{\ell+i-1}\operatorname{Rc}_k\right|$, $0 \le \ell \le j$, and several $\left|\nabla^\ell g_k\right|$, $1 \le \ell \le j$. By the assumption of Lemma 3.11, the induction assumption and the equivalence of $|\cdot|$ and $|\cdot|_k$, we get
$$\left|\nabla^N \operatorname{Rc}_k\right| \le C_N'' \left|\nabla^N g_k\right| + C_N'''.$$

Now we turn to bounding $\left|\nabla^N g_k\right|$. Since g does not depend on t,
$$\frac{\partial}{\partial t}\nabla^N g_k = -2\nabla^N \operatorname{Rc}_k$$
and
$$\frac{\partial}{\partial t}\left|\nabla^N g_k\right|^2 = 2\left\langle \frac{\partial}{\partial t}\nabla^N g_k, \nabla^N g_k \right\rangle \le \left|\frac{\partial}{\partial t}\nabla^N g_k\right|^2 + \left|\nabla^N g_k\right|^2$$
$$= 4\left|\nabla^N \operatorname{Rc}_k\right|^2 + \left|\nabla^N g_k\right|^2 \le \left(1 + 8(C_N'')^2\right)\left|\nabla^N g_k\right|^2 + 8(C_N''')^2.$$

Integrating the above differential inequality of $\left|\nabla^N g_k\right|^2$, we get (compare with (7.47))
$$\left|\nabla^N g_k\right|^2(t) \le e^{\left(1 + 8(C_N'')^2\right)(t-t_0)}\left(\left|\nabla^N g_k\right|^2(t_0) + \frac{8(C_N''')^2}{1 + 8(C_N'')^2}\right).$$

This implies
$$\left|\nabla^N g_k(t)\right| \le \tilde{C}_{N,0},$$
and the induction proof is complete, as well as (3.4) for the $q = 0$ case. Note that the above proof of bounding $\left|\nabla^N \operatorname{Rc}_k\right|$ can be used to show that $\left|\nabla^p \nabla_k^q \operatorname{Rc}_k\right|$, $\left|\nabla^p \nabla_k^q R_k\right|$, and $\left|\nabla^p \nabla_k^q \operatorname{Rm}_k\right|$ are bounded independent of k.

When $q \ge 1$, then $\frac{\partial^q}{\partial t^q}\nabla^p g_k(t) = \nabla^p \frac{\partial^{q-1}}{\partial t^{q-1}}(-2\operatorname{Rc}_k(t))$. Using the evolution equation of the curvature $\operatorname{Rm}_k(t)$, we know that $\left|\frac{\partial^q}{\partial t^q}\nabla^p g_k(t)\right|$ is bounded by a sum of terms which are products of
$$\left|\nabla^{p_1}\nabla_k^{q_1}\operatorname{Rm}_k\right|(t), \quad \left|\nabla^p \nabla_k^q \operatorname{Rc}_k\right|, \quad \text{and} \quad \left|\nabla^p \nabla_k^q R_k\right|.$$
Hence we get $\left|\frac{\partial^q}{\partial t^q}\nabla^p g_k(t)\right| \le \tilde{C}_{p,q}$. □

2.2. Convergence at all times from convergence at one time.

2.2.1. *The Arzela–Ascoli theorem.* With uniform derivative bounds on the metrics in the sequence, the compactness theorem will follow from the **Arzela–Ascoli theorem**.

LEMMA 3.14 (Arzela–Ascoli). *Let X be a σ-compact, locally compact Hausdorff space. If $\{f_k\}_{k \in \mathbb{N}}$ is an equicontinuous, pointwise bounded sequence of continuous functions $f_k : X \to \mathbb{R}$, then there exists a subsequence which converges uniformly on compact sets to a continuous function $f_\infty : X \to \mathbb{R}$.*

The reader is reminded that σ-compact simply means that the space is a countable union of compact sets, and hence any complete Riemannian manifold satisfies the assumption.

COROLLARY 3.15 (Metrics with bounded derivatives preconverge). *Let $(\mathcal{M}^n, g)$ be a Riemannian manifold and let $K \subset \mathcal{M}^n$ be compact. Furthermore, let p be a nonnegative integer. If $\{g_k\}_{k \in \mathbb{N}}$ is a sequence of Riemannian metrics on K such that*

$$\sup_{0 \leq \alpha \leq p+1} \sup_{x \in K} |\nabla^\alpha g_k| \leq C < \infty$$

and if there exists $\delta > 0$ such that $g_k(V, V) \geq \delta g(V, V)$ for all $V \in T\mathcal{M}$, then there exists a subsequence $\{g_k\}$ and a Riemannian metric g_∞ on K such that g_k converges in C^p to g_∞ as $k \to \infty$.

PROOF (SKETCH). We need to show that $\{(g_k)_{bc}\}_{k \in \mathbb{N}}$ form an equicontinuous family. We use the fact that in a coordinate patch

$$\nabla_a (g_k)_{bc} = \frac{\partial}{\partial x^a} (g_k)_{bc} - \Gamma_{ab}^d (g_k)_{dc} - \Gamma_{ac}^d (g_k)_{bd}.$$

Thus if $|\nabla g_k|$ is bounded, then $\left|\frac{\partial}{\partial x^a}(g_k)_{bc}\right|$ is bounded for each a, b, c in each coordinate patch. Hence, by the mean value theorem, the $(g_k)_{bc}$ form an equicontinuous family in the patch and there is a subsequence which converges to $(g_\infty)_{bc}$. Since K is compact, we may take a finite covering by coordinate patches and a subsequence which converges for each coordinate patch. We have thus constructed a limit metric. Note that the uniform upper and lower bounds on the metrics g_k ensure that g_∞ is positive definite.

Similarly, we can use the bound on $|\nabla^2 g_k|$ to get bounds on second derivatives of the metrics $\left|\frac{\partial^2}{\partial x^a \partial x^d}(g_k)_{bc}\right|$ in each coordinate patch and thus show the first derivatives are an equicontinuous family. Taking a further subsequence, we get convergence in C^1. Higher derivatives are similar. $\square$

2.2.2. *Proof of the compactness theorem for solutions assuming the compactness theorem for metrics.* We will now use Corollary 3.15 together with Lemma 3.11 to find a subsequence which converges and complete the proof of Theorem 3.10. Recall that we have assumed Theorem 3.9 and hence there is a subsequence $\{(\mathcal{M}_k, g_k(0), O_k)\}$ which converges to $(\mathcal{M}_\infty^n, g_\infty, O_\infty)$.

We shall show that there are metrics $g_\infty(t)$, for $t \in (\alpha, \omega)$, such that $g_\infty(0) = g_\infty$ and $\{(\mathcal{M}_k, g_k(t), O_k)\}$ converges to $(\mathcal{M}_\infty, g_\infty(t), O_\infty)$ in C^∞.

Since $\{(\mathcal{M}_k, g_k(0), O_k)\}$ converges to $(\mathcal{M}_\infty, g_\infty, O_\infty)$, there are maps $\Phi_k : U_k \to V_k$ such that $\Phi_k^* g_k(0) \to g_\infty$ uniformly on compact sets. We shall apply Lemma 3.11 with $t_0 = 0$, g_∞ as the background metric, and $\Phi_k^* g_k(t)$ as the sequence of metrics. Note that the assumptions of the lemma are satisfied at $t_0 = 0$ because convergence assures us that the $\Phi_k^* g_k(0)$ converge to g_∞ uniformly on compact subsets. We can now apply Corollary 3.15 with $g = g_\infty + dt^2$ on $\mathcal{M}_\infty \times (\alpha, \omega)$ to find a subsequence of $\Phi_k^* g_k(t) + dt^2$ which converges to $g_\infty(t) + dt^2$ in C^∞ on compact subsets. (Since $\frac{\partial}{\partial t}$ is orthogonal to vectors on $\mathcal{M}_\infty$ in each metric $\Phi_k^* g_k(t) + dt^2$, it is orthogonal to them in the limit metric.) Hence there is a subsequence $\{(\mathcal{M}_k, g_k(t), O_k)\}$ which converges to $(\mathcal{M}_\infty, g_\infty(t), O_\infty)$, where $g_\infty(t)$ is defined to be the limit of $\Phi_k^* g_k(t)$. Since all derivatives of the metric converge, the Ricci curvature of $g_k(t)$ converges to the Ricci curvature of $g_\infty(t)$ and hence the limit is a solution of the Ricci flow. This concludes the proof of Theorem 3.10.

3. Extensions of Hamilton's compactness theorem

In this section we give several variations of Theorem 3.10.

3.1. Local compactness theorems. From the proof of the compactness Theorem 3.9 given in the next chapter and the proof of Theorem 3.10 given above, without too much difficulty one sees that a local version of Theorem 3.10 holds. In particular, we have the following.

THEOREM 3.16 (Compactness, local version). *Let $\{(\mathcal{M}_k^n, g_k(t), O_k)\}_{k \in \mathbb{N}}$, $t \in (\alpha, \omega) \ni 0$, be a sequence of complete pointed solutions to the Ricci flow. If there exist $\rho > 0$, $C_0 < \infty$, and $\iota_0 > 0$ independent of k such that*

$$|\mathrm{Rm}_k|_k \leq C_0 \quad \text{in } B_{g_k(0)}(O_k, \rho) \times (\alpha, \omega)$$

and

$$\mathrm{inj}_{g_k(0)}(O_k) \geq \iota_0,$$

then there exists a subsequence such that $\{(B_{g_k(0)}(O_k, \rho), g_k(t), O_k)\}_{k \in \mathbb{N}}$ converges as $k \to \infty$ to a pointed solution $(\mathcal{B}_\infty^n, g_\infty(t), O_\infty)$, $t \in (\alpha, \omega)$, in C^∞ on any compact subset of $\mathcal{B}_\infty \times (\alpha, \omega)$. Furthermore $\mathcal{B}_\infty$ is an open manifold which is complete on the closed ball $\overline{B_{g_\infty(0)}(O_\infty, r)}$ for all $r < \rho$.

EXERCISE 3.17. Prove Theorem 3.16.

A simple consequence of Theorem 3.16 is (see [**186**]) the following corollary.

COROLLARY 3.18 (Compactness theorem yielding complete limits). *Let $\{(\mathcal{M}_k^n, g_k(t), O_k)\}_{k \in \mathbb{N}}$, $t \in (\alpha, \omega) \ni 0$, be a sequence of complete pointed solutions to the Ricci flow. Suppose for any $r > 0$ and $\varepsilon > 0$ there exist constants $C_0(r, \varepsilon) < \infty$ such that*

$$|\mathrm{Rm}_k|_k \leq C_0(r, \varepsilon) \quad \text{on } B_{g_k(0)}(O_k, r) \times (\alpha + \varepsilon, \omega - \varepsilon)$$

for all $k \in \mathbb{N}$. We assume $\mathrm{inj}_{g_k(0)}(O_k) \geq \iota_0$ for some $\iota_0 > 0$. Then there exists a subsequence $\{(\mathcal{M}_k, g_k(t), O_k)\}$ which converges to a complete solution to the Ricci flow $(\mathcal{M}_\infty^n, g_\infty(t), O_\infty)$, $t \in (\alpha, \omega)$.

REMARK 3.19. Note that the limit solution $(\mathcal{M}_\infty^n, g_\infty(t), O_\infty)$ may not have bounded curvature.

Without the injectivity radius estimate, we may use the trick of locally pulling back the solutions by their exponential maps (since the pulled-back solutions satisfy an injectivity radius estimate). We have the following.

COROLLARY 3.20 (Local compactness without injectivity radius estimate). Let $\{(\mathcal{M}_k^n, g_k(t), O_k)\}_{k \in \mathbb{N}}$, $t \in (\alpha, \omega) \ni 0$, be a sequence of complete solutions to the Ricci flow with

$$|\mathrm{Rm}_k|_k \leq C_0 \quad \text{in } B_{g_k(0)}(O_k, \rho) \times (\alpha, \omega).$$

Then there exists a subsequence such that

$$\left\{ \left(B_{T_{O_k}\mathcal{M}_k}(\vec{0}, c) , \left(\exp_{O_k}^{g_k(0)} \right)^* g_k(t) \right), \vec{0} \right\}_{k \in \mathbb{N}}, \quad \text{where } c \doteq \min\{\rho, \pi/\sqrt{C_0}\},$$

converges to a pointed solution $(\mathcal{B}_\infty^n, g_\infty(t), O_\infty)$, $t \in (\alpha, \omega)$, on an open manifold which is complete on the closed ball $\overline{B_{g_\infty(0)}(O_\infty, r)}$ for all $r < c$.

REMARK 3.21. There is a similar result for geodesic tubes; see §25 of Hamilton [186].

3.2. Compactness for Kähler metrics and solutions.
Without much difficulty, the compactness theorems apply to Kähler manifolds and solutions of the Kähler–Ricci flow (see also Cao [48] and Theorem 4.1 on pp. 16–17 of Ruan [314]).

THEOREM 3.22 (Compactness for Kähler metrics). Let $\{(\mathcal{M}_k^{2n}, g_k, O_k)\}$ be a sequence of complete pointed Kähler manifolds of complex dimension n. Suppose
$$|\nabla_k^p \mathrm{Rm}_k|_k \leq C_p \quad \text{on } \mathcal{M}_k$$
for all $p \geq 0$ and k, where $C_p < \infty$ is some sequence of constants independent of k, and
$$\mathrm{inj}_{g_k}(O_k) \geq \iota_0$$
for some constant $\iota_0 > 0$. Then there exists a subsequence $\{j_k\}_{k \in \mathbb{N}}$ such that $\{(\mathcal{M}_{j_k}, g_{j_k}, O_{j_k})\}_{k \in \mathbb{N}}$ converges to a complete pointed complex n-dimensional Kähler manifold $(\mathcal{M}_\infty^{2n}, g_\infty, O_\infty)$ as $k \to \infty$. See the ensuing proof for the meaning of the convergence of the complex structures $J_k \to J_\infty$.

PROOF. Since Kähler manifolds are Riemannian manifolds, we can apply the Cheeger–Gromov Compactness Theorem 3.9 to obtain a pointed limit $(\mathcal{M}_\infty^{2n}, g_\infty, O_\infty)$ which is a complete Riemannian manifold. So the only issue is to show that the limit is Kähler. Let J_k denote the complex structure of $(\mathcal{M}_k, g_k)$. We have for each $k \in \mathbb{N}$,

(1) $J_k^2 = -\mathrm{id}_{T\mathcal{M}_k}$,

(2) $(g_k \circ J_k)(X,Y) \doteq g_k(J_kX, J_kY) = g_k(X,Y)$ for all $X, Y \in T\mathcal{M}_k$,
(3) $\nabla_k J_k = 0$.

Since $\{(\mathcal{M}_k, g_k, O_k)\}$ converges to $(\mathcal{M}_\infty, g_\infty, O_\infty)$, there are diffeomorphisms $\Phi_k : U_k \to V_k$ such that $\Phi_k^* g_k \to g_\infty$ uniformly on compact sets. From (1) we know $(\Phi_k^{-1})_* \circ J_k \circ (\Phi_k)_*$, as $(1,1)$-tensors, are uniformly bounded on any compact set $K \subset \mathcal{M}_\infty$ with respect to the metrics $\Phi_k^* g_k$. Since the $\Phi_k^* g_k$ are equivalent to g_∞ on K, we conclude that $(\Phi_k^{-1})_* \circ J_k \circ (\Phi_k)_*$, as $(1,1)$-tensors, are uniformly bounded on any compact set $K \subset \mathcal{M}_\infty$ with respect to the metric g_∞.

Note that (3) is equivalent to $\nabla^{\Phi_k^* g_k}\left((\Phi_k^{-1})_* \circ J_k \circ (\Phi_k)_*\right) = 0$; hence

$$\left(\nabla^{\Phi_k^* g_k}\right)^p \left((\Phi_k^{-1})_* \circ J_k \circ (\Phi_k)_*\right) = 0 \text{ for all } p.$$

From the proof of Corollary 3.15, there exists a subsequence $(\Phi_k^{-1})_* \circ J_k \circ (\Phi_k)_*$ converging in C^∞, as a $(1,1)$-tensor on compact sets, to a smooth map $J_\infty : T\mathcal{M}_\infty \to T\mathcal{M}_\infty$. Since

(1') $0 = \left((\Phi_k^{-1})_* \circ J_k \circ (\Phi_k)_*\right)^2 + \mathrm{id}_{T\mathcal{M}_\infty} \to J_\infty^2 + \mathrm{id}_{T\mathcal{M}_\infty}$,

(2') $0 = (\Phi_k^* g_k) \circ \left((\Phi_k^{-1})_* \circ J_k \circ (\Phi_k)_*\right) - \Phi_k^* g_k \to g_\infty \circ J_\infty - g_\infty$,

(3') $0 = \nabla^{\Phi_k^* g_k}\left((\Phi_k^{-1})_* \circ J_k \circ (\Phi_k)_*\right) \to \nabla_\infty J_\infty$,

we have $J_\infty^2 = -\mathrm{id}_{T\mathcal{M}_\infty}$, $g_\infty \circ J_\infty = g_\infty$ and $\nabla_\infty J_\infty = 0$. We conclude that $(\mathcal{M}_\infty, g_\infty, J_\infty)$ is a Kähler manifold. $\square$

Applying Theorem 3.10, we obtain the following corresponding result for the Kähler–Ricci flow.

THEOREM 3.23 (Compactness theorem for the Kähler–Ricci flow). *Let $\{(\mathcal{M}_k^{2n}, g_k(t), O_k)\}$, $t \in (\alpha, \omega) \ni 0$, be a sequence of complete pointed solutions to the Kähler–Ricci flow of complex dimension n. Suppose*

$$|\mathrm{Rm}_k|_k \leq C_0 \quad \text{on } \mathcal{M}_k \times (\alpha, \omega)$$

for some constant $C_0 < \infty$ independent of k and that

$$\mathrm{inj}_{g_k(0)}(O_k) \geq \iota_0$$

for some constant $\iota_0 > 0$. Then there exists a subsequence of solutions such that $\{(\mathcal{M}_k, g_k(t), O_k)\}$ converges to a complete pointed complex n-dimensional solution to the Kähler–Ricci flow $(\mathcal{M}_\infty^{2n}, g_\infty(t), O_\infty)$, $t \in (\alpha, \omega)$, as $k \to \infty$.

PROOF. By Theorem 3.10, there exists a subsequence which converges to a Riemannian solution $(\mathcal{M}_\infty, g_\infty(t), O_\infty)$, $t \in (\alpha, \omega)$, to the Ricci flow. From the previous theorem, $(\mathcal{M}_k, g_k(0), J_k)$ converges to $(\mathcal{M}_\infty, g_\infty(0), J_\infty)$ as Kähler manifolds for some complex structure J_∞. Now by assumption, $g_k(t)$ remains Kähler with respect to J_k. Hence $g_k(t) \circ J_k = g_k(t)$ and $\nabla_{g_k(t)} J_k = 0$ for all $t \in (\alpha, \omega)$, which implies $g_\infty(t) \circ J_\infty = g_\infty(t)$ and $\nabla_{g_\infty(t)} J_\infty = 0$ for all $t \in (\alpha, \omega)$. That is, $g_\infty(t)$ remains Kähler with respect to J_∞. $\square$

3.3. Compactness for solutions on orbifolds. Note that Definitions 3.5 and 3.6 can be easily generalized to **orbifolds**. (See [**343**] for the definition of orbifold.) We have the following generalizations of Theorems 3.9 and 3.10 to orbifolds. The version for metrics is

THEOREM 3.24 (Compactness theorem for metrics on orbifolds). *Let $\{(\mathcal{M}_k^n, g_k, O_k)\}$ be a sequence of complete pointed Riemannian orbifolds and let Σ_k be the singular set of $\mathcal{M}_k$. Suppose that*

(i) $|\nabla_k^p \operatorname{Rm}_k|_k \leq C_p$ *on $\mathcal{M}_k$ for all $p \geq 0$ and k, where $C_p < \infty$ are constants independent of k, and*

(ii) $\operatorname{Vol}_{g_k} B_{g_k}(O_k, r_0) \geq v_0$ *for all k, where $r_0 > 0$ and $v_0 > 0$ are two constants independent of k.*

Then either of the following hold.

(1) $\varliminf_{k \to \infty} d_{g_k}(O_k, \Sigma_k) > 0$. *In this case there exists a subsequence $\{(\mathcal{M}_{k_j}, g_{k_j}, O_{k_j})\}$ which converges to a complete pointed Riemannian orbifold $(\mathcal{M}_\infty^n, g_\infty, O_\infty)$ with $|\nabla_{g_\infty}^p \operatorname{Rm}_{g_\infty}|_{g_\infty} \leq C_p$ and $\operatorname{Vol}_{g_\infty} B_{g_\infty}(O_\infty, r_0) \geq v_0$. Furthermore O_∞ is a smooth point in $\mathcal{M}_\infty$.*

(2) $\varliminf_{k \to \infty} d_{g_k}(O_k, \Sigma_k) = 0$. *In this case there exists a subsequence $\{(\mathcal{M}_{k_j}, g_{k_j}, O_{k_j})\}$ such that $\lim_{j \to \infty} d(O_{k_j}, \Sigma_{k_j}) = 0$. If we choose $O'_{k_j} \in \Sigma_{k_j}$ with $d_{g_k}(O_{k_j}, O'_{k_j}) = d(O_{k_j}, \Sigma_{k_j})$, then a subsequence of $\{(\mathcal{M}_{k_j}, g_{k_j}, O'_{k_j})\}$ converges to a complete pointed Riemannian orbifold $(\mathcal{M}_\infty, g_\infty, O_\infty)$ with $|\nabla_{g_\infty}^p \operatorname{Rm}_{g_\infty}|_{g_\infty} \leq C_p$ and $\operatorname{Vol}_{g_\infty} B_{g_\infty}(O_\infty, r_0) \geq v_0$. Furthermore O_∞ is a singular point in $\mathcal{M}_\infty$.*

The version for solutions of Ricci flow is the following.

THEOREM 3.25 (Compactness theorem for solutions on orbifolds). *Let $\{(\mathcal{M}_k^n, g_k(t), O_k)\}$, $t \in (\alpha, \omega)$, be a sequence of complete pointed orbifold solutions of the Ricci flow. Let Σ_k be the singular set of $\mathcal{M}_k$. Suppose that*

(i) $|\operatorname{Rm}_k|_k \leq C_0$ *on $\mathcal{M}_k \times (\alpha, \omega)$ for all k, where $C_0 < \infty$ is a constant independent of k, and*

(ii) $\operatorname{Vol}_{g_k(0)} B_{g_k(0)}(O_k, r_0) \geq v_0$ *for all k, where $r_0 > 0$ and $v_0 > 0$ are two constants independent of k.*

Then either of the following hold.

(1) $\varliminf_{k \to \infty} d_{g_k(0)}(O_k, \Sigma_k) > 0$. *In this case there exists a subsequence $\{(\mathcal{M}_{k_j}, g_{k_j}(t), O_{k_j})\}$ which converges to a complete pointed orbifold solution of the Ricci flow $(\mathcal{M}_\infty^n, g_\infty(t), O_\infty)$ with $|\operatorname{Rm}_{g_\infty}|_{g_\infty} \leq C_0$ on $\mathcal{M}_\infty \times (\alpha, \omega)$ and $\operatorname{Vol}_{g_\infty(0)} B_{g_\infty(0)}(O_\infty, r_0) \geq v_0$. Furthermore O_∞ is a smooth point in $\mathcal{M}_\infty$.*

(2) $\varliminf_{k \to \infty} d_{g_k(0)}(O_k, \Sigma_k) = 0$. *In this case there exists a subsequence $\{(\mathcal{M}_{k_j}, g_{k_j}(t), O_{k_j})\}$ such that $\lim_{j \to \infty} d(O_{k_j}, \Sigma_{k_j}) = 0$. Furthermore if we choose $O'_{k_j} \in \Sigma_{k_j}$ with $d_{g_k(0)}(O_{k_j}, O'_{k_j}) = d_{g_k(0)}(O_{k_j}, \Sigma_{k_j})$, then there is a subsequence of $\{(\mathcal{M}_{k_j}, g_{k_j}(t), O'_{k_j})\}$ which converges to a complete pointed orbifold solution of the Ricci flow $(\mathcal{M}_\infty^n, g_\infty(t), O_\infty)$ with $|\operatorname{Rm}_{g_\infty}|_{g_\infty} \leq C_0$*

on $\mathcal{M}_\infty \times (\alpha, \omega)$ and $\operatorname{Vol}_{g_\infty(0)} B_{g_\infty(0)}(O_\infty, r_0) \geq v_0$. Furthermore O_∞ is a singular point in $\mathcal{M}_\infty$.

Idea of the proofs. Theorem 3.25 can be proved from Theorem 3.24 in the same way as we have proved Theorem 3.10 from Theorem 3.9. On the other hand, Theorem 3.24 can be proved with some modification of the proof of Theorem 3.9 to handle the singularity (see [**257**]). Note that the Bishop–Gromov volume comparison theorem holds for orbifolds. Fix $r > 0$; for any k and $q_k \in \mathcal{M}_k$ with $d_{g_k(0)}(O_k, q_k) \leq r$, we have $\operatorname{Vol}_{g_k(0)} B_{g_k(0)}(q_k, r_0) \geq v_1$, where v_1 is a positive constant independent of k but depending on v_0, r, r_0, n, C_0. This implies that there exists r_1 independent of k such that $B_{g_k(0)}(q_k, r_1)$ has the orbifold topological model $B^n / G(q_k)$, where B^n is the unit ball in Euclidean space centered at the origin and $G(q_k) \subset O(n)$ is a discrete subgroup with rank $|G(q_k)|$ bounded independent of k. The existence of r_1 implies that we can modify the choice of λ^α in Definition 4.26 and Proposition 4.22 so that the ball $\tilde{B}_k^\alpha \doteqdot B(x_k^\alpha, \lambda^\alpha/2)$ has the orbifold topological model $B^n / G(x_k^\alpha)$. The key observation in the proof of Theorem 3.24 is that we can choose a subsequence of orbifolds so that the groups $G(x_k^\alpha)$ and their actions on B^n are independent of k. We can then use the balls $\tilde{B}_k^\alpha, B_k^\alpha, \bar{B}_k^\alpha$ to build the limit orbifold.

4. Applications of Hamilton's compactness theorem

In this section we discuss some applications of Theorems 3.10 and 3.16. We will see more applications of the compactness theorems later in this volume.

4.1. Singularity models. Theorem 3.16 may be applied to study singular, nonsingular, and ancient solutions of the Ricci flow. For example, let $(\mathcal{M}^n, g(t))$, $t \in [0, T)$, where $T \in (0, \infty]$, be a complete solution to the Ricci flow. Given a sequence of points and times $\{(x_k, t_k)\}_{k \in \mathbb{N}}$, let $K_k \doteqdot |\operatorname{Rm}(x_k, t_k)|$. We say that the sequence $\{(x_k, t_k)\}$ satisfies an **injectivity radius estimate** if there exists $\iota_0 > 0$ independent of k such that $\operatorname{inj}_{g(t_k)}(x_k) \geq \iota_0 K_k^{-1/2}$. Given a complete solution of the Ricci flow, we can obtain a local limit of dilations provided we have an injectivity radius estimate and a local bound on the curvatures after dilations.

COROLLARY 3.26 (Existence of singularity models). *Let $(\mathcal{M}^n, g(t))$, $t \in (\alpha, \omega)$, be a complete solution to the Ricci flow. Given a sequence of points and times $\{(x_k, t_k)\}_{k \in \mathbb{N}}$, let $K_k \doteqdot |\operatorname{Rm}(x_k, t_k)| > 0$ and*

$$g_k(t) \doteqdot K_k g\left(t_k + K_k^{-1} t\right).$$

Suppose that the sequence $\{(x_k, t_k)\}$ satisfies an injectivity radius estimate, i.e., $\operatorname{inj}_{g(t_k)}(x_k) \geq \iota_0 K_k^{-1/2}$, for some $\iota_0 > 0$, and suppose that $\alpha_k, \omega_k, \alpha_\infty, \omega_\infty \geq 0$ with $\alpha_k \to \alpha_\infty > 0$, $\omega_k \to \omega_\infty$, and $\left[t_k - \frac{\alpha_k}{K_k}, t_k + \frac{\omega_k}{K_k}\right] \subset (\alpha, \omega)$ are such

that there exist positive constants $\rho \leq \infty$ and $C < \infty$ where
(3.12)
$$\sup_{g_k} \left\{ \left| \mathrm{Rm} \right| (x,t) : (x,t) \in B_{g(t_k)}\left(x_k, \frac{\rho}{\sqrt{K_k}}\right) \times \left[t_k - \frac{\alpha_k}{K_k}, t_k + \frac{\omega_k}{K_k}\right] \right\} \leq C K_k.$$
Then there exists a subsequence of the dilated solutions
$$\left(B_{g(t_k)}\left(x_k, \rho K_k^{-1/2}\right), g_k(t), x_k \right)$$
which converges to a solution $(\mathcal{B}_\infty^n, g_\infty(t), x_\infty)$ on an open manifold on the time interval $(-\alpha_\infty, \omega_\infty]$, which is complete on the closed ball $\overline{B_{g_\infty(0)}(x_\infty, r)}$ for all $r < \rho$. In particular, if $\rho = \infty$, then the solution $(\mathcal{B}_\infty, g_\infty(t), x_\infty)$ is complete.

REMARK 3.27. By definition, we let $B_{g(t_k)}(x_k, \infty) = \mathcal{M}$.

In Chapter 6 we shall show that the injectivity radius estimate in the corollary above for the solution $(\mathcal{M}^n, g(t))$, $t \in [0, T)$, on a closed manifold with $T < \infty$, is a consequence of Perelman's no local collapsing theorem.

THEOREM 3.28 (Local injectivity radius estimate for finite time singular solutions). Let $(\mathcal{M}^n, g(t))$, $t \in [0, T)$, $T < \infty$, be a solution to the Ricci flow on a closed Riemannian manifold. There exists $\delta > 0$ such that if $(x_0, t_0) \in \mathcal{M} \times [0, T)$ is a point and time satisfying
$$|\mathrm{Rm}(x, t_0)| \leq \frac{1}{\rho^2} \quad \text{in } B_{g(t_0)}(x_0, \rho)$$
for some $\rho > 0$, then
$$\inj_{g(t_0)}(x_0) \geq \delta \rho.$$

For finite time singular solutions on closed manifolds we have the following.

COROLLARY 3.29 (Local singularity models for finite time singular solutions). Suppose $(\mathcal{M}^n, g(t))$, $t \in [0, T)$, $T < \infty$, is a solution on a closed Riemannian manifold. If there exist positive constants $\rho \leq \infty$ and $C < \infty$ such that (3.12) holds, where $\alpha_k, \beta_k \geq 0$, $\alpha_k \to \alpha > 0$, and $\beta_k \to \beta$, then there exists a subsequence such that $\left(B_{g(t_k)}\left(x_k, \rho K_k^{-1/2}\right), g_k(t), x_k \right)$ converges to a solution $(\mathcal{B}_\infty^n, g_\infty(t), x_\infty)$, $t \in (-\alpha, \beta)$, to the Ricci flow on an open manifold which is complete on the closed ball $\overline{B_{g_\infty(0)}(x_\infty, r)}$ for all $r < \rho$. In particular, if $\rho = \infty$, then $(\mathcal{B}_\infty, g_\infty(t), x_\infty)$ is complete.

4.2. 3-manifolds with positive Ricci curvature revisited. As an application of Theorem 3.10 we give a proof of the following result of Hamilton, which is a consequence of Theorem 6.3 on p. 173 of Volume One.

THEOREM 3.30 (Closed 3d Rc > 0 manifolds are diffeomorphic to space forms). If $(\mathcal{M}^3, g_0)$ is a closed Riemannian 3-manifold with positive Ricci curvature, then $\mathcal{M}$ admits a metric with positive constant sectional curvature.

PROOF. Let $(\mathcal{M}, g(t))$, $t \in [0, T)$, be the maximal solution to the Ricci flow with $g(0) = g_0$. Recall that $T \leq \frac{3}{2}(R_{\min}(0))^{-1} < \infty$. The proof relies on three main estimates.

(1) (*Positivity of Ricci is preserved.*) $\operatorname{Rc}(g(t)) > 0$ for all $t \geq 0$.

(2) (*Strong curvature pinching.*) There exist $C < \infty$ and $\delta > 0$ depending only on g_0 such that
$$\frac{\left|\operatorname{Rc} - \frac{1}{3} R g\right|}{R} \leq C R^{-\delta}$$
on $\mathcal{M} \times [0, T)$ (see inequality (6.37) on p. 190 of Volume One).

(3) (*Injectivity radius estimate.*) There exists $\iota_0 > 0$ depending on n, T, g_0 such that if $(x, t) \in \mathcal{M} \times [0, T)$ and $r \in (0, 1]$ are such that
$$|\operatorname{Rm}(\cdot, t)| \leq r^{-2} \quad \text{in } B_{g(t)}(x, r),$$
then $\operatorname{inj}_{g(t)}(x) \geq \iota_0 r$ (see Theorem 3.28 or Corollary 6.62).

The main technique is to dilate and apply the compactness theorem. Choose (x_k, t_k) with $x_k \in \mathcal{M}$ and $t_k \to T$ such that
$$R_k \doteqdot R(x_k, t_k) = \max_{\mathcal{M}^3 \times [0, t_k]} R \geq \max_{\mathcal{M}^3 \times [0, t_k]} |\operatorname{Rm}| \to \infty$$
as $k \to \infty$. Consider the sequence of solutions $(\mathcal{M}, g_k(t), x_k)$, $t \in (-t_k R_k, 0]$, where
$$g_k(t) \doteqdot R_k g\left(t_k + R_k^{-1} t\right).$$
Let $r_k \doteqdot R_k^{-1/2}$, which is bounded above by 1 for k large enough. We have
$$|\operatorname{Rm}(\cdot, t_k)| \leq r_k^{-2} \quad \text{in } B_{g(t_k)}(x_k, r_k).$$
Thus by (3), $\operatorname{inj}_{g(t_k)}(x_k) \geq \iota_0 r_k$, which is equivalent to $\operatorname{inj}_{g_k(0)}(x_k) \geq \iota_0$. Hence Hamilton's compactness theorem implies that, for a subsequence, $(\mathcal{M}, g_k(t), x_k)$ converges to $(\mathcal{M}_\infty^3, g_\infty(t), x_\infty)$, a complete solution defined for $t \in (-\infty, 0]$. We claim that $(\mathcal{M}_\infty, g_\infty(t))$ has constant positive sectional curvature. In particular, $\mathcal{M}_\infty$ has bounded diameter and hence is diffeomorphic to $\mathcal{M}$. So the theorem follows.

First note that $R_{g_\infty}(x_\infty, 0) = \lim_{k \to \infty} R_{g_k}(x_k, 0) = \lim_{k \to \infty} 1 = 1$. Hence $R_{g_\infty}(x, 0) > 0$ for x contained in a neighborhood of x_∞. Estimate (2) says that
$$\frac{\left|\operatorname{Rc}(g_k) - \frac{1}{3} R(g_k) g_k\right|}{R(g_k)}(t) \leq C R_k^{-\delta} R(g_k)^{-\delta}.$$
Since the convergence of $(\mathcal{M}, g_k(t), x_k)$ is in C^∞ on compact subsets, we have on the subset of $\mathcal{M}_\infty$ where $R(g_\infty) > 0$,
$$\frac{\left|\operatorname{Rc}(g_k) - \frac{1}{3} R(g_k) g_k\right|}{R(g_k)} \to \frac{\left|\operatorname{Rc}(g_\infty) - \frac{1}{3} R(g_\infty) g_\infty\right|}{R(g_\infty)}$$
(we have swept under the rug the fact that in Cheeger–Gromov convergence one must pull back by appropriate diffeomorphisms from an exhaustion of $\mathcal{M}_\infty$ to $\mathcal{M}$; we leave it to the reader to justify the arguments in this proof).

On the other hand, $CR_k^{-\delta} R(g_k)^{-\delta} \to 0 \cdot R(g_\infty)^{-\delta} = 0$. Hence we conclude that $\mathrm{Rc}(g_\infty) = \frac{1}{3} R(g_\infty) g_\infty$ on the subset of $\mathcal{M}_\infty$ where $R(g_\infty) > 0$. On the other hand, the contracted second Bianchi identity implies $R(g_\infty) = \mathrm{const}$ in any connected subset of the set where $R(g_\infty) > 0$. Hence we conclude $R(g_\infty) \equiv 1$ on all of $\mathcal{M}_\infty$, so that $\mathrm{Rc}(g_\infty) \equiv \frac{1}{3} g_\infty$ on $\mathcal{M}_\infty$. □

4.3. Ricci flow on closed surfaces with $\chi > 0$. Now we give various proofs, which are variations on a theme, of the following consequence of Theorem 5.77 on p. 156 of Volume One.

THEOREM 3.31. *If $(\mathcal{M}^2, g_0)$ is a closed Riemannian surface with positive Euler characteristic, then a smooth solution $g(t)$ of the Ricci flow with $g(0) = g_0$ exists on a maximal time interval $[0, T)$ with $T < \infty$. Moreover, there exists a sequence $\{(x_k, t_k)\}$ with $t_k \to T$ such that $g_k(t) \doteqdot R_k g(t_k + R_k^{-1} t)$, with $R_k = R(x_k, t_k)$, converges to a solution $(\mathcal{M}^2, g_\infty(t))$ with constant positive curvature.*

First we recall the ideas of some proofs from Volume One. The first proof relies on the monotonicity of the quantity $\left| \nabla_i \nabla_j f - \frac{1}{2} \Delta f g_{ij} \right|^2$.

PROOF #I. In this proof, which is the original proof of Hamilton, we actually recall the exponential convergence (not just sequential convergence) in C^∞ of the normalized flow. Consider the normalized flow $\frac{\partial}{\partial t} g = (r - R) g$ on the maximal time interval $[0, T)$. In the rest of this proof we abuse notation by using $g(t), t \in [0, T)$, to stand for the normalized solution rather than the unnormalized solution in the statement of the theorem. One can prove that $T = \infty$. If $R(g(t_0)) > 0$ for some $t_0 < \infty$, then we may combine the entropy and Harnack (or Bernstein–Bando–Shi derivative) estimates to show that there exist $c > 0$ and $C < \infty$ such that

$$0 < c \leq R \leq C$$

on $\mathcal{M} \times [t_0, \infty)$. Let $M_{ij} \doteqdot \nabla_i \nabla_j f - \frac{1}{2} \Delta f g_{ij}$, where $\Delta f \doteqdot R - r$. From the equation (see Corollary 5.35 on p. 130 of Volume One)

$$\frac{\partial}{\partial t} |M|^2 = \Delta |M|^2 - 2 |\nabla M|^2 - 2R |M|^2$$
$$\leq \Delta |M|^2 - 2c |M|^2$$

(using the lower bound for R), we have

(3.13) $$|M| \leq C_1 e^{-ct}$$

for some $C_1 < \infty$. We also have for $p \in \mathbb{N}$ (see Corollary 5.63 on p. 149 of Volume One)

(3.14) $$|\nabla^p M| \leq C_p e^{-c_p t}$$

for some $c_p > 0$ and $C_p < \infty$. By the diffeomorphism invariance of the estimates (3.13) and (3.14), this implies that the modified equation

$$\frac{\partial}{\partial t} g = (r - R) g + \mathcal{L}_{\nabla f} g$$

converges exponentially fast in C^∞ to a gradient shrinker. The nonexistence of nontrivial gradient shrinkers (Proposition 5.21 on p. 118 of Volume One) then implies that under the original equation $\frac{\partial}{\partial t} g = (r - R) g$, the solution converges exponentially fast in C^∞ to a constant curvature metric. Finally, for any g_0 we may modify the entropy estimate to show that there exists $t_0 < \infty$ such that $R(g(t_0)) > 0$. □

The next proof uses Hamilton's isoperimetric estimate.

PROOF #IIA. Suppose that $\mathcal{M}^2$ is diffeomorphic to the 2-sphere. Given an embedded loop γ separating $\mathcal{M}$ into two connected components $\mathcal{M}_1$ and $\mathcal{M}_2$, the isoperimetric ratio of γ is defined by

$$C_H(\gamma) \doteqdot L(\gamma)^2 \left(\frac{1}{\text{Area}(\mathcal{M}_1)} + \frac{1}{\text{Area}(\mathcal{M}_2)} \right)$$

and the isoperimetric constant of $(\mathcal{M}, g)$ is

$$C_H(\mathcal{M}, g) \doteqdot \inf_\gamma C_H(\gamma) \leq 4\pi.$$

Then (see Theorem 5.88 on p. 162 of Volume One) under the Ricci flow

$$\frac{d}{dt} C_H(\mathcal{M}, g(t)) \geq 0,$$

so that

$$C_H(\mathcal{M}, g(t)) \geq C_H(\mathcal{M}, g_0) > 0.$$

On the other hand, in the presence of a curvature bound, the isoperimetric constant bounds the injectivity radius by

$$\operatorname{inj}(\mathcal{M}, g) \geq \left(\frac{\pi}{4 K_{\max}} C_H(\mathcal{M}, g) \right)^{1/2},$$

where $K_{\max} \doteqdot \max_\mathcal{M} K$ (K is the Gauss curvature). Hence we may dilate about a sequence $\{(x_k, t_k)\}$ approaching the singularity of the unnormalized flow $g(t)$, as in the proof of Theorem 3.30, and apply Hamilton's compactness theorem to obtain a limit solution $(\mathcal{M}_\infty^2, g_\infty(t))$. This limit solution is a complete ancient solution with bounded positive curvature. In the case of a Type IIa singularity, by choosing $\{(x_k, t_k)\}$ suitably, the limit is an eternal solution (attaining the supremum of R in space-time), which must be the cigar soliton $\left(\mathbb{R}^2, \frac{dx^2 + dy^2}{1 + x^2 + y^2} \right)$. However, the isoperimetric estimate is preserved in the limit,[2] which contradicts the existence of such a limit. Hence the singularity is Type I. In this case the limit $(\mathcal{M}_\infty, g_\infty(t))$ is compact (see [186] or Proposition 9.16 of [111]) and has constant entropy, which implies that it is a gradient shrinker and hence is a constant curvature solution. □

---

[2]More precisely, the limit being a cigar soliton implies that $\lim_{i \to \infty} C_H(\mathcal{M}^2, g(t_i)) = 0$, which leads to a contradiction.

REMARK 3.32. In the above proof, we could have replaced Hamilton's isoperimetric estimate by Perelman's no local collapsing theorem in Chapter 6, which enables the application of the compactness theorem and at the same time rules out the formation of the cigar soliton singularity model.

Now we give a proof that Type I limits are round 2-spheres using Perelman's entropy (see Chapter 6 for properties used in the proof below; the reader may wish to come back to this part after reading that chapter).

PROOF #IIB, USING PERELMAN'S ENTROPY FOR LAST STEP. For the last step in the above proof, we may use Perelman's entropy instead of Hamilton's entropy. This has the advantage that the entropy is defined for solutions with curvature changing sign, so that we may apply it to the original solution $g(t)$, $t \in [0, T)$, rather than the limit solution. We assume that $g(t)$ forms a *Type I* singularity. Let $\mathcal{W}(g(t), f(t), \tau(t))$ denote the entropy with $\tau \doteqdot T - t$, which is defined for $\tau \in (0, T]$. Taking f to be the constant $f_1(t) \doteqdot -\log \frac{4\pi\tau}{\mathrm{Vol}_{g(t)}(\mathcal{M})}$, so that it satisfies the constraint $\int_{\mathcal{M}} (4\pi\tau)^{-1} e^{-f_1} d\mu = 1$, we see that

$$\mu(g(t), \tau(t)) \leq \mathcal{W}(g, f_1, \tau) \leq \tau R_{\max}(t) - \log \frac{4\pi\tau}{\mathrm{Vol}_{g(t)}(\mathcal{M})} - 2.$$

In particular, by the long-time existence theorem for the Ricci flow on the 2-sphere, we have $\mathrm{Vol}_{g(t)}(\mathcal{M}) = 8\pi(T-t)$. Hence we have an upper bound for the μ-invariant:

$$\mu(g(t), \tau(t)) \leq C - 2 + \log 2,$$

where we have used the Type I assumption $(T-t) R_{\max}(t) \leq C$. In particular, by the monotonicity of μ, $\frac{d}{dt} \mu(g(t), \tau(t)) \geq 0$, the limit

$$\mu_T \doteqdot \lim_{t \to T} \mu(g(t), \tau(t))$$

exists. Dilate the solution about (x_k, t_k) with $g_k(t) = R_k g(t_k + R_k^{-1} t)$ as in the proof of Theorem 3.30. By the scaling property of μ we have

$$\mu(g_k(t), R_k(T - t_k) - t) = \mu(g(t_k + R_k^{-1}t), T - t_k - R_k^{-1}t).$$

Thus for each $t \in (-\infty, \omega_\infty)$, the (maximal) time interval of existence of the limit solution $(\mathcal{M}_\infty^2, g_\infty(t))$,

$$\mu_T \equiv \lim_{k \to \infty} \mu(g_k(t), R_k(T - t_k) - t) = \mu(g_\infty(t), \omega_\infty - t).$$

(We may assume $R_k(T - t_k) \to \omega_\infty$ converges. Here we also used the continuity of $\mu(g, \tau)$ in g and the fact that the convergence, after the pull-back by diffeomorphisms, of $g_k(t)$ to $g_\infty(t)$ is globally pointwise in C^∞ since $\mathcal{M}_\infty^2 \cong \mathcal{M}$ is compact.) Now the theorem follows from the result that a solution having constant μ is a gradient shrinker. □

5. Notes and commentary

Some basic references for compactness theorems for Riemannian metrics are Cheeger [**70**], Greene and Wu [**165**], Gromov [**169**], and Peters [**300**]. A survey of compactness theorems in Riemannian geometry has been given in Petersen [**301**].

The compactness theorem for Ricci flow in this chapter was proven by Hamilton in [**187**] and was used to classify singularities and nonsingular solutions in [**186**], [**190**] and [**297**]. Cheeger–Gromov theory was also directly used to study the Ricci flow in Carfora and Marzuoli [**57**]. Further compactness theorems on the Ricci flow which extend Hamilton's results can be found in [**257**] and [**156**].

It should be noted that there has been much work to ensure the injectivity radius bound for dilations of singularities, most notably by Hamilton [**186**] and Perelman [**297**]. Additional work on injectivity radius estimates has been done by Wu [**372**] and by the authors of [**112**] in the case of 2-dimensional orbifolds and [**109**] for sequences of solutions with almost nonnegative curvature operator.

CHAPTER 4

Proof of the Compactness Theorem

> We think in generalities, but we live in details.
> – Alfred North Whitehead

> There is no royal road to geometry. – Euclid

1. Outline of the proof

We now prove the compactness Theorem 3.9. This is a fundamental result in Riemannian geometry and does not require the Ricci flow. The compactness theorem is in the spirit of Cheeger [**70**] and Gromov [**169**] (see also Greene and Wu [**165**], Peters [**300**], and the book [**37**]). We follow the proof for pointed sequences converging in C^∞ given by Hamilton [**187**]; as Hamilton notes there, things are easier because we can assume bounds on all covariant derivatives of the curvature.

Theorem 3.9 will be proved in several steps. It is outlined as follows.

STEP A: *Construct a sequence of coverings of each manifold $\mathcal{M}_k^n$ which we can compare to each other.* The covers should consist of balls $B_k^\alpha \subset \mathcal{M}_k$ with a number of properties, most notably:

- they are diffeomorphic to Euclidean balls, and for each fixed α they have the same radii for all sufficiently large k,
- they are numbered sequentially in α starting from balls centered at the origin to balls with centers further and further away from the origin,
- if we take smaller radii ($\tilde{B}_k^\alpha \subset B_k^\alpha$), they are disjoint, and if we take larger radii ($B_k^\alpha \subset \bar{B}_k^\alpha \subset \vec{B}_k^\alpha$), they contain their neighbors,
- we can bound the number of balls intersecting a given ball (the most is $I(n, C_0)$, where C_0 is the curvature bound), and
- we can bound the number of these balls that it takes to cover a large ball in $\mathcal{M}_k$, independent of k for k large (it takes fewer than $A(r)$ balls to cover $B(O_k, r)$ if $k \geq K(r)$).

The specifics of this are contained in Lemma 4.18. This process is carried out in Section 3 of this chapter.

STEP B: *Use our nice covering to construct maps $F_{k\ell}^\alpha : \bar{B}_k^\alpha \to \mathcal{M}_\ell$.* We do this by taking the inverse of the exponential map on the ball $\bar{B}_k^\alpha \subset \mathcal{M}_k$, identifying the tangent space of $\mathcal{M}_k$ at the center of the ball B_k^α with Euclidean space and with the tangent space of $\mathcal{M}_\ell$ at the center of $\bar{B}_\ell^\alpha$ and then mapping by the exponential map to $\mathcal{M}_\ell$. We do this for each ball. This is done in subsection 4.1 below.

STEP C: *Use nonlinear averaging to glue together the maps $F_{k\ell}^\alpha$ to obtain maps $F_{k\ell} : B(O_k, 2^k) \to \mathcal{M}_\ell$ which take O_k to O_ℓ* (done in subsection 4.2 below). By taking subsequences, we can ensure that compositions of $F_{k,k+1}$ are approximate isometries which are getting closer to isometries as k goes to ∞.

STEP D: *We form the limit manifold $\mathcal{M}_\infty^n$ as the direct limit of the directed system $\{F_{k,k+1} : B(O_k, 2^k) \to B(O_{k+1}, 2^{k+1})\}$.* The coordinates of $B(O_k, 2^k)$ then form coordinates for the limit $\mathcal{M}_\infty$, i.e., for each coordinate $H_k^\alpha : E^\alpha \to B(O_k, 2^k)$ there is a coordinate for the limit manifold defined as

$$H_{\infty,k}^\alpha = I_k \circ H_k^\alpha : E^\alpha \to \mathcal{M}_\infty,$$

where I_k is the inclusion of $B(O_k, 2^k)$ into $\mathcal{M}_\infty$. Furthermore, for each coordinate E^α of $B(O_k, 2^k)$ there are Riemannian metrics $g_{k,\ell}^\alpha$ which are obtained by the pullbacks $F_{k\ell}^* g_\ell$. Since $F_{k\ell}$ are approximate isometries, the sequence is equicontinuous and thus by the Arzela–Ascoli theorem we can find a convergent subsequence and get local metrics $g_{\infty,k}^\alpha$. It is not hard to see that these metrics form a Riemannian metric g_∞ on $\mathcal{M}_\infty$ via the coordinate charts $H_{\infty,k}^\alpha$. *We can then show that the limit metric is complete* and $(\mathcal{M}_\infty, g_\infty, O_\infty)$ satisfies the theorem, where O_∞ is the equivalence class of the base points in the direct limit. This is done in subsection 4.4 below.

At many points in this construction we will take a subsequence; to simplify notation, at each stage the sequence will be re-indexed to continue to be k.

2. Approximate isometries, compactness of maps, and direct limits

In this section we shall introduce some basic concepts which are essential to the construction of the limit manifold $(\mathcal{M}_\infty^n, g_\infty, O_\infty)$.

2.1. Approximate isometries. In the following, the notation $|T|_g$ means the length at a point of the tensor T with respect to the Riemannian metric g.

DEFINITION 4.1 (Approximate isometry). For any $0 < \varepsilon < 1$ and $p \in \mathbb{N} \cup \{0\}$, a smooth map $\Phi : (\mathcal{M}^n, g) \to (\mathcal{N}^n, h)$ is an (ε, p)-**pre-approximate isometry** if

$$\sup_{x \in \mathcal{M}} |\Phi^* h - g|_g \leq \varepsilon, \qquad \sup_{1 \leq \alpha \leq p} \sup_{x \in \mathcal{M}} \left|\nabla_g^\alpha (\Phi^* h)\right|_g \leq \varepsilon.$$

An (ε, p)-pre-approximate isometry is an (ε, p)-**approximate isometry** if it is a diffeomorphism and

$$\sup_{x \in \mathcal{N}} \left|(\Phi^{-1})^* g - h\right|_h \leq \varepsilon, \qquad \sup_{1 \leq \alpha \leq p} \sup_{x \in \mathcal{N}} \left|\nabla_h^\alpha \left[(\Phi^{-1})^* g\right]\right|_h \leq \varepsilon,$$

i.e., $\Phi^{-1} : (\mathcal{N}, h) \to (\mathcal{M}, g)$ is also an (ε, p)-pre-approximate isometry.

Note the condition $\left|\left(\Phi^{-1}\right)^* g - h\right|_h \leq \varepsilon$ is equivalent to $|g - \Phi^* h|_{\Phi^* h} \leq \varepsilon$ and $\left|\nabla_h^\alpha \left[\left(\Phi^{-1}\right)^* g\right]\right|_h \leq \varepsilon$ is equivalent to $|\nabla_{\Phi^* h}^\alpha g|_{\Phi^* h} \leq \varepsilon$. Another way to express the condition $\sup_{x \in \mathcal{M}} |\Phi^* h - g|_g \leq \varepsilon$ is

$$\left| h_{ab} \frac{\partial \Phi^a}{\partial x^i} \frac{\partial \Phi^b}{\partial x^j} - g_{ij} \right|_g^2 = \left(\left[h_{ab} g^{ik} \frac{\partial \Phi^a}{\partial x^i} \right] \frac{\partial \Phi^b}{\partial x^j} - I_j^k \right) \left(\frac{\partial \Phi^c}{\partial x^k} \left[h_{cd} g^{j\ell} \frac{\partial \Phi^d}{\partial x^\ell} \right] - I_k^j \right)$$

$$= \left| (d\Phi)^T (d\Phi) - \mathrm{id} \right|^2,$$

where id is the identity map on $T\mathcal{M}$, the transpose comes from the two metrics h and g, and $|F|^2 = \mathrm{trace}\left(F^2\right)$. Approximate isometries allow pointwise bounding of metric tensors as follows.

PROPOSITION 4.2 (Approximate isometries and norms). *Let $\varepsilon \in (0, 1)$.*

(i) *If $\Phi : (\mathcal{M}^n, g) \to (\mathcal{N}^n, h)$ is an $(\varepsilon, 0)$-pre-approximate isometry and X is a vector field on $\mathcal{M}$, then*

$$|X|_{\Phi^* h}^2 \leq (1 + \varepsilon) |X|_g^2.$$

(ii) *If $\Phi : (\mathcal{M}^n, g) \to (\mathcal{N}^n, h)$ is an $(\varepsilon, 0)$-approximate isometry and X is a vector field on $\mathcal{M}$, then*

$$\frac{1}{1+\varepsilon} |X|_{\Phi^* h}^2 \leq |X|_g^2 \leq (1 + \varepsilon) |X|_{\Phi^* h}^2.$$

PROOF. (i) Using the Cauchy–Schwarz inequality on the tensor space, we find that

$$(\Phi^* h)_{ij} X^i X^j = \left((\Phi^* h)_{ij} - g_{ij} \right) X^i X^j + g_{ij} X^i X^j$$

$$\leq |\Phi^* h - g|_g \cdot g_{ij} X^i X^j + g_{ij} X^i X^j$$

$$\leq (1 + \varepsilon) g_{ij} X^i X^j.$$

(ii) Inequality $|X|_g^2 \leq (1 + \varepsilon) |X|_{\Phi^* h}^2$ follows from (i) and the fact that $\Phi^{-1} : (\mathcal{N}, h) \to (\mathcal{M}, g)$ is also an (ε, p)-pre-approximate isometry. □

The following now immediately follows from Lemma 3.13.

COROLLARY 4.3 (Norms of tensors). *If $\Phi : (\mathcal{M}^n, g) \to (\mathcal{N}^n, h)$ is an $(\varepsilon, 0)$-approximate isometry, then for any (p, q)-tensor field T on $\mathcal{M}$ we have*

(4.1) $$(1 + \varepsilon)^{-(p+q)/2} |T|_{\Phi^* h} \leq |T|_g \leq (1 + \varepsilon)^{(p+q)/2} |T|_{\Phi^* h}.$$

The following proposition shows how $(\varepsilon, 0)$-approximate isometries deform distances by small amounts.

PROPOSITION 4.4 (Distances). *If $\Phi : (\mathcal{M}^n, g) \to (\mathcal{N}^n, h)$ is an $(\varepsilon, 0)$-pre-approximate isometry, then*

$$\Phi(B_g(x_0, r)) \subset B_h\left(\Phi(x_0), (1 + \varepsilon)^{1/2} r\right).$$

PROOF. Suppose $x \in B_g(x_0, r)$. Then

$$d_h(\Phi(x), \Phi(x_0)) \leq \inf_{\alpha:[a,b] \to \mathcal{M}} \int_a^b h(\Phi_*(\dot{\alpha}), \Phi_*(\dot{\alpha}))^{1/2} dt$$

$$\leq \inf_{\alpha:[a,b] \to \mathcal{M}} \int_a^b ((1+\varepsilon) g(\dot{\alpha}, \dot{\alpha}))^{1/2} dt$$

$$\leq (1+\varepsilon)^{1/2} d_g(x, x_0),$$

where $\alpha(a) = x$ and $\alpha(b) = x_0$. □

We shall need the following propositions about how approximate isometries affect tensors and how to compose approximate isometries.

LEMMA 4.5 (Norms of covariant derivatives of tensors, I). *Given $p \in \mathbb{N}$ and $q_1, q_2 \in \mathbb{N} \cup \{0\}$, there exists a positive constant $C_{p,q_1,q_2} < \infty$ such that if $\Phi : (\mathcal{M}^n, g) \to (\mathcal{N}^n, h)$ is an (ε, p)-approximate isometry with $\varepsilon < 1$, then for any (q_1, q_2)-tensor field T on $\mathcal{M}$ we have*

$$(4.2) \qquad \left|\nabla_g^r T\right|_g \leq \left|\nabla_{\Phi^* h}^r T\right|_g + \varepsilon C_{p,q_1,q_2} \sum_{k=0}^{r-1} \left|\nabla_{\Phi^* h}^k T\right|_g$$

for all $0 < r \leq p$.

PROOF. We begin by proving (4.2) for $p = 1$. Note that $\Gamma_g - \Gamma_{\Phi^* h}$ is a global $(1, 2)$-tensor field since it is the difference of two connections. By (3.8) and (4.1) with $C = 1 + \varepsilon$, we find

$$|\Gamma_g - \Gamma_{\Phi^* h}|_g \leq (1+\varepsilon)^{3/2} |\Gamma_g - \Gamma_{\Phi^* h}|_{\Phi^* h} \leq \frac{3}{2}(1+\varepsilon)^{3/2} |\nabla_g \Phi^* h|_{\Phi^* h}$$

$$(4.3) \qquad \leq \frac{3}{2}(1+\varepsilon)^3 |\nabla_g \Phi^* h|_g \leq 12\varepsilon$$

since $\varepsilon < 1$. We have

$$(\nabla_g)_i T^{\ell_1 \ell_2 \cdots \ell_{q_1}}_{j_1 j_2 \cdots j_{q_2}} = (\nabla_{\Phi^* h})_i T^{\ell_1 \ell_2 \cdots \ell_{q_1}}_{j_1 j_2 \cdots j_{q_2}} - \sum_{s=1}^{q_2} (\Gamma_g - \Gamma_{\Phi^* h})^m_{ij_s} T^{\ell_1 \ell_2 \cdots \ell_{q_1}}_{j_1 \cdots m \cdots j_{q_1}}$$

$$+ \sum_{s=1}^{q_1} (\Gamma_g - \Gamma_{\Phi^* h})^{\ell_s}_{im} T^{\ell_1 \cdots m \cdots \ell_{q_1}}_{j_1 j_2 \cdots j_{q_2}}.$$

The above expression has one term of the form $\nabla_{\Phi^* h} T$ and $q_1 + q_2$ terms of the form $(\Gamma_g - \Gamma_{\Phi^* h}) * T$. Hence we get the estimate

$$|\nabla_g T|_g \leq |\nabla_{\Phi^* h} T|_g + (q_1 + q_2) |\Gamma_g - \Gamma_{\Phi^* h}|_g |T|_g$$

$$\leq |\nabla_{\Phi^* h} T|_g + 12\varepsilon (q_1 + q_2) |T|_g,$$

using (4.3).

We now induct on p. Suppose (4.2) is true for any tensor and $p \leq \rho$. Certainly we may choose $C_{\rho+1,q_1,q_2}$ to be greater than C_{ρ,q_1,q_2} and so we need only prove inequality (4.2) for $r = \rho + 1$. Then

$$|\nabla_g^{\rho+1}T|_g = |\nabla_g^\rho \nabla_g T|_g$$
$$\leq |\nabla_{\Phi^*h}^\rho \nabla_g T|_g + \varepsilon C_{\rho,q_1,q_2+1} \sum_{k=0}^{\rho-1} |\nabla_{\Phi^*h}^k \nabla_g T|_g$$

by the inductive hypothesis. For the first term on the RHS,

$$|\nabla_{\Phi^*h}^\rho \nabla_g T|_g \leq |\nabla_{\Phi^*h}^\rho (\nabla_g - \nabla_{\Phi^*h}) T|_g + |(\nabla_{\Phi^*h})^{\rho+1} T|_g.$$

In turn, a sum of terms of the form

$$\left|(\nabla_{\Phi^*h})^k (\Gamma_g - \Gamma_{\Phi^*h})\right|_g \left|(\nabla_{\Phi^*h})^{\rho-k} T\right|_g,$$

where $0 \leq k \leq \rho$, bounds the first term. As was shown in the proof of Lemma 3.11,

$$\left|\nabla_{\Phi^*h}^k (\Gamma_g - \Gamma_{\Phi^*h})\right|_g \leq C_k \sum_{j=1}^{k+1} \left|\nabla_{\Phi^*h}^j g\right|_g$$
$$\leq C_k \sum_{j=1}^{k+1} (1+\varepsilon)^{(2+j)/2} \left|\nabla_{\Phi^*h}^j g\right|_{\Phi^*h}$$
$$\leq C_k \frac{2^{(k+4)/2}}{\sqrt{2}-1} \varepsilon$$

for some constant C_k depending on k, where we have used that Φ is an $(\varepsilon, \rho+1)$-approximate isometry and $0 \leq k \leq \rho$. Hence we have bounded the term $|\nabla_{\Phi^*h}^\rho \nabla_g T|_g$. Similarly, we can bound the terms $|\nabla_{\Phi^*h}^k \nabla_g T|$, where $0 \leq k \leq \rho - 1$.

Putting it all together, we see that in the estimate of $|\nabla_g^{\rho+1}T|_g$, the coefficients of $|\nabla_{\Phi^*h}^k T|_g$ are all bounded and all have an ε in them except $|\nabla_{\Phi^*h}^{\rho+1} T|_g$. Thus we get

$$|\nabla_g^{\rho+1}T|_g \leq |\nabla_{\Phi^*h}^{\rho+1}T|_g + \varepsilon C_{\rho+1,q_1,q_2} \sum_{k=0}^{\rho} |\nabla_{\Phi^*h}^k T|_g.$$

This completes the proof of the induction. □

COROLLARY 4.6 (Norms of covariant derivatives of tensors, II). *There exist $C_{p,0,q_2}$ for $p, q_2 \in \mathbb{N}$, such that if $\Phi : (\mathcal{M}^n, g) \to (\mathcal{N}^n, h)$ is an (ε, p)-approximate isometry, then for any $(0, q_2)$-tensor T field on $\mathcal{N}$*

$$\left|\nabla_g^r (\Phi^*T)\right|_g \leq (1+\varepsilon)^{(r+q_2)/2} \left(|\nabla_h^r T|_h + \varepsilon C_{p,0,q_2} \sum_{k=0}^{r-1} |\nabla_h^k T|_h \right)$$

for each $1 \leq r \leq p$.

The following proposition about the composition of approximate isometries will be used in the construction of the *directed system* (see subsection 2.3 of this chapter for the definition) when proving Theorem 3.9.

PROPOSITION 4.7 (Composition of approximate isometries, I). *There exist C_p for $p \in \mathbb{N}$ such that if $\Phi_i : (\mathcal{M}_i^n, g_i) \to (\mathcal{M}_{i+1}^n, g_{i+1})$ are (ε_i, p)-approximate isometries for $i = 0, 1$ with $\varepsilon_i \leq 1$, then*

$$\sup_{0 \leq r \leq p} \sup_{x \in \mathcal{M}_1} \left| \nabla_0^r \left(\Phi_0^* \Phi_1^* g_2 - g_0 \right) \right|_{g_0} \leq \varepsilon_0 + \varepsilon_1 C_p,$$

$$\sup_{0 \leq r \leq p} \sup_{x \in \mathcal{M}_2} \left| \nabla_2^r \left(\left(\Phi_1^{-1} \right)^* \left(\Phi_0^{-1} \right)^* g_0 - g_2 \right) \right|_{g_2} \leq \varepsilon_1 + \varepsilon_0 C_p,$$

where we have denoted $\nabla_i \doteqdot \nabla_{g_i}$.

PROOF. Using (4.1), we estimate

$$\begin{aligned}|\Phi_0^* \Phi_1^* g_2 - g_0|_{g_0} &\leq |\Phi_0^* \Phi_1^* g_2 - \Phi_0^* g_1|_{g_0} + |\Phi_0^* g_1 - g_0|_{g_0} \\ &\leq (1 + \varepsilon_0) |\Phi_1^* g_2 - g_1|_{g_1} + \varepsilon_0 \\ &\leq (1 + \varepsilon_0) \varepsilon_1 + \varepsilon_0.\end{aligned}$$

By Corollary 4.6 for $1 \leq r \leq p$

$$\begin{aligned}\left| \nabla_0^r \left(\Phi_0^* \Phi_1^* g_2 \right) \right|_{g_0} &= \left| \nabla_0^r \left(\Phi_0^* \Phi_1^* g_2 - g_0 \right) \right|_{g_0} \\ &\leq \left| \nabla_0^r \left(\Phi_0^* \Phi_1^* g_2 - \Phi_0^* g_1 \right) \right|_{g_0} + \left| \nabla_0^r \left(\Phi_0^* g_1 - g_0 \right) \right|_{g_0} \\ &\leq (1 + \varepsilon_0)^{(r+2)/2} \left(\left| \nabla_1^r \left(\Phi_1^* g_2 - g_1 \right) \right|_{g_1} \right. \\ &\quad \left. + \varepsilon_0 C_{p,0,2} \sum_{k=0}^{r-1} \left| \nabla_1^k \left(\Phi_1^* g_2 - g_1 \right) \right|_{g_1} \right) + \varepsilon_0 \\ &\leq (1 + \varepsilon_0)^{(r+2)/2} \left(\varepsilon_1 + \varepsilon_0 C_{r,0,2} r \varepsilon_1 \right) + \varepsilon_0.\end{aligned}$$

By symmetry we have

$$\left| \left(\Phi_1^{-1} \right)^* \left(\Phi_0^{-1} \right)^* g_0 - g_2 \right|_{g_2} \leq (1 + \varepsilon_1) \varepsilon_0 + \varepsilon_1,$$

$$\left| \nabla_0^r \left(\left(\Phi_1^{-1} \right)^* \left(\Phi_0^{-1} \right)^* g_0 \right) \right|_{g_2} \leq (1 + \varepsilon_1)^{(r+2)/2} \left(\varepsilon_0 + \varepsilon_1 C_{r,0,2} r \varepsilon_0 \right) + \varepsilon_1.$$

The proposition is proved for $C_p = 2^{(p+2)/2} \left(1 + p C_{p,0,2} \right)$. □

COROLLARY 4.8 (Composition of approximate isometries, II). *If*

$$\Phi_i : (\mathcal{M}_i^n, g_i) \to (\mathcal{M}_{i+1}^n, g_{i+1})$$

are (ε_i, p)-approximate isometries for $i = 0, 1, \ldots k$, then $\Phi_k \circ \cdots \circ \Phi_1 \circ \Phi_0 : (\mathcal{M}_0^n, g_0) \to (\mathcal{M}_{k+1}^n, g_{k+1})$ is a $\left(C_p \sum_{i=0}^k \varepsilon_i, p \right)$-approximate isometry.

PROOF. We shall induct on the number of compositions. By induction, $\Phi_{k-1} \circ \cdots \circ \Phi_0$ is a $\left(C_p \sum_{i=0}^{k-1} \varepsilon_i, p \right)$-approximate isometry. By Proposition

4.7, we have for $0 \leq r \leq p$

$$\left|\nabla_0^r \left((\Phi_k \circ \Phi_{k-1} \circ \cdots \circ \Phi_0)^* g_{k+1} - g_0\right)\right|_{g_0}$$
$$= \left|\nabla_0^r \left((\Phi_{k-1} \circ \cdots \circ \Phi_0)^* \Phi_k^* g_{k+1} - g_0\right)\right|_{g_0}$$
$$\leq C_p \sum_{i=0}^{k-1} \varepsilon_i + \varepsilon_k C_p \leq C_p \sum_{i=0}^{k} \varepsilon_i.$$

Similarly, by induction, $\Phi_k \circ \cdots \circ \Phi_1$ is a $\left(C_p \sum_{i=1}^{k} \varepsilon_i, p\right)$-approximate isometry. By Proposition 4.7,

$$\left|\nabla_{k+1}^r \left(\left((\Phi_k \circ \cdots \circ \Phi_1 \circ \Phi_0)^{-1}\right)^* g_0 - g_{k+1}\right)\right|_{g_{k+1}}$$
$$= \left|\nabla_{k+1}^r \left(\left[\left((\Phi_k \circ \cdots \circ \Phi_1)^{-1}\right)^*\right] (\Phi_0^{-1})^* g_0 - g_{k+1}\right)\right|_{g_{k+1}}$$
$$\leq C_p \sum_{i=1}^{k} \varepsilon_i + \varepsilon_0 C_p \leq C_p \sum_{i=0}^{k} \varepsilon_i.$$

Thus $\Phi_k \circ \cdots \circ \Phi_0$ is a $\left(C_p \sum_{i=0}^{k} \varepsilon_i, p\right)$-approximate isometry. $\square$

2.2. Compactness of maps. We shall need a version of the Arzela–Ascoli theorem which applies to maps. We define C^∞-convergence on compact sets for maps in Euclidean space as follows.

DEFINITION 4.9 (C^p-convergence of maps). Let $\mathcal{U}$ and $\mathcal{V}$ be two open sets in $\mathbb{R}^n$ and let $K \subset \mathcal{U}$ be a compact set. We say that a sequence of maps $\Phi_k : \mathcal{U} \to \mathcal{V}$ **converges to a map** $\Phi_\infty : \mathcal{U} \to \mathcal{V}$ **in** C^p **on** K if for every $\varepsilon > 0$ there exists $k_0 = k_0(\varepsilon, p)$ such that

$$\sup_{0 \leq r \leq p} \sup_{x \in K} |\nabla^r (\Phi_k(x) - \Phi_\infty(x))| \leq \varepsilon \quad \text{for } k \geq k_0.$$

Note that the norm given is the Euclidean norm and ∇ is the gradient with respect to the Euclidean metric.

DEFINITION 4.10 (C^∞-convergence of maps uniformly on compact sets). Let $\mathcal{U}$ and $\mathcal{V}$ are two open sets in $\mathbb{R}^n$. A sequence of maps $\Phi_k : \mathcal{U} \to \mathcal{V}$ **converges to a map** $\Phi_\infty : \mathcal{U} \to \mathcal{V}$ **in** C^∞ **uniformly on compact sets** if for any compact set $K \subset \mathcal{U}$ and any $p > 0$ there exists $k_1 = k_1(K, p)$ such that $\{\Phi_k\}_{k \geq k_1}$ converges to Φ_∞ in C^p on K.

The following is a corollary to the Arzela–Ascoli theorem, Lemma 3.14.

COROLLARY 4.11 (Compactness of sequence of isometries). *Let $\mathcal{U}$ and $\mathcal{V}$ be two bounded open sets in $\mathbb{R}^n$. Let $\{g_k\}_{k \in \mathbb{N}}$ and $\{h_k\}_{k \in \mathbb{N}}$ be Riemannian metrics on $\mathcal{U}$ and $\mathcal{V}$, respectively, such that the g_k and h_k are all uniformly equivalent to the Euclidean metric and all of their derivatives (covariant derivatives with respect to the Euclidean metric) are bounded. If the $\Phi_k : (\mathcal{U}, g_k) \to (\mathcal{V}, h_k)$ are isometries, then there is a subsequence of Φ_k which* converges in C^∞ uniformly on compact sets to a C^∞ diffeomorphism $\Phi_\infty : \mathcal{U} \to \mathcal{V}$.

PROOF (SKETCH). Let $\{x^a\}$ and $\{y^\alpha\}$ be the standard Euclidean coordinates on $\mathcal{U}$ and $\mathcal{V}$, respectively. Since Φ_k are isometries, we have

$$(4.4) \qquad (g_k)_{ab} = \frac{\partial (\Phi_k)^\alpha}{\partial x^a} \frac{\partial (\Phi_k)^\beta}{\partial x^b} (h_k)_{\alpha\beta}.$$

Thus

$$n = \frac{\partial (\Phi_k)^\alpha}{\partial x^a} \frac{\partial (\Phi_k)^\beta}{\partial x^b} (h_k)_{\alpha\beta} (g_k)^{ab}$$

and the partial derivatives $\frac{\partial(\Phi_k)^\alpha}{\partial x^a}$ are all bounded. Thus there is a subsequence of $\{\Phi_k\}$ which converges to a map Φ_∞. By symmetry the same argument applies to $\{\Phi_k^{-1}\}$; hence Φ_∞ is invertible.

Taking the derivatives of both sides of (4.4), we get that

$$\frac{\partial}{\partial x^c} (g_k)_{ab} = \frac{\partial^2 (\Phi_k)^\alpha}{\partial x^c \partial x^a} \frac{\partial (\Phi_k)^\beta}{\partial x^b} (h_k)_{\alpha\beta} + \frac{\partial (\Phi_k)^\alpha}{\partial x^a} \frac{\partial^2 (\Phi_k)^\beta}{\partial x^c \partial x^b} (h_k)_{\alpha\beta}$$
$$+ \frac{\partial (\Phi_k)^\alpha}{\partial x^a} \frac{\partial (\Phi_k)^\beta}{\partial x^b} \frac{\partial (\Phi_k)^\gamma}{\partial x^c} \frac{\partial}{\partial y^\gamma} (h_k)_{\alpha\beta}.$$

From this equation we can express $\frac{\partial^2(\Phi_k)^\alpha}{\partial x^c \partial x^a}$ as a polynomial function of $(g_k)_{ab}$, $(g_k^{-1})^{ab}$, $(h_k)_{\alpha\beta}$, $(h_k^{-1})^{\alpha\beta}$, $\frac{\partial}{\partial x^c}(g_k)_{ab}$, $\frac{\partial}{\partial y^\gamma}(h_k)_{\alpha\beta}$ and $\frac{\partial(\Phi_k)^\alpha}{\partial x^a}$ using symmetry in the usual way (see §5 of [**187**] for the explicit formula). Thus $\left|\frac{\partial^2(\Phi_k)^\alpha}{\partial x^c \partial x^a}\right|$ can be bounded. By differentiating the formula for $\frac{\partial^2(\Phi_k)^\alpha}{\partial x^c \partial x^a}$ and using induction, we can bound all higher derivatives of $(\Phi_k)^\alpha$. This implies the corollary. □

2.3. Review of direct limits. Let $\{(A_k, f_k)\}_{k \in \mathbb{N}}$ be a sequence of topological spaces and open embeddings:

$$A_1 \xrightarrow{f_1} A_2 \xrightarrow{f_2} \cdots \to A_k \xrightarrow{f_k} A_{k+1} \to \cdots.$$

Consider the compositions

$$f_{k\ell} \doteq f_{\ell-1} \circ f_{\ell-2} \circ \cdots \circ f_{k+1} \circ f_k : A_k \to A_\ell$$

defined for $k \leq \ell$, where $f_{kk} \doteq \operatorname{id}_{A_k} : A_k \to A_k$, the identity map. Clearly

$$f_{\ell m} \circ f_{k\ell} = f_{km}$$

for all $k \leq \ell \leq m$. That is, $\left(\{A_k\}_{k \in \mathbb{N}}, \{f_{k\ell}\}_{k \leq \ell}\right)$ is a **directed system** of topological spaces (see Definition 15.1 in [**159**]). We will use $\amalg$ to denote disjoint union.

DEFINITION 4.12 (Direct limit). The **direct limit** is

$$\varinjlim A_k = (\amalg_k A_k)/\sim,$$

where $x \sim y$ if $x \in A_k$ and $y \in A_\ell$ for some $k, \ell \in \mathbb{N}$ and either $f_{k\ell}(x) = y$ (if $k \leq \ell$) or $f_{\ell k}(y) = x$ (if $\ell \leq k$). The relation $\sim$ is an equivalence relation. The topology on $\varinjlim A_k$ is the quotient topology.

Note that direct limits can be defined for more general directed systems, but this is sufficient for our needs.

Let $\iota_\ell : A_\ell \hookrightarrow \amalg_k A_k$ denote the inclusion map and $\pi : \amalg_k A_k \to \varinjlim A_k$ denote the quotient map and define

(4.5) $$I_\ell \doteq \pi \circ \iota_\ell : A_\ell \to \varinjlim A_k,$$

the inclusion map into the direct limit. The topology on $\varinjlim A_k$ is the finest topology such that the maps $I_\ell : A_\ell \to \varinjlim A_k$ are continuous for all $\ell \in \mathbb{N}$. Since the maps f_k are one-to-one for all $k \in \mathbb{N}$, the maps $f_{k\ell}$ are one-to-one for all $k \leq \ell$. This implies the following.

LEMMA 4.13. *The maps I_ℓ are one-to-one for all $\ell \in \mathbb{N}$, and for all $m \geq \ell$ we have*
$$I_\ell = I_m \circ f_{\ell m}.$$

PROOF. The first statement is obvious. The second statement follows from $\iota_\ell(x_\ell) \sim \iota_m(f_{\ell m}(x_\ell))$ (in Definition 4.12, we have suppressed the identification of $x \in A_\ell$ with its image $\iota_\ell(x) \in \amalg_k A_k$). $\square$

The following facts about direct limits are elementary.

LEMMA 4.14 (An open cover for the direct limit). *If $U_\ell \subset A_\ell$ is an open set, then $I_\ell(U_\ell) \subset \varinjlim A_k$ is open, i.e., I_ℓ are open maps. Thus $\{I_\ell(A_\ell)\}_{\ell \in \mathbb{N}}$ forms an open cover of $\varinjlim A_k$.*

PROOF. Since the f_k are open maps for all k, we have that
$$\pi^{-1}[I_\ell(U_\ell)] = \bigcup_{m \geq \ell} f_{\ell m}(U_\ell) \cup \bigcup_{m < \ell} (f_{m\ell})^{-1}(U_\ell)$$
is open in $\amalg_k A_k$. Hence $I_\ell(U_\ell) \subset \varinjlim A_k$ is open. $\square$

This has implications about the structure of compact sets in the limit.

COROLLARY 4.15 (Compact sets in the direct limit). *If $K \subset \varinjlim A_k$ is compact, then for k large enough, $K = I_k(K_k)$ for some compact set $K_k \subset A_k$.*

PROOF. K is covered by $\{I_\ell(A_\ell)\}_{\ell \in \mathbb{N}}$ and since it is compact, it is also covered by a finite number of these. Since $I_\ell(A_\ell) \subset I_m(A_m)$ if $\ell \leq m$, we see that K is in the image of I_k for large enough k. Since I_k is a homeomorphism onto its image, $K_k = I_k^{-1}(K)$ is compact. $\square$

Recall that a topological space is called **second-countable** if its topology has a countable base.

COROLLARY 4.16 (Second-countable direct limits). *If each A_k is the countable union of compact sets, then so is $\varinjlim A_k$. In particular, $\varinjlim A_k$ is second-countable.*

PROOF. If $A_k = \bigcup_{\ell=1}^{\infty} K_{k\ell}$ where $K_{k\ell}$ is compact, then

$$\varinjlim A_k = \bigcup_{k=1}^{\infty} I_k(A_k) = \bigcup_{k=1}^{\infty} \bigcup_{\ell=1}^{\infty} I_k(K_{k\ell})$$

which is a countable union of compact sets. □

LEMMA 4.17 (Direct limit of Hausdorff spaces is Hausdorff).
(1) *If $x \in \varinjlim A_k$, there exist ℓ and $x_\ell \in A_\ell$ such that $I_\ell(x_\ell) = x$.*
(2) *If each A_k is Hausdorff, then $\varinjlim A_k$ is Hausdorff.*

PROOF. (1) Since $x \in \varinjlim A_k$, there must be $x_\ell \in \amalg_k A_k$ such that $\pi(x_\ell) = x$, and thus $x_\ell \in A_\ell$ for some ℓ.

(2) Given $x \neq y \in \varinjlim A_k$, there exists $x_k \in A_k$ and $y_\ell \in A_\ell$ for some $k, \ell \in \mathbb{N}$ such that $I_k(x_k) = x$ and $I_\ell(y_\ell) = y$. Assume without loss of generality that $\ell \geq k$. Define $x_\ell \doteq f_{k\ell}(x_k) \in A_\ell$. Since $x \neq y$, we have $x_\ell \neq y_\ell$. Since A_ℓ is Hausdorff, there exist disjoint open neighborhoods N_x and N_y of x_ℓ and y_ℓ, respectively. Since I_ℓ is one-to-one and open, we conclude that $I_\ell(N_x)$ and $I_\ell(N_y)$ are disjoint open neighborhoods of x and y, respectively, in $\varinjlim A_k$. □

The direct limit satisfies the following universal property (see Proposition 15.3 in [**159**]). For any space X and maps $\psi_k : A_k \to X$ such that

$$\psi_\ell \circ f_{k\ell} = \psi_k,$$

there exists a unique map $\Psi : \varinjlim A_k \to X$ such that

$$\psi_k = \Psi \circ I_k.$$

3. Construction of good coverings by balls

3.1. Overview. In this section we shall prove the following lemma, which we will need in order to construct maps between different manifolds. The reader is warned that in this section we will be using superscripts which usually do not represent exponents. We shall need five radii of balls so that the smallest are disjoint and so that all others cover and are successively larger to allow for maps between the intersections. This section is quite technical and the proofs may be skipped in the first reading.

LEMMA 4.18 (Existence of good coverings by balls). *There exist a subsequence of $\{(\mathcal{M}_k^n, g_k, O_k)\}$, convex geodesic balls $\tilde{B}_k^\alpha \subset \hat{B}_k^\alpha \subset B_k^\alpha \subset \bar{B}_k^\alpha \subset$*

$\vec{B}_k^\alpha \subset (\mathcal{M}_k, g_k)$, and functions $A(r)$, $K(r)$, $I(n, C_0)$, where C_0 is the curvature bound in Theorem 3.9 and A and K are nondecreasing in r, so that the following hold.

(1) $\tilde{B}_k^\alpha$, $\hat{B}_k^\alpha$, B_k^α, $\bar{B}_k^\alpha$, and $\vec{B}_k^\alpha$ are concentric, i.e., they have the same center, with center denoted as x_k^α such that $x_k^0 = O_k$.

(2) $\tilde{B}_k^\alpha$ and $\tilde{B}_k^\beta$ are disjoint for $\alpha \neq \beta$.

(3) The exponential map $\exp_{x_k^\alpha} \circ L_k^\alpha : \vec{B}^\alpha \to \vec{B}_k^\alpha$ is a diffeomorphism, where $\vec{B}^\alpha$ is a ball in $\mathbb{R}^n$ and $L_k^\alpha : \mathbb{R}^n \to T_{x_k^\alpha} \mathcal{M}_k$ is a linear isometry defined using an orthonormal frame at x_k^α. Moreover, for each α the ball $\vec{B}_k^\alpha$ is geodesically convex for k large enough (depending on α).

(4) We have the containment
$$B(O_k, r) \subset \bigcup_{\alpha \leq A(r)} \hat{B}_k^\alpha$$
if $k \geq K(r)$.

(5) The number of β such that $B_k^\beta \cap B_k^\alpha \neq \emptyset$ is fewer than $I(n, C_0)$.

(6) If $\alpha, \beta < A(r)$, then $B_k^\alpha \cap B_k^\beta$ is either empty for all $k \geq K(r)$ or nonempty for all $k \geq K(r)$.

(7) If $B_k^\alpha \cap B_k^\beta \neq \emptyset$, where $\alpha, \beta \leq A(r)$ and $k \geq K(r)$, then $B_k^\alpha \subset \bar{B}_k^\beta$ and $\bar{B}_k^\alpha \subset \vec{B}_k^\beta$.

The proof of the above lemma will occupy the rest of the section. In order to prove the lemma, we will need the following result on how the injectivity radius can decay in relation to distance. On a complete manifold with bounded curvature the injectivity radius at a point can decay at most exponentially in distance. A strong partial result in this direction was first obtained by Cheng, Li, and Yau [**93**]. Later, Cheeger, Gromov, and Taylor [**75**] obtained the following stronger estimate using different techniques. Recall that $\operatorname{inj}(x)$ denotes the injectivity radius at x.

PROPOSITION 4.19 (Injectivity radius decay estimate). *Let $(\mathcal{M}^n, g)$ be a complete Riemannian manifold with sectional curvatures $|K| \leq C_0$ and injectivity radius $\operatorname{inj}(O) \geq \iota_0 > 0$. Then there exist constants $a = a(n, C_0) > 0$ and $C = C(n, C_0) < \infty$ such that for any $x \in \mathcal{M}$*
$$\operatorname{inj}(x) \geq \mu[d(x, O), \iota_0],$$
where
$$(4.6) \qquad \mu[r, \iota_0] \doteq a \cdot \min\{\iota_0, 1\}^n \cdot e^{-Cr}.$$

REMARK 4.20. The reason for the unnatural looking exponent n in the estimate is that the result is proved by using relative volume comparisons, and the injectivity radius inj bounds the volume comparable to inj^n whereas the volume V bounds the injectivity radius comparable only to V. That is, the exponent n arises from the conversion from injectivity radius to volume and then back again to injectivity radius.

3.2. Choice of ball centers.

In this subsection we will find centers for the balls which will make up the necessary covers. Define $\lambda[r]$ as

$$(4.7) \qquad \lambda[r] \doteqdot \frac{1}{D}\mu[r,\iota_0] = \frac{a}{D} \cdot \min\{\iota_0, 1\}^n \cdot e^{-Cr},$$

where $D = D(n, \iota_0)$ is a large constant, to be chosen later, depending only on n and the uniform lower bound ι_0 for $\operatorname{inj}(O_k)$. Note that λ is a decreasing function of r. We shall choose D large enough so that $\lambda[0] \leq 1$.

We shall work on an individual manifold $(\mathcal{M}^n, g, O)$ and then apply the result to $(\mathcal{M}^n_k, g_k, O_k)$. Choose a sequence of points $\{x^\alpha\}_{\alpha=0}^N$ in $\mathcal{M}$ with $N \in \mathbb{N} \cup \{0, \infty\}$, which we call a **net**, as follows. Let $x^0 = O$ and $r^0 \doteqdot d(x^0, O) = 0$. Let

$$\mathcal{S}^1 \doteqdot \{x \in \mathcal{M} : B(x, \lambda[d(x,O)]) \cap B(x^0, \lambda[r^0]) = \varnothing\}.$$

If $\mathcal{S}^1$ is empty, then $B(x^0, 2\lambda[r^0]) = \mathcal{M}$ and we take $N = 0$ and stop choosing points. Since the balls in the definition of $\mathcal{S}^1$ are open, $\mathcal{S}^1$ is a closed set. Hence, if $\mathcal{S}^1$ is nonempty, then there exists a point $x^1 \in \mathcal{S}^1$ such that $r^1 \doteqdot d(x^1, O) = d(\mathcal{S}^1, O)$. Since $x^1 \in \mathcal{S}^1$, we have

$$B(x^1, \lambda[r^1]) \cap B(x^0, \lambda[r^0]) = \varnothing.$$

Next let

$$\mathcal{S}^2 \doteqdot \left\{x \in \mathcal{M} : B(x, \lambda[d(x,O)]) \cap B\left(x^\beta, \lambda\left[r^\beta\right]\right) = \varnothing \text{ for } \beta = 0, 1\right\}.$$

Again, since $\mathcal{S}^2 \subset \mathcal{S}^1$ is a closed set, if $\mathcal{S}^2$ is nonempty (if $\mathcal{S}^2$ is empty, we take $N = 1$ and stop), we may choose $x^2 \in \mathcal{S}^2$ such that $r^2 \doteqdot d(x^2, O) = d(\mathcal{S}^2, O)$. We have

$$B\left(x^\beta, \lambda\left[r^\beta\right]\right) \cap B(x^\gamma, \lambda[r^\gamma]) = \varnothing \quad \text{for } \beta \neq \gamma \in \{0, 1, 2\}.$$

By induction, assuming that the points $x^0, x^1, x^2, \ldots, x^{\alpha-1}$ have been chosen, let

$$\mathcal{S}^\alpha \doteqdot \left\{x \in \mathcal{M} : B(x, \lambda[d(x,O)]) \cap B\left(x^\beta, \lambda\left[r^\beta\right]\right) = \varnothing, \beta = 0, \ldots, \alpha-1\right\}.$$

Note that $\mathcal{S}^\alpha \subset \mathcal{S}^{\alpha-1}$. If $\mathcal{S}^\alpha$ is nonempty (if $\mathcal{S}^\alpha$ is empty, we take $N = \alpha - 1$ and stop), choose $x^\alpha \in \mathcal{S}^\alpha$ such that $r^\alpha \doteqdot d(x^\alpha, O) = d(\mathcal{S}^\alpha, O)$. Then

$$B\left(x^\beta, \lambda\left[r^\beta\right]\right) \cap B(x^\gamma, \lambda[r^\gamma]) = \varnothing \quad \text{for } \beta \neq \gamma \in \{0, 1, 2, \ldots, \alpha\}.$$

LEMMA 4.21 (Existence of covers with bounds on the number of balls). *Let $(\mathcal{M}^n, g)$ be a complete Riemannian manifold with sectional curvatures $|K| \leq C_0$ and injectivity radius $\operatorname{inj}(O) \geq \iota_0 > 0$. For each $r > 0$ there exists a nondecreasing function $A(r)$ such that the finite collection*

$$\{B(x^\alpha, 2\lambda[r^\alpha]) : 0 \leq \alpha \leq A(r)\}$$

forms a cover for $B(O, r)$ and $r^\alpha > r$ if $\alpha > A(r)$. Furthermore, we can choose $A(r)$ to depend only on n, r, C_0 and ι_0 (in particular, not depend on the manifold $\mathcal{M}$).

3. CONSTRUCTION OF GOOD COVERINGS BY BALLS

PROOF. There are two steps to the argument. The first is that the balls $\{B(x^\alpha, 2\lambda[r^\alpha]) : 0 \leq \alpha \leq A'\}$ cover for some A'. The second is that we can find a bound A for A' which is independent of the manifold.

Let $A'(r) \doteq \max\{\alpha : r^\alpha \leq r\}$. Note that this definition ensures that $r^\alpha > r$ if $\alpha > A'(r)$. We will now show that $\{B(x^\alpha, 2\lambda[r^\alpha]) : 0 \leq \alpha \leq A'(r)\}$ form a cover of $B(O, r)$. Consider $p \in B(O, r)$ and let $s \doteq d(p, O)$. If p is not covered by these balls, then $d(p, x^\alpha) \geq 2\lambda[r^\alpha]$ for all $\alpha \leq A'(r)$. Hence $B(p, \lambda[s])$ is disjoint from the balls $\{B(x^\alpha, \lambda[r^\alpha]) : 0 \leq \alpha \leq A'(s)\}$ since if $q \in B(p, \lambda[s])$, then for $\alpha \leq A'(s) < A'(r)$,

$$d(q, x^\alpha) \geq d(p, x^\alpha) - d(p, q) \geq 2\lambda[r^\alpha] - \lambda[s] \geq \lambda[r^\alpha]$$

since λ is decreasing. This implies that $p \in \mathcal{S}^{A'(s)+1}$. This is a contradiction because $r^{A'(s)+1}$ must be the minimal distance from O to a point in $\mathcal{S}^{A'(s)+1}$, but since $r^{A'(s)+1} > s$, the minimum should have been $s = d(O, p)$.

In order to estimate $A'(r)$, we shall use the curvature bound to get volume estimates. The Rauch Comparison Theorem [72] and our injectivity assumption imply that there is a number $\varepsilon = \varepsilon(n, C_0)$ depending only on the dimension and the upper curvature bound of C_0 such that

$$\operatorname{Vol}(B(x^\alpha, \lambda[r^\alpha])) \geq \varepsilon \lambda[r^\alpha]^n$$

for all $\alpha \leq A'(r)$. By the Bishop Volume Comparison Theorem, there is a number $M = M(n, C_0)$ depending only on the dimension and the lower curvature bound of $-C_0$ such that

$$\operatorname{Vol}(B(O, r)) \leq M \exp\left[(n-1)\sqrt{C_0}\, r\right].$$

Since $\lambda[r] \leq \lambda[r^\alpha]$ for $\alpha \leq A'(r)$, we have

$$\sum_{\alpha=0}^{A'(r)} \operatorname{Vol}(B(x^\alpha, \lambda[r^\alpha])) \geq \left(A'(r) + 1\right) \varepsilon \lambda[r]^n.$$

Since the balls $B(x^\alpha, \lambda[r^\alpha])$ are disjoint and are contained in $B(O, r + \lambda[0])$, we also have that

$$\sum_{\alpha=0}^{A'(r)} \operatorname{Vol}(B(x^\alpha, \lambda[r^\alpha])) \leq \operatorname{Vol}(B(O, r + \lambda[0]))$$

$$\leq M \exp\left[(n-1)\sqrt{C_0}\,(r + \lambda[0])\right].$$

Thus we get the bound

$$\left(A'(r) + 1\right)\varepsilon\lambda[r]^n \leq M \exp\left[(n-1)\sqrt{C_0}\,(r + \lambda[0])\right]$$

or

$$A'(r) \leq \frac{M \exp\left[(n-1)\sqrt{C_0}\,(r + \lambda[0])\right]}{\varepsilon\lambda[r]^n} - 1.$$

We can take $A(r)$ to be the least integer greater than this number. Since r^α is increases as α increases, we see that if $\alpha > A(r)$, then $r^\alpha \geq r$. □

In summary we have constructed a sequence of balls such that the following holds:

PROPOSITION 4.22 (Good cover of a Riemannian manifold). *Let $(\mathcal{M}^n, g)$ be a complete Riemannian manifold with sectional curvatures $|K| \leq C_0$ and injectivity radius $\operatorname{inj}(O) \geq \iota_0 > 0$. Then there exists a net of points $\{x^\alpha\}_{\alpha=0}^N$, where $N \in \mathbb{N} \cup \{\infty\}$, and a nondecreasing function $A(r)$ (which depends only on n, C_0 and ι_0) such that*

(1) $r^\alpha = d(x^\alpha, O)$ *is a nondecreasing function of α,*
(2) *the balls $\{B(x^\alpha, \lambda[r^\alpha])\}_{\alpha=0}^N$ are disjoint, and*
(3) $B(O, r) \subset \bigcup_{\alpha \leq A(r)} B(x^\alpha, 2\lambda[r^\alpha])$.

3.3. Application to the sequence of manifolds. We are now ready to apply our construction of the coverings by balls to the sequence $\mathcal{M}_k^n$. For each $\mathcal{M}_k$ we can construct nets $\{x_k^\alpha\}$ as in Proposition 4.22. Let $r_k^\alpha \doteq d(x_k^\alpha, O_k)$. We next show that the $\lambda[r_k^\alpha]$ are bounded for each α, so we can find a subsequence which converges.

PROPOSITION 4.23 (Bounds on the distance of the centers to the origins). $r_k^\alpha \leq 2\alpha\lambda[0]$ *and* $r_k^\alpha \geq \lambda[0]$ *for all $\alpha \neq 0$.*

PROOF. The worst case scenario is if the construction is a string of balls such that the centers are on a distance-minimizing geodesic. In this case, the distance $r_k^\alpha \leq \lambda[0] + \sum_{\beta=1}^{\alpha-1} 2\lambda\left[r_k^\beta\right] + \lambda[r_k^\alpha] \leq 2\alpha\lambda[0]$. □

COROLLARY 4.24 (Convergence of the distance of the centers to the origins). *There exists a subsequence $\{(\mathcal{M}_k^n, g_k, O_k)\}_{k \in \mathbb{N}}$ and positive numbers $\{r_\infty^\alpha\}_{\alpha \in \mathbb{N}}$ such that, for each α, $r_k^\alpha \to r_\infty^\alpha$ as $k \to \infty$. Hence there is a function $K(\alpha)$ such that if $k \geq K(\alpha)$, then*

$$\frac{1}{2}\lambda[r_\infty^\alpha] \leq \lambda[r_k^\alpha] \leq 2\lambda[r_\infty^\alpha].$$

PROOF. This is a simple diagonalization argument. □

COROLLARY 4.25. *Let $K'(r) \doteq \max\{K(\alpha) : \alpha \leq A(r)\}$. Then if $\alpha \leq A(r)$ and $k \geq K'(r)$, then*

$$\frac{1}{2}\lambda[r_\infty^\alpha] \leq \lambda[r_k^\alpha] \leq 2\lambda[r_\infty^\alpha].$$

PROOF. If $k \geq K'(r)$ and $\alpha \leq A(r)$, then $k \geq K(\alpha)$. □

We will denote $\lambda[r_\infty^\alpha]$ by λ^α. Now define the following collection of balls:

DEFINITION 4.26 (Various size balls). Define

$$\tilde{B}_k^\alpha \doteq B(x_k^\alpha, \lambda^\alpha/2), \qquad \hat{B}_k^\alpha \doteq B(x_k^\alpha, 4\lambda^\alpha),$$
$$B_k^\alpha \doteq B(x_k^\alpha, 5\lambda^\alpha), \qquad \bar{B}_k^\alpha \doteq B(x_k^\alpha, 45e^{10cC}\lambda^\alpha),$$
$$\vec{B}_k^\alpha \doteq B(x_k^\alpha, 205e^{20cC}\lambda^\alpha),$$

where c and C are defined by $\lambda[r] \doteq ce^{-Cr}$ as in (4.7), i.e.,

$$(4.8) \qquad c \doteq \frac{a}{D} \cdot \min\{\iota_0, 1\}^n.$$

The strange radii is to ensure that Lemma 4.18 is true and it will later be clear why these numbers were chosen. We easily see the following properties:

PROPOSITION 4.27 (Disjointness of smaller balls and covering of larger balls). *If $k \geq K'(r)$ and $\alpha \leq A(r)$, then $\tilde{B}_k^\alpha$ are disjoint and $B(O_k, r) \subset \bigcup_{\alpha \leq A(r)} \hat{B}_k^\alpha$.*

PROOF. It follows from Corollary 4.25 that

$$\tilde{B}_k^\alpha \subset B(x_k^\alpha, \lambda[r_k^\alpha]), \quad B(x_k^\alpha, 2\lambda[r_k^\alpha]) \subset \hat{B}_k^\alpha.$$

The proposition then follows immediately from Proposition 4.22. □

PROPOSITION 4.28 (Bound on index of intersecting balls). *For any $\alpha \geq 0$ there exists an integer $I(\alpha, n)$ (independent of k) such that if*

$$B_k^\beta \cap B_k^\alpha \neq \varnothing$$

for some β and k, then $\beta \leq I(\alpha, n)$.

PROOF. This follows easily from Lemma 4.21. If $y \in B_k^\beta \cap B_k^\alpha$, then

$$r_k^\beta \leq d(O_k, x_k^\alpha) + d(x_k^\alpha, y) + d(y, x_k^\beta)$$
$$\leq r_k^\alpha + 5\lambda^\alpha + 5\lambda^\beta$$
$$(4.9) \qquad \leq r_k^\alpha + 10\lambda[0],$$

so by Proposition 4.23,

$$r_k^\beta \leq (2\alpha + 10)\lambda[0]$$

since $\lambda[0] \leq 1$. By Lemma 4.21 we have $\beta \leq A((2\alpha+10)\lambda[0])$. Now just take $I(\alpha, n) \doteq A((2\alpha+10)\lambda[0])$. □

PROPOSITION 4.29 (Stability of the intersections of balls). *There is a subsequence $\{\mathcal{M}_k^n, g_k\}$ such that for every pair (α, β) there is a number $K(\alpha, \beta)$ such that either $B_k^\alpha \cap B_k^\beta$ is empty for all $k \geq K(\alpha, \beta)$ or $B_k^\alpha \cap B_k^\beta$ is nonempty for all $k \geq K(\alpha, \beta)$.*

PROOF. Let $\{(\alpha_\ell, \beta_\ell)\}_{\ell \in \mathbb{N} \cup \{0\}}$ be an ordering of the elements of the countable set $(\mathbb{N} \cup \{0\}) \times (\mathbb{N} \cup \{0\})$. We construct inductively a nesting sequence of subsets $K_\ell \subset \mathbb{N}$. Let K_0 be an infinite subset such that $B_k^{\alpha_0} \cap B_k^{\beta_0}$ intersect always or never for all $k \in K_0$. This can be done since either the balls intersect for infinitely many k, and we take K_0 to be the set of such k, or they do not intersect for infinitely many k', and we take K_0 to be the set of these k'. Now for each subsequent ℓ we can construct an infinite subset K_ℓ of $K_{\ell-1}$ such that $B_k^{\alpha_\ell} \cap B_k^{\beta_\ell}$ intersect always or never for all $k \in K_\ell$. We now consider the collection of $\{K_\ell\}_{\ell \in \mathbb{N} \cup \{0\}}$ as nested subsequences of $\mathbb{N}$ and take a diagonal subsequence. Define a function $K : (\mathbb{N} \cup \{0\}) \times (\mathbb{N} \cup \{0\}) \to \mathbb{N}$

such that $K(\alpha_\ell, \beta_\ell) \in K_\ell$ is a monotone function of ℓ. If $k \geq K(\alpha_\ell, \beta_\ell)$, then $k \in K_\ell$. Hence either $B_k^{\alpha_\ell} \cap B_k^{\beta_\ell}$ is empty for all $k \geq K(\alpha_\ell, \beta_\ell)$ or it is nonempty for all $k \geq K(\alpha_\ell, \beta_\ell)$. □

DEFINITION 4.30. Let $K(r) \doteq \max(K'(r), \{K(\alpha, \beta) : \alpha, \beta \leq A(r)\})$.

By definition, we have that if $\alpha, \beta \leq A(r)$, then $K(\alpha, \beta) \leq K(r)$. Note also that $K(r)$ is increasing. This proves statement (6) of Lemma 4.18.

3.4. Estimates on inclusions of intersecting balls.
Now to complete the proof of Lemma 4.18 (only statements (5) and (7) remain to be proved), we recall Definition 4.26 and prove the following proposition.

PROPOSITION 4.31. If $B_k^\alpha \cap B_k^\beta \neq \varnothing$, where $k \geq K\left(\max\left\{r_k^\alpha, r_k^\beta\right\}\right)$, then we have $B_k^\alpha \subset \bar{B}_k^\beta$ and $\bar{B}_k^\alpha \subset \vec{B}_k^\beta$.

PROOF. Recall estimate (4.9), which gives
$$r_k^\beta \leq r_k^\alpha + 10\lambda[0].$$
Now, we can estimate λ^β by
$$\lambda^\beta \geq \frac{1}{2}\lambda\left[r_k^\beta\right] \geq \frac{1}{2}\lambda[r_k^\alpha + 10\lambda[0]].$$
Recall that the definition of λ is
$$\lambda[r] = ce^{-Cr}.$$
So $\lambda[0] = c$ and $\lambda[r_k^\alpha + 10\lambda[0]] = e^{-10cC}\lambda[r_k^\alpha]$. Hence
$$\lambda^\beta \geq \frac{1}{2}e^{-10cC}\lambda[r_k^\alpha] \geq \frac{1}{4}e^{-10cC}\lambda^\alpha$$
or
$$\lambda^\alpha \leq 4e^{10cC}\lambda^\beta.$$
Choose a $y \in B_k^\alpha \cap B_k^\beta$. Then for each $x \in B_k^\alpha$ we get
$$d\left(x, x_k^\beta\right) \leq d(x, x_k^\alpha) + d(x_k^\alpha, y) + d\left(y, x_k^\beta\right)$$
$$< 5\lambda^\alpha + 5\lambda^\alpha + 5\lambda^\beta$$
$$\leq 20e^{10cC}\lambda^\beta + 20e^{10cC}\lambda^\beta + 5\lambda^\beta$$
$$\leq 45e^{10cC}\lambda^\beta.$$
Similarly, if $x \in \bar{B}_k^\alpha$, then
$$d\left(x, x_k^\beta\right) \leq d(x, x_k^\alpha) + d(x_k^\alpha, y) + d\left(y, x_k^\beta\right)$$
$$< 45e^{10cC}\lambda^\alpha + 5\lambda^\alpha + 5\lambda^\beta$$
$$\leq 45e^{10cC}\left(4e^{10cC}\lambda^\beta\right) + 5\left(4e^{10cC}\lambda^\beta\right) + 5\lambda^\beta$$
$$\leq 205e^{20cC}\lambda^\beta.$$

□

Finally, we want to ensure that the $\vec{B}_k^\alpha$ are embedded geodesic balls and to ensure that the requirements of Proposition 4.53 are satisfied, so we need that
$$205e^{20cC}\lambda^\alpha \leq 410e^{20cC}\lambda\,[r_k^\alpha] \leq \frac{1}{3}\operatorname{inj}(x_k^\alpha),$$
which is ensured if
$$1230e^{20cC}\lambda\,[r_k^\alpha] \leq \mu\,[r_k^\alpha, \iota_0]$$
or by (4.6), equivalently,
$$1230e^{20cC}/D \leq 1, \quad \text{i.e.,} \quad 20c'C \leq D\log(D/1230),$$
where $c' \doteq a \cdot \min\{\iota_0, 1\}^n$ (so that by (4.8) we have $c \doteq c'/D$). Since $D\log D$ goes to infinity as $D \to \infty$, we can choose D large enough to satisfy this inequality and also such that $c = c'/D$ is less than 1. Finally, we can make D large enough so that the balls have radius less than $\pi/\left(6\sqrt{C_0}\right)$, and hence are convex by Corollary 4.47, and so that the balls have radius less than $c_1/\sqrt{C_0}$ as in Proposition 4.32.

To complete the proof of statement (5) of Lemma 4.18, we need to show that the number of β such that $B_k^\beta \cap B_k^\alpha \neq \varnothing$ is fewer than $I(n, C_0)$. For any β such that $B_k^\beta \cap B_k^\alpha \neq \varnothing$, $B_k^\beta \subset \bar{B}_k^\alpha$. As in the proof Lemma 4.21, we can estimate the volume of $\bar{B}_k^\alpha$ from above by a multiple of $(\lambda_\alpha)^n$ and the volume of each B_k^β from below by a multiple of $(\lambda_\alpha)^n$; this will give the bound $I(n, C_0)$.

4. The limit manifold $(\mathcal{M}_\infty^n, g_\infty)$

We can now construct limit overlap maps of balls and limit metrics; we shall use these to take a direct limit and find the limit Riemannian manifold $(\mathcal{M}_\infty^n, g_\infty)$. Let B_k^α, $\bar{B}_k^\alpha$, $\vec{B}_k^\alpha$, $A(r)$, $K(r)$, and $I(n, C_0)$ be as in Lemma 4.18.

4.1. Local metrics on balls, transition functions, and their limits.
Given $r > 0$, consider the ball $B(O_k, r)$. For this section we shall always assume that $k \geq K(r)$. Note that by Lemma 4.18, $B(O_k, r)$ is covered by the finite collection of balls $\hat{B}_k^\alpha$, where $\alpha \leq A(r)$. For each $\alpha \leq A(r)$ we shall construct maps
$$F_{k\ell}^\alpha : \bar{B}_k^\alpha \to \mathcal{M}_\ell$$
using the exponential maps of $\mathcal{M}_k$ and $\mathcal{M}_\ell$. In addition, we wish to average these maps in such a way that the maps limit to the identity, in a sense, as $k, \ell \to \infty$.

We first look at the metrics in normal coordinates in balls and obtain convergence of the metrics locally. Choose linear isometries
$$L_k^\alpha : \mathbb{R}^n \to T_{x_k^\alpha}\mathcal{M}_k.$$
We can then define all of the diffeomorphisms
$$H_k^\alpha : E^\alpha \to B_k^\alpha, \qquad \bar{H}_k^\alpha : \bar{E}^\alpha \to \bar{B}_k^\alpha, \qquad \vec{H}_k^\alpha : \vec{E}^\alpha \to \vec{B}_k^\alpha$$

as restrictions of the map
$$\exp_{x_k^\alpha} \circ L_k^\alpha,$$
where E^α, $\bar{E}^\alpha$, and $\vec{E}^\alpha$ are the appropriately sized Euclidean balls centered at the origin (the three maps are the same, but they are defined on different domains).[1]

We have the following standard result (for the proof see Corollary 4.12 in [**187**]).

PROPOSITION 4.32 (The $\left|\nabla^\ell \operatorname{Rm}\right| \leq C_\ell$ imply $|\partial^m g| \leq \tilde{C}_m$ in normal coordinates). *Let $(\mathcal{M}^n, g)$ be a Riemannian manifold. Let $p \in \mathcal{M}$ and $r_0 \in \left(0, \frac{1}{4} \operatorname{inj}(p)\right)$. Assume that for all $\ell \geq 0$ there are constants $C_\ell < \infty$ such that*
$$\left|\nabla^\ell \operatorname{Rm}\right| \leq C_\ell \quad \text{in } B(p, r_0).$$
Then in the normal coordinates $\{x^i\}$ on $B(p, r_0)$ there are constants $\tilde{C}_\ell$ depending on $n, \operatorname{inj}(p), C_0, \ldots, C_\ell$ and a constant c_1 depending only on n such that for any multi-index α with $|\alpha| \geq 1$
$$\frac{1}{2}(\delta_{ij}) \leq (g_{ij}) \leq 2(\delta_{ij}) \quad \text{and} \quad \left|\frac{\partial^\alpha g_{ij}}{\partial x^\alpha}\right| \leq \tilde{C}_{|\alpha|}$$
in $B\left(p, \min\left\{c_1/\sqrt{C_0}, r_0\right\}\right)$.

The $\left\{\left(B_k^\alpha, (H_k^\alpha)^{-1}\right)\right\}_{\alpha \leq A(r)}$ form coordinate charts covering $B(O_k, r) \subset \mathcal{M}_k$. Since
$$\vec{g}_k^\beta \doteq \left(\vec{H}_k^\beta\right)^* g_k$$
are Riemannian metrics on $\vec{E}^\beta$ in normal coordinates with uniformly bounded curvatures, by Proposition 4.32, all partial derivatives of the metrics are uniformly bounded. Hence by the Arzela–Ascoli theorem there is a subsequence so that the $\vec{g}_k^\beta$ converge uniformly in C^∞ on compact sets to a **limit Riemannian metric $\vec{g}_\infty^\beta$ defined locally on $\vec{E}^\beta$**. We use our convention that the subsequence is still indexed by k.

We have **transition maps** on $\mathcal{M}_k$ defined as follows. Recall that for $\alpha, \beta \leq A(r)$, if $B_k^\alpha \cap B_k^\beta \neq \varnothing$ for some k, then it is true for all $k \geq K(r)$ by Lemma 4.18. If $B_k^\alpha \cap B_k^\beta \neq \varnothing$, then it makes sense to define the maps
$$J_k^{\alpha\beta} : E^\alpha \to \bar{E}^\beta, \qquad \bar{J}_k^{\alpha\beta} : \bar{E}^\alpha \to \vec{E}^\beta$$
by
$$J_k^{\alpha\beta} \doteq \left(\bar{H}_k^\beta\right)^{-1} \circ H_k^\alpha, \qquad \bar{J}_k^{\alpha\beta} \doteq \left(\vec{H}_k^\beta\right)^{-1} \circ \bar{H}_k^\alpha.$$
The maps $J_k^{\alpha\beta}$ and $\bar{J}_k^{\alpha\beta}$ are embeddings; since for all γ, $E^\gamma \subset \bar{E}^\gamma \subset \vec{E}^\gamma$, $J_k^{\alpha\beta}$ is a restriction of $\bar{J}_k^{\alpha\beta}$.

---

[1]The balls B_k^α, $\bar{B}_k^\alpha$, and $\vec{B}_k^\alpha$ are given by Definition 4.26; consequently the radii of E^α, $\bar{E}^\alpha$, and $\vec{E}^\alpha$ are equal to $5\lambda^\alpha$, $45e^{10cC}\lambda^\alpha$, and $205e^{20cC}\lambda^\alpha$, respectively.

To obtain the local convergence of $J_k^{\alpha\beta}$ as $k \to \infty$, we take some further subsequences. Since the $\bar{J}_k^{\alpha\beta}$ are Riemannian isometries between $\bar{g}_k^\alpha \doteqdot (\bar{H}_k^\alpha)^* g_k$ and $\bar{g}_k^\beta$ for each k and since the derivatives of the metrics are bounded, we have that for each pair $\alpha, \beta \leq A(r)$ such that $B_k^\alpha \cap B_k^\beta \neq \varnothing$ for all $k \geq K(r)$, there exists a subsequence such that the $\bar{J}_k^{\alpha\beta}$ converge to a **limit transition map**
$$\bar{J}_\infty^{\alpha\beta} : \bar{E}^\alpha \to \bar{E}^\beta$$
in C^∞ uniformly on compact sets (by Corollary 4.11). In fact, $\bar{J}_\infty^{\alpha\beta}$ is a Riemannian isometry between $\bar{g}_\infty^\alpha$ and $\bar{g}_\infty^\beta$. Since $J_k^{\alpha\beta}$ is a restriction of $\bar{J}_k^{\alpha\beta}$, we also have that $J_k^{\alpha\beta}$ converges to a map $J_\infty^{\alpha\beta}$. We then diagonalize the sequences so that $\bar{J}_k^{\alpha\beta}$ converges for every α and β. Notice that since $\bar{J}_k^{\beta\alpha} \circ J_k^{\alpha\beta} = \mathrm{id}_\beta : E^\beta \to \vec{E}^\beta$, the identity embedding, we must have that
$$\bar{J}_\infty^{\beta\alpha} \circ J_\infty^{\alpha\beta} = \mathrm{id}_\beta.$$

4.2. Constructing approximate isometries $F_{k\ell;r}$ of large balls in $\mathcal{M}_k$ into $\mathcal{M}_\ell$. We can now construct approximate isometries $F_{k\ell;r}$ between the ball $B(O_k, r) \subset \mathcal{M}_k$ and an open set in $\mathcal{M}_\ell$ for sufficiently large k and ℓ. (We shall use some results about the center of mass given in Section 5 of this chapter.) The following is the main result of this subsection.

PROPOSITION 4.33 (Existence of an approximate isometry on a large ball). *For every $r > 0$, $\varepsilon > 0$, and $p > 0$ there exists $k_0 = k_0(r, \varepsilon, p) > 0$ such that for $k, \ell > k_0$ there is a diffeomorphism*
$$F_{k\ell;r} : B(O_k, r) \to F_{k\ell;r}(B(O_k, r)) \subset \mathcal{M}_\ell$$
which is an (ε, p)-approximate isometry.

Let $F_{k\ell}^\alpha : B_k^\alpha \to B_\ell^\alpha$ (from a ball in $\mathcal{M}_k$ to a ball in $\mathcal{M}_\ell$) be defined by
$$F_{k\ell}^\alpha \doteqdot H_\ell^\alpha \circ (H_k^\alpha)^{-1}.$$
Roughly speaking, we construct the desired map $F_{k\ell;r}$ by averaging the local maps $F_{k\ell}^\alpha$ for $\alpha \leq A(r)$. Notice that in terms of the (inverse) coordinate systems $\left(E^\beta, H_k^\beta\right)$ and $\left(\vec{E}^\beta, \vec{H}_\ell^\beta\right)$, where β satisfies $B_k^\beta \cap B_k^\alpha \neq \varnothing$, the map $F_{k\ell}^\alpha$ corresponds to the map
$$F_{k\ell,\beta}^\alpha : E^\beta \to \vec{E}^\beta$$
between Euclidean balls defined by
$$F_{k\ell,\beta}^\alpha \doteqdot \left(\vec{H}_\ell^\beta\right)^{-1} \circ F_{k\ell}^\alpha \circ H_k^\beta$$
$$= \left(\vec{H}_\ell^\beta\right)^{-1} \circ \bar{H}_\ell^\alpha \circ \left(\bar{H}_k^\alpha\right)^{-1} \circ H_k^\beta$$
$$\tag{4.10} = \bar{J}_\ell^{\alpha\beta} \circ J_k^{\beta\alpha}.$$

Hence we have the following local property which is a key step to Proposition 4.33.

PROPOSITION 4.34 (Local maps converge to the identity in a sense). *If α and β are such that $B_k^\alpha \cap B_k^\beta \neq \varnothing$ for k sufficiently large, then the maps $F_{k\ell,\beta}^\alpha : E^\beta \to \vec{E}^\beta$ converge to the identity (inclusion) map id_β as $k, \ell \to \infty$.*

Now we proceed to average the local maps to construct a map on a large ball. To apply Proposition 4.53 on averaging maps, we need to construct a partition of unity subordinate to the covering $\{B_k^\alpha\}_{\alpha \leq A(r)}$ of $B(O_k, r)$ as follows. For $\alpha \leq A(r)$ let ψ^α be a smooth function which is 1 on $\hat{E}^\alpha \subset E^\alpha$ and 0 outside of E^α ($\hat{E}^\alpha = (H_k^\alpha)^{-1} \hat{B}_k^\alpha$). We construct a partition of unity on $B(O_k, r)$ by letting

$$\varphi_k^\alpha(x) \doteq \begin{cases} \dfrac{\psi^\alpha \circ (H_k^\alpha)^{-1}(x)}{\sum_{\gamma \leq A(r)} \psi^\gamma \circ (H_k^\gamma)^{-1}(x)} & \text{if } x \in B_k^\alpha, \\ 0 & \text{if } x \notin B_k^\alpha, \end{cases}$$

where $\alpha \leq A(r)$. By Lemma 4.18(4) the denominator is no less than 1 and by Lemma 4.18(5) the number of terms in the denominator which are well defined is bounded by $I(n, C_0)$, independent of k. To ensure that the basepoint is preserved, we need the partition of unity function (ϕ_k^α below) indexed by $\alpha \neq 0$ to vanish in a neighborhood of O_k, so we introduce a C^∞ function $\chi : E^0 \to \mathbb{R}$ such that $\chi = 0$ in a neighborhood of the origin and $\chi = 1$ outside $(H_k^0)^{-1} \tilde{B}_k^0$ (note that this set is a Euclidean ball independent of k). Define $\chi_k : \mathcal{M}_k \to \mathbb{R}$ by $\chi_k(x) \doteq \chi \circ (H_k^0)^{-1}(x)$ if $x \in B_k^0$ and $\chi_k(x) \doteq 1$ otherwise. Then we take

$$\phi_k^\alpha : \mathcal{M}_k \to \mathbb{R}$$

to be, for $0 < \alpha \leq A(r)$,

$$\phi_k^\alpha(x) \doteq \begin{cases} \dfrac{\chi_k(x) \cdot \psi^\alpha \circ (H_k^\alpha)^{-1}(x)}{\psi^0 \circ (H_k^0)^{-1}(x) + \sum_{0 < \gamma \leq A(r)} \chi_k(x) \cdot \psi^\gamma \circ (H_k^\gamma)^{-1}(x)} & \text{if } x \in B_k^\alpha, \\ 0 & \text{if } x \notin B_k^\alpha, \end{cases}$$

while for $\alpha = 0$,

$$\phi_k^0(x) \doteq \begin{cases} \dfrac{\psi^0 \circ (H_k^0)^{-1}(x)}{\psi^0 \circ (H_k^0)^{-1}(x) + \sum_{0 < \gamma \leq A(r)} \chi_k(x) \cdot \psi^\gamma \circ (H_k^\gamma)^{-1}(x)} & \text{if } x \in B_k^0, \\ 0 & \text{if } x \notin B_k^0. \end{cases}$$

The collection of functions $\{\phi_k^\alpha\}_{\alpha \leq A(r)}$ is a partition of unity subordinate to the covering $\{B_k^\alpha\}_{\alpha \leq A(r)}$ of $B(O_k, r)$.

Notice that with respect to the coordinates $\left(E^\beta, H_k^\beta\right)$ the map $\phi_k^\alpha(x)$, where $\alpha \neq 0$, can be expressed as

$$\phi_{k,\beta}^\alpha \doteqdot \phi_k^\alpha \circ H_k^\beta = \frac{\chi \circ J_k^{\beta 0} \cdot \psi^\alpha \circ J_k^{\beta \alpha}}{\psi^0 \circ J_k^{\beta 0} + \sum_{0 < \gamma \leq A(r)} \chi \circ J_k^{\beta 0} \cdot \psi^\gamma \circ J_k^{\beta \gamma}},$$

so with respect to the coordinates $\left(E^\beta, H_k^\beta\right)$, we have that $\phi_{k,\beta}^\alpha$ converges to a function $\phi_{\infty,\beta}^\alpha$ defined by

$$\phi_{\infty,\beta}^\alpha \doteqdot \frac{\chi \circ J_\infty^{\beta 0} \cdot \psi^\alpha \circ J_\infty^{\beta \alpha}}{\psi^0 \circ J_\infty^{\beta 0} + \sum_{0 < \gamma \leq A(r)} \chi \circ J_\infty^{\beta 0} \cdot \psi^\gamma \circ J_\infty^{\beta \gamma}}.$$

When $\alpha = 0$, we have

$$\phi_{k,\beta}^0 \doteqdot \phi_k^0 \circ H_k^\beta = \frac{\psi^0 \circ J_k^{\beta 0}}{\psi^0 \circ J_k^{\beta 0} + \sum_{0 < \gamma \leq A(r)} \chi \circ J_k^{\beta 0} \cdot \psi^\gamma \circ J_k^{\beta \gamma}},$$

$$\phi_{k,\beta}^0 \to \phi_{\infty,\beta}^0 \doteqdot \frac{\psi^0 \circ J_\infty^{\beta 0}}{\psi^0 \circ J_\infty^{\beta 0} + \sum_{0 < \gamma \leq A(r)} \chi \circ J_\infty^{\beta 0} \cdot \psi^\gamma \circ J_\infty^{\beta \gamma}}.$$

The definition of $F_{k\ell;r}$ is as follows. For $x \in B(O_k, r)$, we define

(4.11) $$F_{k\ell;r}(x) \doteqdot \mathrm{cm}\left\{F_{k\ell}^0(x), F_{k\ell}^1(x), \ldots, F_{k\ell}^{A(r)}(x)\right\} \in \mathcal{M}_\ell$$

to be the center of mass using the weights $\phi_k^\alpha(x)$; by the choice of balls in Lemma 4.18 we can apply Proposition 4.53 and conclude the existence of $F_{k\ell;r}$ when k, ℓ are large enough. The map $F_{k\ell;r}$ is smooth with all its derivatives $|\nabla^p F_{k\ell;r}|$ bounded by constants $\tilde{C}_{p+1}$ independent of k. From the construction of the weights ϕ_k^α we have $F_{k\ell;r}(O_k) = O_\ell$. Note by Proposition 4.53 and the definition of the center of mass, $F_{k\ell;r}$ satisfies

(i) $F_{k\ell;r}(x)$ is the minimizer $y \in \mathcal{M}_\ell$ of

$$f_x(y) = \sum_{\alpha=0}^{A(r)} \phi_k^\alpha(x) d_{g_\ell}^2(y, F_{k\ell}^\alpha(x)),$$

(ii) $F_{k\ell;r}(x)$ is the solution $y \in \mathcal{M}_\ell$ of

(4.12) $$\sum_{\alpha=0}^{A(r)} \phi_k^\alpha(x) \exp_y^{-1} F_{k\ell}^\alpha(x) = 0,$$

where the exponential map is with respect to g_ℓ. With respect to the coordinates $\left(E^\beta, H_k^\beta\right)$, equation (4.12) can be written in E^β as

$$\sum_{\alpha \leq A(r)} \phi_k^\alpha \circ H_k^\beta(X) \exp_{F_{k\ell;r} \circ H_k^\beta(X)}^{-1} F_{k\ell}^\alpha \circ H_k^\beta(X) = 0.$$

Define the local versions of $F_{k\ell;r}$ by

$$G^{\beta}_{k\ell;r} \doteq \left(\vec{H}^{\beta}_{\ell}\right)^{-1} \circ F_{k\ell;r} \circ H^{\beta}_{k}.$$

We may pull back to $\vec{E}^{\beta}$ via the map $\vec{H}^{\beta}_{\ell}$ to get

$$\sum_{\alpha \leq A(r)} \phi^{\alpha}_{k} \circ H^{\beta}_{k}(X) \exp^{-1}_{G^{\beta}_{k\ell;r}(X)} F^{\alpha}_{k\ell,\beta}(X) = 0,$$

where now the map exp is with respect to the metric $g^{\beta}_{\ell} = \left(\vec{H}^{\beta}_{\ell}\right)^{*} g_{\ell}$.

To see the limiting behavior of $F_{k\ell;r}$ when k, ℓ are large, we note that since Proposition 4.34 implies for each β that we have

$$F^{\alpha}_{k\ell,\beta} \to \mathrm{id}_{\beta} \quad \text{when } k, \ell \to \infty,$$

then by Proposition 4.54, we have $G^{\beta}_{k\ell;r} \to \mathrm{id}_{\beta}$ on any compact subset of E^{β} in C^{∞}. We have proved the following.

PROPOSITION 4.35 (The maps $F_{k\ell;r}$ converge to id in a sense). *For every $r > 0$, $\varepsilon > 0$, and $p \in \mathbb{N} \cup \{0\}$ there exists $k_0 = k_0(r, \varepsilon, p)$ such that for $\beta \leq A(r)$,*

$$\left|\nabla^{p}\left(G^{\beta}_{k\ell;r} - \mathrm{id}_{\beta}\right)\right| \leq \varepsilon$$

for all $k, \ell \geq k_0$, where ∇ and $|\cdot|$ are the covariant derivative and norm with respect to the Euclidean metric on E^{β}.

As a corollary we have the following.

COROLLARY 4.36 ($F_{k\ell;r}$ is a local diffeomorphism). *There exists $k_0 = k_0(r)$ such that if $k, \ell \geq k_0$, then $F_{k\ell;r}|_{B^{\beta}_{k}}$ is a diffeomorphism for each $\beta \leq A(r)$.*

PROOF. If $G^{\beta}_{k\ell;r}$ is sufficiently close to the identity map, then it must be injective since its derivative is nonsingular. □

Now we turn to proving that given (ε, p) and r, for k, ℓ large enough, $F_{k\ell,r}$ is an (ε, p)-pre-approximate isometry. First we have the following general result.

LEMMA 4.37 (Limit of almost-identity pullbacks). *Let $\phi_k : U \to U \subset \mathbb{R}^n$ be diffeomorphisms, let $\mathrm{id} : U \to U$ be the identity map, and let $\{h_k\}_{k \in \mathbb{N}}$ and h_{∞} be Riemannian metrics on U. Suppose h_k and h_{∞} are uniformly equivalent to the Euclidean metric for all $k \in \mathbb{N}$ and their derivatives (covariant derivatives with respect to the Euclidean metric) are uniformly bounded. If $\phi_k \to \mathrm{id}$ and $h_k \to h_{\infty}$ in C^{∞} uniformly on compact sets, then for every $\varepsilon > 0$, $p \in \mathbb{N}$, and compact set $K \subset U$, there exists $k_0 = k_0(\varepsilon, p, K)$ such that if $k \geq k_0$, then*

$$\sup_{0 \leq r \leq p} \sup_{x \in K} |\nabla^{r}(\phi^{*}_{k} h_k - h_{\infty})|(x) \leq \varepsilon,$$

4. THE LIMIT MANIFOLD $(\mathcal{M}_\infty^n, g_\infty)$

where $|\cdot|$ is the Euclidean norm and ∇ is the Euclidean covariant derivative (i.e., partial derivative).

PROOF. Let $x = \{x^i\}$ be the standard Euclidean coordinates on U and let $\phi_k(x) = (\phi_k^a(x))_{a=1}^n$. Since $\phi_k \to \text{id}$ in C^∞ uniformly compact sets, we have that $\frac{\partial \phi_k^a}{\partial x^i} \to \delta_i^a$ and $\frac{\partial^\alpha \phi_k^a}{(\partial x)^\alpha} \to 0$ uniformly on K where α is multi-index with $|\alpha| \geq 2$. Since $h_k \to h_\infty$ in C^∞ uniformly compact sets, we have that $\frac{\partial^\alpha (h_k(x))_{ab}}{(\partial x)^\alpha} \to \frac{\partial^\alpha (h_\infty(x))_{ab}}{(\partial x)^\alpha}$ uniformly on K for any α. Now

$$\phi_k^* h_k - h_\infty = \left((h_k)_{ab} \frac{\partial \phi_k^a}{\partial x^i} \frac{\partial \phi_k^b}{\partial x^j} - (h_\infty)_{ij} \right) \to 0$$

uniformly on K. So the lemma holds for $p = 0$.

For any $r > 0$,

$$\nabla^r (\phi_k^* h_k - h_\infty) = \left(\frac{\partial^\alpha}{(\partial x)^\alpha} \left((h_k)_{ab} \frac{\partial \phi_k^a}{\partial x^i} \frac{\partial \phi_k^b}{\partial x^j} - (h_\infty)_{ij} \right) \right)_{|\alpha| \leq r}$$

$$= \left(\frac{\partial^\alpha (h_k)_{ab}}{(\partial x)^\alpha} \cdot \frac{\partial \phi_k^a}{\partial x^i} \frac{\partial \phi_k^b}{\partial x^j} - \frac{\partial^\alpha (h_\infty)_{ij}}{(\partial x^\ell)^\alpha} + \Theta_{\alpha,i,j} \right),$$

where $\Theta_{\alpha,i,j}$ is a sum of terms of the form

$$\frac{\partial^{\alpha_1} (h_k)_{ab}}{(\partial x)^{\alpha_1}} \cdot \frac{\partial^{\alpha_2} \phi_k^a}{(\partial x)^{\alpha_2} \partial x^i} \cdot \frac{\partial^{\alpha_3} \phi_k^b}{(\partial x)^{\alpha_3} \partial x^j}$$

with $|\alpha_1| + |\alpha_2| + |\alpha_3| = r$ and $|\alpha_2| \geq 1$. Hence $\nabla^r (\phi_k^* h_k - g_\infty) \to 0$ uniformly on K. The lemma is proved. □

With this we are ready to prove the following.

LEMMA 4.38 ($F_{k\ell,r}$ is an (ε, p)-pre-approximate isometry). *For any $\varepsilon > 0$ and $p > 0$ there exists $\kappa = \kappa(\varepsilon, p)$ such that*

$$\left| \nabla_{g_k}^q (F_{k\ell;r}^* g_\ell - g_k) \right|_{g_k} \leq \varepsilon$$

for all $q \leq p$ if $k, \ell \geq \kappa(\varepsilon, p)$. Hence $F_{k\ell,r}$ is an (ε, p)-pre-approximate isometry.

PROOF. We work in a coordinate chart $\left(E^\beta, H_k^\beta \right)$. By Proposition 4.35, for any $\varepsilon > 0$ there exists $k_0 = k_0(r, \varepsilon)$ such that $\left| \nabla^q \left(G_{k\ell;r}^\beta - \text{id}_\beta \right) \right| < \varepsilon$ if $k, \ell \geq k_0$.

By Proposition 4.32, the metrics $g_k^\beta = \left(H_k^\beta \right)^* g_k$ are uniformly equivalent in the C^∞-norm to the Euclidean metric on E^β. Thus it suffices to estimate the partial derivatives of $F_{k\ell;r}^* g_\ell - g_k$ using the Euclidean metric. Since $G_{k\ell;r}^\beta \to \text{id}_\beta$ and $g_\ell^\beta \to g_\infty^\beta$, we may use Lemma 4.37 to conclude that $\left(G_{k\ell;r}^\beta \right)^* g_\ell^\beta \to g_\infty^\beta$ in the C^∞-Euclidean norm as $k, \ell \to \infty$. The desired estimates now follow from the fact that $g_k^\beta \to g_\infty^\beta$ in the C^∞-Euclidean norm as $k \to \infty$. □

Next we turn to prove that $F_{k\ell;r}$ is a diffeomorphism. Since $F_{\ell k}^\alpha$ is the inverse of $F_{k\ell}^\alpha$, by Proposition 4.34, for each β we have $F_{k\ell,\beta}^\alpha \to \mathrm{id}_\beta$ and $F_{\ell k,\beta}^\alpha \to \mathrm{id}_\beta$ when $k,\ell \to \infty$. Then by a simple argument using Proposition 4.54 we conclude that $F_{\ell k,r} \circ F_{k\ell,r}$ and $F_{k\ell,r} \circ F_{\ell k,r}$ both approach the identity map when $k,\ell \to \infty$. It follows that $F_{k\ell,r}$ is invertible.

Now it follows from the inverse function theorem and Proposition 4.35 that $F_{k\ell;r}^{-1}$ is an (ε', p)-pre-approximate isometry. Hence, given (ε, p) and r, there is a k_0 such that $F_{k\ell;r}$ is an (ε, p)-approximate isometry for $k, \ell \geq k_0$. Proposition 4.33 is proved.

4.3. The directed system. We are now in a position to construct a directed system whose direct limit will give us the limit manifold $(\mathcal{M}_\infty^n, g_\infty)$. We first show that, after passing to a subsequence, the existence of approximate isometries whose compositions are also approximate isometries, as close as we like to isometries.

PROPOSITION 4.39 (Metrics are almost isometric on large balls). *There exists a subsequence $\left\{\left(\mathcal{M}_{k_j}^n, g_{k_j}\right)\right\}_{j \in \mathbb{N}}$ such that for any $\varepsilon > 0$ and $p \in \mathbb{N}$ there exists $j_0 = j_0(\varepsilon, p) \in \mathbb{N}$ such that if $j > j_0$, then there exist maps*

$$\Psi_j : B\left(O_{k_j}, 2^j\right) \to B\left(O_{k_{j+1}}, 2^{j+1}\right)$$

with

$$\Psi_j\left(O_{k_j}\right) = O_{k_{j+1}}$$

such that for any $\ell \in \mathbb{N}$ the composition map

$$\Psi_{j,\ell} \doteqdot \Psi_{j+\ell-1} \circ \cdots \circ \Psi_{j+1} \circ \Psi_j : \left(B\left(O_{k_j}, 2^j\right), g_{k_j}\right) \to \left(B\left(O_{k_{j+\ell}}, 2^{j+\ell}\right), g_{k_{j+\ell}}\right)$$

is an (into) (ε, p)-approximate isometry.

PROOF. We may assume that C_j is increasing as j increases and that $C_0 \geq 1$. We shall inductively define the subsequence $\left\{\left(\mathcal{M}_{k_j}^n, g_{k_j}\right)\right\}_{j \in \mathbb{N}}$. It is sufficient to construct a sequence $\{\Psi_j\}_{j \in \mathbb{N}}$ such that Ψ_j is a $\left(C_j^{-1} 2^{-j}, j\right)$-approximate isometry. In this case we can use Corollary 4.8 to see that $\Psi_{r,\ell}$ is a $\left(C_r \sum_{i=r}^{r+\ell-1} C_i^{-1} 2^{-i}, r\right)$-approximate isometry. In fact, since C_j is increasing in j, we have

$$C_r \sum_{i=r}^{r+\ell-1} C_i^{-1} 2^{-i} \leq C_r \sum_{i=r}^{\infty} C_i^{-1} 2^{-i} \leq \sum_{i=r}^{\infty} 2^{-i} = 2^{1-r},$$

which implies that $\Psi_{r,\ell}$ is a $\left(2^{1-r}, r\right)$-approximate isometry. We also have by Proposition 4.4 that $\Psi_0\left(B\left(O_{k_0}, 1\right)\right) \subset B\left(O_{k_1}, \left(1 + C_0^{-1}\right)^{1/2}\right) \subset B\left(O_{k_1}, 2\right)$ since

$$1 + C_0^{-1} < 4.$$

Similarly, we have

$$\Psi_r\left(B\left(O_{k_r}, 2^r\right)\right) \subset B\left(O_{r+1}, \left(1 + C_r^{-1} 2^{-r}\right)^{1/2} 2^r\right) \subset B\left(O_{k_{r+1}}, 2^{r+1}\right)$$

again since $C_r \geq 1$. Hence, given $\varepsilon > 0$ and $p > 0$, we can take $j_0 \doteq \max(1 - \log_2 \varepsilon, p)$.

For $j = 0$, make k_0 large enough so that $F_{k_0 \ell; 1}$ is a $\left(C_0^{-1}, 0\right)$-approximate isometry for any $\ell \geq k_0$ (we can do this by Proposition 4.33). By induction, suppose we can do this up to k_r. We then make k_{r+1} large enough so that $F_{k_{r+1}\ell; 2^{r+1}}$ is a $\left(C_{r+1}^{-1} 2^{-(r+1)}, r+1\right)$-approximate isometry for any $\ell \geq k_{r+1}$. Now choose $\Psi_r \doteq F_{k_r k_{r+1}; 2^r}$. $\square$

Let us re-index the subsequence taken in the previous proposition so that it is once again indexed by k and the index of Ψ coincides with the index of $\mathcal{M}$. We may now take the final subsequence to get metrics on $B\left(O_k, 2^k\right) \subset \mathcal{M}_k$ which will become the limit metric. Since $\Psi_{j,\ell}$ are approximate isometries, we may consider the sequence $\left\{\Psi_{j,\ell}^* g_{j+\ell}\right\}_{\ell=0}^\infty$ of Riemannian metrics on $B\left(O_j, 2^j\right)$. Since $\Psi_{j,\ell}$ are (ε, p)-approximate isometries independent of ℓ, the metrics $\Psi_{j,\ell}^* g_{j+\ell}$ are uniformly bounded together with its derivatives and so there is a subsequence in ℓ so that they converge to a limit metric $g_{j,\infty}$ on $B\left(O_j, 2^j\right)$. We can use this argument diagonally to get the following proposition.

PROPOSITION 4.40 (Existence of almost-isometric limiting metrics on large balls). *There exist a subsequence $\{k_j\}_{j=1}^\infty$ and Riemannian metrics $g_{k_j, \infty}$ on $B\left(O_{k_j}, 2^{k_j}\right)$ such that for every $\varepsilon > 0$ and $p \geq 0$ there exists $j_0 = j_0(\varepsilon, p)$ such that*

$$\left|\nabla_{g_{k_j,\infty}}^r \left(\Psi_{k_j}^* \Psi_{k_j+1}^* \Psi_{k_j+2}^* \cdots \Psi_{k_j+\ell-1}^* g_{k_j+\ell} - g_{k_j,\infty}\right)\right|_{g_{k_j,\infty}} \leq \varepsilon$$

for all $r \leq p$ and $\ell \geq 0$ if $j \geq j_0$.

PROOF. This essentially follows from Lemma 4.37 again. $\square$

We again re-index, replacing Ψ_{k_j} with Ψ_j, which equals $\Psi_{k_{j+1}} \circ \cdots \circ \Psi_{k_j+2} \circ \Psi_{k_j+1} \circ \Psi_{k_j}$ in the old notation, so that we have a sequence of maps $\Psi_j : B\left(O_j, 2^j\right) \to B\left(O_{j+1}, 2^{j+1}\right)$ (note that if j corresponds to k_j, then we have shrunk the ball of radius 2^{k_j} to the ball of radius 2^j).

We note that $\Psi_j : \left(B\left(O_j, 2^j\right), g_{j,\infty}\right) \to \left(B\left(O_{j+1}, 2^{j+1}\right), g_{j+1,\infty}\right)$ is an isometry since

$$\left|\Psi_j^* g_{j+1,\infty} - g_{j,\infty}\right| \leq \left|\Psi_j^* \left(g_{j+1,\infty} - \Psi_{j+1}^* \cdots \Psi_{j+\ell-1}^* g_{j+\ell}\right)\right|$$
$$+ \left|\Psi_j^* \Psi_{j+1}^* \cdots \Psi_{j+\ell-1}^* g_{j+\ell} - g_{j,\infty}\right|$$

and both terms on the RHS go to zero as $\ell \to \infty$.

4.4. Construction of the limit.

We are now ready to construct the limit manifold $(\mathcal{M}_\infty^n, g_\infty)$. Topologically, we take the direct limit

$$\mathcal{M}_\infty^n \doteqdot \varinjlim B\left(O_k, 2^k\right),$$

where the directed system comes from the maps Ψ_k. Note that since Ψ_k are approximate isometries, they must be open embeddings. Hence $\mathcal{M}_\infty$ is a Hausdorff space by Lemma 4.17.

We recall the embeddings $I_k : B\left(O_k, 2^k\right) \to \mathcal{M}_\infty$ defined in (4.5). The coordinate maps $H_k^\alpha : E^\alpha \to B\left(O_k, 2^k\right)$ induce coordinate maps $H_{\infty,k}^\alpha \doteqdot I_k \circ H_k^\alpha : E^\alpha \to \mathcal{M}_\infty$. Note that the transition maps

$$\left(H_{\infty,k}^\beta\right)^{-1} \circ H_{\infty,k+r}^\alpha = \left(I_k \circ H_k^\beta\right)^{-1} \circ I_{k+r} \circ H_{k+r}^\alpha$$

(4.13)
$$= \left(H_\ell^\beta\right)^{-1} \circ \Psi_{k,r} \circ H_k^\alpha$$

are C^∞ diffeomorphisms (when the domain and range are suitably restricted) and hence they induce a C^∞ structure on $\mathcal{M}_\infty$. Furthermore, since $\Psi_{k,r}$ are isometries between $g_{k,\infty}$ and $g_{k+r,\infty}$, we easily see that the transition maps are isometries and there exists a **metric g_∞ on $\mathcal{M}_\infty$** such that

$$I_k^* g_\infty = g_{k,\infty}.$$

We now show that $\{(\mathcal{M}_k^n, g_k, O_k)\}$ **converges to** $(\mathcal{M}_\infty^n, g_\infty, O_\infty)$. Given a compact set $K \subset \mathcal{M}_\infty$, it must be contained in $I_k \left[B\left(O_k, 2^k\right)\right]$ for some $k > 0$ and hence must also be contained in $I_\ell \left[B\left(O_\ell, 2^\ell\right)\right]$ for all $\ell \geq k$. We now claim that for every p there exists $k_0 = k_0(K, p)$ such that for any $\varepsilon > 0$

$$\sup_{x \in K} \left| \nabla^\alpha \left(g_\infty - \left(I_k^{-1}\right)^* g_k \right) \right|_{g_\infty} < \varepsilon$$

for all $\alpha \leq p$, $k \geq k_0$. This follows by pulling this expression back by I_k to get

$$\left| \nabla_{g_\infty}^\alpha \left(g_\infty - \left(I_k^{-1}\right)^* g_k \right)\right|_{g_\infty} = \left| I_k^* \left[\nabla_{g_\infty}^\alpha \left(g_\infty - \left(I_k^{-1}\right)^* g_k \right)\right]\right|_{I_k^* g_\infty}$$

$$= \left| \nabla_{g_{k,\infty}}^\alpha \left(g_{k,\infty} - g_k \right)\right|_{g_{k,\infty}}$$

and by using Proposition 4.40. Hence the maps I_k^{-1} satisfy the requirements of Definition 3.5 and we have shown the following.

PROPOSITION 4.41 (Convergence to a limit). $\{(\mathcal{M}_k^n, g_k, O_k)\}$ *converges to* $(\mathcal{M}_\infty^n, g_\infty, O_\infty)$.

Furthermore, we can show that the limit metric is complete.

PROPOSITION 4.42 (The limit is complete). *The metric g_∞ is complete.*

PROOF. Any closed geodesic ball $\bar{B} \subset \mathcal{M}_\infty$ is contained in the image $I_k \left[B\left(O_k, 2^k\right)\right]$ for some $k \in \mathbb{N}$. Recall that I_k is an open embedding, which implies $I_k^{-1}(\bar{B})$ is closed and bounded, and hence compact since g_k is complete. Therefore $\bar{B}$ is the image of a compact set, and hence compact. We are done because if closed metric balls are compact, then the metric is complete (see, for instance, [**72**]). □

5. Center of mass and nonlinear averages

In this section we review some standard work on the convexity of the distance function and properties of the center of mass. The treatment here follows mostly that in Buser and Karcher [**40**], although we address additional issues related to proving C^∞-convergence. In this section we adopt the convention $\pi/\left(2\sqrt{K}\right) \doteqdot \infty$ when $K \leq 0$ and we assume that geodesics have constant speed, i.e., they are parametrized proportional to arc length.

5.1. Derivatives of the distance function and $\exp^{-1}$. Let $(\mathcal{M}^n, g)$ be a Riemannian manifold and let $d(x,y)$ denote the distance between x and y. Fix $x \in \mathcal{M}$ and consider the function $f(y) \doteqdot \frac{1}{2}d^2(x,y)$. We can write f as an integral by letting $\gamma(r)$ be a minimal geodesic from $\gamma(0) = x$ to $\gamma(1) = y$, so that

$$f(y) = \frac{1}{2}\int_0^1 g(\dot{\gamma}, \dot{\gamma})\, dr$$

since geodesics have constant speed. The quantity $g(\dot{\gamma}, \dot{\gamma})$ is constant and equal to the square of the length of the geodesic. The gradient of f can be expressed as follows.

LEMMA 4.43 (Gradient of the distance squared function). *If y is not in the cut locus of x, then*

$$\operatorname{grad} f(y) = -\exp_y^{-1} x \in T_y\mathcal{M}.$$

PROOF. Let $\gamma(r)$ be the unique minimal geodesic from $\gamma(0) = x$ to $\gamma(1) = y$; then by the Gauss lemma, $\operatorname{grad} f(y) = \dot{\gamma}(1)$. It follows easily from the uniqueness of solutions of the geodesic equation that $\exp_y(-\dot{\gamma}(1)) = x$, so that $-\dot{\gamma}(1) = \exp_y^{-1} x$. The lemma is proved. □

Given $Y \in T_y\mathcal{M}$, let $\alpha : (-\varepsilon, \varepsilon) \to \mathcal{M}$ be the geodesic with $\alpha(0) = y$ and $\dot{\alpha}(0) = Y$ (where $\varepsilon > 0$ is sufficiently small for later purposes). Since y is not in the cut locus of x, there exists a smooth family of unique minimal geodesics $\gamma_s : [0,1] \to \mathcal{M}$, such that $\gamma_s(0) = x$ and $\gamma_s(1) = \alpha(s)$ for $s \in (-\varepsilon, \varepsilon)$. Then $\gamma_0 = \gamma$ is the minimal geodesic joining x and y. Define $\sigma : (-\varepsilon, \varepsilon) \times [0,1] \to \mathcal{M}$ by

$$\sigma(s,r) \doteqdot \gamma_s(r) = \exp_x\left[r \exp_x^{-1} \alpha(s)\right] = \exp_{\alpha(s)}\left((1-r)\exp_{\alpha(s)}^{-1} x\right),$$

where we are considering the curves γ_s both in terms of geodesics from x and geodesics from $\alpha(s)$. Note that

$$\frac{\partial \sigma}{\partial s}(s,0) = 0, \quad \frac{\partial \sigma}{\partial s}(s,1) = \dot{\alpha}(s),$$
$$\frac{\partial \sigma}{\partial r}(s,1) = -\exp_{\alpha(s)}^{-1} x.$$

The second derivative of f has the following expression.

LEMMA 4.44 (Hessian of the distance squared function). *The Hessian of f is given by*

$$(\nabla_Y \operatorname{grad} f)(y) = -\nabla_Y \exp_y^{-1} x = \nabla_{\partial/\partial r} J(1),$$

where $J(r)$ is the Jacobi field along the geodesic between x and y parametrized on $r \in [0,1]$ such that $J(0) = 0$ and $J(1) = Y \in T_y\mathcal{M}$.

PROOF. Given $Y \in T_y\mathcal{M}$, define γ_s and σ as above. Let $J_s(r) \doteq \frac{\partial \sigma}{\partial s}(s,r)$ be the Jacobi field along γ_s; then $J_s(0) = 0$ and $J_s(1) = \dot{\alpha}(s)$. Using $\frac{\partial \sigma}{\partial r}(s,1) = -\exp_{\alpha(s)}^{-1} x$, we compute

$$\nabla_Y \exp_y^{-1} x = -\nabla_{\partial/\partial s} \frac{\partial \sigma}{\partial r}(s,r)\bigg|_{(s,r)=(0,1)} = -(\nabla_{\partial/\partial r} J_0)(1).$$

□

The following is essentially the Hessian comparison theorem.

LEMMA 4.45 (Hessian comparison). *If the sectional curvature of $(\mathcal{M}^n, g)$ is bounded above by K, then there exists a constant $C = C(K) > 0$ such that for any $y \in B\left(x, \pi/\left(2\sqrt{K}\right)\right)$ not in the cut locus we have*

(4.14) $\quad (\operatorname{Hess} f)(Y,Y) = -g\left(\nabla_Y \exp_y^{-1} x, Y\right) \geq C|Y|^2, \quad Y \in T_y\mathcal{M}.$

PROOF. By Lemma 4.44 we need to estimate

$$g\left((\nabla_{\partial/\partial r} J)(1), J(1)\right) = \frac{1}{2}\frac{d}{dr}[g(J(r), J(r))]\bigg|_{r=1} = |J|\frac{d}{dr}|J|\bigg|_{r=1}.$$

Note that we can write $Y = Y^\perp + c\dot{\gamma}(1)$, where $Y^\perp$ is perpendicular to $\dot{\gamma}(1)$ and $c \in \mathbb{R}$. Then

$$J(r) = J^\perp(r) + cr\dot{\gamma}(r),$$

where $J^\perp(r)$ is a Jacobi field satisfying $J^\perp(0) = 0$, $J^\perp(1) = Y^\perp$ and $J^\perp(r) \perp \dot{\gamma}(r)$. Now it is clear that we only need to estimate $|J|\frac{d}{dr}|J|\big|_{r=1}$ assuming that Y is orthogonal to $\dot{\gamma}(1)$.

We compute using the Jacobi equation

$$\frac{d^2}{dr^2}|J| = \frac{d}{dr}\left(\frac{g\left(\nabla_{\partial/\partial r}J, J\right)}{|J|}\right)$$

$$= -\frac{g\left(\nabla_{\partial/\partial r}J, J\right)^2}{|J|^3} + \frac{g\left(\nabla_{\partial/\partial r}\nabla_{\partial/\partial r}J, J\right)}{|J|} + \frac{|\nabla_{\partial/\partial r}J|^2}{|J|}$$

$$= -\frac{g\left(\nabla_{\partial/\partial r}J, J\right)^2}{|J|^3} - \frac{g\left(R\left(J,\dot{\gamma}\right)\dot{\gamma}, J\right)}{|J|} + \frac{|\nabla_{\partial/\partial r}J|^2}{|J|},$$

so

$$\frac{d^2}{dr^2}|J| + K|\dot{\gamma}|^2|J| = |J|^{-1}\left(|J|^2|\dot{\gamma}|^2 K - g\left(R\left(J,\dot{\gamma}\right)\dot{\gamma}, J\right)\right)$$
$$+ |J|^{-3}\left(|J|^2|\nabla_{\partial/\partial r}J|^2 - g\left(\nabla_{\partial/\partial r}J, J\right)^2\right)$$
$$\geq 0,$$

where we used $J(r) \perp \dot{\gamma}(r)$ to conclude that $|J|^2|\dot{\gamma}|^2 K - g\left(R\left(J,\dot{\gamma}\right)\dot{\gamma}, J\right) \geq 0$.

The corresponding ODE for $\phi(r)$ is

$$\phi'' + K|\dot{\gamma}|^2\phi = 0$$

($|\dot{\gamma}| = d(x,y)$ is a constant since γ is a geodesic) and has solutions

$$\phi(r) = \phi(0)\operatorname{cs}_{K|\dot{\gamma}|^2}(r) + \phi'(0)\operatorname{sn}_{K|\dot{\gamma}|^2}(r),$$

where

$$\operatorname{sn}_\kappa(r) = \begin{cases} \frac{1}{\sqrt{\kappa}}\sin(\sqrt{\kappa}r) & \text{if } \kappa > 0, \\ r & \text{if } \kappa = 0, \\ \frac{1}{\sqrt{-\kappa}}\sinh(\sqrt{-\kappa}r) & \text{if } \kappa < 0, \end{cases}$$

and

$$\operatorname{cs}_\kappa(r) = \begin{cases} \cos(\sqrt{\kappa}r) & \text{if } \kappa > 0, \\ 1 & \text{if } \kappa = 0, \\ \cosh(\sqrt{-\kappa}r) & \text{if } \kappa < 0. \end{cases}$$

The functions $\operatorname{sn}_\kappa(r)$ and $\operatorname{cs}_\kappa(r)$ are the solutions to $\phi'' + \kappa\phi = 0$ with $\operatorname{sn}_\kappa(0) = 0$, $\operatorname{sn}'_\kappa(0) = 1$, $\operatorname{cs}_\kappa(0) = 1$, and $\operatorname{cs}'_\kappa(0) = 0$.

Note that $J/|J|$ is a unit vector (and has a limit as $r \to 0$) and that $\nabla_{\partial/\partial r}J(0)$ is well defined. From $\frac{d}{dr}|J| = g\left(\nabla_{\partial/\partial r}J, \frac{J}{|J|}\right)$ we know that the limit $\lim_{r\to 0+}\frac{d}{dr}|J|$ is also well defined. Now we compare $|J|(r)$ with the solution $\phi(r)$ which satisfies $\phi(0) = |J|(0) = 0$ and $\phi'(0) = \lim_{r\to 0+}\frac{d}{dr}|J|$. Note that

$$\phi(r) = \phi'(0)\operatorname{sn}_{K|\dot{\gamma}|^2}(r)$$

and that $\phi(r)$ is nonnegative for $r \in [0,1]$ when $K \leq 0$ or when $K > 0$ and $\sqrt{K}\,|\dot\gamma| \leq \pi$. Assuming $\sqrt{K}\,|\dot\gamma| < \pi$ if $K > 0$, we compute

$$\left(|J|'\phi - |J|\phi'\right)' = |J|''\phi - |J|\phi''$$
$$= \left(\frac{d^2}{dr^2}|J| + K|\dot\gamma|^2 |J|\right)\phi \geq 0$$

for all $r \in [0,1]$. Integrating this from 0 to r gives us

$$|J|'(r)\,\phi(r) - |J(r)|\,\phi'(r) \geq |J|'(0)\,\phi(0) + |J(0)|\,\phi'(0) = 0,$$

that is, for $r \in (0,1]$,

(4.15) $$|J|'(r) \geq |J(r)|\,\frac{\phi'(r)}{\phi(r)}.$$

Hence

$$\left.|J|\frac{d}{dr}|J|\right|_{r=1} \geq |J|^2(1)\,\frac{\phi'(1)}{\phi(1)}$$

provided $\sqrt{K}\,|\dot\gamma| < \pi$ when $K > 0$. This proves

$$-g\left(\nabla_Y \exp_y^{-1} x, Y\right) = \left.|J|\frac{d}{dr}|J|\right|_{r=1} \geq \frac{\operatorname{cs}_{K|\dot\gamma|^2}(1)}{\operatorname{sn}_{K|\dot\gamma|^2}(1)}|Y|^2.$$

Note that $\frac{\operatorname{cs}_{K|\dot\gamma|^2}(1)}{\operatorname{sn}_{K|\dot\gamma|^2}(1)}$ is positive either when $K \leq 0$ or when $K > 0$ and $\sqrt{K}\,|\dot\gamma_0| < \pi/2$. □

Recall that a C^2 function ϕ is (strictly) **convex** if its Hessian is positive definite: $\nabla\nabla\phi > 0$.

COROLLARY 4.46 (Local convexity of the distance squared function). *Suppose the sectional curvatures of $(\mathcal{M}^n, g)$ are bounded above by K. Then the function $f(y) \doteq \frac{1}{2}d^2(x,y)$ is convex for any $y \in B\left(x, \pi/\left(2\sqrt{K}\right)\right)$ not in the cut locus of x.*

PROOF. This follows directly from (4.14). □

We also have

COROLLARY 4.47 (Convexity of small enough balls). *Suppose the sectional curvatures of $(\mathcal{M}^n, g)$ are bounded above by K. Then the ball $B(O, r)$ is convex if $r \leq \min\left\{\operatorname{inj} O, \pi/\left(2\sqrt{K}\right)\right\}$.*

PROOF. Suppose $x, y \in B(O, r)$. Let $\gamma(t)$ be the constant speed minimal geodesic between x and y. We simply need to show that $d(\gamma(t), O) < r$ for every t. Consider the function $f(z) = \frac{1}{2}d(O, z)^2$. By Corollary 4.46, we have that

$$\nabla_t \frac{d}{dt} f(\gamma(t)) = (\nabla^2 f)\left(\frac{d\gamma}{dt}, \frac{d\gamma}{dt}\right) > 0,$$

which implies that the maximum of $f(\gamma(t))$ occurs at the endpoints. Hence
$$d(O, \gamma(t)) \leq \max\{d(O, x), d(O, y)\} < r.$$
□

The next lemma will be used in proving the smooth dependence of the center of mass in the next subsection.

PROPOSITION 4.48 (On the derivatives of $\exp^{-1}$). *Let $(\mathcal{M}^n, g)$ be a Riemannian manifold such that all derivatives of the curvature are bounded:*
$$\left|\nabla^\ell \operatorname{Rm}\right| \leq C_\ell \quad \text{for } \ell = 0, 1, 2, \ldots.$$
There is a constant $c(n) > 0$ such that for any $p \in \mathcal{M}$ and $x, y \in B(p, r_1)$, where $r_1 \leq \min\{\frac{1}{4} \operatorname{inj}(p), c/\sqrt{C_0}\}$, if x is not in the cut locus of y, then

(i) *we have*

(4.16) $$\left|\nabla_y^{\ell_1} \nabla_x^{\ell_2} \exp_y^{-1} x\right| \leq \tilde{C}_{\ell_1 + \ell_2 + 1} \quad \text{for } \ell_1, \ell_2 = 0, 1, 2, \ldots,$$

where $\tilde{C}_\ell = \tilde{C}_\ell(n, \operatorname{inj}(p), \ell, C_0, \ldots, C_\ell) > 0$ are constants independent of x and y, and ∇_y and ∇_x are the covariant derivatives with respect to y and x, respectively;

(ii) *when $x, y \to p_* \in B(p, r_1)$, we have*

(4.17) $$(\nabla_x \exp_y^{-1} x : T_x\mathcal{M} \to T_y\mathcal{M}) \to (\operatorname{id} : T_{p_*}\mathcal{M} \to T_{p_*}\mathcal{M}),$$
$$(\nabla_y \exp_y^{-1} x : T_y\mathcal{M} \to T_y\mathcal{M}) \to (-\operatorname{id} : T_{p_*}\mathcal{M} \to T_{p_*}\mathcal{M}),$$

where we use parallel translation to identify $T_x\mathcal{M}$ and $T_y\mathcal{M}$ with $T_{p_*}\mathcal{M}$ and to define the convergences above.

PROOF. (i) Let $w = \{w^k\}$ be normal coordinates on $B(p, r_1)$. By Proposition 4.32 (Corollary 4.12 in [**187**]) we have in the coordinates w,

(4.18) $$\frac{1}{2}(\delta_{ij}) \leq (g_{ij}) \leq 2(\delta_{ij}) \quad \text{and} \quad \left|\frac{\partial^\alpha}{(\partial w)^\alpha} g_{ij}\right| \leq \tilde{C}_{|\alpha|},$$

where α is a multi-index. In particular the Christoffel symbols Γ_{ij}^k satisfy

(4.19) $$\left|\frac{\partial^\alpha}{(\partial w)^\alpha} \Gamma_{ij}^k\right| \leq \tilde{C}_{|\alpha|+1}.$$

Now we consider the exponential map $\exp : T\mathcal{M} \to \mathcal{M}$ with $\exp_y z = x$ for $z \in T\mathcal{M}$ using the coordinates w; here we abuse notation in that $x = (x^k)$ stands for both a point in $\mathcal{M}$ and its coordinates in the coordinate system w (and the same for y). Define $f(r, y, z)$, $0 \leq r \leq 1$, by

(4.20) $$\frac{d^2 f^k}{dr^2} + \Gamma_{ij}^k(f) \frac{df^i}{dr} \frac{df^j}{dr} = 0,$$
$$f^k(0, y, z) = y^k,$$
$$\frac{df^k}{dr}(0, y, z) = z^k.$$

Then $\left(f^k\left(1,y,z\right)\right) = \left(x^k\right) = \exp_y z$. We will apply the implicit function theorem to
$$F\left(x,y,z\right) \doteq f\left(1,y,z\right) - x$$
to prove that $z = \exp_y^{-1} x$ is a smooth function of (x,y) and that $\exp_y^{-1} x$ has the required derivative estimates. Consider the boundary value problem for the first-order ODE

$$\frac{df^k}{dr} = h^k,$$
$$\frac{dh^k}{dr} + \Gamma_{ij}^k(f) h^i h^j = 0,$$
$$f^k(0,y,z) = y^k,$$
$$h^k(0,y,z) = z^k.$$

From $f(r,y,z) \in B(p,r_1)$, $|h(r,y,z)|_g = |z|_{g(p)} = d(y,x)$, and $\frac{1}{2}(I_{ij}) \leq (g_{ij})$, we have $\left|f^k(r,y,z)\right| \leq \sqrt{2}r_1$ and $\left|h^k(r,y,z)\right| \leq \sqrt{2}d(y,x)$. From the smooth dependence property for ODE (see Theorem 4.1 on p. 100 of Hartman [**196**]), $f^k(r,y,z)$ is a smooth function of (r,y,z). Using the proof of Theorem 3.1 on p. 95 of [**196**] and (4.19), it follows from an induction argument on the order of derivatives that for $r \in [0,1]$,

(4.21)
$$\left|\frac{\partial^{\alpha+\beta}}{(\partial y)^\alpha (\partial z)^\beta} f^k(r,y,z)\right| \leq \tilde{C}_{|\alpha|+|\beta|+1},$$
$$\left|\frac{\partial^{\alpha+\beta}}{(\partial y)^\alpha (\partial z)^\beta} h^k(r,y,z)\right| \leq \tilde{C}_{|\alpha|+|\beta|+1}.$$

Actually the proof of Theorem 3.1 on p. 95 of [**196**] implies the estimate above for $|\alpha| + |\beta| = 1$.

Let $\gamma_{y,z}$ be the geodesic with $\gamma_{y,z}(0) = y$ and $\left(\frac{d}{dr}\gamma_{y,z}\right)(0) = z$. Let $J_{y,z,\tilde{z}}(r)$ denote the Jacobi field along the geodesic $\gamma_{y,z}$ with $J_{y,z,\tilde{z}}(0) = 0$ and $\left(\frac{d}{dr}J_{y,z,\tilde{z}}\right)(0) = \tilde{z} \in T_y\mathcal{M}$. Then the covariant partial derivative in the direction $\tilde{z}$ is

$$D_z f(1,y,z)(\tilde{z}) = \left.\frac{d}{ds} f(1,y,z+s\tilde{z})\right|_{s=0} = J_{y,z,\tilde{z}}(1).$$

To show $D_z f(1,y,z) : T_y\mathcal{M} \to T_{\exp_y z}\mathcal{M}$ is invertible, we prove that there is a constant $c_0 > 0$ such that $|J_{y,z,\tilde{z}}(1)| \geq c_0 |\tilde{z}|$. This follows from the Rauch comparison theorem; here we give a proof using (4.15). As in the proof of Lemma 4.45, it suffices to prove that $|J_{y,z,\tilde{z}}(1)| \geq c_0 |\tilde{z}|$ for those $\tilde{z}$ which are orthogonal to z in $T_y\mathcal{M}$. From (4.15), we have $\left(\frac{|J_{y,z,\tilde{z}}(r)|}{\phi(r)}\right)' \geq 0$, where $\phi(r) = |\tilde{z}| \operatorname{sn}_{C_0|z|^2}(r)$. Since $\lim_{r \to 0} \frac{|J_{y,z,\tilde{z}}(r)|}{\phi(r)} \to 1$, we have $\frac{|J_{y,z,\tilde{z}}(r)|}{\phi(r)} \geq 1$ and

$$|J_{y,z,\tilde{z}}|(1) \geq \phi(1) = |\tilde{z}| \operatorname{sn}_{C_0|z|^2}(1) \geq c_0 |\tilde{z}|.$$

Using $|z| \leq 2c/\sqrt{C_0}$, we can choose $c_0 \doteq \operatorname{sn}_{4c^2}(1)$. We have proved

(4.22) $$\left|(D_z f(1, y, z))^{-1}\right| \leq c_0^{-1}.$$

Now we can apply the implicit function theorem to $F(x, y, z) \doteq f(1, y, z) - x$. From

$$D_y f(1, y, z) + D_z f(1, y, z) \frac{\partial z}{\partial y} = 0,$$

$$D_z f(1, y, z) \frac{\partial z}{\partial x} - \operatorname{id} = 0,$$

we can take higher-order derivatives of the equations above to get formulas for $\frac{\partial^{\alpha+\beta} z}{(\partial y)^\alpha (\partial x)^\beta}$ in terms of the partial derivatives of $\frac{\partial^{\alpha+\beta}}{(\partial y)^\alpha (\partial z)^\beta} f^k(1, y, z)$ and $(D_z f(1, y, z))^{-1}$. From (4.21) and (4.22) we can estimate

$$\left|\frac{\partial^{\alpha+\beta} z}{(\partial y)^\alpha (\partial x)^\beta}\right| \leq \tilde{C}_{|\alpha|+|\beta|+1}$$

by induction on the order of derivatives. From (4.18) we know that the bounds of the covariant derivatives $|\nabla_y^{\ell_1} \nabla_x^{\ell_2} z|$ follow from the bounds of $\left|\frac{\partial^{\alpha+\beta} z}{(\partial y)^\alpha (\partial x)^\beta}\right|$.

(ii) Let $\tilde{w}$ be normal coordinates on $B(p_*, r_*)$ for sufficiently small r_* and let $\tilde{g}_{ij} \doteq g\left(\frac{\partial}{\partial \tilde{w}^i}, \frac{\partial}{\partial \tilde{w}^j}\right)$. From Theorem 4.10 in [**187**] we have $(\tilde{g}_{ij}) \to (\delta_{ij})$ on $B(p_*, r_*)$ as $r_* \to 0$. Hence the geodesic equation (4.20) in the coordinate chart $\tilde{w}$ has a solution $\tilde{f}(r, y, z)$ which converges in C^1 to $\tilde{f}_\infty(r, y, z) \doteq y+rz$ as $r_* \to 0$. So for $x, y \in B(p_*, r_*)$, $\exp_y^{-1} x$ converges in C^1 to $\exp_{\infty y}^{-1} x \doteq x - y$ as $r_* \to 0$. The estimate (4.17) follows from $(\tilde{g}_{ij}) \to (\delta_{ij})$ on $B(p_*, r_*)$ and

$$\frac{\partial}{\partial x^i} \exp_{\infty y}^{-1} x = e_i \quad \text{and} \quad \frac{\partial}{\partial y^i} \exp_{\infty y}^{-1} x = -e_i,$$

where $e_i = (0, \cdots, 1, \cdots, 0)$ is the unit vector in i-th direction. The lemma now is proved. $\square$

REMARK 4.49. (i) Under the assumptions of Proposition 4.48, let $p_* \in B(p, r_1)$ and let $\tilde{w}$ be normal coordinates on $B(p_*, r_*)$ for sufficiently small r_*. Then when $x, y \to p_*$, we have that $\exp_y^{-1} x$ converges in C^1 to the map $x - y$, where $x = \{x^k\}$ stands both for a point in $\mathcal{M}$ and its coordinates in the coordinate system $\tilde{w}$ (and the same for $y = \{y^k\}$).

(ii) Suppose h is another metric on $\mathcal{M}$. From the proof of Proposition 4.48 it is not difficult to see that the map $(\exp^g)^{-1} : \mathcal{M} \times \mathcal{M} \to T\mathcal{M}$ is close to $(\exp^h)^{-1} : \mathcal{M} \times \mathcal{M} \to T\mathcal{M}$ on any compact set in C^k when g is very close to h on any compact set in C^k, for any $k \in \mathbb{N}$.

5.2. Nonlinear averages. Let $(\mathcal{M}^n, g)$ be a complete Riemannian manifold with sectional curvatures bounded above by K. Let $p \in \mathcal{M}$ and $q_1, \ldots, q_k \in B(p, r)$, where $r < \min\left\{\frac{1}{3} \operatorname{inj}(p), \frac{\pi}{6\sqrt{K}}\right\}$. Let $\mu_1, \ldots, \mu_k$ be nonnegative numbers with $\mu_1 + \cdots + \mu_k > 0$. We define the **center of mass with weights** $\mu_1, \ldots, \mu_k$,

$$\operatorname{cm}\{q_1, \ldots, q_k\} = \operatorname{cm}_{(\mu_1, \ldots, \mu_k)}\{q_1, \ldots, q_k\},$$

as the minimizer of

$$\phi : \mathcal{M} \to \mathbb{R},$$

(4.23) $$\phi(q) \doteq \frac{1}{2} \sum_{i=1}^{k} \mu_i d^2(q, q_i).$$

LEMMA 4.50 (Existence of center of mass). *Let $p \in \mathcal{M}$ and $q_1, \ldots, q_k \in B(p, r)$ for some $r < \frac{\pi}{6\sqrt{K}}$. Suppose $\operatorname{inj}(q) > 3r$ for all $q \in B(p, r)$. Then there exists a unique minimizer $\operatorname{cm}\{q_1, \ldots, q_k\}$ of ϕ in $\mathcal{M}$. Furthermore we have $\operatorname{cm}\{q_1, \ldots, q_k\} \in B(p, 2r)$ and*

$$\lim_{q_1, \ldots, q_k \to q_*} \operatorname{cm}_{(\mu_1, \ldots, \mu_k)}\{q_1, \ldots, q_k\} = q_*$$

uniformly in $\mu_1, \ldots, \mu_k$.

PROOF. It is clear that for any $q \in \mathcal{M} \setminus B(p, 2r)$, we have $\phi(q) > \phi(p)$. Hence the minimizer of ϕ exists and must be contained in $B(p, 2r)$. Note that if $q \in B(p, 2r)$, then $q \in B(q_i, 3r)$. Since $B(q_i, 3r) \subset B\left(q_i, \pi/(2\sqrt{K})\right)$, by Lemma 4.43, the functions $q \mapsto \frac{1}{2}d^2(q, q_i)$ are strictly convex in $B(p, 2r)$. Since the weights μ_i are nonnegative and $\mu_1 + \cdots + \mu_k > 0$, ϕ is strictly convex in $B(p, 2r)$. Hence the minimizer must be unique.

To see the last statement, we apply the first part of the statement to $B(q_*, r_*)$ in place of $B(p, r)$, where r_* is small. We get that when $q_1, \ldots, q_k \in B(q_*, r_*)$, we have $\operatorname{cm}_{(\mu_1, \ldots, \mu_k)}\{q_1, \ldots, q_k\} \in B(q_*, 2r_*)$ for all $(\mu_1, \ldots, \mu_k)$. □

By Lemma 4.43 we have

$$\operatorname{grad} \phi(q) = -\sum_{i=1}^{k} \mu_i \exp_q^{-1} q_i.$$

The minimizer occurs at a point q where the gradient of ϕ is zero, so that

$$\sum_{i=1}^{k} \mu_i \exp_q^{-1} q_i = 0.$$

The following proposition tells us about the derivatives of the center of mass.

PROPOSITION 4.51 (Dependence of cm on weights and points). *Suppose $(\mathcal{M}^n, g)$ is a Riemannian manifold such that all of the derivatives of the curvature are bounded:*

$$\left|\nabla^\ell \operatorname{Rm}\right| \leq C_\ell \quad \text{for } \ell = 0, 1, 2, \ldots.$$

Let $\mu_1, \ldots, \mu_k$ be nonnegative weights with $\mu_1 + \cdots + \mu_k > 0$. There exists a constant $c(n) \in \left(0, \frac{\pi}{6}\right)$ such that for any $p \in \mathcal{M}$, if $\operatorname{inj}(q) > 3r$ for all $q \in B(p, r)$, where $r < \frac{c(n)}{\sqrt{C_0}}$. Then we have the following.

(i) (Bounds on the derivatives of cm) *The unique center of mass*

$$\operatorname{cm}_{(\mu_1, \ldots, \mu_k)}\{q_1, \ldots, q_k\}$$

is a smooth function of $q_1, \ldots, q_k \in B(p, r)$ and $\mu_1, \ldots, \mu_k$. The $\nabla_q^\alpha \nabla_\mu^\beta$-covariant derivatives of $\operatorname{cm}_{(\mu_1, \ldots, \mu_k)}\{q_1, \ldots, q_k\}$, with respect to $q_1, \ldots, q_k$ and $\mu_1, \ldots, \mu_k$, satisfy

(4.24) $$\left|\nabla_q^\alpha \nabla_\mu^\beta \operatorname{cm}_{(\mu_1, \ldots, \mu_k)}\{q_1, \ldots, q_k\}\right| \leq \tilde{C}_{|\alpha|+|\beta|+1},$$

where $\nabla_q = (\nabla_{q_1}, \ldots, \nabla_{q_k})$ and $\nabla_\mu = \left(\frac{\partial}{\partial \mu_1}, \ldots, \frac{\partial}{\partial \mu_k}\right)$ and $\tilde{C}_{|\alpha|+|\beta|+1}$ are constants depending on n, $\operatorname{inj}(p)$, $|\alpha|+|\beta|$, and $C_0, \ldots, C_{|\alpha|+|\beta|+1}$.

(ii) *For $q_1, \ldots, q_k \in B(p, r)$ such that $q_1, \ldots, q_k \to q_* \in \overline{B}(p, r)$ (i.e., the points tend to each other), we have*

(a) (change in a weight has negligible effect on cm)

$$\left|\nabla_{\mu_i} \operatorname{cm}_{(\mu_1, \ldots, \mu_k)}\{q_1, \ldots, q_k\}\right| \to 0,$$

(b) (effect of the change in a point on cm)

$$\left(\nabla_{q_i} \operatorname{cm}_{(\mu_1, \ldots, \mu_k)}\{q_1, \ldots, q_k\} : T_{q_i}\mathcal{M} \to T_{\operatorname{cm}_{(\mu_1, \ldots, \mu_k)}\{q_1, \ldots, q_k\}}\mathcal{M}\right)$$

$$\to \left(\frac{\mu_i}{\sum_{j=1}^k \mu_j} \operatorname{id} : T_{q_*}\mathcal{M} \to T_{q_*}\mathcal{M}\right),$$

(c) (effect of the change in a weight and point on cm)

$$\left(\nabla_{q_j} \frac{\partial}{\partial \mu_j} \operatorname{cm}_{(\mu_1, \ldots, \mu_k)}\{q_1, \ldots, q_k\} : T_{q_i}\mathcal{M} \to T_{\operatorname{cm}_{(\mu_1, \ldots, \mu_k)}\{q_1, \ldots, q_k\}}\mathcal{M}\right)$$

$$\to \left(\frac{1}{\sum_{i=1}^k \mu_i} \operatorname{id} : T_{q_*}\mathcal{M} \to T_{q_*}\mathcal{M}\right).$$

The convergences above are defined using parallel translation to identify $T_{q_i}\mathcal{M}$ with $T_{q_*}\mathcal{M}$.

PROOF. (i) We apply the implicit function theorem to the family of maps

$$G_{q_1, \ldots, q_k, \mu_1, \ldots, \mu_k} : \mathcal{M} \to T\mathcal{M}$$

defined by

$$G_{q_1, \ldots, q_k, \mu_1, \ldots, \mu_k}(q) \doteqdot \sum_{i=1}^k \mu_i \exp_q^{-1} q_i.$$

By the previous lemma, $\operatorname{cm}_{(\mu_1,\ldots,\mu_k)}\{q_1,\ldots,q_k\}$ is the unique solution of the equation
$$G(q) \doteq G_{q_1,\ldots,q_k,\mu_1,\ldots,\mu_k}(q) = 0.$$
Consider the partial derivative
$$\nabla_q G = \sum_{i=1}^k \mu_i \nabla_q \exp_q^{-1} q_i : T_q\mathcal{M} \to T_q\mathcal{M}.$$

By Lemma 4.45, $\nabla_q G$ is positive definite with smallest eigenvalue being bounded from below by a constant depending only on C_0 and $\mu_1,\ldots,\mu_k$. It follows from the implicit function theorem that the unique solution
$$\operatorname{cm}_{(\mu_1,\ldots,\mu_k)}\{q_1,\ldots,q_k\}$$
is continuous in $q_1,\ldots,q_k$ and $\mu_1,\ldots,\mu_k$.

To see that the $\nabla_q^\alpha \nabla_\mu^\beta$-covariant derivatives of $\operatorname{cm}_{(\mu_1,\ldots,\mu_k)}\{q_1,\ldots,q_k\}$ are bounded, we compute the other partial derivatives of G:
$$\nabla_{q_i} G = \mu_i \nabla_{q_i} \exp_q^{-1} q_i, \qquad \frac{\partial}{\partial \mu_i} G = \exp_q^{-1} q_i.$$

Hence
(4.25)
$$\left(\nabla_{q_i} G, \frac{\partial}{\partial \mu_j} G\right) + \nabla_q G \cdot \left(\nabla_{q_i} \operatorname{cm}\{q_1,\ldots,q_k\}, \nabla_{\mu_j} \operatorname{cm}\{q_1,\ldots,q_k\}\right) = 0,$$
where $q = \operatorname{cm}\{q_1,\ldots,q_k\}$. Thus
$$\left(\nabla_{q_i} \operatorname{cm}\{q_1,\ldots,q_k\}, \nabla_{\mu_j} \operatorname{cm}\{q_1,\ldots,q_k\}\right) = -(\nabla_q G)^{-1}\left(\nabla_{q_i} G, \frac{\partial}{\partial \mu_j} G\right).$$

This and Proposition 4.48(i) implies (4.24) when $|\alpha| + |\beta| = 1$.

To bound the higher derivatives of $\operatorname{cm}\{q_1,\ldots,q_k\}$, we argue inductively on the order of the derivative $|\alpha| + |\beta|$. We take the appropriate derivatives of (4.25) of order $|\alpha| + |\beta| - 1$ with respect to $q_1,\ldots,q_k$ and $\mu_1,\ldots,\mu_k$ so that $\nabla_q G \cdot \nabla_q^\alpha \nabla_\mu^\beta \operatorname{cm}_{(\mu_1,\ldots,\mu_k)}\{q_1,\ldots,q_k\}$ appears in the resulting equality. Then
$$\nabla_q^\alpha \nabla_\mu^\beta \operatorname{cm}_{(\mu_1,\ldots,\mu_k)}\{q_1,\ldots,q_k\}$$
can be expressed in terms of $\nabla_{q_i}^{\ell_1} \exp_q^{-1} q_i$ with $\ell_1 \leq |\alpha| + |\beta|$, $\nabla_{q_i}^{\ell_2} \nabla_q \exp_q^{-1} q_i$ with $\ell_2 \leq |\alpha| + |\beta|$, $\nabla_q^{\alpha_1} \nabla_\mu^{\beta_1} \operatorname{cm}_{(\mu_1,\ldots,\mu_k)}\{q_1,\ldots,q_k\}$ with $|\alpha_1| + |\beta_1| \leq |\alpha| + |\beta| - 1$, and $(\nabla_q G)^{-1}$. Now it is easy to see from Proposition 4.48(i) that $\left|\nabla_q^\alpha \nabla_\mu^\beta \operatorname{cm}_{(\mu_1,\ldots,\mu_k)}\{q_1,\ldots,q_k\}\right|$ are bounded by constants $\tilde{C}_{|\alpha|+|\beta|+1}$ depending on n, $\operatorname{inj}(p)$, $|\alpha| + |\beta|$, and $C_0,\ldots,C_{|\alpha|+|\beta|+1}$.

(ii) When $q_1,\ldots,q_k \to q_*$, by Lemma 4.50, we have $\frac{\partial}{\partial \mu_i} G \to 0$. By Proposition 4.48(ii) we have $\nabla_{q_i} G \to \mu_i \operatorname{id}$ and $\nabla_q G \to -\left(\sum_{i=1}^k \mu_i\right) \operatorname{id}$.

This proves the first two convergences. Next we estimate

$$\nabla_{q_i} \frac{\partial}{\partial \mu_i} \mathrm{cm}_{(\mu_1,\ldots,\mu_k)} \{q_1,\ldots,q_k\}.$$

Since we have $\nabla_{q_i} \frac{\partial}{\partial \mu_i} G = \nabla_{q_i} \exp_q^{-1} q_i \to \mathrm{id}$, by taking ∇_{q_j}-derivative of

$$\frac{\partial}{\partial \mu_j} G + \nabla_q G \cdot \frac{\partial}{\partial \mu_j} \mathrm{cm}\{q_1,\ldots,q_k\} = 0$$

in (4.25) and taking the limit, we get

$$\mathrm{id} - \left(\sum_{i=1}^k \mu_i\right) \nabla_{q_j} \frac{\partial}{\partial \mu_j} \mathrm{cm}_{(\mu_1,\ldots,\mu_k)} \{q_1,\ldots,q_k\} = 0.$$

This proves the third convergence. □

REMARK 4.52. (i) Note that in Euclidean space $\mathbb{R}^n$, we have the following formula for the center of mass

$$\mathrm{cm}_{(\mu_1,\ldots,\mu_k)}\{q_1,\ldots,q_k\} = \frac{1}{\mu_1 + \cdots + \mu_k} (\mu_1 q_1 + \cdots + \mu_k q_k).$$

It is clear that there are many derivatives of the form

$$\nabla_q^\alpha \nabla_\mu^\beta \mathrm{cm}_{(\mu_1,\ldots,\mu_k)}\{q_1,\ldots,q_k\}, \quad \text{where } |\alpha| + |\beta| \geq 2,$$

whose lengths do not approach 0 as $q_1,\ldots,q_k \to q_* \in \mathbb{R}^n$.

(ii) Suppose h is another metric on $\mathcal{M}$. From the proof of Proposition 4.51 and Remark 4.49(ii), it is not difficult to see that, as a function of $(\mu_1,\ldots,\mu_k,q_1,\ldots,q_k)$, the center of mass map $\mathrm{cm}^g_{(\mu_1,\ldots,\mu_k)}\{q_1,\ldots,q_k\}$ is close to $\mathrm{cm}^h_{(\mu_1,\ldots,\mu_k)}\{q_1,\ldots,q_k\}$ on any compact set in C^∞ when g is very close to h on any compact set in C^∞.

We can use the center of mass to average maps. We have the following.

PROPOSITION 4.53 (Averaging maps). *Let $(\mathcal{N}^n, h)$ and $(\mathcal{M}^n, g)$ be Riemannian manifolds such that all derivatives of the curvature of $\mathcal{M}$ are bounded:*

$$\left|\nabla^\ell \mathrm{Rm}\right| \leq C_\ell \quad \text{for } \ell = 0,1,2,\ldots.$$

Let $\mu_i(x)$, $i = 1,\cdots,k$, be a finite sequence of smooth nonnegative functions on $\mathcal{N}$ with compact support in $U_i \subset \mathcal{N}$ and with bounded derivatives. Let $\mathcal{W}$ be an open set with closure $\overline{\mathcal{W}} \subset \bigcup_i \mu_i^{-1}(0,\infty)$. Let $F_i : U_i \to \mathcal{M}$ be a finite sequence of smooth functions with bounded derivatives. Suppose that, for any $x_0 \in \mathcal{W}$, there exist i_0 and $r_0 \in \left(0, \frac{\pi}{6\sqrt{C_0}}\right)$ such that for any j with $x_0 \in \overline{\mu_j^{-1}(0,\infty)}$ we have $F_j(x_0) \in B\left(F_{i_0}(x_0), \frac{1}{2}r_0\right)$ and $\mathrm{inj}(q) > 3r_0$ for all $q \in B(F_{i_0}(x_0), r_0)$. Then there is a function $F : \mathcal{W} \to \mathcal{M}$ defined uniquely by minimizing $\frac{1}{2}\sum_{i=1}^k \mu_i(x) d^2(F(x), F_i(x))$. In particular, $F(x)$ satisfies

$$(4.26) \qquad \sum_{i=1}^k \mu_i(x) \exp_{F(x)}^{-1} F_i(x) = 0.$$

Furthermore $F(x)$ is smooth and its derivatives $\nabla^\ell F(x)$ are bounded by constants depending on the bounds for $\left|\nabla^{\ell_i}\mu_i(x)\right|$ with $\ell_i \leq \ell$, the bounds for $\left|\nabla^{\ell_i}F_i(x)\right|$ with $\ell_i \leq \ell$, and n, ℓ, and $C_0, \ldots, C_{\ell+1}$.

PROOF. Fix $x_0 \in \mathcal{W}$. By assumption, there exists a small neighborhood $V_{x_0} \subset \mathcal{W}$ of x_0, i_0, r_0, and $j_1, \ldots, j_m$ such that $F_{j_k}(x) \in B(F_{i_0}(x_0), r_0)$ for $k = 1, \ldots, m$ and all $x \in V_{x_0}$ and such that $\mu_j(x) = 0$ for all $j \neq j_1, \ldots, j_m$ and $x \in V_{x_0}$. By Lemma 4.50, for $x \in V$ we can define $F_{x_0}(x)$ on V_{x_0} to be

$$\operatorname{cm}_{(\mu_{j_1}(x), \ldots, \mu_{j_m})} \{F_{j_1}(x), \ldots, F_{j_m}(x)\}$$

which is the composition of $\operatorname{cm}_{(\mu_{j_1}, \ldots, \mu_{j_m})}\{q_{j_1}, \ldots, q_{j_m}\}$ and $\mu_{j_k} = \mu_{j_k}(x)$, $q_{j_k} = F_{j_k}(x)$. From the uniqueness of the center of mass it is easy to see that for any two points $x_1, x_2 \in \mathcal{W}$, $F_{x_1}(x) = F_{x_2}(x)$ on $V_{x_1} \cap V_{x_2}$. Hence this defines $F : \mathcal{W} \to \mathcal{M}$.

We now prove the second part of the lemma. Let $x_0 \in \mathcal{W}$; then, on V_{x_0}, $F(x)$ is the composition of $\operatorname{cm}_{(\mu_{j_1}, \ldots, \mu_{j_m})}\{q_{j_1}, \ldots, q_{j_m}\}$ and $\mu_{j_k} = \mu_{j_k}(x)$, $q_{j_k} = F_{j_k}(x)$. It follows from the chain rule and Proposition 4.51(i) that $F(x)$ is smooth on V_{x_0} and that $F(x)$ on V_{x_0} has the required derivative bounds. □

We also used the following convergence property.

PROPOSITION 4.54 (Average by cm of maps limiting to id limits to id). *Let $B_1 \subset B_2$ be two open subsets of $\mathbb{R}^n$ and let g_k, where $k \in \mathbb{N}$, be a family of Riemannian metrics on B_2. Assume all derivatives of the curvatures of g_k are uniformly bounded and $g_k \to g_\infty$ on any compact set in B_2 in C^∞. Suppose $F_k^\alpha : B_1 \to B_2$, for $\alpha = 1, \ldots, A$, are sequences of smooth maps such that $F_k^\alpha \to \operatorname{id}$ uniformly on compact sets in C^1 for each α as $k \to \infty$. Let μ_k^α be partitions of unities for each k on B_1. For any compact set $K \subset B_1$ we can define $F_k : K \to B_2$ for k sufficiently large by letting $F_k(x)$ be the center of mass of $F_k^\alpha(x)$ with weight $\mu_k^\alpha(x)$ with respect to metric g_k, i.e., $F_k(x)$ is defined by*

$$\sum_{\alpha=1}^{A} \mu_k^\alpha(x) \exp^{-1}_{F_k(x)} F_k^\alpha(x) \doteq 0,$$

where the exponential map $\exp_{F_k(x)}$ is with respect to g_k. Then F_k converges to id as $k \to \infty$ uniformly on any compact set in B_1 in C^1.

PROOF. Because some of $\left|\nabla_q^\alpha \nabla_\mu^\beta \operatorname{cm}_{(\mu_1, \ldots, \mu_k)}\{q_1, \ldots, q_k\}\right|$, $|\alpha| + |\beta| \geq 2$, do not approach 0 as $q_1, \ldots, q_k \to q_*$, we will not prove this proposition using the composition employed in the proof of Proposition 4.53. By Remark 4.52(ii), $\operatorname{cm}^{g_k}_{(\mu_k^1(x), \ldots, \mu_k^A(x))}\{F_k^1(x), \ldots, F_k^A(x)\}$ can be made arbitrarily close to $\operatorname{cm}^{g_\infty}_{(\mu_k^1(x), \ldots, \mu_k^A(x))}\{F_k^1(x), \ldots, F_k^A(x)\}$ on any compact set $x \in K \subset B_1$ in C^∞ when we choose k large enough. On the other hand, fix $x_0 \in B_1$ and let $\tilde{w}$ be normal coordinates centered at x_0 in (B_2, g_∞). It follows from Remark

4.49(ii) that by choosing k large and x to be in a very small neighborhood of x_0, we can make $\operatorname{cm}^{g_\infty}_{(\mu_k^1(x),\ldots,\mu_k^A(x))}\{F_k^1(x),\ldots,F_k^A(x)\}$ arbitrarily close in the C^1-topology to the Euclidean center of mass

$$\frac{1}{\mu_k^1(x)+\cdots+\mu_k^A(x)}\left(\mu_k^1(x)F_k^1(x)+\cdots+\mu_k^A(x)F_k^A(x)\right),$$

where we have identified the point $F_k^\alpha(x)\in B_1$ with its coordinates in the coordinate chart $\tilde{w}$. Since $\mu_k^1(x)+\cdots+\mu_k^A(x)\equiv 1$, we have

$$\frac{1}{\mu_k^1(x)+\cdots+\mu_k^A(x)}\left(\mu_k^1(x)F_k^1(x)+\cdots+\mu_k^A(x)F_k^A(x)\right)-x$$
$$=\mu_k^1(x)\left(F_k^1(x)-x\right)+\cdots+\mu_k^A(x)\left(F_k^A(x)-x\right),$$

which clearly converges to 0 on any compact set within the coordinate chart $\tilde{w}$ in C^1 when $k\to\infty$. Now the proposition is proved. $\square$

6. Notes and commentary

For some additional references on compactness theorems not cited in the previous chapter, see Cheeger and Gromov [**73**], [**74**], Gao [**152**], Yang [**374**], [**375**], Anderson [**4**] and Anderson and Cheeger [**6**].

CHAPTER 5

Energy, Monotonicity, and Breathers

> Truth is ever to be found in the simplicity, and not in the multiplicity and confusion of things. – Sir Isaac Newton
>
> The most beautiful thing we can experience is the mysterious. It is the source of all true art and science. – Albert Einstein

Much of the 'classical' study of the Ricci flow is based on the maximum principle. In large part, this is the point of view we have taken in Volume One. As we have seen in Section 8 in Chapter 5 of Volume One, a notable exception to this is Hamilton's entropy estimate, which holds for closed surfaces with positive curvature.[1] Even in this case, the time-derivative of the entropy is the space integral of Hamilton's trace Harnack quantity, which satisfies a partial differential inequality amenable to the maximum principle.[2] Indeed, this fact is the basis for Hamilton's original proof by contradiction of the entropy estimate which uses the global in time existence of the Ricci flow on surfaces.[3] Originally, Hamilton's entropy was a crucial component of the proofs for the convergence of the Ricci flow on surfaces and the classification of ancient solutions on surfaces. Via dimension reduction, the latter result has applications to singularity analysis in Hamilton's program on 3-manifolds.

An interesting direction is that of finding monotonicity formulas for integrals of local geometric quantities. Beautiful recent examples of this are Perelman's energy and entropy estimates in all dimensions. We briefly touched upon these estimates in Section 8 of Chapter 1 (Theorems 1.72 and 1.73) to motivate the study of gradient Ricci solitons. Perelman's energy is the time-derivative of a classical entropy ((5.64) in Section 4 below). Observe how the resulting calculation in Perelman's proof of the upper bound for the maximum time interval of existence of the gradient flow (Proposition 5.34) is reminiscent of Hamilton's proof of his entropy formula. In fact this upper bound says that a modified classical entropy is increasing (see (5.67)).

Monotonicity formulas usually have geometric applications. In particular, Perelman proved that any breather on a closed manifold is a Ricci soliton of the same type. This statement includes the shrinking case which remained open until his work; previously, we have seen the proofs of the

---

[1]See [**108**], Proposition 5.44, for the case of curvature changing sign.
[2]See (5.70).
[3]See Theorem 5.38.

expanding and steady cases in Proposition 1.13. To prove the nonexistence of nontrivial breathers, Perelman needed to do a separate study of each type of breather. However, in each case, the method is the same: introduce a new functional, study its properties, and apply them to the proof that there are no nontrivial breathers of each type. All such functionals have three basic characteristics:

- they are nondecreasing along systems of equations including the Ricci flow,
- they are invariant under diffeomorphisms and/or homotheties,
- their critical points are gradient Ricci solitons (of a different type in each case).

Moreover, Perelman's functionals are successive modifications of his initial functional $\mathcal{F}$ and are motivated by the consideration of gradient Ricci solitons of each type. So it is important to study the cases of the proofs successively in order to see how the evolutions of the functionals are used and how to modify the functionals gradually to define the entropy functional, which is the key to proving the shrinking case and where the proof follows essentially the same steps as the other two cases but uses the new functional.

In this chapter, we shall discuss in detail the energy functional, its geometric applications and its relation with classical entropy; in the next chapter we study Perelman's entropy and some of its geometric applications. The style of this chapter is that of filling in the details of §§1–2 of Perelman [**297**] in the hopes of aiding the reader in their perusal of [**297**]. Throughout this chapter $\mathcal{M}^n$ is a closed n-manifold.

1. Energy, its first variation, and the gradient flow

The Ricci flow is not a gradient flow of a functional on the space $\mathfrak{Met}$ of smooth metrics on a manifold $\mathcal{M}^n$ with respect to the standard L^2-inner product.[4] On the other hand, variational methods have played major roles in geometric analysis, partial differential equations, and mathematical physics. It was unusual that the Ricci flow, a natural geometric partial differential equation, should appear to be an exception to this. Perelman's introduction of the $\mathcal{F}$ functional (defined below) solved the important question of whether the Ricci flow can be seen as a gradient flow. More precisely, as we shall see in this and the following section, the Ricci flow is a *gradient-like flow*; it is a gradient flow when we enlarge the system. The key to solving the question above is to look for functionals whose critical points are Ricci solitons, that is, fixed points of the Ricci flow modulo diffeomorphisms and homotheties (so that the ambient space in which we consider Ricci flow is $\mathfrak{Met}/\mathfrak{Diff} \times \mathbb{R}_+$ instead of $\mathfrak{Met}$). This is consistent with the point of view we adopted in Chapter 1 on Ricci solitons.

---

[4]An exception is when $n = 2$ (see Appendix B of [**111**]), and more generally, for the Kähler-Ricci flow.

1. ENERGY, ITS FIRST VARIATION, AND THE GRADIENT FLOW

1.1. The energy functional $\mathcal{F}$. Let $C^\infty(\mathcal{M})$ denote the set of all smooth functions on a closed manifold $\mathcal{M}^n$. We define the **energy functional** $\mathcal{F}: \mathfrak{Met} \times C^\infty(\mathcal{M}) \to \mathbb{R}$ by

$$(5.1) \qquad \mathcal{F}(g, f) \doteqdot \int_{\mathcal{M}} \left(R + |\nabla f|^2\right) e^{-f} d\mu.$$

Note, in addition to the metric, the introduction of a function f. This embeds the space of metrics in a larger space. We shall sometimes follow the physics literature and call f the **dilaton**.

Since $\Delta\left(e^{-f}\right) = \left(-\Delta f + |\nabla f|^2\right) e^{-f}$, we see from $\int_{\mathcal{M}} \Delta\left(e^{-f}\right) d\mu = 0$ that

$$(5.2) \qquad \int_{\mathcal{M}} |\nabla f|^2 e^{-f} d\mu = \int_{\mathcal{M}} \Delta f e^{-f} d\mu.$$

So we have two other expressions for the energy:

$$(5.3) \qquad \mathcal{F}(g, f) = \int_{\mathcal{M}} (R + \Delta f) e^{-f} d\mu$$

$$(5.4) \qquad = \int_{\mathcal{M}} (R + 2\Delta f - |\nabla f|^2) e^{-f} d\mu.$$

The second way of expressing the energy is motivated by the pointwise formula (5.43) in subsection 2.3.2 below.

LEMMA 5.1 (Elementary properties of $\mathcal{F}$).

(1) Dirichlet-type energy. *The geometric aspect of $\mathcal{F}$ is reflected by $\mathcal{F}(g, 0) = \int_{\mathcal{M}} R\, d\mu$ being the total scalar curvature and the function theory aspect of $\mathcal{F}$ is reflected by expressing it as*

$$(5.5) \qquad \mathcal{F}(g, f) = \int_{\mathcal{M}} \left(4|\nabla w|^2 + Rw^2\right) d\mu \doteqdot \mathcal{G}(g, w),$$

where $w = e^{-f/2}$, which is a Dirichlet energy with a potential term.

(2) Diffeomorphism invariance. *For any diffeomorphism φ of $\mathcal{M}$, we have*

$$\mathcal{F}(\varphi^* g, f \circ \varphi) = \mathcal{F}(g, f).$$

(3) Scaling. *For any $c > 0$ and b*

$$\mathcal{F}(c^2 g, f + b) = c^{n-2} e^{-b} \mathcal{F}(g, f).$$

EXERCISE 5.2. Prove the properties for the energy in the lemma above.

1.2. The first variation of $\mathcal{F}$. We use the symbol δ to denote the variation of a tensor. We shall denote the variations of the metric and dilaton as $\delta g = v \in C^\infty(T^*\mathcal{M} \otimes_S T^*\mathcal{M})$ and $\delta f = h \in C^\infty(\mathcal{M})$, and we define $V \doteq g^{ij}v_{ij}$. Routine calculations give

(5.6) $$\delta_v \Gamma_{ij}^k(g) = \frac{1}{2}g^{kl}\left(\nabla_i v_{jl} + \nabla_j v_{il} - \nabla_l v_{ij}\right),$$

(5.7) $$\delta_v \Gamma_{pj}^p = \frac{1}{2}\nabla_j V,$$

(5.8) $$\delta_{(v,h)}\left(e^{-f}d\mu\right) = \left(\frac{V}{2} - h\right)e^{-f}d\mu.$$

We calculate the last one, for example,

(5.9) $$\delta_{(v,h)}\left(e^{-f}d\mu\right) = -e^{-f}h\,d\mu + e^{-f}\frac{1}{2}g^{ij}v_{ij}\,d\mu = \left(\frac{V}{2} - h\right)e^{-f}d\mu.$$

LEMMA 5.3 (First variation of $\mathcal{F}$). *Then the **first variation of $\mathcal{F}$** can be expressed as*

(5.10) $$\delta_{(v,h)}\mathcal{F}(g,f) = -\int_{\mathcal{M}} v_{ij}(R_{ij} + \nabla_i\nabla_j f)e^{-f}d\mu$$
$$+ \int_{\mathcal{M}}\left(\frac{V}{2} - h\right)\left(2\Delta f - |\nabla f|^2 + R\right)e^{-f}d\mu,$$

where $\delta_{(v,h)}\mathcal{F}(g,f)$ denotes the variation of $\mathcal{F}$ at (g,f) in the direction (v,h), i.e.,

$$\delta_{(v,h)}\mathcal{F}(g,f) \doteq \left.\frac{d}{ds}\right|_{s=0} \mathcal{F}(g + sv, f + sh).$$

PROOF. Recall (V1-p. 92), i.e.,

$$R_{ij} = R_{pij}^p = \partial_p\Gamma_{ij}^p - \partial_i\Gamma_{pj}^p + \Gamma_{ij}^q\Gamma_{pq}^p - \Gamma_{pj}^q\Gamma_{iq}^p,$$

so that

$$\delta R_{ij} = \nabla_p\left(\delta\Gamma_{ij}^p\right) - \nabla_i\left(\delta\Gamma_{pj}^p\right).$$

Since $\nabla_i\nabla_j = \partial_i\partial_j - \Gamma_{ij}^k\partial_k$ as an operator acting on functions, we have

$$\delta\left(\nabla_i\nabla_j f\right) = \nabla_i\nabla_j\left(\delta f\right) - \left(\delta\Gamma_{ij}^p\right)\nabla_p f.$$

Hence, using (5.7),

$$\delta\left(R_{ij} + \nabla_i\nabla_j f\right) = \nabla_p\left(\delta\Gamma_{ij}^p\right) - \left(\delta\Gamma_{ij}^p\right)\nabla_p f + \nabla_i\left(\nabla_j\left(\delta f\right) - \delta\Gamma_{pj}^p\right)$$
$$= e^f\nabla_p\left(e^{-f}\delta\Gamma_{ij}^p\right) + \nabla_i\nabla_j\left(h - \frac{V}{2}\right).$$

We then compute

(5.11) $$\delta\left[\left(R_{ij} + \nabla_i\nabla_j f\right)e^{-f}d\mu\right]$$
$$= \left[\begin{array}{c}\nabla_p\left(e^{-f}\delta\Gamma_{ij}^p\right) + e^{-f}\nabla_i\nabla_j\left(h - \frac{V}{2}\right) \\ + \left(R_{ij} + \nabla_i\nabla_j f\right)e^{-f}\left(\frac{V}{2} - h\right)\end{array}\right]d\mu.$$

1. ENERGY, ITS FIRST VARIATION, AND THE GRADIENT FLOW

So using (5.8),

$$\delta \left[(R + \Delta f) e^{-f} d\mu \right]$$
$$= g^{ij} \delta \left[(R_{ij} + \nabla_i \nabla_j f) e^{-f} d\mu \right] - \delta g_{ij} \cdot (R_{ij} + \nabla_i \nabla_j f) e^{-f} d\mu$$
$$= \left[\nabla_p \left(e^{-f} g^{ij} \delta \Gamma_{ij}^p \right) + e^{-f} \Delta \left(h - \frac{V}{2} \right) + (R + \Delta f) e^{-f} \left(\frac{V}{2} - h \right) \right] d\mu$$
$$- v_{ij} \cdot (R_{ij} + \nabla_i \nabla_j f) e^{-f} d\mu.$$

Note that $\delta \Gamma_{ij}^p$ is a tensor and we do not need an explicit formula for it in the rest of the proof.

By the Divergence Theorem, we have

$$\delta_{(v,h)} \mathcal{F}(g,f) = \int_{\mathcal{M}} \delta \left[(R + \Delta f) e^{-f} d\mu \right]$$
$$= \int_{\mathcal{M}} \left(-\Delta \left(e^{-f} \right) + (R + \Delta f) e^{-f} \right) \left(\frac{V}{2} - h \right) d\mu$$
$$- \int_{\mathcal{M}} v_{ij} \cdot (R_{ij} + \nabla_i \nabla_j f) e^{-f} d\mu,$$

from which the lemma follows. $\square$

REMARK 5.4. By (5.11), the variation of $(R_{ij} + \nabla_i \nabla_j f) e^{-f} d\mu$ is a divergence when $h = \frac{V}{2}$:

$$\delta \left[(R_{ij} + \nabla_i \nabla_j f) e^{-f} d\mu \right] = \nabla_p \left(e^{-f} \delta \Gamma_{ij}^p \right) d\mu.$$

Note also the factor $\frac{V}{2} - h$ in front of the second term in the RHS of (5.10). The significance of when this factor vanishes will be seen in subsection 1.4 below. By (5.8) we have

LEMMA 5.5. *Define the measure*

$$dm \doteqdot e^{-f} d\mu.$$

If the variations of g and f keep the measure dm fixed, that is, $\delta_{(v,h)}(dm) = 0$, then

(5.12) $$V = 2h.$$

As a consequence of Lemma 5.3, we have

COROLLARY 5.6 (Measure-preserving first variation of $\mathcal{F}$). *For variations (v,h) with $\delta_{(v,h)} (e^{-f} d\mu) = 0$, we have*

(5.13) $$\delta_{(v,h)} \mathcal{F}(g,f) = -\int_{\mathcal{M}} v_{ij} (R_{ij} + \nabla_i \nabla_j f) e^{-f} d\mu.$$

Notice in formula (5.10) for $\delta_{(v,h)} \mathcal{F}(g,f)$ the occurrence of the terms

(5.14) $$R_{ij}^m \doteqdot (\mathrm{Rc}^m)_{ij} \doteqdot R_{ij} + \nabla_i \nabla_j f,$$
(5.15) $$R^m \doteqdot R + 2\Delta f - |\nabla f|^2.$$

The first quantity vanishes on steady gradient solitons flowing along ∇f, whereas the second appeared in (5.4).[5] We call R_{ij}^m and R^m the **modified Ricci curvature and modified scalar curvature**, respectively; they are natural quantities from the perspective of the Ricci flow. We can rewrite

$$\mathcal{F}(g,f) = \int_{\mathcal{M}} g^{ij} R_{ij}^m e^{-f} d\mu = \int_{\mathcal{M}} R^m e^{-f} d\mu$$

and

$$\delta_{(v,h)} \mathcal{F}(g,f) = -\int_{\mathcal{M}} v_{ij} R_{ij}^m e^{-f} d\mu$$

when $V = 2h$.

1.3. The modified Ricci and scalar curvatures. In this subsection we digress by showing R_{ij}^m and R^m are natural quantities. Consider a closed Riemannian manifold $(\mathcal{M}^n, g)$ and a metric $\bar{g} = e^{-\frac{2}{n}f} g$ conformal to g. Let $\bar{R}_{ij} = \mathrm{Rc}(\bar{g})_{ij}$, $R_{ij} = \mathrm{Rc}(g)_{ij}$, $\bar{R} = R(\bar{g})$, and $R = R(g)$. The Ricci and scalar curvatures are related by (see for example subsection 7.2 of Chapter 1 in [**111**] or (A.2) and (A.3) in this volume)

$$\bar{R}_{ij} = R_{ij} + \left(1 - \frac{2}{n}\right) \nabla_i \nabla_j f + \frac{1}{n} \Delta f g_{ij} + \frac{n-2}{n^2} \nabla_i f \nabla_j f - \frac{n-2}{n^2} |\nabla f|^2 g_{ij}. \tag{5.16}$$

Tracing this yields

$$\bar{R} = e^{\frac{2}{n}f} \left(R + \frac{2(n-1)}{n} \Delta f - \frac{(n-1)(n-2)}{n^2} |\nabla f|^2 \right). \tag{5.17}$$

The volume forms are related by $d\mu_{\bar{g}} = e^{-f} d\mu$ and the total scalar curvature of $\bar{g}$ is given by

$$\int_{\mathcal{M}} \bar{R} d\mu_{\bar{g}} = \int_{\mathcal{M}} e^{-\frac{n-2}{n}f} \left(R + \frac{(n-1)(n-2)}{n^2} |\nabla f|^2 \right) d\mu,$$

where we integrated by parts, i.e., we used

$$\int_{\mathcal{M}} e^{-\frac{n-2}{n}f} \Delta f d\mu = \frac{n-2}{n} \int_{\mathcal{M}} e^{-\frac{n-2}{n}f} |\nabla f|^2 d\mu.$$

Now consider the Riemannian product $(\mathcal{M}^n, g) \times (T^q, h_q)$, where (T^q, h_q) is a flat unit volume q-dimensional torus. The formulas for the Ricci curvature and scalar curvature of metric $e^{-\frac{2}{n+q}f}(g + h_q)$ are given by (5.16) and (5.17), respectively, where we replace n by $n + q$. If we take the limit as $q \to \infty$ while fixing $(\mathcal{M}^n, g)$, then we obtain Perelman's modified Ricci tensor:

$$\lim_{q \to \infty} \mathrm{Rc}\left(e^{-\frac{2}{n+q}f}(g + h_q)\right) = \mathrm{Rc} + \nabla \nabla f \tag{5.18}$$

and Perelman's modified scalar curvature:

$$\lim_{q \to \infty} R\left(e^{-\frac{2}{n+q}f}(g + h_q)\right) = R + 2\Delta f - |\nabla f|^2, \tag{5.19}$$

---

[5]Earlier we also encountered these quantities in Chapter 1.

where we think of $\operatorname{Rc}\left(e^{-\frac{2}{n+q}f}(g+h_q)\right)$ and $R\left(e^{-\frac{2}{n+q}f}(g+h_q)\right)$ as quantities on $\mathcal{M}$ since they are independent of the point in T^q. The total scalar curvatures of $\left(\mathcal{M}\times T^q, e^{-\frac{2}{n+q}f}(g+h_q)\right)$ limit to Perelman's $\mathcal{F}$ functional:

$$\lim_{q\to\infty}\int_{\mathcal{M}\times T^q} R\left(e^{-\frac{2}{n+q}f}(g+h_q)\right) d\mu_{e^{-\frac{2}{n+q}f}(g+h_q)}$$
$$= \lim_{q\to\infty}\int_\mathcal{M}\int_{T^q} R\left(e^{-\frac{2}{n+q}f}(g+h_q)\right) e^{-2f} d\mu_{h_q} d\mu_g$$
$$= \int_\mathcal{M} \left(R+|\nabla f|^2\right) e^{-f} d\mu$$
$$= \mathcal{F}(g,f).$$

Note that
$$g^{ij} R_{ij}^m = R + \Delta f = R^m - \Delta f + |\nabla f|^2.$$
There is an analogue of the contracted second Bianchi identity for R_{ij}^m and R^m. In particular we compute
$$\nabla_i R_{ij}^m = \nabla_i R_{ij} + \nabla_i \nabla_j \nabla_i f = \frac{1}{2}\nabla_j R + \nabla_j \Delta f + R_{jk}\nabla_k f$$
and
$$\frac{1}{2}\nabla_j R^m = \nabla_j \Delta f - \frac{1}{2}\nabla_j |\nabla f|^2 + \frac{1}{2}\nabla_j R = \frac{1}{2}\nabla_j R + \nabla_j \Delta f - \nabla_j \nabla_k f \nabla_k f,$$
which imply

(5.20) $$\nabla_i R_{ij}^m = \frac{1}{2}\nabla_j R^m + R_{jk}^m \nabla_k f.$$

To understand this formula further, we define
$$\nabla^{*m} : C^\infty(T^*\mathcal{M} \otimes_S T^*\mathcal{M}) \to C^\infty(T^*\mathcal{M})$$
by
$$(\nabla^{*m} a)_j \doteqdot \nabla_i a_{ij} - a_{ji}\nabla_i f.$$

LEMMA 5.7. *The operator ∇^{*m} is the adjoint of $-\nabla$ with respect to the measure $dm = e^{-f}d\mu$.*

PROOF. For any symmetric 2-tensor a_{ij} and 1-form b_i,
$$\int_\mathcal{M} a_{ij}(-\nabla_i) b_j e^{-f} d\mu = \int_\mathcal{M} b_j \nabla_i\left(a_{ij} e^{-f}\right) d\mu$$
$$= \int_\mathcal{M} b_j (\nabla_i a_{ij} - a_{ij}\nabla_i f) e^{-f} d\mu$$
$$= \int_\mathcal{M} b_j (\nabla^{*m} a)_j e^{-f} d\mu.$$

□

Thus (5.20) implies the following, which is the analogue of the contracted second Bianchi identity.

LEMMA 5.8 (Modified contracted second Bianchi identity).

$$\nabla_i^{*m} R_{ij}^m \doteq (\nabla^{*m} \operatorname{Rc}^m)_j = \frac{1}{2} \nabla_j R^m. \tag{5.21}$$

1.4. The functional $\mathcal{F}^m$ and its gradient flow. Unlike $\mathcal{F}(g, f)$, we can obtain a functional of just the metric g by fixing a measure dm on a closed manifold $\mathcal{M}^n$; by a **measure** we mean a positive n-form on $\mathcal{M}$.[6] Define $\mathcal{F}^m : \mathfrak{Met} \to \mathbb{R}$ by

$$\mathcal{F}^m(g) \doteq \mathcal{F}(g, f) = \int_{\mathcal{M}} (R + |\nabla f|^2) dm, \tag{5.22}$$

where

$$f \doteq \log\left(\frac{d\mu}{dm}\right). \tag{5.23}$$

REMARK 5.9. The expression (5.23) makes sense because, given a fixed measure dm on $\mathcal{M}^n$, we can define the bijection

$$C^\infty(\Lambda^n T^*\mathcal{M}) \to C^\infty(\mathcal{M}),$$
$$\omega \mapsto \varphi,$$

where φ is defined so that $\omega = \varphi dm$ (here we have used the fact that $\Lambda^n T_x^*\mathcal{M} \cong \mathbb{R}$). Thanks to this, it is possible to define the quotient of two n-forms; e.g., if $\omega_1 = \varphi_1 dm$ and $\omega_2 = \varphi_2 dm$, where $\varphi_2 > 0$, then we set

$$\frac{\omega_1}{\omega_2} \doteq \frac{\varphi_1}{\varphi_2}.$$

Without using the notation f, we can write the energy of the metric g as

$$\mathcal{F}^m(g) = \int_{\mathcal{M}} \left(R + \left|\nabla \log\left(\frac{d\mu}{dm}\right)\right|^2\right) dm.$$

Using the modified Ricci and scalar curvatures, we can rewrite

$$\mathcal{F}^m(g) = \int_{\mathcal{M}} g^{ij} R_{ij}^m dm = \int_{\mathcal{M}} R^m dm.$$

REMARK 5.10. Let $\varphi : \mathcal{M} \to \mathcal{M}$ be a diffeomorphism. Note that in general

$$\mathcal{F}^m(\varphi^* g) \neq \mathcal{F}^m(g).$$

That is, by fixing the measure dm, we get $\mathcal{F}^m(g)$, which breaks the diffeomorphism invariance of $\mathcal{F}(g, f)$. In subsection 3.1 of this chapter we shall solve this problem by considering a functional $\lambda(g)$ which is diffeomorphism-invariant.

---

[6]For a calculational motivation for fixing the measure, see the notes and commentary at the end of this chapter.

From (5.13) we have

(5.24) $$\delta_v \mathcal{F}^m(g) = -\int_\mathcal{M} v_{ij}(R_{ij} + \nabla_i \nabla_j f)\, dm,$$

where f is given by (5.23). The L^2-inner product on $\mathfrak{Met}$, using the metric g and the measure dm, is defined by

$$\langle a_{ij}, b_{ij}\rangle_m(g) \doteq \int_\mathcal{M} \langle a_{ij}, b_{ij}\rangle_g\, dm.$$

Then by (5.24) we have

$$\nabla \mathcal{F}^m(g) = -(R_{ij} + \nabla_i \nabla_j f),$$

where f is given by (5.23). Hence (twice) the positive **gradient flow** of $\mathcal{F}^m$ is

(5.25) $$\frac{\partial}{\partial t} g_{ij} = -2(R_{ij} + \nabla_i \nabla_j f),$$

(5.26) $$f = \log\left(\frac{d\mu}{dm}\right).$$

We can also write the above system as

(5.27) $$\frac{\partial}{\partial t} g_{ij} = -2\left[R_{ij} + \nabla_i \nabla_j \log\left(\frac{d\mu}{dm}\right)\right].$$

We shall call an equation of the form (5.25) by itself, for some function f, a **modified Ricci flow**.

It is clear from taking $v_{ij} = -2(R_{ij} + \nabla_i \nabla_j f)$ in (5.13) that we obtain the following.

PROPOSITION 5.11 ($\mathcal{F}^m$ evolution under modified Ricci flow). *Suppose $g(t)$ is a solution of (5.25)–(5.26). Then*

(5.28) $$\frac{d}{dt}\mathcal{F}^m(g(t)) = 2\int_\mathcal{M} |R_{ij} + \nabla_i \nabla_j f|^2 e^{-f}\, d\mu.$$

This is Perelman's **monotonicity formula for the gradient flow** of $\mathcal{F}^m$. We may rewrite (5.28) as

$$\frac{d}{dt}\mathcal{F}^m = \frac{d}{dt}\int_\mathcal{M} R^m dm = 2\int_\mathcal{M} |R_{ij}^m|^2\, dm.$$

Note that for a general measure dm, solutions to the initial-value problem for the gradient flow may **not** exist even for a short time; however, as we shall see, this will not cause us problems in applications.

2. Monotonicity of energy for the Ricci flow

For monotonicity formula (5.28) to be useful, we need a corresponding version for solutions of the Ricci flow. In this section we show that solutions to equations (5.25) and (5.26), if they exist, differ from solutions of the Ricci flow by the pullback by time-dependent diffeomorphisms. Thus this gives a monotonicity formula for the energy of the Ricci flow.

2.1. A coupled system equivalent to the gradient flow of $\mathcal{F}^m$.

There is a coupled system, i.e., (5.29)–(5.30), induced from the gradient flow (5.25)–(5.26) obtained simply by computing the evolution equation for $f = \log(d\mu/dm)$. As we shall see, this coupled system is equivalent to the gradient flow.

LEMMA 5.12 (Measure-preserving evolution of f under modified RF). *The function $f(t)$ in a solution $(g(t), f(t))$ of the gradient flow of $\mathcal{F}^m$ (5.25) and (5.26) satisfies the following equation:*
$$\frac{\partial f}{\partial t} = -\Delta f - R.$$

PROOF. We calculate
$$\frac{\partial f}{\partial t} = \frac{\partial}{\partial t} \log\left(\frac{d\mu}{dm}\right) = \frac{1}{2} g^{ij} \frac{\partial g_{ij}}{\partial t} = -g^{ij}(R_{ij} + \nabla_i \nabla_j f).$$
$\square$

Related to the above calculation, we have the following.

EXERCISE 5.13. Show that if $\omega_1(t)$ and $\omega_2(t)$ are time-dependent n-forms, then
$$\frac{\partial}{\partial t} \log\left(\frac{\omega_1}{\omega_2}\right) = \frac{\frac{\partial}{\partial t}\omega_1}{\omega_1} - \frac{\frac{\partial}{\partial t}\omega_2}{\omega_2},$$
where the quotient of two n-forms is defined as in Remark 5.9.

Hence we consider the **coupled modified Ricci flow**

(5.29) $$\frac{\partial}{\partial t} g_{ij} = -2(R_{ij} + \nabla_i \nabla_j f),$$

(5.30) $$\frac{\partial f}{\partial t} = -\Delta f - R.$$

Note that the first equation is a modified Ricci flow equation whereas the second equation is a **backward heat equation**.

LEMMA 5.14. *The coupled modified Ricci flow equations (5.29)–(5.30) are equivalent to the gradient flow (5.27).*

PROOF. If $g(t)$ is a solution to (5.27), then by Lemma 5.12, $(g(t), f(t))$, where $f = \log(d\mu/dm)$, is a solution to the system (5.29)–(5.30).

Conversely, if $(g(t), f(t))$ is a solution to the system (5.29)–(5.30), then $dm \doteqdot e^{-f} d\mu$ satisfies
$$\frac{\partial}{\partial t}(dm) = \left(-\frac{\partial f}{\partial t} - R - \Delta f\right) e^{-f} d\mu = 0;$$
that is, $g(t)$ is a solution to (5.27) with dm as defined above. $\square$

Hence, by (5.28), if $(g(t), f(t))$ is a solution to (5.29)–(5.30), then

(5.31) $$\frac{d}{dt} \mathcal{F}(g(t), f(t)) = 2 \int_{\mathcal{M}} |R_{ij} + \nabla_i \nabla_j f|^2 e^{-f} d\mu.$$

2.2. Correspondence between solutions of the gradient flow and solutions of the Ricci flow.

2.2.1. *Converting a solution of the gradient flow to a solution of Ricci flow.* We first show that solutions of the gradient flow, if they exist, give rise to solutions of the Ricci flow with the same initial data (Lemma 5.15). In particular, suppose we have a solution $(\bar{g}(t), \bar{f}(t))$ of the flow (5.25) and (5.26) on $[0, T]$; then we can obtain a solution $g(t)$ of the Ricci flow on $[0, T]$ by modifying $\bar{g}(t)$ by diffeomorphisms generated by the gradient of $\bar{f}(t)$.

LEMMA 5.15 (Perelman's coupling for Ricci flow). *Let $(\bar{g}(t), \bar{f}(t))$ be a solution of (5.25) and (5.26) on $[0, T]$. We define a 1-parameter family of diffeomorphisms $\Psi(t) : \mathcal{M} \to \mathcal{M}$ by*

$$\frac{d}{dt}\Psi(t) = \nabla_{\bar{g}(t)}\bar{f}(t), \tag{5.32}$$

$$\Psi(0) = \mathrm{id}_{\mathcal{M}}. \tag{5.33}$$

Then the pullback metric $g(t) = \Psi(t)^*\bar{g}(t)$ and the dilaton $f(t) = \bar{f} \circ \Psi(t)$ satisfy the following system:

$$\frac{\partial g}{\partial t} = -2\operatorname{Rc}, \tag{5.34}$$

$$\frac{\partial f}{\partial t} = -\Delta f + |\nabla f|^2 - R. \tag{5.35}$$

REMARK 5.16. Basically we can see this from the facts that $L_{\nabla f} g = 2\nabla\nabla f$ and $L_{\nabla f} f = |\nabla f|^2$. For the sake of completeness we give the detailed calculations below.

PROOF. First note that by Lemma 3.15 of Volume One the system of ODE (5.32)–(5.33) is always solvable. We compute

$$\frac{\partial g}{\partial t} = \Psi^*\left(\frac{\partial \bar{g}}{\partial t}\right) + \Psi^*\left(L_{\nabla_{\bar{g}}\bar{f}}\bar{g}\right) = -2\Psi^*\left(\operatorname{Rc}(\bar{g})\right) = -2\operatorname{Rc}(g).$$

To obtain the equation for $\frac{\partial f}{\partial t}$, we compute

$$\frac{\partial f}{\partial t} = \frac{\partial(\bar{f} \circ \Psi)}{\partial t} = \frac{\partial \bar{f}}{\partial t} \circ \Psi + \left\langle (\bar{\nabla}\bar{f}) \circ \Psi, \frac{\partial \Psi}{\partial t}\right\rangle_{\bar{g}}$$

$$= (-\bar{\Delta}\bar{f} - \bar{R}) \circ \Psi + |(\bar{\nabla}\bar{f}) \circ \Psi|^2_{\bar{g}}$$

$$= -\Delta f - R + |\nabla f|^2,$$

where barring a quantity indicates that it corresponds to $\bar{g}(t)$. □

So a solution to the gradient flow (5.25)–(5.26) yields a solution to the Ricci flow-backward heat equation system (5.34)–(5.35). Note that we can first solve the Ricci flow (5.34) forward in time and then solve (5.35) backward in time to get a solution of (5.34)–(5.35); this will be useful in applications.

2.2.2. Converting a solution of Ricci flow to a solution of the gradient flow.
Now we show the converse of Lemma 5.15 by reversing the procedure of the last subsection. Given a solution $g(t)$ of the Ricci flow (5.34) on $[0, T]$, we can construct a solution $(\bar{g}(t), \bar{f}(t))$ of the gradient flow (5.25) and (5.26) on $[0, T]$ by modifying the solution $g(t)$ by diffeomorphisms. In doing so, we also need to solve a backward heat equation with initial data at time T.

LEMMA 5.17. *Let $g(t)$ be a solution of the Ricci flow $\frac{\partial g}{\partial t} = -2\operatorname{Rc}$ on $[0, T]$ and let f_T be a function on $\mathcal{M}$.*

(i) We can solve the backward heat equation backwards in time
$$\frac{\partial f}{\partial t} = -\Delta f + |\nabla f|^2 - R, \qquad t \in [0, T],$$
$$f(T) = f_T.$$

(ii) Given a solution $f(t)$ to the equation above, define the 1-parameter family of diffeomorphisms $\Phi(t) : \mathcal{M} \to \mathcal{M}$ by

(5.36) $$\frac{d}{dt}\Phi(t) = -\nabla_{g(t)} f(t), \qquad \Phi(0) = \operatorname{id}_{\mathcal{M}},$$

which is a system of ODE and hence is solvable on $[0, T]$.[7] Then the pulled-back metrics $\bar{g}(t) = \Phi(t)^* g(t)$ and the pulled-back dilaton $\bar{f}(t) = f \circ \Phi(t)$ satisfy (5.29) and (5.30).

PROOF. (i) Let $\tau = T - t$. To get the existence of solutions to equation (5.35), we simply set

(5.37) $$u \doteqdot e^{-f}$$

and compute that

(5.38) $$\frac{\partial u}{\partial \tau} = \Delta u - Ru,$$

which is a linear parabolic equation and has a solution on $[0, T]$ with initial data at $\tau = 0$. Indeed, (5.38) follows from
$$\frac{\partial u}{\partial \tau} = -\frac{\partial u}{\partial t} = u\frac{\partial f}{\partial t} = u\left(-\Delta f + |\nabla f|^2 - R\right) = \Delta u - Ru.$$

(ii) Let $g(t)$ be a solution of the Ricci flow and let $f(t)$ be a solution of equation (5.35). One can verify that they satisfy (5.29) and (5.30) as in the proof of Lemma 5.15. □

2.2.3. The adjoint heat equation.
Let $g(t)$ be a solution of Ricci flow and let $\square \doteqdot \frac{\partial}{\partial t} - \Delta$ be the heat operator acting on functions on $\mathcal{M} \times [0, T]$, where $\mathcal{M} \times [0, T]$ is endowed with the volume form $d\mu dt$. Its adjoint is

(5.39) $$\square^* \doteqdot -\frac{\partial}{\partial t} - \Delta + R$$

---
[7] Again see Lemma 3.15 of Volume One.

since

$$\int_0^T \int_\mathcal{M} b\Box a\, d\mu dt = \int_0^T \int_\mathcal{M} b\left(\frac{\partial}{\partial t} - \Delta\right) a\, d\mu dt$$
$$= \int_0^T \int_\mathcal{M} \left[a\left(-\frac{\partial}{\partial t} - \Delta\right) b\, d\mu - ab\frac{\partial}{\partial t} d\mu\right] dt$$
$$= \int_0^T \int_\mathcal{M} a\Box^* b\, d\mu dt$$

for C^2 functions a and b on $\mathcal{M} \times [0,T]$ with compact support in $\mathcal{M} \times (0,T)$, where we used $\frac{\partial}{\partial t} d\mu = -R d\mu$.

By (5.38), if $(g(t), f(t))$ is a solution to (5.34)–(5.35), then $u = e^{-f}$ satisfies the **adjoint heat equation** (also known as the **conjugate heat equation**)

(5.40) $$\Box^* u = \left(-\frac{\partial}{\partial t} - \Delta + R\right) u = 0.$$

It is often better to think in terms of u than in terms of f since u satisfies the adjoint heat equation. In particular, the fundamental solution to the adjoint heat equation is important.

2.3. Monotonicity of $\mathcal{F}$ for the Ricci flow. In this subsection we give two proofs of the monotonicity of energy for Ricci flow. In the next section we give an application of this formula to the nonexistence of nontrivial breather solutions.

2.3.1. *Deriving the monotonicity of $\mathcal{F}$ from the monotonicity of $\mathcal{F}^m$.* By the diffeomorphism invariance of all the quantities under consideration, the monotonicity formula for the gradient flow implies a **monotonicity formula for the Ricci flow**. This involves a function $f(t)$ obtained by solving the backward heat equation (5.35).

LEMMA 5.18 ($\mathcal{F}$ energy monotonicity). *If $(g(t), f(t))$ is a solution to (5.34)–(5.35) on a closed manifold $\mathcal{M}^n$, then*

(5.41) $$\frac{d}{dt}\mathcal{F}(g(t), f(t)) = 2\int_\mathcal{M} |R_{ij} + \nabla_i \nabla_j f|^2 e^{-f} d\mu.$$

PROOF. Since $(g(t), f(t))$ is a solution to (5.34)–(5.35), $(\bar{g}(t), \bar{f}(t))$, defined by $\bar{g}(t) \doteqdot \Phi^*(t) g(t)$ and $\bar{f}(t) = f(t) \circ \Phi(t)$, where $\Phi(t)$ satisfies (5.36), is a solution to (5.29)–(5.30). Now $\mathcal{F}(g, f) = \mathcal{F}(\bar{g}, \bar{f})$, so that by (5.31), we have

$$\frac{d}{dt}\mathcal{F}(g(t), f(t)) = \frac{d}{dt}\mathcal{F}(\bar{g}(t), \bar{f}(t))$$
$$= 2\int_\mathcal{M} \left|\bar{R}_{ij} + \bar{\nabla}_i \bar{\nabla}_j \bar{f}\right|^2_{\bar{g}} e^{-\bar{f}} d\bar{\mu}$$
$$= 2\int_\mathcal{M} |R_{ij} + \nabla_i \nabla_j f|^2 e^{-f} d\mu.$$

2.3.2. *Deriving the monotonicity of $\mathcal{F}$ from a pointwise estimate.* This second approach to the energy monotonicity formula is based on the pointwise formula (5.43), which is a simpler version of the evolution equation for Perelman's backward Harnack quantity (6.22).

Let $(g(t), f(t))$ be a solution to (5.34)–(5.35). Let $u = e^{-f}$ and

(5.42) $$V \doteqdot (2\Delta f - |\nabla f|^2 + R)u = R^m u,$$

where R^m is the modified scalar curvature defined by (5.15),[8] so that

$$\mathcal{F} = \int_{\mathcal{M}} V d\mu.$$

LEMMA 5.19 (Bochner-type formula for V). *If $(g(t), f(t))$ is a solution to (5.34)–(5.35) and if $u = e^{-f}$, then we have the pointwise differential equality:*

(5.43) $$\square^* V = -2|R_{ij} + \nabla_i \nabla_j f|^2 u.$$

This calculation, which we carry out below, is in a similar spirit to that of the calculations for the differential Harnack quantities considered in §10 of Chapter 5 in Volume One and Part II of this volume. To obtain (5.41) from the lemma, we compute

$$\frac{d}{dt} \mathcal{F}(g(t), f(t)) = \frac{d}{dt} \int_{\mathcal{M}} V d\mu$$
$$= \int_{\mathcal{M}} \left(\frac{\partial}{\partial t} V - RV \right) d\mu$$
$$= \int_{\mathcal{M}} 2|R_{ij} + \nabla_i \nabla_j f|^2 u \, d\mu.$$

PROOF OF THE LEMMA. Using definition (5.42) and $g^{ij} \frac{\partial}{\partial t} \Gamma_{ij}^k = 0$, a direct calculation shows that

$$\frac{\partial}{\partial t} R^m = \frac{\partial}{\partial t}(2\Delta f - |\nabla f|^2 + R)$$
$$= 4R_{ij}\nabla_i\nabla_j f + 2\Delta\left(\frac{\partial f}{\partial t}\right) - 2R_{ij}\nabla_i f \nabla_j f - 2\nabla\left(\frac{\partial f}{\partial t}\right) \cdot \nabla f + \frac{\partial R}{\partial t}$$
$$= 4R_{ij}\nabla_i\nabla_j f - \Delta(2\Delta f - |\nabla f|^2 + R) + \Delta|\nabla f|^2 + 2\nabla\Delta f \cdot \nabla f$$
$$\quad - 2R_{ij}\nabla_i f \nabla_j f - 2\nabla(|\nabla f|^2 - R)\nabla f + \frac{\partial R}{\partial t} - \Delta R.$$

From the above we have

$$\left(\frac{\partial}{\partial t} + \Delta \right) R^m = 2|R_{ij} + \nabla_i \nabla_j f|^2 + 2\nabla R^m \cdot \nabla f.$$

---

[8]The above V is not to be confused with our earlier V, which was the trace of the variation v of g.

On the other hand,

$$\frac{\partial V}{\partial t} + \Delta V - RV = \left(\frac{\partial R^m}{\partial t} + \Delta R^m\right) u + \left(\frac{\partial u}{\partial t} + \Delta u - Ru\right) R^m + 2\nabla R^m \cdot \nabla u.$$

Plugging in the equation for $\left(\frac{\partial}{\partial t} + \Delta\right) R^m$ and using (5.40), we have

$$\frac{\partial V}{\partial t} + \Delta V - RV = 2|R_{ij} + \nabla_i \nabla_j f|^2 u + 2u\nabla R^m \cdot \nabla f + 2\nabla R^m \cdot \nabla u.$$

The last two terms cancel each other since $\nabla f = -\nabla u/u$, which yields the lemma. □

REMARK 5.20 (Backward heat-type equation for modified scalar curvature). From the proof of the lemma, we have

(5.44) $$\frac{\partial}{\partial t} R^m = -\Delta R^m + 2\nabla R^m \cdot \nabla f + 2|R^m_{ij}|^2.$$

Note the similarity to the equation $\frac{\partial R}{\partial t} = \Delta R + 2\,|\mathrm{Rc}|^2$, except now we have a backward heat-type equation.

3. Steady and expanding breather solutions revisited

A solution $g(t)$ of the Ricci flow on a manifold $\mathcal{M}^n$ is called a **Ricci breather** if there exist times $t_1 < t_2$, a constant $\alpha > 0$ and a diffeomorphism $\varphi : \mathcal{M} \to \mathcal{M}$ such that

$$g(t_2) = \alpha \varphi^* g(t_1).$$

When $\alpha = 1$, $\alpha < 1$, or $\alpha > 1$, we call $g(t)$ a **steady**, **shrinking**, or **expanding Ricci breather**, respectively. Recall that $g(t)$ is a Ricci soliton (or **trivial Ricci breather**) if for *each* pair of times $t_1 < t_2$ there exist $\alpha > 0$ and a diffeomorphism $\varphi : \mathcal{M} \to \mathcal{M}$ (α and φ will in general depend on t_1 and t_2) such that $g(t_2) = \alpha \varphi^* g(t_1)$.

Note that if we consider the Ricci flow as a dynamical system on the space of Riemannian metrics modulo diffeomorphisms and homotheties, the Ricci breathers correspond to the periodic orbits whereas the Ricci solitons correspond to the fixed points. Since the Ricci flow is a heat-type equation, we expect that there are no periodic orbits except fixed points.

A nice application of the energy monotonicity formula is the nonexistence of nontrivial steady or expanding breather solutions on closed manifolds (§2 of [**297**]). This was first proved by one of the authors in [**218**] (see Proposition 1.66 in this volume). In the next chapter we shall see the application of Perelman's entropy formula to prove shrinking breather solutions on closed manifolds are gradient Ricci solitons (§3 of [**297**]). Hence we confirm the above expectation.

3.1. The infimum λ of $\mathcal{F}$. Suppose we have a steady breather solution to the Ricci flow with $g(t_2) = \varphi^* g(t_1)$ for some $t_1 < t_2$ and diffeomorphism φ. One drawback of the energy monotonicity formula is that in general the solution f to (5.35) has $f(t_2) \neq f(t_1) \circ \varphi$, so that in general, $\mathcal{F}(g(t_2), f(t_2)) \neq \mathcal{F}(g(t_1), f(t_1))$. By taking the infimum of $\mathcal{F}$ among f, we obtain an invariant of the Riemannian metric g which avoids this trouble.

DEFINITION 5.21 (λ-invariant). Given a metric g on a closed manifold $\mathcal{M}^n$, we define the functional $\lambda : \mathfrak{Met} \to \mathbb{R}$ by

$$(5.45) \qquad \lambda(g) \doteqdot \inf\left\{ \mathcal{F}(g, f) : f \in C^\infty(\mathcal{M}), \int_\mathcal{M} e^{-f} d\mu = 1 \right\}.$$

Taking $w = e^{-f/2}$, we have

$$(5.46) \qquad \lambda(g) = \inf\left\{ \mathcal{G}(g, w) : \int_\mathcal{M} w^2 d\mu = 1,\ w > 0 \right\},$$

where, as in (5.5),[9]

$$(5.47) \qquad \mathcal{G}(g, w) \doteqdot \int_\mathcal{M} \left(4|\nabla w|^2 + Rw^2 \right) d\mu.$$

Thus, when we fix g and minimize $\mathcal{F}(g, f)$ among f, we are minimizing a Dirichlet-type functional and we get an eigenfunction-type equation for w. Aspects of this point of view are discussed in the next two lemmas.

Note that the variation of $\mathcal{G}(g, \cdot)$ is given by

$$\frac{1}{2}\delta_{(0,h)}\mathcal{G}(g, w) = \int_\mathcal{M} (4\nabla w \cdot \nabla h + Rwh)\, d\mu = \int_\mathcal{M} (-4\Delta w + Rw)\, h\, d\mu,$$

where $h = \delta w$. Hence the Euler–Lagrange equation for (note that we dropped the positivity condition on w)

$$\lambda(g) \doteqdot \inf\left\{ \mathcal{G}(g, w) : \int_\mathcal{M} w^2 d\mu = 1 \right\}$$

is

$$(5.48) \qquad Lw \doteqdot -4\Delta w + Rw = \lambda(g)\, w.$$

LEMMA 5.22 (Existence and regularity of minimizer of $\mathcal{G}$). *There exists a unique minimizer w_0 (up to a change in sign) of*

$$(5.49) \qquad \inf\left\{ \mathcal{G}(g, w) : \int_\mathcal{M} w^2 d\mu = 1 \right\}.$$

The minimizer w_0 is positive and smooth. Moreover,

---

[9]In view of Lemma 5.1(1), the monotonicity of $\mathcal{F}$ exhibits a dichotomy, it is analogous to both the monotonicty of the total scalar curvature under its gradient flow, $\frac{\partial}{\partial t}g = -2\left(\text{Rc} - \frac{1}{2}g\right)$, and the monotonicity of the Dirichlet energy under its gradient flow, the backward heat equation $\frac{\partial}{\partial t}w = -\Delta w$. In this sense, the monotonicity of $\mathcal{F}$ exhibits a beautiful synthesis of geometry and analysis.

3. STEADY AND EXPANDING BREATHER SOLUTIONS REVISITED

(1) *the minimum value $\lambda(g)$ of $\mathcal{G}(g,w)$ is equal to $\lambda_1(g)$, where $\lambda_1(g)$ is the lowest eigenvalue of the elliptic operator $-4\Delta + R$, and*
(2) *w_0 is the unique positive eigenfunction of*

(5.50) $$-4\Delta w_0 + R w_0 = \lambda_1(g) w_0$$

with L^2-norm equal to 1.

PROOF. To establish the existence of a minimizer w_0 of (5.46), one takes a minimizing sequence $\{w_i\}_{i=1}^\infty$ of (5.46) in $W^{1,2}(\mathcal{M})$. There then exists a subsequence $\{w_i\}_{i=1}^\infty$ which converges to $w_0 \in W^{1,2}(\mathcal{M})$ weakly in $W^{1,2}(\mathcal{M})$ and strongly in $L^2(\mathcal{M})$ (by the Sobolev embedding theorem). Since

$$0 \leq \int_{\mathcal{M}} |\nabla(w_i - w_0)|^2 d\mu$$
$$= \int_{\mathcal{M}} |\nabla w_i|^2 d\mu + \int_{\mathcal{M}} |\nabla w_0|^2 d\mu - 2\int_{\mathcal{M}} \langle \nabla w_i, \nabla w_0 \rangle d\mu,$$

by the weak convergence in $W^{1,2}$, we have $\lim_{i\to\infty} \int_{\mathcal{M}} \langle \nabla w_i, \nabla w_0 \rangle d\mu = \int_{\mathcal{M}} |\nabla w_0|^2 d\mu$ exists, hence

$$\int_{\mathcal{M}} |\nabla w_0|^2 d\mu \leq \liminf_{i\to\infty} \int_{\mathcal{M}} |\nabla w_i|^2 d\mu.$$

On the other hand, by the strong convergence of $\{w_i\}_{i=1}^\infty$ in L^2, we have

$$\lim_{i\to\infty} \int_{\mathcal{M}} R w_i^2 d\mu = \int_{\mathcal{M}} R w_0^2 d\mu,$$
$$\int_{\mathcal{M}} w_0^2 d\mu = \lim_{i\to\infty} \int_{\mathcal{M}} w_i^2 d\mu = 1.$$

Hence w_0 is a minimizer of (5.46) in $W^{1,2}(\mathcal{M})$, and w_0 is a weak solution to the eigenfunction equation (5.48). By standard regularity theory, $w_0 \in C^\infty$. We also have that any minimizer is either nonnegative or nonpositive, since otherwise $\pm |w_0|$ is a distinct smooth minimizer which agrees with w_0 on an open set, contradicting the unique continuation property of solutions to second-order linear elliptic equations.

We now prove w_0 is unique up to a sign. Without loss of generality, we may assume below that w_0 is nonnegative. Call a minimizer w of $\mathcal{G}$ with $\int_{\mathcal{M}} w^2 d\mu = 1$ a *normalized* minimizer. If the nonnegative normalized minimizer is not unique, then there exist two normalized minimizers $w_0 \geq 0$ and $w_1 \geq 0$ with $\int_{\mathcal{M}} w_0 w_1 d\mu = 0$. Then $w_2 = aw_0 + bw_1$ is also a normalized minimizer for all $a, b \in \mathbb{R}$ such that $a^2 + b^2 = 1$. Indeed, since w_0 and w_1 satisfy the linear equation (5.50), so does $w_2 = aw_0 + bw_1$, and $\int_{\mathcal{M}} w_2^2 d\mu = 1$.

Now it not hard to see that there exist a and b such that w_2 changes sign. In particular, if there are points x and y such that $w_1(x) = cw_0(x)$ and $w_1(y) = dw_0(y)$, where $c \neq d$ and $w_0(x) > 0 < w_0(y)$, then by choosing a and b with $a^2 + b^2 = 1$ such that $a + bc$ and $a + bd$ have opposite signs, we have that $w_2(x) = (a + bc)w_0(x)$ and $w_2(y) = (a + bd)w_0(y)$ have opposite signs, which is a contradiction. Hence w_0 is unique.

Finally we show $w_0 > 0$. By the Hopf boundary point lemma (see Lemma 3.4 of Gilbarg and Trudinger [**155**]), if $w_0 = 0$ somewhere, then there exists a point $x_0 \in \partial\Omega$, where $\Omega = \{x \in \mathcal{M} : w_0(x) > 0\}$, such that $\partial\Omega$ satisfies the interior sphere condition at x_0, so that $w(x_0) = 0$ and $|\nabla w(x_0)| \neq 0$, which is a contradiction to $w_0 \geq 0$.

Finally, properties (1) and (2) follow easily. $\square$

The existence of a unique positive smooth minimizer w_0 of $\mathcal{G}(g, w)$ under the constraint $\int_{\mathcal{M}} w^2 d\mu = 1$ implies the existence of a unique smooth minimizer f_0 of $\mathcal{F}(g, \cdot)$ under the constraint $\int_{\mathcal{M}} e^{-f} d\mu = 1$. From (5.50) we see the following.

LEMMA 5.23 (Euler–Lagrange equation for minimizer of $\mathcal{F}$). *The minimizer $f_0 = -2 \log w_0$ of $\mathcal{F}(g, \cdot)$ is unique, C^∞, and a solution to*

(5.51) $$\lambda(g) = 2\Delta f_0 - |\nabla f_0|^2 + R.$$

That is, the modified scalar curvature is a constant, i.e., $R^m \equiv \lambda(g)$. Note that from setting $v = 0$ in (5.10), for the minimizer f of (5.45), we have

$$\delta_{(0,h)} \mathcal{F}(g, f) = -\int_{\mathcal{M}} h \left(2\Delta f - |\nabla f|^2 + R\right) e^{-f} d\mu$$

for all h such that $\int_{\mathcal{M}} h e^{-f} d\mu_g = 0$. We can also obtain (5.51) directly from this.

We summarize the properties of the functional λ on a closed manifold $\mathcal{M}^n$.

(i) (*Lower bound for λ*) $\lambda(g)$ is well defined (i.e., finite) since

$$\mathcal{F}(g, f) \geq \min_{x \in \mathcal{M}} R(x) \cdot \int_{\mathcal{M}} e^{-f} d\mu = \min_{x \in \mathcal{M}} R(x) \doteqdot R_{\min}.$$

In particular,

$$\lambda(g) \geq R_{\min}.$$

(ii) (*Diffeomorphism invariance*) If $\varphi : \mathcal{M} \to \mathcal{M}$ is a diffeomorphism, then

$$\lambda(\varphi^* g) = \lambda(g).$$

(iii) (*Existence of a smooth minimizer*) There exists $f \in C^\infty(\mathcal{M})$ with $\int_{\mathcal{M}} e^{-f} d\mu = 1$ such that $\lambda(g) = \mathcal{F}(g, f)$, i.e.,

(5.52) $$\lambda(g) = \int_{\mathcal{M}} (R + |\nabla f|^2) e^{-f} d\mu.$$

(iv) (*Upper bound for λ*) We have

(5.53) $$\lambda(g) \leq \frac{1}{\operatorname{Vol}(\mathcal{M})} \int_{\mathcal{M}} R \, d\mu.$$

This can be seen by choosing $f = \log \operatorname{Vol}(\mathcal{M})$, which satisfies

$$\int_{\mathcal{M}} e^{-f} d\mu_g = 1 \quad \text{and} \quad \lambda(g) \leq \int_{\mathcal{M}} (R + |\nabla f|^2) e^{-f} d\mu.$$

(v) *(Scaling)*
$$\lambda(cg) = c^{-1}\lambda(g).$$

3.2. The monotonicity of λ. Let $(\mathcal{M}^n, g(t))$, $t \in [0, T]$, be a solution of the Ricci flow on a closed manifold. In this subsection we discuss some properties related to the continuity and monotonicity of $\lambda(g(t))$. Such properties are key to the proof of the nonexistence of nontrivial expanding or steady breathers. First we show that $\lambda(g(t))$ is a continuous function on $[t_1, t_2]$. This is a consequence of the following elementary result (see also Craioveanu, Puta, and Rassias [118] or Chapter XII of Reed and Simon [310]).[10]

LEMMA 5.24 (Effective estimate for continuous dependence of λ on g). *If g_1 and g_2 are two metrics on $\mathcal{M}$ which satisfy*
$$\frac{1}{1+\varepsilon}g_1 \leq g_2 \leq (1+\varepsilon)g_1 \quad \text{and} \quad R(g_1) - \varepsilon \leq R(g_2) \leq R(g_1) + \varepsilon,$$
then[11]
$$\lambda(g_2) - \lambda(g_1)$$
$$\leq \left((1+\varepsilon)^{\frac{n}{2}+1} - (1+\varepsilon)^{-n/2}\right)(1+\varepsilon)^{n/2}\left(\lambda(g_1) - \min R_{g_1}\right)$$
$$+ \left((1+\delta)\max |R_{g_2} - R_{g_1}| + 2\delta \max |R_{g_1}|\right)(1+\varepsilon)^{n/2},$$
where $\delta \to 0$ as $\varepsilon \to 0$.[12] *In particular, $\lambda : \mathfrak{Met} \to \mathbb{R}$ is a continuous function with respect to the C^2-topology.*

PROOF. The proof is straightforward but slightly tedious. First note that $(1+\varepsilon)^{-n/2} d\mu_{g_1} \leq d\mu_{g_2} \leq (1+\varepsilon)^{n/2} d\mu_{g_1}$. If w is a positive function on $\mathcal{M}$, then in view of (5.46), we compute (writing $a \cdot b - c \cdot d = a(b-d) + (a-c)d$)

$$\int_{\mathcal{M}} w^2 d\mu_{g_1} \mathcal{G}(g_2, w) - \int_{\mathcal{M}} w^2 d\mu_{g_2} \mathcal{G}(g_1, w)$$
$$= 4 \int_{\mathcal{M}} w^2 d\mu_{g_1} \left(\int_{\mathcal{M}} |\nabla w|^2_{g_2} d\mu_{g_2} - \int_{\mathcal{M}} |\nabla w|^2_{g_1} d\mu_{g_1} \right)$$
$$+ 4 \left(\int_{\mathcal{M}} w^2 d\mu_{g_1} - \int_{\mathcal{M}} w^2 d\mu_{g_2} \right) \int_{\mathcal{M}} |\nabla w|^2_{g_1} d\mu_{g_1}$$
$$+ \int_{\mathcal{M}} w^2 d\mu_{g_1} \left(\int_{\mathcal{M}} R_{g_2} w^2 d\mu_{g_2} - \int_{\mathcal{M}} R_{g_1} w^2 d\mu_{g_1} \right)$$
$$+ \left(\int_{\mathcal{M}} w^2 d\mu_{g_1} - \int_{\mathcal{M}} w^2 d\mu_{g_2} \right) \int_{\mathcal{M}} R_{g_1} w^2 d\mu_{g_1},$$

---

[10]Thanks to [231] for this last reference.
[11]To denote the dependence on g_i, we use the subscript $_{g_i}$ instead of (g_i). So $R_{g_1} = R(g_1)$.
[12]See the proof for an explicit dependence of δ on ε.

so that

$$\int_{\mathcal{M}} w^2 d\mu_{g_1} \mathcal{G}(g_2, w) - \int_{\mathcal{M}} w^2 d\mu_{g_2} \mathcal{G}(g_1, w)$$
$$\leq 4\left((1+\varepsilon)^{\frac{n}{2}+1} - 1\right) \int_{\mathcal{M}} w^2 d\mu_{g_1} \int_{\mathcal{M}} |\nabla w|^2_{g_1} d\mu_{g_1}$$
$$+ 4\left(1 - (1+\varepsilon)^{-n/2}\right) \int_{\mathcal{M}} w^2 d\mu_{g_1} \int_{\mathcal{M}} |\nabla w|^2_{g_1} d\mu_{g_1}$$
$$+ \int_{\mathcal{M}} w^2 d\mu_{g_1} \int_{\mathcal{M}} w^2 \left(\left|(R_{g_2} - R_{g_1}) \frac{d\mu_{g_2}}{d\mu_{g_1}}\right| + \left|\left(\frac{d\mu_{g_2}}{d\mu_{g_1}} - 1\right) R_{g_1}\right|\right) d\mu_{g_1}$$
$$+ \left|\int_{\mathcal{M}} w^2 \left(1 - \frac{d\mu_{g_2}}{d\mu_{g_1}}\right) d\mu_{g_1}\right| \left|\int_{\mathcal{M}} R_{g_1} w^2 d\mu_{g_1}\right|.$$

(In the above estimates we took into account that R may change sign.) Let $\delta \doteq \max\left\{(1+\varepsilon)^{n/2} - 1, 1 - (1+\varepsilon)^{-n/2}\right\}$, so that $\delta \to 0$ as $\varepsilon \to 0$. Since $\left|1 - \frac{d\mu_{g_2}}{d\mu_{g_1}}\right| \leq \delta$, we have

$$\int_{\mathcal{M}} w^2 d\mu_{g_1} \int_{\mathcal{M}} w^2 d\mu_{g_2} \left(\frac{\mathcal{G}(g_2, w)}{\int_{\mathcal{M}} w^2 d\mu_{g_2}} - \frac{\mathcal{G}(g_1, w)}{\int_{\mathcal{M}} w^2 d\mu_{g_1}}\right)$$
$$\leq 4\left((1+\varepsilon)^{\frac{n}{2}+1} - (1+\varepsilon)^{-n/2}\right) \int_{\mathcal{M}} w^2 d\mu_{g_1} \int_{\mathcal{M}} |\nabla w|^2_{g_1} d\mu_{g_1}$$
$$+ ((1+\delta) \max |R_{g_2} - R_{g_1}| + \delta \max |R_{g_1}|) \left(\int_{\mathcal{M}} w^2 d\mu_{g_1}\right)^2$$
$$+ \delta \max |R_{g_1}| \left(\int_{\mathcal{M}} w^2 d\mu_{g_1}\right)^2.$$

Hence

$$\frac{\mathcal{G}(g_2, w)}{\int_{\mathcal{M}} w^2 d\mu_{g_2}} - \frac{\mathcal{G}(g_1, w)}{\int_{\mathcal{M}} w^2 d\mu_{g_1}}$$
$$\leq 4\left((1+\varepsilon)^{\frac{n}{2}+1} - (1+\varepsilon)^{-n/2}\right) \frac{\int_{\mathcal{M}} |\nabla w|^2_{g_1} d\mu_{g_1}}{\int_{\mathcal{M}} w^2 d\mu_{g_2}}$$
$$+ ((1+\delta) \max |R_{g_2} - R_{g_1}| + 2\delta \max |R_{g_1}|) (1+\varepsilon)^{n/2}.$$

Taking w to be a minimizer for $\mathcal{G}(g_1, \cdot)$, we have

$$\lambda(g_2) - \lambda(g_1)$$
$$\leq 4\left((1+\varepsilon)^{\frac{n}{2}+1} - (1+\varepsilon)^{-n/2}\right) (1+\varepsilon)^{n/2} \frac{\int_{\mathcal{M}} |\nabla w|^2_{g_1} d\mu_{g_1}}{\int_{\mathcal{M}} w^2 d\mu_{g_1}}$$
$$+ ((1+\delta) \max |R_{g_2} - R_{g_1}| + 2\delta \max |R_{g_1}|) (1+\varepsilon)^{n/2}.$$

The result now follows from

$$4\frac{\int_{\mathcal{M}} |\nabla w|^2_{g_1} d\mu_{g_1}}{\int_{\mathcal{M}} w^2 d\mu_{g_1}} = \frac{\mathcal{G}(g_1, w)}{\int_{\mathcal{M}} w^2 d\mu_{g_1}} - \frac{\int_{\mathcal{M}} R_{g_1} w^2 d\mu_{g_1}}{\int_{\mathcal{M}} w^2 d\mu_{g_1}}$$
$$\leq \lambda(g_1) - \min R_{g_1}.$$

$\square$

The monotonicity of $\mathcal{F}(g(t), f(t))$ under the system (5.34)–(5.35) implies the monotonicity of $\lambda(g(t))$ under the Ricci flow.

LEMMA 5.25 (λ monotonicity). *If $g(t)$, $t \in [0, T]$, is a solution to the Ricci flow, then*

$$\frac{d}{dt}\lambda(g(t)) \geq \frac{2}{n}\lambda^2(g(t)),$$

and $\lambda(g(t))$ is nondecreasing in $t \in [0, T]$. Here the derivative $\frac{d}{dt}$ is in the sense of the lim inf of backward difference quotients.

REMARK 5.26. See the next subsection for the case where $\lambda(g(t))$ is not strictly increasing.

PROOF. Given $t_0 \in [0, T]$, let f_0 be the minimizer of $\mathcal{F}(g(t_0), f)$, so that $\lambda(g(t_0)) = \mathcal{F}(g(t_0), f(t_0))$. Solve

(5.54) $$\frac{\partial}{\partial t} f = -R - \Delta f + |\nabla f|^2, \qquad f(t_0) = f_0,$$

backward in time on $[0, t_0]$. Then $\frac{d}{dt}\mathcal{F}(g(t), f(t)) \geq 0$ for all $t \leq t_0$. Since the constraint $\int_{\mathcal{M}} e^{-f} d\mu$ is preserved under (5.54), we have $\lambda(g(t)) \leq \mathcal{F}(g(t), f(t))$ for $t \leq t_0$. This, (5.41), and $\lambda(g(t_0)) = \mathcal{F}(g(t_0), f(t_0))$ imply both

(5.55) $$\lambda(g(t)) \leq \mathcal{F}(g(t), f(t)) \leq \mathcal{F}(g(t_0), f(t_0)) = \lambda(g(t_0))$$

and the following:

(5.56)
$$\begin{aligned}
\frac{d}{dt}\lambda(g(t))\Big|_{t=t_0} &\geq \frac{d}{dt}\mathcal{F}(g(t), f(t))\Big|_{t=t_0} \\
&= 2\int_{\mathcal{M}} |R_{ij} + \nabla_i \nabla_j f|^2 e^{-f} d\mu_{g(t_0)} \\
&\geq 2\int_{\mathcal{M}} \frac{1}{n}(R + \Delta f)^2 e^{-f} d\mu_{g(t_0)} \\
&\geq \frac{2}{n}\left(\int_{\mathcal{M}} (R + \Delta f) e^{-f} d\mu_{g(t_0)}\right)^2 \\
&= \frac{2}{n}\lambda^2(g(t_0)),
\end{aligned}$$

where $f = f_0$ is the minimizer. Hence, from either (5.55) or (5.56), we see that $\lambda(g(t))$ is nondecreasing under the Ricci flow. $\square$

EXERCISE 5.27. Prove (5.56).

SOLUTION TO EXERCISE 5.27. We compute[13]

$$\frac{d}{dt_-}\lambda\left(g\left(t\right)\right)\Big|_{t=t_0} \doteqdot \liminf_{h\to 0_+} \frac{\lambda\left(g\left(t_0\right)\right) - \lambda\left(g\left(t_0 - h\right)\right)}{h}$$

$$\geq \liminf_{h\to 0_+} \frac{\mathcal{F}\left(g\left(t_0\right), f_0\right) - \mathcal{F}\left(g\left(t_0 - h\right), f\left(t_0 - h\right)\right)}{h},$$

where f_0 is the minimizer for $\mathcal{F}\left(g\left(t_0\right), \cdot\right)$ and $f(t)$ is the solution to (5.54). On the other hand, we conclude by (5.41) that the last expression is equal to $2\int_{\mathcal{M}} |R_{ij} + \nabla_i\nabla_j f_0|^2 e^{-f_0} d\mu_{g(t_0)}$.

3.3. There are no nontrivial steady breathers. As an application of the monotonicity of the diffeomorphism-invariant functional λ we prove the nonexistence of nontrivial steady breathers.

LEMMA 5.28 (No nontrivial steady breathers on closed manifolds). *If $(\mathcal{M}^n, g(t))$ is a solution to the Ricci flow on a closed manifold such that there exist $t_1 < t_2$ with $\lambda\left(g\left(t_1\right)\right) = \lambda\left(g\left(t_2\right)\right)$, then $g(t)$ is a steady gradient Ricci soliton, which must be Ricci flat. In particular, a steady Ricci breather on a closed manifold is Ricci flat.*

PROOF. Note that if $g(t)$ is a steady Ricci breather with $g(t_2) = \varphi^* g(t_1)$ for some $t_1 < t_2$ and diffeomorphism $\varphi : \mathcal{M} \to \mathcal{M}$, then $\lambda(g(t_2)) = \lambda(g(t_1))$. Hence we only need to prove the first part of the lemma.

Suppose that for a solution $g(t)$ to the Ricci flow there exist times $t_1 < t_2$ such that $\lambda(g(t_2)) = \lambda(g(t_1))$. Let f_2 be the minimizer for $\mathcal{F}$ at time t_2 so that $\mathcal{F}\left(g\left(t_2\right), f_2\right) = \lambda\left(g\left(t_2\right)\right)$. Take $f(t)$ to be the solution to the backward heat equation (5.35) on the time interval $[t_1, t_2]$ with the initial data $f(t_2) = f_2$. By the monotonicity formula (5.41) and the definition of λ we have[14]

$$\lambda\left(g\left(t_1\right)\right) \leq \mathcal{F}\left(g\left(t_1\right), f\left(t_1\right)\right) \leq \mathcal{F}\left(g\left(t\right), f\left(t\right)\right) \leq \mathcal{F}\left(g\left(t_2\right), f_2\right) = \lambda\left(g\left(t_2\right)\right)$$

for all $t \in [t_1, t_2]$. Since $\lambda\left(g\left(t_1\right)\right) = \lambda\left(g\left(t_2\right)\right)$ and $\lambda\left(g\left(t\right)\right)$ is monotone, we have

$$\mathcal{F}\left(g\left(t\right), f\left(t\right)\right) = \lambda\left(g\left(t\right)\right) \equiv \text{const}$$

for $t \in [t_1, t_2]$. Therefore the solution $f(t)$ is the minimizer for $\mathcal{F}\left(g\left(t\right), \cdot\right)$ and $\frac{d}{dt}\mathcal{F}\left(g\left(t\right), f\left(t\right)\right) \equiv 0$, so by (5.41) we have

$$\int_{\mathcal{M}} |R_{ij} + \nabla_i \nabla_j f|^2 e^{-f} d\mu(t) \equiv 0$$

for all $t \in [t_1, t_2]$. Thus

(5.57) $\qquad R_{ij} + \nabla_i \nabla_j f = 0$ for $t \in [t_1, t_2]$.

In particular, $g(t)$ is a steady gradient Ricci soliton flowing along $\nabla f(t)$.[15]

---

[13] Here $\frac{d}{dt_-}$ denotes the lim inf of backward difference quotients.

[14] This is the same as (5.55).

[15] See (1.9), where a gradient soliton is steady if $\varepsilon = 0$.

Note by (5.51) that f satisfies the equation

$$2\Delta f - |\nabla f|^2 + R = \lambda(g).$$

On the other hand, $R + \Delta f = 0$, so that

$$|\nabla f|^2 + R = -\lambda(g).$$

However, integrating, we have

$$-\lambda(g) = \int_{\mathcal{M}} \left(|\nabla f|^2 + R\right) e^{-f} d\mu = \lambda(g),$$

so that $\lambda(g) = 0$ and $\Delta f = |\nabla f|^2 = -R$. Note that then

$$0 = \int_{\mathcal{M}} \left(\Delta f - |\nabla f|^2\right) e^f d\mu = -2 \int_{\mathcal{M}} |\nabla f|^2 e^f d\mu$$

implies that f is constant and hence g is Ricci flat by (5.57). Alternatively, we could have argued that since $\Delta f = |\nabla f|^2 \geq 0$, f is subharmonic and hence constant. $\square$

REMARK 5.29. Even when $\mathcal{M}$ is noncompact, we have $|\nabla f|^2 + R$ is constant for gradient Ricci solitons; see Proposition 1.15.

3.4. Nonexistence of nontrivial expanding breathers.
Recall that $\lambda(g)$ is not scale-invariant, e.g., $\lambda(cg) = c^{-1}\lambda(g)$. Thus we define the **normalized λ-invariant**:

(5.58) $$\bar{\lambda}(g) \doteq \lambda(g) \cdot \mathrm{Vol}(\mathcal{M})^{2/n}.$$

It is easy to see that $\bar{\lambda}(cg) = \bar{\lambda}(g)$ for any $c > 0$, so the invariant $\bar{\lambda}$ is potentially useful for expanding and shrinking breathers. *We shall prove the monotonicity of $\bar{\lambda}(g(t))$ under Ricci flow when it is nonpositive.* For this reason it is most useful for expanding breathers.

Recall that by (5.56), we have

(5.59) $$\frac{d}{dt}\lambda(g(t)) \geq 2 \int_{\mathcal{M}} |R_{ij} + \nabla_i \nabla_j f|^2 e^{-f} d\mu,$$

where $\frac{d}{dt}\lambda(g(t))$ is defined as the lim inf of backward difference quotients.[16] Let $V \doteq V(t) \doteq \mathrm{Vol}_{g(t)}(\mathcal{M})$. From (5.59), we compute

$$\frac{d}{dt}\bar{\lambda}(g(t)) = \frac{d}{dt}\left[\lambda(g(t)) \cdot V(t)^{2/n}\right]$$

$$= V^{2/n}\frac{d\lambda}{dt} + \frac{2}{n}V^{\frac{2}{n}-1}\lambda \frac{dV}{dt}$$

$$\geq 2V^{2/n} \int_{\mathcal{M}} |R_{ij} + \nabla_i \nabla_j f|^2 e^{-f} d\mu + \frac{2}{n}\lambda V^{\frac{2}{n}-1} \int_{\mathcal{M}} (-R) d\mu,$$

---

[16]This also applies to the time derivatives below in this argument.

where $f = f(t)$ is the minimizer of $\mathcal{F}(g(t), \cdot)$. From this we obtain

$$\frac{1}{2} V^{-2/n} \frac{d}{dt} \bar{\lambda}(g(t)) \geq \int_{\mathcal{M}} |R_{ij} + \nabla_i \nabla_j f|^2 e^{-f} d\mu$$
$$- \frac{1}{n} \int_{\mathcal{M}} (R + \Delta f) e^{-f} d\mu \cdot \frac{1}{V} \int_{\mathcal{M}} R d\mu.$$

Hence

$$\frac{1}{2} V^{-2/n} \frac{d}{dt} \bar{\lambda}(g(t)) \geq \int_{\mathcal{M}} \left| R_{ij} + \nabla_i \nabla_j f - \frac{1}{n}(R + \Delta f) g_{ij} \right|^2 e^{-f} d\mu$$
$$+ \int_{\mathcal{M}} \frac{1}{n} (R + \Delta f)^2 e^{-f} d\mu$$
$$- \frac{1}{n} \int_{\mathcal{M}} (R + \Delta f) e^{-f} d\mu \cdot \frac{1}{V} \int_{\mathcal{M}} R d\mu.$$

Recall from (5.53) that

$$\int_{\mathcal{M}} (R + \Delta f) e^{-f} d\mu \leq \frac{\int R d\mu}{V}.$$

Assuming $\lambda(t) \leq 0$, so that $\int_{\mathcal{M}} (R + \Delta f) e^{-f} d\mu \leq 0$, we have

(5.60) $\quad \frac{1}{2} V^{-2/n} \frac{d}{dt} \bar{\lambda}(g(t)) - \int_{\mathcal{M}} \left| R_{ij} + \nabla_i \nabla_j f - \frac{1}{n}(R + \Delta f) g_{ij} \right|^2 e^{-f} d\mu$

$$\geq \frac{1}{n} \int_{\mathcal{M}} (R + \Delta f)^2 e^{-f} d\mu - \frac{1}{n} \left(\int_{\mathcal{M}} (R + \Delta f) e^{-f} d\mu \right)^2 \geq 0$$

since $\int_{\mathcal{M}} e^{-f} d\mu = 1$. Hence

LEMMA 5.30. *Let $g(t)$ be a solution to the Ricci flow on a closed manifold $\mathcal{M}^n$. If at some time t, $\bar{\lambda}(t) \leq 0$, then*

(5.61) $\quad \frac{d}{dt} \bar{\lambda}(g(t))$

$$\geq 2 V^{2/n} \int_{\mathcal{M}} \left| R_{ij} + \nabla_i \nabla_j f - \frac{1}{n}(R + \Delta f) g_{ij} \right|^2 e^{-f} d\mu \geq 0,$$

where $V = \mathrm{Vol}_{g(t)}(\mathcal{M})$, $f(t)$ is the minimizer for $\mathcal{F}(g(t), \cdot)$, and the time-derivative is defined as the liminf *of backward difference quotients. By (5.61), if $\frac{d}{dt} \bar{\lambda}(g(t)) = 0$, then $g(t)$ is a gradient Ricci soliton.*

This is reminiscent of the fact that under the normalized Ricci flow, the minimum scalar curvature is nondecreasing as long as it is nonpositive, whereas under the unnormalized Ricci flow, the minimum scalar curvature is always nondecreasing (see Lemma A.20). However these two facts appear to be quite different in nature.

To apply the above monotonicity to the expanding breather case, we need to produce a time t_0 where $\bar{\lambda}(g(t_0)) < 0$. This is accomplished by looking at the evolution of the volume. Below we also give another proof of Lemma 5.28 using $\bar{\lambda}(g(t))$.

3. STEADY AND EXPANDING BREATHER SOLUTIONS REVISITED

LEMMA 5.31. *Expanding or steady breathers on closed manifolds are Einstein.*

PROOF. Let $(\mathcal{M}^n, g(t))$ be an expanding or steady breather with $g(t_2) = \alpha \varphi^* g(t_1)$ for some $t_1 < t_2$ and $\alpha \geq 1$. We have $\bar{\lambda}(g(t_2)) = \bar{\lambda}(g(t_1))$. Let $V(t) \doteqdot \operatorname{Vol}_{g(t)}(\mathcal{M})$. Since $V(t_2) \geq V(t_1)$, we have for some $t_0 \in (t_1, t_2)$,

$$0 \leq \left.\frac{d}{dt}\right|_{t=t_0} \log V(t) = -\frac{\int_{\mathcal{M}} R \, d\mu}{V(t_0)}(t_0) \leq -\lambda(g(t_0)).$$

By Lemma 5.30, if $g(t)$ is not a gradient Ricci soliton, then $\frac{d}{dt}\bar{\lambda}(g(t_0)) > 0$ and we have $\bar{\lambda}(g(t'_0)) < 0$ for some $t'_0 < t_0$. Now since $\bar{\lambda}(g(t))$ is increasing whenever it is negative, we have

$$\bar{\lambda}(g(t_2)) = \bar{\lambda}(g(t_1)) \leq \bar{\lambda}(g(t'_0)) < 0,$$

which implies $\lambda(g(t)) \leq \lambda(g(t_2)) < 0$ for all $t \in [t_1, t_2]$. Hence $\bar{\lambda}(g(t))$ is nondecreasing, which implies $\bar{\lambda}(g(t))$ is constant. By (5.61), we have

$$R_{ij} + \nabla_i \nabla_j f - \frac{1}{n}(R + \Delta f) g_{ij} \equiv 0,$$

and since we are in the equality case of (5.60), we also have

(5.62) $\quad R + \Delta f = C_1(t) = \text{const}$ (depending on time).

That is, we still conclude that $g(t)$ is an expanding or steady gradient Ricci soliton.

Now let $(\mathcal{M}^n, g(t))$ be an expanding or steady gradient Ricci soliton. Recall

$$2\Delta f + R - |\nabla f|^2 = C_2(t) = \text{const}.$$

This, combined with (5.62), implies

$$\Delta f - |\nabla f|^2 = \text{const}.$$

Since

$$\int_{\mathcal{M}} \left(\Delta f - |\nabla f|^2 \right) e^{-f} d\mu = 0,$$

we have $\Delta f - |\nabla f|^2 \equiv 0$. Thus, by the strong maximum principle (or since now $0 = \int_{\mathcal{M}} \left(\Delta f - |\nabla f|^2 \right) e^f d\mu = -2 \int_{\mathcal{M}} |\nabla f|^2 e^f d\mu$), we conclude that $f \equiv \text{const}$. Hence $R_{ij} - \frac{1}{n} R g_{ij} \equiv 0$ and g_{ij} is Einstein. (When $n = 2$, our conclusion is vacuous.) □

REMARK 5.32. As a corollary of the above result, we again see that expanding or steady *solitons* on closed manifolds are Einstein. In the case of shrinking solitons on closed manifolds, using the entropy functional, we shall see in the next chapter that they are necessarily *gradient* shrinking solitons.

Note that on a shrinking breather we have $V(t_2) < V(t_1)$ for $t_2 > t_1$. In particular, it is possible that $\lambda(g(t)) > 0$ for all $t \in [t_1, t_2]$ (on the other hand, if $\lambda(g(t'_0)) < 0$ for some $t'_0 \in [t_1, t_2]$, then the proof above implies that a shrinking breather is Einstein), which causes difficulty in extending the proof above to the shrinking case; in the next chapter this problem is solved by the introduction of Perelman's entropy. (Note that for an Einstein manifold with $R \equiv r = \text{const}$, under the constraint $\int e^{-f} d\mu = 1$ we have

$$\mathcal{F}(g, f) = r + \int_{\mathcal{M}} |\nabla f|^2 e^{-f} d\mu \geq r$$

with equality if and only if $f \equiv \log \text{Vol}(g) = \text{const}$. Hence, if $r > 0$, then $\bar{\lambda}(g) = r \text{Vol}(g)^{2/n} > 0$.)

EXERCISE 5.33 (Behavior of $\bar{\lambda}$ on products). Compute $\bar{\lambda}$ of spheres and products of spheres. Show that $\bar{\lambda}(t)$ of a shrinking $S^2 \times S^1$ under the Ricci flow approaches ∞ as t approaches the singularity time. What happens if we start with $S^2 \times S^2$, where the S^2's have different radii? What is the behavior of $\bar{\lambda}$ for the product of Einstein spaces (or Ricci solitons)?

4. Classical entropy and Perelman's energy

Define the **classical entropy** on a closed manifold $\mathcal{M}^n$ by

(5.63) $$\mathcal{N} \doteqdot \int_{\mathcal{M}} f e^{-f} d\mu = -\int_{\mathcal{M}} u \log u \, d\mu,$$

where $u \doteqdot e^{-f}$. Under the gradient flow (5.29)–(5.30), we have

$$\frac{d\mathcal{N}}{dt} = \int_{\mathcal{M}} \frac{\partial f}{\partial t} e^{-f} d\mu = -\int_{\mathcal{M}} (R + \Delta f) e^{-f} d\mu$$

(5.64) $$= -\mathcal{F}.$$

That is, *the classical entropy is the anti-derivative of the negative of Perelman's energy.*

In this section we show that, by an upper bound for $\mathcal{F}$, a modification of $\mathcal{N}$ is monotone. For comparison, we discuss Hamilton's original proof of surface entropy monotonicity, the entropy formula for Hamilton's surface entropy, the fact that the gradient of Hamilton's surface entropy is the matrix Harnack, and Bakry–Emery's logarithmic Sobolev-type inequality.

4.1. Monotonicity of the classical entropy.
The following gives us an upper bound for the time interval of existence of the Ricci flow in terms of $\int_{\mathcal{M}} dm$ and the initial value of $\mathcal{F}^m$. Equivalently, it also implies the monotonicity of the classical entropy (see also [**356**], pp. 74–75).

PROPOSITION 5.34 (Upper bound for $\mathcal{F}$ in terms of time to blow up). *Suppose that $(g(t), f(t))$ is a solution on a closed manifold $\mathcal{M}^n$ of the gradient flow for $\mathcal{F}^m$, (5.25)–(5.26), for $t \in [0, T)$. Then we have*

(5.65) $$\mathcal{F}^m(g(0)) \leq \frac{n}{2T} \int_{\mathcal{M}} dm,$$

4. CLASSICAL ENTROPY AND PERELMAN'S ENERGY

that is,
$$T \leq \frac{n}{2\mathcal{F}^m(g(0))} \int_\mathcal{M} dm.$$

The proposition is a consequence of the following.

LEMMA 5.35 (Monotonicity formula for the classical entropy $\mathcal{N}$). *If $(g(t), f(t))$, $t \in [0, T)$, is a solution of the gradient flow (5.29)–(5.30) on a closed manifold $\mathcal{M}^n$, then*

$$\frac{d}{dt}\mathcal{F}^m(g(t)) \geq \frac{2}{n}\left(\int_\mathcal{M} dm\right)^{-1}\mathcal{F}^m(g(t))^2,$$

(5.66) $$\mathcal{F}^m(g(t)) \leq \frac{n}{2(T-t)}\int_\mathcal{M} e^{-f} d\mu.$$

By (5.64), this implies the following entropy monotonicity formula:

(5.67) $$\frac{d}{dt}\left(\mathcal{N} - \left(\frac{n}{2}\int_\mathcal{M} e^{-f} d\mu\right)\log(T-t)\right) \geq 0.$$

REMARK 5.36. Following §6.5 of [**356**], we may adjust the entropy quantity on the LHS of (5.67) by adding a constant and define

$$\widetilde{\mathcal{N}} \doteqdot \mathcal{N} - \left(\frac{n}{2}\int_\mathcal{M} e^{-f} d\mu\right)(\log[4\pi(T-t)] + 1).$$

Then we still have $\frac{d\widetilde{\mathcal{N}}}{dt} \geq 0$, whereas $\widetilde{\mathcal{N}}$ has the property that for a fundamental solution $u = e^{-f}$ limiting to a δ-function as $t \to T$, we have $\widetilde{\mathcal{N}} \to 0$ as $t \to T$.

PROOF OF THE LEMMA. From (5.28), we have

$$\frac{d}{dt}\mathcal{F}^m(g(t)) = 2\int_\mathcal{M} |R_{ij} + \nabla_i\nabla_j f|^2 dm \geq \frac{2}{n}\int_\mathcal{M}(R + \Delta f)^2 dm$$

$$\geq \frac{2}{n}\left(\int_\mathcal{M}(R + \Delta f) dm\right)^2 \bigg/ \int_\mathcal{M} dm$$

$$= \frac{2}{n}\left(\int_\mathcal{M} dm\right)^{-1}\mathcal{F}^m(g(t))^2.$$

The solution of the ODE
$$\frac{dx}{dt} = cx^2$$
with $\lim_{t \to T} x(t) = \infty$ is
$$x(t) = \frac{1}{c(T-t)}.$$

Hence, taking $c = \frac{2}{n}\left(\int_\mathcal{M} dm\right)^{-1}$, we get
$$\mathcal{F}^m(g(t)) \leq \frac{n}{2(T-t)}\int_\mathcal{M} dm.$$

□

REMARK 5.37.

(1) The formula above for $\frac{d}{dt}\mathcal{F}^m$ is somewhat reminiscent of Hamilton's formula for the evolution of the time-derivative dN/dt of his entropy $N(g) \doteqdot \int_{\mathcal{M}^2} R \log R \, dA$ on a positively curved surface evolving by Ricci flow (see [**180**]).
(2) We can rewrite (5.66) as

$$\int_{\mathcal{M}} \left(2\Delta f - |\nabla f|^2 + R - \frac{n}{2\tau}\right) e^{-f} d\mu \leq 0,$$

where $\tau \doteqdot T - t$.

4.2. Hamilton's surface entropy. Recall that the *normalized* surface entropy for a closed surface $(\mathcal{M}^2, g)$ with positive curvature is defined by

$$N(g) \doteqdot \int_{\mathcal{M}} \log(RA) \, R d\mu,$$

where A is the area. Let $(\mathcal{M}^2, g(t))$, $t \in [0, T)$, be a solution, on a maximal time interval of existence, of the Ricci flow on a closed surface with $R > 0$. In this subsection we give two proofs of the monotonicity of $N(g(t)) \doteqdot N(t)$.

4.2.1. *Hamilton's original proof of surface entropy monotonicity.* The time-derivative of $N(t)$ is given by

$$\tag{5.68} \frac{dN}{dt} = \int_{\mathcal{M}} Q R d\mu,$$

where

$$Q \doteqdot \Delta \log R + R - r$$

and r is the average scalar curvature. On the other hand, since $\frac{dr}{dt} = r^2$ and by a similar computation to (V1-5.38),

$$\frac{\partial}{\partial t} Q \geq \Delta Q + 2 \langle \nabla \log R, \nabla Q \rangle + Q^2 + 2rQ.$$

By the long-time existence theorem (Proposition 5.19 of Volume One), $r = \frac{1}{T-t}$ and $\text{Area}(g(t)) = 4\pi\chi(T-t)$, where χ denotes the Euler characteristic of $\mathcal{M}$. Differentiating (5.68) with respect to time, we compute that $Z \doteqdot \frac{dN}{dt}$ satisfies

$$\frac{dZ}{dt} = \int_{\mathcal{M}} \left(\frac{\partial}{\partial t} Q\right) R d\mu + \int_{\mathcal{M}} Q \frac{\partial}{\partial t} (R d\mu)$$

$$\geq \int_{\mathcal{M}} (Q^2 + 2rQ) R d\mu$$

$$\geq \frac{1}{\int_{\mathcal{M}} R d\mu} \left(\int_{\mathcal{M}} Q R d\mu\right)^2 + 2rZ$$

$$= \frac{1}{4\pi\chi} Z^2 + \frac{2}{T-t} Z,$$

where we integrated by parts and used Hölder's inequality and the Gauss-Bonnet formula. Thus

$$\text{(5.69)} \qquad \frac{d}{ds}\left(s^{-2}Z\right) \geq \frac{1}{4\pi\chi}\left(s^{-2}Z\right)^2,$$

where $s \doteq \frac{1}{T-t}$. From this we conclude that if $s_0^{-2} Z(s_0) > 0$ for some $s_0 < \infty$, then $s^{-2} Z(s) \to \infty$ as $s \to s_1$ for some $s_1 < \infty$. In other words, if $Z(t_0) > 0$ for some $t_0 < T$, then $Z(t) \to \infty$ as $t \to t_1$ for some $t_1 < T$. This contradicts our assumption that the solution exists on $[0, T)$. Hence $Z(t) \leq 0$ for all t and we have proved the following.

THEOREM 5.38 (Hamilton's surface entropy monotonicity). *For a solution of the Ricci flow on a closed surface with $R > 0$, we have*

$$\frac{dN}{dt}(t) \leq 0$$

for all $t \in [0, T)$.

Note that, from (5.69), we have $t \mapsto (T-t)^2 Z(t)$ is nondecreasing (since $\chi > 0$) and hence there is a constant $C > 0$ such that $(T-t)^2 Z(t) \geq -C$ for all $t \in [0, T)$. By (5.68),

$$Z = \frac{dN}{dt} = -\int_{\mathcal{M}} \frac{|\nabla R|^2}{R} dA + \int_{\mathcal{M}} (R - r)^2 dA,$$

and we have

$$\int_{\mathcal{M}} \frac{|\nabla R|^2}{R} dA \leq \int_{\mathcal{M}} (R - r)^2 dA + C(T - t)^{-2}.$$

REMARK 5.39. An inequality of the above type is often referred to as a reverse Poincaré inequality.

4.2.2. *Entropy formula for Hamilton's surface entropy.* Define the potential function f (up to an additive constant) by $\Delta f = r - R$. In [**97**] the monotonicity of the entropy was proved by relating its time-derivative to Ricci solitons via an integration by parts using the potential function (Proposition 5.39 in Volume One). In particular, we have

$$\frac{dN}{dt}(t) = -2\int_{\mathcal{M}} \left| R_{ij} + \nabla_i \nabla_j f - \frac{1}{2(T-t)} g_{ij} \right|^2 dA$$
$$- \int_{\mathcal{M}} \left| \nabla \log\left(R \cdot e^{-f}\right) \right|^2 R dA.$$

Note that $R_{ij} = \frac{1}{2} R g_{ij}$ and $r = \frac{1}{A} \int_{\mathcal{M}} R dA = (T - t)^{-1}$. We have purposely written this formula to more resemble Perelman's formulas (5.41) and (6.17).

4.2.3. The gradient of Hamilton's surface entropy is the matrix Harnack quantity.
A less well-known fact is that the gradient of Hamilton's entropy in the space of all metrics with the L^2-metric is the matrix Harnack quantity:

$$(5.70) \qquad \delta_v N(g) = \int_{\mathcal{M}} v_{ij} \left(-\Delta \log R \cdot g_{ij} + \nabla_i \nabla_j \log R - \frac{1}{2} R g_{ij} \right) dA,$$

where $\delta g = v$ (see Lemma 10.23 of [**111**] and use $N(g) - E(g)$ is a constant). In the space of metrics in a fixed conformal class, the gradient is the trace Harnack quantity. Note that the same relation is true relating the entropy and the trace Harnack quantity for the Gauss curvature flow of convex hypersurfaces in Euclidean space [**96**].

4.3. Bakry–Emery's logarithmic Sobolev-type inequality.
The proofs of Hamilton's surface entropy formula and Perelman's energy formulas are formally similar to the proof of Bakry and Emery of their logarithmic Sobolev-type inequality [**18**].

PROPOSITION 5.40. *Let $(\mathcal{M}^n, g)$ be a closed Riemannian manifold with $\mathrm{Rc} \geq K$ for some constant $K > 0$. If u is a positive function on $\mathcal{M}$, then*

$$\int_{\mathcal{M}} u \log u \, d\mu \leq \frac{1}{2K} \int_{\mathcal{M}} u |\nabla \log u|^2 \, d\mu + \log \left(\frac{1}{\mathrm{Vol}(\mathcal{M})} \int_{\mathcal{M}} u \, d\mu \right) \int_{\mathcal{M}} u \, d\mu.$$

PROOF. (See [**104**] for more details of the computations.) Consider the solution v to the heat equation $\frac{\partial v}{\partial t} = \Delta v$ with $v(0) = u$. The solution v exists for all time and

$$\lim_{t \to \infty} v = \frac{1}{\mathrm{Vol}(\mathcal{M})} \int_{\mathcal{M}} u \, d\mu.$$

Define $E(t) \doteq \int_{\mathcal{M}} v \log v \, d\mu$. Then

$$(5.71) \qquad \lim_{t \to \infty} E(t) = \int_{\mathcal{M}} u \, d\mu \cdot \log \left(\frac{1}{\mathrm{Vol}(\mathcal{M})} \int_{\mathcal{M}} u \, d\mu \right).$$

We have

$$\frac{dE}{dt} = -\int_{\mathcal{M}} \langle \nabla v, \nabla \log v \rangle \, d\mu = -\int_{\mathcal{M}} v |\nabla \log v|^2 \, d\mu \leq 0.$$

Note that $\lim_{t \to \infty} \frac{dE}{dt}(t) = 0$.

Using $\frac{\partial}{\partial t} \log v = \Delta \log v + |\nabla \log v|^2$, we compute

$$\frac{d^2 E}{dt^2} = 2 \int_{\mathcal{M}} v \left(|\nabla \nabla \log v|^2 + \mathrm{Rc}(\nabla \log v, \nabla \log v) \right) d\mu.$$

Using our assumption $\mathrm{Rc} \geq K$, we find

$$\frac{d^2 E}{dt^2} \geq -2K \frac{dE}{dt}.$$

By $\lim_{t\to\infty}\frac{dE}{dt}(t)=0$ and (5.71), we have

$$\frac{dE}{dt}(0)=-\int_0^\infty \frac{d^2E}{dt^2}(t)\,dt \le 2K\int_0^\infty \frac{dE}{dt}(t)\,dt$$

$$=2K\log\left(\frac{1}{\mathrm{Vol}(\mathcal{M})}\int_\mathcal{M} u\,d\mu\right)\int_\mathcal{M} u\,d\mu - 2KE(0).$$

Hence

$$-\int_\mathcal{M} u\,|\nabla\log u|^2\,d\mu \le 2K\log\left(\frac{1}{\mathrm{Vol}(\mathcal{M})}\int_\mathcal{M} u\,d\mu\right)\int_\mathcal{M} u\,d\mu$$

$$-2K\int_\mathcal{M} u\log u\,d\mu$$

and the proposition follows. $\square$

5. Notes and commentary

Subsection 1.1. As we remarked earlier, the function f is also known as the dilaton; in the physics literature there are numerous references to Perelman's energy functional (see Green, Schwarz, and Witten [162], Polchinksi [307], Strominger and Vafa [341] for example), although Perelman is the first to consider it in the context of Ricci flow. The Ricci flow is the 1-loop approximation of the renormalization group flow (see Friedan [145]).

Subsection 1.2. For a computational motivation for fixing the measure, see also §4 in Chapter 2 of [111], where Perelman's functional is motivated starting from the total scalar curvature functional. In particular, let $\delta g = v$. The variation of the total scalar curvature is

$$\delta \int_\mathcal{M} R\,d\mu = \int_\mathcal{M} \left(\mathrm{div}(\mathrm{div}\,v) - \Delta V - \mathrm{Rc}\cdot v + R\frac{V}{2}\right)d\mu$$

$$= -\int_\mathcal{M}\left(\mathrm{Rc}-\frac{R}{2}g\right)\cdot v\,d\mu.$$

This says that $\nabla\left(\int_\mathcal{M} R\,d\mu\right) = -\mathrm{Rc} + \frac{R}{2}g$, where the gradient is calculated with respect to the standard L^2-metric. To try to find a functional $\mathcal{F}$ with $\nabla\mathcal{F} = -\mathrm{Rc}$, we want to get rid of the $\frac{R}{2}g$ term. Now this term is due to the variation of $d\mu$. So we consider the distorted volume form $e^{-f}d\mu$ and assume its variation is 0. Hence

$$\delta\int_\mathcal{M} Re^{-f}d\mu = \int_\mathcal{M}(\delta R)e^{-f}d\mu = \int_\mathcal{M}(\mathrm{div}(\mathrm{div}\,v)-\Delta V-\mathrm{Rc}\cdot v)e^{-f}d\mu$$

and now we have the extra terms $\int_\mathcal{M}(\mathrm{div}(\mathrm{div}\,v)-\Delta V)e^{-f}d\mu$. We compensate for this by considering

$$\delta\int_\mathcal{M}|\nabla f|^2 e^{-f}d\mu = \int_\mathcal{M}\left(\delta|\nabla f|^2\right)e^{-f}d\mu$$

$$= \int_\mathcal{M}(-v(\nabla f,\nabla f)+\nabla f\cdot\nabla V)e^{-f}d\mu,$$

using $\frac{V}{2} = h \doteq \delta f$. Integrating by parts yields

$$\delta \mathcal{F} = \delta \int_{\mathcal{M}} R e^{-f} d\mu + \delta \int_{\mathcal{M}} |\nabla f|^2 e^{-f} d\mu = -\int_{\mathcal{M}} v_{ij} \left(R_{ij} + \nabla_i \nabla_j f\right) e^{-f} d\mu.$$

Although the Ricci tensor is not strictly elliptic in g, one can ask if the RHS of equation (5.27)

$$\frac{\partial}{\partial t} g_{ij} = -2\left[R_{ij} + \nabla_i \nabla_j \log\left(\frac{d\mu}{dm}\right)\right]$$

is elliptic in g. The answer is still no. In particular, if $\delta g = v$, then

$$\delta \left[\log\left(\frac{d\mu}{dm}\right)\right] = \frac{V}{2}.$$

Hence

$$\delta \left[\nabla_i \nabla_j \log\left(\frac{d\mu}{dm}\right)\right] = \frac{1}{2} \nabla_i \nabla_j v - \delta\left(\Gamma_{ij}^k\right) \cdot \nabla_k \log\left(\frac{d\mu}{dm}\right).$$

Since

$$\delta\left(-2R_{ij}\right) = -\nabla_i \nabla^k v_{jk} - \nabla_j \nabla^k v_{ik} + \nabla_i \nabla_j v + \Delta v_{ij},$$

we have

$$\delta\left(-2\left[R_{ij} + \nabla_i \nabla_j \log\left(\frac{d\mu}{dm}\right)\right]\right) = \Delta v_{ij} - \nabla_i \nabla^k v_{jk} - \nabla_j \nabla^k v_{ik}$$
$$+ \text{ lower-order terms},$$

where the last pair of terms form a Lie derivative of the metric term. However the second-order operator on the RHS is still not elliptic in v.

CHAPTER 6

Entropy and No Local Collapsing

Everything should be made as simple as possible, but not simpler.
– Albert Einstein

Disorder increases with time because we measure time in the direction in which disorder increases. – Stephen Hawking

Close, but no cigar. – Unknown origin

By combining Perelman's energy and the classical entropy in a suitable way, we obtain the entropy functional $\mathcal{W}$, which we shall discuss in this chapter. This is implemented with the introduction of a positive scale-factor τ. The advantage which the addition of this scale-factor yields is that from the functional $\mathcal{W}$ we can understand aspects of the local geometry of the manifold, e.g., volume ratios of balls with radius on the order of $\sqrt{\tau}$. Perelman's entropy is also the integral of his Harnack quantity for fundamental solutions of the adjoint heat equation.[1] As such, one can integrate in space the Harnack partial differential inequality to give a proof of the monotonicity formula for $\mathcal{W}$. Note that here Perelman's Harnack quantity is directly related to the entropy whereas in Hamilton's earlier work on surfaces, Hamilton's Harnack quantity is related to the time-derivative of his entropy.[2] This monotonicity formula can be used to prove that shrinking breathers must be shrinking gradient Ricci solitons. More importantly, this monotonicity formula will be fundamental in proving Hamilton's little loop conjecture or what Perelman calls the no local collapsing theorem.

In this chapter, we shall discuss in detail the entropy estimates, the two functionals μ and ν associated to $\mathcal{W}$, and their geometric applications. We discuss the logarithmic Sobolev inequality, which is related to the entropy functional. We will also give different versions/proofs of the no local collapsing theorem. In the last part of this chapter we shall discuss some interesting calculations related to entropy.

Throughout this chapter $\mathcal{M}^n$ denotes a closed n-dimensional manifold.

1. The entropy functional $\mathcal{W}$ and its monotonicity

Let $(\mathcal{M}^n, g(t))$, $t \in [0, T]$, be a solution of the Ricci flow on a closed manifold. Note that by the proof of Lemma 5.31, we have that when

---
[1]See (6.21).
[2]Equation (6.21) as compared to (5.70).

$\lambda(g(t)) \leq 0$ for some $t \in [t_1, t_2]$, even *shrinking* breathers, i.e., solutions with $g(t_2) = \alpha \Phi^* g(t_1)$ and $\alpha < 1$, are Einstein solutions (trivial Ricci solitons). In order to handle shrinking breathers when $\lambda(g(t)) > 0$ for all $t \in [t_1, t_2]$, we need to generalize the monotonicity formula for the functional $\mathcal{F}$ to a monotonicity formula for a functional related to shrinking breathers. This is the monotonicity formula for the entropy $\mathcal{W}$.

In this section we introduce the entropy $\mathcal{W}$ and discuss its monotonicity. We also give a unified treatment of energy and entropy in the last subsection.

1.1. The entropy $\mathcal{W}$, its first variation and the gradient flow.

1.1.1. *The entropy $\mathcal{W}$.* Let $\mathfrak{Met}$ denote the space of smooth Riemannian metrics on a closed manifold $\mathcal{M}^n$. We define Perelman's **entropy functional** $\mathcal{W} : \mathfrak{Met} \times C^\infty(\mathcal{M}) \times \mathbb{R}^+ \to \mathbb{R}$ by

$$(6.1) \qquad \mathcal{W}(g, f, \tau) \doteqdot \int_\mathcal{M} \left[\tau \left(R + |\nabla f|^2 \right) + f - n \right] (4\pi\tau)^{-n/2} e^{-f} d\mu$$

$$(6.2) \qquad = \int_\mathcal{M} \left[\tau \left(R + |\nabla f|^2 \right) + f - n \right] u d\mu,$$

where[3]

$$(6.3) \qquad u \doteqdot (4\pi\tau)^{-n/2} e^{-f}.$$

This is a modification of the energy functional $\mathcal{F}(g, f)$, which we considered in the last chapter, where we have now introduced the positive parameter τ. By (5.1), we have

$$(6.4) \qquad \mathcal{W}(g, f, \tau) = (4\pi\tau)^{-n/2} \left(\tau \mathcal{F}(g, f) + \int_\mathcal{M} (f - n) e^{-f} d\mu \right)$$

$$(6.5) \qquad = (4\pi\tau)^{-n/2} (\tau \mathcal{F}(g, f) + \mathcal{N}(f)) - n \int_\mathcal{M} u \, d\mu,$$

where the second equality is obtained using definition (5.63). As we shall see, τ plays the dual roles of understanding the geometry of $(\mathcal{M}, g)$ at the *distance* scale $\sqrt{\tau}$ and representing a constant minus *time* for solutions $(\mathcal{M}, g(t))$ of the Ricci flow.[4]

The functional $\mathcal{W}$ has the following elementary properties.

(i) (*Scale invariance*) $\mathcal{W}$ is invariant under the scalings $\tau \mapsto c\tau$ and $g \mapsto cg$, i.e.,

$$(6.6) \qquad \mathcal{W}(cg, f, c\tau) = \mathcal{W}(g, f, \tau).$$

(ii) (*Diffeomorphism invariance*) If $\Phi : \mathcal{M} \to \mathcal{M}$ is a diffeomorphism, then

$$\mathcal{W}(g, f, \tau) = \mathcal{W}(\Phi^* g, \Phi^* f, \tau),$$

where $\Phi^* g$ is the pulled-back metric and $\Phi^* f = f \circ \Phi$.

---

[3] This u is not to be confused with the $u = e^{-f}$ in Chapter 5.
[4] We may think of τ as physically representing temperature (see §5 of [297]).

1.1.2. The first variation of $\mathcal{W}$.
Let $\delta g = v \in C^2(\mathcal{M}, T^*\mathcal{M} \otimes T^*\mathcal{M})$, let $\delta f = h$, and let $\delta \tau = \zeta$. Since

$$\delta(u\, d\mu) = \delta\left((4\pi\tau)^{-n/2} e^{-f} d\mu\right) = \left(-\frac{n}{2\tau}\zeta - h + \frac{V}{2}\right) u\, d\mu, \tag{6.7}$$

where $V \doteq \operatorname{tr}_g v = g^{ij} v_{ij}$, the measure $(4\pi\tau)^{-n/2} e^{-f} d\mu$ preserving variations satisfy

$$-\frac{n}{2\tau}\zeta - h + \frac{V}{2} = 0. \tag{6.8}$$

We find it convenient to write the variation of $\mathcal{W}$ so that this quantity is one of the factors. In particular, we have

LEMMA 6.1 (Entropy first variation formula). *The first variation of $\mathcal{W}$ at (g, f, τ) can be expressed as follows:*

$$\begin{aligned}
&\delta \mathcal{W}_{(v,h,\zeta)}(g, f, \tau) \\
&= \int_{\mathcal{M}} (-\tau v_{ij} + \zeta g_{ij})\left(R_{ij} + \nabla_i \nabla_j f - \frac{1}{2\tau} g_{ij}\right) u\, d\mu \\
&\quad + \int_{\mathcal{M}} \tau\left(\frac{V}{2} - h - \frac{n\zeta}{2\tau}\right)\left(R + 2\Delta f - |\nabla f|^2 + \frac{f - n - 1}{\tau}\right) u\, d\mu.
\end{aligned} \tag{6.9}$$

PROOF. It follows from the first variation formula (5.10) of $\mathcal{F}$ with respect to g and f (keeping τ fixed) that

$$\begin{aligned}
&\delta_{(v,h,0)}\left(\tau(4\pi\tau)^{-n/2} \mathcal{F}(g, f)\right) \\
&= -\int_{\mathcal{M}} \tau v_{ij}(R_{ij} + \nabla_i \nabla_j f) u\, d\mu \\
&\quad + \int_{\mathcal{M}} \tau\left(\frac{V}{2} - h\right)\left(2\Delta f - |\nabla f|^2 + R\right) u\, d\mu.
\end{aligned}$$

Next we calculate the first variation of the remaining term of $\mathcal{W}$ with respect to g and f (again keeping τ fixed),

$$\begin{aligned}
&\delta_{(v,h,0)}\left((4\pi\tau)^{-n/2} \int_{\mathcal{M}} (f - n) e^{-f} d\mu\right) \\
&= \int_{\mathcal{M}} \left[(1 + n - f) h + \frac{V}{2}(f - n)\right] u\, d\mu.
\end{aligned}$$

Now the term from the variation of $\mathcal{W}$ with respect to τ is

$$\begin{aligned}
&\delta_{(0,0,\zeta)}\left(\int_{\mathcal{M}} \left[\tau(R + |\nabla f|^2) + f - n\right](4\pi\tau)^{-n/2} e^{-f} d\mu\right) \\
&= \int_{\mathcal{M}} \left[\left(1 - \frac{n}{2}\right)\zeta(R + |\nabla f|^2) - \frac{n\zeta}{2\tau}(f - n)\right] u\, d\mu.
\end{aligned}$$

Combining the above three formulas and simplifying a little, we get

$$\delta_{(v,h,\zeta)}\mathcal{W}(g,f,\tau)$$
$$= -\int_{\mathcal{M}} \tau v_{ij}(R_{ij} + \nabla_i\nabla_j f)u\,d\mu$$
$$+ \int_{\mathcal{M}} \tau\left(\frac{V}{2} - h\right)\left(2\Delta f - |\nabla f|^2 + R + \frac{f-n}{\tau}\right)u\,d\mu$$
$$+ \int_{\mathcal{M}} \left[h + \left(1 - \frac{n}{2}\right)\zeta(R + |\nabla f|^2) - \frac{n\zeta}{2\tau}(f-n)\right]u\,d\mu.$$

We rewrite the above expression as

$$\delta_{(v,h,\zeta)}\mathcal{W}(g,f,\tau)$$
$$= \int_{\mathcal{M}} (-\tau v_{ij} + \zeta g_{ij})(R_{ij} + \nabla_i\nabla_j f)u\,d\mu$$
$$+ \int_{\mathcal{M}} \tau\left(\frac{V}{2} - h - \frac{n\zeta}{2\tau}\right)\left(2\Delta f - |\nabla f|^2 + R + \frac{f-n}{\tau}\right)u\,d\mu$$
$$+ \int_{\mathcal{M}} -\zeta(R + \Delta f)u\,d\mu + \int_{\mathcal{M}} \frac{n\zeta}{2}\left(2\Delta f - |\nabla f|^2 + R\right)u\,d\mu$$
$$+ \int_{\mathcal{M}} \left(1 - \frac{n}{2}\right)\zeta(R + |\nabla f|^2)u\,d\mu + \int_{\mathcal{M}} hu\,d\mu,$$

which, by combining terms, we further simplify to

$$\delta_{(v,h,\zeta)}\mathcal{W}(g,f,\tau)$$
$$= \int_{\mathcal{M}} (-\tau v_{ij} + \zeta g_{ij})(R_{ij} + \nabla_i\nabla_j f)u\,d\mu$$
$$+ \int_{\mathcal{M}} \tau\left(\frac{V}{2} - h - \frac{n\zeta}{2\tau}\right)\left(2\Delta f - |\nabla f|^2 + R + \frac{f-n}{\tau}\right)u\,d\mu$$
$$+ \int_{\mathcal{M}} (n-1)\zeta\left(\Delta f - |\nabla f|^2\right)u\,d\mu + \int_{\mathcal{M}} hu\,d\mu.$$

Since ζ is a constant, (6.9) follows from a rearrangement and the integration by parts identity: $\int_{\mathcal{M}} \left(\Delta f - |\nabla f|^2\right)e^{-f}d\mu = 0$. □

REMARK 6.2. Analogous to (5.14) and (5.15), the terms

$$R_{ij} + \nabla_i\nabla_j f - \frac{1}{2\tau}g_{ij} \quad \text{and} \quad R + 2\Delta f - |\nabla f|^2 + \frac{f-n}{\tau}$$

in (6.9) are natural quantities vanishing/constant on shrinking gradient Ricci solitons.

1.1.3. *The gradient flow of* $\mathcal{W}$. When we require that the variation (v, h, ζ) satisfies

$$\zeta = -1 \quad \text{and} \quad \frac{V}{2} - h - \frac{n}{2\tau}\zeta = 0,$$

i.e., the variation preserves the measure $(4\pi\tau)^{-n/2}e^{-f}d\mu$ on $\mathcal{M}$, we obtain from (6.9) the gradient flow (assuming $\frac{d\tau}{dt} = -1$)

$$\frac{\partial}{\partial t}g_{ij} = -2(R_{ij} + \nabla_i\nabla_j f), \tag{6.10}$$

$$\frac{\partial f}{\partial t} = -\Delta f - R + \frac{n}{2\tau}, \tag{6.11}$$

$$\frac{d\tau}{dt} = -1. \tag{6.12}$$

LEMMA 6.3 (Entropy monotonicity for gradient flow). *If $(g(t), f(t), \tau(t))$ is a solution to the system (6.10)–(6.12), then*

$$\frac{d}{dt}\mathcal{W}(g(t), f(t), \tau(t))$$
$$= \int_{\mathcal{M}} 2\tau \left| R_{ij} + \nabla_i\nabla_j f - \frac{1}{2\tau}g_{ij} \right|^2 u\,d\mu \geq 0. \tag{6.13}$$

PROOF. This follows directly from substituting

$$v_{ij} = -2(R_{ij} + \nabla_i\nabla_j f),$$
$$h = -\Delta f - R + \frac{n}{2\tau},$$
$$\zeta = -1$$

into (6.9) and using the facts

$$\frac{V}{2} - h - \frac{n}{2\tau}\zeta = 0 \quad \text{and} \quad \int_{\mathcal{M}} \left(\Delta f - |\nabla f|^2\right) e^{-f}d\mu = 0.$$

□

1.2. Coupled evolution equations associated to $\mathcal{W}$ and monotonicity of $\mathcal{W}$.

1.2.1. *The coupled evolution equations associated to $\mathcal{W}$.* As in Chapter 5, there is a system of evolution equations for the triple (g, f, τ) (see (1.3) of [**297**])

$$\frac{\partial}{\partial t}g_{ij} = -2R_{ij}, \tag{6.14}$$

$$\frac{\partial f}{\partial t} = -\Delta f + |\nabla f|^2 - R + \frac{n}{2\tau}, \tag{6.15}$$

$$\frac{d\tau}{dt} = -1, \tag{6.16}$$

whose solution differs from the solution to (6.10)–(6.12) by diffeomorphisms. This leads to the following theorem, which says that

$$\frac{d}{dt}\mathcal{W}(g(t), f(t), \tau(t)) \geq 0.$$

Another motivation for studying this system of equations, from considering gradient Ricci solitons, was discussed in Section 8 of Chapter 1.

THEOREM 6.4 (Entropy monotonicity for Ricci flow). *Let $(g(t), f(t), \tau(t))$, $t \in [0, T]$, be a solution of the modified evolution equations (6.14), (6.15), and (6.16). Then the first variation of $\mathcal{W}$ along this solution is given by the following:*

$$\frac{d}{dt}\mathcal{W}(g(t), f(t), \tau(t))$$

(6.17) $$= \int_{\mathcal{M}} 2\tau \left| R_{ij} + \nabla_i \nabla_j f - \frac{1}{2\tau} g_{ij} \right|^2 \cdot u \, d\mu$$

(6.18) $$= 2\tau \int_{\mathcal{M}} \left| R_{ij} - \frac{1}{u}\left(\nabla_i \nabla_j u - \frac{\nabla_i u \nabla_j u}{u} + \frac{1}{2\tau} u g_{ij} \right) \right|^2 \cdot u \, d\mu \geq 0.$$

EXERCISE 6.5. Show that the line of reasoning of deriving Theorem 6.4 from Lemma 6.3 is rigorous.

REMARK 6.6. The expression

(6.19) $$\nabla_i \nabla_j u - \frac{\nabla_i u \nabla_j u}{u} + \frac{1}{2\tau} u g_{ij} = u \left(\nabla_i \nabla_j \log u + \frac{1}{2\tau} g_{ij} \right)$$

is exactly the matrix Harnack quantity for a solution u of the backward heat equation. In particular, this same expression for positive solutions of the backward heat equation appeared in Hamilton's derivation of the monotonicity formulae for the harmonic map heat flow, mean curvature flow, as well as the Yang–Mills flow; see [**184**].[5]

Equation (6.17) is Perelman's **entropy monotonicity formula** and implies that $\mathcal{W}(g(t), f(t), \tau(t))$ is strictly increasing along a solution of the modified coupled flow except when $g(t)$ is a shrinking gradient Ricci soliton (since $\tau > 0$ in (6.17)), where it must flow along ∇f and where $\mathcal{W}(g(t), f(t), \tau(t))$ is constant. This monotonicity is also fundamental in understanding the local geometry of the solution $g(t)$ to the Ricci flow as we shall see in the proof of the no local collapsing theorem. The function f allows us to localize and the parameter τ tells us at what distance scale to localize ($\sqrt{\tau}$).

Since we do not give the details of how to transform between the systems (6.10)–(6.12) and (6.14)–(6.16), we also compute (6.17) directly through the following exercise.

EXERCISE 6.7 (Deriving $\frac{d\mathcal{W}}{dt}$ from $\frac{d\mathcal{F}}{dt}$). Use the equation for $\frac{d\mathcal{F}}{dt}$ to derive the formula for $\frac{d\mathcal{W}}{dt}$.

SOLUTION TO EXERCISE 6.7. The effect of the extra term $+\frac{n}{2\tau}$ in (6.11) as compared to (5.30) is to add

$$-\frac{n}{2\tau} \int_{\mathcal{M}} \left(R + |\nabla f|^2 \right) e^{-f} d\mu$$

---

[5]For an exposition of the matrix Harnack estimate asssociated to (6.19), see [**104**].

1. THE ENTROPY FUNCTIONAL $\mathcal{W}$ AND ITS MONOTONICITY

to (5.31), so that we get

$$\frac{d\mathcal{F}}{dt} = 2\int_{\mathcal{M}} |R_{ij} + \nabla_i \nabla_j f|^2 e^{-f} d\mu - \frac{n}{2\tau}\mathcal{F}.$$

Similarly, (5.64) becomes

$$\frac{d\mathcal{N}}{dt} = -\mathcal{F} + \frac{n}{2\tau}\int_{\mathcal{M}} e^{-f} d\mu - \frac{n}{2\tau}\mathcal{N}$$

under (6.10)–(6.11). Hence, by (6.5), i.e.,

$$\mathcal{W} = (4\pi\tau)^{-n/2}\left(\tau\mathcal{F}(g,f) + \mathcal{N}\right) - n\int_{\mathcal{M}} u\, d\mu$$

and $\int_{\mathcal{M}} u\, d\mu = $ const, we have

$$\frac{d\mathcal{W}}{dt} = \frac{d}{dt}\left[(4\pi\tau)^{-n/2}(\tau\mathcal{F} + \mathcal{N})\right]$$

$$= (4\pi\tau)^{-n/2}\left(\frac{n}{2\tau}(\tau\mathcal{F} + \mathcal{N}) - \mathcal{F} + \tau\frac{d\mathcal{F}}{dt} + \frac{d\mathcal{N}}{dt}\right),$$

so that

$$(4\pi\tau)^{n/2}\frac{d\mathcal{W}}{dt} = 2\tau\int_{\mathcal{M}} |R_{ij} + \nabla_i\nabla_j f|^2 e^{-f} d\mu - 2\mathcal{F} + \frac{n}{2\tau}\int_{\mathcal{M}} e^{-f} d\mu$$

$$= 2\tau\int_{\mathcal{M}}\left|R_{ij} + \nabla_i\nabla_j f - \frac{1}{2\tau}g_{ij}\right|^2 e^{-f} d\mu.$$

1.2.2. Second proof of the monotonicity of $\mathcal{W}$ from a pointwise estimate.
Analogously to subsection 2.3.2 of Chapter 5 we again derive (6.17) using a pointwise evolution formula. Let[6]

(6.20) $$v \doteqdot \left[\tau\left(R + 2\Delta f - |\nabla f|^2\right) + f - n\right]u.$$

In Part II of this volume we shall see that v is nonpositive when u is a fundamental solution (for this reason v is also called **Perelman's Harnack quantity**). Note that

(6.21) $$\mathcal{W}(g,f,\tau) = \int_{\mathcal{M}} v\, d\mu.$$

We shall show the following below.

LEMMA 6.8 (Perelman's Harnack quantity satisfies adjoint heat-type equation). *Under* (6.14)–(6.16)

(6.22) $$\square^* v = -2\tau\left|R_{ij} + \nabla_i\nabla_j f - \frac{1}{2\tau}g_{ij}\right|^2 u,$$

where $\square^* = -\frac{\partial}{\partial t} - \Delta + R$ is the adjoint heat operator defined in (5.39).

---

[6]This quantity v is not to be confused with $v = \delta g$.

Theorem 6.4 then follows from

$$\frac{d\mathcal{W}}{dt} = \int_\mathcal{M} \left(\frac{\partial}{\partial t} - R\right) v d\mu = \int_\mathcal{M} \left(-\square^* - \Delta\right) v d\mu$$

$$= 2\tau \int_\mathcal{M} \left|R_{ij} + \nabla_i \nabla_j f - \frac{1}{2\tau} g_{ij}\right|^2 u d\mu$$

$$\geq 0.$$

Before we prove (6.22), we need the following lemma concerning the function $u = (4\pi\tau)^{-n/2} e^{-f}$ as defined in (6.3) (compare with (5.40)).

LEMMA 6.9 (u is a solution to the adjoint heat equation). *The evolution equation (6.15) of f is equivalent to the following evolution equation of u:*

(6.23) $$\square^* u = 0.$$

PROOF. We calculate

$$\square^* u = \left(-\frac{\partial}{\partial t} - \Delta + R\right)\left((4\pi\tau)^{-n/2} e^{-f}\right)$$

$$= (4\pi\tau)^{-n/2} e^{-f}\left(-\frac{n}{2\tau} + \frac{\partial f}{\partial t} + R\right) - (4\pi\tau)^{-n/2} \Delta e^{-f} = 0.$$

$\square$

We now show that we can apply the computation in subsection 2.3.2 of Chapter 5 to derive the evolution equation (6.22) for v.

PROOF OF (6.22). Let $\bar{f}$ be defined by $e^{-\bar{f}} \doteqdot u$. Then $\bar{f} = f + \frac{n}{2} \log(4\pi\tau)$ and $(g, \bar{f})$ satisfies (5.34) and (5.35). By (5.43), we have that the quantity

$$V \doteqdot \left(R + 2\Delta \bar{f} - |\nabla \bar{f}|^2\right) u$$

$$= \left(-2\Delta \log u - |\nabla \log u|^2 + R\right) u$$

satisfies the equation

$$\square^* V = -2|R_{ij} - \nabla_i \nabla_j \log u|^2 u.$$

Note that

$$v = \left[\tau\left(R + 2\Delta \bar{f} - |\nabla \bar{f}|^2\right) - \log u - \frac{n}{2} \log(4\pi\tau) - n\right] u$$

(6.24) $$= \tau V - \left(\log u + \frac{n}{2} \log(4\pi\tau) + n\right) u.$$

We compute using $\square^* u = 0$ and the general formula

$$\square^* (ab) = b\square^* a + a\square^* b - 2\langle \nabla a, \nabla b\rangle - Rab$$

that
$$\square^* \left[\left(\log u + \frac{n}{2} \log(4\pi\tau) + n \right) u \right]$$
$$= u \square^* \log u + \log u \, \square^* u - 2 \langle \nabla \log u, \nabla u \rangle - Ru \log u + \frac{n}{2\tau} u$$
$$= -Ru - |\nabla \log u|^2 u + \frac{n}{2\tau} u,$$

since
$$\square^* \log u = \frac{1}{u} \left(-\frac{\partial}{\partial t} - \Delta \right) u + |\nabla \log u|^2 + R \log u$$
$$= -R + |\nabla \log u|^2 + R \log u.$$

From (6.24), we compute
$$\square^* v = \tau \square^* V + V + |\nabla \log u|^2 u + Ru - \frac{n}{2\tau} u$$
$$= -2\tau |R_{ij} - \nabla_i \nabla_j \log u|^2 u$$
$$+ \left(-2\Delta \log u - |\nabla \log u|^2 + R \right) u$$
$$+ |\nabla \log u|^2 u + Ru - \frac{n}{2\tau} u$$
$$= -2\tau |R_{ij} - \nabla_i \nabla_j \log u|^2 u + 2 \left(-\Delta \log u + R \right) u - \frac{n}{2\tau} u.$$

Completing the square, we obtain (6.22). □

1.3. A unified treatment of energy $\mathcal{F}$ and entropy $\mathcal{W}$. We finish this section with several exercises. In total, this unifies part of the discussion of shrinking, steady, and expanding gradient Ricci solitons, which correspond to entropy, energy, and expander entropy on a closed manifold $\hat{\mathcal{M}}^n$ (see the definition below), respectively.[7] In this subsection we use the following convention: $\varepsilon \in \mathbb{R}$, and if $\varepsilon \neq 0$, we take $\tau(t) = \varepsilon t$, whereas if $\varepsilon = 0$, we take $\tau(t) \equiv 1$. We only consider all $t \in \mathbb{R}$ such that $\tau(t) > 0$.

Let
$$(6.25) \qquad V_\varepsilon \doteqdot V_\varepsilon(g, f, \tau) \doteqdot \tau \left(R + 2\Delta f - |\nabla f|^2 \right) - \varepsilon (f - n).$$

Define the ε**-entropy** by
$$(6.26) \quad \mathcal{W}_\varepsilon(g, f, \tau) \doteqdot \int_{\hat{\mathcal{M}}} \left(\tau \left(R + |\nabla f|^2 \right) - \varepsilon (f - n) \right) (4\pi\tau)^{-n/2} e^{-f} d\mu$$
$$= \int_{\hat{\mathcal{M}}} V_\varepsilon u \, d\mu,$$

---

[7] Notation: For the most part we use a hat on $\mathcal{M}$ to emphasize that we are considering a fixed metric instead of a solution to the Ricci flow.

where u is defined in (6.3). When $\varepsilon < 0$, this is Perelman's entropy; when $\varepsilon = 0$, this is Perelman's energy; and when $\varepsilon > 0$, this is called the **expander entropy**.[8] The definition of V_ε is motivated by the following exercise.

EXERCISE 6.10 (Harnack V_ε as an integrand for $\mathcal{W}_\varepsilon$). Let $(\mathcal{M}^n, g(t), f(t))$ be a gradient Ricci soliton in canonical form. By Proposition 1.7, the pair $(g(t), f(t))$ satisfies

$$(6.27) \qquad S^\varepsilon_{ij} \doteqdot R_{ij} + \nabla_i \nabla_j f + \frac{\varepsilon}{2\tau} g_{ij} = 0,$$

where $\varepsilon \in \mathbb{R}$. Note that if $\varepsilon \neq 0$, then $g(t)$ and $f(t)$ are defined for all t such that $\tau(t) > 0$, whereas if $\varepsilon = 0$, then $g(t)$ and $f(t)$ are defined for all $t \in (-\infty, \infty)$. Show that $V_\varepsilon(g(t), f(t), \tau(t))$ is constant in space.

EXERCISE 6.11 ($\mathcal{W}_\varepsilon$ and the Gaussian soliton). Consider the Gaussian soliton $(\mathbb{R}^n, g_\mathbb{E}, f_\varepsilon)$, where $g_\mathbb{E} = \sum_{i=1}^n (dx^i)^2$ and

$$f_\varepsilon(x,t) \doteqdot \begin{cases} -\frac{|x|^2}{4t} & \text{for } t > 0 \text{ if } \varepsilon < 0, \\ 0 & \text{for } t \in \mathbb{R} \text{ if } \varepsilon = 0, \\ -\frac{|x|^2}{4t} & \text{for } t < 0 \text{ if } \varepsilon > 0. \end{cases}$$

Check that

$$\frac{\partial^2 f_\varepsilon}{\partial x^i \partial x^j} + \frac{\varepsilon}{2\tau} \delta_{ij} = 0$$

for all t such that $\tau(t) = \varepsilon t > 0$, and thus $S^\varepsilon_{ij} \equiv 0$. Show $V_\varepsilon = 0$, so that $\mathcal{W}_\varepsilon = 0$. It is useful to keep this example in mind, which reflects the Euclidean heat kernel, when considering the function theory aspects of the material in this chapter and especially the chapter on Perelman's differential Harnack estimate in the second part of this volume.

EXERCISE 6.12 (Basic properties of $\mathcal{W}_\varepsilon$). Show that on a closed Riemannian manifold $(\hat{\mathcal{M}}^n, \hat{g})$,

(1)
$$\mathcal{W}_\varepsilon(\hat{g}, f, \tau) = (4\pi)^{-n/2} \int_\mathcal{M} \left(R_{\tilde{g}} + |\nabla f|^2_{\tilde{g}} - \varepsilon(f - n) \right) e^{-f} d\mu_{\tilde{g}},$$

where $\tilde{g} \doteqdot \tau^{-1} \hat{g}$,

(2) for any constant $c > 0$
$$\mathcal{W}_\varepsilon(c\hat{g}, f, c\tau) = \mathcal{W}_\varepsilon(\hat{g}, f, \tau),$$

and

(3) for any diffeomorphism $\varphi : \hat{\mathcal{M}} \to \hat{\mathcal{M}}$,
$$\mathcal{W}_\varepsilon(\varphi^* \hat{g}, f \circ \varphi, \tau) = \mathcal{W}_\varepsilon(\hat{g}, f, \tau).$$

---

[8]These identifications are true up to constant factors.

1. THE ENTROPY FUNCTIONAL $\mathcal{W}$ AND ITS MONOTONICITY

EXERCISE 6.13 (First variation formula for $\mathcal{W}_\varepsilon$). Show that, on a closed manifold $\mathcal{M}^n$, if

(6.28) $$\delta g = v, \quad \delta f = h, \quad \delta \tau = \zeta$$

at (g, f, τ), then

(6.29) $$\delta_{(v,h,\zeta)} \mathcal{W}_\varepsilon(g, f, \tau) = \int_{\mathcal{M}} \left(\begin{array}{c} (-\tau v_{ij} + \zeta g_{ij}) S_{ij}^\varepsilon \\ + \left(\frac{V}{2} - h - \frac{n\zeta}{2\tau} \right) (V_\varepsilon + \varepsilon) \end{array} \right) u d\mu,$$

where S_{ij}^ε is defined by (6.27). In particular, if

$$\delta_{(v,h,\zeta)} \left((4\pi\tau)^{-\frac{n}{2}} e^{-f} d\mu \right) = \left(\frac{V}{2} - h - \frac{n\zeta}{2\tau} \right) (4\pi\tau)^{-\frac{n}{2}} e^{-f} d\mu = 0,$$

then

(6.30) $$\delta_{(v,h,\zeta)} \mathcal{W}_\varepsilon(g, f, \tau) = \int_{\mathcal{M}} (-\tau v_{ij} + \zeta g_{ij}) S_{ij}^\varepsilon u d\mu.$$

SOLUTION TO EXERCISE 6.13. We compute

$$\delta_v R_{ij} = \nabla_p \left(\delta \Gamma_{ij}^p \right) - \nabla_i \left(\delta \Gamma_{pj}^p \right)$$

and

$$\delta_{(v,h)} \left(\nabla_i \nabla_j f \right) = \nabla_i \nabla_j (\delta f) - \left(\delta \Gamma_{ij}^p \right) \nabla_p f.$$

Adding these two formulas together, we get

$$\delta_{(v,h)} \left(R_{ij} + \nabla_i \nabla_j f \right)$$
$$= \nabla_p \left(\delta \Gamma_{ij}^p \right) - \left(\delta \Gamma_{ij}^p \right) \nabla_p f + \nabla_i \left(\nabla_j (\delta f) - \delta \Gamma_{pj}^p \right)$$
(6.31) $$= e^f \nabla_p \left(e^{-f} \delta \Gamma_{ij}^p \right) + \nabla_i \nabla_j \left(h - \frac{V}{2} \right).$$

Tracing this formula, remembering to take the variation of g^{ij}, and using $\frac{\partial}{\partial s} \tau = \zeta$ imply

$$\delta_{(v,h,\zeta)} \left[\tau (R + \Delta f) (4\pi\tau)^{-\frac{n}{2}} e^{-f} d\mu \right]$$
$$= \tau \left(-\delta g_{ij} \cdot (R_{ij} + \nabla_i \nabla_j f) + g^{ij} \cdot \delta (R_{ij} + \nabla_i \nabla_j f) \right) (4\pi\tau)^{-\frac{n}{2}} e^{-f} d\mu$$
$$+ \tau (R + \Delta f) (4\pi\tau)^{-\frac{n}{2}} \left(\left(1 - \frac{n}{2} \right) \frac{\zeta}{\tau} - h + \frac{V}{2} \right) e^{-f} d\mu$$
$$= \tau \left[\begin{array}{c} -v_{ij} (R_{ij} + \nabla_i \nabla_j f) e^{-f} + \nabla_p (e^{-f} \frac{\partial}{\partial s} \Gamma_{ii}^p) \\ + e^{-f} \Delta \left(h - \frac{V}{2} \right) + (R + \Delta f) e^{-f} \left(\frac{V}{2} - h - \frac{n-2}{2} \frac{\zeta}{\tau} \right) \end{array} \right] (4\pi\tau)^{-\frac{n}{2}} d\mu.$$

Since $\mathcal{W}_\varepsilon(g,f,\tau) = \int_{\mathcal{M}} (\tau(R+\Delta f) - \varepsilon(f-n))(4\pi\tau)^{-n/2} e^{-f} d\mu$, integrating this by parts and applying the divergence theorem, we have

$$\frac{d}{ds}\mathcal{W}_\varepsilon(g,f,\tau)$$

$$= \int_{\mathcal{M}} \tau \begin{bmatrix} -v_{ij}(R_{ij} + \nabla_i\nabla_j f) e^{-f} \\ +(R+\Delta f) e^{-f}\left(\frac{V}{2} - h - \frac{n-2}{2}\frac{\zeta}{\tau}\right) \\ +\Delta(e^{-f})(h - \frac{V}{2}) \\ +\frac{1}{\tau}\left(-\varepsilon h - \varepsilon(f-n)\left(\frac{V}{2} - h - \frac{n\zeta}{2\tau}\right)\right) e^{-f} \end{bmatrix} (4\pi\tau)^{-\frac{n}{2}} d\mu$$

$$= \int_{\mathcal{M}} \begin{bmatrix} -\tau v_{ij}\left(R_{ij} + \nabla_i\nabla_j f + \frac{\varepsilon}{2\tau}g_{ij}\right) e^{-f} \\ +\tau\left(R + 2\Delta f - |\nabla f|^2\right) e^{-f}\left(\frac{V}{2} - h - \frac{n\zeta}{2\tau}\right) \\ -\varepsilon(f-n-1) e^{-f}\left(\frac{V}{2} - h - \frac{n\zeta}{2\tau}\right) \\ +\zeta(R+\Delta f) e^{-f} + \zeta\frac{n}{2}\left(\Delta f - |\nabla f|^2 + \frac{\varepsilon}{\tau}\right) e^{-f} \end{bmatrix} (4\pi\tau)^{-\frac{n}{2}} d\mu.$$

Now, integrating by parts tells us again that the terms on the last line are

$$\int_{\hat{\mathcal{M}}} \zeta\left((R+\Delta f) + \frac{n}{2}\left(\Delta f - |\nabla f|^2 + \frac{\varepsilon}{\tau}\right)\right) e^{-f} (4\pi\tau)^{-\frac{n}{2}} d\mu$$
$$= \int_{\hat{\mathcal{M}}} \zeta\left(R + \Delta f + \frac{\varepsilon n}{2\tau}\right) u d\mu.$$

Substituting this into the above formula yields (6.9).

In the next two exercises we give a unified proof of the monotonicity formulas for entropy, energy, and expander entropy.

EXERCISE 6.14 (Monotonicity of $\mathcal{W}_\varepsilon$ from the first variation formula). When we require that the variation (v, h, ζ) satisfies $\zeta = \varepsilon$ and $\frac{V}{2} - h - \frac{n}{2\tau}\zeta = 0$ in (6.30) (e.g., preserves the measure $(4\pi\tau)^{-n/2} e^{-f} d\mu$), this leads to the following gradient flow:

(6.32) $$\frac{\partial}{\partial t} g_{ij} = -2(R_{ij} + \nabla_i\nabla_j f),$$

(6.33) $$\frac{\partial f}{\partial t} = -\Delta f - R - \frac{n\varepsilon}{2\tau},$$

(6.34) $$\frac{d\tau}{dt} = \varepsilon.$$

Show that if $(g(t), f(t), \tau(t))$ is a solution of the above system, then

$$\frac{d}{dt}\mathcal{W}_\varepsilon(g(t), f(t), \tau(t)) = 2\tau \int_{\mathcal{M}} |S_{ij}^\varepsilon|^2 u d\mu.$$

EXERCISE 6.15 (Evolution of v_ε and monotonicity of $\mathcal{W}_\varepsilon$). Consider the gauge transformed version of (6.32)–(6.34) on a closed manifold $\mathcal{M}^n$:[9]

$$\frac{\partial}{\partial t} g_{ij} = -2R_{ij}, \tag{6.35}$$

$$\frac{\partial f}{\partial t} = -\Delta f - R + |\nabla f|^2 - \frac{n\varepsilon}{2\tau}, \tag{6.36}$$

$$\frac{d\tau}{dt} = \varepsilon. \tag{6.37}$$

Let $v_\varepsilon \doteqdot V_\varepsilon u$. Show that if $(g(t), f(t), \tau(t))$ is a solution of the system above, then

$$\square^* v_\varepsilon = -2\tau \left| R_{ij} + \nabla_i \nabla_j f + \frac{\varepsilon}{2\tau} g_{ij} \right|^2 u = -2\tau \left| S_{ij}^\varepsilon \right|^2 u. \tag{6.38}$$

Also show that this implies

$$\frac{d}{dt} \mathcal{W}_\varepsilon (g(t), f(t), \tau(t)) = 2\tau \int_\mathcal{M} \left| R_{ij} + \nabla_i \nabla_j f + \frac{\varepsilon}{2\tau} g_{ij} \right|^2 u \, d\mu$$

$$= 2\tau \int_\mathcal{M} \left| S_{ij}^\varepsilon \right|^2 u \, d\mu \geq 0.$$

The next exercise relates the first variation of $\mathcal{W}_\varepsilon$ to the linear trace Harnack quantity.

EXERCISE 6.16 (Variation of $\mathcal{W}_\varepsilon$ and linear trace Harnack). Show that, for any symmetric 2-tensor w on a closed manifold $\mathcal{M}^n$, we have:

$$\int_\mathcal{M} w_{ij} S_{ij}^\varepsilon u \, d\mu = \int_\mathcal{M} Z(w, \nabla f) u \, d\mu, \tag{6.39}$$

where Z is the linear trace Harnack inequality defined in (A.27)

$$Z(w, X) \doteqdot \nabla_j \nabla_i w_{ij} + R_{ij} w_{ij} + \frac{\varepsilon}{2\tau} W - 2\nabla_i w_{ij} X_j + w_{ij} X_i X_j \tag{6.40}$$

and $W \doteqdot g^{ij} w_{ij}$. In particular, if $\delta_{(v,h,\varsigma)} \left((4\pi\tau)^{-n/2} e^{-f} d\mu \right) = 0$, then

$$\delta_{(v,h,\varsigma)} \mathcal{W}_\varepsilon(g, f, \tau) = \int_\mathcal{M} Z(\tilde{w}, \nabla f) u \, d\mu,$$

where

$$\tilde{w}_{ij} \doteqdot -\tau v_{ij} + \varsigma g_{ij} = -\tau^2 \delta_{(v,\varsigma)} \left(\tau^{-1} g \right).$$

HINT. Use the identity

$$\int_\mathcal{M} \nabla_j \nabla_i w_{ij} e^{-f} d\mu = \int_\mathcal{M} \nabla_i w_{ij} \nabla_j f e^{-f} d\mu$$

$$= \int_\mathcal{M} w_{ij} \left(\nabla_i f \nabla_j f - \nabla_i \nabla_j f \right) e^{-f} d\mu.$$

In the last two exercises in this subsection, we first rewrite $\mathcal{W}_\varepsilon(g, f, \tau)$ and then we use the new formula to give a lower bound for $\mathcal{W}_\varepsilon(g, f, \tau)$.

---

[9] These equations are equivalent to (6.35)–(6.37) after pulling back by diffeomorphisms generated by the vector fields $\nabla f(t)$.

EXERCISE 6.17. Let

(6.41) $$w \doteqdot (4\pi\tau)^{-n/4} e^{-f/2},$$

so that $w^2 = u$. Show that

(6.42) $$\mathcal{W}_\varepsilon(g, f, \tau) = \int_\mathcal{M} \left(\begin{array}{c} \tau\left(Rw^2 + 4|\nabla w|^2\right) \\ +\varepsilon\left(\log(w^2) + \frac{n}{2}\log(4\pi\tau) + n\right)w^2 \end{array} \right) d\mu$$

(6.43) $$\doteqdot \mathcal{K}_\varepsilon(g, w, \tau).$$

SOLUTION TO EXERCISE 6.17. We obtain (6.42) from substituting the definition of w,

$$f = -2\log w - \frac{n}{2}\log(4\pi\tau), \quad \text{and} \quad \nabla w = -\frac{1}{2}w\nabla f$$

into (6.26).

EXERCISE 6.18 (Lower bounds for $\mathcal{W}_\varepsilon$). Let $(\mathcal{M}^n, g)$ be a closed Riemannian manifold and let $R_{\min} \doteqdot \inf_{x \in \mathcal{M}} R(x)$. Suppose that (g, f, τ) satisfies the constraint $\int_\mathcal{M} (4\pi\tau)^{-n/2} e^{-f} d\mu = 1$. Show that for $\tau > 0$ the following hold.

(1) If $\varepsilon > 0$, then

(6.44) $$\mathcal{W}_\varepsilon(g, f, \tau) \geq \tau R_{\min} - \frac{\varepsilon}{e} \text{Vol}(g) + \varepsilon \left(\frac{n}{2}\log(4\pi\tau) + n\right) > -\infty.$$

(2) If $\varepsilon < 0$, then

$$\mathcal{W}_\varepsilon(g, f, \tau) \geq -2C|\varepsilon| + \tau R_{\min} + \varepsilon\left(\frac{n}{2}\log(4\pi\tau) + n\right) > -\infty,$$

where

$$C \doteqdot \frac{2\tau}{|\varepsilon|} \text{Vol}(g)^{-2/n} + \frac{|\varepsilon|}{2\tau C_s(\mathcal{M}, g)},$$

and $C_s(\mathcal{M}, g)$ is the constant in the Sobolev inequality (6.66). Hence we conclude that when $\varepsilon < 0$, for any $A < \infty$, there exists a constant $C(g, \varepsilon, A) < \infty$ such that

$$\mathcal{W}_\varepsilon(g, f, \tau) \geq -C(g, \varepsilon, A)$$

for $\tau \in [A^{-1}, A]$ and $f \in C^\infty(\mathcal{M})$ with $\int_\mathcal{M}(4\pi\tau)^{-\frac{n}{2}} e^{-f} d\mu = 1$.

SOLUTION TO EXERCISE 6.18. (1) If $\varepsilon > 0$, which corresponds to the expanding case, then (6.44) follows from

(6.45) $$\int_\mathcal{M} u\log u \, d\mu \geq -\frac{1}{e}\text{Vol}(g) > -\infty.$$

(2) If $\varepsilon < 0$, which corresponds to the shrinking case, then *the logarithmic Sobolev inequality implies that the entropy has a lower bound* (see Section 4

of this chapter). In particular, by taking $a = \frac{2\tau}{|\varepsilon|}$ in (6.65) below, we have for $\varepsilon < 0$,

$$\mathcal{W}_\varepsilon(g, f, \tau) \geq \int_M \left(4\tau |\nabla w|^2 + \varepsilon w^2 \log(w^2) \right) d\mu$$
$$+ \tau R_{\min} + \varepsilon \left(\frac{n}{2} \log(4\pi\tau) + n \right)$$
(6.46)
$$\geq -2|\varepsilon| C \left(\frac{2\tau}{|\varepsilon|}, g \right) + \tau R_{\min} + \varepsilon \left(\frac{n}{2} \log(4\pi\tau) + n \right) > -\infty,$$

where

$$C \left(\frac{2\tau}{|\varepsilon|}, g \right) = \frac{2\tau}{|\varepsilon|} \operatorname{Vol}(g)^{-2/n} + \frac{4|\varepsilon|}{2\tau n^2 e^2 C_s(\mathcal{M}, g)}$$

is as in Lemma 6.36.

REMARK 6.19. On a Riemannian manifold $\left(\hat{\mathcal{M}}^n, \hat{g} \right)$, Jensen's inequality says that if φ is convex on $\mathbb{R}$ and $u \in L^1 \left(\hat{\mathcal{M}} \right)$, then

$$\frac{1}{\operatorname{Vol}\left(\hat{\mathcal{M}} \right)} \int_{\hat{\mathcal{M}}} \varphi \circ u \, d\mu \geq \varphi \left(\frac{1}{\operatorname{Vol}\left(\hat{\mathcal{M}} \right)} \int_{\hat{\mathcal{M}}} u \, d\mu \right).$$

Applying Jensen's inequality with $\varphi(x) = x \log x$, we see that for any positive function u with $\int_{\hat{\mathcal{M}}} u \, d\mu = 1$,

(6.47)
$$\int_{\hat{\mathcal{M}}} u \log u \, d\mu \geq -\log \left(\operatorname{Vol}\left(\hat{\mathcal{M}} \right) \right).$$

This is an alternative estimate to (6.45).

2. The functionals μ and ν

Similarly to defining $\lambda(g)$ using the energy $\mathcal{F}(g, f)$, we define two functionals μ and ν using the entropy $\mathcal{W}(g, f, \tau)$. In this section we first discuss the elementary properties of μ and ν obtained directly from the properties of the entropy $\mathcal{W}$. We show using the logarithmic Sobolev inequality that $\mu(g, \tau)$ is finite and we show that a constrained minimizer f of $\mathcal{W}(g, f, \tau)$ exists. We end this section with the monotonicity of μ.

2.1. The diffeomorphism-invariant functionals μ and ν. Let $u \doteq (4\pi\tau)^{-n/2} e^{-f}$ as in (6.3). We define a subset $\mathcal{X}$ of $\mathfrak{Met} \times C^\infty(\mathcal{M}) \times \mathbb{R}^+$ by

(6.48)
$$\mathcal{X} = \left\{ (g, f, \tau) : \int_M u \, d\mu = 1 \right\}.$$

Note that if $(g, f, \tau) \in \mathcal{X}$, then $(cg, f, c\tau) \in \mathcal{X}$ for all $c > 0$, and $(\Phi^* g, \Phi^* f, \tau) \in \mathcal{X}$ for any diffeomorphism $\Phi : \mathcal{M} \to \mathcal{M}$. We consider the restriction of $\mathcal{W}$ to $\mathcal{X}$. Given (g, τ), we first minimize $\mathcal{W}(g, f, \tau)$ over f with $(g, f, \tau) \in \mathcal{X}$ to get $\mu(g, \tau)$, and then we minimize $\mu(g, \tau)$ among $\tau > 0$ to get $\nu(g)$.

DEFINITION 6.20 (Infimum invariants $\mu(g,\tau)$ and $\nu(g)$). The functionals $\mu : \mathfrak{Met} \times \mathbb{R}^+ \to \mathbb{R}$ and $\nu : \mathfrak{Met} \to \mathbb{R}$ are defined by

(6.49) $\mu(g,\tau) \doteqdot \inf \{\mathcal{W}(g,f,\tau) : f \in C^\infty(\mathcal{M}) \text{ satisfies } (g,f,\tau) \in \mathcal{X}\},$

(6.50) $\nu(g) \doteqdot \inf \{\mu(g,\tau) : \tau \in \mathbb{R}^+\}.$

Note that we do not assume that the two functionals $\mu(g,\tau)$ and $\nu(g)$ take finite values. We will see later that $\mu(g,\tau)$ is always finite for any given g and τ, and in the important case where $\lambda(g) > 0$ (where the proof of the nonexistence of nontrivial expanding breathers cannot be applied to the shrinking ones), we have that $\nu(g)$ is finite. We have the following elementary properties of these two functionals.

(i) (*Continuous dependence of μ on g and τ*) $\mu(g_s, \tau)$ is a continuous function of (s, τ) for any C^2 family g_s.[10]

(ii) (*Continuous dependence of ν on g*) $\nu(g_s)$ is a continuous function for any C^2 family g_s.

(iii) (*Scale invariance*) It follows from the scaling property of $\mathcal{W}$ that we have
$$\mu(g, \tau) = \mu(cg, c\tau),$$
$$\nu(g) = \nu(cg).$$

(iv) (*Diffeomorphism invariance*) Since $\mathcal{W}$ is invariant under a diffeomorphism $\Phi : \mathcal{M} \to \mathcal{M}$, we have
$$\mu(g,\tau) = \mu(\Phi^* g, \tau),$$
$$\nu(g) = \nu(\Phi^* g).$$

Compare (i) and (ii) with Lemma 5.24.

EXERCISE 6.21. Prove properties (i)–(iv) above.

Using the fact that the variation of $\mathcal{W} : \mathcal{X} \to \mathbb{R}$ with respect to f is

$$\delta_{(0,h,0)} \mathcal{W}(g,f,\tau) = -\int_{\mathcal{M}} \tau h \left(2\Delta f + \frac{f-(n+1)}{\tau} - |\nabla f|^2 + R \right) u d\mu,$$

where h satisfies $\int_{\mathcal{M}} h u d\mu = 0$, we have the Euler–Lagrange equation of (6.49),

$$\tau \left(2\Delta f - |\nabla f|^2 + R \right) + f - n = C$$

for some constant C. If f_τ is a minimizer of (6.49) (we will see the existence of f_τ in the next subsection), then it follows that

$$\mu(g,\tau) = \int_{\mathcal{M}} \left[\tau \left(2\Delta f_\tau - |\nabla f_\tau|^2 + R \right) + f_\tau - n \right] (4\pi\tau)^{-n/2} e^{-f_\tau} d\mu,$$

and hence $C = \mu(g,\tau)$ for a minimizer. Therefore we have the following.

---

[10]It is easy to see that $\mu(g,\tau)$ is semi-continuous in g. This is a consequence of the fact that if a function $h(x,y)$ is continuous, then $\inf_{y \in Y} h(x,y)$ is upper semi-continuous in x.

LEMMA 6.22 (Euler–Lagrange for minimizer). *The Euler–Lagrange equation of* (6.49) *is*

(6.51) $$\tau\left(2\Delta f - |\nabla f|^2 + R\right) + f - n = C.$$

For the minimizer f_τ of (6.49),

(6.52) $$\tau\left(2\Delta f_\tau - |\nabla f_\tau|^2 + R\right) + f_\tau - n = \mu(g,\tau).$$

Compare (6.52) to the equation $2\Delta f - |\nabla f|^2 + R = \lambda(g)$ for the minimizer f of $\mathcal{F}(g, \cdot)$.

In terms of $w \doteq (4\pi\tau)^{-n/4} e^{-f/2}$ as in (6.41), a simple computation shows that μ is the lowest eigenvalue of the nonlinear operator:

(6.53)
$$N(w) \doteq -4\tau\Delta w + \tau R w - \left(\frac{n}{2}\log(4\pi\tau) + n\right) w - 2w \log w = \mu(g,\tau) w.$$

2.2. The finiteness of μ and the existence of a minimizer f.

As a consequence of Exercise 6.18 we have the following.

LEMMA 6.23 (Finiteness of μ). *For any given g and $\tau > 0$ on a closed manifold $\mathcal{M}^n$,*

(6.54) $$\mu(g,\tau) > -\infty$$

is finite.

PROOF. Since $\mu(\tau g, \tau) = \mu(g, 1)$, we may assume without loss of generality that $\tau = 1$. We need to show that for any metric g there exists a constant $c = c(g)$ such that

(6.55) $$\mathcal{W}(g, f, 1) = \int_{\mathcal{M}} \left(R + |\nabla f|^2 + f - n\right)(4\pi)^{-n/2} e^{-f} d\mu \geq c$$

for any smooth function f on $\mathcal{M}$ satisfying $(4\pi)^{-n/2}\int_{\mathcal{M}} e^{-f} d\mu = 1$. As in (6.41) with $\tau = 1$, let $w = (4\pi)^{-n/4} e^{-f/2}$. By (6.42), the lemma is equivalent to showing that

$$\mathcal{W}(g, f, 1) = \int_{\mathcal{M}} \left(4|\nabla w|^2 + \left(R - 2\log w - \frac{n}{2}\log(4\pi) - n\right) w^2\right) d\mu$$
$$\doteq \mathcal{H}(g, w) \geq c$$

for any $w > 0$ such that $\int_{\mathcal{M}} w^2 d\mu = 1$.

Since $R - n \geq \inf_{x \in \mathcal{M}} R(x) - n > -\infty$, it suffices to show that there exists $C < \infty$ such that

$$\int_{\mathcal{M}} w^2 \log w \, d\mu \leq 2\int_{\mathcal{M}} |\nabla w|^2 d\mu + C$$

for all $w > 0$ with $\int_{\mathcal{M}} w^2 d\mu = 1$. This follows from the logarithmic Sobolev inequality (6.65),[11] which we state in the next subsection. □

---

[11]With $a = 2$.

Next we prove the existence of a smooth minimizer for (6.49); compare the proof below with the proof of Lemma 5.22.

LEMMA 6.24 (Existence of a smooth minimizer for $\mathcal{W}$). *For any metric g on a closed manifold $\mathcal{M}^n$ and $\tau > 0$, there exists a smooth minimizer f_τ of $\mathcal{W}(g,\cdot,\tau)$ over $\mathcal{X}$.*

PROOF. Again assume $\tau = 1$. The lemma will follow from showing that there is a smooth positive minimizer w_1 for $\mathcal{H}(g,w)$ under the constraint $\int_\mathcal{M} w^2 d\mu_g = 1$. A smooth minimizer f_1 of $\mathcal{W}(g,\cdot,1)$ is then given by $f_1 = -2\log w_1 - \frac{n}{2}\log(4\pi)$ (see Rothaus [**312**]). We give a sketch of the proof.

Suppose w is such that $\mathcal{H}(g,w) \leq C_1$. Then the above considerations imply that since $\int_\mathcal{M} w^2 d\mu_g = 1$,

$$C_1 \geq \mathcal{H}(g,w) = \int_\mathcal{M} \left(4|\nabla w|^2 + \left(R - 2\log w - \frac{n}{2}\log(4\pi) - n\right)w^2\right) d\mu$$

$$\geq 2\int_\mathcal{M} |\nabla w|^2 d\mu - C_2,$$

where we used (6.65) below with $a = 1$. Hence any minimizing sequence for $\mathcal{H}(g,\cdot)$ is bounded in $W^{1,2}(\mathcal{M})$. We get a minimizer w_1 in $W^{1,2}(\mathcal{M})$ and by (6.53), w_1 is a weak solution to

$$-4\Delta w_1 + Rw_1 - 2w_1 \log w_1 - \left(\frac{n}{2}\log(4\pi) + n\right)w_1 = \mu(g,1)w_1.$$

By elliptic regularity theory, we have $w_1 \in C^\infty$ (see Gilbarg and Trudinger [**155**] for a general treatise on second-order elliptic PDE). Finally, one can prove that $w_1 > 0$; see [**312**] for more details. □

REMARK 6.25. We also have

(6.56) $$\mu(g,\tau) \doteqdot \inf\left\{\mathcal{W}(g,f,\tau) : f \in W^{1,2}(\mathcal{M}), \int_\mathcal{M} u d\mu = 1\right\},$$

where C^∞ is replaced by $W^{1,2}$ in (6.49) and u is defined in (6.3).

2.3. Monotonicity of μ. Let $(g(t),\tau(t))$, $t \in [0,T]$, be a solution of (6.14) and (6.16) with $\tau(t) > 0$. For any $t_0 \in (0,T]$, let $f(t_0)$ be the minimizer of

$$\{\mathcal{W}(g(t_0),f,\tau(t_0)) : f \in C^\infty(\mathcal{M}) \text{ satisfies } (g,f,\tau) \in \mathcal{X}\}$$

and solve (6.15) for $f(t)$ backwards in time on $[0,t_0]$. By the monotonicity formula, we have

$$\frac{d}{dt}\mathcal{W}(g(t),f(t),\tau(t)) \geq 0$$

for $t \in [0,t_0]$. Note that the integral constraint (6.48) is preserved by the modified coupled equations (6.14)–(6.16). This can be seen from the following calculation:
(6.57)
$$\frac{d}{dt}\int_\mathcal{M} (4\pi\tau)^{-n/2} e^{-f} d\mu = \int_\mathcal{M} \left(\frac{\partial u}{\partial t} - Ru\right) d\mu = \int_\mathcal{M} (-\square^* u - \Delta u) d\mu = 0.$$

Hence we have

(6.58) $\mu(g(t), \tau(t)) \leq \mathcal{W}(g(t), f(t), \tau(t))$
$\leq \mathcal{W}(g(t_0), f(t_0), \tau(t_0)) = \mu(g(t_0), \tau(t_0))$

for $t \in [0, t_0]$.

The above inequality implies

(6.59) $\left.\dfrac{d}{dt}\right|_{t=t_0} \mu(g(t), \tau(t)) \geq \left.\dfrac{d}{dt}\right|_{t=t_0} \mathcal{W}(g(t), f(t), \tau(t)) \geq 0$

in the sense of the lim inf of backward difference quotients. This inequality holds for all $t_0 \in [0, T]$. Actually we have

$$\left.\dfrac{d}{dt}\right|_{t=t_0} \mu(g(t), \tau(t))$$
$$\geq \int_{\mathcal{M}} 2\tau(t_0) \left| R_{ij}(t_0) + \nabla_i \nabla_j f(t_0) - \dfrac{1}{2\tau(t_0)} g(t_0)_{ij} \right|^2$$
$$\times (4\pi\tau(t_0))^{-\frac{n}{2}} e^{-f(t_0)} d\mu_{g(t_0)}$$

for the minimizer $f(t_0)$ of $\{\mathcal{W}(g(t_0), \cdot, \tau(t_0)) : (g(t_0), \cdot, \tau(t_0)) \in \mathcal{X}\}$. Hence, from either (6.58) or (6.59), we have the following monotonicity formula for μ.

LEMMA 6.26 (μ-invariant monotonicity). *Let $(g(t), \tau(t))$, $t \in [0, T]$, be a solution of (6.14) and (6.16) on a closed manifold $\mathcal{M}^n$ with $\tau(t) > 0$. For all $0 \leq t_1 \leq t_2 \leq T$, we have*

(6.60) $\mu(g(t_2), \tau(t_2)) \geq \mu(g(t_1), \tau(t_1))$.

In particular,

(6.61) $\mu(g(t), r^2) \geq \mu(g(0), r^2 + t)$

for $t \in [0, T]$ and $r > 0$.

The following exercise continues our discussion in subsection 1.3 of this chapter. Again $u \doteqdot (4\pi\tau)^{-\frac{n}{2}} e^{-f}$.

EXERCISE 6.27 (μ_ε-invariant monotonicity). Define the μ_ε-**invariant** on a closed manifold $\mathcal{M}^n$:

$$\mu_\varepsilon(g, \tau) \doteqdot \inf\left\{\mathcal{W}_\varepsilon(g, f, \tau) : f \in C^\infty(\mathcal{M}), \int_{\mathcal{M}} u d\mu = 1\right\}.$$

(1) Show that for any $c > 0$,
$$\mu_\varepsilon(cg, c\tau) = \mu_\varepsilon(g, \tau).$$

(2) Show that if $(\mathcal{M}, g(t))$ is a solution to the Ricci flow on a time interval $\mathcal{I} \subset \mathbb{R}$, $\varepsilon \in \mathbb{R}$, and $\frac{d\tau}{dt} = \varepsilon$ with $\tau(t) > 0$, then $\mu_\varepsilon(g(t), \tau(t))$ is monotonically nondecreasing on $\mathcal{I}$. That is, for $t_2 \geq t_1$,
$$\mu_\varepsilon(g(t_2), \tau(t_2)) \geq \mu_\varepsilon(g(t_1), \tau(t_1)).$$

(3) Improving on the previous part, show that for any times $t_1, t_2 \in \mathcal{I}$ with $t_1 \leq t_2$,

$$\mu_\varepsilon\left(g\left(t_2\right), \tau\left(t_2\right)\right) - \mu_\varepsilon\left(g\left(t_1\right), \tau\left(t_1\right)\right)$$
$$\geq \int_{t_1}^{t_2} 2\tau(t) \int_\mathcal{M} \left| R_{ij}(t) + \nabla_i \nabla_j f(t) + \frac{\varepsilon g(t)_{ij}}{2\tau(t)} \right|^2 u(t) \, d\mu_{g(t)} dt,$$

where $f(t)$ is the minimizer of $\mathcal{W}_\varepsilon\left(g(t), \cdot, \tau(t)\right)$ for each $t \in [t_1, t_2]$.

2.4. The behavior of μ under Cheeger–Gromov convergence. If $(\mathcal{N}^n, g)$ is a complete, noncompact Riemannian manifold, we generalize the definition of the functional μ to noncompact manifolds by

$$\mu(g,\tau) \doteq \inf \left\{ \mathcal{W}(g, f, \tau) : e^{-f/2} \in C_c^\infty(\mathcal{N}), \int_\mathcal{N} (4\pi\tau)^{-n/2} e^{-f} d\mu = 1 \right\},$$

where $\mathcal{W}(g, f, \tau)$ is defined as in (6.1) and the infimum is taken over smooth functions with compact support satisfying the constraint.[12]

LEMMA 6.28 (μ under Cheeger–Gromov convergence). *Suppose that we have $(\mathcal{N}_k^n, g_k, x_k) \to (\mathcal{N}_\infty^n, g_\infty, x_\infty)$ in the C^∞ Cheeger–Gromov sense. Then for any $\tau > 0$,*

$$\mu(g_\infty, \tau) \geq \limsup_{k \to \infty} \mu(g_k, \tau).$$

PROOF. By definition, there exists an exhaustion $\{U_k\}_{k \in \mathbb{N}}$ of $\mathcal{N}_\infty$ by open sets with $x_\infty \in U_k$ and diffeomorphisms $\Phi_k : U_k \to V_k \doteq \Phi_k(U_k) \subset \mathcal{N}_k$ with $\Phi_k(x_\infty) = x_k$ such that $\left(U_k, \Phi_k^*\left[g_k|_{V_k}\right]\right) \to (\mathcal{N}_\infty, g_\infty)$ in C^∞ on compact sets. Now let $e^{-f/2} \in C_c^\infty(\mathcal{N}_\infty)$ with

$$\int_{\mathcal{N}_\infty} (4\pi\tau)^{-n/2} e^{-f} d\mu_{g_\infty} = 1.$$

Then by the diffeomorphism invariance of $\mathcal{W}$, for all $k \in \mathbb{N}$ large enough, we have $\int_{\mathcal{N}_k} (4\pi\tau)^{-n/2} e^{-f \circ \Phi_k^{-1}} d\mu_{(\Phi_k^{-1})^* g_\infty} = 1$ and

$$\mathcal{W}\left(\Phi_k^*\left[g_k|_{V_k}\right], f, \tau\right) = \mathcal{W}\left(g_k, f \circ \Phi_k^{-1}, \tau\right).$$

Note that although $\Phi_k^*\left[g_k|_{V_k}\right]$ is not defined on the whole $\mathcal{N}_\infty$, the entropy $\mathcal{W}\left(\Phi_k^*\left[g_k|_{V_k}\right], f, \tau\right)$ makes sense since $U_k \supset \overline{\mathrm{supp}\left(e^{-f/2}\right)}$ when k is large.

---

[12]Implicitly it is understood in the discussion here that we are considering $\mathcal{W}$ as a function of $w \doteq (4\pi\tau)^{-n/4} e^{-f/2} \in C_c^\infty(\mathcal{N})$ so that there is no problem with $f = \infty$. As usual, we use the convention that $w^2 \log w = 0$ when $w = 0$.

Now

$$\left\| g_k|_{V_k} - (\Phi_k^{-1})^* g_\infty \right\|_{C^\ell(\Phi_k(\overline{\operatorname{supp}(f)}),(\Phi_k^{-1})^* g_\infty)}$$
$$= \left\| \Phi_k^* \left[g_k|_{V_k} \right] - g_\infty \right\|_{C^\ell(\overline{\operatorname{supp}(f)},g_\infty)}$$
$$\to 0.$$

Hence

$$\left| \int_{\mathcal{N}_k} (4\pi\tau)^{-n/2} e^{-f \circ \Phi_k^{-1}} d\mu_{(\Phi_k^{-1})^* g_\infty} - \int_{\mathcal{N}_k} (4\pi\tau)^{-n/2} e^{-f \circ \Phi_k^{-1}} d\mu_{g_k} \right|$$
$$\leq (4\pi\tau)^{-n/2} \left| \int_{\mathcal{N}_k} e^{-f \circ \Phi_k^{-1}} \left(1 - \frac{d\mu_{g_k}}{d\mu_{(\Phi_k^{-1})^* g_\infty}} \right) d\mu_{(\Phi_k^{-1})^* g_\infty} \right|$$
$$\leq \left\| g_k|_{V_k} - (\Phi_k^{-1})^* g_\infty \right\|_{C^0(\Phi_k(\overline{\operatorname{supp}(f)}),(\Phi_k^{-1})^* g_\infty)}^{n/2}$$
$$\times \int_{\mathcal{N}_k} (4\pi\tau)^{-n/2} e^{-f \circ \Phi_k^{-1}} d\mu_{(\Phi_k^{-1})^* g_\infty}$$
$$\to 0,$$

which implies

$$C_k \doteqdot \int_{\mathcal{N}_k} (4\pi\tau)^{-n/2} e^{-f \circ \Phi_k^{-1}} d\mu_{g_k} \to 1.$$

We conclude that

$$\mathcal{W}(\mathcal{N}_\infty, g_\infty, f, \tau) = \lim_{k \to \infty} \mathcal{W}\left(U_k, \Phi_k^* \left[g_k|_{V_k} \right], f, \tau\right)$$
$$= \lim_{k \to \infty} \mathcal{W}(\mathcal{N}_k, g_k, f \circ \Phi_k^{-1}, \tau)$$
$$\geq \limsup_{k \to \infty} \mu(g_k, \tau).$$

To see why the last inequality is true, note that the functions $f_k \doteqdot f \circ \Phi_k^{-1} + \log C_k$ satisfy $C_k \to 1$ and the constraints

$$\int_{\mathcal{N}_k} (4\pi\tau)^{-n/2} e^{-f_k} d\mu_{g_k} = 1$$

and

$$\mathcal{W}(\mathcal{N}_k, g_k, f \circ \Phi_k^{-1}, \tau) = C_k \mathcal{W}(\mathcal{N}_k, g_k, f_k, \tau) - C_k \log C_k$$
$$\geq C_k \mu(g_k, \tau) - C_k \log C_k.$$

Since $e^{-f/2}$ is an arbitrary nonnegative function with compact support and since it satisfies the constraint with respect to g_∞, we obtain

$$\mu(g_\infty, \tau) \geq \limsup_{k \to \infty} \mu(g_k, \tau).$$

$\square$

3. Shrinking breathers are shrinking gradient Ricci solitons

Let $(\mathcal{M}^n, g(t))$, $t \in [0, T)$, be a shrinking Ricci breather on a closed manifold with $g(t_2) = \alpha \Phi^* g(t_1)$, where $t_2 > t_1$ and $\alpha \in (0, 1)$. As discussed at the beginning of Section 1 of this chapter, we only need to show that when $\lambda(g(t)) > 0$ for all $t \in [t_1, t_2]$, the shrinking breather is a shrinking Ricci soliton. By Koiso's examples (subsection 7.2 of Chapter 2), such solutions need not be Einstein.

In this section we give two variations on the proof that shrinking breathers on closed manifolds are shrinking gradient solitons. The first proof, which also appears in Hsu [**208**], involves fewer technicalities in that it uses μ instead of ν. The first proof also does not use the assumption $\lambda(g(t)) > 0$. On the other hand, the second proof requires some knowledge of the asymptotic behavior of μ.

3.1. First proof using functional μ.
The following result of Perelman rules out periodic orbits for the Ricci flow in the space of metrics modulo diffeomorphisms and scalings.

THEOREM 6.29 (Shrinking breathers are gradient solitons). *A shrinking breather for the Ricci flow on a closed manifold must be a gradient shrinking Ricci soliton.*

FIRST PROOF. Let $(\mathcal{M}, g(t))$ be a shrinking Ricci breather with $g(t_2) = \alpha \Phi^* g(t_1)$, where $t_2 > t_1$ and $\alpha \in (0, 1)$. Define

$$\tau(t) \doteqdot \frac{t_2 - \alpha t_1}{1 - \alpha} - t,$$

so that $\frac{d\tau}{dt} = -1$,

$$\tau(t_1) = \frac{t_2 - t_1}{1 - \alpha}, \quad \tau(t_2) = \alpha \frac{t_2 - t_1}{1 - \alpha},$$

and $\tau(t_2) = \alpha \tau(t_1)$. By Lemma 6.24, there is a minimizer f_2 for

$$\{\mathcal{W}(g(t_2), f, \tau(t_2)) : f \in C^\infty(\mathcal{M}) \text{ satisfies } (g, f, \tau) \in \mathcal{X}\},$$

so that $\mathcal{W}(g(t_2), f_2, \tau(t_2)) = \mu(g(t_2), \tau(t_2))$. Define $f(t)$ to solve (6.15) on $[t_1, t_2]$ with $f(t_2) = f_2$. By the monotonicity formula (6.17) and the definition of μ, we have

$$\mu(g(t_1), \tau(t_1)) \leq \mathcal{W}(g(t_1), f(t_1), \tau(t_1)) \leq \mathcal{W}(g(t), f(t), \tau(t))$$
$$\leq \mathcal{W}(g(t_2), f_2, \tau(t_2)) = \mu(g(t_2), \tau(t_2))$$

for all $t \in [t_1, t_2]$. Since $g(t_1) = \alpha \Phi^* g(t_2)$ and $\tau(t_2) = \alpha \tau(t_1)$, by the diffeomorphism and scale invariance of μ, we have

$$\mu(g(t_1), \tau(t_1)) = \mu(g(t_2), \tau(t_2)).$$

This and the fact that $\mathcal{W}(g(t), f(t), \tau(t))$ is monotone implies

$$\mathcal{W}(g(t), f(t), \tau(t)) = \mu(g(t), \tau(t)) \equiv \text{const}$$

for $t \in [t_1, t_2]$. Thus $f(t)$ is the minimizer for $\mathcal{W}(g(t), f(t), \tau(t))$ and $\frac{d}{dt}\mathcal{W}(g(t), f(t), \tau(t)) \equiv 0$, so by (6.17), we have

$$\int_{\mathcal{M}} \left| R_{ij} + \nabla_i \nabla_j f - \frac{g_{ij}}{2\tau} \right|^2 e^{-f} d\mu \equiv 0$$

for all $t \in [t_1, t_2]$. We conclude that

(6.62) $\qquad R_{ij} + \nabla_i \nabla_j f - \frac{g_{ij}}{2\tau} = 0 \text{ for } t \in [t_1, t_2].$

Since a breather is a periodic solution of the Ricci flow (modulo diffeomorphisms and homotheties), by the uniqueness of solutions to the Ricci flow on closed manifolds, the behavior of $g(t)$ on $[t_1, t_2]$ determines completely the behavior of $g(t)$ on its whole time interval of existence. This is why, from (6.62), which is valid on $[t_1, t_2]$, one can deduce that $g(t)$ is a breather on $[0, T)$. □

3.2. Asymptotic behavior of μ and finiteness of ν. In the second proof of Theorem 6.29 given below we need the finiteness of ν, which in turn depends on $\lambda(g(t)) > 0$ and the following asymptotic behavior of μ.

We have shown that for each g and $\tau > 0$, $\mu(g, \tau)$ is finite. However we have yet to study the behavior of $\mu(g, \tau)$ as $\tau \to \infty$ or $\tau \to 0$. Recall that

$$\lambda(g) = \lambda_1(-4\Delta + R) = \inf\left\{ \int_{\mathcal{M}} (R + |\nabla f|^2) e^{-f} d\mu : \int_{\mathcal{M}} e^{-f} d\mu = 1 \right\}.$$

Since μ and $\mathcal{W}$ are modifications of λ and $\mathcal{F}$, we can prove the following.

LEMMA 6.30 ($\mu \to \infty$ as $\tau \to \infty$ when $\lambda > 0$). *If $\lambda(g) > 0$, then*

$$\lim_{\tau \to \infty} \mu(g, \tau) = +\infty.$$

REMARK 6.31. The idea of the proof is that when $\tau \to \infty$, the $\mathcal{F}$ term in the expression (6.4) for $\mathcal{W}$ dominates, so if $\inf \mathcal{F} > 0$, then $\inf \mathcal{W} \to \infty$ as $\tau \to \infty$.

PROOF. By Lemma 6.24, for any $\tau > 0$, there exists a C^∞ function f_τ with $\int_{\mathcal{M}} (4\pi\tau)^{-n/2} e^{-f_\tau} d\mu = 1$ such that

$$\mu(g, \tau) = \mathcal{W}(g, f_\tau, \tau) = \int_{\mathcal{M}} \left[\tau \left(R + |\nabla f_\tau|^2 \right) + f_\tau - n \right] (4\pi\tau)^{-n/2} e^{-f_\tau} d\mu.$$

We add a constant to f_τ so that it satisfies the constraint for $\mathcal{F}(g, \cdot)$ (instead of $\mathcal{W}(g, \cdot, \tau)$) and we define $\hat{f} \doteqdot f_\tau + \frac{n}{2} \log(4\pi\tau)$ so that $\int_{\mathcal{M}} e^{-\hat{f}} d\mu = 1$. Then by the logarithmic Sobolev inequality (e.g., Corollary 6.38 with $b = 1$), we have

$$\mu(g, \tau) = \int_{\mathcal{M}} \left[\tau \left(R + |\nabla \hat{f}|^2 \right) + \hat{f} - \frac{n}{2} \log(4\pi\tau) - n \right] e^{-\hat{f}} d\mu$$

$$\geq \int_{\mathcal{M}} \left(\tau R + (\tau - 1) |\nabla \hat{f}|^2 \right) e^{-\hat{f}} d\mu - \frac{n}{2} \log(4\pi\tau) - n - C_1(g)$$

$$\geq (\tau - 1) \int_{\mathcal{M}} \left(R + |\nabla \hat{f}|^2 \right) e^{-\hat{f}} d\mu + R_{\min} - \frac{n}{2} \log(4\pi\tau) - n - C_1(g).$$

Hence, if $\tau \geq 1$, we have

(6.63) $$\mu(g,\tau) \geq (\tau-1)\lambda(g) - \frac{n}{2}\log\tau - C_2(g).$$

Since $\lambda(g) > 0$, we have $\lim_{\tau\to\infty}\mu(g,\tau) = +\infty$. □

EXERCISE 6.32. Show that if $\lambda(g) < 0$, then $\lim_{\tau\to\infty}\mu(g,\tau) = -\infty$. In particular, if $\lambda(g) < 0$, then $\nu(g) = -\infty$.

SOLUTION TO EXERCISE 6.32. Since $\lambda(g) < 0$, there exists f_0 with $\int_{\mathcal{M}} e^{-f_0} d\mu = 1$ and

$$a \doteq \int_{\mathcal{M}} \left(R + |\nabla f_0|^2\right) e^{-f_0} d\mu < 0.$$

Define $\bar{f} \doteq f_0 - \frac{n}{2}\log(4\pi\tau)$ so that $\int_{\mathcal{M}}(4\pi\tau)^{-n/2} e^{-\bar{f}} d\mu = 1$. We have for all $\tau > 0$

$$\mu(g,\tau) \leq \mathcal{W}(g,\bar{f},\tau) = \int_{\mathcal{M}}\left[\tau\left(R+|\nabla\bar{f}|^2\right) + \bar{f} - n\right](4\pi\tau)^{-n/2}e^{-\bar{f}}d\mu$$

$$\leq \tau \int_{\mathcal{M}} \left(R+|\nabla f_0|^2\right)e^{-f_0}d\mu + \frac{1}{e}\int_{\mathcal{M}}(4\pi\tau)^{-n/2}d\mu$$

$$= a\tau + \frac{(4\pi\tau)^{-n/2}}{e}\operatorname{Vol}(g),$$

since $xe^{-x} \leq \frac{1}{e}$ for $x > 0$. The result follows from $a < 0$.

When $\tau \to 0_+$, we have

LEMMA 6.33 (Behavior of $\mu(g,\tau)$ for τ small). *Suppose $(\mathcal{M}^n, g)$ is a closed Riemannian manifold.*

(i) There exists $\bar{\tau} > 0$ such that

$$\mu(g,\tau) < 0 \quad \text{for all } \tau \in (0,\bar{\tau}).$$

(ii)

$$\lim_{\tau\to 0_+}\mu(g,\tau) = 0.$$

The proof of Lemma 6.33 will be given elsewhere.
From Lemmas 6.30 and 6.33, we have

COROLLARY 6.34. *If $\lambda(g) > 0$, then $\nu(g)$ is well defined and finite. Also, there exists $\tau > 0$ such that $\nu(g) = \mu(g,\tau)$.*

3.3. Monotonicity of ν and the second proof. The following lemma is the monotonicity property of $\nu(g(t))$ along the Ricci flow $g(t)$.

LEMMA 6.35 (ν-invariant monotonicity). *Let $(\mathcal{M}^n, g(t))$, $t \in [0,T)$, be a solution to the Ricci flow on a closed manifold.*

(1) *The invariant $\nu(g(t))$ is nondecreasing on $[0,T)$, as long as $\nu(g(t))$ is well defined and finite.*

3. SHRINKING BREATHERS ARE SHRINKING GRADIENT RICCI SOLITONS 245

(2) *Furthermore, if $\lambda(g(t)) > 0$ and if $\nu(g(t))$ is not strictly increasing on some interval, then $g(t)$ is a gradient shrinking Ricci soliton.*

(3) *If $\nu(g(t_0)) = -\infty$ for some t_0, then $\nu(g(t)) = -\infty$ for all $t \in [0, t_0]$.*

PROOF. (1) Given any $0 \leq t_1 < t_2 < T$, we shall show that

$$\nu(g(t_1)) \leq \nu(g(t_2)). \tag{6.64}$$

Since by assumption, $\nu(g(t_2)) > -\infty$, for any $\varepsilon > 0$ there exist f_2 and τ_2 such that

$$\mathcal{W}(g(t_2), f_2, \tau_2) \leq \nu(g(t_2)) + \varepsilon.$$

Let $(f(t), \tau(t))$, $t \in [0, t_2]$, be a solution of the backward heat-type equation (6.15) with $f(t_2) = f_2$ and $\tau(t_2) = \tau_2$. By the monotonicity formula (6.17), we have[13]

$$\mathcal{W}(g(t_2), f(t_2), \tau(t_2)) \geq \mathcal{W}(g(t_1), f(t_1), \tau(t_1)),$$

where equality holds if and only if

$$R_{ij} + \nabla_i \nabla_j f - \frac{1}{2\tau} g_{ij} = 0 \text{ for all } t \in (t_1, t_2).$$

This implies

$$\nu(g(t_2)) + \varepsilon \geq \mathcal{W}(g(t_2), f(t_2), \tau(t_2)) \geq \mathcal{W}(g(t_1), f(t_1), \tau(t_1)) \geq \nu(g(t_1)).$$

The result follows since $\varepsilon > 0$ is arbitrary.

(2) Suppose $\nu(g(t_1)) = \nu(g(t_2))$ for some $t_1 < t_2$. Since $\lambda(g(t)) > 0$, by Corollary 6.34, there exist f_2 and τ_2 such that

$$\mathcal{W}(g(t_2), f_2, \tau_2) = \nu(g(t_2)).$$

In this case, by repeating the argument in (1), we obtain

$$\mathcal{W}(g(t), f(t), \tau(t)) = \nu(g(t)) \equiv \text{const}$$

for all $t \in [t_1, t_2]$. As in the proof of Theorem 6.29, we can conclude that $g(t)$ is a gradient shrinking Ricci soliton.

(3) If $\nu(g(t_0)) = -\infty$, then for any $N > -\infty$ there exist f_0 and τ_0 such that $\mathcal{W}(g(t_0), f_0, \tau_0) \leq N$. Let $(f(t), \tau(t))$, $t \in [0, t_0]$, be the solution of (6.15) with $f(t_0) = f_0$ and $\tau(t_0) = \tau_0$. For all $t \in [0, t_0]$,

$$\nu(g(t)) \leq \mathcal{W}(g(t), f(t), \tau(t)) \leq \mathcal{W}(g(t_0), f(t_0), \tau(t_0)) \leq N.$$

Since $N > -\infty$ is arbitrary, we conclude $\nu(g(t)) = -\infty$ for all $t \in [0, t_0]$. □

Using the ν-invariant instead of the μ-invariant, we can give a

SECOND PROOF OF THEOREM 6.29. As we stated at the beginning of this section, we only need to consider a shrinking breather $g(t)$ with $g(t_1) = \alpha \Phi^* g(t_2)$ and $\lambda(g(t)) > 0$, $t \in [t_1, t_2]$. From the elementary properties (iii) and (iv) of μ and ν in subsection 2.1 of this chapter, we have $\nu(g(t_1)) = $

---

[13]Note that $\tau(t_1) = \tau(t_2) + t_2 - t_1 > 0$.

4. Logarithmic Sobolev inequality

In this section we give a proof of the logarithmic Sobolev inequality which we have used earlier. The logarithmic Sobolev inequality is related to the usual Sobolev inequality and has the advantage of being dimensionless.

4.1. Logarithmic Sobolev inequality on manifolds.

LEMMA 6.36 (Log Sobolev inequality, version 1). *Let $(\mathcal{M}^n, g)$ be a closed Riemannian manifold. For any $a > 0$, there exists a constant $C(a,g)$ (given by (6.67)) such that if $\varphi > 0$ satisfies $\int_\mathcal{M} \varphi^2 d\mu = 1$, then*

$$(6.65) \qquad \int_\mathcal{M} \varphi^2 \log \varphi \, d\mu \le a \int_\mathcal{M} |\nabla \varphi|^2 \, d\mu + C(a, g).$$

PROOF. Recall that the Sobolev inequality (see Lemma 2 in [**245**]) that if $\int_\mathcal{M} \varphi^2 d\mu = 1$, then (assume $n > 2$)

$$(6.66) \qquad \int_\mathcal{M} |\nabla \varphi|^2 \, d\mu \ge C_s(\mathcal{M}, g) \left[\int_\mathcal{M} \varphi^{\frac{2n}{n-2}} d\mu \right]^{\frac{n-2}{n}} - V^{-2/n},$$

where $V = \operatorname{Vol}_g(\mathcal{M})$. Note that for $c_n = \frac{ne}{2}$ we have $c_n \log x \le x^{2/n}$ for all $x > 0$, so that

$$c_n \int_\mathcal{M} \varphi^2 \log \varphi \, d\mu \le \int_\mathcal{M} \varphi^{2 + \frac{2}{n}} d\mu \le \varepsilon \int_\mathcal{M} \varphi^{2 + \frac{4}{n}} d\mu + \frac{1}{\varepsilon} \int_\mathcal{M} \varphi^2 d\mu,$$

for any $\varepsilon > 0$, since $\varphi^{1 + \frac{2}{n}} \varphi \le \varepsilon \varphi^{2(1 + \frac{2}{n})} + \frac{1}{\varepsilon} \varphi^2$. By Hölder's inequality,

$$\int_\mathcal{M} \varphi^2 \varphi^{\frac{4}{n}} d\mu \le \left(\int_\mathcal{M} \varphi^{\frac{2n}{n-2}} d\mu \right)^{\frac{n-2}{n}} \left(\int_\mathcal{M} \varphi^2 d\mu \right)^{\frac{2}{n}}.$$

Hence, using $\int_\mathcal{M} \varphi^2 d\mu = 1$, we have

$$c_n \int_\mathcal{M} \varphi^2 \log \varphi \, d\mu \le \varepsilon \left(\int_\mathcal{M} \varphi^{\frac{2n}{n-2}} d\mu \right)^{\frac{n-2}{n}} + \frac{1}{\varepsilon}$$

$$\le \frac{\varepsilon}{C_s(\mathcal{M}, g)} \left(\int_\mathcal{M} |\nabla \varphi|^2 \, d\mu + V^{-2/n} \right) + \frac{1}{\varepsilon}.$$

Inequality (6.65) follows by choosing

$$(6.67) \qquad C(a, g) = aV^{-2/n} + \frac{4}{an^2 e^2 C_s(\mathcal{M}, g)}.$$

Now we have proved the lemma when $n > 2$. We leave the $n = 2$ case as an exercise. □

EXERCISE 6.37. Prove the above lemma when $n = 2$.

Making the substitution $\varphi = e^{-\phi/2}$ in (6.65), we have the following.

COROLLARY 6.38 (Log Sobolev inequality, version 2). *For any $b > 0$, there exists a constant $C(b, g)$ such that if a function ϕ satisfies $\int_{\mathcal{M}} e^{-\phi} d\mu = 1$, then*

$$(6.68) \qquad -\int_{\mathcal{M}} \phi e^{-\phi} d\mu \leq b \int_{\mathcal{M}} |\nabla \phi|^2 e^{-\phi} d\mu + C(b, g).$$

4.2. Logarithmic Sobolev inequality on Euclidean space. We give a proof of Gross's logarithmic Sobolev inequality on Euclidean space [**170**]. Although this result will not be used elsewhere in Part I of this volume, we include it here since it is both fundamental and elegant.

THEOREM 6.39. *For any nonnegative function $\phi \in W^{1,2}(\mathbb{R}^n)$, we have*

$$\int_{\mathbb{R}^n} \phi^2 \log \phi \, dv \leq \int_{\mathbb{R}^n} |\nabla \phi|^2 \, dv + \frac{1}{2} \int_{\mathbb{R}^n} \phi^2 dv \cdot \log\left(\int_{\mathbb{R}^n} \phi^2 dv\right),$$

where $dv \doteqdot (2\pi)^{-n/2} e^{-|x|^2/2} dx$.

Note that the above inequality is scale-invariant, that is, the inequality is preserved under multiplication of ϕ by a positive constant. Also, if $\int_{\mathbb{R}^n} \phi^2 dx = 1$, then the inequality says that $\int_{\mathbb{R}^n} \phi^2 \log \phi \, dx \leq \int_{\mathbb{R}^n} |\nabla \phi|^2 \, dx$. The following consequence of Gross's logarithmic Sobolev inequality is actually equivalent to it. (We leave the proof of the equivalence to the reader.)

COROLLARY 6.40. *If $\int_{\mathbb{R}^n} (4\pi\tau)^{-n/2} e^{-f} dx = 1$, then*

$$(6.69) \qquad \int_{\mathbb{R}^n} \left(\tau |\nabla f|^2 + f - n\right)(4\pi\tau)^{-n/2} e^{-f} dx \geq 0.$$

In particular, taking $\tau = 1/2$, we have

$$(6.70) \qquad \int_{\mathbb{R}^n} \left(\frac{1}{2}|\nabla f|^2 + f - n\right)(2\pi)^{-n/2} e^{-f} dx \geq 0$$

provided $\int_{\mathbb{R}^n} (2\pi)^{-n/2} e^{-f} dx = 1$. Moreover, if we can perform an integration by parts, then we may rewrite (6.69) as

$$(6.71) \qquad \int_{\mathbb{R}^n} \left(\tau\left(2\Delta f - |\nabla f|^2\right) + f - n\right)(4\pi\tau)^{-n/2} e^{-f} dx \geq 0.$$

REMARK 6.41. Compare the LHS of (6.69) with the entropy (6.1) and compare the integrand on the LHS of (6.71) with Perelman's differential Harnack quantity (6.20).

PROOF OF THE COROLLARY. We shall prove just the case where $\tau = 1/2$ since the general case follows from making the change of variables $\tilde{x} \doteqdot (2\tau)^{-1/2} x$. Let ϕ be defined by $f = \frac{|x|^2}{2} - 2\log\phi$, so that $e^{-f} = e^{-|x|^2/2} \cdot \phi^2$

and $\nabla f = x - 2\frac{\nabla \phi}{\phi}$. We compute

$$\int_{\mathbb{R}^n} \left(\frac{1}{2} |\nabla f|^2 + f - n \right) (2\pi)^{-n/2} e^{-f} dx$$
$$= 2 \int_{\mathbb{R}^n} \left(\frac{1}{2} |x|^2 \phi^2 - \phi\, x \cdot \nabla \phi + |\nabla \phi|^2 - \phi^2 \log \phi - \frac{n}{2} \phi^2 \right) dv,$$

where $dv = (2\pi)^{-n/2} e^{-|x|^2/2} dx$. Now integrating by parts yields

$$\int_{\mathbb{R}^n} -\phi\, x \cdot \nabla \phi\, dv = \frac{1}{2} \int_{\mathbb{R}^n} n\phi^2 dv - \frac{1}{2} \int_{\mathbb{R}^n} |x|^2 \phi^2 dv,$$

so that we have the identity

$$\int_{\mathbb{R}^n} \left(-\phi\, x \cdot \nabla \phi - \frac{n}{2} \phi^2 + \frac{1}{2} |x|^2 \phi^2 \right) dv = 0.$$

Hence
(6.72)
$$\int_{\mathbb{R}^n} \left(\frac{1}{2} |\nabla f|^2 + f - n \right) (2\pi)^{-n/2} e^{-f} dx = 2 \int_{\mathbb{R}^n} \left(|\nabla \phi|^2 - \phi^2 \log \phi \right) dv,$$

with the constraint

$$1 = \int_{\mathbb{R}^n} (2\pi)^{-n/2} e^{-f} dx = \int_{\mathbb{R}^n} \phi^2 dv.$$

Since $\log \left(\int_{\mathbb{R}^n} \phi^2 dv \right) = 0$, by (6.72) and Gross's logarithmic Sobolev inequality, we have

$$\int_{\mathbb{R}^n} \left(\frac{1}{2} |\nabla f|^2 + f - n \right) (2\pi)^{-n/2} e^{-f} dx \geq 0.$$

□

EXERCISE 6.42. Show that Gross's logarithmic Sobolev inequality for Euclidean space implies that Euclidean space $(\mathbb{R}^n, g_{\mathbb{E}})$ has nonnegative entropy:

$$\mathcal{W}(g_{\mathbb{E}}, f, \tau) \geq 0 \quad \text{and} \quad \mu(g_{\mathbb{E}}, \tau) = 0.$$

Now we give Beckner and Pearson's proof of Gross's logarithmic Sobolev inequality, which is a consequence of the following [**23**].

PROPOSITION 6.43. *If $\int_{\mathbb{R}^n} \psi(x)^2 dx = 1$, then*

(6.73) $$\frac{n}{4} \log \left(\frac{2}{\pi e n} \int_{\mathbb{R}^n} |\nabla \psi(x)|^2 dx \right) \geq \int_{\mathbb{R}^n} (\log |\psi(x)|) \psi(x)^2 dx.$$

Note that this inequality is scale-invariant.

We first show that (6.73) implies Gross's logarithmic Sobolev inequality.

PROOF OF THEOREM 6.39 FROM THE PROPOSITION. Given f such that

(6.74) $$\int_{\mathbb{R}^n} (2\pi)^{-n/2} e^{-f} dx = 1,$$

let $\psi \doteq (2\pi)^{-n/4} e^{-f/2}$, so that $\log \psi = -\frac{f}{2} - \frac{n}{4} \log(2\pi)$ and $\int_{\mathbb{R}^n} \psi^2 dx = 1$. Then (6.73) implies

(6.75) $$\frac{n}{4} \log\left(\frac{2}{\pi e n} \int_{\mathbb{R}^n} \frac{1}{4} |\nabla f|^2 e^{-f} (2\pi)^{-n/2} dx\right)$$
$$\geq -\int_{\mathbb{R}^n} \left(\frac{f}{2} + \frac{n}{4} \log(2\pi)\right) e^{-f} (2\pi)^{-n/2} dx,$$

so that

$$\frac{1}{2} \int_{\mathbb{R}^n} |\nabla f|^2 e^{-f} (2\pi)^{-n/2} dx \geq \frac{en}{2} \exp\left\{-\frac{2}{n} \int_{\mathbb{R}^n} f e^{-f} (2\pi)^{-n/2} dx\right\}.$$

We claim

(6.76) $$\frac{en}{2} \exp\left\{-\frac{2}{n} \int_{\mathbb{R}^n} f e^{-f} (2\pi)^{-n/2} dx\right\} \geq \int_{\mathbb{R}^n} (n-f) e^{-f} (2\pi)^{-n/2} dx,$$

which implies the $\tau = 1/2$ case of (6.69):

(6.77) $$\int_{\mathbb{R}^n} \left(\frac{1}{2} |\nabla f|^2 + f - n\right) e^{-f} (2\pi)^{-n/2} dx \geq 0.$$

Since $\int_{\mathbb{R}^n} (2\pi)^{-n/2} e^{-f} dx = 1$, inequality (6.76) is equivalent to

$$\frac{n}{2} \exp\left\{\frac{2}{n} \int_{\mathbb{R}^n} \left(\frac{n}{2} - f\right) e^{-f} (2\pi)^{-n/2} dx\right\} \geq \frac{n}{2} + \int_{\mathbb{R}^n} \left(\frac{n}{2} - f\right) e^{-f} (2\pi)^{-n/2} dx.$$

If we let $du = e^{-f} (2\pi)^{-n/2} dx$ and $g = \frac{n}{2} - f$, then the above inequality becomes

$$\exp\left\{\frac{2}{n} \int_{\mathbb{R}^n} g \, du\right\} \geq 1 + \frac{2}{n} \int_{\mathbb{R}^n} g \, du.$$

This follows from $e^a \geq 1 + a$ for all $a \in \mathbb{R}$. $\square$

Now we present the

PROOF OF PROPOSITION 6.43. By Jensen's inequality, if $\int_{\mathbb{R}^N} |F|^r dx = 1$, then

$$(p-r) \int_{\mathbb{R}^N} (\log |F|) |F|^r dx = \int_{\mathbb{R}^N} \log\left(|F|^{p-r}\right) |F|^r dx$$
$$\leq \log\left(\int_{\mathbb{R}^N} |F|^{p-r} |F|^r dx\right) = \log\left(\int_{\mathbb{R}^N} |F|^p dx\right)$$

for $p \geq r > 0$. The L^2-Sobolev inequality says

$$\|F\|_{L^{N*}}^2 \leq A_N \int_{\mathbb{R}^N} |\nabla F|^2 dx,$$

where $N_* = \frac{2N}{N-2}$ and $A_N = \left(\frac{\Gamma(N)}{\Gamma(N/2)}\right)^{2/N} \frac{1}{\pi N(N-2)}$. By Sterling's formula, $\Gamma(N) \sim \sqrt{2\pi} N^{N-\frac{1}{2}} e^{-N}$, so that $A_N \sim \frac{2}{\pi e N}$ for N large, where Γ is the Gamma function. Hence, if $\int_{\mathbb{R}^N} |F|^2 \, dx = 1$, then by above two inequalities

$$\log\left(A_N \int_{\mathbb{R}^N} |\nabla F|^2 \, dx\right) \geq \frac{2}{N_*} \log\left(\int_{\mathbb{R}^N} |F|^{N_*} \, dx\right)$$

$$\geq \frac{4}{N} \int_{\mathbb{R}^N} (\log|F|) |F|^2 \, dx,$$

where we used $\frac{2}{N_*}(N_* - 2) = \frac{4}{N}$.

Given $f : \mathbb{R}^n \to \mathbb{R}$ with $\int_{\mathbb{R}^n} f^2 dx = 1$, let $N = n\ell$ and define $F : \mathbb{R}^N \to \mathbb{R}$ by

$$F(x) \doteq \prod_{k=1}^{\ell} f(x_k),$$

where $x = (x_1, \ldots, x_\ell)$, $x_k \in \mathbb{R}^n$ for $k = 1, \ldots, \ell$. Now

$$\frac{\nabla F(x)}{F(x)} = \left(\frac{\nabla f(x_1)}{f(x_1)}, \ldots, \frac{\nabla f(x_\ell)}{f(x_\ell)}\right).$$

Hence

$$|\nabla F(x)|^2 = F(x)^2 \sum_{k=1}^{\ell} \frac{|\nabla f(x_k)|^2}{f(x_k)^2}$$

and

$$\int_{\mathbb{R}^{n\ell}} |\nabla F(x)|^2 \, dx = \sum_{k=1}^{\ell} \int_{\mathbb{R}^{n\ell}} \frac{F(x)^2}{f(x_k)^2} |\nabla f(x_k)|^2 \, dx_1 \cdots dx_\ell$$

$$= \sum_{k=1}^{\ell} \left(\prod_{i \neq k} \int_{\mathbb{R}^n} f(x_i)^2 \, dx_i\right) \int_{\mathbb{R}^n} |\nabla f(x_k)|^2 \, dx_k$$

$$= \ell \int_{\mathbb{R}^n} |\nabla f(x)|^2 \, dx,$$

since $\int_{\mathbb{R}^n} f(x)^2 \, dx = 1$. Using $\int_{\mathbb{R}^{n\ell}} F(x)^2 \, dx = \prod_{k=1}^{\ell} \int_{\mathbb{R}^n} f(x_k)^2 \, dx_k = 1$, we have

$$\log\left(A_{n\ell} \ell \int_{\mathbb{R}^n} |\nabla f(x)|^2 \, dx\right) = \log\left(A_{n\ell} \int_{\mathbb{R}^{n\ell}} |\nabla F|^2 \, dx\right)$$

$$\geq \frac{4}{n\ell} \int_{\mathbb{R}^{n\ell}} (\log|F|) |F|^2 \, dx.$$

Now

$$\int_{\mathbb{R}^{n\ell}} (\log|F(x)|) |F(x)|^2 \, dx$$

$$= \int_{\mathbb{R}^{n\ell}} \sum_{k=1}^{\ell} \left(f(x_k)^2 \log|f_k(x_k)| \prod_{i \neq k} f(x_i)^2 \right) dx_1 \cdots dx_\ell$$

$$= \ell \int_{\mathbb{R}^n} (\log|f(x)|) f(x)^2 \, dx.$$

Hence

$$\log \left(A_{n\ell} \int_{\mathbb{R}^n} |\nabla f(x)|^2 \, dx \right) \geq \frac{4}{n} \int_{\mathbb{R}^n} f(x)^2 \log|f(x)| \, dx$$

for all $\ell \in \mathbb{N}$. Recall that $A_{n\ell} \sim \frac{2}{\pi e n \ell}$, which by taking the limit as $\ell \to \infty$, implies

$$\log \left(\frac{2}{\pi e n} \int_{\mathbb{R}^n} |\nabla f(x)|^2 \, dx \right) \geq \frac{4}{n} \int_{\mathbb{R}^n} f(x)^2 \log|f(x)| \, dx.$$

This completes the proof of the proposition. □

5. No finite time local collapsing: A proof of Hamilton's little loop conjecture

In this section we first define the notion of κ-noncollapsed at scale r and show its equivalence to the injectivity radius estimate. We then prove Perelman's celebrated no local collapsing theorem and indicate its equivalence to Hamilton's little loop conjecture. We end this section by showing the *existence of singularity models for solutions of the Ricci flow on closed manifolds developing finite time singularities* corresponding to sequences of points and times with curvatures comparable to their spatial maximums.

Perelman's no local collapsing theorem solves a major stumbling block in Hamilton's program for the Ricci flow on 3-manifolds. In particular, it provides a local injectivity radius estimate which enables one to obtain singularity models when dilating about finite time singular solutions of the Ricci flow on closed manifolds of any dimension. The no local collapsing theorem also rules out the formation of the cigar soliton as a singularity model.[14] The above two consequences of the no local collapsing theorem, together with Hamilton's singularity theory in dimension 3, imply that necks exist in all finite time singular solutions on closed 3-manifolds. This, together with Hamilton's analysis of nonsingular solutions, leads one to hope/expect that

---

[14]More precisely, in dimension 3 it rules out singularity models which are quotients of the product of the cigar soliton and the real line.

Ricci flow with surgery may lead to the resolution of Thurston's geometrization conjecture. Perelman's deeper analysis of 3-dimensional singularity formation greatly strengthens this expectation.[15]

5.1. κ-noncollapsing and injectivity radius lower bound.

5.1.1. *κ-noncollapsing on a Riemannian manifold.* Let $\left(\hat{\mathcal{M}}^n, \hat{g}\right)$ be a complete Riemannian manifold.

DEFINITION 6.44 (κ-noncollapsed). Given $\rho \in (0, \infty]$ and $\kappa > 0$, we say that the metric $\hat{g}$ is **κ-noncollapsed below the scale** ρ if for any metric ball $B(x, r)$ with $r < \rho$ satisfying $|\operatorname{Rm}(y)| \leq r^{-2}$ for all $y \in B(x, r)$, we have

$$(6.78) \qquad \frac{\operatorname{Vol} B(x, r)}{r^n} \geq \kappa.$$

If $\hat{g}$ is κ-noncollapsed below the scale ∞, we say that g is **κ-noncollapsed at all scales**.

Complementarily, we give the following.

DEFINITION 6.45 (κ-collapsed). We say that $\hat{g}$ is **κ-collapsed at the scale** r **at the point** x if $|\operatorname{Rm}(y)| \leq r^{-2}$ for all $y \in B(x, r)$ and

$$(6.79) \qquad \frac{\operatorname{Vol} B(x, r)}{r^n} < \kappa.$$

The metric $\hat{g}$ is said to be **κ-collapsed at the scale** r if there exists $x \in \hat{\mathcal{M}}$ such that $\hat{g}$ is κ-collapsed at the scale r at the point x.

Thus $\hat{g}$ is *not* κ-noncollapsed below the scale ρ if and only if there exists $r < \rho$ and $x \in \hat{\mathcal{M}}$ such that g is κ-collapsed at the scale r at the point x.

REMARK 6.46.
(1) If $\hat{\mathcal{M}}^n$ is closed and flat, then $\hat{g}$ cannot be κ-noncollapsed at all scales since $|\operatorname{Rm}| = 0 \leq r^{-2}$ for all r and $\operatorname{Vol} B(x, r) \leq \operatorname{Vol}\left(\hat{\mathcal{M}}\right)$ so that $\lim_{r \to \infty} \frac{\operatorname{Vol} B(x,r)}{r^n} = 0$ for all $x \in \hat{\mathcal{M}}$.
(2) If $\left(\hat{\mathcal{M}}^n, \hat{g}\right)$ is a closed Riemannian manifold, then for any $\rho > 0$ there exists $\kappa > 0$ such that $\hat{g}$ is κ-noncollapsed below the scale ρ.

We have the following elementary scaling property for κ-noncollapsed metrics.

LEMMA 6.47 (Scaling property of κ-noncollapsed). *If a metric $\hat{g}$ is κ-noncollapsed below the scale ρ, then for any $\alpha > 0$ the metric $\alpha^2 g$ is κ-noncollapsed below the scale $\alpha\rho$.*

---

[15]Some aspects of Perelman's singularity theory are discussed in Chapter 8 of Part I and also in Part II of this volume.

PROOF. We leave it as an exercise to trace through the definition of κ-noncollapsed and verify that the lemma follows from the scaling properties: $B_{\hat{g}}(x,r) = B_{\alpha^2 \hat{g}}(x,\alpha r)$, $|\operatorname{Rm}_{\alpha^2 \hat{g}}(y)| = \alpha^{-2}|\operatorname{Rm}_{\hat{g}}(y)|$, and $\operatorname{Vol}_{\alpha^2 \hat{g}} B_{\alpha^2 \hat{g}}(x,\alpha r) = \alpha^n \operatorname{Vol}_{\hat{g}} B_{\hat{g}}(x,r)$. □

The next lemma says the property of being κ-noncollapsed below the scale ρ is preserved (stable) under pointed Cheeger–Gromov limits.

LEMMA 6.48 (κ-noncollapsed preserved under limits). *Let $\{(\hat{\mathcal{M}}_k^n, \hat{g}_k, O_k)\}$ be a sequence of pointed complete Riemannian manifolds. Suppose that there exist $\kappa > 0$ and $\rho > 0$ so that each $\left(\hat{\mathcal{M}}_k, \hat{g}_k\right)$ is κ-noncollapsed below the scale ρ. Furthermore assume that $(\hat{\mathcal{M}}_k, \hat{g}_k, O_k)$ converges to $(\hat{\mathcal{M}}_\infty^n, \hat{g}_\infty, O_\infty)$ in the pointed Cheeger–Gromov C^2-topology. Then the limit $\left(\hat{\mathcal{M}}_\infty, \hat{g}_\infty\right)$ is κ-noncollapsed below the scale ρ.*

PROOF. This is because the distance function, the curvature, and the volume all converge under the limit. In particular, suppose $x \in \hat{\mathcal{M}}_\infty$ and $r < \rho$ are such that $|\operatorname{Rm}_{\hat{g}_\infty}(y)| \leq r^{-2}$ for all $y \in B_{\hat{g}_\infty}(x,r)$. Then for every $\varepsilon \in (0,r)$, there exists $k(\varepsilon) \in \mathbb{N}$ such that $|\operatorname{Rm}_{\hat{g}_k}(y)| \leq (r-\varepsilon)^{-2}$ for all $y \in B_{\hat{g}_k}(x, r-\varepsilon)$ and for all $k \geq k(\varepsilon)$. Since each $\hat{g}_k$ is κ-noncollapsed below the scale ρ, we have $\operatorname{Vol}_{\hat{g}_k}(B_{\hat{g}_k}(x, r-\varepsilon)) \geq \kappa(r-\varepsilon)^n$ for all $k \geq k(\varepsilon)$. Taking the limit as $k \to \infty$, we have $\operatorname{Vol}_{\hat{g}_\infty}(B_{\hat{g}_\infty}(x, r-\varepsilon)) \geq \kappa(r-\varepsilon)^n$. Letting $\varepsilon \to 0$, we then conclude that $\operatorname{Vol}_{\hat{g}_\infty}(B_{\hat{g}_\infty}(x,r)) \geq \kappa r^n$ as desired. □

Recall that if $\operatorname{Rc} \geq 0$ on a complete Riemannian manifold $\left(\hat{\mathcal{M}}^n, \hat{g}\right)$, then for $p \in \hat{\mathcal{M}}$ fixed, $\frac{\operatorname{Vol} B(p,r)}{r^n}$ is a nonincreasing function of r. When $\hat{\mathcal{M}}$ is noncompact, the notion of κ-noncollapsed at all scales is closely related to another invariant of the geometry of infinity called the asymptotic volume ratio, which we now define.

DEFINITION 6.49 (Asymptotic volume ratio). Let $\left(\hat{\mathcal{M}}^n, \hat{g}\right)$ be a complete noncompact Riemannian manifold with nonnegative Ricci curvature. The **asymptotic volume ratio** is defined as the limit of volume ratios by

(6.80) $$\operatorname{AVR}(\hat{g}) \doteqdot \lim_{r \to \infty} \frac{\operatorname{Vol} B(p,r)}{\omega_n r^n} < \infty,$$

where ω_n is the volume of the unit ball in $\mathbb{R}^n$. We say that $\left(\hat{\mathcal{M}}, \hat{g}\right)$ has **maximum volume growth** if $\operatorname{AVR}(\hat{g}) > 0$.

REMARK 6.50. The asymptotic volume ratio is independent of the choice of basepoint $p \in \hat{\mathcal{M}}$.

EXERCISE 6.51 (AVR > 0 implies noncollapsed). Show that if $\left(\hat{\mathcal{M}}^n, \hat{g}\right)$ is a complete noncompact Riemannian manifold with $\operatorname{Rc} \geq 0$ and $\operatorname{AVR}(\hat{g}) > 0$, then $\hat{g}$ is κ-noncollapsed on all scales for $\kappa = \omega_n \operatorname{AVR}(\hat{g})$.

The next exercise shows that on small enough scales $\hat{g}$ is κ-noncollapsed for some κ.

EXERCISE 6.52. Show that for any Riemannian manifold $\left(\hat{\mathcal{M}}^n, \hat{g}\right)$, $x \in \hat{\mathcal{M}}$, and $\kappa < \omega_n$, there exists $\rho(x) > 0$ such that for every $r \in (0, \rho(x)]$, we have
$$|\mathrm{Rm}| \leq r^{-2} \quad \text{in } B(x, r)$$
and
$$\frac{\mathrm{Vol}\, B(x, r)}{r^n} \geq \kappa.$$

SOLUTION TO EXERCISE 6.52. This follows from the facts that
$$\lim_{r \to 0} r^2 \sup_{B(x,r)} |\mathrm{Rm}| = 0$$
and
$$\lim_{r \to 0} \frac{\mathrm{Vol}\, B(x, r)}{r^n} = \omega_n.$$

REMARK 6.53. In some sense Exercise 6.52 is a local version of Remark 6.46(2).

5.1.2. κ-noncollapsing and injectivity radius lower bound. We now show that κ-noncollapsing and a lower bound of the injectivity radius are equivalent.

LEMMA 6.54. Let $\left(\hat{\mathcal{M}}^n, \hat{g}\right)$ be a complete Riemannian manifold and fix $\rho \in (0, \infty]$.
  (i) If the metric $\hat{g}$ is not κ-collapsed below the scale ρ for some $\kappa > 0$, then there exists a constant $\delta = \delta(n, \kappa)$ which is independent of ρ and $\hat{g}$ such that for any $x \in \hat{\mathcal{M}}$ and $r < \rho$, if $|\mathrm{Rm}| \leq r^{-2}$ in $B(x, r)$, then $\mathrm{inj}(x) \geq \delta r$.
  (ii) Suppose that for any $x \in \hat{\mathcal{M}}$ and $r < \rho$ with $|\mathrm{Rm}| \leq r^{-2}$ in $B(x, r)$ we have $\mathrm{inj}(x) \geq \delta r$ for some $\delta > 0$. Then there exists a constant $\kappa = \kappa(n, \delta)$, independent of ρ and $\hat{g}$, such that $\hat{g}$ is not κ-collapsed below the scale ρ.

PROOF. (i) Let $B(x, r)$ be a ball satisfying $|\mathrm{Rm}| \leq r^{-2}$ in $B(x, r)$ for some $r \leq \rho$. Consider the metric $r^{-2}\hat{g}$ on $B(x, r) = B_{r^{-2}\hat{g}}(x, 1)$. Since $\hat{g}$ is not κ-collapsed on $B(x, r)$, we have $|\mathrm{Rm}_{r^{-2}\hat{g}}| \leq 1$ in $B_{r^{-2}\hat{g}}(x, 1)$ and
$$\mathrm{Vol}_{r^{-2}\hat{g}}\, B_{r^{-2}\hat{g}}(x, 1) = \frac{\mathrm{Vol}_{\hat{g}}\, B(x, r)}{r^n} \geq \kappa.$$

By a result of Cheeger, Gromov, and Taylor (see Theorem A.7), there exists $\delta = \delta(n, \kappa)$ such that $\mathrm{inj}_{r^{-2}\hat{g}}(x) \geq \delta$. Hence $\mathrm{inj}(x) \geq \delta r$.

(ii) Again let $B(x, r)$ be a ball satisfying $|\mathrm{Rm}| \leq r^{-2}$ in $B(x, r)$ for some $r \leq \rho$, and consider the metric $r^{-2}\hat{g}$ on $B(x, r) = B_{r^{-2}\hat{g}}(x, 1)$. We have $|\mathrm{Rm}_{r^{-2}\hat{g}}| \leq 1$ and $\mathrm{inj}_{r^{-2}\hat{g}}(x) \geq \delta$. By the Bishop–Gromov volume (or

Rauch) comparison theorem (comparing $\left(B_{r^{-2}\hat{g}}(x,1), r^{-2}\hat{g}\right)$ with the ball of radius δ in the unit sphere $S^n(1)$), there exists $\kappa = \kappa(n, \delta)$ such that

$$\frac{\operatorname{Vol}_{\hat{g}} B(x, r)}{r^n} = \operatorname{Vol}_{r^{-2}\hat{g}} B_{r^{-2}\hat{g}}(x, 1) \geq \kappa.$$

$\square$

5.1.3. κ-noncollapsing in Ricci flow.

DEFINITION 6.55. We say that a complete solution $(\mathcal{M}^n, g(t))$, $t \in [0, T)$, to the Ricci flow, where $T \in (0, \infty]$, is κ-**noncollapsed below the scale** ρ if for every $t \in [0, T)$, $g(t)$ is κ-noncollapsed below the scale ρ.

If $\mathcal{M}$ is closed, $T_1 < \infty$, and $C_0 \doteqdot \sup_{\mathcal{M} \times [0, T_1)} |\mathrm{Rm}| < \infty$, then, using the metric equivalence $e^{-2(n-1)C_0} g(0) \leq g(t) \leq e^{2(n-1)C_0} g(0)$ for $t \in [0, T_1)$,[16] we see that for every $\rho \in (0, \infty)$ there exists $\kappa = \kappa(n, C_0, g(0), T_1, \rho) > 0$ such that the solution $g(t)$ is κ-noncollapsed below the scale ρ. Hence we are interested in κ-noncollapsing near T when the solution forms a singularity at time T. We shall see that when $T < \infty$ and $\mathcal{M}$ is closed, Perelman's monotonicity of entropy implies that for all $\rho > 0$ the solution is κ-noncollapsed below the scale ρ for some $\kappa = \kappa(n, g(0), T, \rho) > 0$.

In §4.1 of [**297**] Perelman also gave the following.

DEFINITION 6.56 (Locally collapsing solution). Let $(\mathcal{M}^n, g(t))$, $t \in [0, T)$, be a complete solution to the Ricci flow, where $T \in (0, \infty]$. The solution $g(t)$ is said to be **locally collapsing** at T if there exists a sequence of points $x_k \in \mathcal{M}$, times $t_k \to T$, and radii $r_k \in (0, \infty)$ with r_k^2/t_k uniformly bounded (from above) such that the balls $B_{g(t_k)}(x_k, r_k)$ satisfy

(1) (*curvature bound comparable to the radius of the ball*)

$$|\mathrm{Rm}[g(t_k)]| \leq r_k^{-2} \quad \text{in } B_{g(t_k)}(x_k, r_k),$$

(2) (*volume collapse of the ball*)

$$\lim_{k \to \infty} \frac{\operatorname{Vol}_{g(t_k)} B_{g(t_k)}(x_k, r_k)}{r_k^n} = 0.$$

EXERCISE 6.57. It is interesting to consider solutions to the Ricci flow which are defined on a time interval of the form $(0, T)$ with the curvature becoming unbounded as $t \to 0_+$. For example, consider an initial metric g_0 on a surface which is C^∞ except for a conical singularity. We expect a smooth solution $g(t)$ of the Ricci flow to exist on some time interval $(0, T)$ with $g(t) \to g_0$ as $t \to 0_+$. It is interesting to ask if solutions on closed manifolds can locally collapse as $t \to 0_+$. In view of this, for a solution defined on $(0, T)$, formulate the notion of locally collapsing at time 0.

---

[16] For the proof of this metric equivalence, see Corollary 6.50 on p. 204 of Volume One. The argument there is essentially repeated in the proof of the inequalities in (3.3) of this volume.

5.2. The no local collapsing theorem and its proof.

5.2.1. *No local collapsing theorem and little loop conjecture.* One of the major breakthroughs in Ricci flow is the following.

THEOREM 6.58 (No local collapsing—A). *Let $g(t), t \in [0, T)$, be a smooth solution to the Ricci flow on a closed manifold $\mathcal{M}^n$. If $T < \infty$, then for any $\rho \in (0, \infty)$ there exists $\kappa = \kappa(n, g(0), T, \rho) > 0$ such that $g(t)$ is κ-noncollapsed below the scale ρ for all $t \in [0, T)$.*

We shall prove this theorem in the next subsection. Actually Theorem 4.1 of [297] states the result a bit differently.

THEOREM 6.59 (No local collapsing—B). *If $\mathcal{M}$ is closed and $g(t)$ is any solution on $[0, T)$ with $T < \infty$, then $g(t)$ is not locally collapsing at T.*

REMARK 6.60. We leave it as an exercise to show that Theorems 6.58 and 6.59 are equivalent.

Hamilton's little loop conjecture says the following (see §15 of [186]). Let $(\mathcal{M}^n, g(t))$, $t \in [0, T)$, be a smooth solution to the Ricci flow on a closed manifold. There exists $\delta = \delta(n, g(0)) > 0$ such that for any point $(x, t) \in \mathcal{M} \times [0, T)$ where

$$|\operatorname{Rm}(g(t))| \leq \frac{1}{W^2} \quad \text{in } B_{g(t_0)}(x, W)$$

for some $W > 0$, we have

$$\operatorname{inj}_{g(t)}(x) \geq \delta W.$$

Note that the role of the positive number κ in the definition of κ-noncollapsed is similar to the role of δ in the injectivity radius lower bound which is used in the statement of Hamilton's little loop conjecture.

Rephrasing the little loop conjecture (LLC) a little differently, we have the following equivalence between no local collapsing (NLC) at T and the little loop conjecture.

LEMMA 6.61 (NLC and LLC are equivalent). *Let $(\mathcal{M}^n, g(t))$, $t \in [0, T)$, be a smooth complete solution to the Ricci flow where $T \in (0, \infty]$. The following two statements are equivalent.*

(i) *(Little loop conjecture) For any $C > 0$ there exists $\delta > 0$ such that if $(x, t) \in \mathcal{M} \times [0, T)$ and $W \in (0, \sqrt{Ct}]$ satisfy*

$$|\operatorname{Rm}(t)| \leq \frac{1}{W^2} \quad \text{in } B_{g(t)}(x, W),$$

then

(6.81) $$\operatorname{inj}_{g(t)}(x) \geq \delta W.$$

(ii) *(No local collapsing) The solution $g(t)$ is not locally collapsing at T.*

5. NO FINITE TIME LOCAL COLLAPSING

PROOF. (i) $\Longrightarrow$ (ii). We prove (ii) by contradiction. Suppose $g(t)$ is locally collapsing at T. Then there exists a sequence of times $t_k \nearrow T$ and a sequence of metric balls $B_{g(t_k)}(x_k, r_k)$ such that

(1) $\dfrac{r_k^2}{t_k} \leq C$ for some $C < \infty$,

(2) $|\mathrm{Rm}(g(t_k))| \leq r_k^{-2}$ in $B_{g(t_k)}(x_k, r_k)$,

(3) $\dfrac{\mathrm{Vol}_{g(t_k)} B_{g(t_k)}(x_k, r_k)}{r_k^n} \searrow 0$ as $k \to \infty$.

Hence by (i) we have $\mathrm{inj}_{g(t_k)}(x_k) \geq \delta r_k$ for all k, where $\delta > 0$ is independent of k. By Lemma 6.54(ii), the volume collapsing statement (3) above cannot be true, a contradiction.

(ii) $\Longrightarrow$ (i). We also prove (i) by contradiction. If (i) is not true, then there exists $C > 0$ and a sequence of points and times $(x_k, t_k) \in \mathcal{M} \times [0, T)$ and $W_k \in (0, \sqrt{Ct_k}]$ satisfying

$$|\mathrm{Rm}(t_k)| \leq \frac{1}{W_k^2} \quad \text{in } B_{g(t_k)}(x_k, W_k)$$

and

$$\frac{\mathrm{inj}_{g(t_k)}(x_k)}{W_k} \searrow 0.$$

Lemma 6.54(i) implies that $\dfrac{\mathrm{Vol}_{g(t_k)} B_{g(t_k)}(x_k, W_k)}{W_k^n} \searrow 0$ as $k \to \infty$. Thus $g(t)$ is locally collapsing at T and we have a contradiction. The lemma is proved. $\square$

It follows from Theorem 6.59 and Lemma 6.61 that Hamilton's little loop conjecture holds for solutions of the Ricci flow on closed manifolds forming finite time singularities.

COROLLARY 6.62. *Let $g(t)$, $t \in [0, T)$, be a smooth solution to the Ricci flow on a closed manifold $\mathcal{M}^n$. If $T < \infty$, then the little loop conjecture holds. That is, for any $C > 0$ there exists $\delta > 0$ such that if $(x, t) \in \mathcal{M} \times [0, T)$ and $W \in (0, \sqrt{Ct}]$ satisfy*

$$|\mathrm{Rm}(t)| \leq \frac{1}{W^2} \quad \text{in } B_{g(t)}(x, W),$$

then we have $\mathrm{inj}_{g(t)}(x) \geq \delta W$.

The little loop conjecture illustrates the essence of no locally collapsing from the injectivity radius perspective. For convenience we give the following

DEFINITION 6.63 (Local injectivity radius estimate). We say that a complete solution $(\mathcal{M}^n, g(t))$, $t \in [0, T)$, to the Ricci flow satisfies a **local injectivity radius estimate** if for every $\rho \in (0, \infty)$ and $C < \infty$, there exists $c = c(\rho, C, g(t)) > 0$ such that for any $(p, t) \in \mathcal{M} \times [0, T)$ and $r \in (0, \rho]$ which satisfy

$$|\mathrm{Rm}(\cdot, t)| \leq Cr^{-2} \quad \text{in } B_{g(t)}(p, r),$$

we have $\operatorname{inj}_{g(t)}(p) \geq cr$.

Corollary 6.62, i.e., Perelman's no local collapsing theorem, implies that if $(\mathcal{M}^n, g(t))$, $t \in [0, T)$, is a solution of the Ricci flow on a closed manifold with $T < \infty$, then $g(t)$, $t \in [0, T)$, satisfies a local injectivity radius estimate.

5.2.2. *Proof of No Local Collapsing Theorem 6.58.* The idea of the proof is that if a metric g is κ-collapsed at a point x at a distance scale r for κ small and r bounded, then $\mathcal{W}(g, f, r^2)$ is negative and large in magnitude, e.g., on the order of $\log \kappa$, for f concentrated in a ball of radius r centered at x. This contradicts the monotonicity formula for $\mu(g(t), \tau(t))$.

PROOF OF THEOREM 6.58 ASSUMING PROPOSITION 6.64. We shall say that r is the (space) scale of $\mu(g, r^2)$; the justification for this terminology occurs below. Since $T/2 < T$, by the remarks after Definition 6.55, there exists $\kappa_0 = \kappa_0(n, g(0), T, \rho) > 0$ such that $g(t)$ is κ_0-noncollapsed below the scale ρ for all $t \in [0, T/2]$.

On the other hand, if $t \in [T/2, T)$, then for any $0 < r \leq \rho$, we have $t + r^2 \in [T/2, T + \rho^2)$, and by the monotonicity formula (6.61), we have

$$\mu(g(t), r^2) \geq \mu(g(0), t + r^2)$$
$$(6.82) \qquad \geq \inf_{\tau \in [T/2, T+\rho^2]} \mu(g(0), \tau) \doteqdot -C_1(n, g(0), T, \rho) > -\infty$$

since $T < \infty$. In summary, by the monotonicity formula, since the μ-invariant of the initial metric is bounded from below at scales bounded from above and below, the μ-invariant of the solution after a certain amount of time (say $T/2$) is bounded from below at all bounded scales. The theorem will follow from the important observation that *if a Riemannian metric is κ-collapsed at some scale r for κ small, then its μ-invariant is negative and large in magnitude at the time scale r^2* (see Proposition 6.64 below).

If $x \in \mathcal{M}$, $t \in [T/2, T)$, and $r \in (0, \rho]$ are such that $|\operatorname{Rm}_{g(t)}| \leq 1$ in $B_{g(t)}(x, r)$, then $\operatorname{Rc}_{g(t)} \geq -c_1(n) r^{-2}$ and $R_{g(t)} \leq c_1(n) r^{-2}$ in $B_{g(t)}(x, r)$, where $c_1(n) = n(n-1)$. So by (6.82) and (6.83),

$$-C_1(n, g(0), T, \rho) \leq \mu(g(t), r^2) \leq \log \frac{\operatorname{Vol}_{g(t)} B_{g(t)}(x, r)}{r^n} + C_2(n, \rho).$$

We conclude that

$$\frac{\operatorname{Vol}_{g(t)} B_{g(t)}(x, r)}{r^n} \geq \kappa_1(n, g(0), T, \rho) > 0,$$

where $\kappa_1(n, g(0), T, \rho) = e^{-C_1(n, g(0), T, \rho) - C_2(n, \rho)}$. The theorem follows with the choice $\kappa(n, g(0), T, \rho) \doteqdot \min\{\kappa_0, \kappa_1\}$. □

Now we turn to bounding the μ-invariant from above by volume ratios in a Riemannian manifold.

PROPOSITION 6.64 (μ controls volume ratios). *Let $\rho \in (0, \infty)$. There exists a constant $C_2 = C_2(n, \rho) < \infty$ such that if $(\hat{\mathcal{M}}^n, \hat{g})$ is a closed*

Riemannian manifold, $p \in \hat{\mathcal{M}}$ and $r \in (0, \rho]$ are such that $\mathrm{Rc} \geq -c_1(n) r^{-2}$ and $R \leq c_1(n) r^{-2}$ in $B(p, r)$, then

$$\mu(\hat{g}, r^2) \leq \log \frac{\mathrm{Vol}\, B(p, r)}{r^n} + C_2(n, \rho). \tag{6.83}$$

That is,

$$\frac{\mathrm{Vol}\, B(p, r)}{r^n} \geq e^{-C_2(n,\rho)} e^{\mu(\hat{g}, r^2)}.$$

In particular, if for some $\kappa > 0$ and $r \in (0, \rho]$ the metric $\hat{g}$ is κ-collapsed at the scale r, then

$$\mu(\hat{g}, r^2) \leq \log \kappa + C_2(n, \rho).$$

Proof. As in (6.41) with $\tau = r^2$, define the positive function w by

$$w^2 = (4\pi r^2)^{-n/2} e^{-f}. \tag{6.84}$$

From the definition (6.56) of μ as an infimum of $\mathcal{W}$, we have by rewriting $\mathcal{W}(\hat{g}, f, r^2)$ in terms of w (compare to (6.42)),

$$\mu(\hat{g}, r^2) \leq \int_{\hat{\mathcal{M}}} r^2 \left(4 |\nabla w|^2 + R w^2\right) d\mu + \int_{\hat{\mathcal{M}}} (f - n) w^2 d\mu \tag{6.85}$$
$$\doteqdot \mathcal{K}(\hat{g}, w, r^2),$$

where $\int_{\hat{\mathcal{M}}} w^2 d\mu = 1$ and $f = -2 \log w - \frac{n}{2} \log(4\pi r^2)$. While making the convention that $f(y) w^2(y) = 0$ when $w(y) = 0$, we claim that (6.85) holds for nonnegative Lipschitz functions w satisfying $\int_{\hat{\mathcal{M}}} w^2 d\mu = 1$. To see this claim, first by (6.56) we know that (6.85) holds for positive Lipschitz functions w satisfying $\int_{\hat{\mathcal{M}}} w^2 d\mu = 1$. Now given any nonnegative Lipschitz function w satisfying $\int_{\hat{\mathcal{M}}} w^2 d\mu = 1$, define for $\varepsilon \in (0, 1)$,

$$w_\varepsilon \doteqdot C_\varepsilon (w + \varepsilon),$$

where the constant C_ε is defined by $\int_{\hat{\mathcal{M}}} w_\varepsilon^2 d\mu = 1$. Clearly $\lim_{\varepsilon \to 0} C_\varepsilon = 1$, w_ε is a positive Lipschitz function, and hence $\mu(\hat{g}, r^2) \leq \mathcal{K}(\hat{g}, w_\varepsilon, r^2)$ for each $\varepsilon \in (0, 1)$. Using $\lim_{\varepsilon \to 0} \varepsilon \log \varepsilon = 0$ and $f_\varepsilon = -2 \log w_\varepsilon - \frac{n}{2} \log(4\pi r^2)$ in the definition of $\mathcal{K}(\hat{g}, w_\varepsilon, r^2)$, we have $\lim_{\varepsilon \to 0} \mathcal{K}(\hat{g}, w_\varepsilon, r^2) = \mathcal{K}(\hat{g}, w, r^2)$. The claim is proved.

Now let $\phi : [0, \infty) \to [0, 1]$ be a standard cut-off function with $\phi = 1$ on $[0, 1/2]$, $\phi = 0$ on $[1, \infty)$, and $|\phi'| \leq 3$. Assume $p \in \hat{\mathcal{M}}$ and $r \in (0, \rho]$ are such that we have the curvature bounds $\mathrm{Rc} \geq -c_1(n) r^{-2}$ and $R \leq c_1(n) r^{-2}$ in $B(p, r)$ for $c_1(n) = n(n-1)$. We make a judicious choice for f so that the RHS of (6.85) reflects the local geometry at p with respect to the metric $\hat{g}$. In particular, let

$$w^2(x) \doteqdot (4\pi r^2)^{-n/2} \phi \left(\frac{d_{\hat{g}}(x, p)}{r}\right)^2 e^{-c}, \tag{6.86}$$

where $d_{\hat{g}}(x,p)$ is the distance function and the constant $c = c(n,\hat{g},x,r)$ is chosen so that $\int_{\hat{\mathcal{M}}} w^2 d\mu = 1$. Note that w is a Lipschitz function and definition (6.84) implies

(6.87)
$$f = c - \log(\phi^2).$$

(We abuse notation and write $\phi(x) = \phi\left(\frac{d_{\hat{g}}(x,p)}{r}\right)$.) The constant c is related to the volume ratio $\frac{\operatorname{Vol} B(p,r)}{r^n}$ by the following.

LEMMA 6.65. *There exists $C_3(n,\rho) < \infty$ such that for any $r \in (0,\rho)$,*

(6.88)
$$\log \frac{\operatorname{Vol} B(p,r)}{r^n} - C_3(n,\rho) \le c \le \log \frac{\operatorname{Vol} B(p,r)}{r^n}.$$

Proof of the lemma. (1) Since $\int_{\hat{\mathcal{M}}} w^2 d\mu = 1$,

(6.89)
$$e^c = (4\pi r^2)^{-n/2} \int_{\hat{\mathcal{M}}} \phi\left(\frac{d(x,p)}{r}\right)^2 d\mu(x).$$

Applying $\phi \le 1$ and $\operatorname{supp}(w) \subset B(p,r)$, we have

$$e^c \le (4\pi r^2)^{-n/2} \operatorname{Vol} B(p,r),$$

which implies

(6.90)
$$c \le \log \frac{\operatorname{Vol} B(p,r)}{r^n}.$$

(2) On the other hand, since $\phi = 1$ on $[0, 1/2]$, by (6.89) we have

$$c \ge -\frac{n}{2} \log(4\pi) + \log \frac{\operatorname{Vol} B(p,r/2)}{r^n}.$$

Since $\operatorname{Rc} \ge -c_1(n) r^{-2}$ in $B(p,r)$ and $r \le \rho$, by the Bishop–Gromov relative volume comparison theorem, there exists $C_4(n,\rho) < \infty$ such that

$$\operatorname{Vol} B(p,r) \le C_4(n,\rho) \operatorname{Vol} B(p,r/2).$$

Thus

(6.91)
$$c \ge -C_3(n,\rho) + \log \frac{\operatorname{Vol} B(p,r)}{r^n}.$$

This completes the proof of the lemma.

To prove the proposition, we estimate the two terms

$$\int_{\hat{\mathcal{M}}} r^2 \left(4|\nabla w|^2 + Rw^2\right) d\mu + \int_{\hat{\mathcal{M}}} fw^2 d\mu = \mathcal{K}(\hat{g}, w, r^2) + n$$

on the RHS of (6.85) separately. First we have

$$\int_{\hat{\mathcal{M}}} r^2 R w^2 d\mu \le c_1(n)$$

5. NO FINITE TIME LOCAL COLLAPSING

since $r^2 R \leq c_1(n)$ in $B(p,r)$, $\mathrm{supp}(w) \subset B(p,r)$, and $\int_{\hat{\mathcal{M}}} w^2 d\mu = 1$. By (6.86), we have a.e.,

$$r^2 |\nabla w|^2 \leq \left(4\pi r^2\right)^{-n/2} e^{-c} |\phi'|^2 \leq 9 \left(4\pi\right)^{-n/2} \frac{e^{-c}}{r^n}$$

$$\leq \frac{9 \left(4\pi\right)^{-n/2} e^{C_3(n,\rho)}}{\mathrm{Vol}\, B(p,r)},$$

using $|\phi'| \leq 3$, $|\nabla d| = 1$ a.e., and (6.88). Since $|\nabla w|$ has support in $B(p,r)$, we have

$$\int_{\hat{\mathcal{M}}} 4r^2 |\nabla w|^2 d\mu \leq 9 \left(4\pi\right)^{-n/2} e^{C_3(n,\rho)}.$$

We conclude that the energy part of the RHS of (6.85) is bounded from above:

$$\int_{\hat{\mathcal{M}}} r^2 \left(4 |\nabla w|^2 + R w^2\right) d\mu \leq c_1(n) + 9 \left(4\pi\right)^{-n/2} e^{C_3(n,\rho)}.$$

Now we consider the entropy part. Since $f = c - \log(\phi^2)$, by (6.87), we have

$$\int_{\hat{\mathcal{M}}} f w^2 d\mu \leq \int_{\hat{\mathcal{M}}} \left(c - \log(\phi^2)\right) w^2 d\mu$$

$$= c - \left(4\pi r^2\right)^{-n/2} e^{-c} \int_{\hat{\mathcal{M}}} \log(\phi^2) \phi^2 d\mu.$$

Since $-x \log x \leq 1/e$ and ϕ has support in $B(p,r)$, we have[17]

$$\int_{\hat{\mathcal{M}}} \log(\phi^2) \phi^2 d\mu \geq -\frac{1}{e} \mathrm{Vol}\, B(p,r).$$

Hence

$$\int_{\hat{\mathcal{M}}} f w^2 d\mu \leq c + \frac{1}{e} \left(4\pi\right)^{-n/2} \frac{e^{-c} \mathrm{Vol}\, B(p,r)}{r^n}$$

$$\leq c + \frac{1}{e} \left(4\pi\right)^{-n/2} e^{C_3(n,\rho)}$$

$$\leq \log \frac{\mathrm{Vol}\, B(p,r)}{r^n} + \frac{1}{e} \left(4\pi\right)^{-n/2} e^{C_3(n,\rho)}.$$

We conclude that

$$\mu(\hat{g}, r^2) \leq \log \frac{\mathrm{Vol}\, B(p,r)}{r^n} + C_2(n, \rho),$$

where $C_2(n,\rho) \doteqdot \left(9 + \frac{1}{e}\right) \left(4\pi\right)^{-n/2} e^{C_3(n,\rho)} + c_1(n)$. This proves the first part of the proposition.

---

[17]Our convention is $0 \cdot \log 0 = 0$ since $\lim_{\varepsilon \to 0} \varepsilon \log \varepsilon = 0$. Also, $\log(\phi^2) = 0$ in $B(p, r/2)$, so in fact $\int_{\hat{\mathcal{M}}} \log(\phi^2) \phi^2 d\mu \geq -\frac{1}{e} \mathrm{Vol}(B(p,r) - B(p,r/2))$.

The second part follows from that fact that if for some $\kappa > 0$ and $r > 0$ the metric $\hat{g}$ is κ-collapsed at the scale r, then there exists $p \in \hat{\mathcal{M}}$ such that
$$\frac{\operatorname{Vol} B(p,r)}{r^n} \leq \kappa.$$
The proposition is proved. $\square$

EQUIVALENT PROPOSITION. *For every Riemannian manifold $\left(\hat{\mathcal{M}}^n, \hat{g}\right)$, $T \in (0, \infty)$, C_1 and C_2, there exists $c = c(n, T, C_1, C_2) > 0$ such that if for some $r \in (0, \sqrt{T}]$ and $A < \infty$ we have $\mu\left(\hat{g}, r^2\right) \geq -A$, then for any $p \in \hat{\mathcal{M}}$ with $\operatorname{Rc} \geq -C_1 r^{-2}$ in $B(p, r)$ and $R \leq C_2 r^{-2}$ in $B(p, r)$, we have*
$$\operatorname{Vol} B(p, r) \geq \kappa r^n,$$
where $\kappa = c(n, T, C_1, C_2) e^{-A}$.

PROOF. The contrapositive is Proposition 6.64. In particular, the contrapositive is: for every $\left(\hat{\mathcal{M}}^n, \hat{g}\right)$, $T \in (0, \infty)$, C_1 and C_2, there exists a constant $C = C(n, T, C_1, C_2)$ such that if for some $r \in (0, \sqrt{T}]$ and $\kappa > 0$ there exists $p \in \hat{\mathcal{M}}$ with $\operatorname{Rc} \geq -C_1 r^{-2}$ in $B(p, r)$, $R \leq C_2 r^{-2}$ in $B(p, r)$, and $\operatorname{Vol} B(p, r) < \kappa r^n$, then $\mu\left(\hat{g}, r^2\right) < \log \kappa + C$. $\square$

REMARK 6.66. From examining the proof of Proposition 6.64, in the Equivalent Proposition we can replace the condition $\operatorname{Rc} \geq -C_1 r^{-2}$ in $B(p, r)$ by
$$\operatorname{Vol} B(p, r) \leq C_1 \operatorname{Vol} B(p, r/2)$$
since the only place where we used the Ricci curvature lower bound is for the relative volume comparison. That is, there exists $c_0 = c_0(n, T, C_1) > 0$ such that
$$\operatorname{Vol} B(p, r) \geq \kappa_0 r^n,$$
where $\kappa_0 = c_0(n, T, C_1) e^{-A}$.

By assuming a lower bound on $\nu(\hat{g})$ instead of $\mu\left(\hat{g}, r^2\right)$, we may remove the lower bound of Ricci curvature assumption.

PROPOSITION 6.67. *For every Riemannian manifold $\left(\hat{\mathcal{M}}^n, \hat{g}\right)$, $T \in (0, \infty)$, and C_1, there exists $c = c(n, T, C_1) > 0$ such that if for some $r \in (0, \sqrt{T}]$ and $A < \infty$ we have $\nu(g) \geq -A$, then for any $p \in \hat{\mathcal{M}}$ with $R \leq C_1 r^{-2}$ in $B(p, r)$, we have*
$$\operatorname{Vol} B(p, r) \geq \kappa r^n,$$
where $\kappa = c(n, T, C_1) e^{-A}$.

PROOF. If $\operatorname{Vol} B(p, r) \leq 3^n \operatorname{Vol} B(p, r/2)$, then by the above remark, the proposition holds. So we assume that $\operatorname{Vol} B(p, r) > 3^n \operatorname{Vol} B(p, r/2)$. Since $\lim_{s \to 0^+} \frac{\operatorname{Vol} B(p,s)}{\omega_n s^n} = 1$, there exists $k \in \mathbb{N}$ such that $\operatorname{Vol} B\left(p, r/2^k\right) \leq 3^n \operatorname{Vol} B\left(p, r/2^{k+1}\right)$ and $\operatorname{Vol} B\left(p, r/2^i\right) > 3^n \operatorname{Vol} B\left(p, r/2^{i+1}\right)$ for all $0 \leq i <$

k. We can apply Remark 6.66 to the ball $B\left(p, r/2^k\right)$ and get $\operatorname{Vol} B\left(p, r/2^k\right) \geq \kappa_0 \left(r/2^k\right)^n$. Hence

$$\begin{aligned}\operatorname{Vol} B(p, r) &> 3^n \operatorname{Vol} B(p, r/2) \\ &> 3^{n(k-1)} \operatorname{Vol} B\left(p, r/2^k\right) \\ &\geq 3^{n(k-1)} \kappa_0 \left(r/2^k\right)^n \\ &\geq \frac{\kappa_0}{2^n} r^n.\end{aligned}$$

Hence the proposition holds with $c(n, T, C_1) = \frac{c_0(n,T,C_1)}{2^n}$, since by Remark 6.66 we have $\kappa_0 = c_0(n, T, C_1) e^{-A}$. □

5.3. Application of κ-noncollapsing to the analysis of singularities.

We present some applications to end this section. In the next section we give an improvement of the no local collapsing theorem.

5.3.1. *Existence of finite time singularity models.* An injectivity radius estimate (Corollary 6.62) implies that one can apply Hamilton's Cheeger–Gromov-type compactness theorem to obtain the existence of singularity models for singular solutions with finite singularity time. (Recall that the definition of a singularity model is given in Remark 1.29.)

THEOREM 6.68 (Existence of singularity models). *Let $g(t)$, $t \in [0, T)$, be a smooth solution to the Ricci flow on a closed manifold $\mathcal{M}^n$ with $T < \infty$. Suppose that there exists a sequence of times $t_i \nearrow T$, points $p_i \in \mathcal{M}$, and a constant $C < \infty$ such that*

(6.92) $$K_i \doteqdot |\operatorname{Rm}_{g(t_i)}(p_i)| \to \infty,$$

(6.93) $$|\operatorname{Rm}_{g(t)}(x)| \leq C K_i \quad \text{for all } x \in \mathcal{M} \text{ and } t < t_i.$$

Then there exists a subsequence of the sequence of dilated solutions[18]

$$g_i(t) \doteqdot K_i \cdot g\left(t_i + \frac{t}{K_i}\right)$$

such that $(\mathcal{M}^n, g_i(t), p_i)$ converges to a complete ancient solution to the Ricci flow $(\mathcal{M}^n_\infty, g_\infty(t), p_\infty)$ in the sense of C^∞-Cheeger–Gromov convergence. Furthermore there exists $\kappa > 0$ such that $g_\infty(t)$ is κ-noncollapsed on all scales.

PROOF. By Perelman's no local collapsing theorem, we have an injectivity radius estimate at the points p_k with respect to the metrics $g_k(0)$. Hence by Hamilton's compactness theorem (Theorem 3.10), there exists a subsequence such that $(\mathcal{M}^n, g_i(t), p_i)$ converges to a complete ancient solution $(\mathcal{M}^n_\infty, g_\infty(t), p_\infty)$ to the Ricci flow. Since $g(t)$ is κ-noncollapsed on the scale $\sqrt{T}$ for all $t \in [0, T)$, we have $g_i(t)$ is κ-noncollapsed on the scale

---

[18]As usual we denote a subsequence of i still by i rather than i_j to simplify our notation.

$\sqrt{K_i T}$ for all $t \in [-K_i t_i, K_i (T - t_i))$. Since $\lim_{i \to \infty} \sqrt{K_i T} = \infty$, $g_\infty(t)$ is κ-noncollapsed on all scales from Lemma 6.48 (regarding a limit property of sequences of κ-noncollapsed solutions). $\square$

5.3.2. *Ruling out the cigar as a finite time singularity model.* Finally, we observe that Perelman's no local collapsing theorem implies that the cigar (product with any flat solution like $\mathbb{R}^{n-2}$ or a torus T^{n-2}) cannot be a limit of dilations about a finite time singularity as in Theorem 6.68. This is because the cigar (product with any flat solution) is not κ-noncollapsed on all scales for any $\kappa > 0$. An easy way to see this is that the cylinder $\mathcal{S}^1 \times \mathbb{R}$ is a limit of the cigar. Clearly, $\mathcal{S}^1 \times \mathbb{R}$ (product with any flat solution) is not κ-noncollapsed on all scales for any $\kappa > 0$. By the property of κ-noncollapsed being preserved under limits, this implies the same for the cigar (product with any flat solution).

6. Improved version of no local collapsing and diameter control

In this section we give a proof of Perelman's improvement of his no local collapsing theorem to the case where one assumes, in the ball to be shown to be noncollapsed, only the *scalar curvature* has an upper bound. We also present the work of Topping [**357**] on diameter control. We end this section with a variation on the proof of Perelman's no local collapsing theorem.

6.1. Improved version of no local collapsing. We first revisit and revise Proposition 6.64. Let $\left(\hat{\mathcal{M}}^n, \hat{g}\right)$ be a closed Riemannian manifold and $r > 0$. Again we shall consider the inequality (6.85) for $\mu(g, r^2)$ and the test function w defined by (6.86). It is easy to see that the proof of Lemma 6.65 yields the following, where the estimate now involves $\operatorname{Vol} B(p, r/2)$, which we had previously estimated in terms of $\operatorname{Vol} B(p, r)$ under a local Ricci curvature lower bound assumption.

LEMMA 6.69 (c and the volume ratio). *The constant c in (6.86) satisfies the following bounds:*

$$(6.94) \qquad \frac{1}{\operatorname{Vol} B(p, r)} \leq \left(4\pi r^2\right)^{-n/2} e^{-c} \leq \frac{1}{\operatorname{Vol} B(p, r/2)}.$$

Equivalently,

$$(6.95) \quad -\frac{n}{2} \log(4\pi) + \log \frac{\operatorname{Vol} B(p, r/2)}{r^n} \leq c \leq -\frac{n}{2} \log(4\pi) + \log \frac{\operatorname{Vol} B(p, r)}{r^n}.$$

Using the above lemma, we obtain the following (the proof is similar to the proof of Proposition 6.64).

PROPOSITION 6.70 (Bounding μ by the scalar curvature and volume ratio). *The μ-invariant has the following upper bound in terms of local geometric quantities. For any closed Riemannian manifold $\left(\hat{\mathcal{M}}^n, \hat{g}\right)$, point*

$p \in \hat{\mathcal{M}}$, and $r > 0$, we have

$$(6.96) \quad \mu\left(\hat{g}, r^2\right) \leq \log \frac{\operatorname{Vol} B(p,r)}{r^n} + \left(36 + \frac{r^2 \int_{B(p,r)} R_+ d\mu}{\operatorname{Vol} B(p,r)}\right) \frac{\operatorname{Vol} B(p,r)}{\operatorname{Vol} B(p,r/2)},$$

where $R_+ \doteq \max\{R, 0\}$ is the positive part of the scalar curvature.

REMARK 6.71. Proposition 6.64 follows from the above statement and the Bishop–Gromov relative volume comparison theorem.

PROOF. We estimate each of the three terms on the RHS of (6.85) separately, where w is chosen by (6.86). Since $|\phi'| \leq 3$, $|\nabla d| = 1$ a.e., and $\operatorname{supp}(\phi') \subset B(p,r)$, the first term is estimated by

$$4r^2 \int_{\hat{\mathcal{M}}} |\nabla w|^2 d\mu \leq 4 \int_{\hat{\mathcal{M}}} (4\pi r^2)^{-n/2} e^{-c} \left|\phi'\left(\frac{d}{r}\right)\right|^2 d\mu$$

$$\leq 36 \frac{\operatorname{Vol} B(p,r)}{\operatorname{Vol} B(p,r/2)},$$

where we used (6.94) to obtain the last inequality. We also have

$$r^2 \int_{\hat{\mathcal{M}}} Rw^2 d\mu = r^2 \int_{\hat{\mathcal{M}}} R (4\pi r^2)^{-n/2} e^{-c} \phi\left(\frac{d}{r}\right)^2 d\mu$$

$$\leq \frac{r^2}{\operatorname{Vol} B(p,r/2)} \int_{B(p,r)} R_+ d\mu,$$

where we used the fact that $\phi \leq 1$ has support in $B(p,r)$ and (6.94). Finally, by Jensen's inequality (compare with (6.47)),

$$\int_{\hat{\mathcal{M}}} \log(w^2) w^2 d\mu \geq -\log \operatorname{Vol} B(p,r)$$

since $\int_{\hat{\mathcal{M}}} w^2 d\mu = 1$ and $\operatorname{supp}(w^2) \subset B(p,r)$. Hence the third term has the upper bound:

$$\int_{\hat{\mathcal{M}}} fw^2 d\mu = \int_{\hat{\mathcal{M}}} \left(-\frac{n}{2}\log(4\pi r^2) - \log(w^2)\right) w^2 d\mu \leq \log \frac{\operatorname{Vol} B(p,r)}{r^n}.$$

Summing the above three inequalities yields the proposition. □

Motivated by the expression on the RHS of (6.96), we define the **maximal function** on a closed Riemannian manifold $\left(\hat{\mathcal{M}}^n, \hat{g}\right)$ by

$$M_R(p,r) \doteq \sup_{0 < s \leq r} \frac{s^2}{\operatorname{Vol} B(p,s)} \int_{B(p,s)} R_+ d\mu.$$

Since

$$\lim_{s \to 0} \frac{s^2}{\operatorname{Vol} B(p,s)} \int_{B(p,s)} R_+ d\mu = 0,$$

the quantity $M_R(p, r)$ is a well-defined finite number for $0 < r < \infty$. Clearly, if $r_1 \leq r_2$, then $M_R(p, r_1) \leq M_R(p, r_2)$. By (6.96), we have

$$(6.97) \quad \mu\left(\hat{g}, s^2\right) \leq \log \frac{\operatorname{Vol} B(p, s)}{s^n} + (36 + M_R(p, s)) \frac{\operatorname{Vol} B(p, s)}{\operatorname{Vol} B(p, s/2)}.$$

Interestingly, the factor $\frac{\operatorname{Vol} B(p,s)}{\operatorname{Vol} B(p,s/2)}$ on the RHS above does not prevent one from estimating volume ratios by μ and M_R only under a scalar curvature upper bound. We shall prove the following.

PROPOSITION 6.72 (Bounding volume ratios by ν_r and M_R). *If $\left(\hat{\mathcal{M}}^n, \hat{g}\right)$ is a closed Riemannian manifold and $0 < s \leq r$, then*

$$\frac{\operatorname{Vol} B(p, s)}{s^n} \geq e^{-3^n(36 + M_R(p,r))} e^{\nu_r(\hat{g})},$$

where

$$\nu_r(\hat{g}) \doteqdot \inf_{\tau \in (0, r^2]} \mu(\hat{g}, \tau) \geq \nu(\hat{g}).$$

In particular, if $R \leq c_1(n) r^{-2}$ in $B(p, r)$, then $M_R(p, r) \leq c_1(n)$, so that

$$\frac{\operatorname{Vol} B(p, s)}{s^n} \geq e^{-3^n(36 + c_1(n))} e^{\nu_r(\hat{g})}.$$

REMARK 6.73. Given $p \in \hat{\mathcal{M}}$, the function $r \mapsto e^{-3^n 36} e^{\nu_r(\hat{g})} e^{-3^n M_R(p,r)}$ is nonincreasing.

PROOF. If $\frac{\operatorname{Vol} B(p,s)}{\operatorname{Vol} B(p,s/2)} \leq 3^n$ (in this case we say that at p the **volume doubling property** holds at scale s), then (6.97) implies

$$\frac{\operatorname{Vol} B(p, s)}{s^n} \geq e^{-3^n 36} e^{\mu(\hat{g}, s^2)} e^{-3^n M_R(p,s)} \geq e^{-3^n 36} e^{\nu_r(\hat{g})} e^{-3^n M_R(p,r)}$$

and the estimate follows.

If $\frac{\operatorname{Vol} B(p,s)}{\operatorname{Vol} B(p,s/2)} \geq 3^n$, then since $\lim_{k \to \infty} \frac{\operatorname{Vol} B(p,s/2^k)}{\operatorname{Vol} B(p,s/2^{k+1})} = 2^n$, there exists $k \in \mathbb{N}$ such that $\frac{\operatorname{Vol} B(p,s/2^k)}{\operatorname{Vol} B(p,s/2^{k+1})} \leq 3^n$ and $\frac{\operatorname{Vol} B(p,s/2^i)}{\operatorname{Vol} B(p,s/2^{i+1})} > 3^n$ for all $0 \leq i < k$. Applying (6.97) to $B\left(p, s/2^k\right)$, we get

$$\mu\left(\hat{g}, \left(s/2^k\right)^2\right) \leq \log \frac{\operatorname{Vol} B\left(p, s/2^k\right)}{(s/2^k)^n} + \left(36 + M_R\left(p, s/2^k\right)\right) \cdot 3^n.$$

Hence
$$\frac{\operatorname{Vol} B(p,s)}{s^n} \geq \left(\frac{3}{2}\right)^n \frac{\operatorname{Vol} B(p,s/2)}{(s/2)^n}$$
$$\geq \left(\frac{3}{2}\right)^{nk} \frac{\operatorname{Vol} B(p,s/2^k)}{(s/2^k)^n}$$
$$\geq \left(\frac{3}{2}\right)^{nk} e^{-3^n 36} e^{\mu(\hat{g}, s^2/2^{2k})} e^{-3^n M_R(p,s/2^k)}$$
$$\geq \left(\frac{3}{2}\right)^{nk} e^{-3^n 36} e^{\nu_r(\hat{g})} e^{-3^n M_R(p,r)}.$$

The theorem again follows in this case. □

Now we apply Proposition 6.72 to solutions of the Ricci flow and obtain the following improvement of the no local collapsing theorem.

THEOREM 6.74 (No local collapsing theorem improved). *Let* $(\mathcal{M}^n, g(t))$, $t \in [0,T)$, *be a solution to the Ricci flow on a closed manifold with* $T < \infty$ *and let* $\rho \in (0, \infty)$. *There exists a constant* $\kappa = \kappa(n, g(0), T, \rho) > 0$ *such that if* $p \in \mathcal{M}$, $t \in [0,T)$, *and* $r \in (0, \rho]$ *are such that*
$$R \leq r^{-2} \text{ in } B_{g(t)}(p, r),$$
then
$$\frac{\operatorname{Vol}_{g(t)} B_{g(t)}(p, s)}{s^n} \geq \kappa$$
for all $0 < s \leq r$.

PROOF. By (6.61) and the definition of ν_r, we have
$$\nu_r(g(t)) \geq \nu_{\sqrt{\rho^2+T}}(g(0))$$
for $r \in (0, \rho]$ and $t \in [0, T)$. Then the theorem follows from Proposition 6.72. □

6.2. Diameter control. In this subsection we show how ideas related to the previous subsection can be used to obtain a diameter bound for solutions of the Ricci flow in terms of the $L^{(n-1)/2}$-norm of the scalar curvature. This result is due to Topping [357] and our presentation essentially follows his ideas. Recall that Proposition 6.72 implies that if $\left(\hat{\mathcal{M}}^n, \hat{g}\right)$ is a closed Riemannian manifold, then for any $p \in \hat{\mathcal{M}}^n$ and $0 < s < \infty$, we have

(6.98) $$\frac{\operatorname{Vol} B(p,s)}{s^n} \geq e^{-3^n 36} e^{\nu(\hat{g})} e^{-3^n M_R(p,r)},$$

using $\nu(\hat{g}) \doteqdot \inf_{\tau \in [0,\infty)} \mu(\hat{g}, \tau) \leq \nu_r(\hat{g})$ for all r. Recall that by Corollary 6.34, if the lowest eigenvalue is positive, i.e., $\lambda(\hat{g}) \doteqdot \lambda_1(-4\Delta_{\hat{g}} + R_{\hat{g}}) > 0$, then $\nu(\hat{g}) > -\infty$.

THEOREM 6.75 (Topping). *Let $n \geq 3$ and let $\left(\hat{\mathcal{M}}^n, \hat{g}\right)$ be a closed Riemannian manifold with $\nu\left(\hat{g}\right) > -\infty$. Then*

$$\operatorname{diam}\left(\hat{\mathcal{M}}, \hat{g}\right) \leq \max\left\{\frac{12}{\omega_n}, 6e^{3^n 37} e^{-\nu(\hat{g})}\right\} \int_{\hat{\mathcal{M}}} R_+^{\frac{n-1}{2}} d\mu,$$

where ω_n is the volume of the unit n-ball in $\mathbb{R}^n$.

PROOF. Let $\delta\left(\hat{g}\right) \doteq \min\left\{\frac{\omega_n}{2}, \kappa\left(\hat{g}\right) e^{-3^n}\right\}$, where $\kappa\left(\hat{g}\right) \doteq e^{-3^n 36} e^{\nu(\hat{g})}$. Note that $\lim_{s \to 0} \frac{\operatorname{Vol} B(p,s)}{s^n} = \omega_n$ and $\lim_{s \to \infty} \frac{\operatorname{Vol} B(p,s)}{s^n} = 0$ since $\hat{\mathcal{M}}$ is closed. Hence for any point $p \in \hat{\mathcal{M}}$, there exists $s(p) > 0$ such that $\frac{\operatorname{Vol} B(p, s(p))}{s(p)^n} = \delta\left(\hat{g}\right)$ and $\frac{\operatorname{Vol} B(p,s)}{s^n} \geq \delta\left(\hat{g}\right)$ for $s \in (0, s(p)]$.

Applying inequality (6.98), we have

(6.99) $$M_R(p, s(p)) \geq 1.$$

This implies there exists $s'(p) \leq s(p)$ such that

$$\frac{(s'(p))^2}{\operatorname{Vol} B(p, s'(p))} \int_{B(p, s'(p))} R_+ \, d\mu \geq 1.$$

Applying the Hölder inequality, we have

$$\frac{\operatorname{Vol} B(p, s'(p))}{(s'(p))^2} \leq \int_{B(p, s'(p))} R_+ \, d\mu$$

$$\leq \left(\int_{B(p, s'(p))} R_+^{\frac{n-1}{2}} \, d\mu\right)^{\frac{2}{n-1}} \left[\operatorname{Vol} B(p, s'(p))\right]^{\frac{n-3}{n-1}},$$

so that

$$\frac{\operatorname{Vol} B(p, s'(p))}{(s'(p))^{n-1}} \leq \int_{B(p, s'(p))} R_+^{\frac{n-1}{2}} \, d\mu.$$

We have proved that for every $p \in \mathcal{M}$, there exists $s'(p) > 0$ such that

$$\delta\left(\hat{g}\right) s'(p) \leq \frac{\operatorname{Vol} B(p, s'(p))}{(s'(p))^n} \cdot s'(p)$$

$$= \frac{\operatorname{Vol} B(p, s'(p))}{(s'(p))^{n-1}} \leq \int_{B(p, s'(p))} R_+^{\frac{n-1}{2}} \, d\mu,$$

where the first inequality follows from the definition of $s(p)$ and $s'(p) \leq s(p)$.

To finish the proof of the theorem, let γ be a minimal geodesic whose length is the diameter of $\left(\hat{\mathcal{M}}, \hat{g}\right)$. One can show that there exists a countable (possibly finite) number of points $p_i \in \gamma$ such that $B\left(p_i, s'(p_i)\right)$ are disjoint and cover at least $1/3$ of γ (Vitali covering-type theorem). Then

$$\frac{1}{3} \operatorname{diam}\left(\hat{\mathcal{M}}, \hat{g}\right) \leq \sum_i 2 s'(p_i) \leq \frac{2}{\delta\left(\hat{g}\right)} \sum_i \int_{B(p_i, s'(p_i))} R_+^{\frac{n-1}{2}} \, d\mu$$

$$\leq \frac{2}{\delta\left(\hat{g}\right)} \int_{\mathcal{M}} R_+^{\frac{n-1}{2}} \, d\mu.$$

The theorem follows from plugging in the definition of $\delta(\hat{g})$. $\square$

Now we can apply Theorem 6.75 to the Ricci flow and obtain the following.

COROLLARY 6.76. *Let $n \geq 3$ and let $(\mathcal{M}^n, g(t))$, $t \in [0, T)$, be a solution of the Ricci flow on a closed manifold with $T < \infty$. Assume that $\lambda(g(0)) > 0$. Then there exists $C = C(n, g(0)) > 0$ such that*

$$\operatorname{diam}(\mathcal{M}, g(t)) \leq C \int_{\mathcal{M}} R_+(t)^{\frac{n-1}{2}} d\mu_{g(t)}. \tag{6.100}$$

PROOF. Note that by the monotonicity of the λ-invariant we have

$$\lambda(g(t)) \geq \lambda(g(0)) > 0,$$

and hence the theorem is applicable. Now the corollary follows from $\nu(g(t)) \geq \nu(g(0))$. $\square$

6.3. A variation on the proof of no local collapsing. In this subsection we give a modified proof that the no local collapsing theorem follows from entropy monotonicity using a *local* eigenvalue estimate. We also give a heat equation proof of a less sharp form of the *global* version of this eigenvalue estimate.

6.3.1. *Modified proof of no local collapsing theorem.* Recall Cheng's sharp upper bound for the first eigenvalue λ_1 of the Laplacian $-\Delta$ on balls with a lower bound on the Ricci curvature [**92**].

THEOREM 6.77 (Cheng, local eigenvalue comparison). *Let $(\hat{\mathcal{M}}^n, \hat{g})$ be a complete Riemannian manifold with $\operatorname{Rc}(\hat{g}) \geq -(n-1)\hat{g}$. Then for any $p \in \hat{\mathcal{M}}$,*

$$\lambda_1(B(p, 1)) \leq \lambda_1(B_{\mathbb{H}^n}(1)), \tag{6.101}$$

where $B_{\mathbb{H}^n}(1)$ is the open ball of radius 1 in hyperbolic space $\mathbb{H}^n$ of sectional curvature -1. Here λ_1 denotes the first eigenvalue of the Laplacian with the Dirichlet boundary condition.

EXERCISE 6.78. Suppose $(\hat{\mathcal{M}}^n, \hat{g})$ is a complete Riemannian manifold with $\operatorname{Rc}(\hat{g}) \geq (n-1)K\hat{g}$, where $K \leq 0$. Given $r > 0$, determine an upper bound for $\lambda_1(B(p, r))$ in terms of the corresponding model space.

We now give the modified proof of no local collapsing using the above eigenvalue estimate and Jensen's inequality. Given the monotonicity of μ, the first proof we presented relies on inequality (6.83) giving an upper bound for μ in terms of the volume ratio; it is this inequality for which we give a second proof. Recall from (6.85) that we have for a Riemannian manifold

$\left(\hat{\mathcal{M}}^n, \hat{g}\right)$

$$\mu(\hat{g}, r^2) \leq \int_{\hat{\mathcal{M}}} r^2(4|\nabla w|^2 + Rw^2)\, d\mu$$
$$- \int_{\hat{\mathcal{M}}} \left(\log\left(w^2\right) + \frac{n}{2}\log(4\pi r^2) + n\right) w^2 \, d\mu$$

for all w with $\int_{\hat{\mathcal{M}}} w^2 d\mu = 1$. Using (6.47), we have for any w with $\mathrm{supp}(w) \subset B(p, r)$ and $\int_{\hat{\mathcal{M}}} w^2 d\mu = 1$ that

$$- \int_{\hat{\mathcal{M}}} \log(w^2) w^2 \, d\mu \leq \log \mathrm{Vol}\, B(p, r).$$

By assumptions $\mathrm{Rc}(\hat{g}) \geq -(n-1)r^{-2}$ and $R \leq n(n-1)r^{-2}$ in $B(p, r)$,[19] we have for any w with $\mathrm{supp}(w) \subset B(p, r)$ and $\int_{\hat{\mathcal{M}}} w^2 d\mu = 1$,

$$\int_{\hat{\mathcal{M}}} Rw^2 \, d\mu \leq n(n-1).$$

Let $\bar{g} = r^{-2}\hat{g}$; then $B_{\bar{g}}(p, 1) = B(p, r) \doteqdot B_{\hat{g}}(p, r)$. By Theorem 6.77 and the Rayleigh principle for eigenvalues,

$$\inf_{\mathrm{supp}(\bar{w}) \subset B_{\bar{g}}(p,1)} \frac{\int_{\hat{\mathcal{M}}} |\nabla \bar{w}|^2 \, d\mu_{\bar{g}}}{\int_{\hat{\mathcal{M}}} \bar{w}^2 \, d\mu_{\bar{g}}} = \lambda_1\left(B_{\bar{g}}(p, 1)\right) \leq \lambda_1(B_{\mathbb{H}^n}(1)).$$

Hence we have

$$\inf_{\substack{\int_{\hat{\mathcal{M}}} w^2 d\mu = 1, \\ \mathrm{supp}(w) \subset B(p,r)}} \int_{\hat{\mathcal{M}}} r^2 4|\nabla w|^2 \, d\mu \leq 4\lambda_1(B_{\mathbb{H}^n}(1)).$$

Therefore

$$\log \frac{\mathrm{Vol}\, B(p, r)}{r^n} \geq \mu(\hat{g}, r^2) - 4\lambda_1(B_{\mathbb{H}^n}(1)) - n(n-2) + \frac{n}{2} \log(4\pi).$$

This provides the needed estimate to replace (6.83).

6.3.2. A heat equation proof of a global eigenvalue estimate. We now recall a global version of Cheng's Theorem 6.77.

THEOREM 6.79 (Cheng's eigenvalue estimate, global). *If $\left(\hat{\mathcal{M}}^n, \hat{g}\right)$ is a complete noncompact Riemannian manifold with $\mathrm{Rc}(\hat{g}) \geq -(n-1)\hat{g}$, then*

$$\lambda_1(-\Delta) \leq \frac{(n-1)^2}{4}.$$

It turns out that a weaker version of this estimate, i.e., $\lambda_1 \leq \frac{n(n-1)}{4}$, can be proved using the energy/entropy computation of Perelman for the fixed metric case; we give this proof below. First we state a formula which is implicit in [**283**].

---

[19]We choose these constants for our curvature bounds since they are implied by $-r^{-2} \leq \mathrm{sect} \leq r^{-2}$.

LEMMA 6.80. *Let u be a positive solution to the heat equation*

$$\left(\frac{\partial}{\partial t} - \Delta\right) u = 0$$

on a fixed Riemannian manifold $(\hat{\mathcal{M}}^n, \hat{g})$. If $f = -\log u$, that is, $u = e^{-f}$, then

(6.102) $$\frac{d}{dt}\int_{\hat{\mathcal{M}}} |\nabla f|^2 u \, d\mu = -2\int_{\hat{\mathcal{M}}} (|\nabla_i \nabla_j f|^2 + R_{ij}\nabla_i f \nabla_j f) u \, d\mu.$$

PROOF. (1) Using $\frac{\partial}{\partial t} f = \Delta f - |\nabla f|^2$, we calculate

$$\frac{d}{dt}\int_{\hat{\mathcal{M}}} |\nabla f|^2 u \, d\mu = \frac{d}{dt}\int_{\hat{\mathcal{M}}} (\Delta f) u \, d\mu$$

$$= \int_{\hat{\mathcal{M}}} \left(2\Delta f - |\nabla f|^2\right) \Delta u \, d\mu$$

$$= 2\int_{\hat{\mathcal{M}}} (\Delta f \Delta u + \nabla_i \nabla_j f \nabla_i f \nabla_j u) \, d\mu.$$

Now integrating by parts yields

$$\int_{\hat{\mathcal{M}}} \Delta f \Delta u \, d\mu = -\int_{\hat{\mathcal{M}}} \nabla f \cdot \nabla \Delta u \, d\mu$$

$$= -\int_{\hat{\mathcal{M}}} (\nabla f \cdot \Delta \nabla u - R_{ij}\nabla_i f \nabla_j u) \, d\mu$$

$$= \int_{\hat{\mathcal{M}}} (\nabla \nabla f \cdot \nabla \nabla u - u R_{ij}\nabla_i f \nabla_j f) \, d\mu$$

$$= \int_{\hat{\mathcal{M}}} \left(-u|\nabla_i \nabla_j f|^2 - \nabla_i \nabla_j f \nabla_i f \nabla_j u - u R_{ij}\nabla_i f \nabla_j f\right) d\mu.$$

The lemma immediately follows from combining the above two formulas.

(2) Alternatively, one can integrate the following formula to get (6.102) (see (2.1) and Lemma 2.1 in [**283**]):

$$\left(\frac{\partial}{\partial t} - \Delta\right)\left(u\left(2\Delta f - |\nabla f|^2\right)\right) = -2u|\nabla_i \nabla_j f|^2 - 2u R_{ij}\nabla_i f \nabla_j f,$$

which follows directly from the calculation:

$$\frac{\partial}{\partial t}\left(2\Delta f - |\nabla f|^2\right) = 2\Delta\left(\Delta f - |\nabla f|^2\right) - 2\nabla f \cdot \nabla\left(\Delta f - |\nabla f|^2\right)$$

$$= \Delta\left(2\Delta f - |\nabla f|^2\right) - 2\nabla f \cdot \nabla\left(2\Delta f - |\nabla f|^2\right)$$

$$\quad - \Delta |\nabla f|^2 + 2\nabla f \cdot \nabla \Delta f$$

$$= \Delta\left(2\Delta f - |\nabla f|^2\right) - 2\nabla f \cdot \nabla\left(2\Delta f - |\nabla f|^2\right)$$

$$\quad - 2|\nabla_i \nabla_j f|^2 - 2R_{ij}\nabla_i f \nabla_j f.$$

□

Now we can prove the following weaker version of Cheng's Theorem 6.79.

PROPOSITION 6.81 (Weaker version using the heat equation). If $\left(\hat{\mathcal{M}}^n, \hat{g}\right)$ is a complete noncompact Riemannian manifold with $\operatorname{Rc}(\hat{g}) \geq -(n-1)\hat{g}$, then $\lambda_1 \leq \frac{n(n-1)}{4}$.

SKETCH OF PROOF. Assume that ϕ is a normalized first eigenfunction of $-\Delta$; namely, $-\Delta\phi = \lambda_1\phi$ and $\int_{\hat{\mathcal{M}}} \phi^2 = 1$. It is well known that $\phi > 0$ (see [**67**] or [**117**] for example). Now we let $u : \hat{\mathcal{M}} \times [0, \infty) \to \mathbb{R}$ be the solution of the heat equation

$$\left(\frac{\partial}{\partial t} - \Delta\right) u = 0,$$
$$u(0) = \phi^2,$$

and let f be defined as before by $e^{-f} = u$. Since $e^{-f(0)/2} = \phi$ and ϕ is the first eigenfunction, at $t = 0$ we have

$$\lambda_1 = -\frac{\Delta\phi}{\phi} = \frac{1}{4}\left(2\Delta f - |\nabla f|^2\right)$$

and equivalently

(6.103) $$\Delta f = 2\lambda_1 + \frac{1}{2}|\nabla f|^2.$$

Applying the above lemma, we have at $t = 0$,

$$-2\int_{\hat{\mathcal{M}}} (|\nabla_i \nabla_j f|^2 + R_{ij}\nabla_i f \nabla_j f) u \, d\mu$$
$$= \frac{d}{dt}\int_{\hat{\mathcal{M}}} |\nabla f|^2 u \, d\mu$$
$$= \int_{\hat{\mathcal{M}}} \left(2\Delta f - |\nabla f|^2\right) \Delta u \, d\mu$$
$$= \int_{\hat{\mathcal{M}}} 4\lambda_1 \Delta u \, d\mu = 0.$$

Since $\operatorname{Rc} \geq -(n-1)\hat{g}$ and $|\nabla_i \nabla_j f|^2 \geq \frac{1}{n}(\Delta f)^2$, we then have

$$0 \geq \int_{\hat{\mathcal{M}}} \left(\frac{1}{n}(\Delta f)^2 - (n-1)|\nabla f|^2\right) u \, d\mu$$
$$= \int_{\hat{\mathcal{M}}} \left(\frac{1}{n}\left(4\lambda_1^2 + 2\lambda_1|\nabla f|^2 + \frac{1}{4}|\nabla f|^4\right) - (n-1)|\nabla f|^2\right) u \, d\mu.$$

Noting that at $t = 0$,

$$4\lambda_1 = 4\int_{\hat{\mathcal{M}}} |\nabla\phi|^2 \, d\mu = \int_{\hat{\mathcal{M}}} |\nabla f|^2 u \, d\mu,$$

we obtain

(6.104) $$0 \geq \frac{12}{n}\lambda_1^2 - 4(n-1)\lambda_1 + \frac{1}{4n}\int_{\hat{\mathcal{M}}} |\nabla f|^4 u \, d\mu.$$

Applying the Hölder inequality,

$$\int_{\hat{\mathcal{M}}} |\nabla f|^4 u \, d\mu \geq \left(\int_{\hat{\mathcal{M}}} |\nabla f|^2 u \, d\mu \right)^2 = 16\lambda_1^2,$$

which we substitute into (6.104) to conclude

$$\lambda_1 \leq \frac{n(n-1)}{4}.$$

□

7. Some further calculations related to $\mathcal{F}$ and $\mathcal{W}$

In this section we discuss some interesting computations related to energy and entropy including variational formulas for the modified scalar curvature, the second variation of energy and entropy, and a matrix Harnack calculation for the adjoint heat equation.

7.1. Variational structure of the modified scalar curvature.

7.1.1. *Variation of the modified scalar curvature.* We now give yet another proof of the variation formula for the $\mathcal{F}$ functional (5.10) when $\delta f \doteqdot h = \frac{V}{2}$ using the pointwise formula for the variation of the modified scalar curvature. The formulas for $V_0 \doteqdot R + 2\Delta f - |\nabla f|^2$ and $\mathcal{F}$ below should generalize to V_ε and $\mathcal{W}_\varepsilon$ (see (6.25) and (6.26) for the definitions of V_ε and $\mathcal{W}_\varepsilon$).

LEMMA 6.82 (Measure-preserving variations of $R + 2\Delta f - |\nabla f|^2$ and the linear trace Harnack). *If $\delta g = v$ and $\delta f = \frac{V}{2}$, where $V \doteqdot g^{ij} v_{ij}$, on a manifold $\mathcal{M}^n$, then*

(6.105)
$$\delta_{(v, \frac{V}{2})} \left(R + 2\Delta f - |\nabla f|^2 \right)$$
$$= \nabla^i \nabla^j v_{ij} + v_{ij} R_{ij} - 2\nabla_i v_{ik} \nabla_k f + v_{ij} \nabla_i f \nabla_j f - 2v_{ij}(R_{ij} + \nabla_i \nabla_j f).$$

For the proof see the more general Lemma 6.85 below.

Since under our assumptions $\delta \left(e^{-f} d\mu \right) = 0$, we have, using (6.105) and the identity (6.39) with $\varepsilon = 0$, that

$$\frac{\partial \mathcal{F}}{\partial s} = \int_{\mathcal{M}} \left(\delta_{(v, \frac{V}{2})} \left(R + 2\Delta f - |\nabla f|^2 \right) \right) e^{-f} d\mu$$
$$= - \int_{\mathcal{M}} v_{ij} \left(R_{ij} + \nabla_i \nabla_j f \right) e^{-f} d\mu.$$

This is the special case of (5.10) where $h = \frac{V}{2}$.

REMARK 6.83. If $(\mathcal{M}^n, g(t))$, $t \in (-\infty, \infty)$, is a gradient Ricci soliton flowing along ∇f so that $R_{ij} + \nabla_i \nabla_j f = 0$ and if

$$\delta g(t) \doteqdot v(t) \geq 0, \quad t \in (-\infty, \infty),$$

is a bounded nonnegative solution to the Lichnerowicz Laplacian heat equation $\frac{\partial}{\partial t}v = \Delta_L v$, where Δ_L is defined by (VI-3.6), then under measure-preserving variations $(v(t), h(t))$, where $\delta_{(v(t),h(t))}\left(e^{-f(t)}d\mu_{g(t)}\right) = 0$, we have

$$\delta_{(v(t),h(t))}\left(R + 2\Delta f - |\nabla f|^2\right)(t) \geq 0.$$

This follows from the linear trace Harnack estimate for ancient solutions.

EXERCISE 6.84. Show that extending (6.105) to nonmeasure-preserving variations, we have the following. If $\delta g = v$ and $\delta f = h$, then

(6.106)
$$\begin{aligned}\delta_{(v,h)}&\left(R + 2\Delta f - |\nabla f|^2\right) \\ &= \operatorname{div}(\operatorname{div} v) + \langle v, \operatorname{Rc}\rangle - 2(\operatorname{div} v) \cdot \nabla f + v_{ij}\nabla_i f \nabla_j f \\ &\quad + 2(\Delta - \nabla f \cdot \nabla)\left(h - \frac{V}{2}\right) - 2v_{ij}(R_{ij} + \nabla_i \nabla_j f).\end{aligned}$$

Note that (6.105) generalizes to the following, where ∇f is replaced by a vector field X.

LEMMA 6.85. If $\delta g = v$ and $\delta X = \frac{\nabla V}{2}$, then

$$\begin{aligned}\delta_{(v,\frac{\nabla V}{2})}\left(R + 2g^{ij}\nabla_i X_j - |X|^2\right) &= \nabla^i\nabla^j v_{ij} + v_{ij}R_{ij} - 2\nabla_i v_{ik}X_k + v_{ij}X_i X_j \\ &\quad - v_{ij}(2R_{ij} + \nabla_i X_j + \nabla_j X_i).\end{aligned}$$

PROOF. This follows from (VI-p. 69d):

$$\delta_v R = -\Delta v + \nabla^i\nabla^j v_{ij} - v_{ij}R_{ij},$$

$$\delta_{(v,\frac{\nabla V}{2})}\left(-|X|^2\right) = v_{ij}X_i X_j - X \cdot \nabla v,$$

and

$$\begin{aligned}\delta_{(v,\frac{\nabla V}{2})}\left(2g^{ij}\nabla_i X_j\right) &= 2\nabla_i\left(\frac{\partial}{\partial s}X_i\right) - 2\left(\frac{\partial}{\partial s}g_{ij}\right)\nabla_i X_j - 2g^{ij}\left(\frac{\partial}{\partial s}\Gamma^k_{ij}\right)X_k \\ &= \Delta v - 2v_{ij}\nabla_i X_j - 2\nabla_i v_{ik}X_k + \nabla_k v X_k.\end{aligned}$$

□

EXERCISE 6.86 (Generalizing Exercise 6.84). Show that if $\delta g = v$ and $\delta X = Y$, then Lemma 6.85 generalizes to

$$\begin{aligned}\delta_{(v,Y)}\left(R + 2g^{ij}\nabla_i X_j - |X|^2\right) &= \operatorname{div}(\operatorname{div} v) + v \cdot \operatorname{Rc} - 2\operatorname{div}(v) \cdot X + v(X, X) \\ &\quad - v_{ij}(2R_{ij} + \nabla_i X_j + \nabla_j X_i) \\ &\quad + 2\operatorname{div}\left(Y - \frac{1}{2}\nabla V\right) - 2X \cdot \left(Y - \frac{1}{2}\nabla V\right).\end{aligned}$$

A different calculation yields

LEMMA 6.87. *Under the equations $\frac{\partial}{\partial t} g_{ij} = -2R_{ij}$ and*

$$\frac{\partial}{\partial t} X_i = \Delta_d X_i - \nabla_i R + 2 \left(\nabla_X X\right)_i,$$

we have

$$\left(\frac{\partial}{\partial t} + \Delta - 2X \cdot \nabla\right) \left(R + 2g^{ij} \nabla_i X_j - |X|^2\right)$$
$$= 2 \left| R_{ij} + \frac{1}{2} \left(\nabla_i X_j + \nabla_j X_i\right) \right|^2 - \frac{3}{2} |\nabla_i X_j - \nabla_j X_i|^2.$$

This extends the calculation (6.22) of Perelman.

7.1.2. *An extension of the monotonicity formula.* A generalization of Perelman's energy monotonicity formula is given by the following.

LEMMA 6.88. *If $\frac{\partial}{\partial t} g_{ij} = -2R_{ij}$ and*

$$\frac{\partial f}{\partial t} = \Delta f + \alpha \left(R + 2\Delta f - |\nabla f|^2\right),$$

for some $\alpha \in \mathbb{R}$, then

$$\left(\frac{\partial}{\partial t} - (2\alpha + 1) \Delta + 2\alpha \nabla f \cdot \nabla\right) \left(R + 2\Delta f - |\nabla f|^2\right) = 2 \left|R_{ij} + \nabla_i \nabla_j f\right|^2.$$

PROOF. We have by direct calculation,

$$\frac{\partial}{\partial t} |\nabla f|^2 = 2R_{ij} \nabla_i f \nabla_j f + 2\nabla f \cdot \nabla \left[\Delta f + \alpha \left(R + 2\Delta f - |\nabla f|^2\right)\right]$$
$$= (2\alpha + 1) \Delta |\nabla f|^2 - 2(2\alpha + 1) |\nabla_i \nabla_j f|^2 - 4\alpha R_{ij} \nabla_i f \nabla_j f$$
$$+ 2\nabla f \cdot \nabla \left[\alpha \left(R - |\nabla f|^2\right)\right]$$

and

$$\frac{\partial}{\partial t} (\Delta f) = 2R_{ij} \nabla_i \nabla_j f + \Delta \left(\Delta f + \alpha \left(R + 2\Delta f - |\nabla f|^2\right)\right)$$
$$= (2\alpha + 1) \Delta (\Delta f) + 2R_{ij} \nabla_i \nabla_j f + \Delta \left(\alpha \left(R - |\nabla f|^2\right)\right).$$

Hence

$$\left(\frac{\partial}{\partial t} - (2\alpha + 1) \Delta\right) \left(R + 2\Delta f - |\nabla f|^2\right)$$
$$= -2\alpha \Delta R + 2 |R_{ij}|^2$$
$$+ 4 R_{ij} \nabla_i \nabla_j f + 2\Delta \left(\alpha \left(R - |\nabla f|^2\right)\right)$$
$$+ 2 (2\alpha + 1) |\nabla_i \nabla_j f|^2 + 4\alpha R_{ij} \nabla_i f \nabla_j f - 2\nabla f \cdot \nabla \left[\alpha \left(R - |\nabla f|^2\right)\right]$$
$$= 2 |R_{ij}|^2 + 4 R_{ij} \nabla_i \nabla_j f + 2 |\nabla_i \nabla_j f|^2 - 2\alpha \nabla f \cdot \nabla \left(R + 2\Delta f - |\nabla f|^2\right)$$

and the lemma follows. □

From this we deduce the following.

COROLLARY 6.89. If $\frac{\partial}{\partial t} g_{ij} = -2R_{ij}$ and

$$\frac{\partial f}{\partial t} = \Delta f + \alpha \left(R + 2\Delta f - |\nabla f|^2 \right),$$

for some $\alpha \in \mathbb{R}$, then the energy on a closed manifold $\mathcal{M}^n$ satisfies

$$\frac{d\mathcal{F}(g(t), f(t))}{dt} = \frac{d}{dt} \int_{\mathcal{M}} \left(R + 2\Delta f - |\nabla f|^2 \right) e^{-f} d\mu$$

$$= 2 \int_{\mathcal{M}} |R_{ij} + \nabla_i \nabla_j f|^2 e^{-f} d\mu$$

$$- (1 + \alpha) \int_{\mathcal{M}} \left(R + 2\Delta f - |\nabla f|^2 \right)^2 e^{-f} d\mu.$$

So if $\alpha \leq -1$, then

$$\frac{d}{dt} \int_{\mathcal{M}} \left(R + 2\Delta f - |\nabla f|^2 \right) e^{-f} d\mu \geq 0.$$

EXERCISE 6.90. Put a positive constant ε in the above formulas; more precisely, consider the ε-entropy $\mathcal{W}_\varepsilon$ and determine equations for f and τ such that a monotonicity formula for $\mathcal{W}_\varepsilon$ holds. Perhaps one should consider the set of equations

$$\frac{\partial}{\partial t} g_{ij} = -2R_{ij}$$
$$\frac{\partial f}{\partial t} = \Delta f - \frac{\varepsilon}{\tau} \left(f - \frac{n}{2} \right) + \alpha \left(R + 2\Delta f - |\nabla f|^2 - \frac{\varepsilon}{\tau}(f - n) \right)$$
$$\frac{d\tau}{dt} = \varepsilon$$

for $\alpha \leq -1$.

7.2. Second variation of energy and entropy. As is typical for energy-type functionals, we consider the second variation of Perelman's energy and entropy functionals. We are particularly interested in what sense critical points of the entropy functional are stable, i.e., have nonnegative second variation.

Suppose

$$\frac{\partial g}{\partial s} = v \quad \text{and} \quad \frac{\partial f}{\partial s} = h$$

on a closed manifold $\mathcal{M}^n$. Equation (6.31) implies the following:

$$\frac{\partial}{\partial s} \left[(R_{ij} + \nabla_i \nabla_j f) e^{-f} d\mu \right]$$

(6.107)
$$= \nabla_p \left(e^{-f} \frac{\partial}{\partial s} \Gamma_{ij}^p \right) d\mu$$

$$+ \left[\nabla_i \nabla_j \left(h - \frac{V}{2} \right) - (R_{ij} + \nabla_i \nabla_j f) \left(h - \frac{V}{2} \right) \right] e^{-f} d\mu.$$

7. SOME FURTHER CALCULATIONS RELATED TO $\mathcal{F}$ AND $\mathcal{W}$

Differentiating (5.10) again and using (6.106) and (6.107), we have

$$\frac{d^2}{ds^2}\mathcal{F}(g,f)$$

$$= -\int_{\mathcal{M}} \left(\frac{\partial}{\partial s}v_{ij} - 2v_{ik}v_{jk}\right)(R_{ij} + \nabla_i\nabla_j f)e^{-f}d\mu$$

$$- \int_{\mathcal{M}} v_{ij}\nabla_p\left(e^{-f}\frac{\partial}{\partial s}\Gamma^p_{ij}\right)d\mu$$

$$- \int_{\mathcal{M}} v_{ij}\left[\nabla_i\nabla_j\left(h - \frac{V}{2}\right) - (R_{ij} + \nabla_i\nabla_j f)\left(h - \frac{V}{2}\right)\right]e^{-f}d\mu$$

$$+ \int_{\mathcal{M}} \left[\frac{\partial}{\partial s}\left(\frac{V}{2} - h\right)\right]\left(2\Delta f - |\nabla f|^2 + R\right)e^{-f}d\mu$$

$$+ \int_{\mathcal{M}} \left(\frac{V}{2} - h\right)(\operatorname{div}(\operatorname{div} v) + \langle v, \operatorname{Rc}\rangle - 2(\operatorname{div} v)\cdot\nabla f + v_{ij}\nabla_i f\nabla_j f)e^{-f}d\mu$$

$$+ \int_{\mathcal{M}} \left(\frac{V}{2} - h\right)\left(2(\Delta - \nabla f\cdot\nabla)\left(h - \frac{V}{2}\right) - 2v_{ij}(R_{ij} + \nabla_i\nabla_j f)\right)e^{-f}d\mu$$

$$+ \int_{\mathcal{M}} \left(\frac{V}{2} - h\right)^2\left(2\Delta f - |\nabla f|^2 + R\right)e^{-f}d\mu.$$

Integrating by parts, we have

$$\int_{\mathcal{M}} v_{ij}\left[\nabla_i\nabla_j\left(h - \frac{V}{2}\right) - (R_{ij} + \nabla_i\nabla_j f)\left(h - \frac{V}{2}\right)\right]e^{-f}d\mu$$

$$= \int_{\mathcal{M}} \left(h - \frac{V}{2}\right)\left[\nabla_i\nabla_j\left(v_{ij}e^{-f}\right) - v_{ij}(R_{ij} + \nabla_i\nabla_j f)e^{-f}\right]d\mu$$

(6.108) $$= \int_{\mathcal{M}} \left(h - \frac{V}{2}\right)\binom{\operatorname{div}(\operatorname{div} v) + \langle v, \operatorname{Rc}\rangle}{-2(\operatorname{div} v)\cdot\nabla f + v_{ij}\nabla_i f\nabla_j f}e^{-f}d\mu$$

$$- 2\int_{\mathcal{M}} v_{ij}(R_{ij} + \nabla_i\nabla_j f)\left(h - \frac{V}{2}\right)e^{-f}d\mu.$$

Hence $\frac{d^2}{ds^2}\mathcal{F}(g,f)$ is equal to

$$- \int_{\mathcal{M}} \left(\frac{\partial}{\partial s}v_{ij} - 2v_{ik}v_{jk}\right)(R_{ij} + \nabla_i\nabla_j f)e^{-f}d\mu$$

$$+ \int_{\mathcal{M}} \left(\frac{\partial}{\partial s}\Gamma^p_{ij}\right)(\nabla_p v_{ij})e^{-f}d\mu + 2\int_{\mathcal{M}} \left|\nabla\left(h - \frac{V}{2}\right)\right|^2 e^{-f}d\mu$$

$$+ 4\int_{\mathcal{M}} v_{ij}(R_{ij} + \nabla_i\nabla_j f)\left(h - \frac{V}{2}\right)e^{-f}d\mu$$

$$+ \int_{\mathcal{M}} \left[\frac{\partial}{\partial s}\left(\frac{V}{2} - h\right) + \left(\frac{V}{2} - h\right)^2\right]\left(2\Delta f - |\nabla f|^2 + R\right)e^{-f}d\mu$$

$$+ 2\int_{\mathcal{M}} \binom{\operatorname{div}(\operatorname{div} v) + \langle v, \operatorname{Rc}\rangle}{-2(\operatorname{div} v)\cdot\nabla f + v_{ij}\nabla_i f\nabla_j f}\left(\frac{V}{2} - h\right)e^{-f}d\mu.$$

If g is a steady gradient soliton flowing along ∇f, then
$$R_{ij} + \nabla_i \nabla_j f = 0,$$
$$2\Delta f - |\nabla f|^2 + R = 0,$$
so that

(6.109)
$$\begin{aligned}\frac{d^2}{ds^2}\mathcal{F}(g,f) &= \int_{\mathcal{M}} \left(\nabla_i v_{jp} - \frac{1}{2} \nabla_p v_{ij} \right) (\nabla_p v_{ij}) e^{-f} d\mu \\ &\quad + 2 \int_{\mathcal{M}} \left(\begin{array}{c} \operatorname{div}(\operatorname{div} v) + \langle v, \operatorname{Rc}\rangle \\ -2 (\operatorname{div} v) \cdot \nabla f + v_{ij} \nabla_i f \nabla_j f \end{array} \right) \left(\frac{V}{2} - h \right) e^{-f} d\mu \\ &\quad + 2 \int_{\mathcal{M}} \left| \nabla \left(h - \frac{V}{2} \right) \right|^2 e^{-f} d\mu.\end{aligned}$$

Writing the second variation this way, we see the appearance of the linear trace Harnack quadratic.

Note that the first line of (6.109) is independent of h whereas for (pointwise) measure-preserving variations (v, h) of (g, f), the last two lines of (6.109) vanish. This shows that given v_{ij}, the variation (v, h) minimizing $\frac{d^2}{ds^2}\mathcal{F}(g, f)$ while preserving the (integral) constraint $\int_{\mathcal{M}} e^{-f} d\mu = 1$ has the last two lines of (6.109) nonpositive; not surprisingly, the last two lines are less than or equal to a negative norm squared, as we shall show below.

Using (6.108), the second line in (6.109) may be rewritten as
$$-2 \int_{\mathcal{M}} \left[\nabla_i \nabla_j \left(h - \frac{V}{2} \right) \right] v_{ij} e^{-f} d\mu$$
$$= 2 \int_{\mathcal{M}} (\operatorname{div} v - v(\nabla f)) \cdot \left[\nabla \left(h - \frac{V}{2} \right) \right] e^{-f} d\mu.$$

Hence for a steady gradient soliton g flowing along ∇f,
$$\begin{aligned}\frac{d^2}{ds^2}\mathcal{F}(g,f) &= \int_{\mathcal{M}} \left(\nabla_i v_{jp} - \frac{1}{2} \nabla_p v_{ij} \right) (\nabla_p v_{ij}) e^{-f} d\mu \\ &\quad + 2 \int_{\mathcal{M}} (\operatorname{div} v - v(\nabla f)) \cdot \left[\nabla \left(h - \frac{V}{2} \right) \right] e^{-f} d\mu \\ &\quad + 2 \int_{\mathcal{M}} \left| \nabla \left(h - \frac{V}{2} \right) \right|^2 e^{-f} d\mu.\end{aligned}$$

Completing the square, we have
$$\begin{aligned}\frac{d^2}{ds^2}\mathcal{F}(g,f) &= \int_{\mathcal{M}} \left(\nabla_i v_{jp} - \frac{1}{2} \nabla_p v_{ij} \right) (\nabla_p v_{ij}) e^{-f} d\mu \\ &\quad - \frac{1}{2} \int_{\mathcal{M}} |\operatorname{div} v - v(\nabla f)|^2 e^{-f} d\mu \\ &\quad + 2 \int_{\mathcal{M}} \left| \frac{1}{2} (\operatorname{div} v - v(\nabla f)) + \nabla \left(h - \frac{V}{2} \right) \right|^2 e^{-f} d\mu.\end{aligned}$$

If α is a 1-form, then the minimizer of the energy
$$E(h) \doteqdot \int_{\mathcal{M}} |dh + \alpha|^2 \, d\mu$$
is given by
$$\Delta h = -\operatorname{div}(\alpha).$$
Thus, for a steady gradient soliton g flowing along ∇f,

(6.110)
$$\begin{aligned}\frac{d^2}{ds^2}\mathcal{F}(g,f) &\geq \int_{\mathcal{M}} \left(\nabla_i v_{jp} - \frac{1}{2}\nabla_p v_{ij}\right)(\nabla_p v_{ij}) e^{-f} d\mu \\ &\quad - \frac{1}{2}\int_{\mathcal{M}} |\operatorname{div} v - v(\nabla f)|^2 e^{-f} d\mu \\ &\quad + \frac{1}{2}\int_{\mathcal{M}} |\operatorname{div} v - v(\nabla f) + \nabla w|^2 e^{-f} d\mu,\end{aligned}$$

where
$$\Delta w = -\operatorname{div}(\operatorname{div} v - v(\nabla f))$$
and with equality in (6.110) if and only if

(6.111)
$$\Delta\left(h - \frac{V}{2}\right) = -\frac{1}{2}\operatorname{div}(\operatorname{div} v - v(\nabla f)).$$

Since
$$\int_{\mathcal{M}} \langle \operatorname{div} v - v(\nabla f), \nabla w\rangle \, d\mu$$
$$= -\int_{\mathcal{M}} w \operatorname{div}(\operatorname{div} v - v(\nabla f)) \, d\mu$$
$$= \int_{\mathcal{M}} w \Delta w \, d\mu = -\int_{\mathcal{M}} |\nabla w|^2 \, d\mu,$$

we conclude
$$\frac{d^2}{ds^2}\mathcal{F}(g,f) \geq \int_{\mathcal{M}} \left(\nabla_i v_{jp} - \frac{1}{2}\nabla_p v_{ij}\right)(\nabla_p v_{ij}) \, d\mu - \frac{1}{2}\int_{\mathcal{M}} |\nabla w|^2 \, d\mu.$$

Since
$$\int_{\mathcal{M}} \nabla_i v_{jp} \nabla_p v_{ij} \, d\mu = -\int_{\mathcal{M}} v_{jp} \nabla_i \nabla_p v_{ij} \, d\mu$$
$$= \int_{\mathcal{M}} |\operatorname{div} v|^2 \, d\mu + \int_{\mathcal{M}} v_{jp}(-R_{pq}v_{qj} + R_{ipjq}v_{iq}) \, d\mu,$$

we obtain the following.

PROPOSITION 6.91. *Let $(\mathcal{M}^n, g)$ be a steady gradient Ricci soliton flowing along ∇f on a closed manifold. If $\frac{\partial g}{\partial s} = v$ and $\frac{\partial f}{\partial s} = h$, then*

(6.112)
$$\frac{d^2}{ds^2}\mathcal{F}(g,f) \geq \int_{\mathcal{M}} \begin{pmatrix} -\frac{1}{2}|\nabla v|^2 + |\operatorname{div} v|^2 - \frac{1}{2}|\nabla w|^2 \\ -R_{pq}v_{qj}v_{jp} + R_{ipjq}v_{iq}v_{jp} \end{pmatrix} d\mu,$$

with equality if and only if (6.111) *holds.*

REMARK 6.92. Note that

$$(\nabla_i - \nabla_i f)(\nabla_j - \nabla_j f) v_{ij}$$
$$= \operatorname{div}(\operatorname{div} v) + \langle \operatorname{Rc}, v \rangle - 2 \operatorname{div}(v) \cdot \nabla f + v_{ij} \nabla_i f \nabla_j f - (R_{ij} + \nabla_i \nabla_j f) v_{ij}.$$

Also,
$$(\nabla_j - \nabla_j f) v_{ij} = e^f \nabla_j \left(e^{-f} v_{ij} \right)$$

and
$$e^f \nabla_i \nabla_j \left(e^{-f} v_{ij} \right) = (\nabla_i - \nabla_i f)(\nabla_j - \nabla_j f) v_{ij}.$$

For any closed Riemannian manifold $\left(\hat{\mathcal{M}}^n, \hat{g} \right)$ with variation $\frac{\partial}{\partial s} \hat{g} = v$, by (5.10), (5.51), and $\int_{\hat{\mathcal{M}}} \left(\frac{V}{2} - h \right) e^{-f} d\mu = 0$, we have

$$\frac{d}{ds} \lambda(\hat{g}) = - \int_{\hat{\mathcal{M}}} v_{ij} (R_{ij} + \nabla_i \nabla_j f) e^{-f} d\mu,$$

where f is the minimizer of $\mathcal{F}(\hat{g}, \cdot)$. Thus the critical points of $\lambda(\hat{g})$ are the steady gradient Ricci solitons flowing along the gradient of the minimizer. Since, given v, equality holds in (6.112) when h satisfies (6.111), we obtain the following second variation formula proved in [**53**].

THEOREM 6.93 (Cao, Hamilton, and Ilmanen). *Let* $\left(\hat{\mathcal{M}}^n, \hat{g} \right)$ *be a closed Ricci flat manifold and let v be a symmetric 2-tensor. If $g(s) = \hat{g} + sh$, then*

$$\frac{d^2}{ds^2} \bigg|_{s=0} \lambda(g(s)) = \int_{\hat{\mathcal{M}}} \left(-\frac{1}{2} |\nabla v|^2 + |\operatorname{div} v|^2 - \frac{1}{2} |\nabla w|^2 + R_{ipjq} v_{iq} v_{jp} \right) d\mu$$
$$= \int_{\hat{\mathcal{M}}} \langle \mathrm{L} v, v \rangle d\mu,$$

where w is defined by (up to an additive constant)

$$\Delta w \doteq \operatorname{div}(\operatorname{div} v)$$

and

$$\mathrm{L} v \doteq \frac{1}{2} \Delta v - \frac{1}{2} \mathcal{L}_{(\operatorname{div} v)^\sharp} \hat{g} + \operatorname{Rm}(v).$$

Similarly, one can compute the variation of

$$\nu(\hat{g}) \doteq \inf_{\tau \in [0, \infty)} \mu(\hat{g}, \tau) = \inf_{\tau, f} \mathcal{W}(\hat{g}, f, \tau),$$

where $\int_{\hat{\mathcal{M}}} e^{-f} d\mu = (4\pi\tau)^{n/2}$, on an Einstein manifold with positive scalar curvature (see [**53**]).

THEOREM 6.94 (Cao, Hamilton, and Ilmanen). *Let* $\left(\hat{\mathcal{M}}^n, \hat{g} \right)$ *be a closed Einstein manifold with positive scalar curvature, i.e.,* $\operatorname{Rc} = \frac{1}{2\tau} \hat{g}$ *where $\tau > 0$,* and let v be a symmetric 2-tensor. If $g(s) = \hat{g} + sh$, then

$$\left.\frac{d^2}{ds^2}\right|_{s=0} \nu(g(s)) = \frac{\tau}{\operatorname{Vol}(\hat{\mathcal{M}})} \int_{\hat{\mathcal{M}}} \left(-\frac{1}{2}|\nabla v|^2 + |\operatorname{div} v|^2 - \frac{1}{2}|\nabla w|^2\right) d\mu$$

$$+ \frac{\tau}{\operatorname{Vol}(\hat{\mathcal{M}})} \int_{\hat{\mathcal{M}}} \left(R_{ijk\ell} v_{ij} v_{\ell k} + \frac{1}{4\tau} w^2\right) d\mu$$

$$- \frac{1}{2n}\left(\frac{1}{\operatorname{Vol}(\hat{\mathcal{M}})} \int_{\hat{\mathcal{M}}} V d\mu\right)^2,$$

where w is defined uniquely by

$$\Delta w + \frac{w}{2\tau} \doteqdot \operatorname{div}(\operatorname{div} v), \qquad \int_{\hat{\mathcal{M}}} w \, d\mu = 0.$$

7.3. A matrix Harnack calculation for the adjoint heat equation. We also have the following Harnack-type calculation (see [**287**]). We may think of equation (6.113) below as a Bochner-type formula which says that the evolution of the modified Ricci tensor is given by a backward Lichnerowicz Laplacian heat operator with Hamilton's matrix Harnack quadratic as the main term on the RHS.

PROPOSITION 6.95 (Matrix Harnack formula for adjoint heat equation). *Under the system*

$$\frac{\partial}{\partial t} g_{ij} = -2R_{ij},$$

$$\frac{\partial f}{\partial t} = -\Delta f - R + |\nabla f|^2 - \frac{n\varepsilon}{2\tau},$$

$$\frac{d\tau}{dt} = \varepsilon,$$

we have

(6.113)
$$\left(\frac{\partial}{\partial t} + \Delta_L - 2\nabla f \cdot \nabla\right)\left(\nabla_i \nabla_j f + R_{ij} + \frac{\varepsilon}{2\tau} g_{ij}\right)$$
$$= 2\left(\Delta_L R_{ij} - \frac{1}{2}\nabla_i \nabla_j R + R_{ik} R_{jk} + \frac{\varepsilon}{2\tau} R_{ij}\right)$$
$$+ 2\left((P_{i\ell j} + P_{j\ell i})\nabla_\ell f + R_{kij\ell} \nabla_\ell f \nabla_k f\right)$$
$$- \left(\nabla_i \nabla_k f + R_{ik} + \frac{\varepsilon}{2\tau} g_{ik}\right)\left(-\nabla_j \nabla_k f + R_{jk} + \frac{\varepsilon}{2\tau} g_{jk}\right)$$
$$- \left(-\nabla_i \nabla_k f + R_{ik} + \frac{\varepsilon}{2\tau} g_{ik}\right)\left(\nabla_j \nabla_k f + R_{jk} + \frac{\varepsilon}{2\tau} g_{jk}\right).$$

PROOF. We have

$$\left(\frac{\partial}{\partial t} + \Delta_L\right) \nabla_i \nabla_j \phi = \nabla_i \nabla_j \left(\frac{\partial \phi}{\partial t} + \Delta \phi\right) - 2\left(\frac{\partial}{\partial t} \Gamma^\ell_{ij}\right) \nabla_\ell \phi$$

for any function ϕ of space and time. We may also rewrite the evolution of Rc in the following funny way:
$$\left(\frac{\partial}{\partial t}+\Delta_L\right)R_{ij}=2\Delta_L R_{ij}.$$

Hence
$$\left(\frac{\partial}{\partial t}+\Delta_L\right)\left(\nabla_i\nabla_j f + R_{ij} + \frac{\varepsilon}{2\tau}g_{ij}\right)$$
$$= 2\Delta_L R_{ij} + \nabla_i\nabla_j\left(-R+|\nabla f|^2\right) - 2\left(\frac{\partial}{\partial t}\Gamma_{ij}^\ell\right)\nabla_\ell f - \frac{\varepsilon^2}{2\tau^2}g_{ij} - \frac{\varepsilon}{\tau}R_{ij}$$
$$= 2\left(\Delta_L R_{ij} - \frac{1}{2}\nabla_i\nabla_j R + R_{kij\ell}\nabla_k f \nabla_\ell f\right)$$
$$+ 2\nabla_k\nabla_i\nabla_j f \nabla_k f + 2\nabla_i\nabla_k f \nabla_j\nabla_k f$$
$$+ 2\left(\nabla_i R_{j\ell} + \nabla_j R_{i\ell} - \nabla_\ell R_{ij}\right)\nabla_\ell f - \frac{\varepsilon^2}{2\tau^2}g_{ij} - \frac{\varepsilon}{\tau}R_{ij}$$
$$= 2\nabla_k\left(\nabla_i\nabla_j f + R_{ij} + \frac{\varepsilon}{2\tau}g_{ij}\right)\nabla_k f$$
$$+ 2\left(\Delta_L R_{ij} - \frac{1}{2}\nabla_i\nabla_j R + R_{ik}R_{jk} + \frac{\varepsilon}{2\tau}R_{ij}\right)$$
$$+ 2\left((P_{i\ell j} + P_{j\ell i})\nabla_\ell f + R_{kij\ell}\nabla_k f \nabla_\ell f\right)$$
$$+ 2\nabla_i\nabla_k f \nabla_j\nabla_k f - 2R_{ik}R_{jk} - \frac{2\varepsilon}{\tau}R_{ij} - \frac{\varepsilon^2}{2\tau^2}g_{ij}$$
and the result follows from rewriting the last line. $\square$

REMARK 6.96. In the evolution equation for f, we may replace $-\frac{n\varepsilon}{2\tau}$ on the RHS by any function of t.

Since the matrix Harnack quadratic is the space-time Riemann curvature, we ask the following.

PROBLEM 6.97. Is there a space-time interpretation of equation (6.113)?

For convenience, we define
$$T_{ij}^\varepsilon \doteqdot R_{ij} - \nabla_i\nabla_j f + \frac{\varepsilon}{2\tau}g_{ij}$$
and
$$H(X)_{ij} \doteqdot \Delta_L R_{ij} - \frac{1}{2}\nabla_i\nabla_j R + R_{ik}R_{jk} + \frac{\varepsilon}{2\tau}R_{ij}$$
$$+ (P_{i\ell j} + P_{j\ell i})X_\ell + R_{kij\ell}X_k X_\ell.$$

Then we may rewrite (6.113) as
$$\left(\frac{\partial}{\partial t}+\Delta_L - 2\nabla f \cdot \nabla\right)S_{ij}^\varepsilon = 2H(\nabla f)_{ij} - S_{ik}^\varepsilon T_{jk}^\varepsilon - S_{jk}^\varepsilon T_{ik}^\varepsilon,$$
where S_{ik}^ε is defined in (6.27).

7. SOME FURTHER CALCULATIONS RELATED TO $\mathcal{F}$ AND $\mathcal{W}$

EXERCISE 6.98. Under the system of equations

$$\frac{\partial}{\partial t} g_{ij} = -2R_{ij},$$

$$\frac{\partial}{\partial t} X = -\Delta X + \operatorname{Rc}(X) - \nabla R + 2(\nabla_{X^*} X),$$

$$\frac{d\tau}{dt} = \varepsilon,$$

generalize the above calculation to find

$$\left(\frac{\partial}{\partial t} + \Delta_L - X \cdot \nabla - \mathcal{L}_X \right) \left(2R_{ij} + \nabla_i X_j + \nabla_j X_i + \frac{\varepsilon}{\tau} g_{ij} \right).$$

There is also a corresponding formula where ∇f is replaced a closed 1-form X. When X is not closed, there are a couple of extra terms in the calculation which have a dX factor.

A Kähler version of this calculation of (6.113) was obtained by one of the authors.

There is a computation analogous to (6.113) which one can perform for solutions of a forward heat-type equation. In particular (see [**287**]).

LEMMA 6.99. *If on a manifold $\mathcal{M}^n$,*

$$\frac{\partial}{\partial t} g_{ij} = -2R_{ij},$$

$$\frac{\partial f}{\partial t} = \Delta f - R - |\nabla f|^2,$$

then

$$Z_{ij} \doteqdot R_{ij} - \nabla_i \nabla_j f + \frac{1}{2t} g_{ij}$$

satisfies

$$\left(\frac{\partial}{\partial t} - \Delta_L + 2\nabla f \cdot \nabla \right) Z_{ij} = Y_{ij} - Z_{ik} \left(R_{jk} + \nabla_j \nabla_k f + \frac{1}{2t} g_{jk} \right)$$

$$- \left(\nabla_i \nabla_k f + R_{ik} + \frac{1}{2t} g_{ik} \right) Z_{jk},$$

where

$$Y_{ij} \doteqdot \nabla_i \nabla_j R + 2\nabla_k R_{ij} \nabla_k f + 2R_{kij\ell} \nabla_k f \nabla_\ell f + R_{ik} R_{jk} + \frac{1}{t} R_{ij}.$$

One of the authors has made the following conjecture.

CONJECTURE 6.100. *If $(\mathcal{M}^n, g(t))$ and $f(t)$ are solutions to the equations in Lemma 6.99, where $g(t)$ has bounded nonnegative curvature operator, then*

$$Y_{ij} \geq 0.$$

8. Notes and commentary

Section 5. For $T < \infty$ the condition on r_k^2/t_k just says that the r_k are uniformly bounded. In Definition 6.56, condition (1) may be replaced by

(1') For some $C < \infty$, $|\text{Rm}\,[g(t_k)]| \leq C\, r_k^{-2}$ in $B(p_k, r_k)$.

We call the original definition the 'first definition' and the definition with (1') replacing (1) the 'second definition'. If $g(t)$ is locally collapsing according to the first definition, then taking $C = 1$, it is locally collapsing according to the second definition with the same r_k. On the other hand, if $g_{ij}(t)$ is locally collapsing according to the second definition, then the following hold.

(i) If $C \leq 1$, then $g(t)$ is locally collapsing according to the first definition with the same r_k.

(ii) If $C \geq 1$, then set $\bar{r}_k \doteqdot r_k/\sqrt{C} \leq r_k$. Then $|\text{Rm}\,[g(t_k)]| \leq \bar{r}_k^{-2}$ in $B(p_k, \bar{r}_k)$, and we still have $\lim_{k\to\infty} \bar{r}_k^{-n} \text{Vol}\, B(p_k, \bar{r}_k) = 0$, so that $g_{ij}(t)$ is locally collapsing according to the first definition with $\bar{r}_k = r_k/\sqrt{C}$.

CHAPTER 7

The Reduced Distance

> How thoroughly it is ingrained in mathematical science that every real advance goes hand in hand with the invention of sharper tools and simpler methods which, at the same time, assist in understanding earlier theories and in casting aside some more complicated developments. – David Hilbert

> Technical skill is mastery of complexity while creativity is mastery of simplicity.
> – Chris Zeeman

In [**297**] Perelman introduced a new length (energy-like) functional for paths in the space-times of solutions of the Ricci flow, called the $\mathcal{L}$-length. The naturalness of this functional can be justified both by the space-time approach and the various differential inequalities that the quantities associated to the $\mathcal{L}$-length satisfy, which we shall show in this chapter. A fundamental inequality is the monotonicity of the reduced volume. As we shall see in the next chapter, this monotonicity leads to a second proof of (weakened) no local collapsing for finite time singularities. We emphasize that, unlike the entropy proof, *the proof of weakened no local collapsing in Chapter 8 using the reduced volume also holds for complete noncompact solutions with bounded sectional curvature*. Recall that no local collapsing provides a local injectivity radius estimate and at the same time rules out the formation of the cigar soliton singularity model.

Besides bringing comparison geometry and integral monotonicity into Ricci flow, some original aspects of Perelman's work on the reduced distance function are as follows.

(1) A space-time distance-like function which is *not* always nonnegative.
(2) Bochner formulas, i.e., partial differential inequalities, which are *geometry* (e.g., curvature) *independent*, i.e., these formulas and inequalities hold in any dimension and are independent of the initial metric.
(3) Pointwise monotonicity adapted to the space-time geometry. This describes in some sense how, for any solution to the Ricci flow, the geometry improves as time increases.
(4) Using the space-time geometry to understand point-picking and compactness, in particular, understanding the structure of ancient

κ-solutions, finite-time singularity models, and high curvature regions of the solution.

(5) Relating Ricci flow and aspects of function theory and the heat equation, for example, the analysis of the fundamental solution of the adjoint heat equation coupled to the Ricci flow.

In this chapter we discuss the basic properties of the $\mathcal{L}$-length and the associated distance functions for complete (not necessarily compact) solutions to the backward Ricci flow with bounded sectional curvature.[1]

1. The $\mathcal{L}$-length and distance for a static metric

One of the primary antecedents of Perelman's $\mathcal{L}$-length and reduced distance ℓ is the work of Li-Yau on differential Harnack inequalities for the heat equation on a Riemannian manifold with a static metric.[2] With this in mind we start by summarizing the properties of the energy functional for paths in a Riemannian manifold and various monotonicity and comparison results. The purpose for this is to compare properties associated to the linear heat equation with respect to a static metric to properties of the nonlinear case of metrics evolving by Ricci flow and to show a strong analogy between these two cases. The fact that the case of the heat equation is less technical facilitates the presentation of some of the underlying ideas.

Let $(\hat{\mathcal{M}}^n, \hat{g})$ be a complete Riemannian manifold. Given a C^1-path $\gamma : [\tau_1, \tau_2] \to \hat{\mathcal{M}}$, $\tau_1 \geq 0$, joining two points, i.e., $\gamma(\tau_1) = p$ and $\gamma(\tau_2) = q$, we define its **energy** by

$$\mathcal{E}_{\tau_1, \tau_2}(\gamma) \doteqdot \int_{\tau_1}^{\tau_2} \sqrt{\tau} \left| \frac{d\gamma}{d\tau} \right|_{\hat{g}}^2 d\tau.$$

Convention: By saying that a path $\gamma(\tau)$ is C^k, where $k = 1$ or 2, we really mean $\gamma\left(\frac{\sigma^2}{4}\right)$ is a C^k function of $\sigma \doteqdot 2\sqrt{\tau}$.

This energy functional, which is well-suited for studying the heat equation, is equivalent to the usual energy for paths (see for example §12 of Milnor's book [**265**]). Indeed, making the change of variables $\sigma = 2\sqrt{\tau}$ yields

(7.1) $$\mathcal{E}_{\tau_1, \tau_2}(\gamma) = \int_{2\sqrt{\tau_1}}^{2\sqrt{\tau_2}} \left| \frac{d\gamma}{d\sigma} \right|_{\hat{g}}^2 d\sigma.$$

Keeping in mind that we shall be discussing parabolic equations, *we require that the parameter of the path be given by the time variable τ*. In particular, if we want to study the (reversed) concentration process of the fundamental

---

[1] See the last section for some of the notational conventions we use in this chapter.

[2] The later work of Hamilton on the matrix differential Harnack for the Ricci flow is also important in this development.

solution to the heat equation, which is a delta function at $\tau = 0$ at a point $p \in \hat{\mathcal{M}}$, we take $\tau_1 = 0$ so as to define

$$\mathcal{E}(\gamma) \doteqdot \int_0^{\bar{\tau}} \sqrt{\tau} \left| \frac{d\gamma}{d\tau} \right|_{\hat{g}}^2 d\tau \tag{7.2}$$

for C^1-paths $\gamma : [0, \bar{\tau}] \to \hat{\mathcal{M}}$ from p to arbitrary points $q \in \hat{\mathcal{M}}$.

From (7.1) we see that the critical points γ are of the form

$$\gamma(\tau) = \beta(2\sqrt{\tau}),$$

where β is a constant speed geodesic. Hence, in Euclidean space, $\gamma(\tau) = 2\sqrt{\tau}V$ for some $V \in \mathbb{R}^n$, and its graph $(\gamma(\tau), \tau)$ is a *parabola*. This is one justification for the $\sqrt{\tau}$ factor in (7.2): parabolas are more suited to the heat equation. Note that

$$\lim_{\tau \to 0} \sqrt{\tau} \frac{d\gamma}{d\tau}(\tau) = \frac{d\beta}{d\sigma}(0) \in T_p \hat{\mathcal{M}}. \tag{7.3}$$

EXERCISE 7.1. Show that if γ is a critical point of (7.2), then

$$\nabla_X X + \frac{1}{2\tau} X = 0$$

where $X = \frac{d\gamma}{d\tau}$.

SOLUTION. From (7.1) we find that the critical points satisfy

$$\nabla_{\sqrt{\tau}X}(\sqrt{\tau}X) = 0.$$

Given a basepoint p (at time 0) define a **space-time distance** function on $\hat{\mathcal{M}} \times (0, \infty)$ by

$$L(q, \bar{\tau}) \doteqdot \inf_\gamma \mathcal{E}(\gamma),$$

where the infimum is taken over all C^1-paths $\gamma : [0, \bar{\tau}] \to \hat{\mathcal{M}}$ with $\gamma(0) = p$ and $\gamma(\bar{\tau}) = q$. An elementary computation using (7.1) yields

$$L(q, \bar{\tau}) = \frac{d(p,q)^2}{2\sqrt{\bar{\tau}}}, \tag{7.4}$$

where d is the distance with respect to $\hat{g}$ and the infimum of $\mathcal{E}(\gamma)$ is obtained by a minimal geodesic γ from p to q with

$$\left| \frac{d\gamma}{d\tau}(\tau) \right|_{\hat{g}} = \frac{1}{\sqrt{\tau}} \frac{d(p,q)}{2\sqrt{\bar{\tau}}}. \tag{7.5}$$

With the Euclidean heat kernel in mind we define the **reduced distance**:

$$\ell(q, \bar{\tau}) \doteqdot \frac{L(q, \bar{\tau})}{2\sqrt{\bar{\tau}}} = \frac{d(p,q)^2}{4\bar{\tau}}. \tag{7.6}$$

If we wish to remove the time-dependence in (7.6), then we may define the **enlarged distance**:

$$\bar{L}(q, \bar{\tau}) \doteqdot 2\sqrt{\bar{\tau}} L(q, \bar{\tau}) = 4\bar{\tau} \ell(q, \bar{\tau}) = d(p,q)^2. \tag{7.7}$$

EXERCISE 7.2 (*L*-distance and ℓ between two space-time points). Show using (7.1) that $\inf_\gamma \mathcal{E}_{\tau_1,\tau_2}(\gamma)$, where the infimum is taken over all C^1-paths γ with $\gamma(\tau_1) = p$ and $\gamma(\tau_2) = q$, is attained by $\gamma(\tau) = \beta(2\sqrt{\tau})$, where $\beta : [2\sqrt{\tau_1}, 2\sqrt{\tau_2}] \to \hat{\mathcal{M}}$ is a minimal geodesic with constant speed $\left|\frac{d\beta}{d\sigma}\right|_{\hat{g}} \equiv \frac{d(p,q)}{2\sqrt{\tau_2}-2\sqrt{\tau_1}}$. Hence

$$(7.8) \qquad L_{(p,\tau_1)}(q,\tau_2) \doteqdot \inf_\gamma \mathcal{E}_{\tau_1,\tau_2}(\gamma) = \frac{d(p,q)^2}{2\sqrt{\tau_2} - 2\sqrt{\tau_1}}.$$

We may then define the reduced distance by

$$(7.9) \qquad \ell_{(p,\tau_1)}(q,\tau_2) \doteqdot \frac{L_{(p,\tau_1)}(q,\tau_2)}{2\sqrt{\tau_2} + 2\sqrt{\tau_1}} = \frac{d(p,q)^2}{4(\tau_2 - \tau_1)}.$$

We compute under the assumption $\mathrm{Rc}_{\hat{g}} \geq 0$

$$(7.10) \quad \left(\frac{\partial}{\partial \bar\tau} + \Delta\right)(\bar L - 2n\bar\tau) = \Delta(d^2) - 2n = 2\left(d\Delta d + |\nabla d|^2 - n\right) \leq 0$$

in the weak sense,[3] where we used $|\nabla d| = 1$ a.e. and the Laplacian Comparison Theorem (A.9): $d\Delta d \leq n - 1$. That is, $\bar L - 2n\bar\tau$ *is a subsolution of the backward heat equation in the weak sense.*

REMARK 7.3. It is useful to keep in mind the examples of flat tori to see why one cannot prove a stronger statement (see subsection 9.5 of this chapter).

The role of the reduced distance in the study of the heat equation is exhibited by the Li-Yau differential Harnack inequality (A.12), which implies that for any positive solution u of the heat equation on a complete Riemannian manifold with $\mathrm{Rc}_{\hat g} \geq 0$,

$$\frac{u(x_2, \tau_2)}{u(x_1, \tau_1)} \geq \left(\frac{\tau_2}{\tau_1}\right)^{-n/2} \exp\left\{-\ell_{(x_1,\tau_1)}(x_2, \tau_2)\right\},$$

where $\tau_1 < \tau_2$. A similar, but slightly more complicated, statement holds when $\mathrm{Rc}_{\hat g} \geq -K\hat g$, where $K \geq 0$. See [**253**] for details.

2. The $\mathcal{L}$-length and the *L*-distance

Let $\left(\mathcal{N}^n, \tilde h(t)\right)$, $t \in (\alpha, \omega)$, be a solution to the Ricci flow. From this we can easily obtain a solution $(\mathcal{N}^n, h(\tau))$ to the **backward Ricci flow**

$$\frac{\partial}{\partial \tau} h = 2 \mathrm{Rc}$$

---

[3]That is, for any nonnegative C^2 function φ on space-time with compact support,

$$\int_0^\infty \int_{\mathcal{M}} \left(-\frac{\partial \varphi}{\partial \bar\tau} + \Delta\varphi\right)(\bar L - 2n\bar\tau)\, d\mu d\bar\tau \leq 0.$$

See subsection 9.5 of this chapter for a justification of (7.10).

by reversing time. In particular, if $\omega < +\infty$, let $\tau \doteqdot \omega - t$, so that $(\mathcal{N}, h(\tau))$ is a solution to the backward Ricci flow on the time interval $(0, \omega - \alpha)$.[4]

2.1. Space-time motivation for the $\mathcal{L}$-length.

We begin by motivating the definition of the $\mathcal{L}$-length for the Ricci flow as a renormalization of the length with respect to Perelman's potentially infinite Riemannian metric on space-time. Given $N \in \mathbb{N}$, define a metric on $\widetilde{\mathcal{N}} \doteqdot \mathcal{N}^n \times \mathcal{S}^N \times (0, T)$ by

$$\tag{7.11} \tilde{h} \doteqdot h_{ij} dx^i dx^j + \tau h_{\alpha\beta} dy^\alpha dy^\beta + \left(\frac{N}{2\tau} + R \right) d\tau^2,$$

where $h_{\alpha\beta}$ is the metric on $\mathcal{S}^N$ of constant sectional curvature $1/(2N)$ and R denotes the scalar curvature of the evolving metric h on $\mathcal{N}$. Here we have used the convention that $\{x^i\}_{i=1}^n$ will denote coordinates on the $\mathcal{N}$ factor, $\{y^\alpha\}_{\alpha=1}^N$ coordinates on the $\mathcal{S}^N$ factor, and $x^0 \doteqdot \tau$. Latin indices $i, j, k, \ldots$ will be on $\mathcal{N}$, Greek indices $\alpha, \beta, \gamma, \ldots$ will be on $\mathcal{S}^N$, and 0 represents the (minus) time component. Choosing N large enough so that $\frac{N}{2\tau} + R > 0$ implies that the metric $\tilde{h}$ is Riemannian, i.e., positive-definite. In local coordinates,

$$\tag{7.12} \tilde{h}_{ij} = h_{ij},$$

$$\tag{7.13} \tilde{h}_{\alpha\beta} = \tau h_{\alpha\beta},$$

$$\tag{7.14} \tilde{h}_{00} = \frac{N}{2\tau} + R,$$

$$\tag{7.15} \tilde{h}_{i0} = \tilde{h}_{i\alpha} = \tilde{h}_{\alpha 0} = 0.$$

Let $\tilde{\gamma}(s) \doteqdot (x(s), y(s), \tau(s))$ be a shortest geodesic, with respect to the metric $\tilde{h}$, between points $p \doteqdot (x_0, y_0, 0)$ and $q \doteqdot (x_1, y_1, \tau_q) \in \widetilde{\mathcal{N}}$. Since the fibers $\mathcal{S}^N$ pinch to a point as $\tau \to 0$, it is clear that the geodesic $\tilde{\gamma}(s)$ is orthogonal to the fibers $\mathcal{S}^N$. (To see this directly, take a sequence of geodesics from $p_k \doteqdot (x_0, y_1, 1/k)$ to q and pass to the limit as $k \to \infty$.) Therefore it suffices to consider the manifold $\bar{\mathcal{N}} \doteqdot \mathcal{N} \times (0, T)$ endowed with the Riemannian metric:

$$\tag{7.16} \bar{h} \doteqdot h_{ij} dx^i dx^j + \left(\frac{N}{2\tau} + R \right) d\tau^2.$$

(This metric is dual to the metric considered in [100].) For convenience, denote $x(s) \doteqdot \gamma(s)$.

Now we use $s = \tau$ as the parameter of the curve. Let $\dot{\gamma}(\tau) \doteqdot \frac{d\gamma}{d\tau}(\tau)$. The length of a path $\bar{\gamma}(\tau) \doteqdot (\gamma(\tau), \tau)$, with respect to the metric $\bar{h}$, is given by the following:

---

[4] We shall consider the case where $\alpha = -\infty$ (in which case we define $\omega - \alpha \doteqdot +\infty$). On the other hand, if $\omega = +\infty$ and $\alpha = -\infty$, we may simply take $\tau = -t$. However, for the backward Ricci flow we are not as interested in the case where $\omega = +\infty$ and $\alpha > -\infty$.

$$\text{Length}_{\bar{h}}(\bar{\gamma})$$
$$= \int_0^{T_q} \sqrt{\frac{N}{2\tau} + R + |\dot{\gamma}(\tau)|^2} \, d\tau$$
$$= \int_0^{T_q} \sqrt{\frac{N}{2\tau}} \sqrt{1 + \frac{2\tau}{N}\left(R + |\dot{\gamma}(\tau)|^2\right)} \, d\tau$$
$$= \int_0^{T_q} \sqrt{\frac{N}{2\tau}} \left(1 + \frac{\tau}{N}\left(R + |\dot{\gamma}(\tau)|^2\right) + O\left(N^{-2}\right)\right) d\tau$$
$$= \int_0^{T_q} \sqrt{\frac{N}{2\tau}} d\tau + \int_0^{T_q} \sqrt{\frac{\tau}{2N}}\left(R + |\dot{\gamma}(\tau)|^2\right) d\tau + \int_0^{T_q} \sqrt{\frac{1}{2\tau}} O\left(N^{-3/2}\right) d\tau$$
$$= \sqrt{2N T_q} + \frac{1}{\sqrt{2N}} \int_0^{T_q} \sqrt{\tau}\left(R + |\dot{\gamma}(\tau)|^2\right) d\tau + \sqrt{2T_q} O\left(N^{-3/2}\right).$$

The calculation indicates that as $N \to \infty$, a shortest geodesic should approach a minimizer of the $\mathcal{L}$-length functional defined by

$$\mathcal{L}(\gamma) \doteq \int_0^{T_q} \sqrt{\tau}\left(R(\gamma(\tau), \tau) + |\dot{\gamma}(\tau)|^2_{h(\tau)}\right) d\tau.$$

Note that the definition of $\mathcal{L}(\gamma)$ only depends on the data of $(\mathcal{N}, h)$.

EXERCISE 7.4 (Levi-Civita connection of the potentially infinite metric). Consider the metric $\bar{h}$ on $\mathcal{N}^n \times (0, T)$ defined in (7.16) by (7.12), (7.14), and $\bar{h}_{i0} = 0$ (without the $\mathcal{S}^N$ factor). The components of the Levi-Civita connection $^N\bar{\nabla}$ of $\bar{h}$ are defined by

$$^N\bar{\nabla}_{\frac{\partial}{\partial x^a}} \frac{\partial}{\partial x^b} = \sum_{c=0}^n {}^N\bar{\Gamma}^c_{ab} \frac{\partial}{\partial x^c},$$

where $x^0 = \tau$. Show that

$$^N\bar{\Gamma}^k_{ij} = \Gamma^k_{ij},$$
$$^N\bar{\Gamma}^k_{i0} = R^k_i,$$
$$^N\bar{\Gamma}^k_{00} = -\frac{1}{2}\nabla^k R$$

and

$$^N\bar{\Gamma}^0_{ij} = -\left(\frac{N}{2\tau} + R\right)^{-1} R_{ij},$$
$$^N\bar{\Gamma}^0_{i0} = \left(\frac{N}{2\tau} + R\right)^{-1} \frac{1}{2}\nabla_i R,$$
$$^N\bar{\Gamma}^0_{00} = \left(\frac{N}{2\tau} + R\right)^{-1} \frac{1}{2}\left(\frac{\partial R}{\partial \tau} + \frac{R}{\tau}\right) - \frac{1}{2\tau}.$$

In particular, $^N\bar{\Gamma}^k_{ab}$ are independent of N, whereas

$$\lim_{N\to\infty} {}^N\bar{\Gamma}^0_{ij} = 0,$$

$$\lim_{N\to\infty} {}^N\bar{\Gamma}^0_{i0} = 0,$$

$$\lim_{N\to\infty} {}^N\bar{\Gamma}^0_{00} = -\frac{1}{2\tau}.$$

2.2. The $\mathcal{L}$-length. A *natural geometry on space-time* (in the sense of lengths, distances and geodesics) is given by the following.

DEFINITION 7.5 ($\mathcal{L}$-length). Let $(\mathcal{N}^n, h(\tau))$, $\tau \in (A, \Omega)$, be a solution to the backward Ricci flow $\frac{\partial}{\partial \tau} h = 2\operatorname{Rc}$, and let $\gamma : [\tau_1, \tau_2] \to \mathcal{N}$ be a piecewise C^1-path,[5] where $[\tau_1, \tau_2] \subset (A, \Omega)$ and $\tau_1 \geq 0$. The $\mathcal{L}$-**length** of γ is[6]

$$(7.17) \qquad \mathcal{L}(\gamma) \doteqdot \mathcal{L}_h(\gamma) \doteqdot \int_{\tau_1}^{\tau_2} \sqrt{\tau} \left(R(\gamma(\tau), \tau) + \left| \frac{d\gamma}{d\tau}(\tau) \right|^2_{h(\tau)} \right) d\tau.$$

Later we shall take $\tau_1 = 0$ and call $\tau_2 = \bar{\tau}$.

REMARK 7.6. Taking $\tau_1 = 0$, the subsequent degeneracy introduced by the $\sqrt{\tau}$ factor in (7.17) reflects the infinite speed of propagation of the Ricci flow (as a nonlinear heat-type equation for metrics). We also note the formal similarity between $R + \left|\frac{d\gamma}{d\tau}\right|^2$ and the quantity $R + |\nabla f|^2$ which we considered for gradient Ricci solitons and which also appeared in the definitions of energy and entropy; this seems like more than just a coincidence.

The $\mathcal{L}$-length is defined only for paths defined on a *subinterval* of the time interval where the solution to the backward Ricci flow exists. Note that $\mathcal{L}$ may be *negative* since the scalar curvature may be negative somewhere. This is in contrast to the energy defined in Section 1 above for a static metric. Often we shall use the following conventions:

$$(7.18) \qquad \sigma \doteqdot 2\sqrt{\tau} \quad \text{and} \quad \beta(\sigma) \doteqdot \gamma(\sigma^2/4).$$

We may rewrite $\mathcal{L}$ as

$$(7.19) \qquad \mathcal{L}(\gamma) = \int_{2\sqrt{\tau_1}}^{2\sqrt{\tau_2}} \left(\frac{\sigma^2}{4} R(\beta(\sigma), \sigma^2/4) + \left| \frac{d\beta}{d\sigma}(\sigma) \right|^2_{h(\sigma^2/4)} \right) d\sigma.$$

This is especially useful in the case $\tau_1 = 0$.

Because of the $\left|\frac{d\gamma}{d\tau}\right|^2$ term on the RHS of (7.17), $\mathcal{L}(\gamma)$ looks more like an energy than a length. Another way to obtain $\mathcal{L}$, which is related to the

---

[5]That is, $\gamma\left(\frac{\sigma^2}{4}\right)$ is a C^1 function of σ.

[6]$R(\gamma(\tau), \tau)$ is just a notation meaning $R_{h(\tau)}(\gamma(\tau))$, where $R_{h(\tau)}$ is the scalar curvature of $(\mathcal{N}^n, h(\tau))$.

above approach of renormalizing the Riemannian length functional, is as follows. We define the **space-time graph**

$$\tilde{\gamma} : [\tau_1, \tau_2] \to \mathcal{N} \times [\tau_1, \tau_2]$$

of the path γ by $\tilde{\gamma}(\tau) \doteq (\gamma(\tau), \tau)$, so that $\frac{d\tilde{\gamma}}{d\tau}(\tau) = \left(\frac{d\gamma}{d\tau}(\tau), 1\right)$. Note that the parameter τ, of which γ is a function, also serves as time; so it is natural to consider its graph. Define the space-time metric $\check{h} \doteq h + R d\tau^2$. In general, this metric is indefinite since R may be negative somewhere. We easily compute

$$\mathcal{L}(\gamma) = \int_{\tau_1}^{\tau_2} \sqrt{\tau} \left|\frac{d\tilde{\gamma}}{d\tau}(\tau)\right|^2_{\check{h}} d\tau.$$

Using $\sigma = 2\sqrt{\tau}$, we may rewrite the $\mathcal{L}$-length as

$$\mathcal{L}(\gamma) = \int_{\sigma_1}^{\sigma_2} \left|\frac{d}{d\sigma}\left(\tilde{\gamma}\left(\frac{\sigma^2}{4}\right)\right)\right|^2_{\check{h}} d\sigma,$$

where $\sigma_i \doteq 2\sqrt{\tau_i}$, $i = 1, 2$. That is, $\mathcal{L}(\gamma)$ is the *energy* of the space-time path $\tilde{\gamma}$ with respect to the space-time metric $\check{h}$ and the new time parameter σ.

If $\alpha : [\tau_1, \tau_2] \to \mathcal{N}$ and $\beta : [\tau_2, \tau_3] \to \mathcal{N}$ are paths with $\alpha(\tau_2) = \beta(\tau_2)$, then we define the **concatenated path** $\alpha \smile \beta : [\tau_1, \tau_3] \to \mathcal{N}$ by

$$(\alpha \smile \beta)(\tau) = \begin{cases} \alpha(\tau) & \text{if } \tau \in [\tau_1, \tau_2], \\ \beta(\tau) & \text{if } \tau \in [\tau_2, \tau_3]. \end{cases}$$

We have the following additivity property.

LEMMA 7.7 (Additivity of the $\mathcal{L}$-length).

(7.20) $$\mathcal{L}(\alpha \smile \beta) = \mathcal{L}(\alpha) + \mathcal{L}(\beta).$$

However, the $\mathcal{L}$-length of a path γ is not invariant under reparametrizations of γ. We leave it to the reader to make the easy verification of this fact.

The following bound on $\mathcal{L}$ is elementary.

LEMMA 7.8 (Lower bound for the $\mathcal{L}$-length).

(7.21) $$\mathcal{L}(\gamma) \geq \frac{2}{3}\left(\tau_2^{3/2} - \tau_1^{3/2}\right) \inf_{\mathcal{N} \times [\tau_1, \tau_2]} R.$$

This follows directly from $\mathcal{L}(\gamma) \geq \int_{\tau_1}^{\tau_2} \sqrt{\tau} R_{\inf}(\tau) d\tau$, where $R_{\inf}(\tau) \doteq \inf_{\mathcal{N} \times \{\tau\}} R$. The Riemannian counterpart of estimate (7.21) is the obvious fact that the length of a path is nonnegative.

2. THE $\mathcal{L}$-LENGTH AND THE L-DISTANCE

2.3. The L-distance function. Just as for the usual length functional (perhaps it is better to compare with the energy functional), one gives the following definition.

DEFINITION 7.9 (L-distance). Let $(\mathcal{N}^n, h(\tau))$, $\tau \in (A, \Omega)$, be a solution to the backward Ricci flow. Fix a basepoint $p \in \mathcal{N}$. For any $x \in \mathcal{N}$ and $\tau > 0$, define the L-**distance** by

$$L(x, \tau) \doteqdot L_{(p,0)}^h(x, \tau) \doteqdot \inf_\gamma \mathcal{L}(\gamma),$$

where the infimum is taken over all C^1-paths $\gamma : [0, \tau] \to \mathcal{N}$ joining p to x (the graph $\tilde\gamma$ joins $(p, 0)$ to (x, τ)). We call an $\mathcal{L}$-length minimizing path a **minimal $\mathcal{L}$-geodesic**. We also define

(7.22) $$\bar L(x, \tau) \doteqdot \bar L_{(p,0)}^h(x, \tau) \doteqdot 2\sqrt{\tau} L(x, \tau).$$

Note that the L-distance defined above may be negative. To help the reader have a feeling for the L-distance function, we present some exercises.

EXERCISE 7.10 (Scaling properties of $\mathcal{L}$ and L). Let $(\mathcal{N}^n, h(\tau))$ be a solution to the backward Ricci flow, $\gamma : [\tau_1, \tau_2] \to \mathcal{N}$ a C^1-path, and $c > 0$ a constant. Show that for the solution $\hat h(\hat\tau) \doteqdot ch(c^{-1}\hat\tau)$ and the path $\hat\gamma : [c\tau_1, c\tau_2] \to \mathcal{N}$ defined by $\hat\gamma(\hat\tau) \doteqdot \gamma(c^{-1}\hat\tau)$, we have

$$\mathcal{L}_{\hat h}(\hat\gamma) = \sqrt{c}\,\mathcal{L}_h(\gamma).$$

Consequently,

$$L_{(p,0)}^{\hat h}(q, \hat\tau) = \sqrt{c}\, L_{(p,0)}^h(q, c^{-1}\hat\tau).$$

EXERCISE 7.11 ($\mathcal{L}$ and L on Riemannian products). Suppose that we are given a Riemannian product solution $(\mathcal{N}_1^{n_1} \times \mathcal{N}_2^{n_2}, h_1(\tau) + h_2(\tau))$ to the backward Ricci flow and a C^1-path $\gamma = (\alpha, \beta) : [\tau_1, \tau_2] \to \mathcal{N}_1 \times \mathcal{N}_2$. Show that

$$\mathcal{L}_{h_1+h_2}(\gamma) = \mathcal{L}_{h_1}(\alpha) + \mathcal{L}_{h_2}(\beta).$$

Hence

$$L_{(p_1,p_2,0)}^{h_1+h_2}(q_1, q_2, \tau) = L_{(p_1,0)}^{h_1}(q_1, \tau) + L_{(p_2,0)}^{h_2}(q_2, \tau).$$

It is useful to keep in mind Euclidean space as a basic example; more generally we have

EXERCISE 7.12 (L-distance for Ricci flat solutions). Let $(\mathcal{N}^n, h(\tau) = h_0)$ be a static Ricci flat manifold and let $p \in \mathcal{N}$ be the basepoint. Show that given any $q \in \mathcal{N}$ and $\bar\tau > 0$, the $\mathcal{L}$-length of a C^1-path $\gamma : [0, \bar\tau] \to \mathcal{N}$ from p to q is

$$\mathcal{L}(\gamma) = \int_0^{2\sqrt{\bar\tau}} \left|\frac{d\gamma}{d\sigma}(\sigma^2/4)\right|^2 d\sigma,$$

which is the same as (7.2). Hence a minimal $\mathcal{L}$-geodesic γ is of the form

(7.23) $$\gamma(\tau) = \beta(2\sqrt{\tau}),$$

where $\beta : [0, 2\sqrt{\bar{\tau}}] \to \mathcal{N}^n$ is a minimal constant speed geodesic with respect to h_0 joining p to q. Thus

(7.24) $$L(q, \bar{\tau}) = \frac{d(p,q)^2}{2\sqrt{\bar{\tau}}}.$$

For reference below, we have $\bar{L}(q, \bar{\tau}) \doteqdot d(p, q)^2$ and $\ell(q, \bar{\tau}) \doteqdot \frac{1}{2\sqrt{\bar{\tau}}} L(q, \bar{\tau}) = \frac{d(p,q)^2}{4\bar{\tau}}$; the definition of ℓ will be given again in (7.87).

SOLUTION TO EXERCISE 7.12. This exercise is a special case of the discussion in Section 1 above. We also note that

(7.25) $$\tau \left| \frac{d\gamma}{d\tau}(\tau) \right|^2_{g(\tau)} = \left| \frac{d\beta}{d\sigma}(\sigma) \right|^2_{g(\sigma^2/4)} = |V|^2_{g(0)},$$

where $V \doteqdot \lim_{\tau \to 0} \sqrt{\tau} \frac{d\gamma}{d\tau}(\tau) = \lim_{\sigma \to 0} \frac{d\beta}{d\sigma}(\sigma)$. We leave it to the reader to check that for $\mathcal{L}$-geodesics defined on a subinterval $[\tau_1, \tau_2] \subset [0, T]$, we still have

(7.26) $$\tau \left| \frac{d\gamma}{d\tau}(\tau) \right|^2_{g(\tau)} \equiv \text{const}.$$

2.4. Elementary properties of L. In this subsection $(\mathcal{M}^n, g(\tau))$, $\tau \in [0, T]$, shall denote a complete solution to the backward Ricci flow, and $p \in \mathcal{M}$ shall be a basepoint. We will assume the curvature bound

(7.27) $$\max_{(x,\tau) \in \mathcal{M} \times [0,T]} \{|\text{Rm}(x, \tau)|, |\text{Rc}(x, \tau)|\} \leq C_0 < \infty.$$

The curvature bound assumption is written in this way for the convenience of stating later estimates. We prove some elementary C^0-estimates for the L-distance and lengths of $\mathcal{L}$-geodesics, relating them to the Riemannian distance; we shall use these estimates often later.

First recall from (3.3) in Lemma 3.11 that for $\tau_1 < \tau_2$ and $x \in \mathcal{M}$,

$$e^{-2C_0(\tau_2 - \tau_1)} g(\tau_2, x) \leq g(\tau_1, x) \leq e^{2C_0(\tau_2 - \tau_1)} g(\tau_2, x).$$

LEMMA 7.13 ($\mathcal{L}$ and Riemannian distance). *Let $\gamma : [0, \bar{\tau}] \to \mathcal{M}$, $\bar{\tau} \in (0, T]$, be a C^1-path starting at p and ending at q.*

(i) (*Bounding Riemannian distance by $\mathcal{L}$*) *For any $\tau \in [0, \bar{\tau}]$ we have*

$$d^2_{g(0)}(p, \gamma(\tau)) \leq 2\sqrt{\tau} e^{2C_0 \tau} \left(\mathcal{L}(\gamma) + \frac{2nC_0}{3} \bar{\tau}^{3/2} \right).$$

In particular, when $\mathcal{M}$ is noncompact, for any $\bar{\tau} \in (0, T]$, we have

$$\lim_{q \to \infty} \bar{L}(q, \bar{\tau}) = \lim_{q \to \infty} 2\sqrt{\bar{\tau}} L(q, \bar{\tau}) = +\infty.$$

(ii) (*Bounding speed at some time by $\mathcal{L}$*) *There exists $\tau_* \in (0, \bar{\tau})$ such that*

$$\tau_* \left| \frac{d\gamma}{d\tau}(\tau_*) \right|^2_{g(\tau_*)} = \left| \frac{d\beta}{d\sigma}(\sigma_*) \right|^2_{g(\tau_*)} \leq \frac{1}{2\sqrt{\bar{\tau}}} \mathcal{L}(\gamma) + \frac{nC_0}{3} \bar{\tau},$$

where $\beta(\sigma) \doteq \gamma(\tau)$, $\sigma = 2\sqrt{\tau}$, and $\sigma_* \doteq 2\sqrt{\tau_*}$.

(iii) (*Bounding L by Riemannian distance*) *For any $q \in \mathcal{M}$ and $\bar{\tau} > 0$,*

$$L(q, \bar{\tau}) \leq e^{2C_0\bar{\tau}} \frac{d^2_{g(\bar{\tau})}(p,q)}{2\sqrt{\bar{\tau}}} + \frac{2nC_0}{3}\bar{\tau}^{3/2}.$$

REMARK 7.14. In each of the estimates above, on the RHS one may think of the first term as the main term and the second term as an error term. Recall by (7.24) that if $(\mathcal{M}^n, g(\tau) = g_0)$ is Ricci flat and $\gamma : [0, \bar{\tau}] \to \mathcal{M}$ is a minimal $\mathcal{L}$-geodesic from p to q, then

$$d^2_{g_0}(p, \gamma(\tau)) = 2\sqrt{\tau}L(\gamma(\tau), \tau) = 2\sqrt{\tau}\mathcal{L}\left(\gamma|_{[0,\tau]}\right) = \frac{\tau}{\bar{\tau}}d^2_{g_0}(p,q),$$

and for all $\tau_* \in (0, \bar{\tau})$,

$$\tau_* \left|\frac{d\gamma}{d\tau}(\tau_*)\right|^2_{g(\tau_*)} = \frac{d^2_{g_0}(p,q)}{4\bar{\tau}}.$$

Hence for Ricci flat solutions, $\ell(\gamma(\tau), \tau)$ defined in (7.87) is constant ($\equiv \frac{1}{4\bar{\tau}}d^2_{g_0}(p,q)$) along $\mathcal{L}$-geodesics.

PROOF. (i) Let $\tilde{\sigma} = 2\sqrt{\tilde{\tau}}$ and $\beta(\tilde{\sigma}) \doteq \gamma(\tilde{\tau})$. The idea is to first bound the energy of $\beta|_{[0, 2\sqrt{\tau}]}$. By splitting the formula for $\mathcal{L}$ into two time intervals, we see that

$$\int_0^{2\sqrt{\tau}} \left|\frac{d\beta}{d\tilde{\sigma}}(\tilde{\sigma})\right|^2_{g(\tilde{\sigma}^2/4)} d\tilde{\sigma}$$

$$= \mathcal{L}(\gamma) - \int_{2\sqrt{\tau}}^{2\sqrt{\bar{\tau}}} \left|\frac{d\beta}{d\tilde{\sigma}}(\tilde{\sigma})\right|^2_{g(\tilde{\sigma}^2/4)} d\tilde{\sigma} - \int_0^{\bar{\tau}} \sqrt{\tilde{\tau}}R(\gamma(\tilde{\tau}), \tilde{\tau})d\tilde{\tau}$$

(7.28) $$\leq \mathcal{L}(\gamma) + \frac{2nC_0}{3}\bar{\tau}^{3/2},$$

since $R \geq -nC_0$. Hence, since $g(0) \leq e^{2C_0\tau}g(\tilde{\tau})$ for $\tilde{\tau} \in [0, \tau]$, we have

$$d^2_{g(0)}(p, \gamma(\tau)) \leq e^{2C_0\tau}\left(\int_0^{2\sqrt{\tau}} \left|\frac{d\beta}{d\tilde{\sigma}}(\tilde{\sigma})\right|_{g(\tilde{\sigma}^2/4)} d\tilde{\sigma}\right)^2$$

$$\leq e^{2C_0\tau} \cdot 2\sqrt{\tau} \int_0^{2\sqrt{\tau}} \left|\frac{d\beta}{d\tilde{\sigma}}(\tilde{\sigma})\right|^2_{g(\tilde{\sigma}^2/4)} d\tilde{\sigma}$$

$$\leq 2\sqrt{\tau}e^{2C_0\tau}\left(\mathcal{L}(\gamma) + \frac{2nC_0}{3}\bar{\tau}^{3/2}\right).$$

(ii) From the proof of (i) we have (take $\tau = \bar{\tau}$ in (7.28))

$$\frac{1}{2\sqrt{\bar{\tau}}}\int_0^{2\sqrt{\bar{\tau}}} \left|\frac{d\beta}{d\tilde{\sigma}}(\tilde{\sigma})\right|^2_{g(\tilde{\sigma}^2/4)} d\tilde{\sigma} \leq \frac{1}{2\sqrt{\bar{\tau}}}\mathcal{L}(\gamma) + \frac{nC_0}{3}\bar{\tau}.$$

By the mean value theorem for integrals, there exists $\tau_* \in (0, \bar{\tau})$ such that

$$\left|\frac{d\beta}{d\sigma}(\sigma_*)\right|^2_{g(\tau_*)} \leq \frac{1}{2\sqrt{\bar{\tau}}}\mathcal{L}(\gamma) + \frac{nC_0}{3}\bar{\tau}.$$

(iii) Let $\eta : [0, 2\sqrt{\bar{\tau}}] \to \mathcal{M}$ be a minimal geodesic from p to q with respect to the metric $g(\bar{\tau})$. Then

$$L(q, \bar{\tau}) \leq \mathcal{L}(\eta) = \int_0^{2\sqrt{\bar{\tau}}} \left(\frac{\sigma^2}{4} R\left(\eta(\sigma), \sigma^2/4\right) + \left|\frac{d\eta}{d\sigma}\right|^2_{g(\tau)}\right) d\sigma$$

$$\leq \int_0^{2\sqrt{\bar{\tau}}} \left(\frac{nC_0\sigma^2}{4} + e^{2C_0\bar{\tau}}\left|\frac{d\eta}{d\sigma}\right|^2_{g(\bar{\tau})}\right) d\sigma$$

$$\leq \frac{2nC_0}{3}\bar{\tau}^{3/2} + \frac{e^{2C_0\bar{\tau}}}{2\sqrt{\bar{\tau}}}d^2_{g(\bar{\tau})}(p, q).$$

$\square$

3. The first variation of $\mathcal{L}$-length and existence of $\mathcal{L}$-geodesics

Now that we have defined the $\mathcal{L}$-length, we may mimic basic Riemannian comparison geometry in the space-time setting for the Ricci flow. We compute the **first variation** of the $\mathcal{L}$-length and find the equation for the critical points of $\mathcal{L}$ (the $\mathcal{L}$-geodesic equation). We also compare this equation with the geodesic equation for the space-time graph (with respect to a natural space-time connection) and prove two existence theorems for $\mathcal{L}$-geodesics.

3.1. First variation of the $\mathcal{L}$-length. Let $(\mathcal{N}^n, h(\tau))$, $\tau \in (A, \Omega)$, be a solution to the backward Ricci flow. Consider a variation of the C^2-path $\gamma : [\tau_1, \tau_2] \to \mathcal{N}$; that is, let

$$G : [\tau_1, \tau_2] \times (-\varepsilon, \varepsilon) \to \mathcal{N}$$

be a C^2-map such that

$$G|_{[\tau_1, \tau_2] \times \{0\}} = \gamma.$$

Convention: We say that a variation $G(\cdot, \cdot)$ of a C^2-path γ is C^2 if $G\left(\frac{\sigma^2}{4}, s\right)$ is C^2 in (σ, s).

Define $\gamma_s \doteqdot G|_{[\tau_1, \tau_2] \times \{s\}} : [\tau_1, \tau_2] \to \mathcal{N}$ for $-\varepsilon < s < \varepsilon$. Let

$$X(\tau, s) \doteqdot \frac{\partial G}{\partial \tau}(\tau, s) = \frac{\partial \gamma_s}{\partial \tau}(\tau) \text{ and } Y(\tau, s) \doteqdot \frac{\partial G}{\partial s}(\tau, s) = \frac{\partial \gamma_s}{\partial s}(\tau)$$

be the tangent vector field and variation vector field along $\gamma_s(\tau)$, respectively. The first variation formula for $\mathcal{L}$ is given by

3. FIRST VARIATION OF $\mathcal{L}$-LENGTH AND EXISTENCE OF $\mathcal{L}$-GEODESICS

LEMMA 7.15 ($\mathcal{L}$-First Variation Formula). *Given a C^2-family of curves $\gamma_s : [\tau_1, \tau_2] \to \mathcal{N}$, the first variation of its $\mathcal{L}$-length is given by*

$$
\begin{aligned}
\frac{1}{2}(\delta_Y \mathcal{L})(\gamma_s) &\doteqdot \frac{1}{2}\frac{d}{ds}\mathcal{L}(\gamma_s) = \sqrt{\tau}\, Y \cdot X \Big|_{\tau_1}^{\tau_2} \\
&\quad + \int_{\tau_1}^{\tau_2} \sqrt{\tau}\, Y \cdot \left(\frac{1}{2}\nabla R - \frac{1}{2\tau}X - \nabla_X X - 2\operatorname{Rc}(X)\right) d\tau,
\end{aligned}
\tag{7.29}
$$

where the covariant derivative ∇ is with respect to $h(\tau)$.

REMARK 7.16. We use the notation $(\delta_Y \mathcal{L})(\gamma_s)$ since $\frac{d}{ds}\mathcal{L}(\gamma_s)$, at a given value of s, depends only on γ_s and Y along γ_s.

PROOF. We compute in a similar fashion to the usual first variation formula for length (see [**72**], p. 4ff for example)

$$
\begin{aligned}
\frac{d}{ds}\mathcal{L}(\gamma_s) &= \frac{d}{ds}\int_{\tau_1}^{\tau_2}\sqrt{\tau}\left(R(\gamma_s(\tau),\tau) + \left|\frac{\partial \gamma_s}{\partial \tau}(\tau)\right|^2_{h(\tau)}\right) d\tau \\
&= \int_{\tau_1}^{\tau_2} \sqrt{\tau}\left(\langle \nabla R, Y\rangle + 2\langle \nabla_Y X, X\rangle\right) d\tau;
\end{aligned}
\tag{7.30}
$$

here $\langle \cdot, \cdot \rangle = h(\tau)(\cdot, \cdot)$ denotes the inner product with respect to $h(\tau)$. Using $[X, Y] = \left[\frac{\partial G}{\partial \tau}, \frac{\partial G}{\partial s}\right] = 0$ and $\frac{\partial}{\partial \tau} h = 2\operatorname{Rc}$, we have

$$\langle \nabla_Y X, X\rangle = \langle \nabla_X Y, X\rangle = \frac{d}{d\tau}[h(Y, X)] - \langle Y, \nabla_X X\rangle - 2\operatorname{Rc}(Y, X).$$

Hence

$$\frac{1}{2}\frac{d}{ds}\mathcal{L}(\gamma_s) = \int_{\tau_1}^{\tau_2}\sqrt{\tau}\left(\frac{1}{2}\langle \nabla R, Y\rangle + \frac{d}{d\tau}\langle Y, X\rangle - \langle Y, \nabla_X X\rangle - 2\operatorname{Rc}(Y, X)\right) d\tau$$

and integration by parts yields

$$\int_{\tau_1}^{\tau_2} \sqrt{\tau}\frac{d}{d\tau}\langle Y, X\rangle\, d\tau = -\frac{1}{2}\int_{\tau_1}^{\tau_2}\frac{1}{\sqrt{\tau}}\langle Y, X\rangle\, d\tau + \sqrt{\tau}\,\langle Y, X\rangle \Big|_{\tau_1}^{\tau_2}.$$

The lemma follows from the above two equalities. □

REMARK 7.17. In comparison, the Riemannian first variation of arc length formula on $\left(\hat{\mathcal{M}}^n, \hat{g}\right)$ is

$$\frac{d}{du}\operatorname{L}(\gamma_u) = -\int_0^b \langle U, \nabla_T T\rangle\, ds + \langle U, T\rangle \big|_0^b, \tag{7.31}$$

where $\gamma_u : [0, b] \to \hat{\mathcal{M}}$ is a 1-parameter family of paths, $T \doteqdot \frac{\partial \gamma_u}{\partial s} \big/ \left|\frac{\partial \gamma_u}{\partial s}\right|$, $U \doteqdot \frac{\partial}{\partial u}\gamma_u$, and ds is the arc length element.

3.2. The $\mathcal{L}$-geodesic equation.
The $\mathcal{L}$-first variation formula leads us to the following.

DEFINITION 7.18 ($\mathcal{L}$-geodesic). If γ is a critical point of the $\mathcal{L}$-length functional among all C^2-paths with fixed endpoints, then γ is called an **$\mathcal{L}$-geodesic**.

By the $\mathcal{L}$-first variation formula,

COROLLARY 7.19 ($\mathcal{L}$-geodesic equation). *Let $(\mathcal{N}^n, h(\tau))$, $\tau \in (A, \Omega)$, be a solution to the backward Ricci flow. A C^2-path $\gamma : [\tau_1, \tau_2] \to \mathcal{N}$ is an $\mathcal{L}$-geodesic if and only if it satisfies the **$\mathcal{L}$-geodesic equation**:*

$$(7.32) \qquad \nabla_X X - \frac{1}{2}\nabla R + 2\operatorname{Rc}(X) + \frac{1}{2\tau}X = 0,$$

where $X(\tau) \doteqdot \frac{d\gamma}{d\tau}(\tau)$.

For the four terms in (7.32), (1) is the usual term in the geodesic equation, (2) comes from the variation of R in $\mathcal{L}$, (3) comes from $\frac{\partial}{\partial \tau}h$, and (4) comes from $\sqrt{\tau}$ in $\mathcal{L}$ via integration by parts.

In local coordinates, the $\mathcal{L}$-geodesic equation is

$$(7.33) \quad 0 = \frac{d^2\gamma^i}{d\tau^2} + \Gamma^i_{jk}(\gamma(\tau), \tau) \frac{d\gamma^j}{d\tau}\frac{d\gamma^k}{d\tau} - \frac{1}{2}h^{ij}\nabla_j R + 2h^{ij}R_{jk}\frac{d\gamma^k}{d\tau} + \frac{1}{2\tau}\frac{d\gamma^i}{d\tau},$$

where $\gamma^i = x^i \circ \gamma$.

We find it convenient to use the notation $\frac{D}{d\tau}$ for the covariant derivative along the curve γ. Multiplying (7.32) by τ yields

$$(7.34) \qquad \sqrt{\tau}\frac{D}{d\tau}\left(\sqrt{\tau}X\right) - \frac{\tau}{2}\nabla R + 2\sqrt{\tau}\operatorname{Rc}\left(\sqrt{\tau}X\right) = 0.$$

Since the covariant derivative along the curve can be written as

$$\frac{D}{d\tau}V = \nabla_X V,$$

we may write

$$\sqrt{\tau}\frac{D}{d\tau}\left(\sqrt{\tau}X\right) = \sqrt{\tau}\nabla_X\left(\sqrt{\tau}X\right) = \nabla_{\sqrt{\tau}X}\left(\sqrt{\tau}X\right)$$

along $\gamma(\tau)$, where the last two terms require extending $V(\tau) = \sqrt{\tau}X$ to a vector field in a neighborhood of $\gamma(\tau)$. Note that

$$\nabla_{\sqrt{\tau}X}\left(\sqrt{\tau}X\right) = \tau\nabla_X X + \sqrt{\tau}\left(\frac{d}{d\tau}\sqrt{\tau}\right)X$$

$$= \tau\nabla_X X + \frac{1}{2}X,$$

which is different from $\sqrt{\tau}\nabla_{\sqrt{\tau}X}X$ because $\sqrt{\tau}$ must be differentiated along the curve. Using the convention (7.18) and $Z(\sigma) \doteqdot \frac{d\beta(\sigma)}{d\sigma} = \sqrt{\tau}X$, we get

$$(7.35) \qquad \nabla_Z Z - \frac{\sigma^2}{8}\nabla R + \sigma\operatorname{Rc}(Z) = 0.$$

3. FIRST VARIATION OF $\mathcal{L}$-LENGTH AND EXISTENCE OF $\mathcal{L}$-GEODESICS

EXAMPLE 7.20 ($\mathcal{L}$-geodesics on Einstein solutions). Let $(\mathcal{N}_0^n, h_0(\tau))$, $\tau \in (0, \infty)$, be a 'big bang' Einstein solution to the backward Ricci flow with $\mathrm{Rc}_{h_0}(\tau) = \frac{h_0(\tau)}{2\tau}$. Then $R_{h_0}(\tau) = \frac{n}{2\tau}$ and the $\mathcal{L}$-geodesic equation (7.32) is

$$(7.36) \qquad \nabla_X X + \frac{3}{2\tau} X = 0,$$

so that $\nabla_{\frac{d\gamma}{d\tau}}\left(\tau^{3/2} \frac{d\gamma}{d\tau}\right) = 0$. Note that since ∇ is independent of scaling and $h_0(\tau) = \tau h_0(1)$, we have $\nabla^{h_0(\tau)} = \nabla^{h_0(1)}$ is independent of τ. Clearly the constant paths, where $X = 0$, are $\mathcal{L}$-geodesics. More generally, reparametrize γ and define the path β by $\beta(\rho) = \gamma(f(\rho))$, where

$$(7.37) \qquad \tau^{3/2} = f'\left(f^{-1}(\tau)\right).$$

Then $\dot\beta(\rho) \doteqdot \frac{d\beta}{d\rho} = \frac{d\gamma}{d\tau}(f(\rho)) f'(\rho) = \frac{d\gamma}{d\tau}(f(\rho)) f(\rho)^{3/2}$, so that (7.36) implies

$$\nabla_{\dot\beta(\rho)} \dot\beta(\rho) = \nabla_{\tau^{3/2} \frac{d\gamma}{d\tau}(\tau)} \left(\tau^{3/2} \frac{d\gamma}{d\tau}(\tau)\right) = 0;$$

i.e., β is a constant speed geodesic with respect to $h_0(1)$. Since solutions of (7.37) are given by $f(\rho) = \frac{4}{(\rho_0 - \rho)^2}$, the $\mathcal{L}$-geodesics are of the form

$$\gamma(\tau) = \beta\left(\frac{2}{\sqrt{\tau_0}} - \frac{2}{\sqrt{\tau}}\right),$$

defined for $\tau \in (0, \infty)$, where $\beta : (-\infty, \infty) \to \mathcal{N}_0$ is a constant speed geodesic with respect to $h_0(1)$. Note that

$$\left|\frac{d\gamma}{d\tau}(\tau)\right|_{h_0(\tau)} = \sqrt{\tau} \left|\dot\beta\left(\frac{2}{\sqrt{\tau_0}} - \frac{2}{\sqrt{\tau}}\right) \frac{1}{\tau^{3/2}}\right|_{h_0(1)} = \frac{\mathrm{const}}{\tau}.$$

That is,

$$\tau^2 \left|\frac{d\gamma}{d\tau}(\tau)\right|^2_{h_0(\tau)} \equiv \mathrm{const}.$$

(Compare with (7.26) for the Ricci flat case.) In particular, $\left|\frac{d\gamma}{d\tau}(\tau)\right|_{h_0(1)} = \frac{\mathrm{const}}{\tau^{3/2}}$. In any case, the speed of $\frac{d\gamma}{d\tau}(\tau)$ tends to infinity as $\tau \to 0$, whereas the speed of $\frac{d\gamma}{d\tau}(\tau)$ tends to zero as $\tau \to \infty$. Note $\int_{\tau_0}^{\infty} \left|\frac{d\gamma}{d\tau}(\tau)\right|_{h_0(1)} d\tau < \infty$ for all $\tau_0 \in (0, \infty)$, whereas $\int_0^{\tau_0} \left|\frac{d\gamma}{d\tau}(\tau)\right|_{h_0(1)} d\tau = \infty$.

EXERCISE 7.21. Determine the $\mathcal{L}$-geodesics for Einstein solutions with negative scalar curvature. What multiple of $\left|\frac{d\gamma}{d\tau}(\tau)\right|^2_{h(\tau)}$ is constant?

SOLUTION TO EXERCISE 7.21. Consider a maximal Einstein solution to the backward Ricci flow $h(\tau)$, $\tau \in [0,T)$, where $\mathrm{Rc} = -\frac{1}{2(T-\tau)}h$ is negative. In this case $h(\tau) = \frac{T-\tau}{T}h(0)$ (which is easy to see from $\mathrm{Rc}_{h(\tau)}$ being independent of τ). The $\mathcal{L}$-geodesic equation (7.32) for $\gamma(\tau)$ is

$$\nabla_X X + \left(\frac{1}{2\tau} - \frac{1}{T-\tau}\right) X = 0. \tag{7.38}$$

Let $\beta(\rho) \doteqdot \gamma(f(\rho))$, where f is defined by

$$f'\left(f^{-1}(\tau)\right) = \exp\left(\int^\tau \left(\frac{1}{2\bar\tau} - \frac{1}{T-\bar\tau}\right) d\bar\tau\right)$$
$$= \sqrt{\tau}(T-\tau),$$

i.e., $f'(\rho) = \sqrt{f(\rho)}(T - f(\rho))$. Then $\dot\beta(\rho) \doteqdot \frac{d\beta}{d\rho} = \frac{d\gamma}{d\tau}(f(\rho))f'(\rho)$. Let $\tau \doteqdot f(\rho)$. Equation (7.38) and the definition of f imply

$$\nabla_{\dot\beta(\rho)}\dot\beta(\rho) = \nabla_{f'(\rho)\frac{d\gamma}{d\tau}(\tau)}\left(f'(\rho)\frac{d\gamma}{d\tau}(\tau)\right)$$
$$= f''(\rho)\frac{d\gamma}{d\tau}(\tau) - f'(\rho)f'(\rho)\left(\frac{1}{2\tau} - \frac{1}{T-\tau}\right)\frac{d\gamma}{d\tau}(\tau)$$
$$= \frac{d}{d\rho}\left(f'(\rho) - \exp\left(\int^{f(\rho)}\left(\frac{1}{2\bar\tau} - \frac{1}{T-\bar\tau}\right)d\bar\tau\right)\right) \cdot \frac{d\gamma}{d\tau}(\tau)$$
$$= 0.$$

That is, $\beta(\rho)$ is a constant speed geodesic with respect to $h(0)$. We make the rationalizing substitution $x \doteqdot \sqrt{f(\rho)}$, so that

$$\rho = \int \frac{f'(\rho)\,d\rho}{\sqrt{f(\rho)}(T-f(\rho))} = 2\int \frac{dx}{T-x^2} = \frac{1}{\sqrt{T}}\log\left(\frac{\sqrt{T}+x}{\sqrt{T}-x}\right).$$

That is,

$$f(\rho) = x^2 = T\left(\frac{e^{\sqrt{T}\rho}-1}{e^{\sqrt{T}\rho}+1}\right)^2$$

or

$$f^{-1}(\tau) = \frac{1}{\sqrt{T}}\log\left(\frac{\sqrt{T}+\sqrt{\tau}}{\sqrt{T}-\sqrt{\tau}}\right).$$

Using $\gamma(\tau) = \beta(f^{-1}(\tau))$ and the fact that β is constant speed with respect to $h(0)$, we compute

$$\left|\frac{d\gamma}{d\tau}\right|^2_{h(\tau)} = \frac{T-\tau}{T}\left|\dot\beta(f^{-1}(\tau))\right|^2_{h(0)}\left|\frac{d(f^{-1})}{d\tau}\right|^2 = \frac{\mathrm{const}}{T\tau(T-\tau)},$$

since $h(\tau) = \frac{T-\tau}{T}h(0)$ and

$$\frac{d(f^{-1})}{d\tau} = \frac{1}{f'(\rho)} = \frac{1}{\sqrt{\tau}(T-\tau)}.$$

3. FIRST VARIATION OF $\mathcal{L}$-LENGTH AND EXISTENCE OF $\mathcal{L}$-GEODESICS

EXERCISE 7.22. Estimate the speed of an $\mathcal{L}$-geodesic for a solution $(\mathcal{M}^n, g(\tau))$, $\tau \in [0, T)$, to the backward Ricci flow with

$$|\text{Rm}(x, \tau)| \leq \frac{C}{T - \tau} \quad \text{on } \mathcal{M} \times [0, T).$$

Hint: What does the Bernstein–Bando–Shi estimate say about $|\nabla R|$?

3.3. Space-time approach to the $\mathcal{L}$-geodesic equation.

We now compare the $\mathcal{L}$-geodesic equation for γ with the geodesic equation for the graph $\tilde{\gamma}(\tau) = (\gamma(\tau), \tau)$ with respect to the following space-time connection (see also Lemma 4.3 in [**100**]):

(7.39) $$\tilde{\Gamma}^k_{ij} = \Gamma^k_{ij},$$

(7.40) $$\tilde{\Gamma}^k_{i0} = \tilde{\Gamma}^k_{0i} = R^k_i,$$

(7.41) $$\tilde{\Gamma}^k_{00} = -\frac{1}{2}\nabla^k R,$$

(7.42) $$\tilde{\Gamma}^0_{00} = -\frac{1}{2\tau},$$

where $i, j, k \geq 1$ (above and below), and the rest of the components are zero. It is instructive to compare the Christoffel symbols $\tilde{\Gamma}$ above with the symbols ${}^N\tilde{\Gamma}$ of the Levi-Civita connection ${}^N\tilde{\nabla}$ for the metric $\tilde{h}$ introduced in Exercise 7.4. For $k \geq 1$, note that $\tilde{\Gamma}^k_{ab} = {}^N\tilde{\Gamma}^k_{ab}$ is independent of N, whereas $\tilde{\Gamma}^0_{ab} = \lim_{N \to \infty} {}^N\tilde{\Gamma}^0_{ab}$ for all $a, b \geq 0$.

Let $\tau = \tau(\sigma) \doteqdot \sigma^2/4$, i.e., $\sigma \doteqdot 2\sqrt{\tau}$. We look for a geodesic, with respect to the space-time connection defined above, of the form

$$\tilde{\beta}(\sigma) \doteqdot (\gamma(\tau(\sigma)), \sigma^2/4),$$

where $\gamma : [\tau_1, \tau_2] \to \mathcal{M}$ is a path. For convenience, let $\beta(\sigma) \doteqdot \gamma(\tau(\sigma))$, $\tilde{\beta}^i \doteqdot x^i \circ \beta \doteqdot \beta^i$ for $i = 1, \ldots, n$, and $\tilde{\beta}^0 \doteqdot x^0 \circ \tilde{\beta}$ (so that $\tilde{\beta}^0(\sigma) = \sigma^2/4$). By direct computation, we have

$$\frac{d\beta^k}{d\sigma} = \frac{\sigma}{2}\frac{d\gamma^k}{d\tau},$$

$$\frac{d\tilde{\beta}^0}{d\sigma} = \frac{\sigma}{2},$$

and

$$\frac{d^2\beta^k}{d\sigma^2} = \frac{d}{d\sigma}\left(\frac{\sigma}{2}\frac{d\gamma^k}{d\tau}(\tau(\sigma))\right)$$

$$= \left(\frac{\sigma}{2}\right)^2 \frac{d^2\gamma^k}{d\tau^2}(\tau(\sigma)) + \frac{1}{2}\left(\frac{d\gamma^k}{d\tau}(\tau(\sigma))\right).$$

We justify the change of variables from τ to σ via the geodesic equation with respect to $\tilde{\Gamma}$ by showing that the time component of $\tilde{\beta}$ satisfies the geodesic

equation:
$$\frac{d^2\tilde{\beta}^0}{d\sigma^2} + \sum_{0\leq i,j\leq n}\left(\tilde{\Gamma}^0_{ij}\circ\tilde{\beta}\right)\frac{d\tilde{\beta}^i}{d\sigma}\frac{d\tilde{\beta}^j}{d\sigma} = \frac{d^2}{d\sigma^2}\left(\frac{\sigma^2}{4}\right) + \tilde{\Gamma}^0_{00}\left(\tilde{\beta}(\sigma)\right)\left(\frac{\sigma}{2}\right)^2$$
$$= \frac{1}{2} - \frac{1}{2\left(\frac{\sigma^2}{4}\right)}\left(\frac{\sigma}{2}\right)^2 = 0.$$

(This last equation justifies defining the time component of $\tilde{\beta}(\sigma)$ as $\sigma^2/4$, and in particular, the change of variables $\sigma = 2\sqrt{\tau}$.) For the space components, the geodesic equation with respect to $\tilde{\Gamma}$ says that for $k = 1,...,n$,

$$0 = \frac{d^2\tilde{\beta}^k}{d\sigma^2} + \sum_{0\leq i,j\leq n}\tilde{\Gamma}^k_{ij}\frac{d\tilde{\beta}^i}{d\sigma}\frac{d\tilde{\beta}^j}{d\sigma}$$
$$= \frac{d^2\beta^k}{d\sigma^2} + \sum_{1\leq i,j\leq n}\Gamma^k_{ij}\frac{d\beta^i}{d\sigma}\frac{d\beta^j}{d\sigma} + 2\sum_{1\leq i\leq n}\tilde{\Gamma}^k_{i0}\frac{d\beta^i}{d\sigma}\frac{d\tilde{\beta}^0}{d\sigma} + \tilde{\Gamma}^k_{00}\frac{d\tilde{\beta}^0}{d\sigma}\frac{d\tilde{\beta}^0}{d\sigma}.$$

This is equivalent to

$$0 = \left(\frac{\sigma}{2}\right)^2\frac{d^2\gamma^k}{d\tau^2}(\tau(\sigma)) + \sum_{1\leq i,j\leq n}\Gamma^k_{ij}\left(\frac{\sigma}{2}\frac{d\gamma^i}{d\tau}(\tau(\sigma))\right)\left(\frac{\sigma}{2}\frac{d\gamma^j}{d\tau}(\tau(\sigma))\right)$$
$$+ \frac{1}{2}\left(\frac{d\gamma^k}{d\tau}(\tau(\sigma))\right) + 2\sum_{1\leq i\leq n}R^k_i\left(\frac{\sigma}{2}\frac{d\gamma^i}{d\tau}(\tau(\sigma))\right)\left(\frac{\sigma}{2}\right) - \frac{1}{2}\left(\frac{\sigma}{2}\right)^2\nabla^k R,$$

which, after dividing by $\tau = \sigma^2/4$, implies

$$0 = \frac{d^2\gamma^k}{d\tau^2}(\tau(\sigma)) + \sum_{1\leq i,j\leq n}\Gamma^k_{ij}\frac{d\gamma^i}{d\tau}(\tau(\sigma))\frac{d\gamma^j}{d\tau}(\tau(\sigma)) + \frac{1}{2\tau}\left(\frac{d\gamma^k}{d\tau}(\tau(\sigma))\right)$$
$$+ 2\sum_{1\leq i\leq n}R^k_i\frac{d\gamma^i}{d\tau}(\tau(\sigma)) - \frac{1}{2}\nabla^k R.$$

That is, in invariant notation and with $X \doteq \frac{d\gamma}{d\tau}$, we have

$$\nabla_X X - \frac{1}{2}\nabla R + 2\operatorname{Rc}(X) + \frac{1}{2\tau}X = 0,$$

which is the same as (7.32). Thus $\mathcal{L}$-geodesics correspond to geodesics defined with respect to the space-time connection. In particular, $\gamma(\tau)$ is an $\mathcal{L}$-geodesic if and only if $\beta(\sigma) \doteq \gamma(\sigma^2/4)$ is a geodesic with respect to the space-time connection $\tilde{\nabla}$. Since $\tilde{\Gamma}^c_{ab} = \lim_{N\to\infty} {}^N\tilde{\Gamma}^c_{ab}$, we also conclude that the Riemannian geodesic equation for the metric $\tilde{h}$ on $\mathcal{N}^n \times (0,T)$ (defined in Exercise 7.4) limits to the $\sigma = 2\sqrt{\tau}$ reparametrization of the $\mathcal{L}$-geodesic equation as $N \to \infty$.

EXERCISE 7.23 (Motivation for change of time variable). Show that if $\tilde{\beta}: [0,\bar{\sigma}] \to \mathcal{N} \times [0,T]$ is a geodesic, with respect to the connection $\tilde{\nabla}$, with

3. FIRST VARIATION OF $\mathcal{L}$-LENGTH AND EXISTENCE OF $\mathcal{L}$-GEODESICS

$\tilde{\beta}^0(0) = 0$ and $\frac{d\tilde{\beta}^0}{d\sigma}(\sigma) \neq 0$ for $\sigma > 0$, then $\tilde{\beta}^0(\sigma) = A\sigma^2$ for some positive constant A.

SOLUTION TO EXERCISE 7.23. If $\tilde{\beta}^0(\sigma) = \tau(\sigma)$, then the time component of the geodesic equation with respect to $\tilde{\nabla}$ is

$$0 = \frac{d^2\tilde{\beta}^0}{d\sigma^2} + \sum_{0 \leq i,j \leq n} \left(\tilde{\Gamma}^0_{ij} \circ \tilde{\beta}\right) \frac{d\tilde{\beta}^i}{d\sigma} \frac{d\tilde{\beta}^j}{d\sigma}$$

$$= \frac{d^2\tau}{d\sigma^2} - \frac{1}{2\tau}\left(\frac{d\tau}{d\sigma}\right)^2$$

since $\tilde{\Gamma}^0_{ij} = 0$ when $i \geq 1$ or $j \geq 1$, and $\tilde{\Gamma}^0_{00} = -\frac{1}{2\tau}$. Hence, assuming $\tau(\sigma) > 0$ and $\frac{d\tau}{d\sigma}(\sigma) > 0$ for $\sigma > 0$, we have

$$\frac{d}{d\sigma} \log \frac{d\tau}{d\sigma} = \frac{\frac{d^2\tau}{d\sigma^2}}{\frac{d\tau}{d\sigma}} = \frac{\frac{d\tau}{d\sigma}}{2\tau} = \frac{d}{d\sigma} \log \sqrt{\tau},$$

so that

$$\frac{d\tau}{d\sigma} = C\sqrt{\tau}$$

for some constant $C > 0$. Since $\tau(0) = 0$, we conclude

$$\tau(\sigma) = C^2 \frac{\sigma^2}{4}.$$

3.4. Existence of $\mathcal{L}$-geodesics. Our next order of business is to establish the existence of solutions to the initial-value problem for the $\mathcal{L}$-geodesic equation. In this subsection $(\mathcal{M}^n, g(\tau))$, $\tau \in [0,T]$, is a complete solution to the backward Ricci flow with curvature bound $\max\{|\text{Rm}|, |\text{Rc}|\} \leq C_0 < \infty$ on $\mathcal{M} \times [0,T]$. We shall use the following

LEMMA 7.24 (Estimate for speed of $\mathcal{L}$-geodesics). *Let $(\mathcal{M}^n, g(\tau))$, $\tau \in [0,T]$, be a solution to the backward Ricci flow with bounded sectional curvature. There exists a constant $C(n) < \infty$ depending only on n such that given $0 \leq \tau_1 \leq \tau_2 < T$, if $\gamma : [\tau_1, \tau_2] \to \mathcal{M}$ is an $\mathcal{L}$-geodesic with*

$$\lim_{\tau \to \tau_1} \sqrt{\tau} \frac{d\gamma}{d\tau}(\tau) = V \in T_{\gamma(\tau_1)}\mathcal{M},$$

then for any $\tau \in [\tau_1, \tau_2]$,

$$\tau \left|\frac{d\gamma}{d\tau}(\tau)\right|^2_{g(\tau)} \leq e^{6C_0 T} |V|^2 + \frac{C(n)T}{\min\{T - \tau_2, C_0^{-1}\}} \left(e^{6C_0 T} - 1\right),$$

where C_0 is as in (7.27) and $|V|^2 \doteqdot |V|^2_{g(\tau_1)}$.

PROOF. Let $\sigma_i \doteqdot 2\sqrt{\tau_i}$. Define $\beta : [\sigma_1, \sigma_2] \to \mathcal{M}$ by $\beta(\sigma) = \gamma(\sigma^2/4)$, so that

(7.43) $$\lim_{\sigma \to \sigma_1} \left|\frac{d\beta}{d\sigma}\right|^2_{g(\sigma^2/4)} = |V|^2.$$

Since $\tau_2 < T$, by (7.27) and the Bernstein–Bando–Shi derivative estimate (Theorem V1-p. 224), there exists a constant $C(n) < \infty$ such that

$$|\nabla R(x,\tau)| \leq \frac{C(n)C_0}{\sqrt{\min\{T-\tau_2, C_0^{-1}\}}} \doteq C_2 \tag{7.44}$$

for all $(x, \tau) \in \mathcal{M} \times [\tau_1, \tau_2]$. From the $\mathcal{L}$-geodesic equation (7.35), we compute

$$\frac{d}{d\sigma}\left|\frac{d\beta}{d\sigma}\right|^2_{g(\sigma^2/4)} = \frac{\partial g}{\partial \sigma}\left(\frac{d\beta}{d\sigma}, \frac{d\beta}{d\sigma}\right) + 2\left\langle \nabla_{\frac{d\beta}{d\sigma}}\frac{d\beta}{d\sigma}, \frac{d\beta}{d\sigma}\right\rangle$$

$$= -\sigma \operatorname{Rc}\left(\frac{d\beta}{d\sigma}, \frac{d\beta}{d\sigma}\right) + \frac{\sigma^2}{4}\left\langle \nabla R, \frac{d\beta}{d\sigma}\right\rangle. \tag{7.45}$$

Applying the bounds (7.27) and (7.44) on the curvature and its first derivative, we have

$$\frac{d}{d\sigma}\left|\frac{d\beta}{d\sigma}\right|^2_{g(\sigma^2/4)} \leq 2\sqrt{T}C_0 \left|\frac{d\beta}{d\sigma}\right|^2_{g(\sigma^2/4)} + TC_2 \left|\frac{d\beta}{d\sigma}\right|_{g(\sigma^2/4)}$$

$$\leq 3\sqrt{T}C_0 \left|\frac{d\beta}{d\sigma}\right|^2_{g(\sigma^2/4)} + \frac{C_2^2}{4C_0}T^{3/2}. \tag{7.46}$$

In view of the above ordinary differential inequality (7.46), given positive constants c_1 and c_2, consider the ODE

$$\frac{dA}{d\sigma} = c_1 A + c_2. \tag{7.47}$$

Then for positive solutions $A(\sigma)$,

$$\log(c_1 A + c_2)(\sigma) = \log(c_1 A + c_2)(\sigma_1) + c_1(\sigma - \sigma_1),$$

which implies

$$A(\sigma) = e^{c_1(\sigma-\sigma_1)}A(\sigma_1) + \frac{c_2}{c_1}\left(e^{c_1(\sigma-\sigma_1)} - 1\right).$$

Hence, comparing the solution to the ODI (7.46) with (7.43) to the solution to the ODE (7.47) with $A(\sigma_1) = |V|^2$ and taking $c_1 = 3\sqrt{T}C_0$ and $c_2 = \frac{C_2^2}{4C_0}T^{3/2}$, we have

$$\left|\frac{d\beta}{d\sigma}\right|^2_{g(\sigma^2/4)} \leq e^{6C_0\sqrt{T}(\sqrt{\tau}-\sqrt{\tau_1})}|V|^2 + \frac{C_2^2 T}{12C_0^2}\cdot\left(e^{6C_0\sqrt{T}(\sqrt{\tau}-\sqrt{\tau_1})} - 1\right)$$

$$\leq e^{6C_0 T}|V|^2 + \frac{C_2^2 T}{12C_0^2}\left(e^{6C_0 T} - 1\right) \tag{7.48}$$

since $0 \leq \tau_1 \leq \tau \leq \tau_2 < T$. The lemma follows from this and the definition (7.44) of C_2. $\square$

LEMMA 7.25 ($\mathcal{L}$-geodesic IVP—existence). *Let $(\mathcal{M}^n, g(\tau))$, $\tau \in [0, T]$, be a complete solution to the backward Ricci flow with bounded sectional* curvature. Given a space-time point $(p, \tau_1) \in \mathcal{M} \times [0, T]$ and a tangent vector $V \in T_p\mathcal{M}$, there exists a unique $\mathcal{L}$-geodesic $\gamma : [\tau_1, T) \to \mathcal{M}$ with

$$\lim_{\tau \to \tau_1} \sqrt{\tau} \frac{d\gamma}{d\tau}(\tau) = V.$$

NOTATION 7.26. *We shall usually denote γ as γ_V.*

PROOF. *The main idea of the proof is to change variables from τ to σ and to apply standard ODE theory.* In local coordinates, the $\mathcal{L}$-geodesic equation (7.35) for $\beta(\sigma) = \gamma(\sigma^2/4)$ is

(7.49) $$\frac{d^2\beta^i}{d\sigma^2} + \Gamma^i_{jk}\left(\beta(\sigma), \frac{\sigma^2}{4}\right) \frac{d\beta^j}{d\sigma} \frac{d\beta^k}{d\sigma} - \frac{\sigma^2}{8} \nabla^i R + \sigma R^i_k \frac{d\beta^k}{d\sigma} = 0.$$

The above system of ODE is of the form

$$\frac{d^2\beta}{d\sigma^2} = F\left(\frac{d\beta}{d\sigma}, \beta, \sigma\right),$$

where F is a smooth function. From elementary ODE theory, given any point $p \in \mathcal{M}$ and tangent vector $V \in T_p\mathcal{M}$, there exists a path

$$\beta : [2\sqrt{\tau_1}, 2\sqrt{\tau_1} + \varepsilon] \to \mathcal{M}$$

solving (7.49) with $\frac{d\beta}{d\sigma}(2\sqrt{\tau_1}) = V$ for some $\varepsilon > 0$.

We prove by contradiction that β can be extended to an $\mathcal{L}$-geodesic on $[2\sqrt{\tau_1}, 2\sqrt{T})$. Suppose the maximal time interval of existence of β is $[2\sqrt{\tau_1}, 2\sqrt{\tau_2})$ for some $\tau_2 < T$. By Lemma 7.24, we have $\left|\frac{d\beta}{d\sigma}\right|_{g(\sigma^2/4)} \leq C$. This implies $\beta(\sigma)$ converges as $\sigma \to 2\sqrt{\tau_2}$ to some $q \in \mathcal{M}$ using the assumption that the solution $g(\tau)$ is complete. Hence we can extend β beyond $2\sqrt{\tau_2}$. This is a contradiction and the lemma is proved. $\square$

We end this section by proving a lemma about the existence of $\mathcal{L}$-geodesics between any two space-time points (this may also be proved without using the calculus of variations).

LEMMA 7.27 (Existence of minimal $\mathcal{L}$-geodesics). *Let $(\mathcal{M}^n, g(\tau))$, $\tau \in [0, T]$, be a complete solution to the backward Ricci flow with bounded sectional curvature. Given $p, q \in \mathcal{M}$ and $0 \leq \tau_1 < \tau_2 < T$, there exists a smooth path $\gamma(\tau) : [\tau_1, \tau_2] \to \mathcal{M}$ from p to q such that γ has the minimal $\mathcal{L}$-length among all such paths. Furthermore, all $\mathcal{L}$-length minimizing paths are smooth $\mathcal{L}$-geodesics.*

PROOF. Here we sketch a proof using the direct method in the calculus of variations. Let $\gamma_i : [\tau_1, \tau_2] \to \mathcal{M}$, $i \in \mathbb{N}$, be a minimizing sequence for the $\mathcal{L}$-length functional. That is, $\mathcal{L}(\gamma_i) \to \inf_{\hat{\gamma}} \mathcal{L}(\hat{\gamma})$ as $i \to \infty$, where the infimum is taken over all C^1-paths $\hat{\gamma} : [\tau_1, \tau_2] \to \mathcal{M}$ from p to q. We have

$$\int_{\tau_1}^{\tau_2} \sqrt{\tau} \left|\frac{d\gamma_i}{d\tau}(\tau)\right|^2_{g(\tau)} d\tau \leq \mathcal{L}(\gamma_i) - \frac{2}{3}\left(\tau_2^{3/2} - \tau_1^{3/2}\right) \inf_{\mathcal{M} \times [\tau_1, \tau_2]} R(x, \tau) \leq C.$$

Letting $\sigma_j \doteq 2\sqrt{\tau_j}$ and $\beta_i(\sigma) \doteq \gamma_i(\sigma^2/4)$, the above formula says

$$\int_{\sigma_1}^{\sigma_2} \left|\frac{d\beta_i}{d\sigma}\right|^2_{g(\sigma^2/4)} d\sigma \leq C.$$

From standard theory in the calculus of variations, we can conclude that there exists a subsequence such that γ_i converges to a path $\gamma_\infty : [\tau_1, \tau_2] \to \mathcal{M}$ with $\mathcal{L}(\gamma_\infty) = \inf_{\hat\gamma} \mathcal{L}(\hat\gamma)$. All $\mathcal{L}$-length minimizing paths satisfy the $\mathcal{L}$-geodesic equation in the weak sense. By standard theory again, we have that such paths are smooth. □

4. The gradient and time-derivative of the L-distance function

In this section $(\mathcal{M}^n, g(\tau))$, $\tau \in [0, T]$, shall again denote a complete solution to the backward Ricci flow satisfying the pointwise curvature bound $\max\{|\mathrm{Rm}|, |\mathrm{Rc}|\} \leq C_0 < \infty$ on $\mathcal{M} \times [0, T]$, and $p \in \mathcal{M}$ shall denote a basepoint.

4.1. L is locally Lipschitz. Before we study ∇L and $\frac{\partial L}{\partial \tau}$, we prove that, as a consequence of Lemma 7.13, L is locally Lipschitz. As is the case with this and the following one-sided Lipschitz result, we shall prove effective estimates. When we prove L is locally Lipschitz in the space variables, the following one-sided Lipschitz property will be used.

LEMMA 7.28 (One-sided locally Lipschitz in time). *Given* $0 < \tau_0 < T$, *let* $\varepsilon \doteq \min\left\{\frac{\tau_0}{10}, \frac{T-\tau_0}{10}, \frac{1}{10}\right\} > 0$. *For any* $\tau_1 < \tau_2$ *in* $(\tau_0 - \varepsilon, \tau_0 + \varepsilon)$, $q_0 \in \mathcal{M}$, *and* $q \in B_{g(0)}(q_0, \varepsilon)$, *we have*

$$L(q, \tau_1) \leq L(q, \tau_2) + C_1(\tau_2 - \tau_1),$$

where

$$C_1 = C(n, T, C_0, \tau_0) + C(C_0, T)\frac{d^2_{g(\tau_2)}(p, q)}{4\tau_2}.$$

PROOF. Let $\gamma : [0, \tau_2] \to \mathcal{M}$ be a minimal $\mathcal{L}$-geodesic from p to q. We define the piecewise linearly reparametrized path $\eta : [0, \tau_1] \to \mathcal{M}$ by

$$\eta(\tau) \doteq \begin{cases} \gamma(\tau) & \text{if } \tau \in [0, 2\tau_1 - \tau_2], \\ \gamma(\phi(\tau)) & \text{if } \tau \in [2\tau_1 - \tau_2, \tau_1], \end{cases}$$

where

$$\phi(\tau) \doteq 2\tau + \tau_2 - 2\tau_1 \geq \tau$$

for $\tau \in [2\tau_1 - \tau_2, \tau_1]$. Although η is only piecewise smooth, we still have $L(q, \tau_1) \leq \mathcal{L}(\eta)$. (We will use this fact a few times later.) Hence, since

4. GRADIENT AND TIME-DERIVATIVE OF THE L-DISTANCE FUNCTION

$|R| \leq nC_0$,

$$L(q, \tau_1) \leq \mathcal{L}(\gamma) - \int_{2\tau_1-\tau_2}^{\tau_2} \sqrt{\tau} \left(R(\gamma(\tau), \tau) + \left| \frac{d\gamma}{d\tau}(\tau) \right|^2_{g(\tau)} \right) d\tau$$

$$+ \int_{2\tau_1-\tau_2}^{\tau_1} \sqrt{\tau} \left(R(\gamma(\phi(\tau)), \tau) + \left| \frac{d\gamma}{d\tau}(\phi(\tau)) \cdot \dot{\phi}(\tau) \right|^2_{g(\tau)} \right) d\tau$$

$$\leq L(q, \tau_2) + \frac{2nC_0}{3} \left(\tau_2^{3/2} - (2\tau_1 - \tau_2)^{3/2} \right)$$

$$+ \frac{2nC_0}{3} \left(\tau_1^{3/2} - (2\tau_1 - \tau_2)^{3/2} \right)$$

(7.50)
$$+ 2 \int_{2\tau_1-\tau_2}^{\tau_2} \sqrt{\phi^{-1}(\tau)} \left| \frac{d\gamma}{d\tau}(\tau) \right|^2_{g(\phi^{-1}(\tau))} d\tau.$$

Recall that $\beta(\sigma)$ is defined by (7.18). By Lemma 7.13(ii), (iii) there exists $\tau_* \in (0, \tau_2)$ such that

$$\left| \frac{d\beta}{d\sigma}(\sigma_*) \right|^2_{g(\tau_*)} \leq \frac{1}{2\sqrt{\tau_2}} L(q, \tau_2) + \frac{nC_0}{3} \tau_2$$

(7.51)
$$\leq e^{2C_0\tau_2} \frac{d^2_{g(\tau_2)}(p, q)}{4\tau_2} + \frac{2nC_0}{3} \tau_2.$$

Since $\tau_2 \leq T - \varepsilon$, by Shi's derivative estimate we have

$$|\nabla R(x, \tau)| \leq \frac{C(n)C_0}{\sqrt{\min\{T - \tau_2, C_0^{-1}\}}} \doteq C_2$$

for any $(x, \tau) \in \mathcal{M} \times [0, \tau_2]$. From equation (7.46), we have

$$\frac{d}{d\sigma} \left| \frac{d\beta}{d\sigma} \right|^2_{g(\sigma^2/4)} \leq 3\sqrt{T} C_0 \left| \frac{d\beta}{d\sigma} \right|^2_{g(\sigma^2/4)} + \frac{C_2^2}{4C_0} T^{3/2}.$$

Integrating the above inequality over $[\sigma_*, \sigma]$, we get for all $\sigma \in [0, 2\sqrt{\tau_2}]$,

(7.52)
$$\left| \frac{d\beta}{d\sigma} \right|^2_{g(\sigma^2/4)} \leq e^{3\sqrt{T} C_0 |\sigma - \sigma_*|} \left| \frac{d\beta}{d\sigma}(\sigma_*) \right|^2_{g(\tau_*)} + \frac{C_2^2 T}{12 C_0^2} e^{3\sqrt{T} C_0 |\sigma - \sigma_*|}.$$

Noting that $\phi^{-1}(\tau) = \frac{\tau - \tau_2}{2} + \tau_1 \leq \tau$ for $\tau \in [2\tau_1 - \tau_2, \tau_2]$, we can estimate using (7.52) and (7.51)

$$\int_{2\tau_1-\tau_2}^{\tau_2} \sqrt{\phi^{-1}(\tau)} \left| \frac{d\gamma}{d\tau}(\tau) \right|^2_{g(\phi^{-1}(\tau))} d\tau \leq \int_{2\tau_1-\tau_2}^{\tau_2} \sqrt{\tau} \left| \frac{d\gamma}{d\tau}(\tau) \right|^2_{g(\tau)} e^{4C_0\varepsilon} d\tau$$

$$\leq e^{4C_0\varepsilon} \int_{2\tau_1-\tau_2}^{\tau_2} \frac{1}{\sqrt{\tau}} \left| \frac{d\beta}{d\sigma} \right|^2_{g(\tau)} d\tau \leq C_3 \left(\sqrt{\tau_2} - \sqrt{2\tau_1 - \tau_2} \right),$$

where
$$C_3 \doteq 2e^{4C_0\varepsilon+6C_0T}\left(e^{2C_0\tau_2}\frac{d^2_{g(\tau_2)}(p,q)}{4\tau_2} + \frac{2nC_0}{3}\tau_2 + \frac{C_2^2 T}{12C_0^2}\right)$$

(using $|\sigma - \sigma_*| \leq 2\sqrt{T}$). Combining this with (7.50), we obtain[7]

$$\begin{aligned}L(q,\tau_1) - L(q,\tau_2) &\leq \frac{2nC_0}{3}\left(\tau_2^{3/2} - (2\tau_1 - \tau_2)^{3/2}\right) \\ &+ \frac{2nC_0}{3}\left(\tau_1^{3/2} - (2\tau_1 - \tau_2)^{3/2}\right) \\ &+ 2C_3\left(\tau_2^{1/2} - (2\tau_1 - \tau_2)^{1/2}\right) \\ &\leq C_1(\tau_2 - \tau_1),\end{aligned}$$

where
$$C_1 = C(n, T, C_0, \tau_0) + C(C_0, T)\frac{d^2_{g(\tau_2)}(p,q)}{4\tau_2}.$$

□

When combined with the uniqueness of the boundary-value problem for the $\mathcal{L}$-geodesic equation on small time intervals, Lemma 7.13 can also be used to prove the following; we shall give the proof elsewhere.

LEMMA 7.29 (Short $\mathcal{L}$-geodesics are minimizing). *Given $V \in T_p\mathcal{M}$, there exists $\tau_* > 0$ such that $\gamma_V|_{[0,\tau_*]}$ is a minimal $\mathcal{L}$-geodesic.*

The next lemma and Rademacher's Theorem (Lemma 7.110) imply that L is differentiable almost everywhere on $\mathcal{M} \times (0, T)$.

LEMMA 7.30 (L is locally Lipschitz). *The function $L : \mathcal{M} \times (0, T) \to \mathbb{R}$ is Lipschitz with respect to the metric $g(\tau) + d\tau^2$ defined on space-time.*

PROOF. For any $0 < \tau_0 < T$ and $q_0 \in \mathcal{M}$, let $\varepsilon \doteq \min\left\{\frac{\tau_0}{10}, \frac{T-\tau_0}{10}, \frac{1}{10}\right\} > 0$. Then for any $\tau_1 < \tau_2$ in $(\tau_0 - \varepsilon, \tau_0 + \varepsilon)$ and $q_1, q_2 \in B_{g(0)}(q_0, \varepsilon)$,

$$|L(q_1, \tau_1) - L(q_2, \tau_2)| \leq |L(q_1, \tau_1) - L(q_2, \tau_1)| + |L(q_2, \tau_1) - L(q_2, \tau_2)|.$$

To prove that L is Lipschitz near (q_0, τ_0), it suffices to prove (1) and (2) below.

(1) *$L(\cdot, \tau_1)$ is locally Lipschitz in the space variables uniformly in $\tau_1 \in (\tau_0 - \varepsilon, \tau_0 + \varepsilon)$.* Let d_τ denote the distance function with respect to the metric $g(\tau)$ and let $\gamma : [0, \tau_1] \to \mathcal{M}$ be a minimal $\mathcal{L}$-geodesic from p to q_1. Let $\alpha : [\tau_1, \tau_1 + d_0(q_1, q_2)] \to \mathcal{M}$ be a minimal geodesic of constant speed 1, with respect to $g(0)$, joining q_1 to q_2. Then

$$\gamma \smile \alpha : [0, \tau_1 + d_0(q_1, q_2)] \to \mathcal{M}$$

---

[7]Note that $|\tau_2 - (2\tau_1 - \tau_2)| = 2|\tau_2 - \tau_1| = 2|\tau_1 - (2\tau_1 - \tau_2)|$.

4. GRADIENT AND TIME-DERIVATIVE OF THE L-DISTANCE FUNCTION   309

is a piecewise smooth path from p to q_2. We estimate, using $\left|\frac{d\alpha}{d\tau}(\tau)\right|^2_{g(\tau)} \leq e^{2C_0T}\left|\frac{d\alpha}{d\tau}(\tau)\right|^2_{g(0)} = e^{2C_0T}$, that

$$L(q_2, \tau_1 + d_0(q_1, q_2))$$
$$\leq \mathcal{L}(\gamma) + \int_{\tau_1}^{\tau_1 + d_0(q_1, q_2)} \sqrt{\tau}\left(R(\alpha(\tau), \tau) + \left|\frac{d\alpha}{d\tau}(\tau)\right|^2_{g(\tau)}\right) d\tau$$
$$\leq L(q_1, \tau_1) + \frac{2(nC_0 + e^{2C_0T})}{3}\left((\tau_1 + d_0(q_1, q_2))^{3/2} - \tau_1^{3/2}\right)$$
$$\leq L(q_1, \tau_1) + C_1 d_0(q_1, q_2).$$

By Lemma 7.28 we have

$$L(q_2, \tau_1) \leq L(q_2, \tau_1 + d_0(q_1, q_2)) + C_1 d_0(q_1, q_2)$$
$$\leq L(q_1, \tau_1) + C_1 d_0(q_1, q_2)$$
$$\leq L(q_1, \tau_1) + C_1 d_{\tau_1}(q_1, q_2),$$

where we have used $d_0(q_1, q_2) \leq e^{C_0T} d_{\tau_1}(q_1, q_2)$. By the symmetry between q_1 and q_2 we get

$$|L(q_2, \tau_1) - L(q_1, \tau_1)| \leq C_1 d_{\tau_1}(q_1, q_2).$$

(2) $L(q, \cdot)$ *is locally Lipschitz in the time variable uniformly in* $q \in B_{g(0)}(q_0, \varepsilon)$. For any $\tau_1 < \tau_2$ in $(\tau_0 - \varepsilon, \tau_0 + \varepsilon)$, let $\gamma : [0, \tau_1] \to \mathcal{M}$ be a minimal $\mathcal{L}$-geodesic from p to q and let $\beta : [\tau_1, \tau_2] \to \mathcal{M}$ be the constant path $\beta(\tau) = q$. Then $\gamma \smile \beta : [0, \tau_2] \to \mathcal{M}$ is a piecewise smooth path from p to q. Hence

$$L(q, \tau_2) \leq \mathcal{L}(\gamma) + \mathcal{L}(\beta) = \mathcal{L}(\gamma) + \int_{\tau_1}^{\tau_2} \sqrt{\tau} R(q, \tau) d\tau$$
$$\leq L(q, \tau_1) + \frac{2nC_0}{3}\left(\tau_2^{3/2} - \tau_1^{3/2}\right)$$
$$\leq L(q, \tau_1) + C_1(\tau_2 - \tau_1),$$

where C_1 depends only on C_0 and T. Combining this with Lemma 7.28, we obtain

$$|L(q, \tau_2) - L(q, \tau_1)| \leq C_1 |\tau_2 - \tau_1|,$$

where

(7.53) $$C_1 = C'(n, T, C_0, \tau_0) + C(C_0, T)\frac{d^2_{g(\tau_2)}(p, q)}{4\tau_2}.$$

$\square$

COROLLARY 7.31. *L is differentiable almost everywhere on $\mathcal{M} \times (0, T)$ and $L \in W^{1,\infty}_{\text{loc}}(\mathcal{M} \times (0, T))$.*

PROOF. See Lemmas 7.110 and 7.111. $\square$

4.2. Gradient of L.

We compute the gradient of L via the first variation formula for $\mathcal{L}$. Since $L(\cdot, \tau)$ is not smooth in general, the gradient is defined in the barrier sense as described below. Let $\gamma : [0, \bar{\tau}] \to \mathcal{M}$ be a minimal $\mathcal{L}$-geodesic from p to q so that $L(q, \bar{\tau}) = \mathcal{L}(\gamma)$. For any point x in a small neighborhood U of q and any $\tau \in (\bar{\tau} - \varepsilon, \bar{\tau} + \varepsilon)$ with small $\varepsilon > 0$, let $\gamma_{x,\tau} : [0, \tau] \to \mathcal{M}$ be a smooth family of paths with $\gamma_{x,\tau}(0) = p$, $\gamma_{x,\tau}(\tau) = x$ and $\gamma_{q,\bar{\tau}} = \gamma$. (Recall that our definition of a smooth variation says that $\gamma_{x,\tau}\left(\frac{\sigma^2}{4}\right)$ is a smooth function of (σ, x, τ).) Define $\hat{L} : U \times (\bar{\tau} - \varepsilon, \bar{\tau} + \varepsilon) \to \mathbb{R}$ by

$$\hat{L}(x, \tau) = \mathcal{L}(\gamma_{x,\tau}).$$

Then $\hat{L}(x, \tau)$ is a smooth function of (x, τ) when $\tau > 0$, $L(x, \tau) \leq \hat{L}(x, \tau)$ for all $(x, \tau) \in U \times (\bar{\tau} - \varepsilon, \bar{\tau} + \varepsilon)$, and $L(q, \bar{\tau}) = \hat{L}(q, \bar{\tau})$. That is, the function $\hat{L}(\cdot, \cdot)$ is an **upper barrier** for $L(\cdot, \cdot)$ at the point $(q, \bar{\tau})$.

Given a vector $Y(\bar{\tau})$ at q, let $q(s)$ be a smooth path in U with $q(0) = q$ and $\frac{dq}{ds}(0) = Y(\bar{\tau})$. Consider the smooth 1-parameter family of paths $\gamma_s \doteqdot \gamma_{q(s), \bar{\tau}} : [0, \bar{\tau}] \to \mathcal{M}$. Let $Y(\tau) \doteqdot \left.\frac{\partial}{\partial s}\right|_{s=0} \gamma_s(\tau)$ denote the variation vector field along $\gamma(\tau)$. By (7.29), (7.32), and $Y(0) = \vec{0}$, we have

$$\nabla \hat{L}(q, \bar{\tau}) \cdot Y(\bar{\tau}) = \left.\frac{d}{ds}\right|_{s=0} \hat{L}(q(s), \bar{\tau}) = (\delta_Y \mathcal{L})(\gamma) = 2\sqrt{\bar{\tau}} Y(\bar{\tau}) \cdot X(\bar{\tau}).$$

Hence

$$\nabla \hat{L}(q, \bar{\tau}) = 2\sqrt{\bar{\tau}} X(\bar{\tau}).$$

It follows from Lemma 7.30 that $L(\cdot, \bar{\tau})$ is differentiable a.e. on $\mathcal{M}$. Suppose $L(\cdot, \bar{\tau})$ is differentiable at q. Since $\hat{L}(\cdot, \bar{\tau})$ is an upper barrier for $L(\cdot, \bar{\tau})$ at the point q, it is easy to see that

$$\nabla L(q, \bar{\tau}) = \nabla \hat{L}(q, \bar{\tau}) = 2\sqrt{\bar{\tau}} X(\bar{\tau}).$$

Suppose there is another minimal $\mathcal{L}$-geodesic $\gamma' : [0, \bar{\tau}]$ joining p to q. Then we can construct another barrier function $\hat{L}'$ as above; the same proof will imply $\nabla L(q, \bar{\tau}) = 2\sqrt{\bar{\tau}} X'(\bar{\tau})$, where $X'(\bar{\tau}) = \frac{d\gamma'}{d\tau}(\bar{\tau})$. Now both γ and γ' satisfy the same $\mathcal{L}$-geodesic equation and $\gamma(\bar{\tau}) = \gamma'(\bar{\tau}) = q$ and $\frac{d\gamma'}{d\tau}(\bar{\tau}) = \frac{d\gamma}{d\tau}(\bar{\tau}) = \nabla L(q, \bar{\tau})$. By the standard ODE uniqueness theorem, we conclude that $\gamma(\tau) = \gamma'(\tau)$ for $\tau \in [0, \bar{\tau}]$. Hence if $L(\cdot, \bar{\tau})$ is differentiable at q, then the minimal $\mathcal{L}$-geodesic joining $(p, 0)$ to $(q, \bar{\tau})$ is unique.

> **Convention:** If the function $L(\cdot, \bar{\tau})$ is not differentiable at q, then by writing $\nabla L(q, \bar{\tau}) = 2\sqrt{\bar{\tau}} X(\bar{\tau})$,[8] we mean that there is a smooth function $\hat{L}$ satisfying $\hat{L}(x, \bar{\tau}) \geq L(x, \bar{\tau})$ for $x \in U$, $\hat{L}(q, \bar{\tau}) = L(q, \bar{\tau})$, and $\nabla \hat{L}(q, \bar{\tau}) = 2\sqrt{\bar{\tau}} X(\bar{\tau})$.

We have proved the following.

---

[8] Note that $X(\bar{\tau})$ depends on the choice of minimal $\mathcal{L}$-geodesic, which may not be unique.

4. GRADIENT AND TIME-DERIVATIVE OF THE L-DISTANCE FUNCTION

LEMMA 7.32 (Gradient of L formula). *The spatial gradient of the L-distance function is given by*

$$\nabla L(q, \bar{\tau}) = 2\sqrt{\bar{\tau}} X(\bar{\tau}), \tag{7.54}$$

where $X(\bar{\tau}) = \frac{d\gamma}{d\tau}(\bar{\tau})$, for any minimal $\mathcal{L}$-geodesic $\gamma : [0, \bar{\tau}] \to \mathcal{M}$ joining p to q. Furthermore if $L(\cdot, \bar{\tau})$ is differentiable at q, the minimal $\mathcal{L}$-geodesic joining $(p, 0)$ to $(q, \bar{\tau})$ is unique.

REMARK 7.33. The analogy of (7.54) in Riemannian geometry is as follows. Let $d_p(x) \doteqdot d(x, p)$ and suppose d_p is smooth at $q \in (\hat{\mathcal{M}}^n, \hat{g})$. Define $\gamma : [0, b] \to \hat{\mathcal{M}}$ to be the unique unit speed minimal geodesic from p to q. By the first variation formula (7.31), for any $U \in T_q\hat{\mathcal{M}}$,

$$\langle \nabla d_p(q), U \rangle = \left.\frac{d}{du}\right|_{u=0} L(\gamma_u) = \langle \dot{\gamma}(b), U \rangle,$$

provided the $\gamma_u : [0, b] \to \hat{\mathcal{M}}$ satisfy $\gamma_0 = \gamma$, $\gamma_u(0) = p$ and $\left.\frac{\partial}{\partial u}\right|_{u=0} \gamma_u(b) = U$. That is, $\nabla d_p(q) = \dot{\gamma}(b)$.

Taking the norm of (7.54),

$$|\nabla L|^2(q, \bar{\tau}) = 4\bar{\tau} |X(\bar{\tau})|^2 = -4\bar{\tau} R(q, \bar{\tau}) + 4\bar{\tau} \left(R(q, \bar{\tau}) + |X(\bar{\tau})|^2 \right). \tag{7.55}$$

The reason we rewrite this in a seemingly more complicated way is that both R and $R + |X(\bar{\tau})|^2$ are natural quantities.[9]

4.3. Time-derivative of L. Next we compute the time-derivative of L. This time we need to choose $\gamma_{x,\tau}$ used in subsection 4.2 above a little more carefully. Given $(q, \bar{\tau})$, let $\gamma : [0, \bar{\tau}] \to \mathcal{M}$ be a minimal $\mathcal{L}$-geodesic from p to q so that $L(q, \bar{\tau}) = \mathcal{L}(\gamma)$. We first extend γ to a smooth curve $\gamma : [0, \bar{\tau} + \varepsilon] \to \mathcal{M}$ for some $\varepsilon > 0$, and then we choose a smooth family of curves $\gamma_{x,\tau}$ to satisfy $\gamma_{x,\tau}(0) = p$, $\gamma_{\gamma(\tau),\tau} = \gamma|_{[0,\tau]}$ and $\gamma_{x,\tau}(\tau) = x$. Define

$$\hat{L}(x, \tau) \doteqdot \mathcal{L}(\gamma_{x,\tau}) \tag{7.56}$$

for $(x, \tau) \in U \times (\bar{\tau} - \varepsilon, \bar{\tau} + \varepsilon)$.

We compute, using the chain rule and (7.54),

$$\frac{\partial \hat{L}}{\partial \tau}(q, \bar{\tau}) = \left.\frac{\partial \hat{L}(\gamma(\tau), \tau)}{\partial \tau}\right|_{\tau=\bar{\tau}} = \left.\frac{d}{d\tau}\right|_{\tau=\bar{\tau}} \left[\hat{L}(\gamma(\tau), \tau) \right] - \nabla \hat{L} \cdot X$$

$$= \left.\frac{d}{d\tau}\right|_{\tau=\bar{\tau}} \left[\int_0^\tau \sqrt{\tilde{\tau}} \left(R(\gamma(\tilde{\tau}), \tilde{\tau}) + \left|\frac{d\gamma}{d\tau}(\tilde{\tau})\right|^2 \right) d\tilde{\tau} \right] - 2\sqrt{\bar{\tau}} |X(\bar{\tau})|^2$$

$$= \sqrt{\bar{\tau}} \left(R(\gamma(\bar{\tau}), \bar{\tau}) + |X(\bar{\tau})|^2 \right) - 2\sqrt{\bar{\tau}} |X(\bar{\tau})|^2.$$

It follows from Lemma 7.30 that $L(q, \cdot)$ is differentiable a.e. on $(0, T)$. As discussed in subsection 4.2 above, if $L(q, \cdot)$ is differentiable at $\bar{\tau}$, then

---

[9] As it is the integrand of the $\mathcal{L}$-length, $\sqrt{\tau}(R + |X|^2)$ is a natural quantity.

$\frac{\partial L}{\partial \tau}(q, \bar{\tau}) = \frac{\partial \hat{L}}{\partial \tau}(q, \bar{\tau})$. If $L(q, \cdot)$ is not differentiable at $\bar{\tau}$, then by writing $\frac{\partial L}{\partial \tau}(q, \bar{\tau}) = \sqrt{\bar{\tau}} \left(R(\gamma(\bar{\tau}), \bar{\tau}) + |X(\bar{\tau})|^2 \right) - 2\sqrt{\bar{\tau}} |X(\bar{\tau})|^2$, we mean (this is our **convention** below) that there is a smooth function $\hat{L}(x, \tau)$ satisfying $\hat{L}(x, \tau) \geq L(x, \tau)$ for $x \in U$ and $\tau \in (\bar{\tau} - \epsilon, \bar{\tau} + \epsilon)$, $\hat{L}(q, \bar{\tau}) = L(q, \bar{\tau})$ and $\frac{\partial \hat{L}}{\partial \tau}(q, \bar{\tau}) = \sqrt{\bar{\tau}} \left(R(\gamma(\bar{\tau}), \bar{\tau}) + |X(\bar{\tau})|^2 \right) - 2\sqrt{\bar{\tau}} |X(\bar{\tau})|^2$. Now we have proved

LEMMA 7.34 (Time-derivative of L formula). *The time-derivative of the L-distance function is given by*

$$(7.57) \qquad \frac{\partial L}{\partial \tau}(q, \bar{\tau}) = -\sqrt{\bar{\tau}} \left(R(q, \bar{\tau}) + |X(\bar{\tau})|^2 \right) + 2\sqrt{\bar{\tau}} R(q, \bar{\tau}),$$

where $X(\bar{\tau}) = \frac{d\gamma}{d\tau}(\bar{\tau})$, for any minimal $\mathcal{L}$-geodesic $\gamma : [0, \bar{\tau}] \to \mathcal{M}$ joining p to q.

In the case where $(\mathcal{M}^n, g(\tau) \equiv g_0)$ is Ricci flat, we have $\gamma(\tau) = \beta(2\sqrt{\tau})$, where $\beta : [0, 2\sqrt{\bar{\tau}}] \to \mathcal{M}$ is a constant speed Riemannian geodesic with respect to g_0. Thus $\frac{d\gamma}{d\tau} = \frac{1}{\sqrt{\tau}} \dot{\beta}(2\sqrt{\tau})$ and $\left| \dot{\beta}(2\sqrt{\tau}) \right| \equiv \frac{d(q,p)}{2\sqrt{\bar{\tau}}}$. On the other hand $L(q, \tau) = \frac{d(q,p)^2}{2\sqrt{\tau}}$. Hence

$$\frac{\partial L}{\partial \tau}(q, \bar{\tau}) = -\frac{d(q, p)^2}{4\bar{\tau}^{3/2}} = -\sqrt{\bar{\tau}} \left| \frac{d\gamma}{d\tau}(\bar{\tau}) \right|^2,$$

agreeing with (7.57).

5. The second variation formula for $\mathcal{L}$ and the Hessian of L

Recall that the second variation of arc length formula of a geodesic γ is

$$(7.58) \quad \left. \frac{d^2}{du^2} \right|_{u=0} \mathrm{L}(\gamma_u)$$
$$= \int_0^b \left(|\nabla_{\dot{\gamma}} U|^2 - \langle \nabla_{\dot{\gamma}} U, \dot{\gamma} \rangle^2 - \langle \mathrm{Rm}(U, \dot{\gamma})\dot{\gamma}, U \rangle \right) ds + \langle \nabla_U U, \dot{\gamma} \rangle \big|_0^b,$$

where $\gamma_u : [0, b] \to \mathcal{M}$ is parametrized by arc length s and satisfies $\gamma_0 = \gamma$ and $U \doteqdot \frac{\partial}{\partial u} \big|_{u=0} \gamma_u$. This formula is fundamental to Riemannian geometry for a variety of reasons. For example, on a complete Riemannian manifold, any two points can be joined by a minimal geodesic, in which case $\frac{d^2}{du^2} \big|_{u=0} \mathrm{L}(\gamma_u) \geq 0$ for endpoint-preserving variations. The second variation of arc length can also be used to bound from above the Hessian of the distance function. In this section we consider the analogous second variation formula for $\mathcal{L}$-length.

The first variation of $\mathcal{L}$-length determines the $\mathcal{L}$-geodesic equation and is related to the space-time connection. The second variation formula for $\mathcal{L}$-length at an $\mathcal{L}$-geodesic $\gamma : [0, \bar{\tau}] \to \mathcal{M}$ is related to the space-time curvature and hence Hamilton's matrix Harnack quadratic. In the case of a minimal $\mathcal{L}$-geodesic, it also gives an upper bound for the Hessian of the barrier function

$\hat{L}$ defined in subsection 4.2 of this chapter. At a point q where $L(\cdot, \bar{\tau})$ is C^2, the second variation formula gives an upper bound for the Hessian of the L-distance function and tracing this estimate yields an estimate for the Laplacian of L. The Hessian upper bound will be very important in discussing the weak solution formulation later.

In this section $(\mathcal{M}^n, g(\tau))$, $\tau \in [0, T]$, will denote a complete solution to the backward Ricci flow satisfying the pointwise curvature bound $\max\{|\mathrm{Rm}|, |\mathrm{Rc}|\} \leq C_0 < \infty$ on $\mathcal{M} \times [0, T]$, and $p \in \mathcal{M}$ shall denote a basepoint.

5.1. The second variation formula for $\mathcal{L}$. Let $\bar{\tau} \in (0, T)$ and let $\gamma : [0, \bar{\tau}] \to \mathcal{M}$ be an $\mathcal{L}$-geodesic from p to q. Let $\gamma_s : [0, \bar{\tau}] \to \mathcal{M}$, $s \in (-\varepsilon, \varepsilon)$, be a smooth family of paths with $\gamma_0(\tau) = \gamma(\tau)$. Recall that our convention about the smoothness of a variation γ_s of γ is that $\beta_s(\sigma) \doteqdot \gamma_s\left(\frac{\sigma^2}{4}\right)$ is required to be a smooth function of (σ, s). It is easy to see that the nondifferentiability of γ_s at $\tau = 0$ causes no trouble in the following calculation. This would also be clear if we use (7.19) to do the calculation. Define $\frac{\partial \gamma_s}{\partial \tau}(\tau) \doteqdot X(\tau, s)$ and $\frac{\partial \gamma_s}{\partial s}(\tau) \doteqdot Y(\tau, s)$, so that $[X, Y] = 0$.[10] We also write $Y(\tau) \doteqdot Y(\tau, 0)$. Note that $Y\left(\frac{\sigma^2}{4}, s\right)$ is a smooth function of (σ, s).

Recall the first variation formula (7.30)

$$\frac{d}{ds}\mathcal{L}(\gamma_s) = \int_0^{\bar{\tau}} \sqrt{\tau}\left(Y(R) + 2\langle \nabla_Y X, X \rangle\right) d\tau,$$

which holds for all $s \in (-\varepsilon, \varepsilon)$. Differentiating this again, we get

$$(\delta_Y^2 \mathcal{L})(\gamma) \doteqdot \left.\frac{d^2 \mathcal{L}(\gamma_s)}{ds^2}\right|_{s=0}$$
$$= \int_0^{\bar{\tau}} \sqrt{\tau}\left(Y(Y(R)) + 2\langle \nabla_Y \nabla_Y X, X \rangle + 2|\nabla_Y X|^2\right) d\tau.$$

Now since $[X, Y] = 0$,

$$\langle \nabla_Y \nabla_Y X, X \rangle = \langle \nabla_Y \nabla_X Y, X \rangle = \langle R(Y, X) Y, X \rangle + \langle \nabla_X \nabla_Y Y, X \rangle.$$

Hence

$$(7.59) \quad (\delta_Y^2 \mathcal{L})(\gamma) = \int_0^{\bar{\tau}} \sqrt{\tau}\left(\begin{array}{c} Y(Y(R)) + 2\langle R(Y, X) Y, X \rangle \\ +2\langle \nabla_X \nabla_Y Y, X \rangle + 2|\nabla_Y X|^2 \end{array}\right) d\tau.$$

---

[10]Alternately, given any vector field Y along γ, there exists a family of paths γ_s such that $\left.\frac{\partial \gamma_s}{\partial s}\right|_{s=0} = Y$. In this case, we extend Y by defining $\frac{\partial \gamma_s}{\partial s} = Y$ for $s \in (-\varepsilon, \varepsilon)$, so that $[X, Y] = 0$. Technically, X and Y are sections of the bundle $G^*T\mathcal{M}$ on $[0, \bar{\tau}] \times (-\varepsilon, \varepsilon)$, where $G(\tau, s) \doteqdot \gamma_s(\tau)$.

On the other hand, we compute
$$\frac{d}{d\tau}\langle\nabla_Y Y, X\rangle = \langle\nabla_X\nabla_Y Y, X\rangle + \langle\nabla_Y Y, \nabla_X X\rangle$$
$$+ \frac{\partial g}{\partial \tau}(\nabla_Y Y, X) + \left\langle\left(\frac{\partial}{\partial\tau}\nabla\right)_Y Y, X\right\rangle.$$

Now $\frac{\partial g}{\partial \tau} = 2\operatorname{Rc}$ and
$$\left\langle\left(\frac{\partial}{\partial\tau}\nabla\right)_Y Y, X\right\rangle = 2(\nabla_Y \operatorname{Rc})(Y, X) - (\nabla_X \operatorname{Rc})(Y, Y).$$

Hence

(7.60) $\frac{d}{d\tau}\langle\nabla_Y Y, X\rangle = \langle\nabla_X\nabla_Y Y, X\rangle + \langle\nabla_Y Y, \nabla_X X\rangle + 2\operatorname{Rc}(\nabla_Y Y, X)$
$$+ 2(\nabla_Y \operatorname{Rc})(Y, X) - (\nabla_X \operatorname{Rc})(Y, Y).$$

Suppose

(7.61) $$Y(0) = \vec{0}$$

(this and the fact that $\sqrt{\tau}X(\tau)$ has a limit as $\tau \to 0$ are used to get the third equality below). Then applying (7.60) to (7.59) and integrating by parts, we compute

$(\delta_Y^2 \mathcal{L})(\gamma)$
$$= \int_0^{\bar{\tau}} \sqrt{\tau}\left(Y(Y(R)) + 2\langle R(Y, X)Y, X\rangle + 2|\nabla_Y X|^2\right) d\tau$$
$$+ 2\int_0^{\bar{\tau}} \sqrt{\tau}\begin{pmatrix} \frac{d}{d\tau}\langle\nabla_Y Y, X\rangle - \langle\nabla_Y Y, \nabla_X X\rangle - 2\operatorname{Rc}(\nabla_Y Y, X) \\ -2(\nabla_Y \operatorname{Rc})(Y, X) + (\nabla_X \operatorname{Rc})(Y, Y) \end{pmatrix} d\tau$$
$$= \int_0^{\bar{\tau}} \sqrt{\tau}\left(Y(Y(R)) + 2\langle R(Y, X)Y, X\rangle + 2|\nabla_Y X|^2\right) d\tau$$
$$+ 2\int_0^{\bar{\tau}} \sqrt{\tau}\begin{pmatrix} -\langle\nabla_Y Y, \nabla_X X\rangle - 2\operatorname{Rc}(\nabla_Y Y, X) \\ -2(\nabla_Y \operatorname{Rc})(Y, X) + (\nabla_X \operatorname{Rc})(Y, Y) \end{pmatrix} d\tau$$
$$+ 2\sqrt{\tau}\langle\nabla_Y Y, X\rangle\big|_0^{\bar{\tau}} - \int_0^{\bar{\tau}} \frac{1}{\sqrt{\tau}}\langle\nabla_Y Y, X\rangle d\tau$$
$$= 2\sqrt{\bar{\tau}}\langle\nabla_Y Y, X\rangle + \int_0^{\bar{\tau}} \sqrt{\tau}\begin{pmatrix} Y(Y(R)) - \nabla_Y Y \cdot \nabla R \\ +2\langle R(Y, X)Y, X\rangle + 2|\nabla_Y X|^2 \end{pmatrix} d\tau$$
$$+ 2\int_0^{\bar{\tau}} \sqrt{\tau}\begin{pmatrix} -\langle\nabla_Y Y, [\nabla_X X + 2\operatorname{Rc}(X) - \frac{1}{2}\nabla R + \frac{1}{2\tau}X]\rangle \\ -2(\nabla_Y \operatorname{Rc})(Y, X) + (\nabla_X \operatorname{Rc})(Y, Y) \end{pmatrix} d\tau$$
$$= 2\sqrt{\bar{\tau}}\langle\nabla_Y Y, X\rangle + \int_0^{\bar{\tau}} \sqrt{\tau}\left(\nabla_{Y,Y}^2 R + 2\langle R(Y, X)Y, X\rangle + 2|\nabla_Y X|^2\right) d\tau$$
$$+ \int_0^{\bar{\tau}} \sqrt{\tau}\left(-4(\nabla_Y \operatorname{Rc})(Y, X) + 2(\nabla_X \operatorname{Rc})(Y, Y)\right) d\tau,$$

5. THE SECOND VARIATION FORMULA FOR $\mathcal{L}$ AND THE HESSIAN OF L

where we used the $\mathcal{L}$-geodesic equation (7.32) to get the last equality; in the above,

$$\nabla^2_{Y,Y} R \doteqdot Y(Y(R)) - (\nabla_Y Y)(R) = \operatorname{Hess}(R)(Y,Y).$$

That is,

LEMMA 7.35 ($\mathcal{L}$-Second variation — version 1). *Let $\bar{\tau} \in (0,T)$ and let $\gamma : [0, \bar{\tau}] \to \mathcal{M}$ be an $\mathcal{L}$-geodesic from p to q and let $Y \doteqdot \frac{\partial}{\partial s}\gamma_s$ for some smooth variation γ_s of γ with $Y(0) = \vec{0}$. The second variation of $\mathcal{L}$-length is given by*

$$(\delta^2_Y \mathcal{L})(\gamma) = 2\sqrt{\bar{\tau}} \langle \nabla_Y Y, X \rangle (\bar{\tau})$$

(7.62)
$$+ \int_0^{\bar{\tau}} \sqrt{\tau} \left(\begin{array}{c} \nabla^2_{Y,Y} R + 2\langle R(Y,X)Y, X \rangle + 2|\nabla_Y X|^2 \\ -4(\nabla_Y \operatorname{Rc})(Y,X) + 2(\nabla_X \operatorname{Rc})(Y,Y) \end{array} \right) d\tau.$$

REMARK 7.36. Note that by (7.54) we have $(\delta \mathcal{L})(\gamma) = 2\sqrt{\bar{\tau}} X(\bar{\tau})$. Hence

$$(\delta^2_Y \mathcal{L})(\gamma) - 2\sqrt{\bar{\tau}} \langle X, \nabla_Y Y \rangle(\bar{\tau}) = (\delta^2_Y \mathcal{L})(\gamma) - (\delta_{\nabla_Y Y} \mathcal{L})(\gamma),$$

whose value only depends on $Y(\tau)$ defined along $\gamma(\tau)$. This is analogous to considering $(\operatorname{Hess} f)(Y,Y) = YY(f) - \nabla_Y Y \cdot \nabla f$.

We now rewrite the $\mathcal{L}$-second variation formula in a better form, which relates to Hamilton's matrix Harnack quadratic, i.e., the space-time curvature.[11] Since

$$\frac{d}{d\tau}[\operatorname{Rc}(Y(\tau), Y(\tau))] = \left(\frac{\partial}{\partial \tau} \operatorname{Rc}\right)(Y,Y) + (\nabla_X \operatorname{Rc})(Y,Y) + 2\operatorname{Rc}(\nabla_X Y, Y),$$

integrating by parts, we have

$$-\int_0^{\bar{\tau}} \sqrt{\tau} \left(\frac{\partial}{\partial \tau} \operatorname{Rc}\right)(Y,Y) d\tau$$
$$= \int_0^{\bar{\tau}} \sqrt{\tau} \left(\frac{1}{2\tau} \operatorname{Rc}(Y,Y) + (\nabla_X \operatorname{Rc})(Y,Y) + 2\operatorname{Rc}(\nabla_X Y, Y)\right) d\tau$$
$$- \sqrt{\tau} \operatorname{Rc}(Y,Y)\big|_0^{\bar{\tau}}.$$

---

[11] See Section 5 of Chapter 8 for the reason why the space-time curvature is Hamilton's matrix quadratic.

Hence (7.62) and $Y(0) = \vec{0}$ imply

$$\frac{1}{2}\left(\delta_Y^2 \mathcal{L}\right)(\gamma) - \sqrt{\tau}\left\langle \nabla_Y Y, X\right\rangle + \sqrt{\tau}\operatorname{Rc}(Y,Y)\Big|_0^{\bar{\tau}}$$
$$= \int_0^{\bar{\tau}} \sqrt{\tau}\left(\left(\frac{\partial}{\partial \tau}\operatorname{Rc} + \frac{1}{2\tau}\operatorname{Rc}\right)(Y,Y) + \frac{1}{2}\nabla_{Y,Y}^2 R\right) d\tau$$
$$+ \int_0^{\bar{\tau}} \sqrt{\tau}\left(\langle R(Y,X)Y, X\rangle - |\operatorname{Rc}(Y)|^2\right) d\tau$$
$$+ \int_0^{\bar{\tau}} \sqrt{\tau}\left(-2\left(\nabla_Y \operatorname{Rc}\right)(Y,X) + 2\left(\nabla_X \operatorname{Rc}\right)(Y,Y)\right) d\tau$$
$$+ \int_0^{\bar{\tau}} \sqrt{\tau}\left|\nabla_X Y + \operatorname{Rc}(Y)\right|^2 d\tau,$$

where we used $\nabla_X Y = \nabla_Y X$. Let $H(X,Y)$ denote the matrix Harnack expression

(7.63)
$$H(X,Y) \doteqdot -2\left(\frac{\partial}{\partial \tau}\operatorname{Rc}\right)(Y,Y) - \nabla_{Y,Y}^2 R + 2|\operatorname{Rc}(Y)|^2 - \frac{1}{\tau}\operatorname{Rc}(Y,Y)$$
$$- 2\langle R(Y,X)Y, X\rangle - 4\left(\nabla_X \operatorname{Rc}\right)(Y,Y) + 4\left(\nabla_Y \operatorname{Rc}\right)(Y,X).$$

By substituting the definition of $H(X,Y)$ in the above formula, we obtain

(7.64)
$$\left(\delta_Y^2 \mathcal{L}\right)(\gamma) - 2\sqrt{\bar{\tau}}\left\langle \nabla_Y Y, X\right\rangle(\bar{\tau}) + 2\sqrt{\bar{\tau}}\operatorname{Rc}(Y,Y)(\bar{\tau})$$
$$= -\int_0^{\bar{\tau}} \sqrt{\tau} H(X,Y) d\tau + \int_0^{\bar{\tau}} 2\sqrt{\tau}\left|\nabla_X Y + \operatorname{Rc}(Y)\right|^2 d\tau.$$

An even nicer form is

LEMMA 7.37 ($\mathcal{L}$-Second variation — version 2). *Let $\bar{\tau} \in (0,T)$ and let $\gamma : [0, \bar{\tau}] \to \mathcal{M}$ be an $\mathcal{L}$-geodesic. If $Y(\tau) \doteqdot \frac{\partial}{\partial s}\gamma_s(\tau)$, for a smooth variation γ_s of γ, satisfies $Y(0) = \vec{0}$, then*

(7.65)
$$\left(\delta_Y^2 \mathcal{L}\right)(\gamma) - 2\sqrt{\bar{\tau}}\left\langle \nabla_Y Y, X\right\rangle(\bar{\tau}) + 2\sqrt{\bar{\tau}}\operatorname{Rc}(Y,Y)(\bar{\tau}) = \frac{|Y(\bar{\tau})|^2}{\sqrt{\bar{\tau}}}$$
$$- \int_0^{\bar{\tau}} \sqrt{\tau} H(X,Y) d\tau + \int_0^{\bar{\tau}} 2\sqrt{\tau}\left|\nabla_X Y + \operatorname{Rc}(Y) - \frac{1}{2\tau}Y\right|^2 d\tau.$$

PROOF. Since

$$\left|\nabla_X Y + \operatorname{Rc}(Y) - \frac{1}{2\tau}Y\right|^2$$
$$= |\nabla_X Y + \operatorname{Rc}(Y)|^2 - \frac{1}{\tau}\left(\langle \nabla_X Y, Y\rangle + \langle \operatorname{Rc}(Y), Y\rangle\right) + \frac{1}{4\tau^2}|Y|^2$$
$$= |\nabla_X Y + \operatorname{Rc}(Y)|^2 - \frac{1}{2\tau}\frac{d}{d\tau}|Y|^2 + \frac{1}{4\tau^2}|Y|^2,$$

5. THE SECOND VARIATION FORMULA FOR $\mathcal{L}$ AND THE HESSIAN OF L

we have

$$\int_0^{\bar{\tau}} 2\sqrt{\tau} \left| \nabla_X Y + \operatorname{Rc}(Y) \right|^2 d\tau$$
$$= \int_0^{\bar{\tau}} 2\sqrt{\tau} \left| \nabla_X Y + \operatorname{Rc}(Y) - \frac{1}{2\tau} Y \right|^2 d\tau + \frac{1}{\sqrt{\tau}} |Y|^2 \Big|_0^{\bar{\tau}},$$

where we integrated by parts. Note that the assumption of the lemma implies that $Y\left(\frac{\sigma^2}{4}\right)$ is smooth in σ and $\lim_{\tau \to 0} \frac{1}{\sqrt{\tau}} |Y(\tau)|^2 = 0$. The lemma now follows from (7.64). $\square$

We now consider a special case of this formula. As above, let $\gamma : [0, \bar{\tau}] \to \mathcal{M}$ be an $\mathcal{L}$-geodesic. Fix a vector $Y_{\bar{\tau}} \in T_{\gamma(\bar{\tau})}\mathcal{M}$ and define a vector field $Y(\tau)$ along γ by solving the following ODE along $-\gamma$:

(7.66) $$\nabla_X Y = -\operatorname{Rc}(Y) + \frac{1}{2\tau} Y, \quad \tau \in [0, \bar{\tau}],$$
$$Y(\bar{\tau}) = Y_{\bar{\tau}}.$$

Note that any vector field along γ can be considered as a variation vector field. In particular, we may extend $Y(\tau)$ to $Y(\tau, s)$ for some smooth variation of γ.

REMARK 7.38. Equation (7.66) is equivalent to

$$\nabla_X \left(\frac{1}{\sqrt{\tau}} Y \right) = -\operatorname{Rc}\left(\frac{1}{\sqrt{\tau}} Y \right),$$

which essentially says $\frac{1}{\sqrt{\tau}} Y$ is parallel with respect to the space-time connection. Note that if $\gamma_s : [0, \bar{\tau}] \to \mathcal{M}$ is a 1-parameter family of paths such that $X = \frac{\partial}{\partial \tau} \gamma_s$ and $Y = \frac{\partial}{\partial s} \gamma_s$, then $[X, Y] = 0$ and (7.66) may be rewritten as

$$\left(\nabla X + \operatorname{Rc} - \frac{1}{2\tau} g \right)(Y) = 0,$$

which is reminiscent of the gradient shrinker equation.

From (7.66) we compute

(7.67) $$\frac{d}{d\tau} |Y|^2 = \frac{d}{d\tau} [g(Y, Y)] = 2 \langle \nabla_X Y, Y \rangle + 2\operatorname{Rc}(Y, Y) = \frac{1}{\tau} |Y|^2.$$

Solving this ODE, we have

(7.68) $$|Y(\tau)|^2 = \frac{\tau}{\bar{\tau}} |Y(\bar{\tau})|^2.$$

Thus $Y(0) = \vec{0}$. Hence by (7.65) we have the following.

LEMMA 7.39 (*$\mathcal{L}$-Second variation under (7.66)*). *If $\bar{\tau} \in (0, T)$, $Y_{\bar{\tau}} \in T_{\gamma(\bar{\tau})}\mathcal{M}$ and Y is a solution to (7.66) with $Y(\bar{\tau}) = Y_{\bar{\tau}}$, then the second variation of $\mathcal{L}$-length is given by*

$$
\left(\delta_Y^2 \mathcal{L}\right)(\gamma) - 2\sqrt{\bar{\tau}} \left\langle \nabla_Y Y, X \right\rangle (\bar{\tau}) + 2\sqrt{\bar{\tau}} \operatorname{Rc}(Y, Y)(\bar{\tau})
$$

(7.69)
$$
= -\int_0^{\bar{\tau}} \sqrt{\tau} H(X, Y) \, d\tau + \frac{|Y(\bar{\tau})|^2}{\sqrt{\bar{\tau}}}.
$$

5.2. Hessian comparison for L. Corresponding to the $\mathcal{L}$-second variation formula is an upper bound for the Hessian of the L-distance function, which we derive in this subsection. Given any $(q, \bar{\tau})$, let $\gamma : [0, \bar{\tau}] \to \mathcal{M}$ be a minimal $\mathcal{L}$-geodesic from p to q so that $L(q, \bar{\tau}) = \mathcal{L}(\gamma)$. Fix a vector $Y \in T_q \mathcal{M} - \{\vec{0}\}$ and define the vector field $Y(\tau)$ along γ to be the solution to (7.66) with $Y(\bar{\tau}) = Y$. Let $\gamma_s : [0, \bar{\tau}] \to \mathcal{M}$ be a smooth family of curves for $s \in (-\varepsilon, \varepsilon)$ with

$$
\left. \frac{d\gamma_s}{ds} \right|_{s=0} (\tau) = Y(\tau) \quad \text{and} \quad (\nabla_Y Y)(\bar{\tau}) = 0.
$$

Then there exists a small neighborhood U of q, $\delta \in (0, \varepsilon]$, and a smooth family of curves $\gamma_{x,\tau} : [0, \tau] \to \mathcal{M}$ for $(x, \tau) \in U \times (\bar{\tau} - \delta, \bar{\tau} + \delta)$ satisfying

$$
\gamma_{x,\tau}(0) = p, \quad \gamma_{\gamma_s(\bar{\tau}), \bar{\tau}} = \gamma_s, \quad \text{and} \quad \gamma_{x,\tau}(\tau) = x
$$

for $s \in (-\delta, \delta)$. We define $\hat{L}(x, \tau) \doteqdot \mathcal{L}(\gamma_{x,\tau})$. Then $\hat{L}(\gamma_s(\bar{\tau}), \bar{\tau}) = \mathcal{L}(\gamma_s)$.

Since $\hat{L}(\cdot, \cdot)$ is an upper barrier function for the L-distance function $L(\cdot, \cdot)$ at $(q, \bar{\tau})$, we have

$$
\left(\operatorname{Hess}_{(q,\bar{\tau})} L\right)(Y, Y) \leq \left(\operatorname{Hess}_{(q,\bar{\tau})} \hat{L}\right)(Y, Y)
$$

when $L(\cdot, \bar{\tau})$ is C^2 at q. Since $(\nabla_Y Y)(\bar{\tau}) = 0$ and

$$
\left(\operatorname{Hess}_{(q,\bar{\tau})} \hat{L}\right)(Y, Y) = \left. \frac{d^2}{ds^2} \right|_{s=0} \hat{L}(\gamma_s(\bar{\tau}), \bar{\tau}) = \left. \frac{d^2}{ds^2} \right|_{s=0} \mathcal{L}(\gamma_s),
$$

combining this with Lemma 7.39, we get the **Hessian Comparison Theorem for L**.

COROLLARY 7.40 (*Inequality for Hessian of L*). *Given $\bar{\tau} \in (0, T)$, $q \in \mathcal{M}$, and $Y \in T_q \mathcal{M}$, let $\gamma : [0, \bar{\tau}] \to \mathcal{M}$ be a minimal $\mathcal{L}$-geodesic from p to q. The Hessian of the L-distance function $L(\cdot, \bar{\tau})$ at q has the upper bound*

(7.70)
$$
\left(\operatorname{Hess}_{(q,\bar{\tau})} L\right)(Y, Y) \leq -\int_0^{\bar{\tau}} \sqrt{\tau} H(X, Y)(\tau) \, d\tau + \frac{|Y(\bar{\tau})|^2}{\sqrt{\bar{\tau}}}
$$
$$
- 2\sqrt{\bar{\tau}} \operatorname{Rc}(Y, Y)(\bar{\tau}),
$$

where $Y(\tau)$ is a solution to (7.66) with $Y(\bar{\tau}) = Y$ and H is the matrix Harnack expression defined in (7.63). Equality in (7.70) holds when $L(\cdot, \bar{\tau})$ is C^2 at q and $Y(\tau)$ is the variation vector field of a family of minimal $\mathcal{L}$-geodesics.

5. THE SECOND VARIATION FORMULA FOR $\mathcal{L}$ AND THE HESSIAN OF L

If $L(\cdot, \bar{\tau})$ is not C^2 at q, the above inequality is understood in the barrier sense; this is our convention below. More precisely, there is a smooth function $\hat{L}(\cdot, \bar{\tau})$ defined near q such that $\left(\mathrm{Hess}_{(q,\bar{\tau})} \hat{L}\right)(Y,Y)$ satisfies inequality (7.70) and $\hat{L}(\cdot, \bar{\tau})$ is an upper barrier function for $L(\cdot, \bar{\tau})$ with $\hat{L}(q, \bar{\tau}) = L(q, \bar{\tau})$.

LEMMA 7.41 (Upper bound for Hessian of L). *Fix $T_0 \in (0, T)$. Given $\bar{\tau} \in (0, T_0]$, $q \in \mathcal{M}$, and $Y \in T_q \mathcal{M}$, the Hessian of the L-distance function $L(\cdot, \bar{\tau})$ at q has the upper bound*

$$(7.71) \quad \left(\mathrm{Hess}_{(q,\bar{\tau})} L\right)(Y,Y) \leq \left(C_2 + C_2 \frac{d^2_{g(\bar{\tau})}(p,q)}{\sqrt{\bar{\tau}}} + \frac{1}{\sqrt{\bar{\tau}}}\right) |Y|^2,$$

where C_2 is a constant depending only on n, T_0, T, C_0 ($C_0 \geq \sup_{\mathcal{M} \times [0,T]} |\mathrm{Rm}|$ is as in (7.27)).

PROOF. From Shi's derivative estimate and the equation for $\frac{\partial}{\partial \tau} \mathrm{Rc}$, there is a constant C_1 depending on n, $T - T_0$, and C_0 such that $\left|\frac{\partial}{\partial \tau} \mathrm{Rc}\right|$, $|\nabla \nabla R|$, $|\nabla \mathrm{Rc}|$, and $|\nabla \nabla \mathrm{Rm}|$ are all bounded by C_1 on $\mathcal{M} \times [0, T_0]$. From (7.52) and Lemma 7.13(ii) and (iii), we get

$$\begin{aligned}
|\sqrt{\tau} X(\tau)|^2 &\leq e^{6C_0 T} |\sqrt{\tau_*} X(\tau_*)|^2_{g(\tau_*)} + \frac{C_2^2 T}{12 C_0^2} e^{6C_0 T} \\
&\leq e^{6C_0 T} \left(\frac{1}{2\sqrt{\bar{\tau}}} \mathcal{L}(\gamma) + \frac{n C_0}{3} \bar{\tau}\right) + \frac{C_2^2 T}{12 C_0^2} e^{6C_0 T} \\
&\leq e^{6C_0 T} \left(\frac{n C_0}{3} \bar{\tau} + \frac{e^{2C_0 \bar{\tau}}}{4 \bar{\tau}} d^2_{g(\bar{\tau})}(p,q) + \frac{n C_0}{3} \bar{\tau}\right) + \frac{C_2^2 T}{12 C_0^2} e^{6C_0 T} \\
&\leq \left(1 + \frac{d^2_{g(\bar{\tau})}(p,q)}{\bar{\tau}}\right) C_2,
\end{aligned}$$

where C_2 is a constant depending only on n, T_0, T, and C_0.

Hence, from (7.63), we have

$$\begin{aligned}
|H(X,Y)|(\tau) &\leq C_1 |Y(\tau)|^2 + \frac{C_0}{\tau} |Y(\tau)|^2 \\
&\quad + C_0 \frac{1}{\tau} \left(1 + \frac{d^2_{g(\bar{\tau})}(p,q)}{\bar{\tau}}\right) C_2 |Y(\tau)|^2 \\
&\quad + C_1 \sqrt{\frac{1}{\tau} \left(1 + \frac{d^2_{g(\bar{\tau})}(p,q)}{\bar{\tau}}\right) C_2} \cdot |Y(\tau)|^2 \\
&\leq C_2 \left(1 + \frac{1}{\tau} + \frac{d^2_{g(\bar{\tau})}(p,q)}{\tau \bar{\tau}}\right) |Y(\tau)|^2.
\end{aligned}$$

Plugging the above estimate into (7.70) and using $|Y(\tau)|^2 = \frac{\tau}{\bar{\tau}}|Y(\bar{\tau})|^2$, we get

$$\left(\mathrm{Hess}_{(q,\bar{\tau})} L\right)(Y,Y) \le \left(C_2 + C_2 \frac{d^2_{g(\bar{\tau})}(p,q)}{\sqrt{\bar{\tau}}} + \frac{1}{\sqrt{\bar{\tau}}}\right)|Y(\bar{\tau})|^2.$$

□

5.3. Laplacian comparison theorem for L**.** Tracing the Hessian comparison theorem for L in (7.70), we obtain the Laplacian comparison theorem for L. We adopt the notation in Corollary 7.40. Let $\{E_i\}_{i=1}^n$ be an orthonormal basis at $q = \gamma(\bar{\tau})$. For each i, we extend E_i to a vector field $E_i(\tau)$, $\tau \in [0, \bar{\tau}]$, along γ to solve the ODE (7.66) with $E_i(\bar{\tau}) = E_i$. Below, E_i will stand for either $E_i \in T_q\mathcal{M}$ or $E_i(\tau)$, which will be clear from the context. Taking $Y = E_i$ in (7.70) and summing over i, we have

$$\Delta L(q,\bar{\tau}) = \sum_{i=1}^n \mathrm{Hess}_{(q,\bar{\tau})} L(E_i, E_i)$$

$$\le -\int_0^{\bar{\tau}} \sqrt{\tau} \sum_{i=1}^n H(X(\tau), E_i(\tau))\, d\tau + \frac{n}{\sqrt{\bar{\tau}}}$$

$$- 2\sqrt{\bar{\tau}} \sum_{i=1}^n \mathrm{Rc}(E_i, E_i)(\bar{\tau}).$$

We compute, without assuming $\langle E_i(\bar{\tau}), E_j(\bar{\tau})\rangle = \delta_{ij}$, that

$$\frac{d}{d\tau}\langle E_i, E_j\rangle(\tau) = \frac{d}{d\tau}[g(\tau)(E_i(\tau), E_j(\tau))]$$

$$= 2\,\mathrm{Rc}(E_i, E_j) + \langle \nabla_X E_i, E_j\rangle + \langle E_i, \nabla_X E_j\rangle$$

$$= 2\,\mathrm{Rc}(E_i, E_j) + \left\langle -\mathrm{Rc}(E_i) + \frac{1}{2\tau}E_i, E_j\right\rangle$$

$$+ \left\langle E_i, -\mathrm{Rc}(E_j) + \frac{1}{2\tau}E_j\right\rangle$$

$$= \frac{1}{\tau}\langle E_i, E_j\rangle(\tau).$$

Hence

(7.72) $$\langle E_i, E_j\rangle(\tau) = \frac{\tau}{\bar{\tau}}\langle E_i(\bar{\tau}), E_j(\bar{\tau})\rangle.$$

Since $\langle E_i, E_j\rangle(\bar{\tau}) = \delta_{ij}$, we have

$$\langle E_i, E_j\rangle(\tau) = \frac{\tau}{\bar{\tau}}\delta_{ij}$$

for all $\tau \in [0, \bar{\tau}]$. When $i = j$, we recover (7.68), which says $|E_i(\tau)|^2 = \tau/\bar{\tau}$.

5. THE SECOND VARIATION FORMULA FOR $\mathcal{L}$ AND THE HESSIAN OF L

Using the Einstein summation convention (expressions with i repeated are summed from 1 to n), we now simplify $\sum_{i=1}^{n} H(X, E_i)(\tau)$:

$$\sum_{i=1}^{n} H(X, E_i)(\tau)$$
$$= -2\left(\frac{\partial}{\partial \tau} \operatorname{Rc}\right)(E_i, E_i) - \nabla_{E_i} \nabla_{E_i} R + 2|\operatorname{Rc}(E_i)|^2 - \frac{1}{\tau}\operatorname{Rc}(E_i, E_i)$$
$$- 2\langle R(E_i, X) E_i, X\rangle + 4(\nabla_{E_i} \operatorname{Rc})(E_i, X) - 4(\nabla_X \operatorname{Rc})(E_i, E_i)$$
$$= -2\frac{\partial}{\partial \tau}[\operatorname{Rc}(E_i, E_i)] + 4\operatorname{Rc}(\nabla_X E_i, E_i) - \frac{\tau}{\bar\tau}\Delta R + 2\frac{\tau}{\bar\tau}|\operatorname{Rc}|^2$$
$$- \frac{1}{\tau}\frac{\tau}{\bar\tau}R + 2\frac{\tau}{\bar\tau}\operatorname{Rc}(X,X) + 4\frac{\tau}{\bar\tau}\operatorname{div}(\operatorname{Rc})(X) - 4\frac{\tau}{\bar\tau}\nabla_X R$$
$$= -2\frac{\partial}{\partial \tau}\left(\frac{\tau}{\bar\tau}R\right) + 4\operatorname{Rc}\left(-\operatorname{Rc}(E_i) + \frac{1}{2\tau}E_i, E_i\right) - \frac{\tau}{\bar\tau}\Delta R + 2\frac{\tau}{\bar\tau}|\operatorname{Rc}|^2$$
$$- \frac{1}{\bar\tau}R + 2\frac{\tau}{\bar\tau}\operatorname{Rc}(X,X) - 2\frac{\tau}{\bar\tau}\nabla_X R$$
$$= \frac{\tau}{\bar\tau}\left(-2\frac{\partial R}{\partial \tau} - \Delta R - 2|\operatorname{Rc}|^2 - \frac{R}{\tau} + 2\operatorname{Rc}(X,X) - 2\nabla_X R\right).$$

Recall from (V1-p.274) that Hamilton's trace Harnack expression is

(7.73) $\quad H(X) \doteqdot H(X)(\tau) \doteqdot -\frac{\partial R}{\partial \tau} - 2\nabla R \cdot X + 2\operatorname{Rc}(X,X) - \frac{R}{\tau}$

(in (V1-p.274) let $t = -\tau$ and replace X by $-X$). Using $\frac{\partial R}{\partial \tau} = -\Delta R - 2|\operatorname{Rc}|^2$, we get

(7.74) $$\sum_{i=1}^{n} H(X, E_i)(\tau) = \frac{\tau}{\bar\tau} H(X).$$

Hence

$$\Delta L(q, \bar\tau) \leq -\int_0^{\bar\tau} \frac{\tau^{3/2}}{\bar\tau} H(X) \, d\tau + \frac{n}{\sqrt{\bar\tau}} - 2\sqrt{\bar\tau} R(q, \bar\tau)$$
$$= -\frac{1}{\bar\tau} K + \frac{n}{\sqrt{\bar\tau}} - 2\sqrt{\bar\tau} R(q, \bar\tau),$$

where

(7.75) $$K = K(\gamma, \bar\tau) \doteqdot \int_0^{\bar\tau} \tau^{3/2} H(X) \, d\tau.$$

We call K the **trace Harnack integral**. Dropping the bars on the τ's, we have the **Laplacian comparison theorem for L**.

LEMMA 7.42 (Inequality for ΔL). *Given $\tau \in (0, T)$ and $q \in \mathcal{M}$, let $\gamma: [0, \tau] \to \mathcal{M}$ be a minimal $\mathcal{L}$-geodesic joining p to q. Then*

(7.76) $$\Delta L(q, \tau) \leq -\frac{1}{\tau} K + \frac{n}{\sqrt{\tau}} - 2\sqrt{\tau} R.$$

As before, if $L(\cdot, \tau)$ is not C^2 at q, by our convention the above inequality is understood in the barrier sense.

6. Equations and inequalities satisfied by L and ℓ

In this section $(\mathcal{M}^n, g(\tau))$, $\tau \in [0, T]$, will be a complete solution to the backward Ricci flow satisfying the curvature bound $\max\{|\mathrm{Rm}|, |\mathrm{Rc}|\} \leq C_0 < \infty$ on $\mathcal{M} \times [0, T]$, and $p \in \mathcal{M}$ will be a basepoint. Given a point $q \in \mathcal{M}$ and $\bar{\tau} \in (0, T)$, let $\gamma : [0, \bar{\tau}] \to \mathcal{M}$ be a minimal $\mathcal{L}$-geodesic from p to q and let $X(\tau) \doteqdot \frac{d\gamma}{d\tau}$.

6.1. The $\mathcal{L}$-length integrand and the trace Harnack quadratic.
Now using the $\mathcal{L}$-geodesic equation (7.32), we compute the evolution of the $\mathcal{L}$-length integrand as

$$\frac{d}{d\tau}\left(R(\gamma(\tau), \tau) + |X(\tau)|^2_{g(\tau)}\right)$$
$$= \frac{\partial R}{\partial \tau} + \nabla R \cdot X + 2\,\mathrm{Rc}(X, X) + 2\langle \nabla_X X, X \rangle$$
$$= \frac{\partial R}{\partial \tau} + \nabla R \cdot X + 2\,\mathrm{Rc}(X, X) + \left\langle \nabla R - 4\,\mathrm{Rc}(X) - \frac{1}{\tau}X, X \right\rangle$$
$$= \frac{\partial R}{\partial \tau} + 2\nabla R \cdot X - 2\,\mathrm{Rc}(X, X) - \frac{1}{\tau}|X|^2.$$

Hence

(7.77) $$\frac{d}{d\tau}\left(R + |X|^2\right) = -H(X) - \frac{1}{\tau}\left(R + |X|^2\right),$$

where $H(X)$ is defined in (7.73). In another form, we have

$$\frac{d}{d\tau}\left[\tau\left(R + |X|^2\right)\right] = -\tau H(X).$$

Recall that
$$K = K(\gamma, \bar{\tau}) = \int_0^{\bar{\tau}} \tau^{3/2} H(X)\, d\tau$$

defined in (7.75). Multiplying (7.77) by $\tau^{3/2}$ and integrating (by parts for the second equality), we get

$$-K(\gamma, \bar{\tau})$$
$$= \int_0^{\bar{\tau}} \left[\tau^{3/2}\frac{d}{d\tau}\left(R + |X|^2\right) + \tau^{1/2}\left(R + |X|^2\right)\right] d\tau$$
$$= \bar{\tau}^{3/2}\left(R(\gamma(\bar{\tau}), \bar{\tau}) + |X(\bar{\tau})|^2\right) - \frac{1}{2}\int_0^{\bar{\tau}} \tau^{1/2}\left(R + |X|^2\right) d\tau$$
$$= \bar{\tau}^{3/2}\left(R(\gamma(\bar{\tau}), \bar{\tau}) + |X(\bar{\tau})|^2\right) - \frac{1}{2}\mathcal{L}(\gamma).$$

Hence

LEMMA 7.43 (*L*-distance and trace Harnack quadratic). *Let $\bar{\tau} \in (0, T)$. Suppose $\gamma : [0, \bar{\tau}] \to \mathcal{M}$ is a minimal $\mathcal{L}$-geodesic from p to q. Then*

$$\bar{\tau}^{3/2} \left(R(q, \bar{\tau}) + |X(\bar{\tau})|^2 \right) = -K(\gamma, \bar{\tau}) + \frac{1}{2} L(q, \bar{\tau}). \tag{7.78}$$

Using (7.78), we can rewrite (7.57) and (7.55) as follows. For convenience we include (7.76) below as (7.81).

LEMMA 7.44. *Let $\bar{\tau} \in (0, T)$, let $\gamma : [0, \bar{\tau}] \to \mathcal{M}$ be a minimal $\mathcal{L}$-geodesic from p to q, and let $K = K(\gamma, \bar{\tau})$ be given by (7.75). Then, at $(q, \bar{\tau})$,*

$$\frac{\partial L}{\partial \tau} = \frac{1}{\bar{\tau}} K - \frac{1}{2\bar{\tau}} L + 2\sqrt{\bar{\tau}} R, \tag{7.79}$$

$$|\nabla L|^2 = -4\bar{\tau} R - \frac{4}{\sqrt{\bar{\tau}}} K + \frac{2}{\sqrt{\bar{\tau}}} L, \tag{7.80}$$

$$\Delta L \leq -\frac{1}{\bar{\tau}} K + \frac{n}{\sqrt{\bar{\tau}}} - 2\sqrt{\bar{\tau}} R. \tag{7.81}$$

These may seem like strange ways to rewrite equations (7.57) and (7.55). A motivation is given by (7.76), which involves the integral K of Hamilton's trace Harnack quadratic, which naturally arises from the second variation formula for $\mathcal{L}$-length. Note that besides K, these formulas do not explicitly contain the quantity X. When the minimal $\mathcal{L}$-geodesic is not unique, the quantities $\frac{\partial L}{\partial \tau}$, $|\nabla L|^2$, ΔL and K are defined using the choice of γ.

6.2. Inequalities for *L*. Combining (7.79) and (7.80) with (7.81), we get

LEMMA 7.45. *At $(q, \bar{\tau})$ the L-distance function $L(x, \tau)$ satisfies*

$$\frac{\partial L}{\partial \tau} \leq -\Delta L - \frac{1}{2\bar{\tau}} L + \frac{n}{\sqrt{\bar{\tau}}}, \tag{7.82}$$

$$\frac{\partial L}{\partial \tau} - \Delta L + \frac{1}{2\sqrt{\bar{\tau}}} |\nabla L|^2 - \frac{1}{2\bar{\tau}} L - 2\sqrt{\bar{\tau}} R + \frac{n}{\sqrt{\bar{\tau}}} \geq 0. \tag{7.83}$$

Equation (7.82) already exhibits the advantage of the *L*-distance over the usual distance function in Riemannian geometry in the setting of the Ricci flow. It is a subsolution to a backward heat equation.

Recall that $\bar{L}(x, \tau) = 2\sqrt{\tau} L(x, \tau)$. Then at $(q, \bar{\tau})$

$$\frac{\partial \bar{L}}{\partial \tau} \leq 2\sqrt{\bar{\tau}} \left(-\Delta L - \frac{1}{2\bar{\tau}} L + \frac{n}{\sqrt{\bar{\tau}}} \right) + \frac{1}{\sqrt{\bar{\tau}}} L = -\Delta \bar{L} + 2n,$$

so that

LEMMA 7.46 ($\bar{L} - 2n\tau$ supersolution of heat equation). *At $(q, \bar{\tau})$ the function $\bar{L}(x, \tau)$ satisfies*

$$\frac{\partial \bar{L}}{\partial \tau} + \Delta \bar{L} \leq 2n, \tag{7.84}$$

that is,

(7.85) $$\left(\frac{\partial}{\partial \tau} + \Delta\right)\left(\bar{L} - 2n\tau\right) \leq 0.$$

Now we prove a lemma which describes the limiting behavior of $\bar{L}(q, \bar{\tau})$ as $\bar{\tau} \to 0_+$. We will use it to estimate the minimal value of $\frac{1}{2\sqrt{\bar{\tau}}} L(\cdot, \bar{\tau})$ (and likewise $\bar{L}$) over $\mathcal{M}$.

LEMMA 7.47 ($\bar{L}$ tends to its Euclidean value as $\bar{\tau} \to 0$). We have

(7.86) $$\lim_{\bar{\tau} \to 0_+} \bar{L}(q, \bar{\tau}) = \left(d_{g(0)}(p, q)\right)^2.$$

Hence L satisfies

$$\lim_{\bar{\tau} \to 0_+} \frac{L(q, \bar{\tau})}{\left[d_{g(0)}(p, q)\right]^2 / 2\sqrt{\bar{\tau}}} = 1.$$

PROOF. We only need to prove the first equality above; the second equality follows by definition. In Lemma 7.13(i), let $\tau = \tau_2 = \bar{\tau}$ and let $\gamma : [0, \bar{\tau}] \to \mathcal{M}$ be a minimal $\mathcal{L}$-geodesic between p and q. We have

$$d_{g(0)}^2(p, q) \leq e^{2C_0 \bar{\tau}} \left(\bar{L}(q, \bar{\tau}) + \frac{4nC_0}{3} \bar{\tau}^2\right)$$

for any $\bar{\tau} \in (0, T)$. Taking the limit $\lim_{\bar{\tau} \to 0_+}$ of the above inequality, we get $\lim_{\bar{\tau} \to 0_+} \bar{L}(q, \bar{\tau}) \geq \left(d_{g(0)}(p, q)\right)^2$.

To see the other direction of the inequality, choosing $\tau_2 = \bar{\tau}$ in Lemma 7.13(iii), we have

$$\bar{L}(q, \bar{\tau}) \leq \frac{4nC_0}{3} \bar{\tau}^2 + e^{2C_0 \bar{\tau}} d_{g(\bar{\tau})}^2(p, q)$$

for any $\bar{\tau} \in (0, T)$. Taking $\bar{\tau} \to 0_+$, we get

$$\lim_{\bar{\tau} \to 0_+} \bar{L}(q, \bar{\tau}) \leq d_{g(0)}^2(p, q).$$

The lemma is proved. □

Next we estimate the minimum value of $\bar{L}(q, \bar{\tau}) - 2n\bar{\tau}$ over $\mathcal{M}$.

LEMMA 7.48 (Monotonicity and estimate for $\min_{\mathcal{M}} \bar{L}$). Let $(\mathcal{M}^n, g(\tau))$, $\tau \in [0, T]$, be a complete solution to the backward Ricci flow with bounded sectional curvature.

(i) The function
$$\min_{q \in \mathcal{M}} \left(\bar{L}(q, \bar{\tau}) - 2n\bar{\tau}\right)$$
is a nonincreasing function of $\bar{\tau}$.

(ii) For all $\bar{\tau} \in (0, T)$,
$$\min_{q \in \mathcal{M}} \bar{L}(q, \bar{\tau}) \leq 2n\bar{\tau}.$$

6. EQUATIONS AND INEQUALITIES SATISFIED BY L AND ℓ

PROOF. (i) From (7.85) we have

$$\frac{\partial}{\partial \tau}\left(\bar{L}-2n\tau\right)+\Delta\left(\bar{L}-2n\tau\right)\leq 0.$$

If $\bar{L}\left(\cdot,\cdot\right)$ were C^{2}, then the lemma would follow from the maximum principle. Since we only know that $\bar{L}\left(\cdot,\cdot\right)$ is locally Lipschitz, we proceed with the maximum principle argument with some care. Define the function $h:(0,T)\to\mathbb{R}$ by

$$h(\tau)\doteq\min_{x\in\mathcal{M}}\left(\bar{L}\left(x,\tau\right)-2n\tau\right).$$

CLAIM. For any $\tau>0$ there exists $q_\tau\in\mathcal{M}$ such that

$$h(\tau)=\bar{L}\left(q_\tau,\tau\right)-2n\tau,$$

and $h(\tau)$ is a continuous function.

If $\mathcal{M}$ is closed, then the claim follows from $\bar{L}(x,\tau)$ being a continuous function when $\tau>0$. When $\mathcal{M}$ is not closed, the claim follows from Lemma 7.13(i), which says that $\lim_{x\to\infty}\bar{L}\left(x,\tau\right)=+\infty$. This and the local Lipschitz property of $\bar{L}\left(x,\tau\right)$ imply the claim.

Now we estimate the right lim sup derivative of h at $\bar{\tau}\in(0,T)$:

$$\frac{d^{+}h}{d\tau}(\bar{\tau})\doteq\limsup_{s\to 0_{+}}\frac{h(\bar{\tau}+s)-h(\bar{\tau})}{s}$$

$$=\limsup_{s\to 0_{+}}\frac{\bar{L}(q_{\bar{\tau}+s},\bar{\tau}+s)-\bar{L}(q_{\bar{\tau}},\bar{\tau})}{s}-2n$$

$$\leq\limsup_{s\to 0_{+}}\frac{\bar{L}(q_{\bar{\tau}},\bar{\tau}+s)-\bar{L}(q_{\bar{\tau}},\bar{\tau})}{s}-2n,$$

where the last inequality follows from the definition of $q_{\bar{\tau}+s}$. If $\bar{L}$ is C^{2} at $(q_{\bar{\tau}},\bar{\tau})$, then by using (7.84), we have

$$\limsup_{s\to 0_{+}}\frac{\bar{L}(q_{\bar{\tau}},\bar{\tau}+s)-\bar{L}(q_{\bar{\tau}},\bar{\tau})}{s}=\frac{\partial\bar{L}}{\partial\tau}(q_{\bar{\tau}},\bar{\tau})$$

$$\leq 2n-\Delta\bar{L}(q_{\bar{\tau}},\bar{\tau})\leq 2n.$$

The reason why $\Delta\bar{L}(q_{\bar{\tau}},\bar{\tau})\geq 0$ is that $q_{\bar{\tau}}$ is a minimum and smooth point of $\bar{L}(\cdot,\bar{\tau})$. We have proved that $\frac{d^{+}h}{d\tau}(\bar{\tau})\leq 0$ when $\bar{L}$ is C^{2} at $(q_{\bar{\tau}},\bar{\tau})$.

When $\bar{L}$ is not C^{2} at $(q_{\bar{\tau}},\bar{\tau})$, let $\hat{L}$ be a smooth barrier function of L at $(q_{\bar{\tau}},\bar{\tau})$ as in (7.56). Setting $\hat{\bar{L}}(x,\tau)\doteq 2\sqrt{\tau}\hat{L}(x,\tau)$, we see that $q_{\bar{\tau}}$ is a minimum point of the locally defined function $\hat{\bar{L}}(\cdot,\bar{\tau})$ since $\hat{\bar{L}}$ is a barrier function of $\bar{L}$ from above at $(q_{\bar{\tau}},\bar{\tau})$ and $q_{\bar{\tau}}$ is a minimum point of $\bar{L}(\cdot,\bar{\tau})$.

Hence $\Delta \bar{\hat{L}}(q_{\bar\tau}, \bar\tau) \geq 0$. We have

$$\frac{d^+ h}{d\tau}(\bar\tau) \leq \limsup_{s \to 0_+} \frac{\bar{L}(q_{\bar\tau}, \bar\tau + s) - \bar{L}(q_{\bar\tau}, \bar\tau)}{s}$$

$$\leq \limsup_{s \to 0_+} \frac{\bar{\hat{L}}(q_{\bar\tau}, \bar\tau + s) - \bar{\hat{L}}(q_{\bar\tau}, \bar\tau)}{s} = \frac{\partial \bar{\hat{L}}}{\partial \tau}(q_{\bar\tau}, \bar\tau).$$

Then using (7.84), which holds for the barrier function $\bar{\hat{L}}$, we have

$$\frac{\partial \bar{\hat{L}}}{\partial \tau}(q_{\bar\tau}, \bar\tau) \leq 2n - \Delta \bar{\hat{L}}(q_{\bar\tau}, \bar\tau) \leq 2n.$$

We have proved that $\frac{d^+ h}{d\tau}(\bar\tau) \leq 0$ when $\bar{L}$ is not C^2 at $(q_{\bar\tau}, \bar\tau)$. Hence we have proved that $\frac{d^+ h}{d\tau}(\bar\tau) \leq 0$ for all $\bar\tau \in (0, T)$. By the monotonicity principle for Lipschitz functions stated in §3 (Lemma 3.1) of [**179**], $h(\tau)$ is nonincreasing.

(ii) This follows from (i) and

$$\lim_{\bar\tau \to 0_+} h(\bar\tau) = \lim_{\bar\tau \to 0} \min_{q \in \mathcal{M}} \bar{L}(q, \bar\tau)$$

$$\leq \lim_{\bar\tau \to 0} \bar{L}(p, \bar\tau) = \left(d_{g(0)}(p, p)\right)^2$$

$$= 0.$$

□

6.3. The reduced distance function ℓ. To get even better equations than those in Lemma 7.45, we introduce the reduced distance function ℓ.

DEFINITION 7.49. The **reduced distance** ℓ is defined by

(7.87) $$\ell(x, \tau) \doteqdot \frac{1}{2\sqrt{\tau}} L(x, \tau) = \frac{1}{4\tau} \bar{L}(x, \tau).$$

Let $\bar\tau \in (0, T)$ and let $\gamma : [0, \bar\tau] \to \mathcal{M}$ be a minimal $\mathcal{L}$-geodesic from p to q and let $K = K(\gamma, \bar\tau)$ be defined as in (7.75). By (7.79), (7.80), and (7.81), we have at $(q, \bar\tau)$,

(7.88) $$\frac{\partial \ell}{\partial \bar\tau} = \frac{1}{2\bar\tau^{3/2}} K - \frac{\ell}{\bar\tau} + R,$$

(7.89) $$|\nabla \ell|^2 = -R - \frac{1}{\bar\tau^{3/2}} K + \frac{\ell}{\bar\tau},$$

(7.90) $$\Delta \ell \leq -\frac{1}{2\bar\tau^{3/2}} K + \frac{n}{2\bar\tau} - R.$$

From these equations (which involve the trace Harnack integral K), (7.86) and Lemma 7.48, we easily deduce the following which do not involve K.

LEMMA 7.50 (Reduced distance — partial differential inequalities). At $(q, \bar{\tau})$ the reduced distance $\ell(x, \tau)$ satisfies

(7.91) $$\frac{\partial \ell}{\partial \tau} - \Delta \ell + |\nabla \ell|^2 - R + \frac{n}{2\bar{\tau}} \geq 0,$$

(7.92) $$2\Delta \ell - |\nabla \ell|^2 + R + \frac{\ell - n}{\bar{\tau}} \leq 0,$$

(7.93) $$\frac{\partial \ell}{\partial \tau} + \Delta \ell + \frac{\ell}{\bar{\tau}} - \frac{n}{2\bar{\tau}} \leq 0,$$

(7.94) $$2\frac{\partial \ell}{\partial \tau} + |\nabla \ell|^2 - R + \frac{\ell}{\bar{\tau}} = 0,$$

$$\lim_{\bar{\tau} \to 0+} \frac{\ell(q, \bar{\tau})}{[d_{g(0)}(p, q)]^2 / 4\bar{\tau}} = 1,$$

(7.95) $$\min_{q \in \mathcal{M}} \ell(q, \bar{\tau}) = \frac{1}{4\bar{\tau}} \min_{q \in \mathcal{M}} \bar{L}(q, \bar{\tau}) \leq \frac{n}{2}.$$

REMARK 7.51. (i) Note that (7.94) is the only possible equality obtainable from (7.88) and (7.89) which does not involve K.

(ii) In the inequalities above, the direction of the inequality depends on the sign of the coefficient in front of the term $\Delta \ell$. The reason for this is that, analogous to considering $\Delta(d^2)$ on a Riemannian manifold with nonnegative Ricci curvature, $\Delta \ell$ has an upper bound (7.90).

The above equations for ℓ demonstrate, in the context of Ricci flow, the superiority of the reduced distance over the Riemannian distance function. It is a space-time notion of distance which is a subsolution of Laplace-type and heat-type equations. Note that (7.91) is a forward heat superequation whereas (7.93) is a backward heat subequation. One would think that the backward heat subequation is more natural since it is associated to the backward Ricci flow, but we shall find the forward superequation (7.91) very useful.

REMARK 7.52. (i) If we replace the inequality by an equality in (7.91), we obtain equation (6.14) for f used in the study of the entropy $\mathcal{W}$:

$$\frac{\partial}{\partial \tau} f - \Delta f + |\nabla f|^2 - R + \frac{n}{2\tau} = 0.$$

(ii) Compare (7.92) with the equation (6.52):

$$2 \Delta f - |\nabla f|^2 + R + \frac{f - n}{\tau} = \frac{1}{\tau} \mu(g, \tau)$$

for the minimizer f of $\mathcal{W}(g, \cdot, \tau)$.

If we divide by $2\sqrt{\bar{\tau}}$ in the LHS of (7.65), we get $(\nabla \nabla \ell + \operatorname{Rc})(Y, Y)$ at smooth points of ℓ. Motivated by the special case of shrinking solitons, we

may write formula (7.65) at smooth points of ℓ as
(7.96)
$$\left(\nabla\nabla\ell + \mathrm{Rc} - \frac{1}{2\bar{\tau}}g\right)(Y,Y)(\bar{\tau})$$
$$= -\frac{1}{2\sqrt{\bar{\tau}}}\int_0^{\bar{\tau}} \sqrt{\tau} H(X,Y)\, d\tau + \int_0^{\bar{\tau}} \sqrt{\frac{\tau}{\bar{\tau}}}\left|\nabla_X Y + \mathrm{Rc}(Y) - \frac{1}{2\tau}Y\right|^2 d\tau.$$

EXERCISE 7.53. What does the trace of this equality say?

We have the following property of the reduced distance ℓ.

LEMMA 7.54 (Reduced distance as $\tau \to 0$). *Given $V \in T_p\mathcal{M}$, let $\gamma_V : [0,T] \to \mathcal{M}$ be the $\mathcal{L}$-geodesic with $\lim_{\tau \to 0_+} \sqrt{\tau}\frac{d\gamma}{d\tau}(\tau) = V$. Then*
(7.97)
$$\lim_{\bar{\tau} \to 0_+} \ell(\gamma_V(\bar{\tau}), \bar{\tau}) = |V|^2_{g(0)}.$$

PROOF. Since $\lim_{\tau \to 0} \sqrt{\tau}\frac{d\gamma_V}{d\tau}(\tau) = V$ and $\gamma_V|_{[0,\bar{\tau}]}$ is a minimal $\mathcal{L}$-geodesic when $\bar{\tau}$ is small (see Lemma 7.29), we have
$$\lim_{\tau \to 0_+} \ell(\gamma_V(\tau), \tau)$$
$$= \lim_{\tau \to 0_+} \frac{1}{2\sqrt{\tau}} \int_0^{\tau} \sqrt{\tilde{\tau}}\left(R(\gamma_V(\tilde{\tau}), \tilde{\tau}) + \left|\frac{d\gamma_V}{d\tau}(\tilde{\tau})\right|^2_{g(\gamma(\tilde{\tau}),\tilde{\tau})}\right) d\tilde{\tau}$$
$$= \lim_{\tau \to 0_+} \frac{1}{2\sqrt{\tau}} \int_0^{\tau} \sqrt{\tilde{\tau}} \cdot \frac{1}{\tilde{\tau}} |V|^2_{g(0)}\, d\tilde{\tau} = |V|^2_{g(0)}.$$
□

We conclude this subsection with a few exercises concerning the reduced distance ℓ.

EXERCISE 7.55 ($\mathcal{L}$-triangle inequality). Show that for any path $\gamma : [\tau_1, \tau_2] \to \mathcal{M}$,
$$L(\gamma(\tau_2), \tau_2) - L(\gamma(\tau_1), \tau_1)$$
$$= 2\sqrt{\tau_2}\ell(\gamma(\tau_2), \tau_2) - 2\sqrt{\tau_1}\ell(\gamma(\tau_1), \tau_1)$$
$$= \mathcal{L}(\gamma) - \int_{\tau_1}^{\tau_2} \sqrt{\tau}\left|\nabla\ell - \frac{d\gamma}{d\tau}\right|^2 d\tau.$$

REMARK 7.56. Given points $(\bar{q}_1, \bar{\tau}_1)$ and $(\bar{q}_2, \bar{\tau}_2)$, we may define
$$\mathcal{L}d((\bar{q}_1, \bar{\tau}_1), (\bar{q}_2, \bar{\tau}_2)) = L_{(\bar{q}_1, \bar{\tau}_1)}(\bar{q}_2, \bar{\tau}_2)$$
$$\doteq \inf\{\mathcal{L}(\gamma) : \gamma(\bar{\tau}_i) = \bar{q}_i,\ i = 1, 2\}.$$
Then the above formula implies
$$\mathcal{L}d((p,0), (q_1, \tau_1)) + \mathcal{L}d((q_1, \tau_1), (q_2, \tau_2)) \geq \mathcal{L}d((p,0), (q_2, \tau_2)).$$
Equality holds if and only if for some minimal $\mathcal{L}$-geodesic $\gamma : [\tau_1, \tau_2] \to \mathcal{M}$ we have $\frac{d\gamma}{d\tau} = \nabla\ell$ along γ, where ℓ is defined with respect to the basepoint $(p,0)$.

SOLUTION TO EXERCISE 7.55. We compute, using (7.94),

$$\frac{\partial}{\partial \tau}\left(2\sqrt{\tau}\ell\left(\gamma\left(\tau\right),\tau\right)\right) = \sqrt{\tau}\left(2\frac{\partial \ell}{\partial \tau} + \frac{\ell}{\tau} + 2\nabla \ell \cdot \frac{d\gamma}{d\tau}\right)$$

$$= \sqrt{\tau}\left(R + \left|\frac{d\gamma}{d\tau}\right|^2 - \left|\nabla \ell - \frac{d\gamma}{d\tau}\right|^2\right).$$

The next exercise characterizes when an integral curve of ℓ is an $\mathcal{L}$-geodesic.

EXERCISE 7.57 (Integral curves of $\nabla \ell$). Show that for any solution $(\mathcal{M}^n, g(\tau))$, $\tau \in [0, T)$, to the backward Ricci flow, a smooth integral curve γ of $\nabla \ell$ is an $\mathcal{L}$-geodesic if and only if $\frac{\partial}{\partial \tau}(\nabla \ell) = 0$ along γ, where $\nabla \ell$ is the gradient vector field.

SOLUTION TO EXERCISE 7.57. In view of the $\mathcal{L}$-geodesic equation (7.32), we compute

$$\nabla_{\nabla \ell} \nabla \ell - \frac{1}{2}\nabla R + 2\operatorname{Rc}(\nabla \ell) + \frac{1}{2\tau}\nabla \ell$$

$$= \frac{1}{2}\nabla\left(|\nabla \ell|^2 - R + \frac{\ell}{\tau}\right) + 2\operatorname{Rc}(\nabla \ell).$$

Applying the identity (7.94) for ℓ, we obtain

$$\nabla_{\nabla \ell} \nabla \ell - \frac{1}{2}\nabla R + 2\operatorname{Rc}(\nabla \ell) + \frac{1}{2\tau}\nabla \ell$$

$$= -\nabla\left(\frac{\partial \ell}{\partial \tau}\right) + 2\operatorname{Rc}(\nabla \ell)$$

$$= -\frac{\partial}{\partial \tau}(\nabla \ell),$$

where $\nabla \ell$ is the gradient of ℓ (considered as a vector field), i.e., $(\nabla \ell)^i = g^{ij}\nabla_j \ell$.

We also note the following consequence of (7.91) which we shall revisit later when considering the reduced volume monotonicity.

EXERCISE 7.58 (Pointwise monotonicity of reduced volume integrand). Show that for a solution $\frac{\partial}{\partial \tau}g = 2\operatorname{Rc}$ of the backward Ricci flow,

$$\left(\frac{\partial}{\partial \tau} + \mathcal{L}_{\nabla \ell}\right)\left((4\pi\tau)^{-n/2}e^{-\ell}d\mu\right)$$

$$= \left(-\frac{n}{2\tau} - \frac{\partial \ell}{\partial \tau} + R - |\nabla \ell|^2 + \Delta \ell\right)(4\pi\tau)^{-n/2}e^{-\ell}d\mu$$

$$\leq 0,$$

where $\mathcal{L}$ denotes the Lie derivative. In what sense is the inequality true? Note that ℓ is only a Lipschitz function. See Section 9 of this chapter for some hints.

6.4. The growth of the reduced distance function ℓ. The growth of ℓ, in particular the lower bound of ℓ, will be used to justify some technical issues later (for example, the proof of Lemma 7.130). The next lemma follows from Lemma 7.13(i) and (iii).

LEMMA 7.59 (Bounds for the reduced distance). *Let $(\mathcal{M}^n, g(\tau))$, $\tau \in [0, T]$, be a complete solution to the backward Ricci flow with bounded sectional curvature and let p be a basepoint. Then for any $(q, \bar{\tau}) \in \mathcal{M} \times (0, T)$,*

$$\frac{1}{4\bar{\tau}e^{2C_0 T}} d^2_{g(0)}(p, q) - \frac{nC_0\bar{\tau}}{3} \leq \ell(q, \bar{\tau}) \leq \frac{e^{2C_0 T}}{4\bar{\tau}} d^2_{g(\bar{\tau})}(p, q) + \frac{nC_0\bar{\tau}}{3}.$$

Next we bound $|\nabla \ell|^2$ and $\left|\frac{\partial \ell}{\partial \tau}\right|$ by ℓ, and hence we can bound them by $d^2_{g(\bar{\tau})}(p, q)$. These estimates will be improved in Lemmas 7.64 and 7.65 when we assume the curvature operator is nonnegative.

LEMMA 7.60 (Bounds for first derivatives of ℓ). *Suppose $(\mathcal{M}^n, g(\tau))$, $\tau \in [0, T)$, is a complete solution to the backward Ricci flow with bounded sectional curvature. Then for any $\bar{\tau} \in (0, T)$ there exist positive constants $A \geq 1$ and C_1 depending only on $\bar{\tau}, n, T$ and C_0 which satisfy the following properties. For any $q \in \mathcal{M}$ and $\tau \in (0, \bar{\tau}]$, we have $\ell(q, \tau) + A\tau \geq 0$,*

(i)
$$|\nabla \ell|^2 (q, \tau) \leq \frac{C_1}{\tau} (\ell(q, \tau) + A\tau),$$

(ii)
$$\left|\frac{\partial \ell}{\partial \tau}\right|(q, \tau) \leq \frac{C_1}{\tau} (\ell(q, \tau) + A\tau).$$

PROOF. (i) Let $\gamma(\tilde{\tau})$ be a minimal $\mathcal{L}$-geodesic from $(p, 0)$ to (q, τ) and let $X(\tilde{\tau}) \doteqdot \frac{d\gamma}{d\tilde{\tau}}$. By Lemma 7.13(ii) we have

$$|\sqrt{\tau_*}X(\tau_*)|^2 \leq n\hat{C}_1\bar{\tau} + \ell(q, \bar{\tau})$$

for some $\tau_* \in (0, \bar{\tau})$. Here and below $\hat{C}_1$ and $\hat{A}$ are constants depending only on $\bar{\tau}, n, T$ and C_0 and its value may change from line to line. Thus using Shi's derivative estimate and integrating equation (7.45) ($\tilde{\sigma} = 2\sqrt{\tilde{\tau}}$),

$$\frac{d}{d\tilde{\sigma}} \left|\sqrt{\tilde{\tau}}X(\tilde{\tau})\right|^2_{g(\tilde{\sigma}^2/4)} = -\tilde{\sigma}\operatorname{Rc}\left(\sqrt{\tilde{\tau}}X(\tilde{\tau}), \sqrt{\tilde{\tau}}X(\tilde{\tau})\right) + \frac{\tilde{\sigma}^2}{4}\left\langle \nabla R, \sqrt{\tilde{\tau}}X(\tilde{\tau})\right\rangle$$

on $[\tau, \tau_*]$, as in the proof of Lemma 7.28, we get for any $\tau \in [0, \bar{\tau}]$

$$|\sqrt{\tau}X(\tau)|^2 \leq \left(n\hat{C}_1\bar{\tau} + \ell(q, \bar{\tau}) + \hat{C}_1\bar{\tau}\right)e^{\hat{C}_1}.$$

It follows from $\nabla \ell(q, \tau) = X(\tau)$ that

(7.98)
$$|\nabla \ell|^2 (q, \tau) \leq \frac{\hat{C}_1}{\tau}\left(\ell(q, \tau) + \hat{A}\tau\right).$$

In particular $\ell(q, \tau) + \hat{A}\tau \geq 0$ for all (q, τ).

(ii) From (7.94) we have at (q, τ),
$$\left|\frac{\partial \ell}{\partial \tau}\right| \leq \frac{|R|}{2} + \frac{|\ell|}{2\tau} + \frac{1}{2}|\nabla \ell|^2.$$

Using (7.98), we get at (q, τ) that
$$\left|\frac{\partial \ell}{\partial \tau}\right| \leq C_0 + \frac{|\ell|}{2\tau} + \frac{\hat{C}_1}{\tau}\left(\ell + \hat{A}\tau\right)$$
$$\leq C_0 + \frac{\left|\ell + \hat{A}\tau\right| + \hat{A}\tau}{2\tau} + \frac{\hat{C}_1}{\tau}\left(\ell + \hat{A}\tau\right)$$
$$\leq \frac{\left(\hat{C}_1 + \frac{1}{2}\right)}{\tau}\left(\ell + \left(\hat{A} + \frac{C_0 + \frac{1}{2}\hat{A}}{\hat{C}_1 + \frac{1}{2}}\right)\tau\right),$$

where we have used $\ell(q, \tau) + \hat{A}\tau \geq 0$ for all (q, τ) in the last inequality. Thus (ii) follows from taking $C_1 = \hat{C}_1 + \frac{1}{2}$ and $A = \hat{A} + \frac{C_0 + \frac{1}{2}\hat{A}}{\hat{C}_1 + \frac{1}{2}}$. The lemma is proved. □

From Lemma 7.41 we get

LEMMA 7.61 (Hessian of reduced distance). *Suppose $(\mathcal{M}^n, g(\tau))$, $\tau \in [0, T)$, is a complete solution to the backward Ricci flow with bounded sectional curvature. Fix $T_0 \in (0, T)$. Given $\bar{\tau} \in (0, T_0]$ and $q \in \mathcal{M}$, the Hessian of the reduced distance function $\ell(\cdot, \bar{\tau})$ at q has the upper bound*

(7.99) $$\text{Hess}_{(q,\bar{\tau})} \ell \leq \frac{C_2}{2\sqrt{\bar{\tau}}} + \frac{1 + C_2 d_{g(\bar{\tau})}^2(p, q)}{2\bar{\tau}},$$

where C_2 is a constant depending only on $n, T_0, T, \sup_{\mathcal{M} \times [0,T]} |\text{Rm}|$.

NOTATION 7.62. *Let φ be a C^2 function in a Riemannian manifold $\left(\hat{\mathcal{M}}, \hat{g}\right)$. By $\text{Hess}\,\varphi \doteq \nabla\nabla\varphi \leq C$ we actually mean $\nabla\nabla\varphi \leq C\hat{g}$. We shall similarly abuse notation in other parts of this chapter.*

6.5. Estimates for ℓ when $\text{Rm} \geq 0$. As we shall show below, when $(\mathcal{M}^n, g(\tau))$, $\tau \in [0, T)$, has bounded nonnegative curvature operator, there are better estimates for $\text{Hess}_{(q,\bar{\tau})} \ell$, $|\nabla \ell|^2$, and $\left|\frac{\partial \ell}{\partial \bar{\tau}}\right|$. Let $\bar{\tau} \in (0, T)$ and $q \in \mathcal{M}$. Let $\gamma : [0, \bar{\tau}] \to \mathcal{M}$ be a minimal $\mathcal{L}$-geodesic from p to q and let $X(\tau) \doteq \frac{d\gamma}{d\tau}$. Let Y be a solution to (7.66). Hamilton's matrix Harnack estimate implies

$$H(X, Y)(\tau) \geq -\left(\frac{1}{\tau} + \frac{1}{T - \tau}\right) \text{Rc}(Y, Y)(\tau)$$
$$\geq -n\left(\frac{1}{\tau} + \frac{1}{T - \tau}\right) R|Y(\tau)|^2$$
$$\geq -\frac{n}{\bar{\tau}}\left(1 + \frac{\tau}{T - \tau}\right)|Y(\bar{\tau})|^2 R$$

since $|Y(\tau)|^2 = \frac{\tau}{\bar{\tau}} |Y(\bar{\tau})|^2$. Hence by (7.70),

$$\left(\text{Hess}_{(q,\bar{\tau})} L \right) (Y(\bar{\tau}), Y(\bar{\tau}))$$

$$\leq \frac{n}{\bar{\tau}} |Y(\bar{\tau})|^2 \int_0^{\bar{\tau}} \sqrt{\tau} \left(1 + \frac{\tau}{T-\tau} \right) R \, d\tau + \frac{|Y(\bar{\tau})|^2}{\sqrt{\bar{\tau}}}$$

$$\leq \frac{n}{\bar{\tau}} \left(1 + \frac{\bar{\tau}}{T-\bar{\tau}} \right) |Y(\bar{\tau})|^2 \int_0^{\bar{\tau}} \sqrt{\tau} \left(R + \left| \frac{d\gamma}{d\tau} \right|^2 \right) d\tau + \frac{|Y(\bar{\tau})|^2}{\sqrt{\bar{\tau}}}$$

$$= \left[\left(\frac{2n}{\sqrt{\bar{\tau}}} + \frac{2n\sqrt{\bar{\tau}}}{T-\bar{\tau}} \right) \ell(q, \bar{\tau}) + \frac{1}{\sqrt{\bar{\tau}}} \right] |Y(\bar{\tau})|^2.$$

We have proved (compare with Lemma 7.41)

LEMMA 7.63. *Let $(\mathcal{M}^n, g(\tau))$, $\tau \in [0, T)$, be a solution to the backward Ricci flow with bounded nonnegative curvature operator. We have for any $\bar{\tau} \in (0, T)$,*

$$\text{Hess}_{(q,\bar{\tau})} \ell \leq n \left(\frac{1}{\bar{\tau}} + \frac{1}{T-\bar{\tau}} \right) \ell(q, \bar{\tau}) + \frac{1}{2\bar{\tau}}.$$

Recall that Hamilton's trace Harnack estimate says that

$$H(X)(\tau) \geq -\left(\frac{1}{\tau} + \frac{1}{T-\tau} \right) R(\gamma(\tau), \tau).$$

Hence $K(\gamma, \bar{\tau})$ defined in (7.75) satisfies

$$K(\gamma, \bar{\tau}) \geq -\int_0^{\bar{\tau}} \tau^{3/2} \left(\frac{1}{\tau} + \frac{1}{T-\tau} \right) R(\gamma(\tau), \tau) \, d\tau$$

$$\geq -\int_0^{\bar{\tau}} \tau^{1/2} \frac{T}{T-\tau} \left(R(\gamma(\tau), \tau) + \left| \frac{d\gamma}{d\tau} \right|^2_{g(\tau)} \right) d\tau$$

$$\geq -\frac{T}{T-\bar{\tau}} L(\gamma(\bar{\tau}), \bar{\tau}).$$

Therefore, from (7.89), we have at $(q, \bar{\tau})$

$$|\nabla \ell|^2 \leq -R + \frac{1}{\bar{\tau}^{3/2}} \frac{T}{T-\bar{\tau}} L(\gamma(\bar{\tau}), \bar{\tau}) + \frac{\ell}{\bar{\tau}}$$

$$\leq -R + \frac{1}{\bar{\tau}} \left(1 + \frac{2T}{T-\bar{\tau}} \right) \ell.$$

In particular,

LEMMA 7.64. *Let $(\mathcal{M}^n, g(\tau))$, $\tau \in [0, T)$, be a solution to the backward Ricci flow with bounded nonnegative curvature operator. If $\bar{\tau} \in (0, (1-c)T)$ for some $c \in (0, 1)$, then for any $q \in \mathcal{M}$*

$$|\nabla \ell|^2 (q, \bar{\tau}) + R(q, \bar{\tau}) \leq \frac{C}{\bar{\tau}} \ell(q, \bar{\tau}),$$

where $C = 1 + \frac{2}{c}$. If $T = \infty$, then for any $(q, \bar\tau) \in \mathcal{M} \times (0, \infty)$

$$|\nabla \ell|^2 (q, \bar\tau) + R(q, \bar\tau) \leq \frac{3}{\bar\tau} \ell(q, \bar\tau).$$

Hence, for an ancient solution with bounded $\operatorname{Rm} \geq 0$, the reduced distance bounds Rm.

Now we can estimate $\left|\frac{\partial \ell}{\partial \tau}\right|$.

LEMMA 7.65. *Let $(\mathcal{M}^n, g(\tau))$, $\tau \in [0, T]$, be a solution to the backward Ricci flow with bounded nonnegative curvature operator. If $\bar\tau \in (0, (1-c)T)$ for some $c \in (0,1)$, then at any $(q, \bar\tau) \in \mathcal{M} \times (0, \infty)$,*

$$\left|\frac{\partial \ell}{\partial \tau}\right| \leq \frac{C+1}{2} \frac{\ell}{\bar\tau},$$

where $C = 1 + \frac{2}{c}$. If $T = \infty$, then at any $(q, \bar\tau) \in \mathcal{M} \times (0, \infty)$,

(7.100) $$\left|\frac{\partial}{\partial \tau} \log \ell\right| \leq \frac{2}{\bar\tau},$$

and for any $0 < \tau_1 < \tau_2$ and $q \in \mathcal{M}$,

(7.101) $$\left(\frac{\tau_1}{\tau_2}\right)^2 \leq \frac{\ell(q, \tau_2)}{\ell(q, \tau_1)} \leq \left(\frac{\tau_2}{\tau_1}\right)^2.$$

PROOF. From (7.94) we have at $(q, \bar\tau)$ that

$$\frac{\partial \ell}{\partial \tau} + \frac{1}{2}|\nabla \ell|^2 = \frac{1}{2}R - \frac{1}{2\bar\tau}\ell.$$

Hence by Lemma 7.64 and $\ell(x, \tau) > 0$, we have

$$\left|\frac{\partial \ell}{\partial \tau}\right| \leq \frac{1}{2}\left(|\nabla \ell|^2 + R\right) + \frac{1}{2\bar\tau}\ell$$

$$\leq \frac{C+1}{2\bar\tau}\ell.$$

If $T = \infty$, we can choose c arbitrarily close to 1 and get $\left|\frac{\partial}{\partial \tau} \log \ell\right| \leq \frac{2}{\bar\tau}$. Since $\log \ell(x, \tau)$ is a locally Lipschitz function of $\tau > 0$, we have

$$\left|\log \frac{\ell(q, \tau_2)}{\ell(q, \tau_1)}\right| \leq \int_{\tau_1}^{\tau_2} \frac{2}{\tau} d\tau = \log\left(\frac{\tau_2}{\tau_1}\right)^2,$$

and (7.101) follows. □

6.6. Reduced distance under Cheeger–Gromov convergence.
Finally we discuss the convergence of the reduced distance under Cheeger–Gromov convergence. Let $\{(\mathcal{M}_k^n, g_k(\tau), p_k)\}_{k \in \mathbb{N}}$ and $(\mathcal{M}_\infty^n, g_\infty(\tau), p_\infty)$, $\tau \in [0, T]$, be complete pointed solutions to the backward Ricci flow satisfying the curvature bound

$$\max\{|\operatorname{Rm}_{g_k}|, |\operatorname{Rc}_{g_k}|\} \leq C_0 < \infty \quad \text{on } \mathcal{M}_k \times [0, T]$$

for all $k \in \mathbb{N}$. Suppose that $(\mathcal{M}_k, g_k(\tau), p_k) \to (\mathcal{M}_\infty, g_\infty(\tau), p_\infty)$ on the time interval $[0, T]$ in the C^∞ Cheeger–Gromov sense; that is, there exist an exhaustion $\{U_k\}_{k\in\mathbb{N}}$ of $\mathcal{M}_\infty$ by open sets with $p_\infty \in U_k$ and diffeomorphisms $\Phi_k : U_k \to V_k \doteq \Phi_k(U_k) \subset \mathcal{M}_k$ with $\Phi_k(p_\infty) = p_k$ such that $\left(U_k, \Phi_k^*\left(g_k(\tau)|_{V_k}\right)\right) \to (\mathcal{M}_\infty, g_\infty(\tau))$ in C^∞ on compact sets in $\mathcal{M}_\infty \times [0, T]$.

LEMMA 7.66 (ℓ under Cheeger–Gromov convergence). *Under the setup above, for any $(q, \bar\tau) \in \mathcal{M}_\infty \times (0, T)$, we have*

$$\ell^{g_\infty}_{(p_\infty, 0)}(q, \bar\tau) = \lim_{k\to\infty} \ell^{g_k}_{(p_k, 0)}(\Phi_k(q), \bar\tau).$$

The convergence is uniform on compact subsets of $\mathcal{M}_\infty \times (0, T)$. Furthermore, the convergence is uniform in C^∞ on compact subsets of the open set of points at which $\ell^{g_\infty}_{(p_\infty, 0)}$ is C^∞.

PROOF. Let $\breve\gamma : [0, \bar\tau] \to \mathcal{M}_\infty$ be a minimal $\mathcal{L}$-geodesic, with respect to g_∞, joining p_∞ to q. By the convergence of the sequence of solutions $\{g_k(\tau)\}$ in C^2 on compact sets, we have

$$\ell^{g_\infty}_{(p_\infty, 0)}(q, \bar\tau) = \frac{1}{2\sqrt{\bar\tau}} \mathcal{L}_{g_\infty}(\breve\gamma) = \frac{1}{2\sqrt{\bar\tau}} \lim_{k\to\infty} \mathcal{L}_{g_k}(\Phi_k \circ \breve\gamma)$$
$$\geq \limsup_{k\to\infty} \ell^{g_k}_{(p_k, 0)}(\Phi_k(q), \bar\tau).$$

Next we prove the opposite inequality. Since for any compact set $\mathcal{K} \subset \mathcal{M}_\infty$,

(7.102) $$\left\|\Phi_k^*\left(g_k|_{V_k}\right) - g_\infty\right\|_{C^2(\mathcal{K}\times[0,T], g_\infty)} \to 0,$$

we know for every $q \in \mathcal{M}_\infty$ that

$$d_{g_k(\bar\tau)}(p_k, \Phi_k(q)) = d_{\Phi_k^*(g_k|_{V_k})(\bar\tau)}(p_\infty, q) \to d_{g_\infty(\bar\tau)}(p_\infty, q)$$

as $k \to \infty$. For sufficiently large k, let $\gamma_k : [0, \bar\tau] \to \mathcal{M}_k$ be the minimizing $\mathcal{L}$-geodesic with respect to g_k from p_k to $\Phi_k(q)$. By Lemma 7.13(iii) and (i), we have

$$\mathcal{L}(\gamma_k) = L^{g_k}_{(p_k, 0)}(\Phi_k(q), \bar\tau) \leq C_2,$$

and for any $\tau \in [0, \bar\tau]$,

$$d_{g_k(0)}(p_k, \gamma_k(\tau)) \leq C_2$$

for k large enough, where C_2 is a constant independent of k and τ (depending on $n, \bar\tau, T, C_0, d_{g_\infty(\bar\tau)}(p_\infty, q)$). Hence $\Phi_k^{-1}(\gamma_k([0, \bar\tau]))$ stays inside some compact set $\mathcal{K}_1 \subset \mathcal{M}_\infty$ independent of k. For k sufficiently large, we have $L^{g_k}_{(p_k, 0)}(\Phi_k(q), \bar\tau) = \mathcal{L}_{\Phi_k^*(g_k|_{V_k})}\left(\Phi_k^{-1}(\gamma_k)\right)$. By (7.102), we get

$$\liminf_{k\to\infty} \ell^{g_k}_{(p_k, 0)}(\Phi_k(q), \bar\tau) = \frac{1}{2\sqrt{\bar\tau}} \liminf_{k\to\infty} \mathcal{L}_{\Phi_k^*(g_k|_{V_k})}\left(\Phi_k^{-1} \circ \gamma_k\right)$$
$$= \frac{1}{2\sqrt{\bar\tau}} \liminf_{k\to\infty} \mathcal{L}_{g_\infty}\left(\Phi_k^{-1} \circ \gamma_k\right) \geq \ell^{g_\infty}_{(p_\infty, 0)}(q, \bar\tau).$$

The second equality above needs uniform bound $\sqrt{\tau}\left|\frac{d\gamma_k}{d\tau}\right|^2_{g_k}(\tau) \leq C_2$, which follows from Lemma 7.13(ii) and an argument similar to that in the proof of Lemma 7.28. The last inequality above follows from the fact that $\Phi_k^{-1} \circ \gamma_k$ is a curve from p_∞ to q. We have shown the desired opposite inequality and the lemma is proved (we leave it as an exercise to prove the uniform convergence). $\square$

7. The ℓ-function on Einstein solutions and Ricci solitons

7.1. ℓ function on an Einstein solution with positive scalar curvature.
We consider an Einstein solution $(\mathcal{M}^n, g(\tau))$ of the backward Ricci flow with positive scalar curvature defined on a time interval containing 0. We first consider the case that the solution is smooth at $\tau = 0$ and then generalize to the case that the solution becomes singular as $\tau \searrow 0$. From the scalar curvature evolution equation $\frac{dR}{d\tau} = -\frac{2}{n}R^2$, we have

(7.103) $$R(\tau) = \frac{1}{R(0)^{-1} + \frac{2\tau}{n}} = \frac{1}{1 + \frac{2\tau}{n}R(0)} R(0)$$

and

$$g(\tau) = \left(1 + \frac{2\tau}{n}R(0)\right)g(0), \quad \tau \in \left(-\frac{n}{2R(0)}, \infty\right).$$

Given a curve $\gamma : [0, \bar{\tau}] \to \mathcal{M}$ from p to q, we have

$$\mathcal{L}(\gamma) = \int_0^{\bar{\tau}} \sqrt{\tau}\left(R(\gamma(\tau), \tau) + \left|\frac{d\gamma}{d\tau}\right|^2_{g(\tau)}\right) d\tau$$

$$= \int_0^{\bar{\tau}} \sqrt{\tau}\left(\frac{1}{R(0)^{-1} + \frac{2\tau}{n}} + \left(1 + \frac{2\tau}{n}R(0)\right)\left|\frac{d\gamma}{d\tau}\right|^2_{g(0)}\right) d\tau.$$

Let $\sigma = 2\sqrt{\tau}$, so that

$$\int_0^{\bar{\tau}} \frac{\sqrt{\tau}}{R(0)^{-1} + \frac{2\tau}{n}} d\tau = \frac{n}{2}\int_0^{2\sqrt{\bar{\tau}}} \frac{\sigma^2}{2n \cdot R(0)^{-1} + \sigma^2} d\sigma$$

$$= n\sqrt{\bar{\tau}}\left(1 - \frac{\tan^{-1}\left(\sqrt{2\bar{\tau}R(0)/n}\right)}{\sqrt{2\bar{\tau}R(0)/n}}\right).$$

Recall that the minimum of the functional $\int_0^{\bar{\tau}} \phi(\tau) x(\tau)^2 d\tau$ under the constraint $\int_0^{\bar{\tau}} x(\tau) d\tau = d$ is given by $x(\tau) = \frac{1}{\phi(\tau)} \frac{d}{\int_0^{\bar{\tau}} \phi(\tau)^{-1} d\tau}$. Similarly the minimum of

(7.104) $$E(\gamma) \doteqdot \int_0^{\bar{\tau}} \phi(\tau)\left|\frac{d\gamma}{d\tau}\right|^2_{g(0)} d\tau$$

is
$$\frac{d_{g(0)}^2(p,q)}{\int_0^{\bar{\tau}} \phi(\tau)^{-1} d\tau}$$

and the minimizer is given by a minimal geodesic γ with speed

$$\left|\frac{d\gamma}{d\tau}\right|_{g(0)} = \frac{1}{\phi(\tau)} \frac{d_{g(0)}(p,q)}{\int_0^{\bar{\tau}} \phi(\tau)^{-1} d\tau}.$$

Caveat: Here we have assumed that the improper integral $\int_0^{\bar{\tau}} \phi(\tau)^{-1} d\tau$ is convergent.

Hence

$$\inf \int_0^{\bar{\tau}} \sqrt{\tau}\left(1 + \frac{2\tau}{n} R(0)\right) \left|\frac{d\gamma}{d\tau}\right|_{g(0)}^2 d\tau$$

$$= \frac{d_{g(0)}^2(p,q)}{\int_0^{\bar{\tau}} \tau^{-1/2} \left(1 + \frac{2\tau}{n} R(0)\right)^{-1} d\tau}.$$

Again we make the rationalizing substitution $\sigma = 2\sqrt{\tau}$ to get

$$\int_0^{\bar{\tau}} \tau^{-1/2} \left(1 + \frac{2\tau}{n} R(0)\right)^{-1} d\tau = \frac{2n}{R(0)} \int_0^{2\sqrt{\bar{\tau}}} \frac{1}{2nR(0)^{-1} + \sigma^2} d\sigma$$

$$= \sqrt{2n} R(0)^{-1/2} \tan^{-1}\left(\frac{2\sqrt{\bar{\tau}}}{\sqrt{2n} R(0)^{-1/2}}\right).$$

We conclude

LEMMA 7.67. *Let $(\mathcal{M}^n, g(\tau))$ be an Einstein solution of the backward Ricci flow on a time interval containing 0 with $R(0) > 0$. Let $p \in \mathcal{M}$ be the basepoint. For all $\bar{\tau} \in (0, \infty)$, we have*

$$L(q, \bar{\tau}) = n\sqrt{\bar{\tau}} \left(1 - \frac{\tan^{-1}\left(\sqrt{2\bar{\tau} R(0)/n}\right)}{\sqrt{2\bar{\tau} R(0)/n}}\right)$$

$$+ \frac{R(0)^{1/2} d_{g(0)}^2(p,q)}{\sqrt{2n} \tan^{-1}\left(\sqrt{2\bar{\tau} R(0)/n}\right)}.$$

In particular,

(7.105) $$\ell(q, \bar{\tau}) = \frac{n}{2}\left(1 - \frac{\tan^{-1}\left(\sqrt{2\bar{\tau} R(0)/n}\right)}{\sqrt{2\bar{\tau} R(0)/n}}\right)$$

$$+ \frac{\sqrt{2\bar{\tau} R(0)/n}}{\tan^{-1}\left(\sqrt{2\bar{\tau} R(0)/n}\right)} \frac{d_{g(0)}^2(p,q)}{4\bar{\tau}}.$$

Since by (7.103),
$$\frac{2\bar{\tau}R(0)}{n} = \frac{R(0)}{R(\bar{\tau})} - 1,$$
we may rewrite (7.105) as

(7.106) $$\ell(q,\bar{\tau}) = \frac{n}{2}\left(1 - \frac{\tan^{-1}(s(\bar{\tau}))}{s(\bar{\tau})}\right) + \frac{d^2_{g(0)}(p,q)}{4\bar{\tau}}\frac{s(\bar{\tau})}{\tan^{-1}(s(\bar{\tau}))},$$

where
$$s(\bar{\tau}) \doteq \sqrt{\frac{R(0)}{R(\bar{\tau})} - 1}.$$

Now we consider the extreme case: $R(0) = \infty$.

EXERCISE 7.68. Let $(\mathcal{M}^n, g(\tau))$, $\tau > 0$, be an Einstein solution of the backward Ricci flow. Suppose that $\lim_{\tau \to 0} R(\tau) = \infty$ so that $R(\tau) = \frac{n}{2\tau}$, $\mathrm{Rc}(\tau) = \frac{1}{2\tau}g(\tau)$ and $g(\tau) = \tau g(1)$. Although the metric $g(0)$ is not defined, we may still consider $(p, 0)$, $p \in \mathcal{M}$, as the basepoint for defining $\mathcal{L}$, L and ℓ as before. We have for $\gamma : [0, \bar{\tau}] \to \mathcal{M}$ from p to q,
$$\mathcal{L}(\gamma) = n\sqrt{\bar{\tau}} + \int_0^{\bar{\tau}} \tau^{3/2} \left|\frac{d\gamma}{d\tau}\right|^2_{g(1)} d\tau.$$

Show by considering the paths
$$\tilde{\gamma}(\tau) = \begin{cases} \beta(\tau/\eta) & \tau \leq \eta, \\ q & \tau > \eta, \end{cases}$$

where $\beta : [0, 1] \to \mathcal{M}$ is a constant speed geodesic with respect to $g(1)$ joining p to q, and letting $\eta \to 0$, that

(7.107) $$\inf_\gamma \int_0^{\bar{\tau}} \tau^{3/2} \left|\frac{d\gamma}{d\tau}\right|^2_{g(1)} d\tau = 0,$$

where the infimum is taken over $\gamma : [0, \bar{\tau}] \to \mathcal{M}$ joining p to q. That is,

(7.108) $$L(q, \bar{\tau}) = n\sqrt{\bar{\tau}}.$$

Hence for an Einstein solution $g(\tau)$ of the backward Ricci flow with
$$\lim_{\tau \to 0} R(\tau) = \infty,$$
we have
$$\ell(q, \bar{\tau}) \equiv \frac{n}{2}.$$

Using the rule for the ℓ function on product spaces (see Exercise 7.11), if we have a product solution $(\mathcal{M} \times \mathcal{N}, g(\tau) + h(\tau))$, where $(\mathcal{M}^n, g(\tau))$ is an Einstein solution of the backward Ricci flow with $\lim_{\tau \to 0} R_g(\tau) = \infty$ and where $(\mathcal{N}^m, h(\tau))$ is Ricci flat, then

$$\ell^{g+h}_{(p_1, p_2, 0)}(q_1, q_2, \bar{\tau}) = \frac{n}{2} + \frac{d_h(p_2, q_2)^2}{4\bar{\tau}}.$$

For example, we may take $(\mathcal{M}^2, g(\tau))$ to be an evolving 2-sphere and $(\mathcal{N}^1, h(\tau))$ to be the line to get the cylinder $S^2 \times \mathbb{R}$.

EXERCISE 7.69. Assuming the solution $(\mathcal{M}^n, g(\tau))$ is defined only for $\tau > 0$, rewrite equation (7.106) using the metric $g(1)$ instead of $g(0)$. Show that this equivalent form is consistent with (7.108).

We follow up on the above exercise by translating time in our Einstein solution so that $R(0) \to \infty$ in (7.103), and correspondingly, $R(\tau) \to \frac{n}{2\tau}$. Since we then have $d^2_{g(0)}(p,q) \to 0$, we choose to rewrite the formula for ℓ in terms of $d^2_{g(\bar{\tau})}(p,q)$ using

$$\frac{d^2_{g(0)}(p,q)}{d^2_{g(\bar{\tau})}(p,q)} = \frac{R(\bar{\tau})}{R(0)} = \frac{1}{1 + \frac{2\bar{\tau}}{n} R(0)}.$$

In particular, from (7.105), we have

$$\ell(q, \bar{\tau}) = \frac{n}{2}\left(1 - \frac{\tan^{-1}\left(\sqrt{2\bar{\tau} R(0)/n}\right)}{\sqrt{2\bar{\tau} R(0)/n}}\right)$$
$$+ \frac{R(\bar{\tau}) d^2_{g(\bar{\tau})}(p,q)}{\tan^{-1}\left(\sqrt{2\bar{\tau} R(0)/n}\right)} \frac{1}{2\sqrt{2n}\sqrt{\bar{\tau} R(0)}}$$
$$\to \frac{n}{2}$$

as $R(0) \to \infty$.

EXERCISE 7.70. Let $(\mathcal{M}^n, \tilde{g}(t))$, $t \in [0, T)$, be a maximal shrinking Einstein solution of Ricci flow so that $R(t) = \frac{n}{2(T-t)}$. Given any $(x_i, t_i) \in \mathcal{M} \times (0, T)$, we define a solution $g_i(\tau) \doteqdot \tilde{g}(t_i - \tau)$ of the backward Ricci flow and $\tilde{\ell}_{(x_i, t_i)}(x, t) = \ell^{g_i}_{(x_i, 0)}(x, t_i - t)$. Check that

$$\tilde{\ell}^{g_i}_{(x_i, t_i)}(x, t) = \frac{n}{2}\left(1 - \frac{\tan^{-1}\sqrt{\frac{t_i - t}{T - t_i}}}{\sqrt{\frac{t_i - t}{T - t_i}}}\right)$$
$$+ \frac{d^2_{g(t)}(x, x_i)}{4(T-t)\sqrt{\frac{t_i-t}{T-t_i}}\tan^{-1}\sqrt{\frac{t_i-t}{T-t_i}}}.$$

From the above remarks, $\tilde{\ell}^{g_i}_{(x_i, t_i)}(x, t) \to \frac{n}{2}$ as $i \to \infty$ if $t_i \to T$.

7.2. The ℓ function on a steady gradient Ricci soliton. Consider a steady gradient Ricci soliton $g(\tau) = \varphi^*_\tau g_0$ on $\mathcal{M}^n$ where $\operatorname{Rc}(g_0) + \nabla\nabla f_0 = 0$ and φ_τ satisfies

(7.109) $$\frac{\partial \varphi_\tau}{\partial \tau}(x) = -(\operatorname{grad}_{g_0} f_0)(\varphi_\tau(x)),$$
$$\varphi(0) = \operatorname{id}_{\mathcal{M}}.$$

7. THE ℓ-FUNCTION ON EINSTEIN SOLUTIONS AND RICCI SOLITONS

Given a path $\gamma(\tau)$, its $\mathcal{L}$-length is

$$\mathcal{L}(\gamma) = \int_0^{\bar{\tau}} \sqrt{\tau} \left(R(\gamma(\tau), \tau) + \left|\frac{d\gamma}{d\tau}\right|^2_{g(\tau)} \right) d\tau$$

$$= \int_0^{\bar{\tau}} \sqrt{\tau} \left(R(\varphi_\tau(\gamma(\tau)), 0) + \left|(\varphi_\tau)_* \frac{d\gamma}{d\tau}\right|^2_{g(0)} \right) d\tau.$$

Let $\beta(\tau) \doteqdot \varphi_\tau(\gamma(\tau))$. Note that since $g(\tau) = \varphi_\tau^* g_0$, geometrically, a point (x, τ) is the same as the point $(\varphi_\tau(x), 0)$. That is, $(\gamma(\tau), \tau)$ is the same as $(\beta(\tau), 0)$. We have

$$\dot{\beta}(\tau) \doteqdot \frac{d\beta}{d\tau} = (\varphi_\tau)_* \frac{d\gamma}{d\tau} + \frac{\partial \varphi_\tau}{\partial \tau}(\beta(\tau)),$$

so that

$$(\varphi_\tau)_* \frac{d\gamma}{d\tau} = \dot{\beta}(\tau) + (\operatorname{grad}_{g_0} f_0)(\beta(\tau)).$$

Hence

$$\mathcal{L}(\gamma) = \int_0^{\bar{\tau}} \sqrt{\tau} \left(R(\beta(\tau), 0) + \left|\dot{\beta}(\tau) + (\operatorname{grad}_{g_0} f_0)(\beta(\tau))\right|^2_{g(0)} \right) d\tau$$

$$= \int_0^{\bar{\tau}} \sqrt{\tau} \left(R(\beta(\tau), 0) + |(\operatorname{grad}_{g_0} f_0)(\beta(\tau))|^2_{g(0)} \right) d\tau$$

$$+ \int_0^{\bar{\tau}} \sqrt{\tau} \left(\left|\dot{\beta}(\tau)\right|^2_{g(0)} + 2 \left\langle \dot{\beta}(\tau), (\operatorname{grad}_{g_0} f_0)(\beta(\tau)) \right\rangle_{g(0)} \right) d\tau.$$

Now from (1.34) we have

(7.110) $$R(\beta(\tau), 0) + |(\operatorname{grad}_{g_0} f_0)(\beta(\tau))|^2_{g(0)} = \tilde{C}$$

independent of $\beta(\tau)$, and

$$\left\langle \dot{\beta}(\tau), (\operatorname{grad}_{g_0} f_0)(\beta(\tau)) \right\rangle_{g(0)} = \frac{d}{d\tau}(f_0(\beta(\tau))).$$

Hence, letting $\sigma = 2\sqrt{\tau}$ and $\alpha(\sigma) \doteqdot \beta(\sigma^2/4)$ and integrating by parts, we have an alternate formula for the $\mathcal{L}$-length on a steady gradient Ricci soliton.

LEMMA 7.71. *On a steady gradient Ricci soliton* $\operatorname{Rc}(g_0) + \nabla\nabla f_0 = 0$, *we have*

$$\mathcal{L}(\gamma) = \frac{2}{3}\tilde{C}\bar{\tau}^{3/2} + 2\sqrt{\bar{\tau}}f_0\left(\alpha\left(2\sqrt{\bar{\tau}}\right)\right) + \int_0^{2\sqrt{\bar{\tau}}} \left(\left|\frac{d\alpha}{d\sigma}(\sigma)\right|^2_{g(0)} - f_0(\alpha(\sigma)) \right) d\sigma,$$

where $\tilde{C}$ *is given in* (7.110), $\alpha(\sigma) = \varphi_{\sigma^2/4}\left(\gamma\left(\frac{\sigma^2}{4}\right)\right)$ *and* φ_τ *is defined by* (7.109).

Later we shall see that the shrinking case gives us a more explicit formula.

Now we compute the geodesic equation by taking a variation $Y = \delta\alpha$ of α which vanishes at the endpoints. We have

COROLLARY 7.72. *The variation of the $\mathcal{L}$-length on a steady gradient Ricci soliton* $\mathrm{Rc}(g_0) + \nabla\nabla f_0 = 0$ *is given by*

$$\delta\mathcal{L}(\gamma)(Y) = -\int_0^{2\sqrt{\bar\tau}} \left\langle 2\nabla_{\frac{d\alpha}{d\sigma}} \frac{d\alpha}{d\sigma} + \nabla f_0, Y \right\rangle d\sigma,$$

where ∇ *is the Levi-Civita connection for* $g(0)$. *Hence the $\mathcal{L}$-geodesic equation is*

$$\nabla_{\frac{d\alpha}{d\sigma}} \frac{d\alpha}{d\sigma} + \frac{1}{2}\nabla f_0 = 0.$$

This also implies that the following directional derivative vanishes:

$$\frac{d\alpha}{d\sigma}\left(\left|\frac{d\alpha}{d\sigma}\right|^2_{g(0)} + f_0 \right) = 0.$$

That is, along an $\mathcal{L}$-geodesic, the square of its speed with respect to σ plus the potential function is constant.

EXERCISE 7.73. Show that the above equation is equivalent to (7.35):

$$0 = D_Z Z - \frac{\sigma^2}{8}\nabla R + \sigma\,\mathrm{Rc}(Z),$$

where $Z \doteqdot \sqrt{\tau}\frac{d\gamma}{d\tau}$. Note that on a gradient soliton $\frac{1}{2}\nabla R = \mathrm{Rc}(\nabla f)$.

In the rest of this subsection we consider $\mathcal{L}$-geodesics on the cigar on $\mathbb{R}^2$. The scalar curvature of the cigar solution to the backward Ricci flow $\frac{\partial g}{\partial \tau} = 2\,\mathrm{Rc}$,

$$g(x,y,\tau) \doteqdot \frac{dx^2 + dy^2}{e^{-4\tau} + x^2 + y^2},$$

is given by

$$R(x,y,\tau) = \frac{4}{1 + e^{4\tau}(x^2 + y^2)}.$$

Let $r^2 = x^2 + y^2$. The $\mathcal{L}$-length of a *radial path* $\gamma(\tau)$, $0 \leq \tau \leq \bar\tau$, with $r(\tau) \doteqdot r(\gamma(\tau))$, is

$$\mathcal{L}(\gamma) = \int_0^{\bar\tau} \sqrt{\tau}\left(\frac{4}{1 + e^{4\tau}r^2} + \frac{1}{e^{-4\tau} + r^2}\left(\frac{dr}{d\tau}\right)^2 \right) d\tau.$$

Define $s(\tau) \doteqdot \sinh^{-1}(e^{2\tau}r(\tau))$ (which is the distance to the origin with respect to $g(\tau)$) so that

$$\mathcal{L}(\gamma) = \int_0^{\bar\tau} \left(4\,\mathrm{sech}^2 s + \left(\frac{ds}{d\tau} - 2\tanh s\right)^2 \right)\sqrt{\tau}\,d\tau.$$

Setting $\sigma \doteq 2\sqrt{\tau}$, we have

$$\mathcal{L}(\gamma) = \int_0^{2\sqrt{\tau}} \left(\sigma^2 \operatorname{sech}^2 s + \left(\frac{ds}{d\sigma} - \sigma \tanh s \right)^2 \right) d\sigma$$

$$= \int_0^{2\sqrt{\tau}} \left(\sigma^2 + \left(\frac{ds}{d\sigma} \right)^2 - 2\sigma \tanh s \frac{ds}{d\sigma} \right) d\sigma.$$

Those $\mathcal{L}$-geodesics (i.e., the critical points of $\mathcal{L}$) that emanate from the origin are given by radial paths $s = s(\sigma)$ which satisfy

$$\frac{d^2 s}{d\sigma^2} - \tanh s = 0,$$
$$s(0) = 0.$$

Indeed, if $\delta s = v$, then

$$\delta \mathcal{L}(\gamma) = \int_0^{2\sqrt{\tau}} \left(2 \frac{ds}{d\sigma} \frac{dv}{d\sigma} - 2\sigma \left(v \tanh' s \frac{ds}{d\sigma} + \tanh s \frac{dv}{d\sigma} \right) \right) d\sigma$$

$$= \int_0^{2\sqrt{\tau}} \left(-2v \frac{d^2 s}{d\sigma^2} + 2v \tanh s \right) d\sigma.$$

Multiplying the $\mathcal{L}$-geodesic equation above by $\frac{ds}{d\sigma}$ and integrating, we get

$$\left(\frac{ds}{d\sigma} \right)^2 - 2 \log \cosh s = s'(0)^2,$$

or equivalently,

$$\frac{ds}{d\sigma} = \sqrt{a^2 + 2 \log \cosh s},$$

where $a = s'(0)$. Note that $\frac{ds}{d\sigma} \geq a$ and for s large, $\frac{ds}{d\sigma} \approx \sqrt{2s}$. We leave it as an exercise to check that $s \approx \frac{\sigma^2}{2} = 2\tau$. In comparison, the radial $\mathcal{L}$-geodesics on a cylinder of any dimension $S^{n-1} \times \mathbb{R}$ satisfy $s = 2a\sqrt{\tau}$ (see (7.23)).

REMARK 7.74. The solutions of the linearized $\mathcal{L}$-geodesic equation at $s = 0$, i.e.,

$$\frac{d^2 s}{d\sigma^2} - s = 0,$$
$$s(0) = 0,$$

are $s(\sigma) = a \sinh \sigma$.

PROBLEM 7.75. Determine the qualitative properties of the reduced distance ℓ on the Bryant soliton.

7.3. ℓ function on a gradient shrinker. Let $\left(\mathcal{N}^n, \tilde{h}(t)\right)$, $-\infty < t < 1$, be a shrinking gradient Ricci soliton in canonical form as given in Proposition 1.7 with $\varepsilon = -1$. Define $\tau \doteqdot 1 - t$, and let

(7.111) $\quad h(\tau) \doteqdot \tilde{h}(1-\tau) \quad \text{and} \quad f(\tau) \doteqdot \tilde{f}(1-\tau), \quad 0 < \tau < \infty.$

Then $h(\tau)$ is a solution of the backward Ricci flow on the maximal time interval $(0, \infty)$. Since $\tilde{h}(t) = (1-t)\tilde{\varphi}_t^* \tilde{h}(0)$, we have for $\tau > 0$

$$h(\tau) = \tau \varphi_\tau^* h(1),$$

where $\varphi_\tau \doteqdot \tilde{\varphi}_{1-\tau}$. Although the metric $\tilde{h}(0)$ is not well defined, we may still define $\mathcal{L}$, L, and ℓ as before using the basepoint $(p, 0)$. By (1.12) we have

$$f(\tau) = f(1) \circ \varphi_\tau.$$

Note that $R_h(x, \tau) = \frac{1}{\tau} R_{h(1)}(\varphi_\tau(x))$. From

$$\frac{\partial}{\partial t} \tilde{\varphi}_t(x) = \frac{1}{1-t} \left(\operatorname{grad}_{\tilde{h}(0)} \tilde{f}(0) \right) (\tilde{\varphi}_t(x)),$$

we have

(7.112) $\quad \dfrac{\partial}{\partial \tau} \varphi_\tau = -\left(\dfrac{\partial}{\partial t} \tilde{\varphi}_t\right) = -\dfrac{1}{\tau} \left(\operatorname{grad}_{h(1)} f(1) \right) \circ \varphi_\tau.$

REMARK 7.76. Note that if $h(\tau)$ is a shrinking gradient soliton flowing along ∇f, then $\operatorname{Rc} + \nabla \nabla f - \frac{1}{2\tau} h = 0$. Since f satisfies $\frac{\partial f}{\partial \tau} = -|\nabla f|^2$, the gradient vector field ∇f satisfies

$$\frac{\partial}{\partial \tau} (\nabla f) = -2\operatorname{Rc}(\nabla f) + \nabla \left(\frac{\partial f}{\partial \tau} \right) = -2\operatorname{Rc}(\nabla f) - \nabla |\nabla f|^2 = -\frac{1}{\tau} \nabla f.$$

Given a path $\gamma : [0, \bar{\tau}] \to \mathcal{N}$ from p to q, its $\mathcal{L}$-length is

$$\mathcal{L}(\gamma) = \int_0^{\bar{\tau}} \sqrt{\tau} \left(R_{h(\tau)}(\gamma(\tau)) + |\dot{\gamma}(\tau)|^2_{h(\tau)} \right) d\tau$$

$$= \int_0^{\bar{\tau}} \sqrt{\tau} \left(\frac{1}{\tau} R_{h(1)}(\varphi_\tau(\gamma(\tau))) + \tau |(\varphi_\tau)_* \dot{\gamma}(\tau)|^2_{h(1)} \right) d\tau.$$

Let $\beta(\tau) \doteqdot \varphi_\tau(\gamma(\tau))$. Note that the point $(\beta(\tau), 1)$ corresponds geometrically to the point $(\gamma(\tau), \tau)$ since in general the point (x, τ) corresponds to $(\varphi_\tau(x), 1)$. We have

$$\dot{\beta}(\tau) = (\varphi_\tau)_* \dot{\gamma}(\tau) + \frac{\partial \varphi_\tau}{\partial \tau}(\varphi_\tau(\gamma(\tau))),$$

which implies

$$(\varphi_\tau)_* \dot{\gamma}(\tau) = \dot{\beta}(\tau) + \frac{1}{\tau}\left(\operatorname{grad}_{h(1)} f(1)\right)(\beta(\tau)).$$

Hence
$$\mathcal{L}(\gamma) = \int_0^{\bar{\tau}} \sqrt{\tau}\left(\frac{1}{\tau}R_{h(1)}(\beta(\tau)) + \tau\left|\dot{\beta}(\tau) + \frac{1}{\tau}(\operatorname{grad}_{h(1)}f(1))(\beta(\tau))\right|^2_{h(1)}\right)d\tau$$
$$= \int_0^{\bar{\tau}} \tau^{-1/2}\left(R_{h(1)}(\beta(\tau)) + |(\operatorname{grad}_{h(1)}f(1))(\beta(\tau))|^2_{h(1)}\right)d\tau$$
$$+ \int_0^{\bar{\tau}} \tau^{3/2}\left|\dot{\beta}(\tau)\right|^2_{h(1)} d\tau + \int_0^{\bar{\tau}} 2\sqrt{\tau}\frac{d}{d\tau}(f(\beta(\tau),1))\,d\tau$$
$$= \int_0^{\bar{\tau}} \tau^{-1/2}\left(R_{h(1)}(\beta(\tau)) + |(\operatorname{grad}_{h(1)}f)(\beta(\tau),1)|^2_{h(1)} - f(\beta(\tau),1)\right)d\tau$$
$$+ \int_0^{\bar{\tau}} \tau^{3/2}\left|\dot{\beta}(\tau)\right|^2_{h(1)} d\tau + 2\sqrt{\bar{\tau}}f(\beta(\bar{\tau}),1).$$

On a shrinking gradient Ricci soliton we have
(7.113) $\quad R_{h(1)}(\beta(\tau)) + |(\operatorname{grad}_{h(1)}f)(\beta(\tau),1)|^2_{h(1)} - f(\beta(\tau),1) = \hat{C}.$

Hence
$$\mathcal{L}(\gamma) = 2\sqrt{\bar{\tau}}\left(f(\beta(\bar{\tau}),1) + \hat{C}\right) + \int_0^{\bar{\tau}} \tau^{3/2}\left|\dot{\beta}(\tau)\right|^2_{h(1)} d\tau$$

and
$$\frac{1}{2\sqrt{\bar{\tau}}}\mathcal{L}(\gamma) = f(\beta(\bar{\tau}),1) + \hat{C} + \frac{1}{2\sqrt{\bar{\tau}}}\int_0^{\bar{\tau}} \tau^{3/2}\left|\dot{\beta}(\tau)\right|^2_{h(1)} d\tau.$$

We have
$$f(\beta(\bar{\tau}),1) = f\left(\varphi_{\bar{\tau}}^{-1}(\beta(\bar{\tau})),\bar{\tau}\right) = f(\gamma(\bar{\tau}),\bar{\tau}).$$

Note that from (7.107),

(7.114) $\quad\displaystyle\inf_{\beta}\int_0^{\bar{\tau}} \tau^{3/2}\left|\dot{\beta}(\tau)\right|^2_{h(1)} d\tau = 0,$

where the infimum is taken over all $\beta : [0,\bar{\tau}] \to \mathcal{N}$ joining p to $\varphi_{\bar{\tau}}(q)$. Since $\beta(\bar{\tau}) = \varphi_{\bar{\tau}}(q)$ implies $\gamma(\bar{\tau}) = q$, we conclude

LEMMA 7.77 (Reduced distance on shrinker). *For a shrinking gradient Ricci soliton as in Proposition 1.7 with $\varepsilon = -1$,*

(7.115) $\quad\quad\quad\quad\quad \ell(q,\bar{\tau}) = f(q,\bar{\tau}) + \hat{C},$

where f is defined in (7.111) and $\hat{C}$ is from (7.113). That is,
$$\ell(q,\bar{\tau}) = f(\beta(\bar{\tau}),1) = f(\varphi_{\bar{\tau}}(q),1) + \hat{C},$$
where $\beta(\tau) \doteqdot \varphi_{\tau}(\gamma(\tau))$ and φ_{τ} is defined by (7.112).

REMARK 7.78. Note that for flat Euclidean space, thought of as a shrinking (Gaussian) soliton, the potential $f = \frac{|x|^2}{4\tau}$ given in (1.16) is the same as ℓ.

EXERCISE 7.79. Show that for a shrinking gradient Ricci soliton, the paths $\gamma(\tau) = \varphi_{1-\tau}(x) = \tilde{\varphi}_{\tau}(x)$ for $x \in \mathcal{N}$ fixed are $\mathcal{L}$-geodesics.

Following up on details related to the previous exercise, we have

EXERCISE 7.80. In regards to (7.114), show that although for $\varphi_{\bar\tau}(q) \neq p$ a smooth minimizer of $\inf_\beta \int_0^{\bar\tau} \tau^{3/2} \left|\dot\beta(\tau)\right|^2_{h(1)} d\tau$, where the infimum is taken over paths joining p to $\varphi_{\bar\tau}(q)$, does not exist (which is related to $h(0)$ not being well-defined), any minimizing sequence β_i of paths joining p to $\varphi_{\bar\tau}(q)$ limits to the constant path $\tau \mapsto \varphi_{\bar\tau}(q)$. That is, any minimizing sequence γ_i of paths joining p to q for $\frac{1}{2\sqrt{\bar\tau}}\mathcal{L}(\gamma)$ limits to the path $\tau \mapsto \varphi_\tau^{-1}(\varphi_{\bar\tau}(q))$. Note that in general, $q_0 \doteqdot \lim_{\tau \to 0} \varphi_\tau^{-1}(\varphi_{\bar\tau}(q)) \neq p$. We may think of this as saying that the minimal geodesic starting at $(p,0)$ immediately jumps to $(q_0, 0)$ and then becomes a constant path in the geometric sense. **Caveat:** The solution is undefined at the 'big bang' time $\tau = 0$.

We now present another proof of Lemma 7.77, following an original idea of one of the authors [289]. (The proof given above is also inspired by his line of reasoning.) Given a path $\gamma : [0, \bar\tau] \to \mathcal{N}$, we have

$$\frac{d}{d\tau}\left(\sqrt{\tau} f(\gamma(\tau), \tau)\right) = \sqrt{\tau}\left(\frac{f}{2\tau} + \frac{\partial f}{\partial \tau} + \nabla f \cdot \dot\gamma\right)$$

(7.116)
$$= \sqrt{\tau}\left(\frac{f}{2\tau} + \frac{\partial f}{\partial \tau} + \frac{1}{2}|\nabla f|^2 + \frac{1}{2}|\dot\gamma|^2 - \frac{1}{2}|\dot\gamma - \nabla f|^2\right).$$

Now for a gradient shrinker in canonical form, where (1.14) holds, i.e.,

(7.117) $$\frac{\partial f}{\partial \tau} = -|\nabla f|^2,$$

assuming f has a critical point,[12] we may normalize f (by adding the appropriate constant) so that

(7.118) $$R + |\nabla f|^2 - \frac{1}{\tau} f \equiv 0.$$

Substituting this with (7.117) into (7.116), we have

$$\frac{d}{d\tau}\left(\sqrt{\tau} f(\gamma(\tau), \tau)\right) = \frac{1}{2}\sqrt{\tau}\left(R + |\dot\gamma|^2 - |\dot\gamma - \nabla f|^2\right).$$

Hence

$$f(\gamma(\bar\tau), \bar\tau) = \frac{1}{2\sqrt{\bar\tau}}\mathcal{L}(\gamma) - \frac{1}{2\sqrt{\bar\tau}}\int_0^{\bar\tau} \sqrt{\tau}\,|\dot\gamma(\tau) - \nabla f(\gamma(\tau), \tau)|^2_{h(\tau)}\, d\tau.$$

Given a point $q \in \mathcal{N}$, we may take $\gamma : (0, \bar\tau] \to \mathcal{N}$ to be the path with $\dot\gamma(\tau) = \nabla f(\tau)$ for all $\tau \in (0, \bar\tau]$ and $\gamma(\bar\tau) = q$. We have

$$f(\gamma(\bar\tau), \bar\tau) = \frac{1}{2\sqrt{\bar\tau}}\mathcal{L}(\gamma) \geq \ell(\gamma(\bar\tau), \bar\tau).$$

Since $\tau = 0$ is the big bang time, $\ell(\gamma(\bar\tau), \bar\tau)$ is independent of the basepoint chosen.

---

[12]If $\mathcal{N}$ is compact, this assumption is always satisfied (though it is also satisfied for the shrinking Gaussian soliton).

On the other hand, taking γ to be a minimal $\mathcal{L}$-geodesic with $\gamma(\bar{\tau}) = q$,[13] we have

$$f(\gamma(\bar{\tau}), \bar{\tau}) = \ell(\gamma(\bar{\tau}), \bar{\tau}) - \frac{1}{2\sqrt{\bar{\tau}}} \int_0^{\bar{\tau}} \sqrt{\tau} |\dot{\gamma}(\tau) - \nabla f(\gamma(\tau), \tau)|_{h(\tau)}^2 d\tau$$
$$\leq \ell(\gamma(\bar{\tau}), \bar{\tau}).$$

Since $\gamma(\bar{\tau}) = q$ is arbitrary, we conclude that $f = \ell$ on $\mathcal{N} \times (0, \infty)$ for f defined by (7.118).

EXERCISE 7.81. Let $\left(\mathcal{N}^n, \tilde{h}(t)\right)$, $-\infty < t < 1$, be a shrinking gradient Ricci soliton in canonical form and consider $(\mathcal{N}, h(\tau))$, $\tau \in (0, \infty)$, where $h(\tau) \doteqdot \tilde{h}(1-\tau)$. Show that for any $p, q \in \mathcal{N}$ and $\bar{\tau} > 0$, if we take any sequence $\tau_i \to 0$ and minimal $\mathcal{L}$-geodesics $\gamma_i : [\tau_i, \bar{\tau}] \to \mathcal{N}$ with $\gamma_i(\tau_i) = p$ and $\gamma_i(\bar{\tau}) = q$, then a subsequence γ_i converges to a minimal $\mathcal{L}$-geodesic $\gamma : (0, \bar{\tau}] \to \mathcal{N}$ with $\gamma(\bar{\tau}) = q$. In particular, independent of the choice of $p \in \mathcal{N}$, for any path $\beta : (0, \bar{\tau}] \to \mathcal{N}$ with $\gamma(\bar{\tau}) = q$, we have

$$\mathcal{L}(\beta) \geq \mathcal{L}(\gamma).$$

8. $\mathcal{L}$-Jacobi fields and the $\mathcal{L}$-exponential map

Continuing our mimicry of Riemannian comparison geometry, in this section we derive the $\mathcal{L}$-Jacobi equation, which is a linear second-order ODE along an $\mathcal{L}$-geodesic, and we derive an estimate for the norm of an $\mathcal{L}$-Jacobi field. We also discuss the $\mathcal{L}$-exponential map, its Jacobian, called the $\mathcal{L}$-Jacobian, and briefly mention the $\mathcal{L}$-index lemma. These results will be of crucial importance to our discussion of the reduced volume and its applications in the next chapter.

Throughout this section $(\mathcal{M}^n, g(\tau))$, $\tau \in [0, T]$, will denote a complete solution to the backward Ricci flow satisfying the curvature bound $\max\{|\text{Rm}|, |\text{Rc}|\} \leq C_0 < \infty$ on $\mathcal{M} \times [0, T]$, and $p \in \mathcal{M}$ is a basepoint.

Before discussing Ricci flow, we first recall some basic Riemannian geometry that is relevant to the material in this section. A good reference for comparison Riemannian geometry, besides Cheeger and Ebin [**72**], is Milnor's book [**265**]; in our setting, the Riemannian path energy is analogous to the $\mathcal{L}$-length. Let $\gamma : [a, b] \to \left(\hat{\mathcal{M}}, \hat{g}\right)$ be a unit speed geodesic, let $\dot{\gamma} \doteqdot \frac{d\gamma}{ds}$, and let ds denote the arc length element. The **index form** is defined by

$$(7.119) \qquad \text{I}(V, W) \doteqdot \int_a^b \left(\langle \nabla_{\dot{\gamma}} V, \nabla_{\dot{\gamma}} W \rangle - \langle R(V, \dot{\gamma})\dot{\gamma}, W \rangle\right) ds,$$

where V and W are vector fields along γ perpendicular to $\dot{\gamma}$ and vanishing at the endpoints. By the second variation of arc length formula, under these assumptions on V and W, we have

$$\delta_{V,W}^2 \, \text{L}(\gamma) = \text{I}(V, W),$$

---

[13] See the exercise below.

where L(γ) denotes the length of γ. Integrating by parts on (7.119), we may express this as

$$\mathrm{I}(V,W) = -\int_a^b \langle \nabla_{\dot\gamma}\nabla_{\dot\gamma}V + R(V,\dot\gamma)\dot\gamma, W\rangle\, ds.$$

Recall that a vector field J along γ is a **Jacobi field** if

$$\nabla_{\dot\gamma}\nabla_{\dot\gamma}J + R(J,\dot\gamma)\dot\gamma = 0.$$

(If J vanishes at the endpoints of γ, then $\mathrm{I}(J,W)=0$ for all W.) Equivalently, a Jacobi field is the variation vector field of a 1-parameter family of geodesics. Given a unit speed geodesic $\gamma:[a,b]\to\hat{\mathcal{M}}$, the set of Jacobi fields along γ is isomorphic to $T_{\gamma(s_0)}\hat{\mathcal{M}} \times T_{\gamma(s_0)}\hat{\mathcal{M}}$, for any $s_0 \in [a,b]$; each Jacobi field J is determined by the initial data $J(s_0)=J_0$ and $(\nabla_{\dot\gamma}J)(s_0)=J_1$ for $J_0, J_1 \in T_{\gamma(s_0)}\hat{\mathcal{M}}$.

The **Index Lemma** says that if $\gamma:[a,b]\to\hat{\mathcal{M}}$ is a unit speed geodesic *without conjugate points*, then among all vector fields along γ perpendicular to $\dot\gamma$ with prescribed values at the endpoints, the unique such Jacobi field minimizes the index form; i.e., given $A \in T_{\gamma(a)}\hat{\mathcal{M}}$ and $B \in T_{\gamma(b)}\hat{\mathcal{M}}$ perpendicular to $\dot\gamma$, the Jacobi field J with $J(a)=A$ and $J(b)=B$ satisfies

(7.120) $$\mathrm{I}(J,J) \le \mathrm{I}(W,W)$$

for all W perpendicular to $\dot\gamma$ and such that $W(a)=A$ and $W(b)=B$. Equality in (7.120) holds if and only if $W=J$. In particular, for any $W \ne \vec{0}$ perpendicular to $\dot\gamma$ with $W(a)=\vec{0}$ and $W(b)=\vec{0}$, we have $\mathrm{I}(W,W) > 0$. If γ has no conjugate points in the interior, but possibly one at b, then (7.120) still holds although J may not be unique. Furthermore, we have $\mathrm{I}(W,W) \ge 0$ for any W perpendicular to $\dot\gamma$ with $W(a)=\vec{0}$ and $W(b)=\vec{0}$, where equality holds if and only if W is a Jacobi field.

8.1. $\mathcal{L}$-Jacobi fields. Let $\gamma:[0,\bar\tau]\to\mathcal{M}$ be an $\mathcal{L}$-geodesic, where $\bar\tau \in (0,T)$, and let $X(\tau) \doteqdot \frac{d\gamma}{d\tau}$ be its tangent vector field.

DEFINITION 7.82 ($\mathcal{L}$-Jacobi field). An **$\mathcal{L}$-Jacobi field** along an $\mathcal{L}$-geodesic γ is the variation vector field of a smooth 1-parameter family of $\mathcal{L}$-geodesics γ_s, $s\in(-\varepsilon,\varepsilon)$, for some $\varepsilon>0$, all defined on the same time interval as $\gamma_0=\gamma$.

Let $X(\tau,s) \doteqdot \frac{\partial\gamma_s}{\partial\tau}$, $Y(\tau,s) \doteqdot \frac{\partial\gamma_s}{\partial s}$, and let $Y(\tau) \doteqdot Y(\tau,0)$ be an $\mathcal{L}$-Jacobi field along γ. Using the $\mathcal{L}$-geodesic equation (7.32), we compute

$$\nabla_X(\nabla_X Y) = \nabla_X(\nabla_Y X) = R(X,Y)X + \nabla_Y(\nabla_X X)$$
$$= R(X,Y)X + \nabla_Y\left(\frac{1}{2}\nabla R - 2\operatorname{Rc}(X) - \frac{1}{2\tau}X\right).$$

8. $\mathcal{L}$-JACOBI FIELDS AND THE $\mathcal{L}$-EXPONENTIAL MAP

Thus we have a linear second-order ODE for the $\mathcal{L}$-Jacobi field $Y(\tau)$, called the $\mathcal{L}$-**Jacobi equation**:

$$(7.121) \quad \nabla_X(\nabla_X Y) = R(X,Y)X + \frac{1}{2}\nabla_Y(\nabla R) - 2(\nabla_Y \operatorname{Rc})(X)$$
$$- 2\operatorname{Rc}(\nabla_X Y) - \frac{1}{2\tau}\nabla_X Y.$$

Since $\tau = 0$ is a singular point because of the $\frac{1}{\tau}$ factor in the last term, we rewrite the equation as

$$D_{\sqrt{\tau}X}\left(\nabla_{\sqrt{\tau}X} Y\right) = \tau\left(\nabla_X(\nabla_X Y) + \frac{1}{2\tau}\nabla_X Y\right)$$
$$= R(\sqrt{\tau}X, Y)\sqrt{\tau}X + \frac{\tau}{2}\nabla_Y(\nabla R)$$
$$- 2\sqrt{\tau}(\nabla_Y \operatorname{Rc})(\sqrt{\tau}X) - 2\sqrt{\tau}\operatorname{Rc}\left(\nabla_{\sqrt{\tau}X} Y\right).$$

Let $Z(\sigma) \doteqdot \sqrt{\tau}X(\tau)$, where $\sigma = 2\sqrt{\tau}$ and $\beta(\sigma) = \gamma(\sigma^2/4)$. Then $Z(\sigma) = \frac{d\beta}{d\sigma}$ and we can rewrite the $\mathcal{L}$-Jacobi equation for $Y(\tau)$ as

$$\nabla_Z(\nabla_Z Y) = -2\sigma \operatorname{Rc}(\nabla_Z Y) + R(Z,Y)Z$$
$$(7.122) \quad - 2\sigma(\nabla_Y \operatorname{Rc})(Z) + \frac{\sigma^2}{2}\nabla_Y(\nabla R),$$

where we view $Y(\sigma^2/4)$ as a function of σ. Suppose $Z(0) = \lim_{\tau \to 0} \sqrt{\tau}X = V \in T_{\gamma(0)}\mathcal{M}$. We have the following by solving the initial-value problem for (7.122).

LEMMA 7.83. *Given initial data $Y_0, Y_1 \in T_{\gamma(0)}\mathcal{M}$, there exists a unique solution $Y(\tau)$ of (7.121) with $Y(0) = Y_0$ and $(\nabla_Z Y)(0) = Y_1$.*

Since (7.121) is linear, the space of $\mathcal{L}$-Jacobi fields along an $\mathcal{L}$-geodesic γ is a finite-dimensional vector space, isomorphic to $T_{\gamma(0)}\mathcal{M} \times T_{\gamma(0)}\mathcal{M}$.

REMARK 7.84. If the solution $(\mathcal{M}^n, g(\tau) = g_0)$ is Ricci flat, then the $\mathcal{L}$-Jacobi equation (7.121) says

$$\nabla_X(\nabla_X Y) = R(X,Y)X - \frac{1}{2\tau}\nabla_X Y.$$

That is, we obtain the Riemannian Jacobi equation for g_0,

$$D_{\sqrt{\tau}X}\left(\nabla_{\sqrt{\tau}X} Y\right) = R(\sqrt{\tau}X, Y)\sqrt{\tau}X;$$

i.e.,

$$\nabla_Z(\nabla_Z Y) = R(Z,Y)Z.$$

On the other hand, if $g(\tau)$ is Einstein and satisfies $\operatorname{Rc} = \frac{1}{2\tau}g$, then

$$\nabla_X(\nabla_X Y) = R(X,Y)X - \frac{3}{2\tau}\nabla_X Y.$$

We now rewrite the $\mathcal{L}$-Jacobi equation in a more natural way in view of the space-time geometry associated to the Ricci flow. Consider the quantity $\mathrm{Rc}_{g(\tau)}(Y)$. The time-dependent symmetric 2-tensor $\mathrm{Rc}_{g(\tau)}$ is defined on all of $\mathcal{M}$ whereas Y is a vector field along the path $\gamma(\tau)$ in $\mathcal{M}$. In local coordinates, $\mathrm{Rc}_{g(\tau)}(Y)^i = g^{ij} R_{jk} Y^k$, so actually we are considering Rc as a $(1,1)$-tensor. **Caveat:** When we take the time-derivative of Rc, we consider it as a $(2,0)$-tensor and then raise an index to get a $(1,1)$-tensor! By $\nabla_X [\mathrm{Rc}(Y)](\tau_0)$ we simply mean the covariant derivative along $\gamma(\tau)$ of the vector field $\mathrm{Rc}_{g(\tau_0)}(Y(\gamma(\tau)))$ at $\tau = \tau_0$. In this respect the vector field $\mathrm{Rc}_{g(\tau_0)}(Y(\gamma(\tau)))$ along $\gamma(\tau)$ should be distinguished from $\mathrm{Rc}_{g(\tau)}(Y(\gamma(\tau)))$, where in the latter case the Ricci tensor depends on time.

Combining the equations[14]

$$D_{\frac{d}{d\tau}}\left[\mathrm{Rc}_{g(\tau)}(Y)\right] = \left(\frac{\partial}{\partial \tau}\mathrm{Rc}\right)(Y) + (\nabla_X \mathrm{Rc})(Y) + \mathrm{Rc}(\nabla_X Y) - 2\,\mathrm{Rc}^2(Y)$$

and

$$D_{\frac{d}{d\tau}}\left(\nabla_Y^{g(\tau)} X\right) = \nabla_X (\nabla_Y X) + \left(\frac{\partial}{\partial \tau}\nabla\right)_Y X$$
$$= \nabla_X (\nabla_Y X) + (\nabla_Y \mathrm{Rc})(X)$$
$$+ (\nabla_X \mathrm{Rc})(Y) - (\nabla \mathrm{Rc})\left(\underset{2}{Y}, \underset{3}{X}\right),$$

where $(\nabla \mathrm{Rc})\left(\underset{2}{Y}, \underset{3}{X}\right)(Z) \doteqdot (\nabla \mathrm{Rc})(Z, Y, X)$, and commuting derivatives, we have

(7.123) $\quad D_{\frac{d}{d\tau}}\left(\mathrm{Rc}(Y) + \nabla_Y X\right)$

$$= \left(\frac{\partial}{\partial \tau}\mathrm{Rc}\right)(Y) + 2(\nabla_X \mathrm{Rc})(Y) + \mathrm{Rc}(\nabla_X Y) - 2\,\mathrm{Rc}^2(Y)$$
$$+ \nabla_Y (\nabla_X X) + R(X,Y)X + (\nabla_Y \mathrm{Rc})(X) - (\nabla \mathrm{Rc})\left(\underset{2}{Y}, \underset{3}{X}\right),$$

where we used

$$\nabla_X (\nabla_Y X) = \nabla_Y (\nabla_X X) + R(X,Y)X.$$

Substituting the $\mathcal{L}$-geodesic equation

$$0 = \nabla_X X - \frac{1}{2}\nabla R + 2\,\mathrm{Rc}(X) + \frac{1}{2\tau}X$$

in (7.123) and adding this to

$$D_{\frac{d}{d\tau}}\left(-\frac{1}{2\tau}Y\right) = \frac{1}{2\tau^2}Y - \frac{1}{2\tau}\nabla_X Y,$$

---

[14] In accordance with the caveat above, $\frac{\partial}{\partial \tau}\mathrm{Rc}$ denotes the derivative of Rc as a $(2,0)$-tensor and $\left(\frac{\partial}{\partial \tau}\mathrm{Rc}\right)(Y)^i \doteqdot g^{ij}\left(\frac{\partial}{\partial \tau}\mathrm{Rc}\right)_{jk} Y^k$.

we have
$$D_{\frac{d}{d\tau}}\left(\operatorname{Rc}(Y) + \nabla_Y X - \frac{1}{2\tau}Y\right)$$
$$= \left(\frac{\partial}{\partial \tau}\operatorname{Rc}\right)(Y) + 2(\nabla_X \operatorname{Rc})(Y) - 2\operatorname{Rc}^2(Y)$$
$$+ \frac{1}{2}\nabla_Y(\nabla R) - 2(\nabla_Y \operatorname{Rc})(X) - \operatorname{Rc}(\nabla_Y X)$$
$$+ R(X,Y)X + (\nabla_Y \operatorname{Rc})(X) - (\nabla \operatorname{Rc})\left(Y, X\atop 2\ \ 3\right)$$
$$+ \frac{1}{2\tau^2}Y - \frac{1}{\tau}\nabla_Y X.$$

This may be rewritten as
$$D_{\frac{d}{d\tau}}\left(\operatorname{Rc}(Y) + \nabla_Y X - \frac{1}{2\tau}Y\right)$$
$$= \left(\frac{\partial}{\partial \tau}\operatorname{Rc}\right)(Y) + \frac{1}{2}\nabla_Y(\nabla R) - \operatorname{Rc}^2(Y) + \frac{1}{2\tau}\operatorname{Rc}(Y)$$
$$- 2(\nabla_Y \operatorname{Rc})(X) + 2(\nabla_X \operatorname{Rc})(Y) + R(X,Y)X$$
$$+ (\nabla \operatorname{Rc})\left(Y, X\atop 1\ \ 3\right) - (\nabla \operatorname{Rc})\left(Y, X\atop 2\ \ 3\right)$$
$$- \operatorname{Rc}\left(\operatorname{Rc}(Y) + \nabla_Y X - \frac{1}{2\tau}Y\right) - \frac{1}{\tau}\left(\operatorname{Rc}(Y) + \nabla_Y X - \frac{1}{2\tau}Y\right).$$

Define the matrix Harnack expression
$$J(Y) \doteqdot -\left(\frac{\partial}{\partial \tau}\operatorname{Rc}\right)(Y) - \frac{1}{2}\nabla_Y(\nabla R) + \operatorname{Rc}^2(Y) - \frac{1}{2\tau}\operatorname{Rc}(Y)$$
$$+ 2(\nabla_Y \operatorname{Rc})(X) - 2(\nabla_X \operatorname{Rc})(Y) - R(X,Y)X$$
$$- (\nabla \operatorname{Rc})\left(Y, X\atop 1\ \ 3\right) + (\nabla \operatorname{Rc})\left(Y, X\atop 2\ \ 3\right),$$

so that (note $\left(-(\nabla \operatorname{Rc})\left(Y, X\atop 1\ \ 3\right) + (\nabla \operatorname{Rc})\left(Y, X\atop 2\ \ 3\right)\right)(Y) = 0$)
$$\langle J(Y), Y\rangle = \frac{1}{2}H(X,Y).$$

Thus we have the following.

LEMMA 7.85. *The $\mathcal{L}$-Jacobi equation is equivalent to*

(7.124) $$\left(D_{\frac{d}{d\tau}} + \operatorname{Rc} + \frac{1}{\tau}\right)\left(\operatorname{Rc}(Y) + \nabla_X Y - \frac{1}{2\tau}Y\right) = -J(Y),$$

where we have replaced $\nabla_Y X$ by $\nabla_X Y$.

EXERCISE 7.86. Rewrite the above equation using Uhlenbeck's trick.

8.2. Bounds for $\mathcal{L}$-Jacobi fields. Let $\varepsilon > 0$ and let $\gamma_s : [0, \bar{\tau}] \to \mathcal{M}$, $s \in (-\varepsilon, \varepsilon)$, be a smooth 1-parameter family of $\mathcal{L}$-geodesics. In this subsection we adopt the notation of subsection 8.1 above. Assume $Y(0, s) = \vec{0}$ for $s \in (-\varepsilon, \varepsilon)$ (for simplicity we may assume $\gamma_s(0) = \gamma(0)$ for all s). We shall estimate from above the norms of $\mathcal{L}$-Jacobi fields $Y(\tau) = Y(\tau, 0)$.

By the first variation formula for the $\mathcal{L}$-length and the $\mathcal{L}$-geodesic equation, we have for $s \in (-\varepsilon, \varepsilon)$,

$$\delta_Y \mathcal{L}(\gamma_s) = 2\sqrt{\bar{\tau}} \langle X_s, Y_s \rangle (\bar{\tau}).$$

We differentiate this again to get

$$\left(\delta_Y^2 \mathcal{L}\right)(\gamma) = 2\sqrt{\bar{\tau}} \langle \nabla_X Y, Y \rangle (\bar{\tau}) + 2\sqrt{\bar{\tau}} \langle X, \nabla_Y Y \rangle (\bar{\tau}),$$

where we used $\nabla_Y X = \nabla_X Y$.

Now the derivative of the norm squared of the $\mathcal{L}$-Jacobi field is

$$\left.\frac{d}{d\tau}\right|_{\tau = \bar{\tau}} |Y|^2 = \left.\frac{d}{d\tau}\right|_{\tau = \bar{\tau}} |Y(\tau)|^2_{g(\tau)} = 2 \langle \nabla_X Y, Y \rangle (\bar{\tau}) + 2 \operatorname{Rc}(Y, Y)(\bar{\tau})$$

(7.125)
$$= 2 \operatorname{Rc}(Y, Y)(\bar{\tau}) + \frac{1}{\sqrt{\bar{\tau}}} \left(\delta_Y^2 \mathcal{L}\right)(\gamma) - 2 \langle X, \nabla_Y Y \rangle (\bar{\tau}),$$

which is expressed in terms of the second variation of $\mathcal{L}$. Let $\tilde{Y}$ be a vector field along γ which satisfies the ODE

(7.126) $$\left(\nabla_X \tilde{Y}\right)(\tau) = -\operatorname{Rc}\left(\tilde{Y}(\tau)\right) + \frac{1}{2\tau}\tilde{Y}(\tau), \quad \tau \in [0, \bar{\tau}],$$

(7.127) $$\tilde{Y}(\bar{\tau}) = Y(\bar{\tau}).$$

(The first equation is the same as (7.66).) As in (7.68),

(7.128) $$\left|\tilde{Y}(\tau)\right|^2 = \frac{\tau}{\bar{\tau}} |Y(\bar{\tau})|^2.$$

In particular, $\tilde{Y}(0) = \vec{0} = Y(0)$.

Now we further *assume* that the γ_s are *minimal* $\mathcal{L}$-geodesics for each $s \in (-\varepsilon, \varepsilon)$. Let $\tilde{\gamma}_s : [0, \bar{\tau}] \to \mathcal{M}$ be a 1-parameter variation of γ with

$$\left.\frac{\partial}{\partial s}\right|_{s=0} \tilde{\gamma}_s = \tilde{Y}, \quad \tilde{\gamma}_s(\bar{\tau}) = \gamma_s(\bar{\tau}) \quad \text{and} \quad \tilde{\gamma}_s(0) = \gamma_s(0) ;$$

this is possible because $\tilde{Y}(0) = Y(0)$ and $\tilde{Y}(\bar{\tau}) = Y(\bar{\tau})$. Then $\mathcal{L}(\tilde{\gamma}_s) \geq \mathcal{L}(\gamma_s)$ for all s, and equality holds at $s = 0$. Hence

$$\left(\delta_Y^2 \mathcal{L}\right)(\gamma) \leq \left(\delta_{\tilde{Y}}^2 \mathcal{L}\right)(\gamma),$$

where equality holds if $\tilde{Y}$ is an $\mathcal{L}$-Jacobi field. Combining this with (7.125), we get

$$\left.\frac{d}{d\tau}\right|_{\tau = \bar{\tau}} |Y|^2 \leq 2 \operatorname{Rc}(Y, Y)(\bar{\tau}) + \frac{1}{\sqrt{\bar{\tau}}} \left(\delta_{\tilde{Y}}^2 \mathcal{L}\right)(\gamma) - 2 \langle X, \nabla_Y Y \rangle (\bar{\tau}).$$

By (7.69), since (7.126) holds and $Y(0) = \vec{0}$, we have

$$\left(\delta_{\tilde{Y}}^2 \mathcal{L}\right)(\gamma) - 2\sqrt{\bar{\tau}} \left\langle X, \nabla_{\tilde{Y}} \tilde{Y} \right\rangle (\bar{\tau})$$
$$= -\int_0^{\bar{\tau}} \sqrt{\tau} H\left(X, \tilde{Y}\right) d\tau + \frac{|Y(\bar{\tau})|^2}{\sqrt{\bar{\tau}}} - 2\sqrt{\bar{\tau}} \operatorname{Rc}(Y, Y)(\bar{\tau}).$$

Note that $\tilde{\gamma}_s(\bar{\tau}) = \gamma_s(\bar{\tau})$ implies $\nabla_{\tilde{Y}} \tilde{Y}(\bar{\tau}) = \nabla_Y Y(\bar{\tau})$. Hence

LEMMA 7.87 (Differential inequality for length of $\mathcal{L}$-Jacobi field). *Let $\gamma_s : [0, \tau_2] \to \mathcal{M}$, where $\tau_2 \in (0, T)$, $s \in (-\varepsilon, \varepsilon)$, and $\varepsilon > 0$, be a smooth family of minimal $\mathcal{L}$-geodesics with $Y_s(0) = \vec{0}$ for $s \in (-\varepsilon, \varepsilon)$. Then for any $\bar{\tau} \in (0, \tau_2]$ the $\mathcal{L}$-Jacobi field $Y(\tau) = \left.\frac{d\gamma_s}{ds}\right|_{s=0}$ satisfies the estimate*

$$(7.129) \qquad \left.\frac{d}{d\tau}\right|_{\tau=\bar{\tau}} |Y|^2 \leq -\frac{1}{\sqrt{\bar{\tau}}} \int_0^{\bar{\tau}} \sqrt{\tau} H\left(X, \tilde{Y}\right) d\tau + \frac{|Y(\bar{\tau})|^2}{\bar{\tau}},$$

where $\tilde{Y}$ satisfies (7.126) and (7.127) and $H\left(X, \tilde{Y}\right)$ is Hamilton's Harnack quantity defined in (7.63).

Note that the only place where we used an inequality (versus an equality) in our derivation is $\left(\delta_Y^2 \mathcal{L}\right)(\gamma) \leq \left(\delta_{\tilde{Y}}^2 \mathcal{L}\right)(\gamma)$. Hence equality holds in (7.129) if and only if the vector field $\tilde{Y}$ satisfying (7.126) and (7.127) is an $\mathcal{L}$-Jacobi field. Then

$$\left.\frac{d}{d\tau}\right|_{\tau=\bar{\tau}} |Y|^2 = \left.\frac{d}{d\tau}\right|_{\tau=\bar{\tau}} |\tilde{Y}|^2 = \frac{|Y(\bar{\tau})|^2}{\bar{\tau}}$$

in (7.129), and $\int_0^{\bar{\tau}} \sqrt{\tau} H\left(X, \tilde{Y}\right) d\tau = 0$. From (7.125) we get

$$(7.130) \quad \left.\frac{d}{d\tau}\right|_{\tau=\bar{\tau}} |Y|^2 = 2\operatorname{Rc}(Y, Y)(\bar{\tau}) + \frac{1}{\sqrt{\bar{\tau}}} (\operatorname{Hess} L)(Y, Y)(\bar{\tau}) = \frac{|Y(\bar{\tau})|^2}{\bar{\tau}}.$$

Applying Hamilton's matrix Harnack inequality to (7.129), we get

LEMMA 7.88 (Estimate for time-derivative of length of $\mathcal{L}$-Jacobi field). *If the solution $(\mathcal{M}^n, g(\tau))$, $\tau \in [0, T]$, to the backward Ricci flow has bounded nonnegative curvature operator and the $\mathcal{L}$-Jacobi field $Y(\tau)$ along a minimal $\mathcal{L}$-geodesic $\gamma : [0, \tau_2] \to \mathcal{M}$ satisfies $Y(0) = \vec{0}$, then for $c \in (0, 1)$ and $\bar{\tau} \in (0, \min\{\tau_2, (1-c)T\}]$,*

$$\left.\frac{d}{d\tau}\right|_{\tau=\bar{\tau}} \log|Y|^2 \leq \frac{1}{\bar{\tau}} \left(C\ell(\gamma(\bar{\tau}), \bar{\tau}) + 1 \right),$$

where $C = \frac{2}{c}$. If $T = \infty$, then

$$\left.\frac{d}{d\tau}\right|_{\tau=\bar{\tau}} \log|Y|^2 \leq \frac{1}{\bar{\tau}} \left(2\ell(\gamma(\bar{\tau}), \bar{\tau}) + 1 \right).$$

REMARK 7.89. Note that for Euclidean space, $\mathcal{L}$-Jacobi fields satisfy $|Y(\tau)|^2 = \text{const} \cdot \tau$. In particular, $\frac{d}{d\tau} \log |Y(\tau)|^2 = \frac{1}{\tau}$ (and ℓ is constant along $\mathcal{L}$-geodesics).

PROOF. Since $g(\tau)$ has nonnegative curvature operator, Hamilton's matrix inequality holds and we have for any $\tau \in [0, \bar{\tau}]$,

$$H\left(X, \tilde{Y}\right)(\tau) + \left(\frac{1}{\tau} + \frac{1}{T-\tau}\right) \operatorname{Rc}\left(\tilde{Y}, \tilde{Y}\right)(\tau) \geq 0.$$

Since $\operatorname{Rc} \geq 0$ and $\left|\tilde{Y}(\tau)\right|^2 = \frac{\tau}{\bar{\tau}} |Y(\bar{\tau})|^2$, from $\bar{\tau} \leq (1-c)T$, we get for $\tau \in [0, \bar{\tau}]$,

$$H\left(X, \tilde{Y}\right)(\tau) \geq -\left(\frac{1}{\tau} + \frac{1}{T-\tau}\right) R(\gamma(\tau), \tau) \left|\tilde{Y}(\tau)\right|^2$$

$$= -\frac{T}{\bar{\tau}(T-\tau)} R(\gamma(\tau), \tau) |Y(\bar{\tau})|^2$$

$$\geq -\frac{1}{c\bar{\tau}} R(\gamma(\tau), \tau) |Y(\bar{\tau})|^2.$$

Then (7.129) implies

$$\left.\frac{d}{d\tau}\right|_{\tau=\bar{\tau}} |Y|^2 \leq \left(\frac{1}{c\sqrt{\bar{\tau}}} \int_0^{\bar{\tau}} \sqrt{\tau} R(\gamma(\tau), \tau) d\tau + 1\right) \frac{|Y(\bar{\tau})|^2}{\bar{\tau}}$$

$$\leq \left(\frac{2}{c} \ell(\gamma(\bar{\tau}), \bar{\tau}) + 1\right) \frac{|Y(\bar{\tau})|^2}{\bar{\tau}},$$

since γ is a minimal $\mathcal{L}$-geodesic. Hence

$$\left.\frac{d}{d\tau}\right|_{\tau=\bar{\tau}} \log |Y|^2 \leq \frac{1}{\bar{\tau}}\left(\frac{2}{c} \ell(\gamma(\bar{\tau}), \bar{\tau}) + 1\right).$$

Finally, we leave it as an exercise to check that when $T = \infty$, one can in essence take $c = 1$ in the inequality above. □

8.3. The $\mathcal{L}$-exponential map. The $\mathcal{L}$-exponential map

$$\mathcal{L}\exp : T\mathcal{M} \times [0, T) \to \mathcal{M}$$

is defined by

$$\mathcal{L}\exp(V, \bar{\tau}) \doteqdot \mathcal{L}\exp_V(\bar{\tau}) \doteqdot \gamma_V(\bar{\tau}),$$

where $\gamma_V(\tau)$ is the $\mathcal{L}$-geodesic with $\lim_{\tau \to 0} \sqrt{\tau} \frac{d\gamma}{d\tau}(\tau) = V \in T_p\mathcal{M}$ (and $\gamma_V(0) = p$). Given $\bar{\tau}$, define the **$\mathcal{L}$-exponential map at time $\bar{\tau}$**

$$\mathcal{L}_{\bar{\tau}} \exp : T\mathcal{M} \to \mathcal{M}$$

by

$$\mathcal{L}_{\bar{\tau}} \exp(V) \doteqdot \gamma_V(\bar{\tau}).$$

EXAMPLE 7.90 ($\mathcal{L}$-exponential map on a Ricci flat solution). To get a feel for the $\mathcal{L}$-exponential map, we first consider a *Ricci flat* solution $(\mathcal{M}^n, g(\tau) = g_0)$. Here, by (7.23), for $V \in T_p\mathcal{M}$,

$$\mathcal{L}\exp(V, \bar{\tau}) = \exp\left(2\sqrt{\bar{\tau}}V\right),$$

where exp is the usual exponential map of $(\mathcal{M}, g_0)$ with basepoint p. Note that for a Ricci flat solution, the $\mathcal{L}$-exponential map has the scaling property:

$$\mathcal{L}\exp(V, \bar{\tau}) = \mathcal{L}\exp\left(cV, \frac{\bar{\tau}}{c^2}\right)$$

for any $c > 0$. However this is not true for general solutions of the Ricci flow.

The $\mathcal{L}$-exponential map at $\bar{\tau} = 0$ is related to $\exp^{g(0)}$ which is the usual (Riemannian) exponential map with respect to the metric $g(0)$.

LEMMA 7.91 ($\mathcal{L}$-exponential map as $\bar{\tau} \to 0$). *Let $(\mathcal{M}^n, g(\tau))$, $\tau \in [0, T]$, be a complete solution to the backward Ricci flow with bounded sectional curvature. Given $V \in T_p\mathcal{M}$, as $\bar{\tau} \to 0$, the $\mathcal{L}$-exponential map tends to the Riemannian exponential map of $g(0)$ in the following sense:*

$$(7.131) \qquad \lim_{\bar{\tau} \to 0} \mathcal{L}\exp\left(\frac{1}{2\sqrt{\bar{\tau}}}V, \bar{\tau}\right) = \exp^{g(0)}(V).$$

From the proof we can see that the convergence in (7.131) can be made into C^∞-convergence.

PROOF. Motivated by the Ricci flat case, we define the path $\beta : [0, 1] \to \mathcal{M}$ by

$$\beta(\rho) \doteqdot \mathcal{L}\exp\left(\frac{1}{2\sqrt{\bar{\tau}}}V, \rho^2\bar{\tau}\right) = \gamma_{\frac{1}{2\sqrt{\bar{\tau}}}V}(\rho^2\bar{\tau}),$$

so that $\beta(1) = \mathcal{L}\exp\left(\frac{1}{2\sqrt{\bar{\tau}}}V, \bar{\tau}\right)$. (Note that β depends on $\bar{\tau}$ but we do not emphasize this in our notation.) We have

$$(7.132) \qquad \frac{d\beta}{d\rho}(\rho) = \frac{d}{d\rho}\left(\gamma_{\frac{1}{2\sqrt{\bar{\tau}}}V}(\rho^2\bar{\tau})\right) = 2\rho\bar{\tau}\frac{d\gamma_{\frac{1}{2\sqrt{\bar{\tau}}}V}}{d\tau}(\rho^2\bar{\tau}).$$

Hence the $\mathcal{L}$-geodesic equation (7.32) becomes

$$0 = \nabla_{\frac{1}{2\rho\bar{\tau}}\frac{d\beta}{d\rho}}\left(\frac{1}{2\rho\bar{\tau}}\frac{d\beta}{d\rho}\right) - \frac{1}{2}\nabla R + 2\operatorname{Rc}\left(\frac{1}{2\rho\bar{\tau}}\frac{d\beta}{d\rho}\right) + \frac{1}{2\rho^2\bar{\tau}}\frac{1}{2\rho\bar{\tau}}\frac{d\beta}{d\rho}.$$

Multiplying this by $4\rho^2\bar{\tau}^2$ yields for $\rho \in [0, 1]$,

$$(7.133) \qquad \nabla_{\frac{d\beta}{d\rho}}\frac{d\beta}{d\rho} - 2\rho^2\bar{\tau}^2\nabla R + 4\rho\bar{\tau}\operatorname{Rc}\left(\frac{d\beta}{d\rho}\right) = 0.$$

The covariant derivative and Ricci tensor are with respect to $g(\rho^2\bar{\tau})$. Since

$$\lim_{\tau \to 0}\sqrt{\tau}\frac{d\gamma_{\frac{1}{2\sqrt{\bar{\tau}}}V}}{d\tau}(\tau) = \frac{1}{2\sqrt{\bar{\tau}}}V,$$

we have
$$\lim_{\rho \to 0} \frac{d\beta}{d\rho}(\rho) = \lim_{\rho \to 0} 2\rho\bar{\tau} \frac{d\gamma_{\frac{1}{2\sqrt{\bar{\tau}}}V}}{d\tau}(\rho^2\bar{\tau}) = V$$
independent of $\bar{\tau}$. Hence, by taking the limit of (7.133) as $\bar{\tau} \to 0$, we have $\nabla^{g(0)}_{\frac{d\beta}{d\rho}}\frac{d\beta}{d\rho} = 0$ and
$$\rho \mapsto \lim_{\bar{\tau} \to 0} \mathcal{L}\exp\left(\frac{1}{2\sqrt{\bar{\tau}}}V, \rho^2\bar{\tau}\right) = \lim_{\bar{\tau} \to 0} \beta(\rho)$$
is a constant speed geodesic with respect to $g(0)$ and with initial vector V. Thus when we evaluate it at $\rho = 1$, we get that the limit is $\exp^{g(0)}(V)$. □

Next we compute the differential of $\mathcal{L}_{\bar{\tau}}\exp$.

LEMMA 7.92. *For $\bar{\tau} \in (0,T)$, $\mathcal{L}_{\bar{\tau}}\exp$ is differentiable at V and the tangent map*
$$D[\mathcal{L}_{\bar{\tau}}\exp(V)] \doteqdot (\mathcal{L}_{\bar{\tau}}\exp)_* : T_p\mathcal{M} \to T_{\mathcal{L}_{\bar{\tau}}\exp(V)}\mathcal{M}$$
is given by
$$(\mathcal{L}_{\bar{\tau}}\exp)_*(W) = J(\bar{\tau}),$$
where $J(\tau)$ is the $\mathcal{L}$-Jacobi field along $\mathcal{L}\exp(V,\tau)$ with
$$J(0) = 0 \quad \text{and} \quad \left.\frac{d}{d\sigma}\right|_{\sigma=0} J(\frac{\sigma^2}{4}) = W.$$

PROOF. We have a family of $\mathcal{L}$-geodesics $\mathcal{L}\exp(V + sW, \tau)$, $\tau \in [0, \bar{\tau}]$, for $s \in (-\varepsilon, \varepsilon)$. By the definition of $\mathcal{L}$-Jacobi field,
$$J(\tau) \doteqdot \left.\frac{d}{ds}\right|_{s=0} \mathcal{L}\exp(V + sW, \tau)$$
is an $\mathcal{L}$-Jacobi field. Since $\mathcal{L}\exp(V + sW, 0) = p$, we have $J(0) = 0$. From
$$\left.\frac{d}{d\sigma}\right|_{\tau=0} \mathcal{L}\exp(V + sW, \tau) = V + sW,$$
where $\sigma = 2\sqrt{\tau}$, we get by taking $\left.\frac{d}{ds}\right|_{s=0}$,
$$\left.\frac{d}{d\sigma}\right|_{\tau=0} J(\tau) = W.$$
Note that the tangent map of $\mathcal{L}_{\bar{\tau}}\exp$ at V is given by
$$D[\mathcal{L}_{\bar{\tau}}\exp(V)](W) = \left.\frac{d}{ds}\right|_{s=0} \mathcal{L}\exp(V + sW, \bar{\tau}).$$
The lemma follows. □

As a simple corollary of the lemma we have

COROLLARY 7.93. *Fix $\bar{\tau} \in (0,T)$ and consider the $\mathcal{L}$-exponential map $\mathcal{L}_{\bar{\tau}} \exp : T_p\mathcal{M} \to \mathcal{M}$. Then V is a critical point of the map $\mathcal{L}_{\bar{\tau}} \exp$ if and only if there is a nontrivial $\mathcal{L}$-Jacobi field $J(\tau)$ along $\mathcal{L}\exp(V,\tau)$, $\tau \in [0,\bar{\tau}]$, such that $J(0) = 0$ and $J(\bar{\tau}) = 0$.*

PROOF. If V is a critical point of $\mathcal{L}_{\bar{\tau}} \exp$, then there exists W such that $D\left[\mathcal{L}\exp(V,\bar{\tau})\right](W) = 0$. By the lemma, $\frac{d}{ds}\big|_{s=0} \mathcal{L}\exp(V + sW, \tau)$ is the required $\mathcal{L}$-Jacobi field.

On the other hand, if we have a nontrivial $\mathcal{L}$-Jacobi field $J_1(\tau)$ with $J_1(0) = 0$ and $J_1(\bar{\tau}) = 0$, then we define $W \doteqdot \frac{d}{d\sigma}\big|_{\tau=0} J_1(\tau)$. Since $J_1(\tau)$ is nontrivial, we have $W \neq 0$. By the uniqueness of solutions of the initial-value problem for $\mathcal{L}$-Jacobi fields, we know that

$$D\left[\mathcal{L}_{\bar{\tau}} \exp(V)\right](W) = J(\bar{\tau}) = J_1(\bar{\tau}) = 0.$$

We see that V is a critical point of $\mathcal{L}_{\bar{\tau}} \exp$. □

The Hopf-Rinow theorem in Riemannian geometry can be generalized for $\mathcal{L}$-geodesics and the $\mathcal{L}$-exponential map. The proof of this result will appear elsewhere.

LEMMA 7.94 ($\mathcal{L}$-Hopf-Rinow). *Suppose $(\mathcal{N}^n, h(\tau))$, $\tau \in [0,T]$, is a solution to the backward Ricci flow satisfying the curvature bound $|\text{Rm}(x,\tau)| \leq C_0 < \infty$ for $(x,\tau) \in \mathcal{N} \times [0,T]$. The following are equivalent:*

(1) *for every $\tau \in [0,T)$, the metric $h(\tau)$ is complete;*
(2) *$\mathcal{L}\exp$ is defined on all of $T_p\mathcal{N} \times [0,T)$ for some $p \in \mathcal{N}$;*
(3) *$\mathcal{L}\exp$ is defined on all of $T_p\mathcal{N} \times [0,T)$ for all $p \in \mathcal{N}$.*

Moreover,

(4) *any of the above statements implies that given any two points p,q and $0 \leq \tau_1 < \tau_2 < T$, there is a minimal $\mathcal{L}$-geodesic $\gamma : [\tau_1, \tau_2] \to \mathcal{N}$ joining p and q.*

8.4. $\mathcal{L}$-cut locus. We have the following simple lemma which is analogous to the corresponding theorem in Riemannian geometry.

LEMMA 7.95 (When $\mathcal{L}$-geodesics stop minimizing). *Given $V \in T_p\mathcal{M}$, there exists $\tau_V \in (0,T]$ such that*

$$\mathcal{L}\left(\gamma_V|_{[0,\tau]}\right) = L(\gamma_V(\tau), \tau) \quad \text{for all } \tau \in [0, \tau_V)$$

and

$$\mathcal{L}\left(\gamma_V|_{[0,\tau]}\right) > L(\gamma_V(\tau), \tau) \quad \text{for any } \tau \in (\tau_V, T),$$

where $\gamma_V : [0,T) \to \mathcal{M}$ is the $\mathcal{L}$-geodesic with $\lim_{\tau \to 0} \sqrt{\tau}\frac{d\gamma}{d\tau}(\tau) = V$.

PROOF. The existence of τ_V follows from the additivity property for concatenated paths (7.20). The fact that $\tau_V > 0$ follows from Lemma 7.29. □

That is, if $\tau_V < T$, then τ_V is the first time the $\mathcal{L}$-geodesic $\gamma_V : [0, T) \to \mathcal{M}$ stops minimizing. On the other hand, $\tau_V = T$ if and only if γ_V is minimal. The lemma establishes that either γ_V is minimal or there exists a first positive time τ_V past which γ_V does not minimize.

Let $\gamma : [0, T) \to \mathcal{M}$ be an $\mathcal{L}$-geodesic with $\gamma(0) = p$. We say that a point $(\gamma(\bar{\tau}), \bar{\tau})$, $\bar{\tau} \in (0, T)$, is an **$\mathcal{L}$-conjugate point to $(p, 0)$ along** γ if there exists a nontrivial $\mathcal{L}$-Jacobi field along γ which vanishes at the endpoints $(p, 0)$ and $(\gamma(\bar{\tau}), \bar{\tau})$. A point $(q, \bar{\tau})$ is an **$\mathcal{L}$-conjugate point to $(p, 0)$** if $(q, \bar{\tau})$ is $\mathcal{L}$-conjugate to $(p, 0)$ along some minimal $\mathcal{L}$-geodesic $\gamma(\tau)$, $\tau \in [0, \bar{\tau}]$, from p to q. If $\gamma(\tau) = \gamma_V(\tau)$ for some $V \in T_p\mathcal{M}$, then this is equivalent to V being a critical point of the $\mathcal{L}$-exponential map $\mathcal{L}_{\bar{\tau}} \exp$ (see Corollary 7.93).

DEFINITION 7.96. (i) The **$\mathcal{L}$-cut locus of $(p, 0)$ in the tangent space of space-time** is defined by
$$\mathcal{L}C_{(p,0)} \doteqdot \{(V, \tau_V) : V \in T_p\mathcal{M}\},$$
where τ_V is defined above. Since $\tau_V > 0$, we have $\mathcal{L}C_{(p,0)} \subset T_p\mathcal{M} \times (0, T]$.

(ii) The **$\mathcal{L}$-cut locus of $(p, 0)$ at time $\bar{\tau} \in (0, T]$ in the tangent space** is defined by
$$\mathcal{L}C_{(p,0)}(\bar{\tau}) \doteqdot \mathcal{L}C_{(p,0)} \cap (T_p\mathcal{M} \times \{\bar{\tau}\}).$$

(iii) Define
$$\Omega_{(p,0)}(\bar{\tau}) \doteqdot \{V \in T_p\mathcal{M} : \tau_V > \bar{\tau}\}.$$

In words, $\Omega_{(p,0)}(\bar{\tau})$ is the open set of tangent vectors at p for which the corresponding $\mathcal{L}$-geodesic minimizes past time $\bar{\tau}$. Note that $\Omega_{(p,0)}(\bar{\tau})$ is not necessarily star-shaped in the sense that if $V \in \Omega_{(p,0)}(\bar{\tau})$, then $aV \in \Omega_{(p,0)}(\bar{\tau})$ for any $a \in (0, 1)$. However $\Omega_{(p,0)}(\bar{\tau})$ is an open subset in $T_p\mathcal{M}$ and $\Omega(\tau_2) \subset \Omega(\tau_1)$ if $\tau_1 < \tau_2$.

Next we define the $\mathcal{L}$-cut locus of the map $\mathcal{L}_{\bar{\tau}} \exp$ for $\bar{\tau} \in (0, T)$.

DEFINITION 7.97. (i) The **$\mathcal{L}$-cut locus of $(p, 0)$ at time $\bar{\tau}$** is defined by
$$\mathcal{L}\operatorname{Cut}_{(p,0)}(\bar{\tau}) \doteqdot \{\mathcal{L}_{\bar{\tau}} \exp(V) : V \in T_p\mathcal{M} \text{ and } \tau_V = \bar{\tau}\}.$$

(ii) We define $\mathcal{L}\operatorname{Cut}^2_{(p,0)}(\bar{\tau})$ to be the set of points $q \in \mathcal{M}$ such that there are at least two different minimal $\mathcal{L}$-geodesics on $[0, \bar{\tau}]$ from p to q.

(iii) We define $\mathcal{L}\operatorname{Cut}^c_{(p,0)}(\bar{\tau})$ to be the set of points q such that $(q, \bar{\tau})$ is $\mathcal{L}$-conjugate to $(p, 0)$.

EXERCISE 7.98. Show that for $\bar{\tau} \in (0, T)$ we have $q \notin \mathcal{L}\operatorname{Cut}_{(p,0)}(\bar{\tau})$ if and only if for every minimal $\mathcal{L}$-geodesic $\gamma : [0, \bar{\tau}] \to \mathcal{M}$ joining p to q, we may extend γ as a minimal $\mathcal{L}$-geodesic past time $\bar{\tau}$.

There is a characterization of $\mathcal{L}$-cut locus points analogous to the characterization of cut locus points in Riemannian geometry. The first lemma below can be proved using the locally Lipschitz property of the $\mathcal{L}$-distance L, while the second lemma below can be proved via a calculation similar to the proof of the Riemannian index lemma. We will give the details of the proof elsewhere.

LEMMA 7.99 ($\mathcal{L}$-cut locus).
 (i) $\mathcal{L}\operatorname{Cut}_{(p,0)}(\bar{\tau}) = \mathcal{L}\operatorname{Cut}^2_{(p,0)}(\bar{\tau}) \cup \mathcal{L}\operatorname{Cut}^c_{(p,0)}(\bar{\tau})$ and is closed.
 (ii) $\mathcal{L}\operatorname{Cut}^2_{(p,0)}(\bar{\tau})$ has measure 0.
 (iii) $\mathcal{L}\operatorname{Cut}^c_{(p,0)}(\bar{\tau})$ is closed and has measure 0.

The proof of the above lemma depends on an index lemma for $\mathcal{L}$-length. Define the $\mathcal{L}$-**index form** $\mathcal{L}\operatorname{I}(Y,W)$ by

$$(7.134) \quad \mathcal{L}\operatorname{I}(Y,W) \doteqdot \int_{\tau_a}^{\tau_b} \sqrt{\tau} \left[\begin{array}{l} \frac{1}{2}\nabla_Y \nabla_W R + \langle R(Y,X)W, X \rangle \\ + \langle \nabla_X Y, \nabla_X W \rangle - (\nabla_Y \operatorname{Rc})(W,X) \\ - (\nabla_W \operatorname{Rc})(Y,X) + (\nabla_X \operatorname{Rc})(Y,W) \end{array} \right] d\tau.$$

Note that the second variation of $\mathcal{L}$-length (7.62) is related to the $\mathcal{L}$-index form by

$$(7.135) \quad (\delta_Y^2 \mathcal{L})(\gamma) = 2\sqrt{\tau}\langle \nabla_Y Y, X\rangle \big|_{\tau_a}^{\tau_b} + 2\mathcal{L}\operatorname{I}(V,V).$$

On the other hand, the $\mathcal{L}$-index form is related to the $\mathcal{L}$-Jacobi equation by

$$\mathcal{L}\operatorname{I}(Y,W) = \sqrt{\tau}\langle \nabla_X Y, W\rangle \big|_{\tau_a}^{\tau_b}$$
$$- \int_{\tau_a}^{\tau_b} \sqrt{\tau} \left[\left\langle \begin{array}{l} \nabla_X(\nabla_X Y) - \operatorname{Rm}(X,Y)X - \frac{1}{2}\nabla_Y(\nabla R) \\ +2(\nabla_Y \operatorname{Rc})(X) + 2\operatorname{Rc}(\nabla_X Y) + \frac{1}{2\tau}\nabla_X Y \end{array}, W \right\rangle \right] d\tau.$$

This can be proved by integrating by parts on the term $\sqrt{\tau}\langle \nabla_X Y, \nabla_X W\rangle$ in (7.134).

LEMMA 7.100 ($\mathcal{L}$-index lemma). *Let γ be an $\mathcal{L}$-geodesic from (p, τ_a) to (q, τ_b) such that there are no points $\mathcal{L}$-conjugate to (p, τ_a) along γ. For any piecewise smooth vector field W along γ with $W(\tau_a) = 0$, let Y be the unique $\mathcal{L}$-Jacobi field such that $Y(\tau_a) = W(\tau_a) = 0$ and $Y(\tau_b) = W(\tau_b)$. Then*

$$\mathcal{L}\operatorname{I}(Y,Y) \leq \mathcal{L}\operatorname{I}(W,W)$$

and the equality holds if only if $Y = W$. Here we have used the obvious generalization of the definition of $\mathcal{L}$-conjugate point with $(p,0)$ replaced by (p, τ_a).

Lemma 7.99 implies the following.

COROLLARY 7.101 (Differentiability of L away from $\mathcal{L}$-cut locus). *Given $\bar{\tau} \in (0, T)$ and $q \in \mathcal{M}$, suppose that there is only one minimal $\mathcal{L}$-geodesic γ joining $(p, 0)$ and $(q, \bar{\tau})$ and suppose that $(q, \bar{\tau})$ is not an $\mathcal{L}$-conjugate point of $(p, 0)$, i.e., $(q, \bar{\tau})$ is not an $\mathcal{L}$-cut locus point. Then the L-distance $L(\cdot, \cdot)$ and reduced distance ℓ are C^2-differentiable at $(q, \bar{\tau})$.*

PROOF. Suppose $\lim_{\tau \to 0} \sqrt{\tau} \frac{d\gamma}{d\tau}(0) = V$; the hypothesis implies by Lemma 7.92 that there is some $\varepsilon > 0$ and some small neighborhood U_V of $V \in T_p \mathcal{M}$ such that the map

$$(\mathcal{L}\exp, \mathrm{id}) : U_V \times (\bar{\tau} - \varepsilon, \bar{\tau} + \varepsilon) \to \mathcal{M} \times (\bar{\tau} - \varepsilon, \bar{\tau} + \varepsilon),$$
$$(\mathcal{L}\exp, \mathrm{id})(W, \tau) = (\mathcal{L}_\tau \exp(W), \tau)$$

is a local diffeomorphism. For each $W \in U_V$ and $\tau_* \in (\bar{\tau} - \varepsilon, \bar{\tau} + \varepsilon)$, we **claim** the curve $\mathcal{L}\exp(W, \tau)$, $\tau \in [0, \tau_*]$, is a minimal $\mathcal{L}$-geodesic. Hence using the local diffeomorphism property, there are an $\varepsilon_1 > 0$, a small neighborhood U_q of q, and a family of minimal $\mathcal{L}$-geodesics $\gamma_{\hat{q}, \tau_*}$ smoothly depending on the endpoint $\hat{q} \in U_q$ and $\tau_* \in (\bar{\tau} - \varepsilon_1, \bar{\tau} + \varepsilon_1)$. Now $L(\hat{q}, \tau_*) = \mathcal{L}(\gamma_{\hat{q}, \tau_*})$ is a smooth function of $(\hat{q}, \tau_*)$, L is differentiable near $(q, \bar{\tau})$.

We now prove the claim by contradiction. If the claim is false, then there is a sequence of points $(W_i, \tau_i) \to (V, \bar{\tau})$ such that $\mathcal{L}\exp(W_i, \tau)$, $\tau \in [0, \tau_i]$, is not a minimal $\mathcal{L}$-geodesic. Let $\mathcal{L}\exp(\hat{W}_i, \tau)$, $\tau \in [0, \tau_i]$, be a minimal $\mathcal{L}$-geodesic from p to $\mathcal{L}\exp(W_i, \tau_i)$. As in the proof of Lemma 7.28, using Lemma 7.13(ii) and the $\mathcal{L}$-geodesic equation, it is easy to show that $\left|\hat{W}_i\right|_{g(p,0)}$ is bounded. Hence there is a subsequence $\hat{W}_i \to \hat{W}_\infty$. If $\hat{W}_\infty \neq V$, then we get two minimal $\mathcal{L}$-geodesics γ_V and $\gamma_{\hat{W}_\infty}$ joining $(p, 0)$ and $(q, \bar{\tau})$, which contradicts the assumption of the lemma. If $\hat{W}_\infty = V$, then $(\mathcal{L}\exp, \mathrm{id})$ cannot be a local diffeomorphism since $(\mathcal{L}\exp(W_i, \tau_i), \tau_i) = \left(\mathcal{L}\exp(\hat{W}_i, \tau_i), \tau_i\right)$. The claim is proved and the lemma is proved. □

As a simple consequence, we have

(7.136) $$\mathcal{L}_{\bar{\tau}} \exp : \Omega_{(p,0)}(\bar{\tau}) \to \mathcal{M} \backslash \mathcal{L}\mathrm{Cut}_{(p,0)}(\bar{\tau})$$

is a diffeomorphism. Note that $\mathcal{M} \backslash \mathcal{L}\mathrm{Cut}_{(p,0)}(\bar{\tau})$ is open and dense in $\mathcal{M}$.

Now we end this subsection by rewriting the formula for the $\mathcal{L}$-index form in terms of Hamilton's matrix Harnack quadratic. Using

$$\frac{d}{d\tau}[\mathrm{Rc}(Y, W)] = \left(\frac{\partial}{\partial \tau} \mathrm{Rc}\right)(Y, W) + (\nabla_X \mathrm{Rc})(Y, W)$$
$$+ \mathrm{Rc}(\nabla_X Y, W) + \mathrm{Rc}(Y, \nabla_X W),$$

we may write (7.134) as

$$\mathcal{L}\mathrm{I}(Y, W) = -\int_{\tau_a}^{\tau_b} \sqrt{\tau} \frac{d}{d\tau}[\mathrm{Rc}(Y, W)] d\tau$$
$$+ \int_{\tau_a}^{\tau_b} \sqrt{\tau} \begin{bmatrix} \left(\frac{\partial}{\partial \tau} \mathrm{Rc}\right)(Y, W) + \frac{1}{2}\nabla_Y \nabla_W R - \langle \mathrm{Rc}(Y), \mathrm{Rc}(W) \rangle \\ + \langle R(Y, X)W, X \rangle - (\nabla_Y \mathrm{Rc})(W, X) \\ - (\nabla_W \mathrm{Rc})(Y, X) + 2(\nabla_X \mathrm{Rc})(Y, W) \\ + \langle \mathrm{Rc}(Y) + \nabla_X Y, \mathrm{Rc}(W) + \nabla_X W \rangle \end{bmatrix} d\tau.$$

Rearranging terms and integrating by parts, we express this as follows. Define the symmetric 2-tensor Q by

$$Q(Y,W) \doteq \left(\frac{\partial}{\partial \tau} \operatorname{Rc}\right)(Y,W) + \frac{1}{2}\nabla_Y \nabla_W R - \langle \operatorname{Rc}(Y), \operatorname{Rc}(W)\rangle$$
$$+ \frac{1}{2\tau} \operatorname{Rc}(Y,W) + \langle R(Y,X)W, X\rangle - (\nabla_Y \operatorname{Rc})(W,X)$$
$$- (\nabla_W \operatorname{Rc})(Y,X) + 2(\nabla_X \operatorname{Rc})(Y,W)$$

and

$$S(Y) \doteq \operatorname{Rc}(Y) + \nabla_X Y - \frac{1}{2\tau} Y.$$

Then we have

LEMMA 7.102 ($\mathcal{L}$-index form). *The $\mathcal{L}$-index form $\mathcal{L}\operatorname{I}(Y,W)$ can be written as*

$$\mathcal{L}\operatorname{I}(Y,W) = -\sqrt{\tau}\operatorname{Rc}(Y,W)\big|_{\tau_a}^{\tau_b} + \int_{\tau_a}^{\tau_b} \sqrt{\tau} Q(Y,W)\, d\tau$$
$$+ \int_{\tau_a}^{\tau_b} \sqrt{\tau} \left(\begin{array}{c} \langle S(Y), S(W)\rangle + \langle S(Y), \frac{W}{2\tau}\rangle \\ + \langle \frac{Y}{2\tau}, S(W)\rangle + \frac{\langle Y,W\rangle}{4\tau^2} \end{array} \right) d\tau.$$

Note that, assuming $[X,Y] = \vec{0}$ and $[X,W] = \vec{0}$, we have

$$\frac{d}{d\tau}\left(\frac{\langle Y,W\rangle}{\tau}\right) = \frac{2}{\tau}\left(\operatorname{Rc} + \operatorname{Sym}(\nabla X) - \frac{1}{2\tau}g\right)(Y,W),$$

where $\operatorname{Sym}(\nabla X)_{ij} \doteq \frac{1}{2}(\nabla_i X_j + \nabla_j X_i) = \frac{1}{2}\mathcal{L}_X g$.

8.5. $\mathcal{L}$-Jacobian. First we recall the **Jacobian** in Riemannian geometry. Let $(\hat{\mathcal{M}}^n, \hat{g})$ be a Riemannian manifold, let $p \in \hat{\mathcal{M}}$, and given $V \in T_p\hat{\mathcal{M}}$ with $|V| = 1$, let $\gamma_V : [0, s_V) \to \hat{\mathcal{M}}$ be the maximal unit speed minimal geodesic with $\dot{\gamma}(0) = V$. Take $\{E_i\}_{i=1}^{n-1}$ to be an orthonormal frame at p perpendicular to V and define Jacobi fields $\{J_i(s)\}$ along γ_V so that $J_i(0) = \vec{0}$ and $(\nabla_V J_i)(0) = E_i$. The Jacobian J is defined by

$$\operatorname{J}(\gamma_V(s)) \doteq \sqrt{\det(\langle J_i(s), J_j(s)\rangle)},$$

where $(\langle J_i(s), J_j(s)\rangle)$ is an $(n-1) \times (n-1)$ matrix. Note that

(7.137) $$\lim_{s \to 0} \frac{\operatorname{J}(\gamma_V(s))}{s^{n-1}} = 1.$$

Let $d\sigma_{S^{n-1}}$ denote the volume form on the unit $(n-1)$-sphere in $T_p\hat{\mathcal{M}}$, which naturally extends to $T_p\hat{\mathcal{M}} - \{\vec{0}\}$, and let $\operatorname{Cut}(p)$ be the cut locus of p in $\hat{\mathcal{M}}$. We define the $(n-1)$-form $d\sigma$ on $\hat{\mathcal{M}} \backslash (\operatorname{Cut}(p) \cup \{p\})$ by

$$d\sigma \doteq \left(\exp_p^{-1}\right)^* d\sigma_{S^{n-1}},$$

where the exponential map is restricted to inside the cut locus in the tangent space. Then the volume form of $\hat{g}$ on $\hat{\mathcal{M}} \backslash (\operatorname{Cut}(p) \cup \{p\})$ is given by

$$d\mu_{\hat{g}} = \operatorname{J}(\gamma_V(s)) \, dr \wedge d\sigma.$$

The volume forms of the geodesic spheres $S(p, r) \doteqdot \{x \in \hat{\mathcal{M}} : d(x, p) = r\}$, at smooth points, are given by

$$d\sigma_{S(p,r)} = \operatorname{J}(\gamma_V(s)) \, d\sigma.$$

The Jacobian is related to the mean curvatures of the geodesic spheres and the Ricci curvatures of the metric $\hat{g}$ on $\hat{\mathcal{M}}$ by the following formulas:

$$\frac{\partial}{\partial r} \log \operatorname{J} = H$$

and

$$\frac{\partial}{\partial r} H = -\operatorname{Rc}\left(\frac{\partial}{\partial r}, \frac{\partial}{\partial r}\right) - |h|^2$$
$$\leq -\operatorname{Rc}\left(\frac{\partial}{\partial r}, \frac{\partial}{\partial r}\right) - \frac{H^2}{n-1},$$

where h is the second fundamental form of $S(p, r)$. The Bishop–Gromov volume comparison theorem may be proved this way (see [**111**] for example).

Now we turn to the case of Ricci flow. Let $\gamma_V(\tau)$, $\tau \in [0, T)$, be an $\mathcal{L}$-geodesic emanating from p with $\lim_{\tau \to 0} \sqrt{\tau} \dot{\gamma}(\tau) = V$. Let $J_i^V(\tau)$, $i = 1, \ldots, n$, be $\mathcal{L}$-Jacobi fields along γ_V with

$$J_i^V(0) = 0 \quad \text{and} \quad (\nabla_V J_i^V)(0) = E_i^0,$$

where $\{E_i^0\}_{i=1}^n$ is an orthonormal basis for $T_p \mathcal{M}$ with respect to $g(0)$. Note that $J_i^V(\tau)$ is a smooth function of V and $\tau > 0$ since $g(\tau)$ is smooth. Via the orthonormal basis $\{E_i^0\}_{i=1}^n$ we can identify $T_p \mathcal{M}$ with $\mathbb{R}^n$. Since $D(\mathcal{L}\exp(V, \tau))(E_i^0) = J_i^V(\tau)$ (see Lemma 7.92), the **Jacobian of the $\mathcal{L}$-exponential map** $\mathcal{L}\operatorname{J}_V(\tau) \in \mathbb{R}$ (called the $\mathcal{L}$-Jacobian for short) is the square root of the determinant (computed using the inner products on the tangent spaces from the Riemannian metric $g(\tau)$) of the basis of $\mathcal{L}$-Jacobi fields:

$$\left(J_1^V(\tau), \ldots, J_n^V(\tau)\right).$$

That is,

$$\mathcal{L}\operatorname{J}_V(\tau) \doteqdot \sqrt{\det\left(\left\langle J_i^V(\tau), J_j^V(\tau)\right\rangle_{g(\tau)}\right)_{n \times n}}.$$

It is clear that $\mathcal{L}\operatorname{J}_V(\tau)$ is a smooth function of (V, τ) when $\tau > 0$. Another equivalent way of describing $\mathcal{L}\operatorname{J}_V(\tau)$ is to define

$$\mathcal{L}\operatorname{J}_V(\tau) \, dx(V) \doteqdot \left[(\mathcal{L}_\tau \exp(V))^* \, d\mu_{g(\tau, \mathcal{L}_\tau \exp(V))}\right],$$

where dx is the standard Euclidean volume form on $(T_p \mathcal{M}, g(0, p))$.

8. $\mathcal{L}$-JACOBI FIELDS AND THE $\mathcal{L}$-EXPONENTIAL MAP

To get a feeling of the $\mathcal{L}$-Jacobian, we calculate an example.

EXAMPLE 7.103 ($\mathcal{L}$-Jacobian of Ricci flat solution). Recall the fact that if $(\mathcal{M}^n, g(\tau) = g_0)$ is a Ricci flat solution, then an $\mathcal{L}$-geodesic is of the form $\gamma(\tau) = \beta(2\sqrt{\tau})$ where $\beta(\sigma)$ is a constant speed geodesic. Then an $\mathcal{L}$-Jacobi field is of the form
$$J^V(\tau) = K(2\sqrt{\tau})$$
where $K(\sigma)$ is a Riemannian Jacobi field along $\beta(\sigma)$ with respect to g_0. Hence, by choosing $J_n^V(\tau) = 2\sqrt{\tau}\dfrac{\frac{d\beta}{d\sigma}(2\sqrt{\tau})}{\left|\frac{d\beta}{d\sigma}(2\sqrt{\tau})\right|_{g(0)}}$,
$$\mathcal{L}\,\mathrm{J}_V(\tau) = 2\sqrt{\tau}\,\mathrm{J}_V(2\sqrt{\tau}),$$
where J_V is the Jacobian of the Riemannian exponential map of g_0. Since by (7.137),
$$\lim_{\sigma \to 0_+} \frac{\mathrm{J}_V(\sigma)}{\sigma^{n-1}} = 1,$$
we have
$$\lim_{\tau \to 0_+} \frac{\mathcal{L}\,\mathrm{J}_V(\tau)}{\tau^{n/2}} = 2^n.$$
Note that for Euclidean space $\mathbb{R}^n$ we have $\mathcal{L}\,\mathrm{J}_V(\tau) = 2^n \tau^{n/2}$.

As suggested by Example 7.103 and (7.131), we now prove the following lemma.

LEMMA 7.104 ($\mathcal{L}$-Jacobian as $\tau \to 0$). Let $(\mathcal{M}^n, g(\tau))$ be a solution of the backward Ricci flow with bounded sectional curvature. We have the following asymptotics for the $\mathcal{L}$-Jacobian at $\tau = 0$:

(7.138)
$$\lim_{\tau \to 0_+} \frac{\mathcal{L}\,\mathrm{J}_V(\tau)}{\tau^{n/2}} = 2^n.$$

PROOF. Let $\bar{E}_i(\tau)$ denote the parallel translation of E_i^0 along $\gamma_V(\tau)$ with respect to $g(0)$. Since $J_i^V(0) = 0$ and $\left(\nabla_{\frac{d}{d\sigma}} J_i^V\right)(0) = \left(\nabla_V J_i^V\right)(0) = E_i^0$, then by the definition of derivative,
$$\lim_{\tau \to 0_+} \frac{\left|J_i^V(\tau) - 2\sqrt{\tau}\bar{E}_i(\tau)\right|_{g(0)}}{2\sqrt{\tau}} = \lim_{\tau \to 0_+} \frac{\left|J_i^V(\tau) - \sigma \bar{E}_i(\tau)\right|_{g(0)}}{\sigma} = 0.$$
Hence
$$\lim_{\tau \to 0_+} \frac{\mathcal{L}\,\mathrm{J}_V(\tau)}{\tau^{n/2}} = \lim_{\tau \to 0_+} \tau^{-n/2}\sqrt{\det\left(\langle 2\sqrt{\tau}\bar{E}_i(\tau), 2\sqrt{\tau}\bar{E}_j(\tau)\rangle_{g(0)}\right)}$$
$$= 2^n.$$
□

For the proof of the no local collapsing result via the L-distance in Chapter 8 we need the following properties of the $\mathcal{L}$-Jacobians.

PROPOSITION 7.105 (Time-derivative of $\mathcal{L}$-Jacobian). *Let $(\mathcal{M}^n, g(\tau))$, $\tau \in [0,T]$, be a solution of the backward Ricci flow with bounded sectional curvature. Along a minimizing $\mathcal{L}$-geodesic $\gamma_V(\tau)$, $\tau \in [0, \tau_V)$, with $\gamma_V(0) = p$, where τ_V is defined in Lemma 7.95, for $0 < \bar\tau < \tau_V$ the $\mathcal{L}$-Jacobian $\mathcal{L}\,\mathrm{J}_V(\tau)$ satisfies*

$$(7.139) \qquad \left(\frac{d}{d\tau} \log \mathcal{L}\,\mathrm{J}_V\right)(\bar\tau) \leq \frac{n}{2\bar\tau} - \frac{1}{2\bar\tau^{\frac{3}{2}}} K,$$

where $K = K(\gamma_V, \bar\tau)$ is defined by (7.75). Equality in (7.139) holds at the point $\gamma_V(\bar\tau)$ only if

$$(7.140) \qquad \mathrm{Rc}\,(\gamma_V(\bar\tau), \bar\tau) + (\mathrm{Hess}\,\ell)\,(\gamma_V(\bar\tau), \bar\tau) = \frac{g\,(\gamma_V(\bar\tau), \bar\tau)}{2\bar\tau}.$$

REMARK 7.106. The proof of (7.139) is closely modeled on that of the classical Bishop–Gromov volume comparison theorem. Here we follow the derivation using $\mathcal{L}$-Jacobi fields. There are other ways to prove volume comparison such as in Li [**246**].

REMARK 7.107. If we let $\bar\sigma = 2\sqrt{\bar\tau}$ and $\sigma = 2\sqrt{\tau}$, then (7.139) says

$$\left(\frac{d}{d\sigma} \log \mathcal{L}\,\mathrm{J}_V\right)(\bar\tau) \leq -\frac{2}{\bar\sigma^2} K + \frac{n}{\bar\sigma}.$$

Compare this with (7.76).

PROOF OF PROPOSITION 7.105. Choose an orthonormal basis $\{E_i(\bar\tau)\}$ of $T_{\gamma_V(\bar\tau)}\mathcal{M}$. Since there is no point on $\gamma_V(\tau)$, $0 \leq \tau \leq \bar\tau$, which is $\mathcal{L}$-conjugate to $(p, 0)$ along γ_V, we can extend $E_i(\bar\tau)$ to an $\mathcal{L}$-Jacobi field $E_i(\tau)$ along γ_V for $\tau \in [0, \bar\tau]$ with $E_i(0) = 0$. Actually for the same reason, we know that both $\{J_i^V(\tau)\} \in T_{\gamma_V(\tau)}\mathcal{M}$ and $\{E_i(\tau)\} \in T_{\gamma_V(\tau)}\mathcal{M}$ are linearly independent when $\tau \in (0, \bar\tau]$. We can write

$$J_i^V(\bar\tau) = \sum_{j=1}^n A_i^j E_j(\bar\tau)$$

for some matrix $\left(A_i^j\right) \in \mathrm{GL}\,(n, \mathbb{R})$. Then

$$J_i^V(\tau) = \sum_{j=1}^n A_i^j E_j(\tau)$$

for all $\tau \in [0, \bar\tau]$, since we cannot have a nontrivial $\mathcal{L}$-Jacobi field vanishing at the endpoints $\tau = 0, \bar\tau$.

Now we compute that the evolution of the $\mathcal{L}$-Jacobian along γ_V is given by

$$\frac{d}{d\tau}\bigg|_{\tau=\bar{\tau}} \log \mathcal{L} J(\tau) = \frac{d}{d\tau}\bigg|_{\tau=\bar{\tau}} \log \sqrt{\det\left(\left\langle \sum_{k=1}^{n} A_i^k E_k(\tau), \sum_{\ell=1}^{n} A_j^\ell E_\ell(\tau) \right\rangle_{g(\tau)}\right)}$$

$$= \frac{1}{2} \frac{d}{d\tau}\bigg|_{\tau=\bar{\tau}} \log \det\left(\langle E_i, E_j \rangle(\tau)\right) + \frac{1}{2} \frac{d}{d\tau}\bigg|_{\tau=\bar{\tau}} \det\left(A_i^k\right) + \frac{1}{2} \frac{d}{d\tau}\bigg|_{\tau=\bar{\tau}} \det\left(A_j^\ell\right)$$

$$= \frac{1}{2} \sum_{i=1}^{n} \frac{d}{d\tau}\bigg|_{\tau=\bar{\tau}} \langle E_i, E_i \rangle(\tau)$$

$$\leq -\frac{1}{2} \frac{1}{\sqrt{\bar{\tau}}} \int_0^{\bar{\tau}} \sqrt{\tau} \sum_{i=1}^{n} H\left(X, \tilde{E}_i\right) d\tau + \frac{1}{2} \sum_{i=1}^{n} \frac{|E_i(\bar{\tau})|^2}{\bar{\tau}}.$$

The last inequality is due to (7.129). Here the $\tilde{E}_i(\tau)$ are the vector fields along γ_V satisfying

$$\nabla_X \tilde{E}_i = -\operatorname{Rc}\left(\tilde{E}_i\right) + \frac{1}{2\tau} \tilde{E}_i,$$

where $\tilde{E}_i(\bar{\tau}) = E_i(\bar{\tau})$ and $H\left(X, \tilde{E}_i\right)(\tau)$ is the matrix Harnack quadratic given by (7.63). By (7.72), we have $\left\langle \tilde{E}_i, \tilde{E}_j \right\rangle(\tau) = \frac{\tau}{\bar{\tau}} \delta_{ij}$, and by (7.74), we have

$$\sum_{i=1}^{n} H\left(X, \tilde{E}_i\right)(\tau) = \frac{\tau}{\bar{\tau}} H(X)(\tau).$$

So

$$\left(\frac{d}{d\tau} \log \mathcal{L} J\right)(\bar{\tau}) \leq -\frac{1}{2\bar{\tau}^{3/2}} \int_0^{\bar{\tau}} \tau^{3/2} H(X) d\tau + \frac{n}{2\bar{\tau}}$$

(7.141)
$$= -\frac{1}{2\bar{\tau}^{3/2}} K + \frac{n}{2\bar{\tau}}.$$

If equality in (7.139) holds, then we have equality in (7.129) for each $Y(\tau) = \tilde{E}_i(\tau)$, $i = 1, \cdots, n$. By (7.130), we have

$$2 \operatorname{Rc}\left(\tilde{E}_i(\bar{\tau}), \tilde{E}_i(\bar{\tau})\right) + \frac{1}{\sqrt{\bar{\tau}}} (\operatorname{Hess} L)\left(\tilde{E}_i(\bar{\tau}), \tilde{E}_i(\bar{\tau})\right) = \frac{\left|\tilde{E}_i(\bar{\tau})\right|^2}{\bar{\tau}}$$

for each i. Since $\tilde{E}_i(\bar{\tau}) = E_i(\bar{\tau})$ can be chosen arbitrarily, this implies (7.140). □

9. Weak solution formulation

The purpose of this section is to prove the integration by parts inequality (7.148) for the reduced distance ℓ and to give the inequalities we proved for ℓ a weak interpretation. We first recall some of the well-known results in real analysis which we shall need. An excellent reference for properties of

Lipschitz functions and other aspects of real analysis on $\mathbb{R}^n$ is the book by Evans and Gariepy [**139**]. Many of the results in their book easily extend to Riemannian manifolds; when this is the case, we state the extensions without proof. In this section we shall assume that $\left(\hat{\mathcal{M}}^n, \hat{g}\right)$ is a complete Riemannian manifold.

9.1. Locally Lipschitz functions. Recall the definition of differentiability on Riemannian manifolds.

DEFINITION 7.108 (Differentiable function). A function $f : \hat{\mathcal{M}} \to \mathbb{R}$ is **differentiable** at $p \in \hat{\mathcal{M}}$ if there exists a linear map

$$L_p : T_p\hat{\mathcal{M}} \to \mathbb{R}$$

such that

$$\lim_{X \to 0} \frac{\left|f\left(\exp_p(X)\right) - f(p) - L_p(X)\right|}{|X|} = 0.$$

When this is the case, by definition we write

$$\left|f\left(\exp_p(X)\right) - f(p) - L_p(X)\right| = o(|X|) \quad \text{as} \quad X \to 0.$$

This implies for every $X \in T_p\hat{\mathcal{M}}$ that the directional derivative

$$D_X f \doteq \lim_{s \to 0} \frac{f\left(\exp_p(sX)\right) - f(p)}{s} = L_p(X)$$

exists.

REMARK 7.109. Note that differentiability can be defined more generally for differentiable manifolds, but in the definition above we chose to endow the manifold with a Riemannian metric.

Now we list three results in Evans and Gariepy's book. The first is Theorem 2 in §3.1.2 on p. 81 of [**139**].

LEMMA 7.110 (Rademacher's Theorem). *Let $\left(\hat{\mathcal{M}}, \hat{g}\right)$ be a Riemannian manifold. If $f : \hat{\mathcal{M}} \to \mathbb{R}$ is a locally Lipschitz function, then f is differentiable almost everywhere with respect to the Riemannian (Lebesgue) measure.*

Secondly, Theorem 5 in §4.2.3 on p. 131 of [**139**].

LEMMA 7.111 (Locally Lipschitz is equivalent to being in $W_{\text{loc}}^{1,\infty}$). *Let U be an open set in a Riemannian manifold $\left(\hat{\mathcal{M}}, \hat{g}\right)$. Then $f : U \to \mathbb{R}$ is locally Lipschitz if and only if $f \in W_{\text{loc}}^{1,\infty}(U)$.*

Thirdly, Theorem 2 in §2.4.2 on p. 76 of [**139**].

LEMMA 7.112 (Hausdorff dimension of a Lipschitz graph). *Let $\left(\hat{\mathcal{M}}_1^n, \hat{g}_1\right)$ and $\left(\hat{\mathcal{M}}_2^m, \hat{g}_2\right)$ be two Riemannian manifolds. If $f : \hat{\mathcal{M}}_1 \to \hat{\mathcal{M}}_2$ is locally Lipschitz in the sense that for any $p \in \hat{\mathcal{M}}_1$, there is an open neighborhood U_p of p and a constant C_p such that $d_{\hat{g}_2}(f(q_1), f(q_2)) \leq C_p d_{\hat{g}_1}(q_1, q_2)$ for any $q_1, q_2 \in U_p$, then*

$$\mathcal{H}_{\dim}\left\{(x, f(x)) : x \in \hat{\mathcal{M}}_1\right\} = n,$$

where $\mathcal{H}_{\dim}$ denotes the Hausdorff dimension. In particular, the $(n+m)$-dimensional Riemannian measure vanishes:

$$\operatorname{meas}_{\hat{\mathcal{M}}_1 \times \hat{\mathcal{M}}_2}\left\{(x, f(x)) : x \in \hat{\mathcal{M}}_1\right\} = 0.$$

Later we shall recall some more basic results, especially about convex functions, as we need them.

Now we give a proof that integration by parts holds for locally Lipschitz functions. We say a vector field v on $\hat{\mathcal{M}}$ is **locally Lipschitz** if for any $p \in \hat{\mathcal{M}}$ and local coordinates $\{x^i\}$ in a neighborhood of p, we have for each i that the function $v^i(x)$ is locally Lipschitz, where $v(x) \doteq v^i(x) \frac{\partial}{\partial x^i}$. It is well known that integration by parts holds for Lipschitz functions.

LEMMA 7.113 (Integration by parts for Lipschitz functions). *Let f be a locally Lipschitz function on $\hat{\mathcal{M}}$ and let v be a locally Lipschitz vector field on $\hat{\mathcal{M}}$. Suppose that at least one of f and v has compact support. Then*

$$\int_{\hat{\mathcal{M}}} f \operatorname{div} v \, d\mu_{\hat{g}} = -\int_{\hat{\mathcal{M}}} v \cdot \nabla f \, d\mu_{\hat{g}}.$$

Here $\operatorname{div} v$ and $v \cdot \nabla f$ are defined with respect to $\hat{g}$.

PROOF. We prove the lemma in the case where v has compact support; the other case can be proved similarly.[15] Rademacher's Theorem says that both derivatives $\operatorname{div} v$ and ∇f exist almost everywhere. Since v has support in some compact set $\mathcal{K}$, we have $|v|, |\operatorname{div} v| \in L^\infty(\mathcal{K})$, and $f, |\nabla f| \in L^\infty_{\text{loc}}(\hat{\mathcal{M}})$, so the integrals in the lemma make sense. We can choose a smooth partition of unity $\{\phi_\alpha\}_{\alpha=1}^k$ where the support of each ϕ_α is contained in a local coordinate chart. Clearly $\phi_\alpha v$ is locally Lipschitz. To prove the lemma, it suffices to show

$$\int_{\hat{\mathcal{M}}} f \operatorname{div}(\phi_\alpha v) \, d\mu_h = -\int_{\hat{\mathcal{M}}} \phi_\alpha v \cdot \nabla f \, d\mu_h.$$

Hence without loss of generality we may assume the support of v is contained in some local coordinate chart $(U, \{x^i\})$.

---

[15] See the exercise below.

Let U_1 be an open neighborhood of the support of v satisfying $U_1 \subset\subset U$.[16] We can apply Theorem 1 in §6.6.1 on p. 251 of [**139**] to f and v. This tells us that for any $\varepsilon > 0$ there exists a C^1 function f_ε and a C^1 vector field $v_\varepsilon = v_\varepsilon^i \frac{\partial}{\partial x^i}$ defined on U such that v_ε has compact support in U_1 and

$$\operatorname{meas}\{x \in U_1 : f_\varepsilon(x) \neq f(x) \text{ or } \nabla f_\varepsilon(x) \neq \nabla f(x)\} \leq \varepsilon,$$
$$\operatorname{meas}\{x \in U_1 : v_\varepsilon(x) \neq v(x) \text{ or } \nabla_j v_\varepsilon(x) \neq \nabla_j v(x) \text{ for some } j\} \leq \varepsilon,$$
$$\sup_{x \in V}\left|\frac{\partial f_\varepsilon}{\partial x^i}(x)\right| \leq C(n)\operatorname{Lip}(f,U) \quad \text{for all } i,$$
$$\sup_{x \in V}\left|\frac{\partial v_\varepsilon^i}{\partial x^j}(x)\right| \leq C(n)\operatorname{Lip}(v^i,U) \quad \text{for all } i,j,$$

where meas is the n-dimensional Riemannian (Lebesgue) measure on U and $\operatorname{Lip}(f,U)$ is the Lipschitz constant of f on U. By the divergence theorem, which clearly holds for C^1 functions, we have

$$\int_{U_1} f_\varepsilon \operatorname{div} v_\varepsilon d\mu_{\hat{g}} = -\int_{U_1} v_\varepsilon \cdot \nabla f_\varepsilon d\mu_{\hat{g}}.$$

Taking $\varepsilon \searrow 0$, we get $\int_{U_1} f \operatorname{div} v d\mu_{\hat{g}} = -\int_{U_1} v \cdot \nabla f d\mu_{\hat{g}}$ and hence

$$\int_{\hat{\mathcal{M}}} f \operatorname{div} v\, d\mu_{\hat{g}} = -\int_{\hat{\mathcal{M}}} v \cdot \nabla f\, d\mu_{\hat{g}}.$$

$\square$

In summary, the divergence theorem for Lipschitz functions follows from a standard approximation result.

EXERCISE 7.114. Prove Lemma 7.113 in the case where f, instead of v, has compact support.

9.2. Convex functions. Convex functions have nice differentiability properties. Recall the notion of derivatives on $\mathbb{R}^n$.

DEFINITION 7.115 (First and second derivatives). Let $U \subset \mathbb{R}^n$ be an open set. Given a continuous function $f : U \to \mathbb{R}$, we say

(i) f has **first derivative** $Df(x) \in \mathbb{R}^n$ at $x \in U$ if

$$|f(y) - f(x) - Df(x) \cdot (y - x)| = o(|y - x|) \quad \text{as} \quad y \to x.$$

This definition agrees with Definition 7.108.

(ii) f has **second derivative** $D^2 f(x) \in \mathbb{M}^{n \times n}$ at $x \in U$, where $\mathbb{M}^{n \times n}$ is the set of $n \times n$ matrices, if there exists a vector $Df(x) \in \mathbb{R}^n$

---

[16]$U_1 \subset\subset U$ means $\overline{U_1}$ is compact and $\overline{U_1} \subset U$. In this case we say that U_1 is compactly contained in U.

(the first derivative) such that

$$\left| f(y) - f(x) - Df(x) \cdot (y-x) - \frac{1}{2}(y-x)^T \cdot D^2 f(x) \cdot (y-x) \right|$$
$$= o\left(|y-x|^2\right) \quad \text{as} \quad y \to x.$$

We also recall the following related notion (see p. 167 of [**139**] for example).

DEFINITION 7.116 (Locally bounded variation). Let U be an open set in $\mathbb{R}^n$. A function $f \in L^1_{\text{loc}}(U)$ has **locally bounded variation** if for every open set $U_1 \subset\subset U$,

$$\sup \left\{ \int_{U_1} f \operatorname{div} \phi \, dx \, : \, \phi \in C^1_c(U_1; \mathbb{R}^n), \, |\phi| \leq 1 \right\} < \infty,$$

where the lower index c in C^1_c indicates having compact support. In this case we write $f \in BV_{\text{loc}}(U)$.

Clearly if $f \in C^1(U)$, then $f \in BV_{\text{loc}}(U)$ since

$$\int_{U_1} f \operatorname{div} \phi \, dx = -\int_{U_1} \nabla f \cdot \phi \, dx \leq \operatorname{meas}(U_1) \cdot \sup_{\operatorname{supp} \phi} |\nabla f| < \infty$$

for all $U_1 \subset\subset U$. On the other hand, as a partial converse, note that as remarked on p. 166 at the beginning of Chapter 5 in [**139**], "...a BV function is 'measure theoretically C^1.'"

The following lemma is well known; see Theorem 1(i) in §6.3 on p. 236 of [**139**] for the proof of (i), Aleksandrov's Theorem in §6.4 on p. 242 of [**139**] for the proof of (ii), and Theorem 3 in §6.3 on pp. 240–241 of [**139**] for the proof of (iii). Let $B(r) \subset \mathbb{R}^n$ denote the ball of radius r centered at $\vec{0}$.

LEMMA 7.117 (Regularity properties of convex functions). *Let $f : B(r) \to \mathbb{R}$ be a convex function. Then*
  (i) *f is locally Lipschitz,*
  (ii) *(Aleksandrov's Theorem) f has second derivative $D^2 f(x)$ for a.e. $x \in B(r)$,*
  (iii) *$\frac{\partial f}{\partial x^i}$ has locally bounded variation for each i, and*

$$D^2 f \in L^1_{\text{loc}}\left(B(r); \mathbb{M}^{n \times n}\right).$$

REMARK 7.118. Note that it follows easily from the convexity of f that $D^2 f(x) \geq 0$ when it exists.

We consider the approximation of continuous functions, in particular convex functions, by smooth functions via the convolution with mollifiers. A standard **mollifier** is $\eta : \mathbb{R}^n \to \mathbb{R}$ defined by

$$\eta(x) \doteq \begin{cases} a e^{1/(|x|^2 - 1)} & \text{if } |x| < 1, \\ 0 & \text{if } |x| \geq 1, \end{cases}$$

where $a > 0$ is chosen so that $\int_{B(1)} \eta \, dx = 1$. This function is C^∞ with support contained in $B(1) \subset \mathbb{R}^n$ and all derivatives vanishing for $|x| \geq 1$ including $|x| = 1$. Let $f : B(r) \to \mathbb{R}$ be a continuous function. For $\varepsilon \in (0, r)$ define the **mollified function**

$$f_\varepsilon : B(r - \varepsilon) \to \mathbb{R}$$

by

(7.142) $$f_\varepsilon(y) \doteqdot \frac{1}{\varepsilon^n} \int_{B(\varepsilon)} f(y + z) \eta\left(\frac{z}{\varepsilon}\right) dz.$$

It is a standard fact that f_ε is C^∞.

We now prove a lemma about convex functions.

LEMMA 7.119 (Mollifiers and derivatives). *Let $f : B(r) \to \mathbb{R}$ be a continuous function. Let $\frac{\partial}{\partial y^i}$ and $\frac{\partial^2}{\partial y^i \partial y^j}$ denote the standard partial derivatives on $\mathbb{R}^n$.*

(i) *If f has first derivative $Df(x)$ at $x \in B(r)$, then*

$$\lim_{\varepsilon \to 0_+} \left(\frac{\partial f_\varepsilon}{\partial y^i}(x)\right)_{i=1}^n = Df(x).$$

(ii) *If f has second derivative $D^2 f(x)$ at $x \in B(r)$, then*

$$\lim_{\varepsilon \to 0_+} \left(\frac{\partial^2 f_\varepsilon}{\partial y^i \partial y^j}(x)\right)_{i,j=1}^n = D^2 f(x).$$

(iii) *If f is a convex function on $B(r)$, then f_ε is a convex function on $B(r - \varepsilon)$.*

(iv) *If $f \in L^p_{\mathrm{loc}}$, where $p \in [1, \infty)$, then f_ε converges to f in L^p_{loc}.*

PROOF. (i) Define $\tilde{f}$ by

$$f(x + y) \doteqdot f(x) + Df(x) \cdot y + \tilde{f}(y).$$

Then $\tilde{f}(y)$ has first derivative $D\tilde{f}(0) = \vec{0}$. It suffices to prove that the mollified function $\tilde{f}_\varepsilon(y)$ satisfies $\lim_{\varepsilon \to 0_+} \frac{\partial \tilde{f}_\varepsilon}{\partial y^i}(0) = 0$ for each i. Note that

$$\tilde{f}_\varepsilon(y) = \frac{1}{\varepsilon^n} \int_{B(y,\varepsilon)} \tilde{f}(z) \eta\left(\frac{z - y}{\varepsilon}\right) dz$$

and

$$\frac{\partial \tilde{f}_\varepsilon}{\partial y^i}(0) = -\frac{1}{\varepsilon^n} \int_{B(\varepsilon)} \tilde{f}(z) \left(\frac{\partial \eta}{\partial z^i}\right)\left(\frac{z}{\varepsilon}\right) \frac{dz}{\varepsilon}.$$

Since $D\tilde{f}(0) = \vec{0}$, given any $\varepsilon_1 > 0$ there exists $\delta > 0$ such that $\left|\tilde{f}(y)\right| \leq \varepsilon_1 |y|$ whenever $|y| \leq \delta$. Hence, if $\varepsilon \leq \delta$, then

$$\left|\frac{\partial \tilde{f}_\varepsilon}{\partial y^i}(0)\right| \leq \frac{1}{\varepsilon^n} \int_{B(\varepsilon)} \left|\tilde{f}(z)\right| \cdot \left|\frac{\partial \eta}{\partial z^i}\right| \left(\frac{z}{\varepsilon}\right) \frac{dz}{\varepsilon}$$

$$\leq \varepsilon_1 \int_{B(\varepsilon)} \left|\frac{z}{\varepsilon}\right| \cdot \left|\frac{\partial \eta}{\partial z^i}\right| \left(\frac{z}{\varepsilon}\right) d\left(\frac{z}{\varepsilon}\right)$$

$$= c_1 \varepsilon_1,$$

where $c_1 \doteq \int_{B(1)} |y| \cdot \left|\frac{\partial \eta}{\partial y^i}\right|(y)\, dy$ is independent of ε. This implies

$$\lim_{\varepsilon \to 0+} \frac{\partial \tilde{f}_\varepsilon}{\partial y^i}(0) = 0.$$

(ii) Write

$$f(x+y) \doteq f(x) + Df(x) \cdot y + \frac{1}{2} y^T \cdot D^2 f(x) \cdot y + \hat{f}(y).$$

Then $\hat{f}(y)$ has first derivative $D\hat{f}(0) = \vec{0}$ and second derivative $D^2 \hat{f}(0) = 0$. It suffices to prove that for all (i,j) the mollified function $\hat{f}_\varepsilon(y)$ satisfies $\lim_{\varepsilon \to 0+} \frac{\partial^2 \hat{f}_\varepsilon}{\partial y^i \partial y^j}(0) = 0$. Note that

$$\hat{f}_\varepsilon(y) = \frac{1}{\varepsilon^n} \int_{B(y,\varepsilon)} \hat{f}(z) \eta\left(\frac{z-y}{\varepsilon}\right) dz$$

and

$$\frac{\partial^2 \hat{f}_\varepsilon}{\partial y^i \partial y^j}(0) = \frac{1}{\varepsilon^n} \int_{B(\varepsilon)} \hat{f}(z) \left(\frac{\partial^2 \eta}{\partial z^i \partial z^j}\right) \left(\frac{z}{\varepsilon}\right) \frac{dz}{\varepsilon^2}.$$

Since $D\hat{f}(0) = \vec{0}$, given any $\varepsilon_1 > 0$ there is a $\delta > 0$ such that $\left|\hat{f}(y)\right| \leq \varepsilon_1 |y|^2$ when $|y| \leq \delta$. Hence when $\varepsilon \leq \delta$,

$$\left|\frac{\partial^2 \hat{f}_\varepsilon}{\partial y^i \partial y^j}(0)\right| \leq \frac{1}{\varepsilon^n} \int_{B(\varepsilon)} \left|\hat{f}(z)\right| \cdot \left|\frac{\partial^2 \eta}{\partial z^i \partial z^j}\right| \left(\frac{z}{\varepsilon}\right) \frac{dz}{\varepsilon^2}$$

$$\leq \varepsilon_1 \cdot \int_{B(\varepsilon)} \left|\frac{z}{\varepsilon}\right|^2 \left|\frac{\partial^2 \eta}{\partial z^i \partial z^j}\right| \left(\frac{z}{\varepsilon}\right) d\left(\frac{z}{\varepsilon}\right)$$

$$= c_2 \varepsilon_1,$$

where $c_2 \doteq \int_{B(1)} |y|^2 \left|\frac{\partial^2 \eta}{\partial y^i \partial y^j}\right|(y)\, dy$ is independent of ε. This implies that $\frac{\partial^2 \hat{f}_\varepsilon}{\partial y^i \partial y^j}(0) = 0$ as $\varepsilon \to 0+$.

(iii) For $0 \leq \lambda \leq 1$ and $x, y \in B(r - \varepsilon)$, we compute

$$f_\varepsilon(\lambda x + (1-\lambda) y) = \frac{1}{\varepsilon^n} \int_{B(\varepsilon)} f(\lambda(x+z) + (1-\lambda)(y+z)) \eta\left(\frac{z}{\varepsilon}\right) dz$$

$$\leq \frac{1}{\varepsilon^n} \int_{B(\varepsilon)} (\lambda f(x+z) + (1-\lambda) f(y+z)) \eta\left(\frac{z}{\varepsilon}\right) dz$$

$$= \lambda f_\varepsilon(x) + (1-\lambda) f_\varepsilon(y).$$

(iv) See Theorem 6 on p. 630 of Evans [**137**]. □

9.3. Functions with Hessian upper bound. Recall that $\left(\hat{\mathcal{M}}^n, \hat{g}\right)$ is a complete Riemannian manifold. First we give a definition.

DEFINITION 7.120 (Hessian upper bound in support sense). Let $C \in \mathbb{R}$ and $W \subset \hat{\mathcal{M}}$ be an open set. A continuous function $f : W \to \mathbb{R}$ has the **Hessian upper bound** C in the support sense if for any $p \in W$ and any $\varepsilon > 0$ there is a neighborhood U of p and a C^2 local upper barrier function $\varphi : U \to \mathbb{R}$ such that $\varphi(p) = f(p)$, $f(x) \leq \varphi(x)$ for all $x \in U$, and $\nabla\nabla\varphi(p) \leq C + \varepsilon$. We denote this by

$$\operatorname{Hess}_{\sup}(f) \leq C.$$

REMARK 7.121. Clearly we can generalize the above definition to the case where $C : W \to \mathbb{R}$ is a function. However we shall only need the case where C is a constant.

It is easy to see that when f is C^2, $\operatorname{Hess}_{\sup}(f) \leq C$ implies $\nabla\nabla f(p) \leq C$ for each $p \in W$. The following elementary lemma enables us to study the differentiability properties of functions with a Hessian upper bound via the theory of convex functions in Euclidean space. The idea is that we can add a suitable smooth function to a function with a Hessian upper bound to make it concave. Let $B_p(r) \subset T_p\hat{\mathcal{M}}$ denote the open ball of radius r centered at 0, and let $B(p, r) \subset \hat{\mathcal{M}}$ denote the open geodesic ball of radius r centered at $p \in \hat{\mathcal{M}}$.

LEMMA 7.122 (Functions with a Hessian upper bound). *Let $W \subset \hat{\mathcal{M}}$ be an open set, let $f : W \to \mathbb{R}$ be a continuous function, and let $p \in W$.*

(i) *If* $\operatorname{Hess}_{\sup}(f) \leq C$, *then for any $c > 0$ there exists $r > 0$ and a C^2 function ψ on $B(p,r)$ such that* $\operatorname{Hess}_{\sup}(f + \psi) \leq -c$ *on $B(p,r)$.*

(ii) *If* $\operatorname{Hess}_{\sup}(f) \leq 0$ *on W, then $-f$ is a convex function on $(W, \hat{g})$ in the Riemannian sense, i.e., the restriction of $-f$ to each geodesic segment in W is convex.*

(iii) *Let $r \leq \min\left\{1, \frac{\operatorname{inj}(p)}{3}\right\}$ and $x : B(p,r) \to B_p(r) \subset T_p\hat{\mathcal{M}}$ be normal coordinates. If there exist constants $C_1, C_2 < \infty$ such that for each $q \in B(p,r)$ there exists a local upper barrier function φ_q for f at q satisfying*

$$|\nabla\varphi_q|(q) \leq C_1 \quad \text{and} \quad \nabla\nabla\varphi_q(q) \leq C_2 g(q),$$

then there exists a C^2 function $\psi(x)$ defined on $B_p(2r) \subset T_p\hat{\mathcal{M}}$ such that $\left(f \circ \exp_p^{-1} + \psi\right)(x)$ is a concave function on $B_p(r)$ with respect to the Euclidean metric $g(p)$. In particular, f is locally Lipschitz on $B(p, r)$.

PROOF. (i) We may assume $C + c > 0$ since otherwise we are done. Since $\nabla\nabla\left[d(x,p)^2\right](p) = 2g(p)$, where g is the metric, by the continuity of $\nabla\nabla\left[d(x,p)^2\right]$ near p, there exists $r > 0$ (depending only on g and how big of a ball centered at p fits inside W) such that $\nabla\nabla\left[d(x,p)^2\right] \geq g$ in $B(p,r) \subset W$. Let $\psi(x) = -(C+c)d(x,p)^2$. Then
$$\text{Hess}_{\text{supp}}(\psi) = \nabla\nabla\psi \leq -(C+c)g.$$
Since $\text{Hess}_{\text{supp}}(f) \leq C$, we conclude $\text{Hess}_{\text{supp}}(f + \psi) \leq -c$ on $B(p,r)$. Note that we may also assume $|\nabla\psi| \leq C'$ on $B(p,r)$ for some constant $C' < \infty$.

(ii) Let $\gamma : [0, a] \to W$ be any geodesic parametrized by arc length. To verify that $-f$ is a convex function on $(W, \hat{g})$ in the Riemannian sense, we need to show that $f \circ \gamma : [0, a] \to \mathbb{R}$ is a concave function. Since $f \circ \gamma$ also satisfies $\text{Hess}_{\text{supp}}(f \circ \gamma) \leq 0$ on $[0, a]$, this reduces the original problem to a 1-dimensional problem.

For every point $s_0 \in (0, a)$ and $\varepsilon > 0$ there exists a C^2 function $\varphi(s)$ defined on a subinterval $(s_0 - \delta, s_0 + \delta)$ such that
$$f \circ \gamma(s_0) = \varphi(s_0) \quad \text{and} \quad f \circ \gamma(s) \leq \varphi(s)$$
for all $s \in (s_0 - \delta, s_0 + \delta)$, and $\varphi''(s_0) \leq \varepsilon$. Let
$$\tilde{\varphi}(s) \doteqdot \varphi(s_0) + \varphi'(s_0)(s - s_0) + \varepsilon(s - s_0)^2.$$
Note that $\tilde{\varphi}(s_0) = \varphi(s_0)$, $\tilde{\varphi}'(s_0) = \varphi'(s_0)$, and $\tilde{\varphi}''(s) \equiv 2\varepsilon > \varepsilon \geq \varphi''(s_0)$. Hence there exists $\delta_1 \in (0, \delta)$ such that
$$f \circ \gamma(s_0) = \tilde{\varphi}(s_0) \quad \text{and} \quad f \circ \gamma(s) < \tilde{\varphi}(s)$$
for $s \in (s_0 - \delta_1, s_0 + \delta_1) - \{s_0\}$. We claim that $f \circ \gamma(s) \leq \tilde{\varphi}(s)$ for all $s \in [0, a]$. By taking $\varepsilon \to 0$, the claim then implies
$$f \circ \gamma(s) \leq f \circ \gamma(s_0) + \varphi'(s_0)(s - s_0) \quad \text{for all } s_0, s \in [0, a].$$
We conclude $f \circ \gamma(s)$ is concave on $[0, a]$.

Finally, suppose the claim is false; then there exists $s_1 \in [0, a] - \{s_0\}$ such that $f \circ \gamma(s_1) = \tilde{\varphi}(s_1)$. Suppose $s_2 \in (s_0, s_1)$ is a minimum point of $f \circ \gamma(s) - \tilde{\varphi}(s)$ on $[s_0, s_1]$. Then
$$(7.143) \quad f \circ \gamma(s) \geq f \circ \gamma(s_2) + [\varphi'(s_0) + 2\varepsilon(s_2 - s_0)](s - s_2) + \varepsilon(s - s_2)^2$$
on $[s_0, s_1]$. On the other hand, by our hypothesis on $f \circ \gamma$, there exists a C^2 function $\varphi_2(s)$ defined for s near s_2 such that
$$f \circ \gamma(s) \leq f \circ \gamma(s_2) + \varphi'_2(s_2)(s - s_2) + \frac{\varepsilon}{2}(s - s_2)^2,$$

which contradicts (7.143). This completes the proof of the claim and part (ii).

(iii) By hypothesis, for every $q \in B(p,r)$ there exists a C^2 function $\varphi_q : U \to \mathbb{R}$ defined on a neighborhood U of q such that $\varphi_q(q) = f(q)$, $f(x) \leq \varphi_q(x)$ for all $x \in U$, $|\nabla \varphi_q|(q) \leq C_1$, and $\nabla \nabla \varphi_q(q) \leq C_2 g(q)$. In normal coordinates $\{x^i\}$ centered at p, we have

$$\left(\frac{\partial^2 \varphi_q}{\partial x^i \partial x^j} - \Gamma_{ij}^k \frac{\partial \varphi_q}{\partial x^k} \right)(q) = \left(\operatorname{Hess}_g (\varphi_q)_{ij} \right)(q) \leq C_2 \left(g_{ij} \right)(q) \leq C' C_2 \left(\delta_{ij} \right),$$

where $C' < \infty$ is independent of $q \in B(p,r)$. Since $|\nabla \varphi_q|(q) \leq C_1$ and $\left| \Gamma_{ij}^k \right|(q) \leq C''$, where $C'' < \infty$ is independent of q, there exists a constant C_3 independent of q such that the matrix $\left(\frac{\partial^2 \varphi_q}{\partial x^i \partial x^j} - C_3 \delta_{ij} \right)$ is negative definite at q. Choose $\tilde{\psi}(x) = -C_3 d_{g(p)}(x,p)^2$. Then $\varphi_q + \tilde{\psi}$ is a local upper barrier function for $f + \tilde{\psi}$ at q, and since $\frac{\partial^2 \tilde{\psi}}{\partial x^i \partial x^j} = -2C_3 \delta_{ij}$, we have

$$\left(\frac{\partial^2 \left(\varphi_q + \tilde{\psi} \right)}{\partial x^i \partial x^j} \right)(q) < 0.$$

By repeating the proof of (ii), where we replace the geodesic γ in (ii) by any straight line in $B_p(r)$, we see that $\left(f \circ \exp_p^{-1} + \tilde{\psi} \circ \exp_p^{-1} \right)(x)$ is a convex function of $x \in B_p(r)$.

Finally, since $-\left(f \circ \exp_p^{-1} + \psi \right)$ is a convex function on $B_p(r)$, we have $-\left(f \circ \exp_p^{-1} + \psi \right)$ is locally Lipschitz on $B_p(r)$, and hence f is locally Lipschitz on $B(p,r)$. $\square$

REMARK 7.123. Let $f : B(p,r) \to \mathbb{R}$ be a continuous function, where $p \in \hat{\mathcal{M}}$ and $r < \operatorname{inj}(p)$. Suppose there exist $C_1, C_2 < \infty$ such that for each $q \in B(p,r)$ there is a local upper barrier function φ_q for f at q satisfying $|\nabla \varphi_q|(q) \leq C_1$ and $\nabla \nabla \varphi_q(q) \leq C_2 g(q)$.

(i) By Lemma 7.122(iii), we conclude $f(x)$ satisfies (i), (ii), and (iii) in Lemma 7.117 for any choice of local coordinate chart x.[17] In particular, let $q \in B(p,r)$ and let x be normal coordinates centered at q (in particular, $x(q) = 0$). If $D^2 f(0)$ exists, then we can define the Hessian

$$\nabla \nabla f(q) \doteqdot D^2 f(0)$$

and the Laplacian

$$\Delta f(q) \doteqdot \Delta_{\hat{g}} f(q) \doteqdot \operatorname{tr}|_{x=0} \left(D^2 f \right)(x).$$

It is clear that $(\nabla \nabla f)(X,Y), \Delta f \in L_{\mathrm{loc}}^1(B(p,r))$ for any continuous vector fields X and Y on $B(p,r)$.

---

[17]Here we have abused notation. When we write $f(q)$, we treat f as a function on $\hat{\mathcal{M}}$, and when we write $f(x)$, we actually mean $f \circ x^{-1}$.

(ii) Let $q \in B(p,r)$ and let coordinates x be the same as in (i), where $\nabla\nabla f(q) = D^2 f(0)$ exists. For any $\varepsilon > 0$ we define the C^∞ function on $T_q \mathcal{M}$,

$$\tilde{\varphi}_{q,\varepsilon}(x) = f(0) + Df(0) \cdot x + \frac{1}{2}x^T \cdot \nabla\nabla f(q) \cdot x + \frac{\varepsilon}{2} x^T \cdot x.$$

It is clear that $f(0) = \tilde{\varphi}_{q,\varepsilon}(0)$, $f(x) \leq \tilde{\varphi}_{q,\varepsilon}(x)$ in a small neighborhood of 0, and

$$\nabla\nabla \tilde{\varphi}_{q,\varepsilon}(q) = \nabla\nabla f(q) + \varepsilon g(q).$$

Thus if $\nabla\nabla f(q) \leq k_q g(q)$, then $\mathrm{Hess}_{\sup}(f)(q) \leq k_q$.

(iii) Again assuming $\nabla\nabla f(q) = D^2 f(0)$ exists, for any $\varepsilon > 0$, define

$$\bar{\varphi}_{q,\varepsilon}(x) = f(0) + Df(0) \cdot x + \frac{1}{2}x^T \cdot \nabla\nabla f(q) \cdot x - \frac{\varepsilon}{2} x^T \cdot x.$$

We have $f(0) = \bar{\varphi}_{q,\varepsilon}(0)$ and $f(x) \geq \bar{\varphi}_{q,\varepsilon}(x)$ in a small neighborhood of 0. If $\mathrm{Hess}_{\sup}(f)(q) \leq k_q < \infty$, then for any $\delta > 0$ there exists a C^2 local upper barrier function $\varphi: U \to \mathbb{R}$ such that $\nabla\nabla\varphi(q) \leq (k_q + \delta)g(q)$. Since $\bar{\varphi}_{q,\varepsilon} \leq f \leq \varphi$ in a neighborhood of 0 and $\bar{\varphi}_{q,\varepsilon}(0) = \varphi(0)$, we have

$$\nabla\nabla f(q) - \varepsilon g(q) = \nabla\nabla \bar{\varphi}_{q,\varepsilon} \leq \nabla\nabla \varphi(q) \leq (k_q + \delta)g(q).$$

Taking $\delta \to 0$ and $\varepsilon \to 0$, we conclude

$$\nabla\nabla f(q) \leq k_q g(q),$$

and tracing, we have $\Delta f(q) \leq n k_q$.

9.4. The equivalence of notions of supersolution for nonsmooth functions. Note that $\left(\hat{\mathcal{M}}^n, \hat{g}\right)$ is a complete Riemannian manifold. Recall the following definition.

DEFINITION 7.124 (Weak-type notions of supersolution). Let $f: \hat{\mathcal{M}} \to \mathbb{R}$ be a continuous function.

(i) Suppose f satisfies $\mathrm{Hess}_{\sup}(f) \leq C$ in the support sense. Then f is said to satisfy $\Delta f \leq k$ **in the support sense** for some continuous function k if for every $p \in \hat{\mathcal{M}}$ there are a neighborhood U of p and a constant C_1 which have the following property. For any $\varepsilon > 0$ and any $q \in U$ there are a constant $r > 0$ and C^2 function $\varphi: B(q,r) \to \mathbb{R}$ such that $\varphi(q) = f(q)$, $f(x) \leq \varphi(x)$ for all $x \in B(q,r)$, $|\nabla\varphi|(q) \leq C_1$, and

$$\Delta\varphi(q) \leq k + \varepsilon.$$

(ii) A continuous function $f: \hat{\mathcal{M}} \to \mathbb{R}$ is said to satisfy $\Delta f \leq k$ **in the weak sense** for some function $k \in L^1_{\mathrm{loc}}(\hat{\mathcal{M}})$ if for any nonnegative C^2 function φ with compact support, we have

(7.144) $$\int_{\hat{\mathcal{M}}} f \Delta\varphi \, d\mu_{\hat{g}} \leq \int_{\hat{\mathcal{M}}} \varphi k \, d\mu_{\hat{g}}.$$

(iii) A continuous function $f : \hat{\mathcal{M}} \to \mathbb{R}$ is said to satisfy $\Delta f \leq k$ **in the viscosity sense** for some continuous function k if for every $p \in \hat{\mathcal{M}}$ and any C^2 function $\varphi : U \to \mathbb{R}$ on some neighborhood of p satisfying $\varphi(p) = f(p)$, $f(x) \geq \varphi(x)$ for all $x \in U$, we have $\Delta \varphi(p) \leq k$.

(iv) A continuous function $f : \hat{\mathcal{M}} \to \mathbb{R}$ is called a **supersolution** of $\Delta f \leq k$ for some continuous function k if for every $p \in M$, any $r < \frac{1}{3}\operatorname{inj}(p)$, and every C^2 function φ on $\overline{B}(p,r)$ with $\Delta \varphi = k$ and $\varphi|_{\partial \overline{B}(p,r)} = f|_{\partial \overline{B}(p,r)}$ we have $\varphi \leq f$ on $B(p,r)$.

Now we can prove the following.

LEMMA 7.125 (Equivalence of notions of supersolution). *Let $k : \hat{\mathcal{M}} \to \mathbb{R}$ be a continuous function.*

(i) *Let $f : \hat{\mathcal{M}} \to \mathbb{R}$ be a continuous function with $\operatorname{Hess}_{\sup}(f) \leq C_p < \infty$ on $B(p, \frac{1}{2}\operatorname{inj}(p))$ for each $p \in \hat{\mathcal{M}}$. Suppose that for each $q \in B(p, \frac{1}{3}\operatorname{inj}(p))$ there is a local upper barrier function φ_q for f near q satisfying $|\nabla \varphi_q|(q) \leq \tilde{C}_p < \infty$. We have for any $\varphi \in C_c^2(\hat{\mathcal{M}})$*

(7.145) $$\int_{\hat{\mathcal{M}}} f \Delta \varphi \, d\mu_{\hat{g}} \leq \int_{\hat{\mathcal{M}}} \Delta f \cdot \varphi \, d\mu_{\hat{g}}.$$

In particular if $\Delta f \leq k$ in the support sense, then f satisfies $\Delta f \leq k$ in the weak sense.

(ii) *f satisfying $\Delta f \leq k$ in the weak sense is equivalent to f satisfying $\Delta f \leq k$ in the viscosity sense; they are both equivalent to f being a supersolution of $\Delta f \leq k$.*

PROOF. (i) Let ψ be a C^2 function on $\hat{\mathcal{M}}$. Then having $\Delta f \leq k$ in the support sense is equivalent to $\Delta(f + \psi) \leq k + \Delta \psi$ in the support sense, and $\Delta f \leq k$ in the weak sense is equivalent to $\Delta(f + \psi) \leq k + \Delta \psi$ in the weak sense. We use a partition of unity $\{\phi_\alpha\}$ to rewrite $\varphi = \sum \varphi \phi_\alpha$, so that we only need to verify inequality (7.145) and (7.144) when φ has small compact support, say in $B(p, \frac{1}{2}r)$ for some $p \in \hat{\mathcal{M}}$, where $r < \min\left\{1, \frac{\operatorname{inj}(p)}{3}\right\}$ as determined by Lemma 7.122(iii). From Lemma 7.122(iii) and by adding to f another concave C^2 function if necessary, we may assume that $f(q)$ satisfies $\operatorname{Hess}_{\sup}(f) \leq -1$ on $B(p,r)$ and that $f(x)$ is a concave function on $B_p(r) \subset T_p\hat{\mathcal{M}}$ in normal coordinates $\{x^i\}$ on $B(p,r)$.

From Lemma 7.117(ii) and Lemma 7.119(ii), there are smooth functions $f_\varepsilon(x)$ such that $f_\varepsilon(x) \to f(x)$ uniformly on $B_p(\frac{2}{3}r)$, $\left(\frac{\partial^2 f_\varepsilon}{\partial x^i \partial x^j}(x)\right) \to D^2 f(x)$ for x a.e. on $B_p(\frac{2}{3}r)$. By Lemma 7.119(iii), $\left(\frac{\partial^2 f_\varepsilon}{\partial x^i \partial x^j}(x)\right) \leq 0$ for all $x \in B_p(\frac{2}{3}r)$. Hence for any $\delta_1 > 0$ and any C^2-test function φ supported in $B_p(\frac{1}{2}r)$, there exist a sequence $\varepsilon_k \to 0_+$ and a set $W_{\delta_1} \subset B_p(\frac{1}{2}r)$ with $\operatorname{meas}(W_{\delta_1}) \leq \delta_1$ such that $\left(\frac{\partial f_{\varepsilon_k}}{\partial x^i}(x)\right) \to Df(x)$ and $\left(\frac{\partial^2 f_{\varepsilon_k}}{\partial x^i \partial x^j}(x)\right) \to$

$D^2 f(x)$ uniformly on $B_p\left(\frac{1}{2}r\right) \setminus W_{\delta_1}$. This in turn implies that $\Delta_{\hat{g}} f_{\varepsilon_k} \to \Delta_{\hat{g}} f$ uniformly on $B_p\left(\frac{1}{2}r\right) \setminus W_{\delta_1}$. We compute using $\Delta f_{\varepsilon_k} \leq 0$,

$$\int_{\hat{\mathcal{M}}} f_{\varepsilon_k} \Delta \varphi d\mu_{\hat{g}} = \int_{\hat{\mathcal{M}}} \Delta f_{\varepsilon_k} \varphi d\mu_{\hat{g}} = \int_{B_p\left(\frac{1}{2}r\right) \setminus W_{\delta_1}} \Delta f_{\varepsilon_k} \varphi d\mu_{\hat{g}} + \int_{W_{\delta_1}} \Delta f_{\varepsilon_k} \varphi d\mu_{\hat{g}}$$

$$= \int_{B_p\left(\frac{1}{2}r\right) \setminus W_{\delta_1}} \Delta f_{\varepsilon_k} \varphi d\mu_{\hat{g}} + \int_{W_{\delta_1}} \hat{g}^{ij} \left(\frac{\partial^2 f_{\varepsilon_k}}{\partial x^i \partial x^j} - \Gamma_{ij}^l \frac{\partial f_{\varepsilon_k}}{\partial x^l} \right) \varphi d\mu_{\hat{g}}$$

$$\leq \int_{B_p\left(\frac{1}{2}r\right) \setminus W_{\delta_1}} \Delta f_{\varepsilon_k} \varphi d\mu_{\hat{g}} - \int_{W_{\delta_1}} \hat{g}^{ij} \Gamma_{ij}^l \frac{\partial f_{\varepsilon_k}}{\partial x^l} \varphi d\mu_{\hat{g}}.$$

Since $f(x)$ is Lipschitz in $B(r)$, there is a constant $C_2 > 0$ independent of k and δ_1 such that $\left| \hat{g}^{ij} \Gamma_{ij}^l \frac{\partial f_{\varepsilon_k}}{\partial x^l} \right| \leq C_2$. Fix δ_1 and let $k \to \infty$; we get

$$\int_{\hat{\mathcal{M}}} f \Delta \varphi d\mu_{\hat{g}} \leq \int_{B_p\left(\frac{1}{2}r\right) \setminus W_{\delta_1}} \Delta f \cdot \varphi d\mu_{\hat{g}} + C_2 \int_{W_{\delta_1}} \varphi d\mu_{\hat{g}}$$

$$= \int_{B_p\left(\frac{1}{2}r\right)} \Delta f \cdot \varphi d\mu_{\hat{g}} - \int_{W_{\delta_1}} \Delta f \cdot \varphi d\mu_{\hat{g}} + C_2 \int_{W_{\delta_1}} \varphi d\mu_{\hat{g}}.$$

Since $\text{meas}(W_{\delta_1}) \leq \delta_1$ and $\Delta f \in L^1_{\text{loc}}$ by Remark 7.123(i), if we let $\delta_1 \to 0_+$, then $\int_{W_{\delta_1}} \Delta f \cdot \varphi d\mu_{\hat{g}} \to 0$ and

$$\int_{\hat{\mathcal{M}}} f \Delta \varphi d\mu_{\hat{g}} \leq \int_{B_p\left(\frac{1}{2}r\right)} \Delta f \cdot \varphi d\mu_{\hat{g}} = \int_{\hat{\mathcal{M}}} \Delta f \cdot \varphi d\mu_{\hat{g}}.$$

If $\Delta f \leq k$ in the support sense, then by Remark 7.123(iii), f satisfies $\Delta f \leq k$ a.e. on $\hat{\mathcal{M}}$. Combining this with (7.145), we get

$$\int_{\hat{\mathcal{M}}} f \Delta \varphi d\mu_{\hat{g}} \leq \int_{\hat{\mathcal{M}}} \varphi k d\mu_{\hat{g}}.$$

(ii) This equivalence is well known. One can find a proof that f satisfies $\Delta f \leq k$ in the viscosity sense if and only if f is a supersolution of $\Delta f \leq k$ in L. Hörmander's book [205] on p. 147, Proposition 3.2.10. One can find a proof in Juutinen, Lindqvist, and Manfredi [226] that f satisfying $\Delta f \leq k$ in the viscosity sense is equivalent to $\Delta f \leq k$ holding in the weak sense, where they actually prove the equivalence for $\text{div}\left(|\nabla f|^{p-2} \nabla f \right) \leq k$ for $p > 0$. Actually, the equivalence of the notions of weak and viscosity supersolutions was first proved by Ishii and Lions [214]. $\square$

9.5. Comparison theory for Riemannian distance d and reduced distance ℓ. As a simple application of the results in the previous subsection we consider the distance function d on a Riemannian manifold and the reduced distance ℓ of a solution to the backward Ricci flow.

9.5.1. *Weak differentiability of the distance function d on a Riemannian manifold.* Let $\left(\hat{\mathcal{M}}^n, \hat{g}\right)$ be a Riemannian manifold with sect $(\hat{g}) \geq -K$. Let $p \in \hat{\mathcal{M}}$ and define $d_p(x) \doteqdot d(p, x)$. Since $|\nabla d_p(x)| = 1$ for a.e. $x \in \hat{\mathcal{M}}$ and since the Hessian and Laplacian comparison theorems (A.9) hold pointwise a.e., d_p satisfies $\Delta d_p \leq (n-1)\sqrt{K}\coth\left(\sqrt{K}d_p\right)$ in the support sense from Definition 7.124(i). Applying Lemma 7.125 in the previous subsection to $d_p(x)$, we obtain the following (see for example, [**316**], [**246**] or Theorem 1.128 in [**111**]).

LEMMA 7.126 (Laplacian comparison). *Let $\left(\hat{\mathcal{M}}, \hat{g}\right)$ be a complete Riemannian manifold with* sect $(\hat{g}) \geq -K$ *for some* $K \geq 0$.

(i) *If* $K > 0$, *then* $\Delta d_p \leq (n-1)\sqrt{K}\coth\left(\sqrt{K}d_p\right)$ *in the weak sense.*
(ii) *If* $K = 0$, *then* $\Delta d_p \leq \frac{n-1}{d_p}$ *in the weak sense.*
(iii) $\Delta d_p \in L^1_{\mathrm{loc}}\left(\hat{\mathcal{M}}\right)$.
(iv) $\int_{\hat{\mathcal{M}}} d_p \Delta \varphi \, d\mu_{\hat{g}} \leq \int_{\hat{\mathcal{M}}} \Delta d_p \cdot \varphi \, d\mu_{\hat{g}}$ *for any* $\varphi \in C^2_c\left(\hat{\mathcal{M}}\right)$.

REMARK 7.127. It is not clear to us whether $\Delta d_p \in L^r_{\mathrm{loc}}$ for $r > 1$ when dimension $n \geq 3$. When $n = 2$, near the point p, we have $\Delta d_p \notin L^r_{\mathrm{loc}}$ for $r \geq 2$.

Below we give an example to show that in general,

$$\int_{\hat{\mathcal{M}}} \Delta d_p(x) \cdot \varphi \, d\mu_{\hat{g}} \neq \int_{\hat{\mathcal{M}}} d_p(x) \cdot \Delta \varphi \, d\mu_{\hat{g}},$$

for $\varphi \in C^2_c\left(\hat{\mathcal{M}}\right)$. Hence d_p does not belong to the Sobolev space $W^{2,1}\left(\hat{\mathcal{M}}\right)$.

Let $T^n \doteqdot [-1,1]^n / \sim$ be the flat torus and choose $p = 0$. We consider d_p^2 to facilitate our discussion. We have $\Delta d_p^2 = 2n$ outside of the cut locus, which is $\mathrm{Cut}(p) = \partial\left([-1,1]^n\right)/\sim$. For $\varphi = 1$ we have

$$\int_{T^n} \Delta d_p^2 \cdot \varphi \, d\mu = 2^{n+1} n \neq 0 = \int_{T^n} d_p^2 \cdot \Delta \varphi \, d\mu.$$

The underlying reason for this phenomenon is the nonsmoothness of d_p^2 on the cut locus of p. Let $T^n_\varepsilon \doteqdot [-1+\varepsilon, 1-\varepsilon]^n$ for $\varepsilon > 0$ (which has piecewise linear boundary). Let ν denote the outward normal of $\partial T^n_\varepsilon \subset T^n$. Then

$$\int_{T^n_\varepsilon} \Delta d_p^2 \cdot \varphi \, d\mu$$
$$= \int_{T^n_\varepsilon} d_p^2 \cdot \Delta \varphi \, d\mu + \int_{\partial T^n_\varepsilon} \left(\nabla d_p^2 \cdot \nu\right) \varphi \, d\mu - \int_{\partial T^n_\varepsilon} d_p^2 \left(\nabla \varphi \cdot \nu\right) d\mu.$$

Note that $\lim_{\varepsilon \to 0_+} \int_{\partial T_\varepsilon^n} d_p^2 (\nabla \varphi \cdot \nu) \, d\mu = 0$ (since n pairs of opposing boundary terms cancel in the limit) and

$$\int_{\partial T_\varepsilon^n} (\nabla d_p^2 \cdot \nu) \varphi \, d\mu = 2 \int_{\partial T_\varepsilon^n} d_p (\nabla d_p \cdot \nu) \varphi \, d\mu = 2(1-\varepsilon) \int_{\partial T_\varepsilon^n} \varphi \, d\mu$$

since $d_p \nabla d_p = (1-\varepsilon) \nu$. Hence

$$\lim_{\varepsilon \to 0_+} \int_{\partial T_\varepsilon^n} (\nabla d_p^2 \cdot \nu) \varphi \, d\mu = 4 \sum_{i=1}^n \int_{T^{n-1}} \varphi(x^1, \ldots, x^{i-1}, 1, x^{i+1}, \ldots, x^n) \, d\sigma.$$

We conclude that

$$2n \int_{T^n} \varphi \, d\mu = \int_{T^n} \Delta d_p^2 \cdot \varphi \, d\mu = \int_{T^n} d_p^2 \cdot \Delta \varphi \, d\mu + 4 \int_{\text{Cut}(p)} \varphi \, d\sigma.$$

This reflects the fact that, in the sense of weak derivatives, $\Delta d_p^2 \leq 2n$ and the distribution $[\Delta d_p^2]$ has its singular part supported on the cut locus.

9.5.2. *Weak differentiability of the reduced distance function ℓ.* In this subsection $(\mathcal{M}^n, g(\tau))$, $\tau \in [0, T]$, is a solution to the backward Ricci flow with the curvature bound $\max\{|\text{Rc}|, |\text{Rm}|\} \leq C_0 < \infty$ on $\mathcal{M} \times [0, T]$. Let $p \in \mathcal{M}$ and let $\ell(x, \tau)$ be the ℓ-distance function defined using basepoint $(p, 0)$. From the gradient estimate of ℓ in space-time and from the Hessian and Laplacian pointwise estimates of ℓ in space (Lemma 7.61 and (7.90)), $\ell(\cdot, \tau)$ satisfies (7.90) in the support sense from Definition 7.124(i). Lemma 7.125 in the previous subsection applies to $\ell(x, \tau)$ as a function of the space variable. We have the following.

PROPOSITION 7.128 (Regularity properties of the reduced distance). *For every $\tau \in (0, T)$,*

(i) *$\ell(q, \tau)$ is locally Lipschitz in the variable q,*
(ii) *$\Delta \ell(q, \tau) \in L^1_{\text{loc}}(\mathcal{M})$,*
(iii) *for any smooth nonnegative function φ on $\mathcal{M}$ with compact support,*

$$\int_{\mathcal{M}} \ell(\cdot, \tau) \Delta \varphi \, d\mu \leq \int_{\mathcal{M}} \varphi \Delta \ell(\cdot, \tau) \, d\mu,$$

where $d\mu$ denotes the volume form with respect to $g(\tau)$.

Note that (i) gives an abstract proof that $\ell(\cdot, \tau)$ is locally Lipschitz, which is different from the effective proof given earlier.

From Lemma 7.113, we have

$$\int_{\mathcal{M}} \ell(\cdot, \tau) \Delta \varphi \, d\mu = -\int_{\mathcal{M}} \nabla \ell(\cdot, \tau) \cdot \nabla \varphi \, d\mu.$$

Combining this with the proposition, we obtain that the inequalities (7.91) and (7.92) hold in the weak sense.

LEMMA 7.129 (Space-time Laplacian comparison for ℓ holds in weak sense). *Let $(\mathcal{M}^n, g(\tau))$, $\tau \in [0, T]$, be a solution to the backward Ricci flow with bounded sectional curvature. We have the following.*

(i) The inequality
$$\frac{\partial \ell}{\partial \tau} - \Delta \ell + |\nabla \ell|^2 - R + \frac{n}{2\tau} \geq 0$$
holds in the weak sense on $\mathcal{M} \times (0, T)$, i.e.,

(7.146) $$\int_{\tau_1}^{\tau_2} \int_{\mathcal{M}} \left[\nabla \ell \cdot \nabla \varphi + \left(\frac{\partial \ell}{\partial \tau} + |\nabla \ell|^2 - R + \frac{n}{2\tau} \right) \varphi \right] d\mu \, d\tau \geq 0$$

for any nonnegative C^2 function φ on $\mathcal{M} \times [\tau_1, \tau_2]$ with $0 < \tau_1 < \tau_2 < T$ such that $\varphi(\cdot, \tau)$ has compact support for each $\tau \in [\tau_1, \tau_2]$.

(ii) For each $\tau \in (0, T)$,
$$2\Delta \ell - |\nabla \ell|^2 + R + \frac{\ell - n}{\tau} \leq 0$$
holds in the weak sense on $\mathcal{M} \times \{\tau\}$, i.e.,

(7.147) $$\int_{\mathcal{M}} \left[-2\nabla \ell \cdot \nabla \varphi + \varphi \left(-|\nabla \ell|^2 + R + \frac{\ell - n}{\tau} \right) \right] d\mu(\tau) \leq 0$$

for any nonnegative C^2 function φ on $\mathcal{M}$ with compact support.

Let $\tilde{\varphi}$ be any nonnegative locally Lipschitz function which satisfies the decay conditions:
$$\tilde{\varphi}(q, \tau), \, |\nabla \tilde{\varphi}(q, \tau)| \leq \frac{1}{c} e^{-c d_{g(0)}^2(p, q)}$$
for some constant $c > 0$. We now show that (7.146) and (7.147) hold for $\varphi = \tilde{\varphi}$. In the proof of the main theorem in §11.2 of [**297**] (see Chapter 8 of this volume), we need this generalization to be able to take $\varphi = e^{-\ell}$. To see this generalization, by the Bishop–Gromov volume comparison theorem, we have the volume growth bound:
$$\text{Vol}_{g(\tau)} B_{g(\tau)}(p, r) \leq C e^{Cr}$$
for some constant $C > 0$. By Lemmas 7.59 and 7.60 we know that both integrals in (7.146) and (7.147) are finite. Let $\{\phi_\alpha\}$ be a partition of unity on $\mathcal{M}$ with compact support. Because of the finiteness of the integrals it suffices to show that for each α the two inequalities hold for $\varphi = \phi_\alpha \tilde{\varphi}$, which is Lipschitz and has compact support in space. Since $\phi_\alpha \tilde{\varphi}$ can be approximated by smooth functions along with its first derivative outside a set of arbitrary small measure, by a proof similar to that of Lemma 7.113 we conclude that for any α, (7.146) and (7.147) hold for $\varphi = \phi_\alpha \tilde{\varphi}$. Hence we have proved that (7.146) and (7.147) hold for any nonnegative locally Lipschitz function $\tilde{\varphi}$ which satisfies the decay conditions $\tilde{\varphi}(q, \tau), \, |\nabla \tilde{\varphi}(q, \tau)| \leq c^{-1} e^{-c d_{g(0)}^2(p, q)}$ for some constant $c > 0$. Note by Lemmas 7.59 and 7.60 that we can choose $\varphi = e^{-\ell}$ in (7.146) and (7.147).

Similarly we may take $\varphi = e^{-\ell}$ in Proposition 7.128(iii) and we obtain the following.

LEMMA 7.130 (Integration by parts inequality for ℓ). *We have*

$$\int_{\mathcal{M}} \left(\Delta\ell - |\nabla\ell|^2\right) e^{-\ell} d\mu \geq 0. \tag{7.148}$$

PROOF. Since integration by parts holds for Lipschitz functions, we can write Proposition 7.128(iii) as

$$-\int_{\mathcal{M}} \nabla\ell(\cdot,\tau) \cdot \nabla\varphi \, d\mu \leq \int_{\mathcal{M}} \varphi \Delta\ell(\cdot,\tau) \, d\mu, \tag{7.149}$$

where $\varphi \in C_c^2(\mathcal{M})$ is nonnegative. By an approximation argument as discussed above, this inequality holds for any nonnegative locally Lipschitz function φ satisfying $\varphi(q,\tau), |\nabla\varphi(q,\tau)| \leq c^{-1} e^{-cd_{g(0)}^2(p,q)}$ for some constant $c > 0$. Taking $\varphi = e^{-\ell}$, we get

$$\int_{\mathcal{M}} |\nabla\ell|^2 e^{-\ell} d\mu \leq \int_{\mathcal{M}} e^{-\ell} \Delta\ell \, d\mu.$$

□

As a simple consequence of the above lemma and $\frac{\partial\ell}{\partial\tau} - \Delta\ell + |\nabla\ell|^2 - R + \frac{n}{2\tau} \geq 0$ a.e., we obtain the following.

COROLLARY 7.131. *For a solution to the backward Ricci flow* $(\mathcal{M}^n, g(\tau))$, $\tau \in [0,T]$, *with bounded sectional curvature, we have*

$$\int_{\mathcal{M}} \left(\frac{\partial\ell}{\partial\tau} - R + \frac{n}{2\tau}\right)(4\pi\tau)^{n/2} e^{-\ell} d\mu \geq 0 \tag{7.150}$$

and, for $0 < \tau_1 < \tau_2 < T$,

$$\int_{\tau_1}^{\tau_2} \int_{\mathcal{M}} \left(\frac{\partial\ell}{\partial\tau} - R + \frac{n}{2\tau}\right)(4\pi\tau)^{n/2} e^{-\ell} d\mu \, d\tau \geq 0. \tag{7.151}$$

10. Notes and commentary

1. This chapter is a discussion of §7.1 and §7.2 of Perelman's [**297**].

2. The reader may also consult Rugang Ye's notes on the ℓ-function [**382**], which we have partially used as a source. See also the appendix of [**134**] for a brief discussion of reduced distance.

3. Notational conventions. In this chapter we have endeavored to maintain a consistent convention for the notation we have used. In particular we have used the following notation when discussing solutions to the forward and backward Ricci flow:

$\left(\hat{\mathcal{M}}^n, \hat{g}\right)$: a static Riemannian manifold,

$\left(\mathcal{N}^n, \tilde{h}(t)\right)$: an arbitrary (not necessarily complete or with bounded curvature) solution to the Ricci flow,

$\left(\mathcal{N}^n, h(\tau)\right)$, $\tau \in (A, \Omega)$: an arbitrary solution to the backward Ricci flow,

$(\mathcal{M}^n, g(\tau))$: a solution to the backward Ricci flow, usually defined for $\tau \in [0, T]$, complete, and satisfying the pointwise curvature bound $\max\{|\text{Rm}|, |\text{Rc}|\} \leq C_0 < \infty$ on $\mathcal{M} \times [0, T]$.

CHAPTER 8

Applications of the Reduced Distance

> Now ... the basic principle of modern mathematics is to achieve a complete fusion [of] 'geometric' and 'analytic' ideas. – Jean Dieudonné

In this chapter we give some geometric applications of the reduced distance for Ricci flow. We give two proofs of the monotonicity of the reduced volume. This result is the Ricci flow analogue of the Bishop–Gromov volume comparison theorem in Riemannian geometry. A beautiful and striking aspect of this monotonicity formula is that unlike Hamilton's matrix Harnack inequality and most other monotonicity formulas, no curvature assumption is needed. Using the reduced volume monotonicity, we prove a weakened no local collapsing theorem and we also prove that certain backward limits of ancient κ-solutions are gradient shrinkers. All of these results are due to Perelman.

Throughout this chapter we shall use $\left(\hat{\mathcal{M}}^n, \hat{g}\right)$ to denote a complete Riemannian manifold and $(\mathcal{M}^n, g(\tau))$ to denote a solution of the backward Ricci flow.

1. Reduced volume of a static metric

We begin by defining the reduced volume of a static metric since this is technically easier than the Ricci flow case yet it still exhibits many of the ideas.

1.1. The reduced volume for a static metric and its monotonicity when $\mathrm{Rc} \geq 0$. Consider the following functional for a complete Riemannian manifold $\left(\hat{\mathcal{M}}^n, \hat{g}\right)$. Given a point $p \in \hat{\mathcal{M}}$, define the **static reduced volume** by

$$(8.1) \qquad \bar{V}\left(\hat{g}, \tau\right) \doteqdot \int_{\hat{\mathcal{M}}} (4\pi\tau)^{-n/2} e^{-d(x,p)^2/4\tau} d\mu(x).$$

This geometric invariant depends on $\hat{g}$, τ and p. Clearly $\bar{V}$ is positive. In general, we can think of this as the integral of the Euclidean heat kernel transplanted to a manifold via the exponential map. If the Ricci curvatures of $\left(\hat{\mathcal{M}}, \hat{g}\right)$ are bounded below by a constant, then by the Bishop–Gromov volume comparison theorem, which gives an upper bound for the volumes of balls, the integral defining $\bar{V}\left(\hat{g}, \tau\right)$ converges for all $\tau > 0$ even when $\hat{\mathcal{M}}$ is noncompact.

EXERCISE 8.1. Show that if $\left(\hat{\mathcal{M}}^n, \hat{g}\right)$ is a complete noncompact Riemannian manifold with $\mathrm{Rc}_{\hat{g}} \geq -K$ for some $K \in \mathbb{R}$, then the integral defining $\bar{V}(\hat{g}, \tau)$ converges for all $\tau > 0$.

In Euclidean space $\bar{V}$ is the integral of the heat kernel, which is the constant 1. Note also that for any $\left(\hat{\mathcal{M}}^n, \hat{g}\right)$ and $p \in \hat{\mathcal{M}}$,

(8.2) $$\lim_{\tau \to 0} \bar{V}(\hat{g}, \tau) = 1$$

essentially since manifolds are locally Euclidean.

EXERCISE 8.2. Prove (8.2).

Let
$$u(x, \tau) \doteq (4\pi\tau)^{-n/2} e^{-d(x,p)^2/4\tau},$$
which is a Lipschitz function, and let $d(x) \doteq d(x, p)$. We can think of $\bar{V}(\hat{g}, \tau) = \int_{\hat{\mathcal{M}}} u(x, \tau) \, d\mu(x)$ as a weighted volume centered at p with the radial weight function u. As $\tau \to 0$, the weight u concentrates at p and as $\tau \to \infty$, u diffuses throughout $\hat{\mathcal{M}}$.

REMARK 8.3. If $\hat{\mathcal{M}}$ is closed, then the upper bound
$$\bar{V}(\hat{g}, \tau) \leq \int_{\hat{\mathcal{M}}} (4\pi\tau)^{-n/2} d\mu(x) = (4\pi\tau)^{-n/2} \mathrm{Vol}(\hat{g})$$
implies that $\lim_{\tau \to \infty} \bar{V}(\hat{g}, \tau) = 0$.

Now assume that $\left(\hat{\mathcal{M}}^n, \hat{g}\right)$ is complete with nonnegative Ricci curvature. Since $\mathrm{Rc}_{\hat{g}} \geq 0$, the Bishop–Gromov volume comparison theorem says that the **volume ratio** $r^{-n} \mathrm{Vol}\, B(p, r)$ is a nonincreasing function of r. It is thus natural to expect that $\bar{V}(\hat{g}, \tau)$ is a nonincreasing function of τ since as τ increases, the weighting favors larger radii. Indeed we have

LEMMA 8.4 (Static reduced volume monotonicity). *If $\left(\hat{\mathcal{M}}^n, \hat{g}\right)$ is complete with $\mathrm{Rc}_{\hat{g}} \geq 0$, then*

(8.3) $$\frac{d}{d\tau} \bar{V}(\hat{g}, \tau) = \int_{\hat{\mathcal{M}}} \left(\frac{\partial}{\partial \tau} - \Delta\right) u \, d\mu \leq 0.$$

In particular, by (8.2),
$$\bar{V}(\hat{g}, \tau) \leq 1 \quad \text{for all } \tau > 0.$$

REMARK 8.5. Clearly this lemma implies $\lim_{\tau \to \infty} \bar{V}(\hat{g}, \tau) \in [0, 1]$ exists.

PROOF. We compute that u is a subsolution, in the weak sense, to the heat equation:
$$\left(\frac{\partial}{\partial \tau} - \Delta\right) u = u \left(-\frac{n}{2\tau} + \frac{d^2}{4\tau^2} + \frac{(d\Delta d + |\nabla d|^2)}{2\tau} - \frac{d^2}{4\tau^2} |\nabla d|^2\right)$$

(8.4) $$\leq 0,$$

where we used $|\nabla d| = 1$ a.e. and the Laplacian comparison theorem, i.e., $d\Delta d \leq n - 1$. Note that the Laplacian comparison theorem is equivalent to the Bishop–Gromov volume comparison theorem. □

It is useful to keep the following simple examples in mind when pondering Lipschitz continuous sub- and super-solutions of the heat equation.

EXAMPLE 8.6 (Heat equation on $\mathcal{S}^1$). Let $\mathcal{M}^1 = \mathcal{S}^1 = \mathbb{R}/(2\pi\mathbb{Z})$ with the standard metric $g = d\theta^2$. Consider the function

$$f(\theta, t) = t + \frac{\theta^2}{2}, \quad \theta \in (-\pi, \pi] \text{ and } t \in \mathbb{R}.$$

For each fixed θ, $f(\theta, \cdot)$ is a smooth (linear) function of time and, for each fixed t, we have $f(\cdot, t)$ is Lipschitz on $\mathcal{S}^1$ and C^∞ except at $\theta = \pi$. Moreover f is a solution to the heat equation almost everywhere. In particular,

$$\frac{\partial f}{\partial t}(\theta, t) = \frac{\partial^2 f}{\partial \theta^2}(\theta, t) = 1$$

for all $\theta \in \mathcal{S}^1 - \{\pi\}$ and $t \in \mathbb{R}$. On the other hand,

$$\frac{d}{dt} \int_{\mathcal{S}^1} f(\theta, t) \, d\theta = 2\pi > 0.$$

This is consistent with the fact that f is not a subsolution of the heat equation in the weak sense (as we shall now see, f is a supersolution). Note that

$$\frac{\partial f}{\partial \theta}(\theta, t) = \theta$$

for all $\theta \in (-\pi, \pi)$ and $t \in \mathbb{R}$ (and $\frac{\partial f}{\partial \theta}$ is undefined for $\theta = \pi$). In particular, for each $t \in \mathbb{R}$, $\frac{\partial f}{\partial \theta}(\cdot, t)$ has a jump discontinuity at $\theta = \pi$. In the sense of distributions, we have $\frac{\partial f}{\partial \theta}(\theta, t) = \theta$ and

(8.5) $$\frac{\partial^2 f}{\partial \theta^2}(\cdot, t) = 1 - 2\pi \cdot \delta_\pi,$$

where δ_π is the Dirac δ-function centered at $\theta = \pi$. Hence, in the sense of distributions,

$$\frac{\partial f}{\partial t} = \frac{\partial^2 f}{\partial \theta^2} + 2\pi \cdot \delta_\pi \geq \frac{\partial^2 f}{\partial \theta^2}.$$

EXERCISE 8.7. Prove (8.5).

SOLUTION TO EXERCISE 8.7. For any C^2 function $\varphi : \mathcal{S}^1 \to \mathbb{R}$ we have

$$\int_{\mathcal{S}^1} \frac{\theta^2}{2} \varphi''(\theta) \, d\theta = \left. \frac{\theta^2}{2} \varphi'(\theta) \right|_{-\pi}^{\pi} - \int_{\mathcal{S}^1} \theta \varphi'(\theta) \, d\theta$$

$$= \int_{\mathcal{S}^1} \varphi(\theta) \, d\theta - \left. \theta \varphi(\theta) \right|_{-\pi}^{\pi}$$

$$= \int_{\mathcal{S}^1} \varphi(\theta) \, d\theta - 2\pi \varphi(\pi).$$

That is,
$$\frac{\partial^2}{\partial \theta^2}\left(\frac{\theta^2}{2}\right) = 1 - 2\pi \cdot \delta_\pi.$$

The square torus is also a nice concrete example for which we can compute the static reduced volume explicitly.

EXAMPLE 8.8 (Static reduced volume for square torus). Consider the torus $\hat{\mathcal{M}}^n \doteqdot \mathbb{R}^n/(2\mathbb{Z})^n$ with the standard flat metric $\hat{g} = dx_1^2 + \cdots + dx_n^2$. A fundamental domain for the covering $\mathbb{R}^n \to \hat{\mathcal{M}}$ is $D \doteqdot (-1,1]^n$. Let $p = 0$ be the origin so that $d(x,p) = |x|$ for $x \in (-1,1]^n = \hat{\mathcal{M}}$. The static reduced volume is[1]

$$\bar{V}(\hat{g}, \tau) \doteqdot \int_{(-1,1]^n} (4\pi\tau)^{-n/2} e^{-|x|^2/4\tau} dx$$
(8.6)
$$= \pi^{-n/2} \int_{\left(-\frac{1}{2\sqrt{\tau}}, \frac{1}{2\sqrt{\tau}}\right]^n} e^{-|\tilde{x}|^2} d\tilde{x}$$

using the change of variables $\tilde{x} = \frac{x}{2\sqrt{\tau}}$. From (8.6) it is clear that $\bar{V}(\hat{g}, \tau)$ is a decreasing function of τ.

1.2. Static reduced volume and volume ratios. We may think of $\bar{V}$ as the static manifold analogue of **Perelman's reduced volume** for the Ricci flow (see (8.16) defined later in this chapter), which is defined similarly with $d^2/4\tau$ replaced by the reduced distance function ℓ. The monotonicity formula (8.3) is analogous to Perelman's monotonicity of the reduced volume (8.28). In the Ricci flat case, $\bar{V}$ is the same as Perelman's reduced volume (see Exercise 7.12 and (8.16)). Intuitively the static reduced volume $\bar{V}$ says something about volume ratios $r^{-n} \operatorname{Vol} B(p,r)$ at scales $r \sim \sqrt{\tau}$. Motivated by these elementary yet *a posteriori* considerations, we now relate $\bar{V}$ to the volume ratio $r^{-n} \operatorname{Vol} B(p,r)$ under the assumption that the Ricci curvatures of $\hat{g}$ are nonnegative.

Let $\left(\hat{\mathcal{M}}^n, \hat{g}\right)$ be a complete Riemannian manifold with $\operatorname{Rc}_{\hat{g}} \geq 0$. We divide the integral $\bar{V}$ into two parts:

(8.7) $$\bar{V}(\hat{g}, \tau) = \int_{B(p,r)} u\, d\mu + \int_{\hat{\mathcal{M}} - B(p,r)} u\, d\mu.$$

For the first term on the RHS, just using the obvious fact that $e^{-r^2/4\tau} \leq 1$, we have
$$\int_{B(p,r)} u\, d\mu \leq (4\pi\tau)^{-n/2} \operatorname{Vol} B(p,r).$$

---

[1] Since the torus is (Ricci) flat, the static metric reduced volume is the same as the Ricci flow reduced volume of Perelman.

1. REDUCED VOLUME OF A STATIC METRIC

Let $A(s)$ denote the volume of the geodesic $(n-1)$-sphere of radius s centered at p. Since $\mathrm{Rc}_{\hat{g}} \geq 0$, we may apply the Bishop–Gromov volume comparison theorem (see (A.8)), which says that for $s \geq r$,

$$(8.8) \qquad A(s) \leq A(r)\frac{s^{n-1}}{r^{n-1}} \leq n\frac{\operatorname{Vol} B(p,r)}{r^n}s^{n-1},$$

to estimate the second term on the RHS of (8.7):

$$\int_{\hat{\mathcal{M}}-B(p,r)} u\, d\mu = \int_r^\infty (4\pi\tau)^{-n/2} e^{-s^2/4\tau} A(s)\, ds$$

$$\leq n\frac{\operatorname{Vol} B(p,r)}{r^n} \int_r^\infty (4\pi\tau)^{-n/2} e^{-s^2/4\tau} s^{n-1}\, ds$$

$$= \frac{n}{2}\pi^{-n/2}\frac{\operatorname{Vol} B(p,r)}{r^n} \int_{\frac{r^2}{4\tau}}^\infty e^{-\eta}\eta^{\frac{n-2}{2}}\, d\eta,$$

where we made the change of variables $\eta \doteqdot \frac{s^2}{4\tau}$. Hence we have

$$\bar{V}(\hat{g},\tau) \leq (4\pi)^{-n/2}\frac{\operatorname{Vol} B(p,r)}{\tau^{n/2}} + \frac{n}{2}\pi^{-n/2}\frac{\operatorname{Vol} B(p,r)}{r^n}\int_{\frac{r^2}{4\tau}}^\infty e^{-\eta}\eta^{\frac{n-2}{2}}\, d\eta$$

$$(8.9) \qquad = \frac{\operatorname{Vol} B(p,r)}{r^n}\left(\left(\frac{r^2}{4\pi\tau}\right)^{n/2} + \frac{n}{2}\pi^{-n/2}\int_{\frac{r^2}{4\tau}}^\infty e^{-\eta}\eta^{\frac{n-2}{2}}\, d\eta\right).$$

This tells us that a lower bound for $\bar{V}$ yields a lower bound for the volume ratios of balls. Note that

$$1 = \int_{\mathbb{R}^n} \pi^{-n/2} e^{-|x|^2}\, dx = \frac{n\omega_n}{2}\pi^{-n/2}\int_0^\infty e^{-\eta}\eta^{\frac{n-2}{2}}\, d\eta,$$

where ω_n is the volume of the unit Euclidean n-ball. Hence

LEMMA 8.9 (The static reduced volume is bounded by volume ratios). If $\left(\hat{\mathcal{M}}^n,\hat{g}\right)$ is complete with $\mathrm{Rc}_{\hat{g}} \geq 0$, then for all $r > 0$ and $\tau > 0$,

$$(8.10) \qquad \bar{V}(\hat{g},\tau) \leq \frac{\operatorname{Vol} B(p,r)}{r^n}\left(\left(\frac{r^2}{4\pi\tau}\right)^{n/2} + \frac{1}{\omega_n}\right).$$

Thus, for $r \leq \rho$, the static reduced volume controls the volume ratio in the sense that

$$(8.11) \qquad \frac{\operatorname{Vol} B(p,r)}{r^n} \geq c_n^{-1}\bar{V}(\hat{g},\rho^2),$$

where $c_n \doteqdot \left(\frac{1}{2\sqrt{\pi}}\right)^n + \frac{1}{\omega_n}$. Note also that if for some $p \in \hat{\mathcal{M}}$ and $r > 0$ we have

$$(8.12) \qquad \frac{\operatorname{Vol} B(p,r)}{r^n} \leq \kappa,$$

then
$$0 < \bar{V}\left(\hat{g}, \kappa^{\alpha} r^2\right) \le \frac{\kappa^{1-\frac{\alpha n}{2}}}{(4\pi)^{n/2}} + \frac{\kappa}{\omega_n}.$$

Hence if $\alpha < 2/n$, then assumption (8.12), as $\kappa \to 0$, implies $\bar{V}\left(\hat{g}, \kappa^{\alpha} r^2\right) \to 0$.

1.3. The noncompact case and the asymptotic volume ratio.
Consider the case where $\left(\hat{\mathcal{M}}^n, \hat{g}\right)$ is complete and noncompact with $\mathrm{Rc}_{\hat{g}} \ge 0$. It is natural to believe that the limit as $\tau \to \infty$ of the static reduced volume $\bar{V}\left(\hat{g}, \tau\right)$ is related to the limit as $r \to \infty$ of the volume ratios; as we now show, this is indeed the case. Inequality (8.10) implies

$$\lim_{\tau \to \infty} \bar{V}\left(\hat{g}, \tau\right) \le \inf_{r > 0} \frac{\mathrm{Vol}\, B(p, r)}{\omega_n r^n} = \mathrm{AVR}\left(\hat{g}\right),$$

where

$$\mathrm{AVR}\left(\hat{g}\right) \doteqdot \lim_{r \to \infty} \frac{\mathrm{Vol}\, B(p, r)}{\omega_n r^n} = \lim_{s \to \infty} \frac{A(s)}{n \omega_n s^{n-1}}$$

as in (6.80). Next we show the opposite inequality. Since, by (8.8),

$$A(r) \ge n\omega_n \mathrm{AVR}\left(\hat{g}\right) r^{n-1},$$

we have for all $\tau > 0$,

$$\bar{V}\left(\hat{g}, \tau\right) = \int_0^\infty (4\pi\tau)^{-n/2} e^{-s^2/4\tau} A(s)\, ds$$
$$\ge n\omega_n \mathrm{AVR}\left(\hat{g}\right) \int_0^\infty (4\pi\tau)^{-n/2} e^{-s^2/4\tau} s^{n-1}\, ds = \mathrm{AVR}\left(\hat{g}\right).$$

Therefore the limit, as τ tends to infinity, of the static reduced volume is the asymptotic volume ratio.

LEMMA 8.10 (Asymptotic limit of $\bar{V}$ is AVR). *If $\left(\hat{\mathcal{M}}^n, \hat{g}\right)$ is a complete noncompact Riemannian manifold with $\mathrm{Rc}_{\hat{g}} \ge 0$, then*

$$\lim_{\tau \to \infty} \bar{V}\left(\hat{g}, \tau\right) = \mathrm{AVR}\left(\hat{g}\right).$$

REMARK 8.11. When $n = 2$, (8.9) says for any $r > 0$

(8.13) $$\bar{V}\left(\hat{g}, \tau\right) \le \frac{\mathrm{Area}\, B(p, r)}{\pi r^2}\left(\frac{r^2}{4\tau} + e^{-\frac{r^2}{4\tau}}\right).$$

Note that the function $F(x) \doteqdot x + e^{-x}$, $x \ge 0$, is an increasing function and its minimum value of 1 is attained at $x = 0$. In particular, as in (8.10) the upper bound in (8.13) improves as τ increases (for fixed $r > 0$).

2. Reduced volume for Ricci flow

In this section $(\mathcal{M}^n, g(\tau))$, $\tau \in [0, T]$, will be a complete solution to the backward Ricci flow satisfying the curvature bound $|\mathrm{Rm}(x, \tau)| \le C_0 < \infty$ for $(x, \tau) \in \mathcal{M} \times [0, T]$.

2.1. Volumes of geodesic spheres in $\widetilde{\mathcal{M}}$.

We motivate the definition of the reduced volume by computing the volume of geodesic spheres in the potentially infinite-dimensional manifold $\left(\widetilde{\mathcal{M}}, \tilde{g}\right)$ introduced in subsection 2.1 of Chapter 7. In particular, let $p = (x_0, y_0, 0)$, $\bar{\tau} \in (0, T)$, and

$$B_{\tilde{g}}\left(p, \sqrt{2N\bar{\tau}}\right) \subset \widetilde{\mathcal{M}} \doteqdot \mathcal{M} \times \mathcal{S}^N \times (0, T)$$

denote the ball centered at p with radius $\sqrt{2N\bar{\tau}}$ with respect to the metric:

$$\tilde{g} \doteqdot g_{ij} dx^i dx^j + \tau g_{\alpha\beta} dy^\alpha dy^\beta + \left(\frac{N}{2\tau} + R\right) d\tau^2,$$

where $g_{\alpha\beta}$ is the metric on $\mathcal{S}^N$ of constant sectional curvature $1/(2N)$. For any point $w = (x, y, \tau_w) \in \partial B_{\tilde{g}}(p, \sqrt{2N\bar{\tau}})$, because of the factor τ in $\tau g_{\alpha\beta} dy^\alpha dy^\beta$, we have

$$\sqrt{2N\bar{\tau}} = d_{\tilde{g}}(w, p) = d_{\tilde{g}}((x, y, \tau_w), (x_0, y_0, 0))$$
$$= d_{\tilde{g}}((x, y, \tau_w), (x_0, y, 0)).$$

Hence, letting $\gamma(\tau) \doteqdot (\gamma_\mathcal{M}(\tau), y, \tau)$, $\tau \in [0, \tau_w]$, with $\gamma(0) = (x_0, y, 0)$ and $\gamma_\mathcal{M}(\tau_w) = w$, we have

$$\sqrt{2N\bar{\tau}} = \inf_\gamma \operatorname{Length}_{\tilde{g}}(\gamma)$$

$$= \inf_{\gamma_\mathcal{M}} \left(\begin{array}{c} \frac{1}{\sqrt{2N}} \int_0^{\tau_w} \sqrt{\tau} \left(R + |\dot{\gamma}_\mathcal{M}(\tau)|^2\right) d\tau \\ + \sqrt{2N\tau_w} + O\left(N^{-3/2}\right) \end{array} \right)$$

(8.14)
$$= \sqrt{2N\tau_w} + \frac{1}{\sqrt{2N}} L(x, \tau_w) + O\left(N^{-3/2}\right),$$

where

$$L(x, \tau_w) \doteqdot \inf_{\gamma_\mathcal{M}} \int_0^{\tau_w} \sqrt{\tau} \left(R + |\dot{\gamma}_\mathcal{M}(\tau)|^2\right) d\tau$$

and the infimum is taken over $\gamma_\mathcal{M} : [0, \tau_w] \to \mathcal{M}$ with $\gamma_\mathcal{M}(0) = x_0$ and $\gamma_\mathcal{M}(\tau_w) = x$. Therefore for any $w = (x, y, \tau_w) \in \partial B_{\tilde{g}}(p, \sqrt{2N\bar{\tau}})$,

$$\sqrt{\tau_w} = \sqrt{\bar{\tau}} - \frac{1}{2N} L(x, \tau_w) + O\left(N^{-2}\right).$$

This implies that the geodesic sphere $\partial B_{\tilde{g}}\left(p, \sqrt{2N\bar{\tau}}\right)$, with respect to $\tilde{g}$, is $O(N^{-1})$-close to the hypersurface $\mathcal{M} \times \mathcal{S}^N \times \{\bar{\tau}\}$.

EXERCISE 8.12. Justify the equality (8.14).

Note that since the fibers $\mathcal{S}^N$ pinch to a point as $\tau \to 0$, if $w = (x, y, \tau_w) \in \partial B_{\tilde{g}}(p, \sqrt{2N\bar{\tau}})$, then any point in $\{x\} \times \mathcal{S}^N \times \{\tau_w\}$ also lies on the sphere $\partial B_{\tilde{g}}\left(p, \sqrt{2N\bar{\tau}}\right)$. We have that the volume of $\partial B_{\tilde{g}}\left(p, \sqrt{2N\bar{\tau}}\right)$

is roughly (since the sphere has small curvature for N large) the volume of the hypersurface $\mathcal{M} \times \mathcal{S}^N \times \{\bar{\tau}\}$ in $\widetilde{\mathcal{M}}$ and its volume can be computed as

$$\operatorname{Vol}_{\tilde{g}} \partial B_{\tilde{g}}\left(p, \sqrt{2N\bar{\tau}}\right)$$
$$\approx \int_{\partial B_{\tilde{g}}(p,\sqrt{2N\bar{\tau}})} d\mu_{g_{\mathcal{M}}(\tau_w)}(x) \wedge \tau_w^{N/2} d\mu_{\mathcal{S}^N}(y)$$
$$\approx \operatorname{Vol}(\mathcal{S}^N, g_{\mathcal{S}^N}) \int_{\mathcal{M}} \left(\sqrt{\bar{\tau}} - \frac{1}{2N} L(x, \tau_w) + O(N^{-2})\right)^N d\mu_{g_{\mathcal{M}}(\bar{\tau})}$$
$$\approx \omega_N \left(\sqrt{2N\bar{\tau}}\right)^N \int_{\mathcal{M}} \left(1 - \frac{1}{2N\sqrt{\bar{\tau}}} L(x, \bar{\tau}) + O(N^{-2})\right)^N d\mu_{g_{\mathcal{M}}(\bar{\tau})},$$

where ω_N is the volume of the unit sphere $\mathcal{S}^N$ (recall that $g_{\mathcal{S}^N}$ has constant sectional curvature $1/(2N)$, i.e., radius $\sqrt{2N}$). Observing that

$$\lim_{N\to\infty} \left(1 - \frac{1}{2N\sqrt{\bar{\tau}}} L(x, \bar{\tau}) + O(N^{-2})\right)^N$$
$$= \lim_{N\to\infty} \left(1 - \frac{1}{N} \frac{1}{2\sqrt{\bar{\tau}}} L(x, \bar{\tau})\right)^N = e^{-\frac{1}{2\sqrt{\bar{\tau}}} L(x,\bar{\tau})} = e^{-\ell(x,\bar{\tau})},$$

one can prove

(8.15) $$\frac{\operatorname{Vol}_{\tilde{g}}\left(\partial B_{\tilde{g}}\left(p, \sqrt{2N\bar{\tau}}\right)\right)}{\left(\sqrt{2N\bar{\tau}}\right)^{N+n}}$$
$$= (2N)^{-n/2} \omega_N \left(\int_{\mathcal{M}} \bar{\tau}^{-n/2} e^{-\ell(x,\bar{\tau})} d\mu_{g_{\mathcal{M}}(\bar{\tau})} + O(N^{-1})\right).$$

In particular, we obtain the geometric invariant

$$\int_{\mathcal{M}} \bar{\tau}^{-n/2} e^{-\ell(x,\bar{\tau})} d\mu_{g_{\mathcal{M}}(\bar{\tau})}$$

for $\bar{\tau} \in (0, T)$.

EXERCISE 8.13. Make the above arguments rigorous (especially the approximations) and in particular prove (8.15).

2.2. Definition of Perelman's reduced volume. Thus we are led to the following.

DEFINITION 8.14 (Reduced volume for Ricci flow). Let $(\mathcal{M}^n, g(\tau))$, $\tau \in [0, T]$, be a complete solution to the backward Ricci flow with bounded curvature. The **reduced volume** functional is defined by

(8.16) $$\tilde{V}(\tau) \doteqdot \int_{\mathcal{M}} (4\pi\tau)^{-n/2} \exp\left[-\ell(q, \tau)\right] d\mu_{g(\tau)}(q)$$

for $\tau \in (0, T)$.

See Lemma 8.16(ii) below for why $\tilde{V}(\tau)$ is well defined even when $\mathcal{M}$ is noncompact.

In the case of a Ricci flat solution $g(\tau) \equiv \hat{g}$, we have

$$\tilde{V}(\tau) = \int_{\mathcal{M}} (4\pi\tau)^{-n/2} \exp\left\{-\frac{d(p,q)^2}{4\bar{\tau}}\right\} d\mu_{\hat{g}}.$$

(Compare with (8.1).)

We now give heuristic (e.g., unjustified) proofs of the reduced volume monotonicity. *Provided* we can differentiate (8.16) under the integral sign, we obtain

(8.17)
$$\begin{aligned}
\frac{d\tilde{V}}{d\tau}(\tau) &= \frac{d}{d\tau}\left(\int_{\mathcal{M}} (4\pi\tau)^{-n/2} e^{-\ell(\cdot,\tau)} d\mu_{g(\tau)}\right) \\
&\overset{U}{=} \int_{\mathcal{M}} \frac{\partial}{\partial\tau}\left((4\pi\tau)^{-n/2} e^{-\ell(\cdot,\tau)} d\mu_{g(\tau)}\right) \\
&= \int_{\mathcal{M}} \left(-\frac{n}{2\tau} - \frac{\partial\ell}{\partial\tau} + R\right)(4\pi\tau)^{-n/2} e^{-\ell} d\mu,
\end{aligned}$$

where $\overset{U}{=}$ denotes an unjustified inequality.[2] By applying inequalities (7.91) and (7.148), which results in (7.150), we obtain

$$\frac{d\tilde{V}}{d\tau}(\tau) \leq \int_{\mathcal{M}} \left(|\nabla\ell|^2 - \Delta\ell\right)(4\pi\tau)^{-n/2} e^{-\ell} d\mu \leq 0.$$

Recall that for any vector field X, we have $\mathcal{L}_X d\mu = \operatorname{div}(X) d\mu$. In particular, $\mathcal{L}_{\nabla h} d\mu = \Delta h \, d\mu$ for any C^2 function h. Define the first-order differential operator acting on time-dependent tensors and forms:

$$\frac{D}{d\tau} \doteqdot \frac{\partial}{\partial\tau} + \mathcal{L}_X.$$

EXERCISE 8.15. Show that for any C^1-vector field X and C^1 function ϕ, one of them with compact support, under the backward Ricci flow $(\mathcal{M}^n, g(\tau))$, we have

$$\begin{aligned}
\frac{d}{d\tau}\int_{\mathcal{M}}\phi d\mu &= \int_{\mathcal{M}} \frac{D}{d\tau}(\phi d\mu) \\
&= \int_{\mathcal{M}}\left(\frac{\partial}{\partial\tau}\phi + X\cdot\nabla\phi + \phi R + \phi\operatorname{div}(X)\right)d\mu.
\end{aligned}$$

SOLUTION TO EXERCISE 8.15. The result follows from $\frac{\partial}{\partial\tau}d\mu = Rd\mu$ and the integration by parts identity

$$\int_{\mathcal{M}}(X\cdot\nabla\phi + \phi\operatorname{div}(X))d\mu = 0.$$

Now we consider again the time-derivative of the reduced volume of $(\mathcal{M}^n, g(\tau))$ under the backward Ricci flow and we heuristically discuss some

---

[2]For the justification, see the proof of Theorem 8.20 below.

formulas to be considered rigorously later. Let $X = \nabla \ell$. Again, *provided* we can differentiate under the integral sign, we compute

$$\frac{d\tilde{V}}{d\tau} \stackrel{\text{U}}{=} \int_{\mathcal{M}} \frac{D}{d\tau}\left((4\pi\tau)^{-n/2} e^{-\ell(\cdot,\tau)} d\mu_{g(\tau)}\right)$$
$$= \int_{\mathcal{M}} \left(-\frac{n}{2\tau} - \frac{\partial \ell}{\partial \tau} + R - |\nabla \ell|^2 + \Delta \ell\right)(4\pi\tau)^{-n/2} e^{-\ell} d\mu$$
$$\leq 0,$$

where we used (7.91) to obtain the last inequality. Note that we actually have the *pointwise* inequality

$$\frac{D}{d\tau}\left((4\pi\tau)^{-n/2} e^{-\ell} d\mu\right)$$
$$= \left(-\frac{n}{2\tau} - \frac{\partial \ell}{\partial \tau} + R - |\nabla \ell|^2 + \Delta \ell\right)(4\pi\tau)^{-n/2} e^{-\ell} d\mu \leq 0.$$

Pulling this back to the tangent space $T_p\mathcal{M}$, we have

$$\frac{d}{d\tau}\left[(4\pi\tau)^{-n/2} e^{-\ell(\gamma_V(\tau),\tau)} \mathcal{L}\mathrm{J}_V(\tau)\right] \leq 0$$

as in (8.22) below.

2.3. Monotonicity of reduced volume: A proof using the $\mathcal{L}$-Jacobian. In this subsection we give a rigorous proof of the reduced volume monotonicity. Recall that the open set

$$\Omega(\tau) = \Omega_{(p,0)}(\tau) \doteqdot \{V \in T_p\mathcal{M} : \tau_V > \tau\} \subset T_p\mathcal{M}$$

is given by Definition 7.96(iii) and satisfies $\Omega_{(p,0)}(\tau_2) \subset \Omega_{(p,0)}(\tau_1)$ if $\tau_1 < \tau_2$. Recall from (7.136) that the $\mathcal{L}$-exponential map restricted to $\Omega(\tau)$,

$$\mathcal{L}_\tau \exp : \Omega_{(p,0)}(\tau) \to \mathcal{M} \backslash \mathcal{L}\operatorname{Cut}_{(p,0)}(\tau)$$

is a diffeomorphism. If $\Omega_{(p,0)}(\tau) = T_p\mathcal{M}$ for some $\tau > 0$, then $\mathcal{L}\operatorname{Cut}_{(p,0)}(\tau) = \varnothing$ and $\mathcal{M}^n$ is diffeomorphic to Euclidean space.

Since $\mathcal{L}\operatorname{Cut}_{(p,0)}(\tau)$ has measure zero in $(\mathcal{M}, g(\tau))$, we have by the definition of the $\mathcal{L}$-Jacobian and (7.136),

$$\tilde{V}(\tau) = \int_{\mathcal{M}\backslash\mathcal{L}\operatorname{Cut}_{(p,0)}(\tau)} (4\pi\tau)^{-n/2} \exp\left[-\ell(q,\tau)\right] d\mu_{g(\tau)}(q)$$

(8.18)
$$= \int_{\Omega_{(p,0)}(\tau)} (4\pi\tau)^{-n/2} e^{-\ell(\gamma_V(\tau),\tau)} \mathcal{L}\mathrm{J}_V(\tau) dx(V),$$

where γ_V is the $\mathcal{L}$-geodesic emanating from p with $\lim_{\tau \to 0_+} \sqrt{\tau}\dot{\gamma}_V(\tau) = V$ and $\mathcal{L}\mathrm{J}_V(\tau)$ is the $\mathcal{L}$-Jacobian associated to the $\mathcal{L}$-geodesic γ_V. Here dx is the volume form on $T_p\mathcal{M}$ with respect to the Euclidean metric $g(0,p)$, $\mathcal{L}_\tau \exp(V) = \gamma_V(\tau)$, and

$$\mathcal{L}\mathrm{J}_V(\tau) dx(V) = (\mathcal{L}_\tau \exp)^* d\mu_{g(\mathcal{L}_\tau \exp(V),\tau)}.$$

We will use the convention
$$\mathcal{L}\, \mathrm{J}_V(\tau) \doteq 0 \quad \text{for } \tau \geq \tau_V.$$
We can then write the reduced volume as

(8.19) $$\tilde{V}(\tau) = \int_{T_p\mathcal{M}} (4\pi\tau)^{-n/2} e^{-\ell(\gamma_V(\tau),\tau)} \mathcal{L}\, \mathrm{J}_V(\tau)\, dx(V).$$

We compute the evolution of ℓ along a minimal $\mathcal{L}$-geodesic $\gamma_V(\tau)$ for $0 \leq \tau < \tau_V$, where $V \in T_p\mathcal{M}$. For $q = \gamma_V(\tau)$, $\tau \in [0,\tau_V)$, the function $\ell(\cdot,\cdot)$ is smooth in some small neighborhood of (q,τ); hence the following derivatives of ℓ at such (q,τ) exist. Recall from (7.78) that

$$\tau^{3/2}\left(R + |X|^2\right)(\tau) = -K + \frac{1}{2} L(q,\tau),$$

where $K = K(\tau)$ is the trace Harnack integral defined by (7.75). Thus

(8.20) $$\ell(q,\tau) = \tau\left(R + |X|^2\right)(\tau) + \tau^{-1/2} K.$$

Recall equation (7.88):
$$\frac{\partial \ell}{\partial \tau} = \frac{1}{2\tau^{3/2}} K - \frac{1}{\tau}\ell + R,$$
and from (7.54) recall that
$$\nabla \ell(q,\tau) = \dot{\gamma}_V(\tau) = X(\tau).$$

Hence the derivative of the reduced distance along a minimal $\mathcal{L}$-geodesic is given by
$$\frac{d}{d\tau}\left[\ell(\gamma_V(\tau),\tau)\right] = \frac{\partial \ell}{\partial \tau} + \nabla \ell \cdot X$$
$$= \frac{1}{2\tau^{3/2}} K - \frac{\ell}{\tau} + R + |X|^2$$

(8.21) $$= -\frac{1}{2}\tau^{-3/2} K$$

by (8.20).

The following lemma can be viewed as an infinitesimal Bishop–Gromov volume comparison result for the **Ricci flow geometry**. The striking part is that no curvature assumption is needed.

LEMMA 8.16 (Pointwise monotonicity along $\mathcal{L}$-geodesics). *Suppose that $(\mathcal{M}^n, g(\tau))$, $\tau \in [0,T]$, is a complete solution to the backward Ricci flow with bounded curvature.*

(i) *For any $V \in T_p\mathcal{M}$ and $0 < \tau < \tau_V$,*

(8.22) $$\frac{d}{d\tau}\left[(4\pi\tau)^{-n/2} e^{-\ell(\gamma_V(\tau),\tau)} \mathcal{L}\, \mathrm{J}_V(\tau)\right] \leq 0,$$

where equality holds if equality in (7.139) holds.

(ii) *For any $V \in T_pM$ and $0 < \tau < T$,*

(8.23) $$(4\pi\tau)^{-n/2} e^{-\ell(\gamma_V(\tau),\tau)} \mathcal{L}\,\mathrm{J}_V(\tau) \leq \pi^{-n/2} e^{-|V|^2_{g(0,p)}}.$$

Hence, even for a complete solution on a noncompact manifold, the reduced volume is well defined.

PROOF. (i) Recall from (7.139) that

$$\left(\frac{d}{d\tau} \log \mathcal{L}\,\mathrm{J}_V\right)(\tau) \leq \frac{n}{2\tau} - \frac{1}{2}\tau^{-\frac{3}{2}} K.$$

From this and (8.21), we compute

$$\frac{d}{d\tau}\left[(4\pi\tau)^{-n/2} e^{-\ell(\gamma_V(\tau),\tau)} \mathcal{L}\,\mathrm{J}_V(\tau)\right]$$

$$= (4\pi\tau)^{-n/2} e^{-\ell(\gamma_V(\tau),\tau)} \mathcal{L}\,\mathrm{J}_V(\tau)\left(-\frac{n}{2\tau} - \frac{d\ell}{d\tau} + \frac{d}{d\tau}\log\mathcal{L}\,\mathrm{J}_V\right)$$

$$\leq 0.$$

(ii) It follows from (8.22) that for any $0 < \tau < \tau_V$, we have

$$(4\pi\tau)^{-n/2} e^{-\ell(\gamma_V(\tau),\tau)} \mathcal{L}\,\mathrm{J}_V(\tau)$$

$$\leq \lim_{\tau_1 \to 0_+} (4\pi\tau_1)^{-n/2} e^{-\ell(\gamma_V(\tau_1),\tau_1)} \mathcal{L}\,\mathrm{J}_V(\tau_1)$$

$$= \lim_{\tau_1 \to 0_+}\left[(4\pi\tau_1)^{-n/2} \mathcal{L}\,\mathrm{J}_V(\tau_1)\right] e^{-\lim_{\tau_1 \to 0_+} \ell(\gamma_V(\tau_1),\tau_1)}$$

(8.24) $$= \pi^{-n/2} e^{-|V|^2},$$

where in the last equality we have used (7.138) and (7.97). If $\tau \geq \tau_V$, then the statement is obvious. □

An immediate consequence of the above lemma is the following fundamental result: the monotonicity of the reduced volume.

COROLLARY 8.17 (Reduced volume monotonicity). *Suppose $(\mathcal{M}^n, g(\tau))$, $\tau \in [0,T]$, is a complete solution to the backward Ricci flow with the curvature bound $|\mathrm{Rm}(x,\tau)| \leq C_0 < \infty$ for $(x,\tau) \in \mathcal{M} \times [0,T]$. Then*

(i) $\lim_{\tau \to 0_+} \tilde{V}(\tau) = 1.$
(ii) *The reduced volume is nonincreasing:*

(8.25) $$\tilde{V}(\tau_1) \geq \tilde{V}(\tau_2)$$

for any $0 < \tau_1 < \tau_2 < T$, and $\tilde{V}(\tau) \leq 1$ for any $\tau \in (0,T)$.
(iii) *Equality in (8.25) holds if and only if $(\mathcal{M}, g(\tau))$ is isometric to Euclidean space $(\mathbb{R}^n, g_\mathbb{E})$, regarded as the Gaussian soliton.*

PROOF. (i) From equation (7.131) it follows that

$$\lim_{\tau \to 0} \Omega_{(p,0)}(\tau) = T_p \mathcal{M}^n.$$

Since $\Omega_{(p,0)}(\tau_1) = \Omega(\tau_1) \supset \Omega(\tau_2)$ for $\tau_1 < \tau_2$, we have $\lim_{\tau \to 0_+} \chi_{\Omega(\tau)} = 1$, where χ_Ω denotes the characteristic function of the set Ω. We compute

$$\lim_{\tau \to 0_+} \tilde{V}(\tau) = \lim_{\tau \to 0_+} \int_{\Omega(\tau)} (4\pi\tau)^{-n/2} e^{-\ell(\gamma_V(\tau),\tau)} \mathcal{L} \operatorname{J}_V(\tau) \, dx(V)$$

$$= \int_{T_pM} \lim_{\tau \to 0_+} \left[\chi_{\Omega(\tau)} (4\pi\tau)^{-n/2} e^{-\ell(\gamma_V(\tau),\tau)} \mathcal{L} \operatorname{J}_V(\tau) \right] dx(V)$$

$$= \int_{T_pM} 1 \cdot \pi^{-n/2} e^{-|V|^2} dx(V) = 1,$$

where we used (8.24).

(ii) From (8.22), we have for any $0 < \tau_1 < \tau_2$ and $V \in \Omega(\tau_2)$,

$$\tau_1^{-n/2} e^{-\ell(\gamma_V(\tau_1),\tau_1)} \mathcal{L} \operatorname{J}_V(\tau_1) \geq \tau_2^{-n/2} e^{-\ell(\gamma_V(\tau_2),\tau_2)} \mathcal{L} \operatorname{J}_V(\tau_2),$$

so

$$\tilde{V}(\tau_2) = \int_{\Omega(\tau_2)} (4\pi\tau_2)^{-n/2} e^{-\ell(\gamma_V(\tau_2),\tau_2)} \mathcal{L} \operatorname{J}_V(\tau_2) \, dx(V)$$

$$\leq \int_{\Omega(\tau_2)} (4\pi\tau_1)^{-n/2} e^{-\ell(\gamma_V(\tau_1),\tau_1)} \mathcal{L} \operatorname{J}_V(\tau_1) \, dx(V)$$

$$\leq \int_{\Omega(\tau_1)} (4\pi\tau_1)^{-n/2} e^{-\ell(\gamma_V(\tau_1),\tau_1)} \mathcal{L} \operatorname{J}_V(\tau_1) \, dx(V)$$

$$= \tilde{V}(\tau_1),$$

where we used $\Omega(\tau_1) \supset \Omega(\tau_2)$. Note that $\tilde{V}(\tau) \leq 1$ for any $\tau > 0$ follows from (8.25) and (i).

(iii) We prove this statement in two steps by first showing that $g(\tau)$ is a shrinking gradient Ricci soliton and then showing that $(\mathcal{M}, g(\tau))$ is Euclidean space. If $\tilde{V}(\tau_1) = \tilde{V}(\tau_2)$ for a pair of times $0 < \tau_1 < \tau_2$, then for $\tau \in (\tau_1, \tau_2)$ and $V \in \Omega(\tau)$, we have that equality in (8.22) holds:

$$\frac{d}{d\tau} \left[(4\pi\tau)^{-n/2} e^{-\ell(\gamma_V(\tau),\tau)} \mathcal{L} \operatorname{J}_V(\tau) \right] = 0,$$

which, by the proof of Lemma 8.16, implies that we have equality in (7.139). Hence, by (7.140), we get

(8.26) $$\operatorname{Rc}(\gamma_V(\tau), \tau) + (\operatorname{Hess} \ell)(\gamma_V(\tau), \tau) = \frac{g(\gamma_V(\tau), \tau)}{2\tau}$$

for all $V \in \Omega(\tau)$ and $\tau \in [\tau_1, \tau_2]$, and where ℓ is C^∞ at $(\gamma_V(\tau), \tau)$ for all such (V, τ). Since $\tilde{V}(\tau_1) = \tilde{V}(\tau_2)$, we have $\Omega(\tau_1) = \Omega(\tau_2)$.

Suppose there exists $V_1 \in T_pM$ such that $\tau_{V_1} \leq \tau_1$. Since $\Omega(\tau_1) \neq \emptyset$, there exists $V_2 \in T_pM$ such that $\tau_{V_2} > \tau_1$. Since the function $V \mapsto \tau_V$ is a continuous function, there exists $V_3 \in T_pM$ such that $\tau_{V_3} \in (\tau_1, \tau_2)$. Thus $V_3 \in \Omega(\tau_1) - \Omega(\tau_2)$, which is a contradiction. Therefore, for every $V \in T_pM$, we have $\tau_V > \tau_1$, so that $\Omega(\tau_1) = T_pM$ (and hence $\Omega(\tau) = T_pM$

for all $\tau \in [\tau_1, \tau_2]$). This implies ℓ is C^∞ on $\mathcal{M} \times [\tau_1, \tau_2]$ and (8.26) holds on $\mathcal{M} \times [\tau_1, \tau_2]$. Thus $g(\tau)$ is a shrinking gradient Ricci soliton.

Given $\tau_0 \in [\tau_1, \tau_2]$, by Proposition 1.7, the shrinking gradient Ricci soliton structure $\left(\mathcal{M}, g(\tau_0), \nabla \ell(\tau_0), -\frac{1}{\tau_0}\right)$ may be put into a canonical time-dependent form (1.11) defined for all $t < \tau_0$,

$$\tilde{g}(t) = \frac{\tilde{\tau}(t)}{\tau_0} \varphi(t)^* g(\tau_0),$$

where $\tilde{g}(t)$ is a solution of the Ricci flow, by (1.10), i.e., $\tilde{\tau}(t) = \tau_0 - t$ ($\varepsilon = -\frac{1}{\tau_0}$), and $\varphi(t)$ is a 1-parameter family of diffeomorphisms with $\varphi(0) = \mathrm{id}_\mathcal{M}$. By the uniqueness of complete solutions of the Ricci flow with bounded curvature (see Chen and Zhu [82]) and since $\tilde{g}(0) = g(\tau_0)$, we have

$$g(\tau) = \tilde{g}(\tau_0 - \tau),$$

so that

(8.27) $$\varphi(\tau_0 - \tau)^* g(\tau_0) = \frac{\tau_0}{\tau} g(\tau) \quad \text{for } \tau \in (0, \tau_0].$$

Since $|\mathrm{Rm}\,[g(\tau)]| \leq C_0 < \infty$ for $\tau \in [0, T]$ (we just use this for τ small), by (8.27) we have

$$\sup_\mathcal{M} |\mathrm{Rm}\,[g(\tau_0)]| = \sup_\mathcal{M} |\mathrm{Rm}\,[\varphi(\tau_0 - \tau)^* g(\tau_0)]|$$
$$\leq \frac{\tau}{\tau_0} \sup_\mathcal{M} |\mathrm{Rm}\,[g(\tau)]| \leq C_0 \frac{\tau}{\tau_0}$$

for all $\tau \in (0, \tau_0]$. Hence $|\mathrm{Rm}\,[g(\tau_0)]| \equiv 0$. Since $\Omega(\tau_0) = T_p\mathcal{M}$, $\mathcal{M}$ is diffeomorphic to $\mathbb{R}^n$. Part (iii) follows since a flat shrinking gradient Ricci soliton on $\mathbb{R}^n$ must be the Gaussian soliton. □

REMARK 8.18. (i) The Riemannian analogue of Corollary 8.17(i) is

$$\lim_{r \to 0} \frac{\mathrm{Vol}\, B(p, r)}{\omega_n r^n} = 1.$$

(ii) Note that for the shrinking gradient Ricci soliton $g(\tau)$ in subsection 7.3 of Chapter 7, the metric $g(0)$ is not well-defined.

The monotonicity of the reduced volume can be easily generalized to the following. For any fixed measurable subset $A \subset T_p\mathcal{M}$, we can define $D(A, \tau)$ to be the set of vectors $V \in A$ such that $\tau_V > \tau$, i.e.,

$$D(A, \tau) \doteqdot \{V \in A : \tau_V > \tau\} = A \cap \Omega(\tau).$$

It is clear that $D(A, \tau)$ satisfies $D(A, \tau_2) \subset D(A, \tau_1)$ if $\tau_2 > \tau_1$.

COROLLARY 8.19 ($\mathcal{L}$-relative volume comparison). *Suppose $(\mathcal{M}^n, g(\tau))$, $\tau \in [0, T]$, is a complete smooth solution to the backward Ricci flow with the curvature bound $|\mathrm{Rm}\,(x, \tau)| \leq C_0 < \infty$ for $(x, \tau) \in \mathcal{M} \times [0, T]$. Define for any $\tau \in (0, T)$ and any measurable subset $A \subset T_p\mathcal{M}$,*

$$\tilde{V}_A(\tau) \doteqdot \int_{\mathcal{L}_\tau \exp(D(A, \tau))} (4\pi\tau)^{-n/2} \exp[-\ell(q, \tau)]\, d\mu_{g(\tau)}(q).$$

Then for any $\tau_1 < \tau_2$,
$$\tilde{V}_A(\tau_1) \geq \tilde{V}_A(\tau_2).$$

PROOF. By the definition of the $\mathcal{L}$-Jacobian we know that for any L^1 function f on $\mathcal{M}$
$$\int_{\mathcal{L}_\tau \exp(D(A,\tau))} f(y)\, d\mu_{g(\tau)}(y) = \int_{D(A,\tau)} f(\mathcal{L}_\tau \exp(V)) \mathcal{L}\,\mathrm{J}_V(\tau)\, dx(V).$$
(We have used this change of variables formula for $A = T_p\mathcal{M}$ in previous sections.) We have
$$\tilde{V}_A(\tau_2) = \int_{D(A,\tau_2)} (4\pi\tau_2)^{-n/2} e^{-\ell(\gamma_V(\tau_2),\tau_2)} \mathcal{L}\,\mathrm{J}_V(\tau_2)\, dx(V)$$
$$\leq \int_{D(A,\tau_2)} (4\pi\tau_1)^{-n/2} e^{-\ell(\gamma_V(\tau_1),\tau_1)} \mathcal{L}\,\mathrm{J}_V(\tau_1)\, dx(V)$$
$$\leq \int_{D(A,\tau_1)} (4\pi\tau_1)^{-n/2} e^{-\ell(\gamma_V(\tau_1),\tau_1)} \mathcal{L}\,\mathrm{J}_V(\tau_1)\, dx(V)$$
$$= \tilde{V}_A(\tau_1).$$
□

The above can be thought of as a relative volume comparison theorem for the Ricci flow. This is along the lines of the generalization by Shunhui Zhu in [384] (see also Theorem 1.135 in [111] for example).

2.4. Monotonicity of reduced volume revisited. Now we give another proof of the monotonicity of the reduced volume without using the $\mathcal{L}$-Jacobian. Recall that under the evolution equations
$$\frac{\partial}{\partial \tau} g_{ij} = 2R_{ij}, \qquad \frac{\partial}{\partial \tau} f - \Delta f + |\nabla f|^2 - R + \frac{n}{2\tau} = 0,$$
with $\tau > 0$, we have
$$\frac{d}{d\tau} \int_{\mathcal{M}} \tau^{-n/2} e^{-f} d\mu_g = 0.$$
In comparison, by (7.146), the reduced distance ℓ is a subsolution to the above equation for f. We use this fact to give another proof of the monotonicity of reduced volume.

THEOREM 8.20 (Monotonicity of the reduced volume: second proof). *Let $(\mathcal{M}^n, g(\tau))$, $\tau \in [0,T]$, be a complete solution to the backward Ricci flow satisfying the curvature bound $|\mathrm{Rm}(x,\tau)| \leq C_0 < \infty$ for $(x,\tau) \in \mathcal{M} \times [0,T]$. Then for any $\tau \in (0,T)$, the reduced volume $\tilde{V}(\tau)$ is differentiable and nonincreasing:*

(8.28)
$$\frac{d\tilde{V}}{d\tau}(\tau) \leq 0.$$

PROOF. We justify the differentiation under the integral sign in equality (8.17). Consider the difference quotient for the reduced volume integrand:

$$\Phi(q,\tau,h) \doteqdot \frac{(\tau+h)^{-n/2}e^{-\ell(q,\tau+h)}\frac{d\mu_{g(\tau+h)}(q)}{d\mu_{g(\tau)}(q)} - \tau^{-n/2}e^{-\ell(q,\tau)}}{h}.$$

Note that

(8.29) $$\frac{d\tilde{V}}{d\tau}(\tau) = (4\pi)^{-n/2}\lim_{h\to 0}\int_{\mathcal{M}}\Phi(q,\tau,h)d\mu_{g(\tau)}(q),$$

so that the time-derivative of $\tilde{V}(\tau)$ exists if the limit on the RHS exists. At any point (q,τ) where ℓ is differentiable (e.g., for each τ, a.e. on $\mathcal{M}$), we have

$$\lim_{h\to 0}\Phi(q,\tau,h) = \tau^{-n/2}\exp\left[-\ell(q,\tau)\right]\left(-\frac{n}{2\tau} - \frac{\partial\ell}{\partial\tau} + R\right).$$

Recall by Lebesgue's dominated convergence theorem that if we can show there exists a function $\Psi(q,\tau)$ such that for $\tau > 0$ there exists $\varepsilon_\tau > 0$ where

(8.30) $$|\Phi(q,\tau,h)| \leq \Psi(q,\tau) \text{ on } \mathcal{M}$$

for $h \in (-\varepsilon_\tau, \varepsilon_\tau)$, and $\int_{\mathcal{M}}\Psi(q,\tau)d\mu_{g(\tau)}(q) < \infty$, then

$$\lim_{h\to 0}\int_{\mathcal{M}}\Phi(q,\tau,h)d\mu_{g(\tau)}(q) = \int_{\mathcal{M}}\lim_{h\to 0}\Phi(q,\tau,h)d\mu_{g(\tau)}(q).$$

Thus, provided we have (8.30),

$$\frac{d}{d\tau}\tilde{V}(\tau) = (4\pi)^{-n/2}\int_{\mathcal{M}}\lim_{h\to 0}\Phi(q,\tau,h)d\mu_{g(\tau)}(q)$$
$$= (4\pi)^{-n/2}\int_{\mathcal{M}}\tau^{-n/2}\exp\left[-\ell(q,\tau)\right]\left(-\frac{n}{2\tau} - \frac{\partial\ell}{\partial\tau} + R\right)d\mu_{g(\tau)}(q)$$
$$\leq 0,$$

where the last inequality follows from (7.151). This is the reduced volume monotonicity formula.

To see (8.30), we first observe that

$$\Phi_1(q,\tau,h) \doteqdot (\tau+h)^{-n/2}e^{-\ell(q,\tau+h)}\frac{d\mu_{g(\tau+h)}(q)}{d\mu_{g(\tau)}(q)} - \tau^{-n/2}e^{-\ell(q,\tau)}$$

is a locally Lipschitz function of h near $h = 0$, for $\tau > 0$ fixed. (Note that $\Phi_1(q,\tau,0) = 0$.) Hence $\Phi_1(q,\tau,h)$ is an absolutely continuous function of h on the interval $[-\delta,\delta]$, where $\delta \doteqdot \min\left\{\frac{\tau}{10}, \frac{T-\tau}{10}, \frac{1}{10}\right\}$. By elementary real analysis (see Corollary 15 on p. 110 of [**313**]), we have

$$\Phi(q, \tau, \bar{h}) = \frac{1}{\bar{h}} \left(\Phi_1(q, \tau, \bar{h}) - \Phi_1(q, \tau, 0) \right)$$

$$= \frac{1}{\bar{h}} \int_0^{\bar{h}} \frac{\partial}{\partial h} \left[(\tau + h)^{-n/2} e^{-\ell(q, \tau + h)} \frac{d\mu_{g(\tau+h)}(q)}{d\mu_{g(\tau)}(q)} \right] dh$$

(8.31)
$$= \frac{1}{\bar{h}} \int_0^{\bar{h}} \left(-\frac{n}{2(\tau+h)} - \frac{\partial \ell}{\partial \tau}(q, \tau + h) + R(q, \tau + h) \right)$$

$$\times (\tau + h)^{-\frac{n}{2}} e^{-\ell(q, \tau + h)} \frac{d\mu_{g(\tau+h)}(q)}{d\mu_{g(\tau)}(q)} dh.$$

By Lemma 7.59, we have

(8.32)
$$\exp\left[-\ell(q, \bar{\tau})\right] \leq \exp\left(-e^{-2C_0 T} \frac{d_{g(0)}^2(p,q)}{4\bar{\tau}} + \frac{nC_0\bar{\tau}}{3}\right),$$

which decays exponentially quadratically in terms of the distance function. Also we have $\frac{\partial}{\partial \tau} d\mu_{g(\tau)}\big|_{\tau=\bar{\tau}}(q) = R(q, \bar{\tau}) d\mu_{g(\bar{\tau})}(q)$, so that we have the following bounds for the volume form:

$$e^{-nC_0|h|} \leq \frac{d\mu_{g(\tau+h)}(q)}{d\mu_{g(\tau)}(q)} \leq e^{nC_0|h|}.$$

By Lemma 7.59, we also have

$$\ell(q, \bar{\tau}) \leq e^{2C_0 T} \frac{d_{g(\bar{\tau})}^2(p,q)}{4\bar{\tau}} + \frac{nC_0\bar{\tau}}{3},$$

so it follows from Lemma 7.60(ii) that

$$\left|\frac{\partial \ell}{\partial \tau}(q, \tau)\right|_{\tau=\bar{\tau}} \leq \frac{C_1}{\bar{\tau}} \left(\ell(q, \bar{\tau}) + A\bar{\tau} \right)$$

(8.33)
$$\leq \frac{C_1}{\bar{\tau}} \left(e^{2C_0 T} \frac{d_{g(\bar{\tau})}^2(p,q)}{4\bar{\tau}} + \left(\frac{nC_0}{3} + A\right)\bar{\tau} \right),$$

where $A < \infty$. Hence, for $\bar{\tau} = \tau + h$, we have

$$\left|-\frac{n}{2\bar{\tau}} - \frac{\partial \ell}{\partial \tau}(q, \bar{\tau}) + R(q, \bar{\tau})\right| \bar{\tau}^{-\frac{n}{2}} e^{-\ell(q,\bar{\tau})} \frac{d\mu_{g(\bar{\tau})}(q)}{d\mu_{g(\tau)}(q)}$$

(8.34)
$$\leq \left| \frac{n}{2\bar{\tau}} + \frac{C_1}{\bar{\tau}} \left(e^{2C_0 T} \frac{d_{g(\bar{\tau})}^2(p,q)}{4\bar{\tau}} + \left(\frac{nC_0}{3} + A\right)\bar{\tau} \right) + nC_0 \right|$$

$$\times \bar{\tau}^{-\frac{n}{2}} \exp\left(-e^{-2C_0 T} \frac{d_{g(0)}^2(p,q)}{4\bar{\tau}} + \frac{nC_0\bar{\tau}}{3} \right) e^{nC_0|\bar{\tau}-\tau|}.$$

On the other hand, by the curvature lower bound and the Bishop–Gromov volume comparison theorem,

(8.35)
$$\operatorname{Vol} B_{g(\tau)}\left(p, d_{g(\tau)}(p,q)\right) \leq C_4 e^{C_4 d_{g(\tau)}(p,q)}$$

for some constant $C_4 < \infty$. From (8.34), (8.35), and (8.31), it is easy to see that $|\Phi(q, \tau, h)|$ is bounded by an integrable function on $\mathcal{M}$, independent of h small enough. □

EXERCISE 8.21. Let $(\mathcal{M}^n, g)$ be a Riemannian manifold, $p \in \mathcal{M}$, and assume that $\exp_p : T_p\mathcal{M} \to \mathcal{M}$ is a diffeomorphism ($\mathcal{M}$ is then diffeomorphic to $\mathbb{R}^n$). Define
$$\varphi_s : \mathcal{M} \to \mathcal{M}$$
by
$$\varphi_s : \exp_p(V) \mapsto \exp_p(e^s V)$$
for $V \in T_p\mathcal{M}$. Show that $\{\varphi_s\}_{s \in \mathbb{R}}$ is a 1-parameter group of diffeomorphisms and
$$\frac{\partial}{\partial s} \varphi_s = \nabla \left(\frac{r^2}{2}\right) \circ \varphi_s,$$
where $r(x) \doteq d(x, p)$.

SOLUTION TO EXERCISE 8.21. Another way to define φ_s is
$$\varphi_s(x) = \exp_p\left(e^s \exp_p^{-1} x\right).$$
We have
$$\varphi_{s_1}(\varphi_{s_2}(x)) = \exp_p\left(e^{s_1} \exp_p^{-1} \circ \exp_p\left(e^{s_2} \exp_p^{-1} x\right)\right)$$
$$= \exp_p\left(e^{s_1+s_2} \exp_p^{-1} x\right) = \varphi_{s_1+s_2}(x).$$
Note $r(\varphi_s(x)) = e^s \left|\exp_p^{-1} x\right|$. We compute
$$\left(\frac{\partial}{\partial s} \varphi_s\right)(x) = (d\exp_p)_{e^s \exp_p^{-1} x} \left(e^s \exp_p^{-1} x\right)$$
$$= (r\nabla r)(\varphi_s(x)) = \nabla\left(\frac{r^2}{2}\right)(\varphi_s(x)).$$

Now we consider a Ricci flow analogue of the above discussion. Given $p \in \mathcal{M}$ and $\tau_0 > 0$, assume that $\Omega_{(p,0)}(\tau_0) = T_p\mathcal{M}$. For τ such that $\Omega_{(p,0)}(\tau) = T_p\mathcal{M}$, define
$$\phi_\tau : \mathcal{M} \to \mathcal{M}$$
by
$$\phi_\tau : \mathcal{L}_{\tau_0} \exp_p(V) \mapsto \mathcal{L}_\tau \exp_p(V),$$
that is,
$$\phi_\tau(x) = \mathcal{L}_\tau \exp_p\left((\mathcal{L}_{\tau_0} \exp_p)^{-1}(x)\right).$$
We compute
$$\left(\frac{\partial}{\partial \tau} \phi_\tau\right)(x) = \left(\frac{\partial}{\partial \tau} \mathcal{L}_\tau \exp_p\right)\left((\mathcal{L}_{\tau_0} \exp_p)^{-1}(x)\right)$$
$$= X(\phi_\tau(x)) = (\nabla \ell)(\phi_\tau(x)),$$
where $X \doteq \frac{d}{d\tau} \gamma_V(\tau)$ for $V = (\mathcal{L}_{\tau_0} \exp_p)^{-1}(x)$.

3. A weakened no local collapsing theorem via the monotonicity of the reduced volume

In this section, $(\mathcal{M}^n, \tilde{g}(t))$, $t \in [0, T)$, shall denote a complete solution to the Ricci flow with $T < \infty$ and $\sup_{x \in \mathcal{M}, \, t \in [0, t_1]} |\operatorname{Rm}_{\tilde{g}}(x, t)| < \infty$ for any $t_1 < T$ (i.e., the curvatures are bounded, but possibly not uniformly as $t \to T$, as in the case of a singular solution). We fix a time $T_0 \in \left(\frac{T}{2}, T\right)$ and a basepoint $p_0 \in \mathcal{M}$. Let

$$g(\tau) \doteqdot \tilde{g}(T_0 - \tau).$$

Then $(\mathcal{M}^n, g(\tau))$, $\tau \in [0, T_0]$, is a solution to the backward Ricci flow with initial metric $g(0) = \tilde{g}(T_0)$ and bounded sectional curvature. Let $\mathcal{L}(\gamma)$ denote the $\mathcal{L}$-length of a curve γ, let $L : \mathcal{M} \times (0, T_0] \to \mathbb{R}$ denote the L-distance, let $\ell : \mathcal{M} \times (0, T_0] \to \mathbb{R}$ denote the reduced distance, and let $\tilde{V} : (0, T_0] \to (0, \infty)$ denote the reduced volume, all with respect to $g(\tau)$ and the basepoint $(p_0, 0)$.

3.1. A bound of the reduced distance.
The following lower bound of the reduced volume will be used in the proof of the Weakened No Local Collapsing Theorem 8.26.

LEMMA 8.22 (Lower bound for $\tilde{V}$ at initial time).

(i) (ℓ upper bound) *Fix an arbitrary $r_0 > 0$. There exists a constant $C_1 > 0$, depending only on r_0, n, T, and $\sup_{\mathcal{M} \times [0, T/2]} \operatorname{Rc}_{\tilde{g}(t)}$, and there exists $q_0 \in \mathcal{M}$ such that*

$$\ell(q, T_0) \leq C_1 \quad \text{for every } q \in B_{\tilde{g}(0)}(q_0, r_0).$$

(ii) ($\tilde{V}$ lower bound) *Suppose there exist $r_1 > 0$ and $v_1 > 0$ such that*

$$\operatorname{Vol}_{\tilde{g}(0)} B_{\tilde{g}(0)}(w, r_1) \geq v_1$$

for all $w \in \mathcal{M}$. Then there exists a constant $C_2 > 0$, depending only on r_1, v_1, n, T, and $\sup_{\mathcal{M} \times [0, T/2]} \operatorname{Rc}_{\tilde{g}(t)}$, such that

$$\tilde{V}(T_0) \geq C_2.$$

PROOF. (i) By Lemma 7.50 there exists $q_0 \in \mathcal{M}$ such that[3]

$$\ell(q_0, T_0 - T/2) = \min_{q \in \mathcal{M}} \ell(q, T_0 - T/2) \leq \frac{n}{2}.$$

For any $q \in B_{\tilde{g}(0)}(q_0, r_0)$, let $\beta : [T_0 - T/2, T_0] \to \mathcal{M}$ be a constant speed minimal geodesic from q_0 to q with respect to $\tilde{g}(0)$. Defining

$$C_0 \doteqdot \sup_{\mathcal{M} \times [0, T/2]} \operatorname{Rc}_{\tilde{g}(t)},$$

---
[3]This corresponds to time $t = T/2$.

we have $|\cdot|^2_{g(\tau)} \leq e^{C_0 T} |\cdot|^2_{g(T_0)} = e^{C_0 T} |\cdot|^2_{\tilde{g}(0)}$ acting on vector fields for $\tau \in [T_0 - T/2, T_0]$. We can estimate $\mathcal{L}(\beta)$ as follows:

$$\mathcal{L}(\beta)$$
$$\leq \int_{T_0 - T/2}^{T_0} \sqrt{\tau} \left(R_{g(\tau)}(\beta(\tau)) + \left|\frac{d\beta}{d\tau}\right|^2_{g(\tau)} \right) d\tau$$
$$\leq \frac{2}{3}\left(T_0^{3/2} - (T_0 - T/2)^{3/2}\right) \sup_{\mathcal{M} \times [0, T/2]} R_{\tilde{g}(t)} + e^{C_0 T} \int_{T_0 - T/2}^{T_0} \sqrt{\tau} \left|\frac{d\beta}{d\tau}\right|^2_{\tilde{g}(0)} d\tau$$
$$= \frac{2}{3}\left(T_0^{3/2} - (T_0 - T/2)^{3/2}\right) \left(\sup_{\mathcal{M} \times [0, T/2]} R_{\tilde{g}(t)} + 4 e^{C_0 T} \frac{(d_{\tilde{g}(0)}(q, q_0))^2}{T^2} \right)$$
$$\leq \frac{2}{3} T^{3/2} \left(n C_0 + \frac{4 e^{C_0 T} r_0^2}{T^2} \right).$$

Let $\alpha : [0, T_0 - T/2] \to \mathcal{M}$ be a minimal $\mathcal{L}$-geodesic with $\alpha(0) = p_0$ and $\alpha(T_0 - T/2) = q_0$. Then

$$\mathcal{L}(\alpha) = 2\sqrt{T_0 - T/2} \cdot \ell(q_0, T_0 - T/2) \leq n\sqrt{T/2}.$$

Consider the concatenated path:

$$\gamma(\tau) \doteqdot (\alpha \smile \beta)(\tau) = \begin{cases} \alpha(\tau) & \text{if } t \in [0, T_0 - T/2], \\ \beta(\tau) & \text{if } t \in [T_0 - T/2, T_0]. \end{cases}$$

This path is well defined and piecewise smooth. We have

$$\ell(q, T_0) \leq \frac{1}{2\sqrt{T_0}} \mathcal{L}(\gamma) = \frac{1}{2\sqrt{T_0}} [\mathcal{L}(\alpha) + \mathcal{L}(\beta)]$$
$$\leq \frac{1}{\sqrt{2T}} \left[n\sqrt{T/2} + \frac{2}{3} T^{3/2} \left(n C_0 + \frac{4 e^{C_0 T} r_0^2}{T^2} \right) \right]$$
$$\doteqdot C_1 \left(r_0, n, T, \sup_{\mathcal{M} \times [0, T/2]} \operatorname{Rc}_{\tilde{g}(t)} \right).$$

(ii) Choosing $r_0 = r_1$ in (i), we have

$$\tilde{V}(T_0) = \int_{\mathcal{M}} (4\pi T_0)^{-n/2} e^{-\ell(q, T_0)} d\mu_{\tilde{g}(0)}(q)$$
$$\geq \int_{B_{\tilde{g}(0)}(q_0, r_1)} (4\pi T_0)^{-n/2} e^{-\ell(q, T_0)} d\mu_{\tilde{g}(0)}(q)$$
$$\geq \int_{B_{\tilde{g}(0)}(q_0, r_1)} (4\pi T_0)^{-n/2} e^{-C_1} d\mu_{\tilde{g}(0)}(q)$$
$$\geq v_1 (4\pi T)^{-n/2} e^{-C_1} \doteqdot C_2 \left(v_1, r_1, n, T, \sup_{\mathcal{M} \times [0, T/2]} \operatorname{Rc}_{\tilde{g}(t)} \right).$$

$\square$

3. NO LOCAL COLLAPSING VIA REDUCED VOLUME MONOTONICITY

3.2. The weakened no local collapsing theorem. In Lemma 8.9 we saw how the reduced volume of a static metric bounds the volume ratios of balls from below. Similarly, the *reduced volume monotonicity* for solutions of the Ricci flow enables one to prove a weakened form of the no local collapsing theorem, which we first encountered in Chapter 6 using *entropy monotonicity*.

DEFINITION 8.23 (Strongly κ-collapsed). Let $\kappa > 0$ be a constant. We say that a solution $(\mathcal{M}^n, \tilde{g}(t))$, $t \in [0, T)$, to the Ricci flow is **strongly κ-collapsed at** $(q_0, t_0) \in \mathcal{M} \times (0, T)$ **at scale** $r > 0$ if

(1) *(curvature bound in a parabolic cylinder)* $|\mathrm{Rm}_{\tilde{g}}(x, t)| \leq \frac{1}{r^2}$ for all $x \in B_{\tilde{g}(t_0)}(q_0, r)$ and $t \in [\max\{t_0 - r^2, 0\}, t_0]$ and
(2) *(volume of ball is κ-collapsed)*
$$\frac{\mathrm{Vol}_{\tilde{g}(t_0)} B_{\tilde{g}(t_0)}(q_0, r)}{r^n} < \kappa.$$

Given an $r > 0$, if for any $t_0 \in [r^2, T)$ and any $q_0 \in \mathcal{M}$ the solution $\tilde{g}(t)$ is *not* strongly κ-collapsed at (q_0, t_0) at scale r, then we say that $(\mathcal{M}, \tilde{g}(t))$ is **weakly κ-noncollapsed at scale** r.

Recall that the reduced volume $\tilde{V}(\tau)$ has the upper bound 1. When the solution is strongly κ-collapsed, we shall obtain a better upper bound for $\tilde{V}$ which tends to 0 as κ tends to 0.

THEOREM 8.24 (Main estimate for weakened no local collapsing). *Let $(\mathcal{M}^n, \tilde{g}(t))$, $t \in [0, T)$, be a complete solution to the Ricci flow with $T < \infty$ and suppose $\sup_{\mathcal{M} \times [0, t_1]} |\mathrm{Rm}| < \infty$ for any $t_1 < T$. Then there exists $c_1 = c_1(n) \in (0, \frac{1}{2}]$ depending only on n such that if for some $\kappa^{1/n} \leq c_1(n)$, the solution $\tilde{g}(t)$ is strongly κ-collapsed at (p_*, t_*) at scale r, where $t_* > \frac{T}{2}$ and $r < \sqrt{t_*}$, then the reduced volume $\tilde{V}_*$ of $g_*(\tau) \doteqdot \tilde{g}(t_* - \tau)$ with basepoint p_* has the upper bound*
$$\tilde{V}_*(\varepsilon r^2) \leq \phi(\varepsilon, n),$$

where
$$\varepsilon \doteqdot \kappa^{1/n}$$

and
$$\phi(\varepsilon, n) \doteqdot \frac{\exp\left(\frac{1}{6}n(n-1)\right)}{(4\pi)^{n/2}} \varepsilon^{n/2} + \omega_{n-1}(n-2)^{\frac{n-2}{2}} e^{-\frac{n-2}{2}} \exp\left(-\frac{1}{2\sqrt{\varepsilon}}\right).$$

REMARK 8.25. Note that $\lim_{\varepsilon \to 0} \phi(\varepsilon, n) = 0$.

We will prove this theorem in the next subsection. This theorem gives a proof of the following **weakened no local collapsing theorem**.

THEOREM 8.26 (Weakened no local collapsing). *Let $(\mathcal{M}^n, \tilde{g}(t))$, $t \in [0, T)$, be a complete solution to the Ricci flow with $T < \infty$. Suppose*

(1) $\sup_{\mathcal{M} \times [0, t_1]} |\mathrm{Rm}| < \infty$ *for any $t_1 < T$ and*

(2) there exist $r_1 > 0$ and $v_1 > 0$ such that $\text{Vol}_{\tilde{g}(0)} B_{\tilde{g}(0)}(x, r_1) \geq v_1$ for all $x \in \mathcal{M}$.

Then there exists $\kappa > 0$ depending only on r_1, v_1, n, T, and $\sup_{\mathcal{M} \times [0, T/2]} \text{Rc}_{\tilde{g}(t)}$ such that $\tilde{g}(t)$ is weakly κ-noncollapsed at any point $(p_*, t_*) \in \mathcal{M} \times (T/2, T)$ at any scale $r < \sqrt{T/2}$.

Note that if $\mathcal{M}$ is a closed manifold, then assumptions (1) and (2) of the theorem are automatically true.

PROOF OF THEOREM 8.26. Let $c_1(n)$ be as in Theorem 8.24. Suppose $\tilde{g}(t)$ is strongly κ-collapsed at a point $(p_*, t_*) \in \mathcal{M} \times (T/2, T)$ at a scale $r < \sqrt{T/2}$ with $\kappa^{1/n} \leq c_1(n)$. Let $\varepsilon \doteqdot \kappa^{1/n}$. Consider the backward solution of the Ricci flow

$$g_*(\tau) \doteqdot \tilde{g}(t_* - \tau)$$

with basepoint p_*. Since $\varepsilon r^2 \leq r^2 \leq t_*$, by the monotonicity of the reduced volume $\tilde{V}_*$, we have

$$\tilde{V}_*(t_*) \leq \tilde{V}_*(\varepsilon r^2).$$

By Lemma 8.22(ii), choosing $(p_0, T_0) = (p_*, t_*)$, we have $\tilde{V}_*(t_*) \geq C_2 > 0$. Applying Theorem 8.24 yields

$$0 < C_2 \left(r_1, v_1, n, T, \sup_{\mathcal{M} \times [0, T/2]} \text{Rc}_{\tilde{g}(t)} \right)$$
$$\leq \frac{\exp\left(\frac{1}{6} n(n-1)\right)}{(4\pi)^{n/2}} \varepsilon^{n/2} + \omega_{n-1}(n-2)^{\frac{n-2}{2}} e^{-\frac{n-2}{2}} \exp\left(-\frac{1}{2\sqrt{\varepsilon}}\right).$$

This implies a positive lower bound for $\varepsilon = \kappa^{1/n}$ and the theorem is proved. $\square$

REMARK 8.27. Under the assumptions of Theorem 8.26, $(\mathcal{M}^n, \tilde{g}(t))$, $t \in [0, T/2]$, has bounded geometry; so it is easy to see that there exists $\kappa_1 > 0$ depending only r_1, v_1, n, T, and $\sup_{\mathcal{M} \times [0, T/2]} \text{Rc}_{\tilde{g}(t)}$ such that $\tilde{g}(t)$ is weakly κ_1-noncollapsed at $(p_*, t_*) \in \mathcal{M} \times (0, T/2]$ at any scale $r < \sqrt{T/2}$.

3.3. Bounding reduced volume from above when the solution is strongly κ-collapsed. This subsection will be devoted to the proof of Theorem 8.24, which follows directly from Propositions 8.28 and 8.30 below. Note that the assumptions on $\tilde{g}(t)$ translate to the following assumptions on $g_*(\tau)$:

(8.36) $\quad |\text{Rm}_{g_*}(x, \tau)| \leq \dfrac{1}{r^2} \quad$ for all $x \in B_{g_*(0)}(p_*, r)$ and $\tau \in [0, r^2]$,

(8.37) $\quad \dfrac{\text{Vol}_{g_*(0)} B_{g_*(0)}(p_*, r)}{r^n} \leq \kappa.$

3. NO LOCAL COLLAPSING VIA REDUCED VOLUME MONOTONICITY

Shi's local derivative estimate implies that there is a constant $c_2 = c_2(n)$ depending only on n such that

(8.38) $\quad |\nabla_{g_*} R(x, \tau)| \leq \dfrac{c_2}{r^3}$ for $x \in B_{g_*(0)}\left(p_*, \frac{r}{2}\right)$ and $\tau \in \left[0, \frac{1}{2}r^2\right]$.

From (8.19) we can write the reduced volume of $g_*(\tau)$ as

$$\tilde{V}_*(\tau) = \tilde{V}_1(\tau) + \tilde{V}_2(\tau),$$

where

$$\tilde{V}_1(\tau) \doteqdot \int_{|V|_{g_*(0)} \leq \varepsilon^{-1/4}} (4\pi\tau)^{-n/2} e^{-\ell(\gamma_V(\tau), \tau)} \mathcal{L}\mathrm{J}_V(\tau) \, dx(V),$$

$$\tilde{V}_2(\tau) \doteqdot \int_{|V|_{g_*(0)} > \varepsilon^{-1/4}} (4\pi\tau)^{-n/2} e^{-\ell(\gamma_V(\tau), \tau)} \mathcal{L}\mathrm{J}_V(\tau) \, dx(V).$$

Here both the reduced distance $\ell(\gamma_V(\tau), \tau)$ and the $\mathcal{L}$-Jacobian $\mathcal{L}\mathrm{J}_V(\tau)$ are defined with respect to $g_*(\tau)$.

PROPOSITION 8.28. *Under the assumptions of Theorem 8.24, there exists $c_1 = c_1(n) \in (0, \frac{1}{2}]$, depending only on n, such that if $\varepsilon = \kappa^{1/n} \leq c_1(n)$, then*

$$\tilde{V}_1(\varepsilon r^2) \leq \frac{\exp\left(\frac{1}{6} n(n-1)\right)}{(2\pi)^{n/2}} \varepsilon^{n/2}.$$

The idea of the proof is to show that for some choice of c_1, $\gamma_V(\tau)$ is contained in $B_{\tilde{g}(0)}(p_*, r/2)$ and $\ell(\gamma_V(\varepsilon r^2), \varepsilon r^2)$ has a lower bound independent of ε when $|V|_{g_*(0)} \leq \varepsilon^{-1/4}$. The proposition then follows easily.

LEMMA 8.29. *Suppose (p_*, t_*), where $t_* > \frac{T}{2}$, and $r < \sqrt{t_*}$ are such that (8.36) holds, i.e., such that*

$$|\mathrm{Rm}_{g_*}(x, \tau)| \leq \frac{1}{r^2} \quad \text{for all } x \in B_{g_*(0)}(p_*, r) \text{ and } \tau \in [0, r^2].$$

Then there exists $c_1 = c_1(n) \in (0, \frac{1}{2}]$ depending only on n such that if $\varepsilon = \kappa^{1/n} \leq c_1(n)$, then

$$\gamma_V(\tau) \in B_{g_*(0)}(p_*, r/2) \quad \text{for any } V \in B_{g_*(p_*,0)}\left(\vec{0}, \varepsilon^{-1/4}\right) \text{ and } \tau \in \left[0, \varepsilon r^2\right].$$

PROOF. We prove the lemma by contradiction. Suppose

$$V \in B_{g_*(p_*,0)}\left(\vec{0}, \varepsilon^{-1/4}\right)$$

and $\tau' \in (0, \varepsilon r^2]$ is the first time such that $\gamma_V(\tau') \in \partial B_{g_*(0)}(p_*, r/2)$. Let $V(\tau) \doteqdot \sqrt{\tau}\dot{\gamma}_V(\tau) = \sqrt{\tau}X(\tau)$, so that $\lim_{\tau \to 0} V(\tau) = V$. Since $\gamma_V([0, \tau']) \in \overline{B_{g_*(0)}(p_*, r/2)}$ and $\tau' \in \left[0, \frac{1}{2}r^2\right]$, (8.36) and (8.38) imply

$$\mathrm{Rc}_{g_*} \geq -\frac{n-1}{r^2} \quad \text{and} \quad |\nabla_{g_*} R(x, \tau)| \leq \frac{c_2}{r^3}$$

along γ_V, and hence we can use (7.48) to get[4]

$$|V(\tau)|^2_{g_*(\tau)} \leq e^{6(n-1)\varepsilon} \varepsilon^{-1/2} + \frac{c_2^2 \varepsilon}{12(n-1)^2} \left(e^{6(n-1)\varepsilon} - 1\right).$$

Therefore by Hölder's inequality,

$$\left(\int_0^{\tau'} |\dot{\gamma}_V(\tau)|_{g_*(\tau)} d\tau\right)^2$$

$$\leq \int_0^{\tau'} \tau^{-1/2} d\tau \int_0^{\tau'} \sqrt{\tau} |\dot{\gamma}_V(\tau)|^2_{g_*(\tau)} d\tau$$

$$\leq 2\sqrt{\varepsilon} r \int_0^{\tau'} \tau^{-1/2} |V(\tau)|^2_{g_*(\tau)} d\tau$$

$$\leq (2\sqrt{\varepsilon} r)^2 \left(e^{6(n-1)\varepsilon} \varepsilon^{-1/2} + \frac{c_2^2 \varepsilon}{12(n-1)^2} \left(e^{6(n-1)\varepsilon} - 1\right)\right)$$

$$\leq \frac{r^2}{16}.$$

We get the last inequality by choosing $c_1 \leq \frac{1}{2}$ such that

$$4\left(e^{6(n-1)\bar{\varepsilon}} \bar{\varepsilon}^{1/2} + \frac{c_2^2 \bar{\varepsilon}^2}{12(n-1)^2} \left(e^{6(n-1)\bar{\varepsilon}} - 1\right)\right) \leq \frac{1}{16}$$

for all $\bar{\varepsilon} \in [0, c_1]$. Hence

$$\int_0^{\tau'} |\dot{\gamma}_V(\tau)|_{g_*(\tau)} d\tau \leq \frac{r}{4}.$$

If we also require $c_1 \leq \frac{1}{6n}$, then $\tau' \leq \varepsilon r^2 \leq \frac{r^2}{6n}$. Since $|\text{Rm}_{g_*}(x,\tau)| \leq \frac{1}{r^2}$ for all $x \in B_{g_*(0)}(p_*, r)$ and $\tau \in [0, r^2]$, we have $g_*(x,\tau) \geq \frac{2}{3} g_*(x,0)$ for $\tau \in [0, \tau']$ and $x \in B_{g_*(0)}(p_*, r)$. Hence

$$d_{g_*(0)}(\gamma_V(\tau'), p_*) \leq \int_0^{\tau'} |\dot{\gamma}_V(\tau)|_{g_*(0)} d\tau \leq \frac{3}{2} \int_0^{\tau'} |\dot{\gamma}_V(\tau)|_{g_*(\tau)} d\tau$$

$$\leq \frac{3r}{8} < \frac{r}{2}.$$

This contradicts $\gamma_V(\tau') \in \partial B_{g_*(0)}(p_*, r/2)$. The lemma is proved. □

We now give a proof of Proposition 8.28.

---

[4]Note that $\beta(\sigma) \doteqdot \gamma_V(\sigma^2/4)$ satisfies $\frac{d\beta}{d\sigma} = V$. In (7.48) we take $C_0 = \frac{n-1}{r^2}$, $T = \varepsilon r^2$, and $C_2 = \frac{c_2}{r^3}$.

PROOF OF PROPOSITION 8.28. Let c_1 be chosen as in Lemma 8.29 and let $\kappa^{1/n} \leq c_1$. If $V \in B_{g_*(p_*,0)}\left(\vec{0}, \varepsilon^{-1/4}\right)$ and $\gamma_V|_{[0,\varepsilon r^2]}$ is a minimal $\mathcal{L}$-geodesic, then by Lemma 8.29,

$$\ell\left(\gamma_V\left(\varepsilon r^2\right), \varepsilon r^2\right) = \frac{1}{2\sqrt{\varepsilon r^2}} L\left(\gamma_V\left(\varepsilon r^2\right), \varepsilon r^2\right)$$

$$\geq \frac{1}{2\sqrt{\varepsilon r^2}} \int_0^{\varepsilon r^2} \sqrt{\tau} R_{g_*}\left(\gamma_V\left(\tau\right), \tau\right) d\tau$$

$$\geq -\frac{n(n-1)r^{-2}}{2\sqrt{\varepsilon r^2}} \int_0^{\varepsilon r^2} \sqrt{\tau} d\tau$$

$$= -\frac{1}{3} n(n-1)\varepsilon.$$

In the integral defining $\tilde{V}_1(\varepsilon r^2)$, it follows from (8.18) that we only need to consider those vectors $V \in B_{g_*(p_*,0)}\left(\vec{0}, \varepsilon^{-1/4}\right)$ for which $\gamma_V|_{[0,\varepsilon r^2]}$ is a minimal $\mathcal{L}$-geodesic. Hence

$$\tilde{V}_1(\varepsilon r^2) = \int_{|V|_{g_*(0)} \leq \varepsilon^{-1/4}} \left(4\pi\varepsilon r^2\right)^{-n/2} e^{-\ell(\gamma_V(\varepsilon r^2), \varepsilon r^2)} \mathcal{L} \mathrm{J}_V(\varepsilon r^2) \, dx(V)$$

$$\leq \left(4\pi\varepsilon r^2\right)^{-n/2} \exp\left(\frac{1}{3}n(n-1)\varepsilon\right) \int_{|V|_{g_*(0)} \leq \varepsilon^{-1/4}} \mathcal{L} \mathrm{J}_V(\varepsilon r^2) \, dx(V)$$

$$\leq (4\pi)^{-n/2} \varepsilon^{-n/2} \exp\left(\frac{1}{3}n(n-1)\varepsilon\right) \frac{\mathrm{Vol}_{g_*(\varepsilon r^2)} B_{g_*(0)}(p_*, r/2)}{r^n}$$

$$\leq (2\pi)^{-n/2} \varepsilon^{-n/2} \exp\left(\frac{1}{3}n(n-1)\varepsilon\right) \frac{\mathrm{Vol}_{g_*(0)} B_{g_*(0)}(p_*, r/2)}{r^n}$$

$$\leq \frac{\exp\left(\frac{1}{6}n(n-1)\right)}{(2\pi)^{n/2}} \varepsilon^{n/2},$$

where we have used $g_*(\varepsilon r^2) \leq 2g_*(0)$, and also $\frac{\mathrm{Vol} B_{g_*(0)}(p_*, r)}{r^n} \leq \varepsilon^n$ in the last inequality. $\square$

Finally we give an estimate for $\tilde{V}_2(\varepsilon r^2)$ from above.

PROPOSITION 8.30. *Under the assumptions of Theorem 8.24, we have*

$$\tilde{V}_2(\varepsilon r^2) \leq \omega_{n-1}(n-2)^{\frac{n-2}{2}} e^{-\frac{n-2}{2}} \exp\left\{-\frac{1}{2\sqrt{\varepsilon}}\right\}.$$

PROOF. By (8.23), we have

$$\left(4\pi\varepsilon r^2\right)^{-n/2} e^{-\ell(\gamma_V(\varepsilon r^2), \varepsilon r^2)} \mathcal{L} \mathrm{J}_V(\tau) \leq \exp\left(-|V|^2_{g_*(0)}\right).$$

Therefore

$$\tilde{V}_2\left(\varepsilon r^2\right) \leq \int_{|V|_{g_*(0)} \geq \varepsilon^{-1/4}} \exp\left(-|V|^2_{g_*(0)}\right) dx(V)$$

$$= \omega_{n-1} \int_{\varepsilon^{-1/4}}^{\infty} e^{-r^2} r^{n-1} dr.$$

Noting that $r^{n-2} e^{-\frac{1}{2}r^2} \leq (n-2)^{\frac{n-2}{2}} e^{-\frac{n-2}{2}}$, we get

$$\tilde{V}_2\left(\varepsilon r^2\right) \leq \omega_{n-1}(n-2)^{\frac{n-2}{2}} e^{-\frac{n-2}{2}} \exp\left(-\frac{1}{2\sqrt{\varepsilon}}\right).$$

□

4. Backward limit of ancient κ-solution is a shrinker

DEFINITION 8.31 (ancient κ-solution). Let κ be a positive constant. A complete ancient solution $(\mathcal{M}^n, \tilde{g}(t))$, $t \in (-\infty, 0]$, of the Ricci flow is called an **ancient κ-solution** (or κ**-solution** for short) if it satisfies the following three conditions.

(i) $\tilde{g}(t)$ is nonflat and has nonnegative curvature operator for each $t \in (-\infty, 0]$.
(ii) There is a constant $C < \infty$ such that $R_{\tilde{g}}(x, t) \leq C$ for all $(x, t) \in \mathcal{M} \times (-\infty, 0]$.
(iii) $\tilde{g}(t)$ is κ-noncollapsed on all scales for all $t \in (-\infty, 0]$; i.e., for any $\rho > 0$ and for any $(p, t) \in \mathcal{M} \times (-\infty, 0]$, if $|\operatorname{Rm}_{\tilde{g}}(x, t)| \leq \rho^{-2}$ for all $x \in B_{\tilde{g}(t)}(p, \rho)$, then

$$\frac{\operatorname{Vol}_{\tilde{g}(t)} B_{\tilde{g}(t)}(p, \rho)}{\rho^n} \geq \kappa.$$

If the curvature bound condition (ii) in the definition is replaced by the requirement that $\tilde{g}(t)$ satisfies the trace Harnack inequality

$$\frac{\partial R}{\partial t} + 2\nabla R \cdot X + 2\operatorname{Rc}(X, X) \geq 0 \quad \text{for all } X,$$

we say that $\tilde{g}(t)$ is a κ**-solution with Harnack**. In Part II of this volume we will prove that in dimension $n = 3$ the notions of κ-solution with Harnack and ancient κ-solution are equivalent. In this section we prove that in all dimensions *certain backward limits of ancient κ-solutions are nonflat shrinking gradient Ricci solitons* (Theorem 8.32 below). The proof will take several steps, which will be carried out in the following subsections.

4.1. Statement of the theorem. Let $(\mathcal{M}^n, \tilde{g}(t))$, $t \in (-\infty, 0]$, be a κ-solution. For $t_0 \in (-\infty, 0]$ we define a solution to the backward Ricci flow $(\mathcal{M}, g(\tau))$, $\tau \in [0, \infty)$, by

$$g(\tau) \doteqdot \tilde{g}(t_0 - \tau).$$

Given $p_0 \in \mathcal{M}$, we have the reduced length $\ell(q,\tau) \doteqdot \ell^g(q,\tau)$, $\mathcal{L}$-Jacobian $\mathcal{L}\operatorname{J}_V(\tau) \doteqdot \mathcal{L}\operatorname{J}_V^g(\tau)$, and the reduced volume $\tilde{V}(\tau)$ defined with respect to $g(\tau)$ using the basepoint p_0.

For any $\tau > 0$, define dilated backward solutions:

(8.39) $$g_\tau(\theta) \doteqdot \tau^{-1} \cdot g(\tau\theta), \quad \text{for } \theta \in [0,\infty).$$

Let $q_\tau \in \mathcal{M}$ be a point such that

$$\ell(q_\tau, \tau) \leq \frac{n}{2}$$

(by Lemma 7.50 such a point always exists). The following is Proposition 11.2 in [**297**].

THEOREM 8.32.

(1) *For any sequence $\tau_i \to \infty$ and $A > 1$, there exists a subsequence, still denoted by τ_i, such that $(\mathcal{M}^n, g_{\tau_i}(\theta), q_{\tau_i})$, $\theta \in (A^{-1}, A)$, converges in the Cheeger–Gromov sense to a complete nonflat shrinking gradient Ricci soliton $(\mathcal{M}^n_\infty, g_\infty(\theta), q_\infty)$.*

(2) *By choosing a sequence of $A_k \to \infty$ and using a diagonalization argument, we have for any $\tau_i \to \infty$ that there exists a subsequence such that $(\mathcal{M}, g_{\tau_i}(\theta), q_{\tau_i})$, $\theta \in (0,\infty)$, converges in the Cheeger–Gromov sense to a complete nonflat shrinking gradient Ricci soliton, which we also denote by $(\mathcal{M}_\infty, g_\infty(\theta), q_\infty)$. Since the trace Harnack estimate holds for the sequence, it also holds for $g_\infty(\theta)$; hence the limit $(\mathcal{M}_\infty, g_\infty(\theta))$ is a κ-solution with Harnack.*

In dimension $n = 3$, because of the equivalence between κ-solutions with Harnack and ancient κ-solutions, $g_\infty(\theta)$ has bounded curvature.

REMARK 8.33. When $n \geq 4$, it is not clear to us if the limit has bounded sectional curvature.

Before we begin the proof of Theorem 8.32, we end this subsection with some elementary properties about the change of the reduced length ℓ and the $\mathcal{L}$-Jacobian $\mathcal{L}\operatorname{J}_V(\tau)$ under scaling (8.39). Let $\gamma_W^{g_\tau}(\cdot)$ be the $\mathcal{L}$-geodesic, with respect to the solution g_τ, satisfying $\gamma_W^{g_\tau}(0) = p_0 \in \mathcal{M}$ and $\lim_{\theta \to 0} \sqrt{\theta} \dot{\gamma}_W^{g_\tau}(\theta) = W$. The reduced length $\ell^{g_\tau}(q, \theta)$ and the $\mathcal{L}$-Jacobian $\mathcal{L}\operatorname{J}_V^{g_\tau}(\theta)$ shall be defined with respect to $g_\tau(\theta)$ using the basepoint p_0.

LEMMA 8.34 (Elementary scaling properties). *For any $\tau > 0$ and $\theta \in [0,\infty)$, we have*

(8.40) $$\gamma_{\sqrt{\tau}V}^{g_\tau}(\theta) = \gamma_V^g(\tau\theta),$$

(8.41) $$\ell^{g_\tau}(q, \theta) = \ell^g(q, \tau\theta),$$

(8.42) $$\tau^{-n/2} \mathcal{L}\operatorname{J}_{\sqrt{\tau}V}^{g_\tau}(\theta) = \mathcal{L}\operatorname{J}_V^g(\tau\theta).$$

PROOF. Given that we know how curvature changes under scaling, it is easy to check that $\gamma_V^g(\tau\theta)$ satisfies the $\mathcal{L}$-geodesic equation for the solution $g_\tau(\theta)$. We compute using the change of variable $\tilde{\tau} = \tau\theta$ that

$$\lim_{\theta \to 0_+} \sqrt{\theta}\frac{d\gamma_V^g(\tau\theta)}{d\theta} = \lim_{\tilde{\tau} \to 0_+} \sqrt{\tilde{\tau}}\frac{d\gamma_V^g(\tilde{\tau})}{d\tilde{\tau}}\sqrt{\tau} = \sqrt{\tau}V.$$

By the uniqueness of the initial-value problem for the $\mathcal{L}$-geodesic equation, we get (8.40).

Given any curve $\alpha(\tilde{\tau})$, $\tilde{\tau} \in [0, \tau\theta]$, from $(p_0, 0)$ to $(q, \tau\theta)$, the curve $\alpha_\tau(\tilde{\theta}) \doteqdot \gamma(\tau\tilde{\theta})$, $\tilde{\theta} \in [0, \theta]$, joins $(p_0, 0)$ to (q, θ). We compute

$$\frac{1}{2\sqrt{\tau\theta}}\mathcal{L}^g(\alpha) = \frac{1}{2\sqrt{\tau\theta}} \int_0^{\tau\theta} \sqrt{\tilde{\tau}} \left(R_{g(\tilde{\tau})} + \left|\frac{d\alpha}{d\tilde{\tau}}\right|^2_{g(\tilde{\tau})} \right) d\tilde{\tau}$$

$$= \frac{1}{2\sqrt{\tau\theta}} \int_0^{\theta} \sqrt{\tau\tilde{\theta}} \left(\tau^{-1} R_{g_\tau(\tilde{\theta})} + \tau^{-1} \left|\frac{d\alpha_\tau}{d\tilde{\theta}}\right|^2_{g_\tau(\tilde{\theta})} \right) \tau d\tilde{\theta}$$

$$= \frac{1}{2\sqrt{\theta}} \int_0^{\theta} \sqrt{\tilde{\theta}} \left(R_{g_\tau(\tilde{\theta})} + \left|\frac{d\alpha_\tau}{d\tilde{\theta}}\right|^2_{g_\tau(\tilde{\theta})} \right) d\tilde{\theta}$$

$$= \frac{1}{2\sqrt{\theta}} \mathcal{L}^{g_\tau}(\alpha_\tau).$$

Since $\ell^g(q, \tau\theta) = \inf_\alpha \left[\frac{1}{2\sqrt{\tau\theta}}\mathcal{L}^g(\alpha)\right]$ and $\ell^{g_\tau}(q, \theta) = \inf_\beta \left[\frac{1}{2\sqrt{\theta}}\mathcal{L}^{g_\tau}(\beta)\right]$, we conclude that $\ell^g(q, \tau\theta) = \ell^{g_\tau}(q, \theta)$ and the minimizing $\mathcal{L}$-geodesics are related by (8.40).

From (8.40) we get

$\mathcal{L}^g_{\tau\theta}\exp : (T_{p_0}\mathcal{M}, g(0, p_0)) \to (\mathcal{M}, g(\tau\theta))$ where $V \to \gamma_V^g(\tau\theta)$,

$\mathcal{L}^{g_\tau}_\theta\exp : (T_{p_0}\mathcal{M}, \tau^{-1}g(0, p_0)) \to (\mathcal{M}, \tau^{-1}g(\tau\theta))$ where $\sqrt{\tau}V \to \gamma_V^g(\tau\theta)$.

Hence the Jacobian of the above two maps are related by (8.40). □

4.2. The blowdown limit of $g_{\tau_i}(\theta)$. Recall by Lemma 7.64 that the estimate

(8.43) $$|\nabla \ell(q, \tau)|^2 + R(q, \tau) \leq \frac{3\ell(q, \tau)}{\tau}$$

holds for the solution $g(\tau)$. We have the following consequence in regards to the space-time points (q_τ, τ).

LEMMA 8.35 (Estimates for ℓ and R in large neighborhoods of (q_τ, τ)). For any $\varepsilon > 0$ and $A > 1$, there exists $\delta > 0$ such that for any $\tau > 0$,

$$\ell(q, \tilde{\tau}) \leq \delta^{-1} \quad \text{and} \quad \tilde{\tau} R(q, \tilde{\tau}) \leq \delta^{-1}$$

for all $(q, \tilde{\tau}) \in B_{g(\tau)}\left(q_\tau, \sqrt{\varepsilon^{-1}\tau}\right) \times [A^{-1}\tau, A\tau]$.

4. BACKWARD LIMIT OF ANCIENT κ-SOLUTION IS A SHRINKER

PROOF. First we shall prove that there exists $\delta > 0$ such that for any $\tau > 0$,

$$\ell(q, \tilde{\tau}) \leq \frac{1}{3}\delta^{-1} \tag{8.44}$$

for all $(q, \tilde{\tau}) \in B_{g(\tau)}\left(q_\tau, \sqrt{\varepsilon^{-1}\tau}\right) \times [A^{-1}\tau, A\tau]$. From (8.43), since our ancient solution has $\mathrm{Rm} \geq 0$ (in particular, $R \geq 0$)

$$\left|\nabla\sqrt{\ell(q, \tilde{\tau})}\right|_{g(\tilde{\tau})} \leq \sqrt{\frac{3}{4\tilde{\tau}}}.$$

For any $q \in B_{g(\tau)}\left(q_\tau, \sqrt{\varepsilon^{-1}\tau}\right)$, let $\gamma(s)$ be a minimal normal geodesic from q_τ to q with respect to the metric $g(\tau)$. Since ℓ is locally Lipschitz, we have

$$\left|\sqrt{\ell(q, \tau)} - \sqrt{\ell(q_\tau, \tau)}\right| \leq \int_0^{d_{g(\tau)}(q, q(\tau))} \left|\nabla\sqrt{\ell(\gamma(s), \tau)}\right|_{g(\tau)} ds$$

$$\leq \sqrt{\frac{3}{4\tau}} \cdot \sqrt{\varepsilon^{-1}\tau} = \sqrt{\frac{3}{4\varepsilon}}.$$

Since $\ell(q_\tau, \tau) \leq \frac{n}{2}$, by the above estimate, we have

$$\ell(q, \tau) \leq \left(\sqrt{\frac{n}{2}} + \sqrt{\frac{3}{4\varepsilon}}\right)^2 \quad \text{for } q \in B_{g(\tau)}\left(q_\tau, \sqrt{\varepsilon^{-1}\tau}\right). \tag{8.45}$$

From Lemma 7.65, we have for $q \in \mathcal{M}$

$$A^{-2} \leq \frac{\ell(q, \tau)}{\ell(q, \tilde{\tau})} \quad \text{if } \tilde{\tau} \in [A^{-1}\tau, \tau],$$

$$\frac{\ell(q, \tilde{\tau})}{\ell(q, \tau)} \leq A^2 \quad \text{if } \tilde{\tau} \in [\tau, A\tau].$$

(Note that the inequalities in the above two lines have the same form.) Hence for $q \in B_{g(\tau)}\left(q_\tau, \sqrt{\varepsilon^{-1}\tau}\right)$ and $\tilde{\tau} \in [A^{-1}\tau, A\tau]$,

$$\ell(q, \tilde{\tau}) \leq A^2 \left(\sqrt{\frac{n}{2}} + \sqrt{\frac{3}{4\varepsilon}}\right)^2.$$

Now (8.44) is proved by choosing

$$\delta^{-1} = 3A^2 \left(\sqrt{\frac{n}{2}} + \sqrt{\frac{3}{4\varepsilon}}\right)^2.$$

Now $\tilde{\tau}R(q, \tilde{\tau}) \leq \delta^{-1}$ follows directly from (8.43) and (8.44); the lemma is proved. $\square$

For any sequence $\tau_i \to \infty$, consider the sequence of solutions

$$(\mathcal{M}^n, g_{\tau_i}(\theta), q_{\tau_i}), \quad \theta \in [A^{-1}, A].$$

For any $\varepsilon > 0$, Lemma 8.35, after parabolic rescaling $g(\tau)$ by τ_i, yields the curvature bound $\delta^{-1} = \delta(n, \varepsilon, A)^{-1}$ for $g_{\tau_i}(\theta)$ on $B_{g_{\tau_i}(1)}\left(q_{\tau_i}, \sqrt{\varepsilon^{-1}}\right) \times [A^{-1}, A]$. Applying Lemma 8.35 with $\varepsilon = 1$ and $A = 2$, we obtain

$$\left|\operatorname{Rm}_{g_{\tau_i}}(q, 1)\right| \leq \delta^{-1} = \delta(n, 1, 2)^{-1} \quad \text{for } q \in B_{g_{\tau_i}(1)}(q_{\tau_i}, 1).$$

Since $g(\theta)$ is κ-noncollapsed on all scales, we have $g_{\tau_i}(\theta)$ is κ-noncollapsed on $B_{g_{\tau_i}(1)}(q_{\tau_i}, 1)$ and the injectivity radius estimate $\operatorname{inj}_{g_{\tau_i}(1)}(q_{\tau_i}) \geq \delta_1(n, \kappa)$. Now we can apply Hamilton's Cheeger–Gromov-type Compactness Theorem 3.10 to the sequence of solutions $g_{\tau_i}(\theta)$ to the backward Ricci flow to get

$$(\mathcal{M}^n, g_{\tau_i}(\theta), q_{\tau_i}) \longrightarrow (\mathcal{M}^n_\infty, g_\infty(\theta), q_\infty) \quad \text{for } \theta \in [A^{-1}, A].$$

The limit is a complete solution to the backward Ricci flow. Since each $g_{\tau_i}(\theta)$ satisfies the trace Harnack estimate, the limit $g_\infty(\theta)$ satisfies the trace Harnack estimate. Note that $g_\infty(\theta)$ is κ-noncollapsed on all scales, has nonnegative curvature operator, and satisfies $\operatorname{inj}_{g_\infty(1)}(q_\infty) \geq \delta_1(n, \kappa)$.

To finish the proof of Theorem 8.32, we need to show that for each θ, $g_\infty(\theta)$ is a nonflat shrinking gradient Ricci soliton.

4.3. The limit of reduced length $\ell_i(q, \theta)$. Let $\ell_i(q, \theta) \doteq \ell^{g_{\tau_i}}(q, \theta)$, $\theta \in [A^{-1}, A]$, denote the reduced distance of the solution $g_{\tau_i}(\theta)$ with respect to the basepoint p_0.

LEMMA 8.36 (Limit of the reduced distance).

(i) *The limit*

$$\lim_{i \to \infty} \ell_i(q, \theta) \doteq \ell_\infty(q, \theta)$$

exists in the Cheeger–Gromov sense on $\mathcal{M}_\infty \times [A^{-1}, A]$.[5]

(ii) *The limit $\ell_\infty(q, \theta)$ is a locally Lipschitz function on $\mathcal{M}_\infty \times [A^{-1}, A]$ and $\nabla_{g_\infty(\theta)} \ell_\infty(q, \theta)$ and $\frac{\partial \ell_\infty}{\partial \theta}(q, \theta)$ exist a.e. on $\mathcal{M}_\infty \times [A^{-1}, A]$.*

PROOF. (i) Suppose $\Phi_i : U_i \subset \mathcal{M}_\infty \to V_i \doteq \Phi_i(U_i) \subset \mathcal{M}$ are diffeomorphisms which yield the convergence $(\mathcal{M}^n, g_{\tau_i}(\theta), q_{\tau_i}) \longrightarrow (\mathcal{M}^n_\infty, g_\infty(\theta), q_\infty)$ in the sense of Definition 3.6. Given any $\varepsilon > 0$, by Lemma 8.35, (8.43), (7.100), and the scaling property (8.40), we obtain for all i, $\theta \in [A^{-1}, A]$, and $q \in B_{g_{\tau_i}(1)}(q_{\tau_i}, \sqrt{\varepsilon^{-1}}) \subset \mathcal{M}$, the estimates

(8.46)
$$0 \leq \ell_i(q, \theta) \leq \delta^{-1},$$
$$\left|\nabla_{g_{\tau_i}(\theta)} \ell_i(q, \theta)\right|^2_{g_{\tau_i}(\theta)} \leq 3\delta^{-1},$$
$$\left|\frac{\partial \ell_i}{\partial \theta}(q, \theta)\right| \leq 2\delta^{-1}.$$

Hence for i large enough, $\ell_i(\Phi_i(\cdot), \cdot)$ is a sequence of uniformly Lipschitz functions on $B_{g_\infty(1)}\left(q_\infty, \frac{9}{10}\sqrt{\varepsilon^{-1}}\right) \subset \mathcal{M}_\infty$. By the Arzela–Ascoli theorem we get $\ell_i(\Phi_i(q), \theta) \to \ell_\infty(q, \theta)$ on the closed set $\overline{B}_{g_\infty(1)}\left(q_\infty, \frac{4}{5}\sqrt{\varepsilon^{-1}}\right) \times$

---

[5]See the proof below for what we mean by convergence in the Cheeger–Gromov sense.

$[A^{-1}, A]$ for some subsequence. Since ε is arbitrary, (i) then follows from a diagonalization argument. The convergence of the pulled-back reduced distance functions $\ell_i(\Phi_i(q), \theta)$ is what we mean by $\ell_i(q,\theta) \to \ell_\infty(q,\theta)$ on $\mathcal{M}_\infty \times [A^{-1}, A]$ in the Cheeger–Gromov sense.

(ii) It is clear that $\ell_\infty(q, \theta)$ is a locally Lipschitz function on $\mathcal{M}_\infty \times [A^{-1}, A]$. By Rademacher's Theorem (Lemma 7.110) for locally Lipschitz functions, we know $\nabla_{g_\infty(\theta)} \ell_\infty(q,\theta)$ and $\frac{\partial \ell_\infty}{\partial \theta}(q,\theta)$ exist a.e. on $\mathcal{M}_\infty \times [A^{-1}, A]$. □

In the next lemma we show that the equality $2\frac{\partial \ell}{\partial \tau} = R - |\nabla \ell|^2 - \frac{\ell}{\tau}$ is preserved under the limit.

LEMMA 8.37 (Properties of the limit of the reduced distance functions).

(i) We have

(8.47) $$2\frac{\partial \ell_\infty}{\partial \theta} + |\nabla_{g_\infty} \ell_\infty|^2 - R_{g_\infty} + \frac{\ell_\infty}{\theta} \equiv 0.$$

(ii) For any smooth compactly supported $\varphi(q, \theta) \geq 0$ on $\mathcal{M}_\infty \times (A^{-1}, A)$, we have

(8.48) $$\int_{A^{-1}}^{A} \int_{\mathcal{M}_\infty} \left(\begin{array}{c} \frac{\partial \ell_\infty}{\partial \theta} + \nabla_{g_\infty} \ell_\infty \cdot \nabla_{g_\infty} \varphi \\ -R_{g_\infty} + \frac{n}{2\theta} \end{array} \right) e^{-\ell_\infty(q,\theta)} \varphi d\mu_{g_\infty(\theta)}(q) d\theta \geq 0.$$

PROOF. (i) Equation (7.94) tells us that $2\frac{\partial \ell}{\partial \tau} + |\nabla \ell|^2 - R + \frac{\ell}{\tau} = 0$. By scaling, we have

$$2\frac{\partial \ell_i}{\partial \theta} + |\nabla_{g_{\tau_i}(\theta)} \ell_i|^2 - R_{g_{\tau_i}(\theta)} + \frac{\ell_i}{\theta} = 0.$$

It suffices to prove that

$$\frac{\partial \ell_i}{\partial \theta}(\Phi_i(q), \theta) \to \frac{\partial \ell_\infty}{\partial \theta}(q, \theta) \quad \text{and} \quad |\nabla_{g_{\tau_i}(\theta)} \ell_i|^2 (\Phi_i(q), \theta) \to |\nabla_{g_\infty} \ell_\infty|^2 (q, \theta)$$

a.e. on $\mathcal{M}_\infty \times (A^{-1}, A)$. Applying Lemma 7.63 (with $T = \infty$) to $\ell_i(q, \theta) = \ell(q, \tau_i \theta)$ and using the scale-invariance of $(\text{Hess}_{(q,\tau)} \ell)(Y, Y)$, we get for $\theta \in [A^{-1}, A]$ that

$$(\text{Hess}_{(q,\theta)} \ell_i)(Y(\theta), Y(\theta)) \leq \left(\frac{n\ell_i(q,\theta)}{\theta} + \frac{1}{2\theta} \right) |Y(\theta)|^2_{g_{\tau_i}(\theta)}$$

$$\leq A \left(n\delta^{-1} + \frac{1}{2} \right) |Y(\theta)|^2_{g_{\tau_i}(\theta)}.$$

Since $g_{\tau_i}(\Phi_i(q), \theta) \to g_\infty(q, \theta)$ in the C^∞-norm on any compact subset $K \subset \mathcal{M}_\infty \times (A^{-1}, A)$, the Hessian $\text{Hess}_{g_\infty(\theta)} \ell_i(\Phi_i(q), \theta)$ on K is uniformly bounded. From the discussion in Section 9 of Chapter 7, for any $q \in \mathcal{M}_\infty$, there exists a neighborhood $B_{g_\infty(\theta)}\left(q, \frac{1}{4}\mathrm{inj}_{g_\infty(\theta)}(q)\right)$ and a smooth function F on K such that for each $\theta \in (A^{-1}, A)$ the function

$$\varsigma_i(x, \theta) \doteq F(x, \theta) - \ell_i\left(\Phi_i \circ \exp_{g_\infty(1,q)}(x), \theta\right)$$

is convex on $B_{g_\infty(1,q)}\left(0, \frac{1}{16}\mathrm{inj}_{g_\infty(1)}(q)\right) \subset (T_q\mathcal{M}_\infty, g_\infty(1,q))$. Since $\ell_i(\cdot, \theta)$ is differentiable almost everywhere,

$$\nabla_{g_\infty(1,q)}\varsigma_i(x, \theta) \to \nabla_{g_\infty(1,q)}\left(F(x, \theta) - \ell_\infty\left(\exp_{g_\infty(1,q)}(x), \theta\right)\right)$$

a.e. on $B_{g_\infty(1,q)}\left(0, \frac{1}{16}\mathrm{inj}_{g_\infty(1)}(q)\right)$ by Theorem D6.2.7 in [**202**]. Hence

$$\nabla_{g_\infty(1,q)}\ell_i\left(\Phi_i \circ \exp_{g_\infty(1,q)}(x), \theta\right) \text{ converges to } \nabla_{g_\infty(1,q)}\ell_\infty\left(\exp_{g_\infty(1,q)}(x), \theta\right)$$

a.e. on $B_{g_\infty(1,q)}\left(0, \frac{1}{16}\mathrm{inj}_{g_\infty(1)}(q)\right)$ and $\nabla_{g_{\tau_i}(\theta)}\ell_i\left(\Phi_i \circ \exp_{g_\infty(1,q)}(x), \theta\right)$ converges to $\nabla_{g_\infty(\theta)}\ell_\infty\left(\exp_{g_\infty(1,q)}(x), \theta\right)$ a.e. on $B_{g_\infty(1,q)}\left(0, \frac{1}{16}\mathrm{inj}_{g_\infty(1)}(q)\right)$. Because q is an arbitrarily point on $\mathcal{M}_\infty$, we have proved that

(8.49) $$\left|\nabla_{g_{\tau_i}}\ell_i\right|^2_{g_{\tau_i}(\theta)}(\Phi_i(\cdot), \theta) \to \left|\nabla_{g_\infty}\ell_\infty\right|^2_{g_\infty(\theta)}(\cdot, \theta)$$

a.e. on $\mathcal{M}_\infty$ for each $\theta \in (A^{-1}, A)$.

From $2\frac{\partial \ell_i}{\partial \theta} + |\nabla_{g_{\tau_i}(\theta)}\ell_i|^2 - R_{g_{\tau_i}(\theta)} + \frac{\ell_i}{\theta} = 0$ and (8.49) we know $\frac{\partial \ell_i}{\partial \theta}(\Phi_i(\cdot), \theta)$ converges a.e. on $\mathcal{M}_\infty$ for each $\theta \in (A^{-1}, A)$. Since $\ell_i(\Phi_i(\cdot), \theta)$ converges to $\ell_\infty(\cdot, \theta)$ uniformly, we conclude

$$\frac{\partial \ell_i}{\partial \theta}(\Phi_i(\cdot), \theta) \to \frac{\partial \ell_\infty}{\partial \theta}(\cdot, \theta)$$

a.e. on $\mathcal{M}_\infty$ for each $\theta \in (A^{-1}, A)$.

(ii) For any smooth compactly supported function $\varphi(q, \theta) \geq 0$ on $\mathcal{M}_\infty \times (A^{-1}, A)$, for i large enough we can extend $\varphi\left(\Phi_i^{-1}(q_1), \theta\right)$ by 0 to a smooth function, still denoted by $\varphi\left(\Phi_i^{-1}(q_1), \theta\right)$, which has compact support on $\mathcal{M} \times (A^{-1}, A)$. Using the Lipschitz test function $e^{-\ell_i(q_1, \theta)}\varphi\left(\Phi_i^{-1}(q_1), \theta\right)$ in (7.146), we get

$$\int_{A^{-1}}^{A}\int_{\mathcal{M}}\left(\frac{\partial \ell_i}{\partial \theta} + \nabla_{g_{\tau_i}}\ell_i \cdot \nabla_{g_{\tau_i}}\varphi - R_{g_{\tau_i}} + \frac{n}{2\theta}\right)$$
$$\times e^{-\ell_i(q_1, \theta)}\varphi\left(\Phi_i^{-1}(q_1), \theta\right)d\mu_{g_{\tau_i}(\theta)}(q_1)\,d\theta \geq 0$$

and

$$\int_{A^{-1}}^{A}\int_{\mathcal{M}}\left(\frac{\partial \ell_i}{\partial \theta} + \nabla_{g_{\tau_i}}\ell_i \cdot \nabla_{g_{\tau_i}}\varphi - R_{g_{\tau_i}} + \frac{n}{2\theta}\right)$$
$$\times e^{-\ell_i(\Phi_i(q), \theta)}\varphi(q, \theta)\,d\mu_{\Phi_i^* g_{\tau_i}(\theta)}(q)\,d\theta \geq 0.$$

Taking the limit $i \to \infty$, we obtain (8.48). The lemma is proved. $\square$

4. BACKWARD LIMIT OF ANCIENT κ-SOLUTION IS A SHRINKER

4.4. The limit of the reduced volume. Note that the limit $g_\infty(\theta)$ is defined for $\theta \in [A^{-1}, A]$. Instead of considering the reduced volume of this limit, we define the function

$$\hat{V}_\infty(\theta) \doteqdot \int_{\mathcal{M}_\infty} (4\pi\theta)^{-n/2} e^{-\ell_\infty(q,\theta)} d\mu_{g_\infty(\theta)}(q) \quad \text{for } \theta \in [A^{-1}, A],$$

which will play the role of the reduced volume; formally this is the reduced volume using the limit function ℓ_∞. The fact that $\hat{V}_\infty(\theta)$ is finite follows from the following lemma.

LEMMA 8.38.

(i) We have
$$\hat{V}_\infty(\theta) \equiv \lim_{\tau \to \infty} \tilde{V}(\tau).$$

(ii) $\hat{V}_\infty(\theta)$ is a constant contained in $(0, 1)$.

(iii) For any $\psi(\theta)$ which has compact support in (A^{-1}, A),

(8.50)
$$\int_{A^{-1}}^{A} \int_{\mathcal{M}_\infty} (4\pi\theta)^{-n/2} e^{-\ell_\infty(q,\theta)} \psi'(\theta) d\mu_{g_\infty(\theta)}(q) d\theta$$
$$= \int_{A^{-1}}^{A} \hat{V}_\infty(\theta) \psi'(\theta) d\theta = 0,$$

where $\psi'(\theta) = d\psi/d\theta$.

PROOF. (i) Let τ_i be a subsequence such that both g_{τ_i} and $\ell_i(q, \theta)$ converge. By (8.23), we have

$$(4\pi\tau_i\theta)^{-n/2} e^{-\ell(\gamma_V(\tau_i\theta), \tau_i\theta)} \mathcal{L} \mathbf{J}_V(\tau_i\theta) \leq (4\pi)^{-n/2} e^{-|V|^2_{g(0)}}.$$

Then by Lebesgue's dominated convergence theorem and (8.40), we have

$$\lim_{i \to \infty} \tilde{V}(\tau_i \theta) = \int_{\mathbb{R}^n} \lim_{i \to \infty} \left((4\pi\tau_i\theta)^{-n/2} e^{-\ell(\gamma_V(\tau_i\theta), \tau_i\theta)} \mathcal{L} \mathbf{J}_V(\tau_i\theta) d\mu_{g(0)}(V) \right)$$
$$= \int_{\mathcal{M}_\infty} (4\pi\theta)^{-n/2} \lim_{i \to \infty} \left(\tau_i^{-n/2} e^{-\ell_i(\gamma_{\sqrt{\tau_i}V}(\theta), \theta)} d\mu_{g(\tau_i\theta)} \right)$$
$$= \int_{\mathbb{R}^n} (4\pi\theta)^{-n/2} e^{-\ell_\infty(q,\theta)} \lim_{i \to \infty} d\mu_{g_{\tau_i}(\theta)}$$
$$= \int_{\mathcal{M}_\infty} (4\pi\theta)^{-n/2} e^{-\ell_\infty(q,\theta)} d\mu_{g_\infty(\theta)}(q) = \hat{V}_\infty(\theta).$$

Since $\tilde{V}(\tau)$ is a monotone decreasing function, we have

(8.51)
$$\hat{V}_\infty(\theta) = \lim_{\tau \to \infty} \tilde{V}(\tau) \doteqdot \tilde{V}(\infty).$$

In particular, $\hat{V}_\infty(\theta)$ is independent of θ.

(ii) Note that $\tilde{V}(\infty) < 1$ follows from Corollary 8.17(iii). To see $\tilde{V}(\infty) > 0$, we compute using (8.46) and $\theta = 1$ that

$$\tilde{V}(\tau_i) = \int_{\mathcal{M}} (4\pi)^{-n/2} e^{-\ell_i(q,1)} d\mu_{g_{\tau_i}(1)}(q)$$

$$\geq \int_{B_{g_{\tau_i}(1)}(q_{\tau_i}, \varepsilon^{-1/2})} (4\pi)^{-n/2} e^{-\ell_i(q,1)} d\mu_{g_{\tau_i}(1)}(q)$$

$$\geq (4\pi A)^{-n/2} e^{-\delta^{-1}} \operatorname{Vol}_{g_{\tau_i}(1)} B_{g_{\tau_i}(1)} \left(q_{\tau_i}, \varepsilon^{-1/2} \right)$$

$$= (4\pi A)^{-n/2} e^{-\delta^{-1}} \cdot \tau_i^{-n/2} \operatorname{Vol}_{g(\tau_i)} B_{g(\tau_i)} \left(q_{\tau_i}, \tau_i^{1/2} \varepsilon^{-1/2} \right).$$

By Lemma 8.35, we have $R(q, \tau_i) \leq \frac{\delta^{-1}}{\tau_i}$ on the ball $B_{g(\tau_i)}(q_{\tau_i}, \tau_i^{1/2} \varepsilon^{-1/2})$. It follows from $g(\tau)$ being κ-noncollapsed on all scales (choosing the scale $r = \min\left\{ \tau_i^{1/2} \varepsilon^{-1/2}, \tau_i^{1/2} \delta^{1/2} \right\}$) that

$$\tau_i^{-n/2} \operatorname{Vol}_{g(\tau_i)} B_{g(\tau_i)} \left(q_{\tau_i}, \tau_i^{1/2} \varepsilon^{-1/2} \right) \geq \kappa \cdot \left(\min\left\{ \varepsilon^{-1/2}, \delta^{1/2} \right\} \right)^n.$$

Hence

$$\tilde{V}(\tau_i) \geq (4\pi A)^{-n/2} e^{-\delta^{-1}} \kappa \cdot \left(\min\left\{ \varepsilon^{-1/2}, \delta^{1/2} \right\} \right)^n$$

and $\tilde{V}(\infty) > 0$.

(iii) For any $\psi(\theta)$ which has compact support in (A^{-1}, A), we compute

$$\int_{A^{-1}}^{A} \int_{\mathcal{M}_\infty} (4\pi\theta)^{-n/2} e^{-\ell_\infty(q,\theta)} \psi'(\theta) d\mu_{g_\infty(\theta)}(q) d\theta$$

$$= \int_{A^{-1}}^{A} \hat{V}_\infty(\theta) \psi'(\theta) d\theta = \hat{V}_\infty(\theta) \int_{A^{-1}}^{A} \psi'(\theta) d\theta = 0.$$

□

4.5. The limit is a shrinking gradient soliton. Let $\psi(\theta) \geq 0$ be a smooth function only of θ with compact support in (A^{-1}, A). Applying Stokes's theorem for Lipschitz functions, we get

$$\int_{A^{-1}}^{A} \int_{\mathcal{M}_\infty} (4\pi\theta)^{-\frac{n}{2}} \left(\frac{\partial \ell_\infty}{\partial \theta} - R_{g_\infty} + \frac{n}{2\theta} \right) e^{-\ell_\infty(q,\theta)} \psi(\theta) d\mu_{g_\infty(\theta)}(q) d\theta$$

$$= \int_{A^{-1}}^{A} \int_{M} (4\pi\theta)^{-\frac{n}{2}} e^{-\ell_\infty(q,\theta)} \psi'(\theta) d\mu_{g_\infty(\theta)}(q) d\theta$$

$$= 0,$$

where we have used (8.50). Hence, for $\varphi_1(q, \theta) = (4\pi\theta)^{-\frac{n}{2}} \psi(\theta)$,

$$\int_{A^{-1}}^{A} \int_{M} \left(\frac{\partial \ell_\infty}{\partial \theta} + \nabla_{g_\infty} \ell_\infty \cdot \nabla_{g_\infty} \varphi - R_{g_\infty} + \frac{n}{2\theta} \right) e^{-\ell_\infty(q,\theta)} \varphi_1(q, \theta) d\mu_{g_\infty(\theta)}(q) d\theta$$

$$= 0,$$

where $\psi(\theta) \geq 0$ is an arbitrarily smooth function with compact support in (A^{-1}, A).

For any smooth compactly supported $\varphi_2(q, \theta) \geq 0$ on $\mathcal{M}_\infty \times (A^{-1}, A)$, we can choose $\psi(\theta)$ such that $\varphi_1(q, \theta) \geq \varphi_2(q, \theta)$. Plugging $\varphi(q, \theta) \doteqdot \varphi_1(q, \theta) - \varphi_2(q, \theta) \geq 0$ into (8.48), we get

$$0 = \int_{A^{-1}}^{A} \int_{\mathcal{M}_\infty} \left(\frac{\partial \ell_\infty}{\partial \theta} - R_{g_\infty} + \frac{n}{2\theta} \right) e^{-\ell_\infty(q,\theta)} \varphi_1(q, \theta) \, d\mu_{g_\infty(\theta)}(q) \, d\theta$$

$$\geq \int_{A^{-1}}^{A} \int_{\mathcal{M}_\infty} \left(\frac{\partial \ell_\infty}{\partial \theta} - R_{g_\infty} + \frac{n}{2\theta} \right) e^{-\ell_\infty(q,\theta)} \varphi_2(q, \theta) \, d\mu_{g_\infty(\theta)}(q) \, d\theta.$$

Note that $\varphi(q, \theta)$ does not have compact support in $\mathcal{M}_\infty \times (A^{-1}, A)$ if $\mathcal{M}_\infty$ is noncompact. In this case, we choose a partition of unity $\psi_\alpha(q)$ on $\mathcal{M}_\infty$ and use the test functions $\varphi(q, \theta) \psi_\alpha(q)$ to get a sequence of inequalities; the inequality above follows from summing the sequence of inequalities. It also follows from (8.48) that

$$\int_{A^{-1}}^{A} \int_{\mathcal{M}_\infty} \left(\frac{\partial \ell_\infty}{\partial \theta} - R_{g_\infty} + \frac{n}{2\theta} \right) e^{-\ell_\infty(q,\theta)} \varphi_2(q, \theta) \, d\mu_{g_\infty(\theta)}(q) \, d\theta \geq 0.$$

Hence

$$\int_{A^{-1}}^{A} \int_{\mathcal{M}_\infty} \left(\frac{\partial \ell_\infty}{\partial \theta} - R_{g_\infty} + \frac{n}{2\theta} \right) e^{-\ell_\infty(q,\theta)} \varphi_2(q, \theta) \, d\mu_{g_\infty(\theta)}(q) \, d\theta = 0$$

for any smooth compactly supported $\varphi_2(q, \theta) \geq 0$ on $\mathcal{M}_\infty \times (A^{-1}, A)$. This implies that the above equation holds for any smooth compactly supported $\varphi_2(q, \theta)$ on $\mathcal{M}_\infty \times (A^{-1}, A)$, so ℓ_∞ is a weak solution of the parabolic equation

(8.52) $$\frac{\partial \ell_\infty}{\partial \theta} - \Delta_{g_\infty} \ell_\infty + |\nabla_{g_\infty} \ell_\infty|^2 - R_{g_\infty} + \frac{n}{2\theta} = 0.$$

Hence $\ell_\infty(q, \theta)$ is a smooth function by standard regularity theory for parabolic PDE (see G. Lieberman [255], Chapters 5 and 6) and $\ell_\infty(q, \theta)$ satisfies (8.52) in the classical sense.

Let $u_\infty \doteqdot (4\pi\theta)^{-\frac{n}{2}} e^{-\ell_\infty}$, $\square^* \doteqdot \frac{\partial}{\partial \theta} - \Delta + R_{g_\infty}$, and

$$v_\infty \doteqdot \left(\theta(2\Delta \ell_\infty - |\nabla \ell_\infty|^2 + R_{g_\infty}) + \ell_\infty - n \right) u_\infty.$$

Equation (8.52) implies $\square^* u_\infty = 0$. Hence we can apply the same calculation as used to obtain (6.22) to get

(8.53) $$\square^* v_\infty = -2\theta \left| \operatorname{Rc}(g_\infty)_{ij} + \nabla_i \nabla_j \ell_\infty - \frac{1}{2\theta} (g_\infty)_{ij} \right|^2 u_\infty.$$

It follows from (8.47) and (8.52) that $v_\infty \equiv 0$. Thus $\square^* v_\infty \equiv 0$ and hence

(8.54) $$\operatorname{Rc}(g_\infty)_{ij} + \nabla_i \nabla_j \ell_\infty - \frac{1}{2\theta} (g_\infty)_{ij} \equiv 0;$$

i.e., $g_\infty(\theta)$ is a shrinking gradient Ricci soliton on $\mathcal{M}_\infty$ for $\theta \in (A^{-1}, A)$.

4.6. The limit is nonflat. We argue by contradiction. If $g_\infty(\theta)$ is flat for some θ_0, then $R_{g_\infty(\theta_0)} \equiv 0$ and we get from (8.54) that

$$\nabla_i \nabla_j \ell_\infty(q, \theta_0) = \frac{1}{2\theta_0} (g_\infty)_{ij}(\theta_0). \tag{8.55}$$

Taking the trace of the above equation, we get $\Delta_{g_\infty(\theta_0)} \ell_\infty(q, \theta_0) = \frac{n}{2\theta}$. Plugging this into

$$\theta(2\Delta \ell_\infty - |\nabla \ell_\infty|^2 + R_{g_\infty}) + \ell_\infty - n \equiv 0,$$

we obtain

$$|\nabla_{g_\infty(\theta_0)} \ell_\infty|^2_{g_\infty(\theta_0)}(q, \theta_0) = \frac{\ell_\infty(q, \theta_0)}{\theta_0}. \tag{8.56}$$

It follows from Lemma 7.59 that (one estimates ℓ_i and takes the limit as $i \to \infty$)

$$\ell_\infty(q, \theta_0) \geq \frac{1}{4\theta_0 e^{2C_0 \theta_0}} d^2_{g_\infty(0)}(p_0, q) - \frac{nC_0 \theta_0}{3}$$

so that $\ell_\infty(q, \theta_0)$ must have a minimum point $p_1 \in \mathcal{M}_\infty$. On the other hand, (8.55) implies that $\ell_\infty(q, \theta_0)$ is a strictly convex function on $(\mathcal{M}_\infty, g_\infty(\theta_0))$. It is well known that a strictly convex function has at most one critical point.[6] Hence p_1 is the only critical point of $\ell_\infty(\cdot, \theta_0)$ on $\mathcal{M}_\infty$, so that $\mathcal{M}_\infty^n$ is diffeomorphic to $\mathbb{R}^n$. Since, by assumption, $(\mathcal{M}_\infty, g_\infty(\theta_0))$ is a flat manifold, it is then isometric to $(\mathbb{R}^n, g_\mathbb{E})$ under Φ. We can choose Φ so that $\Phi(p_1) = \vec{0}$ and global coordinates $\{x^i\}$ such that $g_\infty(\theta_0) = (dx^1)^2 + \cdots + (dx^n)^2$. Then (8.55) implies

$$\ell_\infty(x, \theta_0) = \frac{1}{4\theta_0}(x^1)^2 + \cdots + \frac{1}{4\theta_0}(x^n)^2 + c_1 x^1 + \cdots + c_n x^n + c_{n+1}.$$

Equation (8.56) implies $(c_1)^2 + \cdots + (c_n)^2 = \theta_0^{-1} c_{n+1}$ and hence

$$\ell_\infty(x, 1) = \frac{1}{4\theta_0}(x^1 + 2\theta_0 c_1)^2 + \cdots + \frac{1}{4\theta_0}(x^n + 2\theta_0 c_n)^2.$$

From the definition of $\hat{V}_\infty(\theta) = \int_{\mathcal{M}_\infty} (4\pi\theta)^{-\frac{n}{2}} e^{-\bar{\ell}_\infty(q,\theta)} d\mu_{g_\infty(\theta)}(q)$, we compute $\hat{V}_\infty(\theta_0) = 1$ by using the formulas for $\ell_\infty(x, \theta_0)$ and $g_\infty(\theta_0)$. This contradicts $\hat{V}_\infty(\theta_0) = \tilde{V}(\infty) < 1$ in (8.51). Hence $g_\infty(\theta)$ is not flat.

PROPOSITION 8.39. *In dimension 3 this shrinking gradient soliton limit has bounded sectional curvature.*

---

[6]Suppose $\ell_\infty(q, \theta_0)$ has two different critical points p and p_1. Let $\gamma(s)$, $0 \leq s \leq s_0$, be a minimal geodesic, with respect to $g_\infty(\theta_0)$, from p to p_1. We have by (8.55) that

$$0 = \frac{d}{ds} \ell_\infty(\gamma(s), \theta_0)\Big|_{s=0}^{s=s_0} = \int_0^{s_0} \frac{d^2}{ds^2} \ell_\infty(\gamma(s), \theta_0) ds$$

$$= \int_0^{s_0} \nabla^2 \ell_\infty(\gamma'(s), \gamma'(s)) ds = \int_0^{s_0} \frac{1}{2\theta} |\gamma'(s)|^2_{\hat{g}_\infty(\theta_0)} ds > 0,$$

which is a contradiction.

PROOF. The sequence $(\mathcal{M}^3, g_{\tau_i}(\theta), q_{\tau_i})$, $\theta \geq 1$, is a sequence of ancient κ-solutions with bounded curvature for each i. The convergence $g_{\tau_i}(\theta) \to g_\infty(\theta)$ for $\theta \in [A^{-1}, A]$ implies that $R_{g_{\tau_i}}(q_{\tau_i}, 1) \to R_{g_\infty}(q_\infty, 1)$. Since the limit $g_\infty(\theta)$ is nonflat, by the strong maximum principle, $R_{g_\infty(1)}(q_\infty) > 0$. By the compactness theorem of the set of 3-dimensional ancient κ-solutions (see Part II of this volume), we know that a subsequence

$$\left(\mathcal{M}, R^{-1}_{g_{\tau_i}(1, q_{\tau_i})} g_{\tau_i}(\theta), q_{\tau_i} \right)$$

converges to a limit $\left(\widetilde{\mathcal{M}}_\infty, \tilde{g}_\infty(\theta), \tilde{q}_\infty \right)$. The limit has bounded sectional curvature. Using the definition of Cheeger–Gromov convergence, it is easy to check that $\left(\mathcal{M}, R^{-1}_{g_\infty(1, q_\infty)} g_{\tau_i}(\theta), q_{\tau_i} \right)$ converges to the limit $\left(\widetilde{\mathcal{M}}_\infty, \tilde{g}_\infty(\theta), \tilde{q}_\infty \right)$. On the other hand it follows from Theorem 8.32 that

$$\left(\mathcal{M}, R^{-1}_{g_\infty(1, q_\infty)} g_{\tau_i}(\theta), q_{\tau_i} \right) \to \left(\mathcal{M}_\infty, R^{-1}_{g_\infty(1, q_\infty)} g_\infty(\theta), q_\infty \right).$$

Hence, by the uniqueness of the Cheeger–Gromov pointed limit, $g_\infty(\theta) = R_{g_\infty(1, q_\infty)} \tilde{g}_\infty(\theta)$ for $\theta \in [1, A]$, which has bounded curvature. We have proved that the shrinking gradient soliton $g_\infty(\theta)$ is complete and has nonflat bounded nonnegative curvature. □

5. Perelman's Riemannian formalism in potentially infinite dimensions

Here we discuss Perelman's potentially infinite Riemannian metric in more detail. We discuss the calculation of its curvature tensor, the Ricci flatness (modulo renormalization) of its metric, and a geometric interpretation of Perelman's entropy formula. Although this section does not belong to the linear flow of this chapter, it provides a unifying viewpoint for various components of this volume. In particular, we have the following.

(1) In Section 2.1 of Chapter 7 we considered the Riemannian metric $\tilde{g}$ on $\widetilde{\mathcal{M}} = \mathcal{M}^n \times \mathcal{S}^N \times (0, T)$ and showed that the renormalization of its Riemannian length is the $\mathcal{L}$-length.[7]

(2) In subsection 3.3 of Chapter 7 we showed that the $\mathcal{L}$-geodesic equation is the same (up to the time reparametrization $\sigma = 2\sqrt{\tau}$) as the geodesic equation of the space-time connection defined by (7.39)–(7.42), which is the limit, as $N \to \infty$, of the Levi-Civita connections of the Riemannian metrics $\tilde{g}$ defined by (7.11) (see Exercise 7.4). Thus the $\mathcal{L}$-geodesic equation is the limit, as $N \to \infty$, of the geodesic equations of the Riemannian metrics $\tilde{g}$.

(3) In subsection 2.1 of this chapter we showed how to obtain the reduced volume functional from a renormalization of the volumes of geodesic balls with respect to the metric $\tilde{g}$ on $\widetilde{\mathcal{M}}$.

---

[7]Here we have switched the notation from $\tilde{h}$ to $\tilde{g}$ and $\widetilde{\mathcal{N}}$ to $\widetilde{\mathcal{M}}$.

Recall from (7.11) that given a solution $(\mathcal{M}^n, g(\tau))$, $\tau \in (0, T)$, of the backward Ricci flow, Perelman [**297**] introduced the manifold $\widetilde{\mathcal{M}} = \mathcal{M} \times \mathcal{S}^N \times (0, T)$ with the following metric:

$$\tilde{g}_{ij} = g_{ij}, \quad \tilde{g}_{\alpha\beta} = \tau g_{\alpha\beta}, \quad \tilde{g}_{00} = \frac{N}{2\tau} + R, \quad \tilde{g}_{i\alpha} = \tilde{g}_{i0} = \tilde{g}_{\alpha 0} = 0,$$

i.e.,

$$\tilde{g} = g_{ij} dx^i dx^j + \tau g_{\alpha\beta} dy^\alpha dy^\beta + \left(R + \frac{N}{2\tau}\right) d\tau^2,$$

where the metric $g_{\alpha\beta}$ on $\mathcal{S}^N$ has constant sectional curvature $\frac{1}{2N}$.

5.1. Riemann curvature tensor of $(\widetilde{\mathcal{M}}, \tilde{g})$. We shall apply the following two steps to compute the Riemann curvature tensor of the manifold $(\widetilde{\mathcal{M}}, \tilde{g})$. First we treat $(\mathcal{M}, g)$ as a hypersurface in the manifold $\overline{\mathcal{M}} = \mathcal{M} \times (0, T)$ with the metric

$$\bar{g} = g_{ij} dx^i dx^j + \left(R + \frac{N}{2\tau}\right) d\tau^2.$$

We compute the curvature of the manifold $(\overline{\mathcal{M}}, \bar{g})$ using the Gauss equations and Koszul's formula. Secondly, we consider the manifold $(\widetilde{\mathcal{M}}, \tilde{g})$ as a warped product with base $(\overline{\mathcal{M}}, \bar{g})$ and fiber $\mathcal{S}^N$. We then use O'Neill's formulas to compute the curvature of $\tilde{g}$. This method of computation essentially follows Guofang Wei [**368**].

Let $\partial_i \doteqdot \frac{\partial}{\partial x^i}$ denote the coordinate vector fields on the $\mathcal{M}$ factor, let $\partial_\tau \doteqdot \frac{\partial}{\partial \tau}$, and let

$$\nu = \frac{1}{\left(R + \frac{N}{2\tau}\right)^{1/2}} \partial_\tau$$

be the unit normal vector field of $\mathcal{M} \times \{\tau\} \subset \overline{\mathcal{M}}$. Direct computation, using $[\partial_\tau, \partial_i] = 0$, gives

(8.57) $$[\nu, \partial_i] = \frac{\nabla_i R}{2\left(R + \frac{N}{2\tau}\right)} \nu.$$

By the formula for the evolution of the metric of a hypersurface evolving in the direction of its normal with speed $|\partial_\tau| = \left(R + \frac{N}{2\tau}\right)^{1/2}$, we have[8]

$$2R_{ij} = \partial_\tau g_{ij} = 2|\partial_\tau| \cdot \text{II}_{ij},$$

where II denotes the second fundamental form of $\mathcal{M} \times \{\tau\} \subset \overline{\mathcal{M}}$.[9] Therefore

$$\text{II}(\partial_i, \partial_j) = \text{II}_{ij} = \frac{1}{\left(R + \frac{N}{2\tau}\right)^{1/2}} R_{ij},$$

and the Levi-Civita connections of $\bar{g}$ and g are related by

---

[8]See the proof of (B.13) with the mean curvature H replaced by $|\partial_\tau|$.

[9]In other parts of this volume we have sometimes used h instead of II to denote the second fundamental form.

$$\bar{\nabla}_{\partial_i}\partial_j = \nabla_{\partial_i}\partial_j - \mathrm{II}_{ij}\nu = \nabla_{\partial_i}\partial_j - \frac{1}{\left(R+\frac{N}{2\tau}\right)^{-1/2}}R_{ij}\nu.$$

By the Gauss equations, we have

$$\langle \bar{R}(\partial_i,\partial_j)\partial_k,\partial_\ell\rangle$$
$$= \langle R(\partial_i,\partial_j)\partial_k,\partial_\ell\rangle - \langle \mathrm{II}(\partial_i,\partial_\ell),\mathrm{II}(\partial_j,\partial_k)\rangle + \langle \mathrm{II}(\partial_j,\partial_\ell),\mathrm{II}(\partial_i,\partial_k)\rangle$$
$$= \langle R(\partial_i,\partial_j)\partial_k,\partial_\ell\rangle - \frac{1}{\left(R+\frac{N}{2\tau}\right)}(R_{i\ell}R_{jk} - R_{j\ell}R_{ik}).$$

By Koszul's formula for the Levi-Civita connection of $\bar{g}$, i.e.,

$$2\langle \bar{\nabla}_X Y, Z\rangle = X\langle Y,Z\rangle + Y\langle X,Z\rangle - Z\langle X,Y\rangle$$
$$+ \langle [X,Y],Z\rangle - \langle [X,Z],Y\rangle - \langle [Y,Z],X\rangle,$$

where the inner products are with respect to $\bar{g}$, we have

(8.58) $$\bar{\nabla}_{\partial_i}\nu = \frac{1}{\left(R+\frac{N}{2\tau}\right)^{1/2}}\mathrm{Rc}\,(\partial_i),$$

(8.59) $$\bar{\nabla}_\nu \nu = -\frac{1}{2\left(R+\frac{N}{2\tau}\right)}\nabla R.$$

To derive the two formulas above, we used $\langle \bar{\nabla}_{\partial_i}\nu,\nu\rangle = 0$, $\langle \bar{\nabla}_\nu\nu,\nu\rangle = 0$, and

$$2\langle \bar{\nabla}_{\partial_i}\nu,\partial_k\rangle = \nu\langle \partial_i,\partial_k\rangle = \left(R+\frac{N}{2\tau}\right)^{-1/2}\partial_\tau g_{ik},$$

$$\langle \bar{\nabla}_\nu\nu,\partial_k\rangle = -\langle [\nu,\partial_k],\nu\rangle = -\frac{\nabla_k R}{2\left(R+\frac{N}{2\tau}\right)}$$

(these follow from Koszul's formula and (8.57)).

Applying another covariant derivative to (8.58) and (8.59), we have

$$\langle \bar{\nabla}_{\partial_i}\bar{\nabla}_\nu \nu,\partial_j\rangle = \frac{1}{2\left(R+\frac{N}{2\tau}\right)^2}\nabla_i R \nabla_j R - \frac{1}{2\left(R+\frac{N}{2\tau}\right)}\nabla_i\nabla_j R,$$

$$-\langle \bar{\nabla}_\nu\bar{\nabla}_{\partial_i}\nu,\partial_j\rangle = \frac{1}{2\left(R+\frac{N}{2\tau}\right)^2}\left(\partial_\tau R - \frac{N}{2\tau^2}\right)R_{ij} + \frac{1}{R+\frac{N}{2\tau}}(R_{i\ell}R_{\ell j} - \partial_\tau R_{ij}),$$

$$-\langle \bar{\nabla}_{[\partial_i,\nu]}\nu,\partial_j\rangle = -\frac{1}{4\left(R+\frac{N}{2\tau}\right)^2}\nabla_i R \nabla_j R,$$

where, to obtain the second formula, we used

$$\langle \bar{\nabla}_{\partial_\tau}(\mathrm{Rc}\,(\partial_i)),\partial_j\rangle = \partial_\tau R_{ij} - \langle \mathrm{Rc}\,(\partial_i),\bar{\nabla}_{\partial_\tau}\partial_j\rangle$$
$$= \partial_\tau R_{ij} - R_{i\ell}R_{\ell j},$$

(note that $\langle \bar{\nabla}_{\partial_\tau}\partial_j,\partial_k\rangle = \frac{1}{2}\partial_\tau \langle \partial_j,\partial_k\rangle = R_{jk}$), and the third formula follows from (8.57) and (8.59).

Hence the curvatures in the normal direction are given by

$$\langle \bar{R}(\partial_i, \nu)\nu, \partial_j\rangle = \frac{1}{2\left(R+\frac{N}{2\tau}\right)^2}\left[\left(\partial_\tau R - \frac{N}{2\tau^2}\right)R_{ij} + \frac{1}{2}\nabla_i R \nabla_j R\right]$$
$$+ \frac{1}{\left(R+\frac{N}{2\tau}\right)}\left(-\partial_\tau R_{ij} - \frac{1}{2}\nabla_i\nabla_j R + R_{i\ell}R_{\ell j}\right),$$
$$\langle \bar{R}(\partial_i, \partial_j)\partial_k, \nu\rangle = \langle \bar{\nabla}_{\partial_i}\bar{\nabla}_{\partial_j}\partial_k - \bar{\nabla}_{\partial_j}\bar{\nabla}_{\partial_i}\partial_k, \nu\rangle$$
$$= \frac{1}{2\left(R+\frac{N}{2\tau}\right)^{3/2}}\left(R_{jk}\nabla_i R - R_{ik}\nabla_j R\right)$$
$$- \frac{1}{\left(R+\frac{N}{2\tau}\right)^{1/2}}\left(\nabla_i R_{jk} - \nabla_j R_{ik}\right).$$

In terms of ∂_i and ∂_τ, this says
$$\langle \bar{R}(\partial_i, \partial_\tau)\partial_\tau, \partial_j\rangle = \frac{1}{2\left(R+\frac{N}{2\tau}\right)}\left[\left(\partial_\tau R - \frac{N}{2\tau^2}\right)R_{ij} + \frac{1}{2}\nabla_i R \cdot \nabla_j R\right]$$
$$+ \left(R_{i\ell}R_{\ell j} - \frac{1}{2}\nabla_i\nabla_j R - \partial_\tau R_{ij}\right),$$
$$\langle \bar{R}(\partial_i, \partial_j)\partial_k, \partial_\tau\rangle = \frac{1}{2\left(R+\frac{N}{2\tau}\right)}\left(R_{jk}\nabla_i R - R_{ik}\nabla_j R\right) - \left(\nabla_i R_{jk} - \nabla_j R_{ik}\right).$$

Note that the curvatures of $\bar{g}$ are the components of Hamilton's matrix Harnack expression in the following sense:
$$\langle \bar{R}(\partial_i, \partial_j)\partial_k, \partial_\ell\rangle \equiv R_{ijk\ell} \mod O(N^{-1}),$$
$$\langle \bar{R}(\partial_i, \partial_\tau)\partial_\tau, \partial_j\rangle \equiv -\partial_\tau R_{ij} - \frac{1}{2}\nabla_i\nabla_j R + R_{i\ell}R_{\ell j} - \frac{1}{2\tau}R_{ij} \mod O(N^{-1})$$
$$\equiv \Delta R_{ij} - \frac{1}{2}\nabla_i\nabla_j R + 2R_{ikj\ell}R_{k\ell}$$
$$- R_{i\ell}R_{\ell j} - \frac{1}{2\tau}R_{ij} \mod O(N^{-1}),$$
$$\langle \bar{R}(\partial_i, \partial_j)\partial_k, \partial_\tau\rangle \equiv \nabla_i R_{jk} - \nabla_j R_{ik} \mod O(N^{-1}).$$

Taking the trace gives the entries of the trace Harnack expression:
$$\overline{\mathrm{Rc}}(\partial_i, \partial_j) \equiv R_{ij} \mod O(N^{-1}),$$
$$\overline{\mathrm{Rc}}(\partial_i, \partial_\tau) \equiv \frac{1}{2}\nabla_j R \mod O(N^{-1}),$$
$$\overline{\mathrm{Rc}}(\partial_\tau, \partial_\tau) \equiv \frac{1}{2}\frac{\partial R}{\partial t} = \frac{1}{2}\Delta R + |\mathrm{Rc}|^2 \mod O(N^{-1}).$$

Now consider $(\widetilde{\mathcal{M}}, \tilde{g})$ as a warped product:
$$\widetilde{\mathcal{M}} = B \times_{f^2} F,$$
where the base manifold is $(B, g_B) = (\overline{\mathcal{M}}, \bar{g})$, the fiber is $(F, g_F) = (\mathcal{S}^N, g_{\alpha\beta})$, and $f = \sqrt{\tau}$. We use O'Neill's formula to compute the curvatures of $(\widetilde{\mathcal{M}}, \tilde{g})$.

5. PERELMAN'S FORMALISM IN POTENTIALLY INFINITE DIMENSIONS

LEMMA 8.40 (O'Neill). *Let X, Y, Z be vector fields on B and let U, V, W be vector fields on F. Then*

(1) $\tilde{R}(U,V)W = K(U,V)W - |G|_{\tilde{g}}^2 \left(g_F(U,W)V - g_F(V,W)U\right),$

(2) $\tilde{R}(X,V)Y = -\frac{1}{f}(\bar{\nabla}_X G, Y)V = -\frac{1}{f}\operatorname{Hess}_{\bar{g}}(f)(X,Y)V,$

(3) $\tilde{R}(X,Y)V = \tilde{R}(V,W)X = 0,$

(4) $\tilde{R}(X,V)W = \tilde{R}(X,W)V = f(V,W)\bar{\nabla}_X G,$

(5) $\tilde{R}(X,Y)Z$ *is the same on either B or $\widetilde{\mathcal{M}}$.*

Here, K denotes the curvature tensor of (F, g_F) and

$$G = \operatorname{grad} f = \frac{1}{2\sqrt{\tau}\left(R + \frac{N}{2\tau}\right)}\partial_\tau.$$

Let $\{\theta_\alpha\}_{\alpha=1}^N$ be a basis of tangent vectors on $\mathcal{S}^N$ and $g_{\alpha\beta} \doteqdot g(\theta_\alpha, \theta_\beta)$. By (3) we have

$$\tilde{R}(\partial_i, \partial_j)\theta_\alpha = 0, \quad \tilde{R}(\partial_i, \partial_\tau)\theta_\alpha = 0, \quad \tilde{R}(\theta_\alpha, \theta_\beta)\partial_i = 0, \quad \tilde{R}(\theta_\alpha, \theta_\beta)\partial_\tau = 0,$$

and hence

$$\left\langle \tilde{R}(\partial_i, \partial_j)\theta_\alpha, \partial_k \right\rangle = 0, \quad \left\langle \tilde{R}(\partial_i, \partial_j)\theta_\alpha, \partial_\tau \right\rangle = 0, \quad \left\langle \tilde{R}(\partial_i, \partial_j)\theta_\alpha, \theta_\beta \right\rangle = 0,$$
$$\left\langle \tilde{R}(\partial_i, \partial_\tau)\theta_\alpha, \partial_j \right\rangle = 0, \quad \left\langle \tilde{R}(\partial_i, \partial_\tau)\theta_\alpha, \partial_\tau \right\rangle = 0, \quad \left\langle \tilde{R}(\partial_i, \partial_\tau)\theta_\alpha, \theta_\beta \right\rangle = 0,$$
$$\left\langle \tilde{R}(\theta_\alpha, \theta_\beta)\partial_i, \theta_\gamma \right\rangle = 0, \quad \left\langle \tilde{R}(\theta_\alpha, \theta_\beta)\partial_\tau, \theta_\gamma \right\rangle = 0.$$

By (1) we have

$$\tilde{R}(\theta_\alpha, \theta_\beta)\theta_\gamma = K(\theta_\alpha, \theta_\beta)\theta_\gamma - |G|_{\tilde{g}}^2 [g(\theta_\alpha, \theta_\gamma)\theta_\beta - g(\theta_\beta, \theta_\gamma)\theta_\alpha],$$

and hence

$$\left\langle \tilde{R}(\theta_\alpha, \theta_\beta)\theta_\gamma, \theta_\delta \right\rangle_{\tilde{g}} = \tau \left\langle K(\theta_\alpha, \theta_\beta)\theta_\gamma, \theta_\delta \right\rangle_g$$
$$- \frac{1}{4\left(R + \frac{N}{2\tau}\right)}[g(\theta_\alpha, \theta_\gamma)g(\theta_\beta, \theta_\delta) - g(\theta_\beta, \theta_\gamma)g(\theta_\alpha, \theta_\delta)]$$
$$= \frac{\tau R}{2N\left(R + \frac{N}{2\tau}\right)}(g_{\beta\gamma}g_{\alpha\delta} - g_{\alpha\gamma}g_{\beta\delta})$$
$$\equiv 0 \mod O(N^{-1}),$$

since $|G|_{\tilde{g}}^2 = \frac{1}{4\tau\left(R + \frac{N}{2\tau}\right)}$, $\tilde{g}(\theta_\beta, \theta_\delta) = \tau g(\theta_\beta, \theta_\delta)$, and

$$\left\langle K(\theta_\alpha, \theta_\beta)\theta_\gamma, \theta_\delta \right\rangle_g = \frac{1}{2N}(g_{\beta\gamma}g_{\alpha\delta} - g_{\alpha\gamma}g_{\beta\delta}).$$

By (2) we have

$$\tilde{R}(\partial_i, \theta_\alpha)\partial_j = -\frac{1}{\sqrt{\tau}} \operatorname{Hess}_{\bar{g}}(\sqrt{\tau})(\partial_i, \partial_j)\theta_\alpha,$$

$$\tilde{R}(\partial_i, \theta_\alpha)\partial_\tau = -\frac{1}{\sqrt{\tau}} \operatorname{Hess}_{\bar{g}}(\sqrt{\tau})(\partial_i, \partial_\tau)\theta_\alpha,$$

$$\tilde{R}(\partial_\tau, \theta_\alpha)\partial_\tau = -\frac{1}{\sqrt{\tau}} \operatorname{Hess}_{\bar{g}}(\sqrt{\tau})(\partial_\tau, \partial_\tau)\theta_\alpha,$$

and hence

$$\left\langle \tilde{R}(\partial_i, \theta_\alpha)\partial_j, \theta_\beta \right\rangle = \frac{1}{2\left(R + \frac{N}{2\tau}\right)} R_{ij} g_{\alpha\beta}$$
$$\equiv 0 \mod O(N^{-1}),$$

$$\left\langle \tilde{R}(\partial_i, \theta_\alpha)\partial_\tau, \theta_\beta \right\rangle = -\frac{1}{4\left(R + \frac{N}{2\tau}\right)} g_{\alpha\beta} \nabla_i R$$
$$\equiv 0 \mod O(N^{-1}),$$

$$\left\langle \tilde{R}(\partial_\tau, \theta_\alpha)\partial_\tau, \theta_\beta \right\rangle = -\frac{1}{4\left(R + \frac{N}{2\tau}\right)} \left(\partial_\tau R + \frac{R}{\tau}\right) g_{\alpha\beta}$$
$$\equiv 0 \mod O(N^{-1}).$$

By (5) we have

$$\tilde{R}(\partial_i, \partial_j)\partial_k = \bar{R}(\partial_i, \partial_j)\partial_k,$$
$$\tilde{R}(\partial_i, \partial_\tau)\partial_\tau = \bar{R}(\partial_i, \partial_\tau)\partial_\tau,$$

and hence

$$\left\langle \tilde{R}(\partial_i, \partial_j)\partial_k, \partial_\ell \right\rangle = \left\langle \bar{R}(\partial_i, \partial_j)\partial_k, \partial_\ell \right\rangle$$
$$= \left\langle R(\partial_i, \partial_j)\partial_k, \partial_\ell \right\rangle - \frac{1}{R + \frac{N}{2\tau}}(R_{i\ell}R_{jk} - R_{j\ell}R_{ik})$$
$$\equiv \left\langle R(\partial_i, \partial_j)\partial_k, \partial_\ell \right\rangle \mod O(N^{-1}),$$

$$\left\langle \tilde{R}(\partial_i, \partial_j)\partial_k, \partial_\tau \right\rangle = \left\langle \bar{R}(\partial_i, \partial_j)\partial_k, \partial_\tau \right\rangle$$
$$= \frac{1}{2\left(R + \frac{N}{2\tau}\right)}(R_{jk}\nabla_i R - R_{ik}\nabla_j R) - (\nabla_i R_{jk} - \nabla_j R_{ik})$$
$$\equiv -(\nabla_i R_{jk} - \nabla_j R_{ik}) \mod O(N^{-1}),$$

5. PERELMAN'S FORMALISM IN POTENTIALLY INFINITE DIMENSIONS

$$\left\langle \tilde{R}(\partial_i, \partial_\tau)\partial_\tau, \partial_j \right\rangle = \left\langle \bar{R}(\partial_i, \partial_\tau)\partial_\tau, \partial_j \right\rangle$$
$$= \frac{1}{2\left(R + \frac{N}{2\tau}\right)} \left(\left(\partial_\tau R - \frac{N}{2\tau^2}\right) R_{ij} + \frac{1}{2}\nabla_i R \cdot \nabla_j R \right)$$
$$+ \left(R_{ij}^2 - \frac{1}{2}\nabla_i \nabla_j R - \partial_\tau R_{ij} \right)$$
$$\equiv -\frac{1}{2\tau} R_{ij} + \triangle_g R_{ij} + 2R_{ikj\ell} R_{k\ell}$$
$$- R_{ij}^2 - \frac{1}{2}\nabla_i \nabla_j R \mod O(N^{-1}).$$

The components of the curvature tensor of this metric coincide (modulo N^{-1}) with the components of Hamilton's matrix Harnack expression.

5.2. Ricci flatness of $(\widetilde{\mathcal{M}}, \tilde{g})$. Taking the trace of the above formulas for the curvature tensor yields

$$\widetilde{\text{Rc}}(\partial_i, \partial_j) = g^{k\ell} \left\langle \tilde{R}(\partial_i, \partial_k)\partial_\ell, \partial_j \right\rangle + \left\langle \tilde{R}(\partial_i, \nu)\nu, \partial_j \right\rangle$$
$$+ \frac{1}{\tau} g^{\alpha\beta} \left\langle \tilde{R}(\partial_i, \theta_\alpha)\theta_\beta, \partial_j \right\rangle$$
$$= \frac{1}{2\left(R + \frac{N}{2\tau}\right)^2} \left[\left(\partial_\tau R - \frac{N}{2\tau^2}\right) \text{Rc}(\partial_i, \partial_j) + \frac{1}{2}\nabla_i R \nabla_j R \right]$$
$$+ \frac{1}{\left(R + \frac{N}{2\tau}\right)} \left(-\partial_\tau R_{ij} - \frac{1}{2}\nabla_i \nabla_j R + 2R_{i\ell} R_{\ell j} \right)$$
$$\equiv 0 \mod O(N^{-1}),$$
$$\widetilde{\text{Rc}}(\partial_i, \partial_\tau) = g^{k\ell} \left\langle \tilde{R}(\partial_i, \partial_k)\partial_\ell, \partial_\tau \right\rangle + \left\langle \tilde{R}(\partial_i, \nu)\nu, \partial_\tau \right\rangle + \frac{1}{\tau} g^{\alpha\beta} \left\langle \tilde{R}(\partial_i, \theta_\alpha)\theta_\beta, \partial_\tau \right\rangle$$
$$= -\frac{1}{2\left(R + \frac{N}{2\tau}\right)} R_i^j \nabla_j R$$
$$\equiv 0 \mod O(N^{-1}),$$
$$\widetilde{\text{Rc}}(\partial_\tau, \partial_\tau) = g^{k\ell} \left\langle \tilde{R}(\partial_\tau, \partial_k)\partial_\ell, \partial_\tau \right\rangle + \left\langle \tilde{R}(\partial_\tau, \nu)\nu, \partial_\tau \right\rangle$$
$$+ \frac{1}{\tau} g^{\alpha\beta} \left\langle \tilde{R}(\partial_\tau, \theta_\alpha)\theta_\beta, \partial_\tau \right\rangle$$
$$= \frac{1}{4\left(R + \frac{N}{2\tau}\right)} |\nabla R|^2$$
$$\equiv 0 \mod O(N^{-1}),$$
$$\widetilde{\text{Rc}}(\partial_i, \theta_\alpha) = g^{k\ell} \left\langle \tilde{R}(\partial_i, \partial_k)\partial_\ell, \theta_\alpha \right\rangle + \left\langle \tilde{R}(\partial_i, \nu)\nu, \theta_\alpha \right\rangle + \frac{1}{\tau} g^{\gamma\beta} \left\langle \tilde{R}(\partial_i, \theta_\gamma)\theta_\beta, \theta_\alpha \right\rangle$$
$$= 0,$$

$$\widetilde{\mathrm{Rc}}(\partial_\tau, \theta_\alpha) = g^{k\ell}\left\langle \tilde{R}(\partial_\tau, \partial_k)\partial_\ell, \theta_\alpha \right\rangle + \left\langle \tilde{R}(\partial_\tau, \nu)\nu, \theta_\alpha \right\rangle$$
$$+ \frac{1}{\tau} g^{\gamma\beta} \left\langle \tilde{R}(\partial_\tau, \theta_\gamma)\theta_\beta, \theta_\alpha \right\rangle$$
$$= 0,$$
$$\widetilde{\mathrm{Rc}}(\theta_\alpha, \theta_\beta) = g^{k\ell}\left\langle \tilde{R}(\theta_\alpha, \partial_k)\partial_\ell, \theta_\beta \right\rangle + \left\langle \tilde{R}(\theta_\alpha, \nu)\nu, \theta_\beta \right\rangle$$
$$+ \frac{1}{\tau} g^{\gamma\delta} \left\langle \tilde{R}(\theta_\alpha, \theta_\gamma)\theta_\delta, \theta_\beta \right\rangle$$
$$= \frac{1}{2\left(R + \frac{N}{2\tau}\right)^2}\left(\partial_\tau R - \frac{2R^2}{N}\right)g_{\alpha\beta}$$
$$\equiv 0 \mod O(N^{-1}).$$

Hence all of the components of the Ricci tensor are equal to zero (modulo N^{-1}).

Finally, taking the trace again yields the scalar curvature of $\tilde{g}$:

$$\tilde{R} = \tilde{g}^{00}\tilde{R}_{00} + \tilde{g}^{ij}\tilde{R}_{ij} + \tilde{g}^{\alpha\beta}\tilde{R}_{\alpha\beta}$$
$$= \frac{1}{2\left(R + \frac{N}{2\tau}\right)^2}\left(-R\partial_\tau R + |\nabla R|^2 - \frac{R^2}{\tau}\right) - \frac{1}{2\left(R + \frac{N}{2\tau}\right)}\left(\triangle R + \frac{R}{\tau}\right).$$

5.3. Geometric interpretation of Perelman's entropy integrand.

Let $(\mathcal{M}^n, g(\tau))$ be a solution to the backward Ricci flow and let f satisfy (6.15). In §6 of [**297**] Perelman also gave a geometric interpretation of the integrand (6.20), i.e.,

$$\tau\left(2\Delta f - |\nabla f|^2 + R\right) + f - n,$$

of his entropy $\mathcal{W}(g, f, \tau)$. In this subsection we discuss this interpretation.

Define the diffeomorphism

$$\tilde{\varphi} = \tilde{\varphi}_{f,N} : \widetilde{\mathcal{M}} \to \widetilde{\mathcal{M}}$$

by

$$\tilde{\varphi}_{f,N} : (x, y, \tau) \mapsto \left(x, y, \left(1 - \frac{2f}{N}\right)\tau\right).$$

Clearly, $\lim_{N\to\infty} \tilde{\varphi}_{f,N} = \mathrm{id}_{\widetilde{\mathcal{M}}}$, i.e., for N large, $\tilde{\varphi}_{f,N}$ is close to the identity.

Consider the pulled-back metric

$$\tilde{g}^m \doteqdot (\tilde{\varphi}_{f,N})^* \tilde{g},$$

which we think of as a perturbation of the metric $\tilde{g}$. By definition,

$$\tilde{g}^m\left(\tilde{X}, \tilde{Y}\right) = \tilde{g}\left((\tilde{\varphi}_{f,N})_* \tilde{X}, (\tilde{\varphi}_{f,N})_* \tilde{Y}\right).$$

We first compute the components of the metric $\tilde{g}^m$.

5. PERELMAN'S FORMALISM IN POTENTIALLY INFINITE DIMENSIONS

LEMMA 8.41.

(1) $\tilde{g}^m_{00} = \tilde{g}_{00} - 2\frac{\partial f}{\partial \tau} - \frac{f}{\tau} + O(N^{-1})$

$\quad = \frac{N}{2\tau} + R - 2\frac{\partial f}{\partial \tau} - \frac{f}{\tau} + O(N^{-1})$,

(2) $\tilde{g}^m_{i0} = -\frac{\partial f}{\partial x^i} + O(N^{-1})$,

(3) $\tilde{g}^m_{ij} = \tilde{g}_{ij} + O(N^{-1}) = g_{ij} + O(N^{-1})$,

(4) $\tilde{g}^m_{\alpha\beta} = \left(1 - \frac{2f}{N}\right)\tilde{g}_{\alpha\beta}$,

(5) $\tilde{g}^m_{i\alpha} = 0$,

(6) $\tilde{g}^m_{\alpha 0} = 0$,

where $\frac{\partial}{\partial \tau} g_{ij} = 2R_{ij}$.

PROOF. Let $\bar{\tau} = \bar{\tau}(x, \tau) \doteqdot \left(1 - \frac{2f}{N}\right)\tau$. In the formulas below, 0 denotes the time index; $i, j, k, \ldots$ denote indices on $\mathcal{M}$; $\alpha, \beta, \gamma, \ldots$ denote indices on $\mathcal{S}^N$; and $a, b, c, \ldots$ denote arbitrary indices. The pulled-back metric is given by

$$\tilde{g}^m_{ab}(x, y, \tau) = \frac{\partial \tilde{\varphi}^c}{\partial z^a}(x, y, \tau) \frac{\partial \tilde{\varphi}^d}{\partial z^b}(x, y, \tau) \tilde{g}_{cd}(\tilde{\varphi}(x, y, \tau)),$$

where $z = x, y$, or τ. Using the formulas for $\tilde{g}_{ij}, \tilde{g}_{\alpha\beta}, \tilde{g}_{00}$, and $\tilde{g}_{i\alpha} = \tilde{g}_{i0} = \tilde{g}_{\alpha 0} = 0$, we obtain the following.

(1)

$\tilde{g}^m_{00}(x, y, \tau) = \left(\frac{\partial \tilde{\varphi}^0}{\partial \tau}(x, y, \tau)\right)^2 \tilde{g}_{00}(\tilde{\varphi}(x, y, \tau))$

$\quad = \left(\frac{\partial \bar{\tau}}{\partial \tau}\right)^2 \tilde{g}_{00}(x, y, \bar{\tau})$

$\quad = \left(1 - \frac{2f}{N} - \frac{2\tau}{N}\frac{\partial f}{\partial \tau}\right)^2 \left(\frac{N}{2\tau}\left(1 - \frac{2f}{N}\right)^{-1} + R\right)$

$\quad = \left(1 + 2\left(-\frac{2f}{N} - \frac{2\tau}{N}\frac{\partial f}{\partial \tau}\right) + O(N^{-2})\right)$

$\quad \quad \times \left(R + \frac{N}{2\tau}\left(1 + \frac{2f}{N}\right) + O(N^{-1})\right)$

$\quad = R + \frac{N}{2\tau}\left(1 + \frac{2f}{N}\right) + \frac{N}{2\tau}2\left(-\frac{2f}{N} - \frac{2\tau}{N}\frac{\partial f}{\partial \tau}\right) + O(N^{-1})$

$\quad = R + \frac{N}{2\tau} - \frac{f}{\tau} - 2\frac{\partial f}{\partial \tau} + O(N^{-1})$

$\quad = \tilde{g}_{00}(x, y, \tau) - \frac{f}{\tau} - 2\frac{\partial f}{\partial \tau} + O(N^{-1})$.

(2)

$$\tilde{g}_{i0}^m(x,y,\tau) = \frac{\partial \tilde{\varphi}^c}{\partial x^i}(x,y,\tau) \frac{\partial \tilde{\varphi}^0}{\partial \tau}(x,y,\tau) \tilde{g}_{c0}(\tilde{\varphi}(x,y,\tau))$$
$$= \frac{\partial \tilde{\varphi}^0}{\partial x^i}(x,y,\tau) \frac{\partial \tilde{\varphi}^0}{\partial \tau}(x,y,\tau) \tilde{g}_{00}(x,y,\bar{\tau})$$
$$= -\frac{2}{N}\frac{\partial f}{\partial x^i}\tau\left(1 - \frac{2f}{N} - \frac{2\tau}{N}\frac{\partial f}{\partial \tau}\right)\left(\frac{N}{2\tau}\left(1 - \frac{2f}{N}\right)^{-1} + R\right)$$
$$= -\frac{2}{N}\frac{\partial f}{\partial x^i}\tau\left(1 + O(N^{-1})\right)\left(\frac{N}{2\tau} + O(1)\right)$$
$$= -\frac{\partial f}{\partial x^i}(x,\tau) + O(N^{-1}).$$

(3)

$$\tilde{g}_{ij}^m(x,y,\tau) = \frac{\partial \tilde{\varphi}^c}{\partial x^i}(x,y,\tau) \frac{\partial \tilde{\varphi}^d}{\partial x^j}(x,y,\tau) \tilde{g}_{cd}(\tilde{\varphi}(x,y,\tau))$$
$$= \frac{\partial \tilde{\varphi}^k}{\partial x^i}(x,y,\tau) \frac{\partial \tilde{\varphi}^\ell}{\partial x^j}(x,y,\tau) \tilde{g}_{k\ell}(x,y,\bar{\tau})$$
$$+ \frac{\partial \tilde{\varphi}^0}{\partial x^i}(x,y,\tau) \frac{\partial \tilde{\varphi}^0}{\partial x^j}(x,y,\tau) \tilde{g}_{00}(x,y,\bar{\tau})$$
$$= \tilde{g}_{ij}(x,y,\bar{\tau}) + \left(\frac{2\tau}{N}\right)^2 \frac{\partial f}{\partial x^i}\frac{\partial f}{\partial x^j}\left(\frac{N}{2\tau}\left(1 - \frac{2f}{N}\right)^{-1} + R\right)$$
$$= g_{ij}(x,\tau) + O(N^{-1}).$$

(4)

$$\tilde{g}_{\alpha\beta}^m(x,y,\tau) = \frac{\partial \tilde{\varphi}^c}{\partial y^\alpha}(x,y,\tau) \frac{\partial \tilde{\varphi}^d}{\partial y^\beta}(x,y,\tau) \tilde{g}_{cd}(\tilde{\varphi}(x,y,\tau))$$
$$= \tilde{g}_{\alpha\beta}(x,y,\bar{\tau}) = \left(1 - \frac{2f(x,\tau)}{N}\right)\tilde{g}_{\alpha\beta}(x,y,\tau).$$

(5)

$$\tilde{g}_{i\alpha}^m(x,y,\tau) = \frac{\partial \tilde{\varphi}^c}{\partial x^i}(x,y,\tau) \frac{\partial \tilde{\varphi}^d}{\partial y^\alpha}(x,y,\tau) \tilde{g}_{cd}(\tilde{\varphi}(x,y,\tau))$$
$$= \tilde{g}_{i\alpha}(x,y,\bar{\tau}) + \frac{\partial \tilde{\varphi}^0}{\partial x^i}(x,y,\tau) \tilde{g}_{0\alpha}(x,y,\bar{\tau})$$
$$= 0.$$

(6)
$$\tilde{g}^m_{\alpha 0} = \frac{\partial \tilde{\varphi}^c}{\partial y^\alpha}(x,y,\tau)\frac{\partial \tilde{\varphi}^d}{\partial \tau}(x,y,\tau)\tilde{g}_{cd}(\tilde{\varphi}(x,y,\tau))$$
$$= \frac{\partial \tilde{\varphi}^0}{\partial \tau}(x,y,\tau)\tilde{g}_{\alpha 0}(x,y,\bar{\tau})$$
$$= 0.$$

$\square$

Now define the 1-parameter family of diffeomorphisms
$$\psi_\tau : \mathcal{M} \to \mathcal{M}$$
by
$$\frac{\partial}{\partial \tau}\psi_\tau = (\nabla f)(\psi_\tau(x),\tau).$$
Consider the diffeomorphism
$$\tilde{\Psi} : \widetilde{\mathcal{M}} \to \widetilde{\mathcal{M}}$$
defined by
$$\tilde{\Psi}(x,y,\tau) \doteqdot (\psi_\tau(x),y,\tau)$$
and the diffeomorphism
$$\bar{\Psi} : \overline{\mathcal{M}} \to \overline{\mathcal{M}}$$
defined by
$$\bar{\Psi}(x,\tau) \doteqdot (\psi_\tau(x),\tau).$$
Let
$$g^m \doteqdot \tilde{\Psi}^* \tilde{g}^m,$$
which is a Riemannian metric on $\widetilde{\mathcal{M}}$.

We compute the components of the metric g^m. First note that
$$2\psi_\tau^*(\operatorname{Hess}_g f) = \psi_\tau^*\left(\mathcal{L}_{\operatorname{grad}_g f} g\right) = \mathcal{L}_{\psi_\tau^*(\operatorname{grad}_g f)}(\psi_\tau^* g)$$
$$= \mathcal{L}_{\operatorname{grad}_{\psi_\tau^* g}(f \circ \bar{\Psi})}(\psi_\tau^* g)$$
$$= 2\operatorname{Hess}_{\psi_\tau^* g}(f \circ \bar{\Psi}).$$

LEMMA 8.42. *Let $(\mathcal{M}^n, g(\tau))$ be a solution of the backward Ricci flow and let f satisfy* (6.15)
$$\frac{\partial f}{\partial \tau} = \Delta f - |\nabla f|^2 + R - \frac{n}{2\tau}.$$
The (spatial) components g^m_{ij} satisfy the following:
$$\frac{\partial}{\partial \tau} g^m_{ij} \equiv (\psi_\tau^*(2\operatorname{Rc}(g) + 2\nabla\nabla f))_{ij} \mod O(N^{-1})$$
$$\equiv 2\operatorname{Rc}(g^m|_{\mathcal{M}\times\{y\}\times\{\tau\}})_{ij} + 2\nabla^m_i \nabla^m_j f \circ \bar{\Psi} \mod O(N^{-1}),$$

where ∇^m is the covariant derivative associated to the metric g_{ij}^m on $\mathcal{M}$. The other components of g^m are given by

$$g_{00}^m = \tilde{g}_{00}^m - |\nabla f|^2$$
$$= \frac{N}{2\tau} - 2(\Delta f - |\nabla f|^2 + R - \frac{n}{2\tau}) - \frac{f}{\tau} - |\nabla f|^2 + O\left(N^{-1}\right)$$
$$= \frac{1}{\tau}\left(\frac{N}{2} - [\tau(2\Delta f - |\nabla f|^2 + R) + f - n]\right) + O\left(N^{-1}\right),$$
$$g_{\alpha\beta}^m = \tilde{g}_{\alpha\beta}^m = \left(1 - \frac{2f}{N}\right)\tilde{g}_{\alpha\beta},$$
$$g_{i0}^m = -\nabla_{(\psi_\tau)_*(\partial_i)} f + O\left(N^{-1}\right),$$
$$g_{\alpha 0}^m = g_{i\alpha}^m = 0.$$

PROOF. We have

$$g_{ij}^m(x, y, \tau) = \tilde{g}^m((\psi_\tau)_*(\partial_i), (\psi_\tau)_*(\partial_j))$$
$$= \tilde{g}_{k\ell}^m(\psi_\tau(x), y, \tau)\frac{\partial \psi_\tau^k}{\partial x^i}\frac{\partial \psi_\tau^\ell}{\partial x^j}$$
$$= [g_{k\ell}(\psi_\tau(x), \tau) + O(N^{-1})]\frac{\partial \psi_\tau^k}{\partial x^i}\frac{\partial \psi_\tau^\ell}{\partial x^j}$$
$$= (\psi_\tau^* g)_{ij}(x, \tau) + O(N^{-1}).$$

Hence

$$\frac{\partial}{\partial \tau} g_{ij}^m = \frac{\partial}{\partial \tau}(\psi_\tau^* g)_{ij} + O(N^{-1})$$
$$= \left(\psi_\tau^*\left(\frac{\partial g}{\partial \tau}\right)\right)_{ij} + \left(\mathcal{L}_{(\psi_\tau)_*(\nabla f)}(\psi_\tau^* g)\right)_{ij} + O(N^{-1})$$
$$\equiv \psi_\tau^*(2\operatorname{Rc}(g))_{ij} + 2\nabla_i^m \nabla_j^m f \circ \bar{\Psi} \mod O(N^{-1})$$
$$\equiv 2\operatorname{Rc}(g^m|_{\mathcal{M}\times\{y\}\times\{\tau\}})_{ij} + 2\nabla_i^m \nabla_j^m f \circ \bar{\Psi} \mod O(N^{-1}).$$

We compute

$$g_{00}^m = \tilde{g}^m\left(\tilde{\Psi}_*(\partial_\tau), \tilde{\Psi}_*(\partial_\tau)\right)$$
$$= \tilde{g}^m((\nabla f, 0, 1), (\nabla f, 0, 1))$$
$$= |\nabla f|^2 - 2|\nabla f|^2 + \tilde{g}_{00}^m$$
$$= \tilde{g}_{00}^m - |\nabla f|^2.$$

Using (6.15), we find that

$$g^m_{00} = \tilde{g}^m_{00} - |\nabla f|^2 = \frac{1}{\tau}\left(\frac{N}{2} - [\tau(2\Delta f - |\nabla f|^2 + R) + f - n]\right) + O(N^{-1}),$$

$$g^m_{\alpha\beta} = \tilde{g}^m\left(\tilde{\Psi}_*(\partial_\alpha), \tilde{\Psi}_*(\partial_\beta)\right) = \tilde{g}^m(\partial_\alpha, \partial_\beta) = \tilde{g}^m_{\alpha\beta}(\psi_\tau(x), y, \tau)$$

$$= \left(1 - \frac{2f \circ \bar{\Psi}}{N}\right)\tilde{g}_{\alpha\beta}(\psi_\tau(x), y, \tau),$$

$$g^m_{i0} = \tilde{g}^m(\tilde{\Psi}_*(\partial_i), \tilde{\Psi}_*(\partial_\tau)) = \tilde{g}^m((\psi_\tau)_*(\partial_i), \partial_\tau)$$

$$= -\nabla_{(\psi_\tau)_*(\partial_i)} f + O(N^{-1}),$$

where $\partial_\alpha \doteq \frac{\partial}{\partial y^\alpha}$ and $\partial_\beta \doteq \frac{\partial}{\partial y^\beta}$. We leave the proofs of the formulas for the rest of components to the reader as exercises. □

Let $\mathcal{H}_c \subset \widetilde{\mathcal{M}}$ denote the hypersurface $\{(x,y,\tau) \in \widetilde{\mathcal{M}} \mid \tau = c\}$ ($\mathcal{H}_c$ is simply a time-slice) with the metric induced by g^m, where c is a constant. By the definition of the metric g^m, we have the following:

$$d\mu_{\mathcal{H}_c} = d\mu_{g^m_{ij}} \wedge d\mu_{g^m_{\alpha\beta}}$$

$$= \left(\psi_\tau^* d\mu_{\tilde{g}^m_{ij}}\right) \wedge \left(1 - \frac{2f \circ \bar{\Psi}}{N}\right)^{\frac{N}{2}} d\mu_{\tilde{g}_{\alpha\beta}}$$

$$= \tau^{\frac{N}{2}}\left(1 - \frac{2f \circ \bar{\Psi}}{N}\right)^{\frac{N}{2}}\left(\psi_\tau^* d\mu_{g_{ij}} + O(N^{-1})\right) \wedge d\mu_{g_{\alpha\beta}}.$$

Using $\left(1 - \frac{2f}{N}\right)^{\frac{N}{2}} = e^{-f} + O(N^{-1})$, we see that the volume form $d\mu_{\mathcal{H}_c}$ of the hypersurface is

$$d\mu_{\mathcal{H}_c} = \tau^{\frac{N}{2}}\left(e^{-f \circ \bar{\Psi}}\left(\psi_\tau^* d\mu_{(\mathcal{M},g)}\right) \wedge d\mu_{\mathcal{S}^N} + O(N^{-1})\right).$$

To find the scalar curvature of the metric on $\mathcal{H}_c$ induced by the metric g^m, first we calculate the Ricci tensor Rc of the metric on $\mathcal{H}_c$ induced by the metric $\tilde{g}^m$. Second, we pull back the tensor Rc by the diffeomorphism induced by $\tilde{\Psi}$.

LEMMA 8.43. *The Christoffel symbols $\hat{\Gamma}^c_{ab}$ corresponding to the metric on $\mathcal{H}_c$ induced by the metric $\tilde{g}^m$ are given by*

$$\hat{\Gamma}^k_{ij} \equiv \Gamma^k_{ij}(\mathcal{M}, g) \mod O(N^{-1}),$$

$$\hat{\Gamma}^\alpha_{ij} = 0, \qquad \hat{\Gamma}^k_{i\alpha} = 0,$$

$$\hat{\Gamma}^\alpha_{i\beta} \equiv -\frac{\delta^\alpha_\beta \nabla_i f}{N} \mod O(N^{-2}),$$

$$\hat{\Gamma}^k_{\alpha\beta} \equiv \tau g_{\alpha\beta}\frac{\nabla^k f}{N} \mod O(N^{-2}),$$

$$\hat{\Gamma}^\gamma_{\alpha\beta} = \Gamma^\gamma_{\alpha\beta}(\mathcal{S}^N),$$

where $\Gamma_{ij}^k(\mathcal{M}, g)$ and $\Gamma_{\alpha\beta}^\gamma(\mathcal{S}^N)$ denote the Christoffel symbols of $(\mathcal{M}, g)$ and $\mathcal{S}^N$, respectively.

PROOF. Recall that the induced metric $\tilde{g}^m \doteqdot \tilde{g}^m|_{\mathcal{H}_c}$ on $\mathcal{H}_c$ is given by

$$\tilde{g}_{ij}^m(x, y, \tau) = \tilde{g}_{ij}(x, y, \tau) + O(N^{-1}) = g_{ij}(x, \tau) + O(N^{-1}),$$

$$\tilde{g}_{\alpha\beta}^m(x, y, \tau) = \left(1 - \frac{2f(x,\tau)}{N}\right)\tilde{g}_{\alpha\beta}(x, y, \tau),$$

$$\tilde{g}_{i\alpha}^m(x, y, \tau) = 0.$$

We compute the Christoffel symbols $\hat{\Gamma}_{ab}^c$ as follows (where p denotes an index in the $\mathcal{M}$ and $\mathcal{S}^N$ directions, i, j, k, ℓ denote indices in the $\mathcal{M}$ direction only, and $\alpha, \beta, \gamma, \delta$ denote indices in the $\mathcal{S}^N$ direction only):

$$\hat{\Gamma}_{ij}^k = \frac{1}{2}(\tilde{g}^m)^{kp}(\partial_i \tilde{g}_{jp}^m + \partial_j \tilde{g}_{ip}^m - \partial_p \tilde{g}_{ij}^m)$$

$$= \frac{1}{2}(\tilde{g}^m)^{k\ell}(\partial_i \tilde{g}_{j\ell}^m + \partial_j \tilde{g}_{i\ell}^m - \partial_\ell \tilde{g}_{ij}^m)$$

$$\equiv \Gamma_{ij}^k(\mathcal{M}, g) \mod O(N^{-1}),$$

$$\hat{\Gamma}_{ij}^\alpha = \frac{1}{2}(\tilde{g}^m)^{\alpha p}(\partial_i \tilde{g}_{jp}^m + \partial_j \tilde{g}_{ip}^m - \partial_p \tilde{g}_{ij}^m)$$

$$= \frac{1}{2}(\tilde{g}^m)^{\alpha\delta}(\partial_i \tilde{g}_{j\delta}^m + \partial_j \tilde{g}_{i\delta}^m - \partial_\delta \tilde{g}_{ij}^m)$$

$$= 0,$$

$$\hat{\Gamma}_{i\beta}^\alpha = \frac{1}{2}(\tilde{g}^m)^{\alpha p}(\partial_i \tilde{g}_{\beta p}^m + \partial_\beta \tilde{g}_{ip}^m - \partial_p \tilde{g}_{i\beta})$$

$$= \frac{1}{2}\tilde{g}^{\alpha\gamma}(\partial_i \tilde{g}_{\beta\gamma}^m + \partial_\beta \tilde{g}_{i\gamma}^m - \partial_\gamma \tilde{g}_{i\beta}^m)$$

$$= \frac{1}{2}\tilde{g}^{\alpha\gamma}\partial_i \tilde{g}_{\beta\gamma}^m \equiv -\frac{\delta_\beta^\alpha \nabla_i f}{N} \mod O(N^{-2}),$$

$$\hat{\Gamma}_{i\alpha}^k = \frac{1}{2}(\tilde{g}^m)^{kp}(\partial_i \tilde{g}_{\alpha p}^m + \partial_\alpha \tilde{g}_{ip}^m - \partial_p \tilde{g}_{i\alpha}^m)$$

$$= \frac{1}{2}(\tilde{g}^m)^{k\ell}(\partial_i \tilde{g}_{\alpha\ell}^m + \partial_\alpha \tilde{g}_{i\ell}^m - \partial_\ell \tilde{g}_{i\alpha}^m)$$

$$= 0,$$

$$\hat{\Gamma}_{\alpha\beta}^k = \frac{1}{2}(\tilde{g}^m)^{kp}(\partial_\alpha \tilde{g}_{\beta p}^m + \partial_\beta \tilde{g}_{\alpha p}^m - \partial_p \tilde{g}_{\alpha\beta}^m)$$

$$= \frac{1}{2}(\tilde{g}^m)^{k\ell}(\partial_\alpha \tilde{g}_{\beta\ell}^m + \partial_\beta \tilde{g}_{\alpha\ell}^m - \partial_\ell \tilde{g}_{\alpha\beta}^m)$$

$$\equiv \tau g_{\alpha\beta}\frac{\nabla^k f}{N} \mod O(N^{-2}),$$

5. PERELMAN'S FORMALISM IN POTENTIALLY INFINITE DIMENSIONS 431

$$\hat{\Gamma}^{\gamma}_{\alpha\beta} = \frac{1}{2}(\tilde{g}^m)^{\gamma p}(\partial_\alpha \tilde{g}^m_{\beta p} + \partial_\beta \tilde{g}^m_{\alpha p} - \partial_p \tilde{g}^m_{\alpha\beta})$$
$$= \frac{1}{2}(\tilde{g}^m)^{\gamma\delta}(\partial_\alpha \tilde{g}^m_{\beta\delta} + \partial_\beta \tilde{g}^m_{\alpha\delta} - \partial_\delta \tilde{g}^m_{\alpha\beta})$$
$$= \Gamma^{\gamma}_{\alpha\beta}(\mathcal{S}^n).$$

□

Therefore the components of the Ricci tensor $\hat{R}_{pq}$ of $(\mathcal{H}_c, \tilde{g}^m|_{\mathcal{H}_c})$ are given by

$$\hat{R}_{jk} = \hat{R}^p_{pjk} = \hat{R}^i_{ijk} + \hat{R}^\alpha_{\alpha jk}$$
$$= \partial_i \hat{\Gamma}^i_{jk} - \partial_j \hat{\Gamma}^i_{ik} + \hat{\Gamma}^i_{ip}\hat{\Gamma}^p_{jk} - \hat{\Gamma}^i_{jp}\hat{\Gamma}^p_{ik}$$
$$+ \partial_\alpha \hat{\Gamma}^\alpha_{jk} - \partial_j \hat{\Gamma}^\alpha_{\alpha k} + \hat{\Gamma}^\alpha_{\alpha p}\hat{\Gamma}^p_{jk} - \hat{\Gamma}^\alpha_{jp}\hat{\Gamma}^p_{\alpha k}$$
$$= \partial_i \hat{\Gamma}^i_{jk} - \partial_j \hat{\Gamma}^i_{ik} + \hat{\Gamma}^i_{i\ell}\hat{\Gamma}^\ell_{jk} - \hat{\Gamma}^i_{j\ell}\hat{\Gamma}^\ell_{ik} + \hat{\Gamma}^i_{i\delta}\hat{\Gamma}^\delta_{jk} - \hat{\Gamma}^i_{j\delta}\hat{\Gamma}^\delta_{ik}$$
$$+ \partial_\alpha \hat{\Gamma}^\alpha_{jk} - \partial_j \hat{\Gamma}^\alpha_{\alpha k} + \hat{\Gamma}^\alpha_{\alpha\delta}\hat{\Gamma}^\delta_{jk} - \hat{\Gamma}^\alpha_{j\delta}\hat{\Gamma}^\delta_{\alpha k} + \hat{\Gamma}^\alpha_{\alpha\ell}\hat{\Gamma}^\ell_{jk} - \hat{\Gamma}^\alpha_{j\ell}\hat{\Gamma}^\ell_{\alpha k}$$
$$\equiv \partial_i \hat{\Gamma}^i_{jk} - \partial_j \hat{\Gamma}^i_{ik} + \hat{\Gamma}^i_{i\ell}\hat{\Gamma}^\ell_{jk} - \hat{\Gamma}^i_{j\ell}\hat{\Gamma}^\ell_{ik} - \partial_j \hat{\Gamma}^\alpha_{\alpha k} + \hat{\Gamma}^\ell_{jk}\hat{\Gamma}^\alpha_{\alpha\ell} \mod O(N^{-1})$$
$$\equiv R_{jk} + \nabla_j \nabla_k f \mod O(N^{-1})$$

and

$$\hat{R}_{\alpha\beta} = \hat{R}^\delta_{\delta\alpha\beta} + \hat{R}^i_{i\alpha\beta}$$
$$= \partial_\delta \hat{\Gamma}^\delta_{\alpha\beta} - \partial_\alpha \hat{\Gamma}^\delta_{\delta\beta} + \hat{\Gamma}^\delta_{\delta p}\hat{\Gamma}^p_{\alpha\beta} - \hat{\Gamma}^\delta_{\alpha p}\hat{\Gamma}^p_{\delta\beta}$$
$$+ \partial_i \hat{\Gamma}^i_{\alpha\beta} - \partial_\alpha \hat{\Gamma}^i_{i\beta} + \hat{\Gamma}^i_{ip}\hat{\Gamma}^p_{\alpha\beta} - \hat{\Gamma}^i_{\alpha p}\hat{\Gamma}^p_{i\beta}$$
$$= \partial_\delta \hat{\Gamma}^\delta_{\alpha\beta} - \partial_\alpha \hat{\Gamma}^\delta_{\delta\beta} + \hat{\Gamma}^\delta_{\delta\gamma}\hat{\Gamma}^\gamma_{\alpha\beta} - \hat{\Gamma}^\delta_{\alpha\gamma}\hat{\Gamma}^\gamma_{\delta\beta} + \hat{\Gamma}^\delta_{\delta\ell}\hat{\Gamma}^\ell_{\alpha\beta} - \hat{\Gamma}^\delta_{\alpha\ell}\hat{\Gamma}^\ell_{\delta\beta}$$
$$+ \partial_i \hat{\Gamma}^i_{\alpha\beta} - \partial_\alpha \hat{\Gamma}^i_{i\beta} + \hat{\Gamma}^i_{i\ell}\hat{\Gamma}^\ell_{\alpha\beta} - \hat{\Gamma}^i_{\alpha\ell}\hat{\Gamma}^\ell_{i\beta} + \hat{\Gamma}^i_{i\delta}\hat{\Gamma}^\delta_{\alpha\beta} - \hat{\Gamma}^i_{\alpha\delta}\hat{\Gamma}^\delta_{i\beta}$$
$$\equiv \partial_\delta \hat{\Gamma}^\delta_{\alpha\beta} - \partial_\alpha \hat{\Gamma}^\delta_{\delta\beta} + \hat{\Gamma}^\delta_{\delta\gamma}\hat{\Gamma}^\gamma_{\alpha\beta} - \hat{\Gamma}^\delta_{\alpha\gamma}\hat{\Gamma}^\gamma_{\delta\beta} + \hat{\Gamma}^\delta_{\delta\ell}\hat{\Gamma}^\ell_{\alpha\beta}$$
$$+ \partial_i \hat{\Gamma}^i_{\alpha\beta} + \hat{\Gamma}^i_{i\ell}\hat{\Gamma}^\ell_{\alpha\beta} \mod O(N^{-2})$$
$$\equiv \operatorname{Rc}(\mathcal{S}^N)_{\alpha\beta} - \frac{\tau g_{\alpha\beta}|\nabla f|^2}{N} + \frac{\tau g_{\alpha\beta}\Delta f}{N} \mod O(N^{-2}).$$

(Note that $\operatorname{Rc}(\mathcal{S}^N)_{\alpha\beta} = \frac{N-1}{2N}g_{\alpha\beta}$.) Thus, the scalar curvature $\hat{R}$ of the hypersurface $(\mathcal{H}_c, g^m|_{\mathcal{H}_c})$, where $g^m = \tilde{\Psi}^*\tilde{g}^m$, is (denote $\tilde{g}^m \doteq \tilde{g}^m|_{\mathcal{H}_c}$)

$$\hat{R} = (g^m)^{ij}(\tilde{\Psi}^*\operatorname{Rc}(\tilde{g}^m))_{ij} + (g^m)^{\alpha\beta}(\tilde{\Psi}^*\operatorname{Rc}(\tilde{g}^m))_{\alpha\beta},$$

so that

$$\hat{R} \equiv g^{k\ell} \frac{\partial x^i}{\partial \psi_\tau^k} \frac{\partial x^j}{\partial \psi_\tau^\ell} \sum_{p,q=1}^n (R_{pq} + \nabla_p \nabla_q f) \frac{\partial \psi_\tau^p}{\partial x^i} \frac{\partial \psi_\tau^q}{\partial x^j}$$

$$+ \left(1 - \frac{2f}{N}\right)^{-1} \frac{g^{\alpha\beta}}{\tau} \left(\operatorname{Rc}\left(\mathcal{S}^N\right)_{\alpha\beta} - \frac{\tau g_{\alpha\beta} |\nabla f|^2}{N} + \frac{\tau g_{\alpha\beta} \Delta f}{N} \right) \mod O(N^{-1})$$

$$\equiv 2\Delta f - |\nabla f|^2 + R + \left(1 - \frac{2f}{N}\right)^{-1} \frac{g^{\alpha\beta}}{\tau} \frac{N-1}{2N} g_{\alpha\beta} \mod O(N^{-1})$$

$$\equiv 2\Delta f - |\nabla f|^2 + R + \frac{1}{2\tau}\left(1 + \frac{2f}{N}\right)(N-1) \mod O(N^{-1})$$

$$\equiv \frac{1}{\tau}\left(\frac{N-1}{2} + \tau(2\Delta f - |\nabla f|^2 + R) + f\right) \mod O(N^{-1}).$$

This is the same as the formula on p. 13 of [**297**] except for the -1 in $\frac{N-1}{2}$. If we instead choose the metric g on $\mathcal{S}^N$ so that $R_{\alpha\beta}(\mathcal{S}^N) = \frac{1}{2} g_{\alpha\beta}$, then we would obtain the exact formula.

The integrand for the entropy $\mathcal{W}(g, f, \tau)$ is related to the scalar curvature of the hypersurface $(\mathcal{H}_c, g^m|_{\mathcal{H}_c})$ by the following formula:

$$\tau\left(2\Delta f - |\nabla f|^2 + R\right) + f - n = \tau \hat{R} - n - \frac{N-1}{2} \mod O(N^{-1}).$$

6. Notes and commentary

Section 2. Corollary 8.17 is from §7.1 of [**297**].

Section 3. Theorem 8.26 is from §7.3 of [**297**]. A localized version of the theorem is given by Perelman in §8.2 of [**297**].

Section 4. Theorem 8.26 is from §11.2 of [**297**].

CHAPTER 9

Basic Topology of 3-Manifolds

The purpose of this chapter is to introduce certain well-known facts in 3-manifold topology which are related to the Ricci flow. It is by no means meant to be a complete survey of the subject, and we have omitted many important results in the field. Due to space limitations, we will not provide proofs and instead will refer the reader to the literature.

Unless mentioned otherwise, we shall assume in this chapter that all 3-manifolds are connected, orientable, and with possibly nonempty boundary. The n-dimensional sphere, n-dimensional ball and n-dimensional Euclidean space are denoted by $\mathcal{S}^n$, B^n, and $\mathbb{R}^n$, respectively.

1. Essential 2-spheres and irreducible 3-manifolds

1.1. Topological, PL and smooth categories. The fundamental work of Moise [**268**] in 1952 shows that any topological homeomorphism of an open set in $\mathbb{R}^3$ into $\mathbb{R}^3$ can be C^0 approximated by **piecewise-linear** (PL) homeomorphisms. As a consequence, he proved that any topological 3-manifold can be triangulated and that there is a unique PL structure on any topological 3-manifold. Different proofs of the Moise theorem can be found in Bing [**28**] and Shalen [**327**]. It is shown in Hirsch [**203**] and Munkres [**279**] that the PL and smooth categories in dimension 3 are equivalent. For the rest of this chapter, we assume that all manifolds and maps between them are smooth.

1.2. Sphere decompositions and irreducibility. The study of 2-spheres in 3-manifolds probably began with the Schoenflies problem. It asks if any smoothly embedded 2-sphere in Euclidean 3-space must bound a 3-ball. The affirmative solution of the problem in dimension 3 by Alexander [**1**] is one of the milestones in the field. We say a 2-sphere in a 3-manifold is **essential** if it does not bound a 3-ball. Essential 2-spheres are closely related to the connected sum decomposition. Indeed, if a 3-manifold $\mathcal{M}$ is a connected sum $\mathcal{M} = \mathcal{M}_1 \# \mathcal{M}_2$, where neither $\mathcal{M}_1$ nor $\mathcal{M}_2$ is $\mathcal{S}^3$ or B^3, then the decomposing 2-sphere is essential. On the other hand, if $\mathcal{S}^2$ is an essential 2-sphere in $\mathcal{M}$ which decomposes the manifold into two pieces $\mathcal{N}_1$ and $\mathcal{N}_2$, i.e., $\mathcal{S}^2$ is **separating**, then this gives a connected sum decomposition $\mathcal{M} = \mathcal{M}_1 \# \mathcal{M}_2$. Namely, we simply take $\mathcal{M}_i$ to be $\mathcal{N}_i$ capped off by a 3-ball. If the essential 2-sphere $\mathcal{S}^2$ does not separate $\mathcal{M}$, i.e., $\mathcal{M} \backslash \mathcal{S}^2$ is connected, then one can again obtain a connected sum decomposition $\mathcal{M} = (\mathcal{S}^2 \times \mathcal{S}^1)\#\mathcal{M}'$. This gives more information than a separating 2-sphere.

Here is a way to see the $\mathcal{S}^2 \times \mathcal{S}^1$ factor in $\mathcal{M}$. Take an embedded arc A whose endpoints lie in $\mathcal{S}^2$ in such a way that A starts from one side of $\mathcal{S}^2$, ends at the other side, and has no other intersection with $\mathcal{S}^2$. We may assume that, after an isotopy, the endpoints of A are the same. Thus there is an embedded $\mathcal{S}^1$ in $\mathcal{M}$ which intersects the 2-sphere transversely in one point. Let $\mathcal{N}$ be a small regular neighborhood of $\mathcal{S}^2 \cup \mathcal{S}^1$. Then it is easy to see that $\mathcal{N}$ is homeomorphic to $(\mathcal{S}^2 \times \mathcal{S}^1) \setminus \{$ open 3-ball $\}$. In particular, this shows that the boundary of $\mathcal{N}$ is a 2-sphere. This gives the connected sum decomposition. (To see the topology of $\mathcal{N}$, the reader may try to consider the corresponding problem in dimension two: replace $\mathcal{S}^2$ by $\mathcal{S}^1$ inside a surface. In this case, we have two simple loops a and b inside an oriented surface so that a intersects b in one point transversely. Then it is an elementary exercise in topology to show that the regular neighborhood $\mathcal{N}(a \cup b)$ is $(\mathcal{S}^1 \times \mathcal{S}^1) \setminus \{$ open 2-disk $\}$. It turns out that this fact holds in all dimensions.)

Thus essential 2-spheres correspond to connected sum decompositions. A 3-manifold is called **irreducible** if each embedded 2-sphere bounds a 3-ball. If a 3-manifold $\mathcal{M}$ is not irreducible, it contains an essential 2-sphere. Using the operations above, one concludes that either $\mathcal{M} = \mathcal{S}^2 \times \mathcal{S}^1$ or $\mathcal{M} = \mathcal{M}_1 \# \mathcal{M}_2$, where neither $\mathcal{M}_1$ nor $\mathcal{M}_2$ is $\mathcal{S}^3$ or B^3. Now one asks if each of the factor 3-manifolds is irreducible or not. Continuing in this way, one bumps into the question of whether this decomposition process for a compact 3-manifold stops after finitely many steps. This was resolved affirmatively by Kneser in 1929 [**233**] for compact triangulated 3-manifolds.

THEOREM 9.1 (Kneser). *Let $\mathcal{M}^3$ be a compact triangulated 3-manifold. Then $\mathcal{M}$ can be decomposed into a connected sum*

$$\mathcal{M}_1^3 \# \mathcal{M}_2^3 \# \cdots \# \mathcal{M}_n^3 \# (\mathcal{S}^2 \times \mathcal{S}^1) \# \cdots \# (\mathcal{S}^2 \times \mathcal{S}^1),$$

where each $\mathcal{M}_i$ is irreducible.

A counterexample to the Poincaré conjecture is usually called a 'fake' 3-sphere. An interesting consequence of Kneser's Theorem is the following statement: if there exists a counterexample to the Poincaré conjecture in dimension 3, then there exists an irreducible fake 3-sphere. This holds because the fundamental group of a connected sum is the free product of the fundamental groups of the factors.

One calls two 2-spheres in a 3-manifold **parallel** if they are disjoint and bound a region homeomorphic to $\mathcal{S}^2 \times [0, 1]$. Kneser's theorem states that for any compact triangulated 3-manifold, there is an integer k such that any collection of more than k disjoint essential 2-spheres in the manifold must contain a pair of parallel 2-spheres. Since Kneser's finiteness theorem is so deeply related to Hamilton's program, we will indicate the basic ideas of its proof here. (See [**201**] for a complete proof. Note too that the proof was generalized by W. Haken to normal surface theory [**176**].) Let us fix a triangulation T of a compact 3-manifold $\mathcal{M}$. Let t be the number of tetrahedra in the triangulation. Kneser proved that if $n > 6t + |H_1(\mathcal{M}; \mathbb{Z}_2)|$, where $H_1(\mathcal{M}; \mathbb{Z}_2)$ is the first homology group of $\mathcal{M}$ with $\mathbb{Z}_2$ coefficients, then any n disjoint essential 2-spheres contain a parallel pair. Here is the basic argument. Suppose $\{\mathcal{S}_1, ..., \mathcal{S}_n\}$ is a collection of n essential disjoint 2-spheres in $\mathcal{M}$. By isotopy and topological surgeries, one may find a new collection of n essential disjoint 2-spheres, still denoted by $\{\mathcal{S}_1, ..., \mathcal{S}_n\}$, that are in 'nice position' with respect to the triangulation. Here, 'nice position' means that the intersection of each 2-sphere with each tetrahedron consists of a disjoint union of geometric triangles and quadrilaterals. These are called **normal surfaces**. There are only seven normally isotopic triangles and quadrilaterals in a tetrahedron. This shows that inside a tetrahedron σ^3, all but at most six components of $\sigma^3 \setminus (\mathcal{S}_1 \cup \cdots \cup \mathcal{S}_n)$ are parallel regions. It follows from this fact by a simple computation that there are two parallel 2-spheres if $n > 6t + |H_1(\mathcal{M}; \mathbb{Z}_2)|$.

1.3. Irreducible 3-manifolds. Irreducible 3-manifolds joined in connected sums may be regarded as building blocks for 3-manifolds. Most familiar 3-manifolds are irreducible. For instance, the complement of a knot in $\mathcal{S}^3$ is irreducible by Alexander's theorem. Also, if a covering space $\mathcal{N}$ of a 3-manifold $\mathcal{M}$ is irreducible, then $\mathcal{M}$ is irreducible. This is due to the fact that 2-spheres are simply connected. Thus any 2-sphere in $\mathcal{M}$ can be lifted to a 2-sphere in $\mathcal{N}$. Now using the irreducibility of $\mathcal{N}$, one produces a 3-ball in $\mathcal{N}$ bounding the lifted 2-sphere. Using the Brouwer fixed point theorem, one then shows that the 3-ball is mapped injectively to $\mathcal{M}$ by the covering map. This proves $\mathcal{M}$ is irreducible. In particular, if the universal cover of a 3-manifold is $\mathbb{R}^3$, then the manifold is irreducible. This shows, for instance, that all flat (e.g., $\mathcal{S}^1 \times \mathcal{S}^1 \times \mathcal{S}^1$) and all hyperbolic 3-manifolds are irreducible. A deep result of Meeks and Yau [**263**] shows that the converse is also true. Namely, if a manifold is irreducible, then all covering spaces of it are irreducible.

In [**264**] Milnor proved that the connected sum decomposition of a compact 3-manifold is unique up to self-homeomorphism of the 3-manifold. But the decomposition in Kneser's theorem is in general not unique up to isotopy of the 3-manifold. This is due to the action of the diffeomorphism group of the 3-manifold on the decomposition. For instance, the manifold $(\mathcal{S}^2 \times \mathcal{S}^1)\#(\mathcal{S}^2 \times \mathcal{S}^1)$ has many nonisotopic essential separating 2-spheres, due to the large diffeomorphism group of the manifold.

2. Incompressible surfaces and the geometrization conjecture

2.1. Incompressible surfaces and Haken manifolds. The success of the study of 2-spheres in 3-manifolds prompted people to look for more general surfaces. Evidently, surfaces that can be contained inside a coordinate chart are not going to be interesting. Haken introduced the following important concept. A compact, connected, properly embedded surface F in a 3-manifold $\mathcal{M}$ is said to be **incompressible** if one of the following conditions holds:

(1) $F \neq \mathcal{S}^2$ or B^2 and the inclusion map induces an injective homomorphism in the fundamental group; or
(2) $F = \mathcal{S}^2$ is an essential 2-sphere; or
(3) $F = B^2$ and ∂F is not null homotopic in $\partial \mathcal{M}$.

An orientable **Haken 3-manifold** is a compact and irreducible 3-manifold $\mathcal{M}$ that admits a two-sided incompressible surface. This definition is the same as saying that $\mathcal{M}$ is a compact, orientable, and irreducible manifold which contains an incompressible surface other than $\mathbb{RP}^2$. The reason is that if a compact, irreducible, orientable 3-manifold $\mathcal{M}$ contains $\mathbb{RP}^2$ as an incompressible surface, then $\mathcal{M} = \mathbb{RP}^3$. Also, if a compact surface is incompressible in $\mathcal{M}$ and is one-sided, then the boundary of a regular neighborhood of the surface is a two-sided incompressible surface. Furthermore, since we assume the 3-manifold is orientable, two-sided surfaces are the same as orientable surfaces.

Haken manifolds constitute a huge portion of all of 3-manifolds. For instance, if a compact orientable 3-manifold $\mathcal{M}$ is irreducible and has nonempty boundary, then $\mathcal{M}$ is Haken. Also, if a closed, irreducible 3-manifold has positive first Betti number, it is Haken. The homeomorphism classification of Haken manifolds is considered to be solved. It is due to the deep work of F. Waldhausen [**363**] in 1968. Among the many results he proved, the following stands out as one of the most striking.

THEOREM 9.2 (Waldhausen). *Two homotopically equivalent closed Haken manifolds are homeomorphic.*

2.2. Torus decompositions and the geometrization conjecture. A **Seifert 3-manifold** (also called a **Seifert space**) is a compact 3-manifold admitting a foliation whose leaves are $\mathcal{S}^1$. Although this was not the original definition by Seifert in 1931, subsequent work of Epstein [**136**] shows that this simpler definition is equivalent to Seifert's original formulation. If a compact 3-manifold admits an $\mathcal{S}^1$ action without global fixed points (i.e., no point is fixed by all elements in $\mathcal{S}^1$), then the manifold is a Seifert space. Seifert 3-manifolds have been classified. In particular, there exist Seifert manifolds which are irreducible, non-Haken 3-manifolds and have infinite fundamental group. In 1976, Thurston constructed closed hyperbolic 3-manifolds that are not Haken.

Suppose $\mathcal{M}$ is a closed, irreducible, and orientable 3-manifold. A natural step after the connected sum decomposition is to look for incompressible tori. This is called the **torus decomposition**. A compact, irreducible 3-manifold is called **geometrically atoroidal** if every incompressible torus is isotopic to a boundary component. (If the manifold is closed, this simply means that there are no incompressible tori.) The torus decomposition theorem of Jaco and Shalen [**224**] and Johannson [**225**] says the following. (For simplicity, we state the result for closed manifolds only.)

THEOREM 9.3 (Jaco and Shalen, Johannson). *Suppose $\mathcal{M}$ is a closed, orientable, and irreducible 3-manifold. Then there exists a possibly empty disjoint union of incompressible tori in $\mathcal{M}$ that decomposes $\mathcal{M}$ into pieces which are either Seifert 3-manifolds or geometrically atoroidal 3-manifolds. Furthermore, the minimal such collection of tori is unique up to isotopies.*

Thurston's work on the geometrization of 3-manifolds addresses the geometries underlying the Seifert pieces and the geometrically atoroidal pieces. Thurston proved that if $\mathcal{N}$ is a non-Seifert, geometrically atoroidal manifold appearing in the decomposition above (so that it has nonempty boundary), then the interior of the manifold $\mathcal{N}$ admits a complete hyperbolic metric of finite volume. Also, it is proved that the interior of any compact Seifert 3-manifold admits a complete, locally homogeneous Riemannian metric of finite volume. Thus, the remaining issue is the geometry of a closed, irreducible, geometrically atoroidal 3-manifold.

The **geometrization conjecture of Thurston** for a closed, irreducible, orientable 3-manifold $\mathcal{M}$ states that there is an embedding of a (possibly empty) disjoint union of incompressible tori in $\mathcal{M}$ such that every component of the complement is either a Seifert space or else admits a complete Riemannian metric of constant curvature and finite volume.

Using Thurston's theorem and the torus decomposition theorem of Jaco and Shalen and of Johannson, one can reduce the geometrization conjecture to the following form. Suppose $\mathcal{M}$ is a closed, irreducible, orientable 3-manifold without any incompressible tori.

Conjecture I: If the fundamental group of $\mathcal{M}$ is infinite and does not contain any subgroup isomorphic to $\mathbb{Z} \oplus \mathbb{Z}$, then $\mathcal{M}$ admits a hyperbolic metric.

Conjecture II: If the fundamental group of $\mathcal{M}$ contains a subgroup isomorphic to $\mathbb{Z} \oplus \mathbb{Z}$, then $\mathcal{M}$ is a Seifert space.

Conjecture III: If the fundamental group of $\mathcal{M}$ is finite, then $\mathcal{M}$ admits a spherical (constant positive curvature) metric.

Topologists have made great progress toward resolving these conjectures. First of all, if the manifold is Haken, Conjecture I was shown to be valid by Thurston. (See also McMullen [**262**] and Otal [**294**].) Conjecture II for Haken manifolds was shown to be valid by the work of Gordon and Heil [**158**], Johannson [**225**], Jaco and Shalen [**224**], Scott [**318**] and Waldhausen [**364**]. Conjecture II for non-Haken manifolds was solved affirmatively by Casson and Jungreis [**61**] and Gabai [**149**] in 1992. Furthermore, Gabai, Myerhoff, and Thurston [**150**] proved that if a closed, irreducible 3-manifold $\mathcal{M}$ is homotopic to a hyperbolic 3-manifold $\mathcal{N}$, then $\mathcal{M}$ is homeomorphic to $\mathcal{N}$. This gives evidence that Conjecture I holds.

Note that Conjecture III implies the **Poincaré conjecture**. Indeed, if $\mathcal{M}$ is a simply-connected 3-manifold, then by Kneser's theorem, we may assume that $\mathcal{M}$ is irreducible. Since $\mathcal{M}$ has a trivial fundamental group, it is geometrically atoroidal. (By definition, if a manifold contains an incompressible torus, its fundamental group must contain $\mathbb{Z} \oplus \mathbb{Z}$, the fundamental group of the torus.) Thus, if Conjecture III holds, $\mathcal{M}$ admits a metric of constant positive sectional curvature. Thus $\mathcal{M}$ must be $\mathcal{S}^3$.

2.3. Examples of 3-manifolds and their geometries. There are eight locally homogeneous Riemannian geometries in dimension 3. Besides the well-known constant curvature metrics $\mathcal{S}^3$, $\mathbb{R}^3$, and $\mathcal{H}^3$, the rest of the five geometries are given by the standard metrics on $\mathcal{S}^2 \times \mathbb{R}$, $\mathcal{H}^2 \times \mathbb{R}$, $\widetilde{SL}(2,\mathbb{R})$, nil, and sol. (The Ricci flow of homogeneous metrics on $\mathcal{S}^3$ is discussed in Section 5 of Chapter 1 in Volume One.)

Here is a description of the three nonproduct geometries. The geometry $\widetilde{SL}(2,\mathbb{R})$ may be regarded as the universal cover of the Lie group $SL(2,\mathbb{R})$ with a metric invariant under left multiplication. The nilpotent 3-dimensional Lie group nil is the Heisenberg group of strictly upper-triangular 3×3 matrices. (The Ricci flow on nil is discussed in Section 7 of Chapter 1 in Volume One.) The geometry sol is that of the 3-dimensional solvable Lie group defined as the semi-direct product of $\mathbb{R}^2$ with $\mathbb{R}$, where the action of $t \in \mathbb{R}$ on $\mathbb{R}^2$ is given by the matrix $\begin{pmatrix} e^t & 0 \\ 0 & e^{-t} \end{pmatrix}$. (The Ricci flow on sol is discussed in Section 7 of Chapter 1 in Volume One.)

A closed, orientable manifold with $\mathcal{S}^2 \times \mathbb{R}$ geometry must be either $\mathcal{S}^2 \times \mathcal{S}^1$ or $\mathbb{RP}^3 \# \mathbb{RP}^3$. It is interesting to note that $\mathbb{RP}^3 \# \mathbb{RP}^3$ is the only connected-sum 3-manifold admitting a locally homogeneous metric. The product of a closed surface of genus at least two with the circle admits an $\mathcal{H}^2 \times \mathcal{S}^1$ metric, i.e., the product of their standard metrics given by the uniformization theorem. The unit tangent bundle over a surface of negative Euler characteristic is a 3-manifold with $\widetilde{SL}(2,\mathbb{R})$ geometry. More generally, any circle bundle over a surface of negative Euler characteristic admits the $\widetilde{SL}(2,\mathbb{R})$ geometry if the Chern class of the bundle is nonzero. A circle bundle over the torus with nonzero Chern class is a 3-manifold with nil geometry. A torus bundle over the circle whose monodromy is a linear map with distinct real eigenvalues has sol geometry. It can be shown that any closed 3-manifold with one of these five geometries is finitely covered by one of the examples just mentioned. For more information, see the excellent reference [**319**].

A closed 3-manifold with sol geometry is not a Seifert 3-manifold. It admits a nontrivial torus decomposition coming from the torus fiber. On the other hand, any manifold admitting one of the six geometries $\mathcal{S}^3$, $\mathbb{R}^3$, $\mathcal{S}^2 \times \mathbb{R}$, $\mathcal{H}^2 \times \mathbb{R}$, $\widetilde{SL}(2,\mathbb{R})$, or nil is a Seifert manifold.

We will say a compact 3-manifold $\mathcal{M}$ is a **topological graph manifold** if it admits a torus decomposition such that each complementary piece is a Seifert manifold. (This is a standard definition in low-dimensional topology.)

In particular, this implies that the boundary of $\mathcal{M}$ is a possibly empty collection of tori. Graph manifolds appear in the work of Cheeger and Gromov as those 3-manifolds admitting an $\mathcal{F}$-structure. Note that the definition of graph manifold that appears in Cheeger and Gromov's work [74] is slightly different from the one above. To be more precise, a graph manifold in the sense of Cheeger–Gromov is a closed 3-manifold admitting a decomposition by (not necessarily incompressible) tori such that each complementary piece is a Seifert space. It can be shown that if $\mathcal{M}$ is a graph manifold in the sense of Cheeger–Gromov, then it is either a topological graph manifold or else a connected sum of topological graph manifolds with $\mathcal{S}^2 \times \mathcal{S}^1$ factors and lens spaces. At any rate, there are no fake 3-spheres or fake 3-balls embedded inside a Cheeger–Gromov graph manifold.

3. Decomposition theorems and the Ricci flow

Recall that a solution $(\mathcal{M}^3, g(t))$, $t \in [0, T)$, to the Ricci flow is said to develop a singularity at time $T \in (0, \infty)$ if the norm of the Riemann curvature tensor becomes infinite at some point or points of the manifold as $t \nearrow T$. (See Corollary 7.2 of Volume One.) A typical situation in which a finite time singularity develops is the **neckpinch**. It is important to note that the formation of neckpinch singularities may be triggered more by the (local) nonlinearity of the Ricci flow PDE than by the (global) topology of the underlying manifold. In any case, here is a heuristic description. (For precise statements, see Section 5 of Chapter 2 in Volume One, as well as the recent papers of Angenent and one of the authors [7, 8].) Suppose a 3-manifold $\mathcal{M}$ contains a separating 2-sphere. Then under the Ricci flow, a region homeomorphic to $\mathcal{S}^2 \times \mathbb{R}$ may develop in $\mathcal{M}$ such that, as $t \nearrow T$, the sectional curvatures become infinite precisely along the hypersurface identified with $\mathcal{S}^2 \times \{0\}$. In this evolution, the geometry of the region identified topologically with $\mathcal{S}^2 \times \mathbb{R}$ asymptotically approaches the cylinder $\mathcal{S}^2 \times \mathbb{R}$ with its standard product metric.

Hamilton developed a program of applying Ricci flow techniques to general 3-manifolds and analyzed the singularities which may arise (see especially [186], [189], and [190]).[1] Some of Hamilton's ideas are as follows; we first consider [186]. Via point-picking arguments and assuming an injectivity radius estimate, one dilates about singularities and takes limits using the Cheeger–Gromov-type compactness theorem for solutions of the Ricci flow to obtain so-called singularity models, which are nonflat ancient solutions of the Ricci flow. In dimension 3 these ancient solutions have nonnegative

---

[1]Some other papers in which Hamilton developed his program to approach Thurston's geometrization conjecture by Ricci flow are as follows: characterizing spherical space forms [178], weak and strong maximum principles for systems [179], ancient 2-dimensional solutions and surface entropy monotonicity [180] (as used in [186]), matrix Harnack estimate [181] and its applications to eternal solutions [182], and the compactness theorem [187]. (These are only partial descriptions that reflect aspects of the papers' relevance to Hamilton's 3-manifold program.)

sectional curvature. By the strong maximum principle, the universal covers of the ancient solutions either split as the product of a surface solution with $\mathbb{R}$ or have positive sectional curvature in which case they are diffeomorphic to either $\mathcal{S}^3$ or $\mathbb{R}^3$.[2] In the case of splitting, Hamilton proved that the ancient surface solution is either a round shrinking $\mathcal{S}^2$ (and the universal cover of the singularity model is hence geometrically a shrinking round product cylinder; by definition we say that in this case a neck singularity forms) or it has a backward limit which is the cigar soliton. Note that Perelman's no local collapsing theorem rules out the last case of a cigar. In the case when the universal cover of the singularity model has positive sectional curvature and is diffeomorphic to $\mathbb{R}^3$, the covering is trivial. This ancient solution is either Type I or has backward limit which is a steady Ricci soliton on a topological $\mathbb{R}^3$.

In the *latter* case, the asymptotic scalar curvature ratio is infinite, and by dimension reduction, there exists a sequence of points tending to spatial infinity for which the corresponding dilations of the solution limit to an ancient product solution, which again must be a shrinking round cylinder.[3] In this case the singularity model is expected to be the positively curved and rotationally symmetric Bryant soliton and the forming singularity is expected to be a degenerate neckpinch. In summary, we should have that at the largest curvature scale the dilations yield the Bryant soliton, whereas at lower scales dilations yield round product cylinders. This agrees with the fact that the dimension reduction of the Bryant soliton, and more generally a 3-dimensional gradient steady soliton with positive curvature which is κ-noncollapsed at all scales, is a round product cylinder.

On the other hand, the *former* case of a Type I ancient solution with positive sectional curvature, if it exists, also dimension reduces to a round cylinder. Thus a consequence of Hamilton's 3-dimensional singularity formation theory and Perelman's no local collapsing theorem is that if a finite time singularity forms on a closed 3-manifold, then either $\mathcal{M}$ is diffeomorphic to a spherical space form or a neckpinch forms.

In [**186**] Hamilton studies 3-dimensional singularity formation by considering regions in the solution where the scalar curvature is comparable to its spatial maximum. He studies the regions where the scalar curvature is *not* comparable to its spatial maximum by the technique of dimension reduction. For example, when a 3-dimensional steady Ricci soliton singularity model forms, Hamilton proved an injectivity radius estimate to obtain a second limit which is either a shrinking round product cylinder or the product of a cigar with $\mathbb{R}$. (Again no local collapsing rules out the latter case.) The

---

[2]When the universal cover of the singularity model has positive sectional curvature and is diffeomorphic to $\mathcal{S}^3$, $\mathcal{M}$ admits a metric with positive sectional curvature (e.g., $g(t)$ for t large enough) and hence is topologically diffeomorphic to a spherical space form. In this case the singularity model must be geometrically a shrinking spherical space form with its underlying manifold diffeomorphic to $\mathcal{M}$.

[3]With the help of no local collapsing.

regions with curvature comparable to their spatial maximum are geometrically close to a 3-dimensional steady Ricci soliton, whereas some regions with curvature not comparable to their spatial maximum are geometrically close to a shrinking round product cylinder.

In [**189**] Hamilton developed surgery theory and formulated a version of Ricci flow with surgery. Although this theory was developed for solutions on closed 4-manifolds with positive isotropic curvature, its higher aim was clearly a surgery theory for 3-manifolds. Indeed, the class of 4-manifolds with positive isotropic curvature is flexible enough to allow for connected sums. Many of the techniques developed in [**189**] applied to the setting of the Ricci flow on closed 3-manifolds. Limiting arguments and the study of ancient solutions were developed by Hamilton with the aim of enabling surgery. A contradiction argument using limiting techniques was proposed to show that for suitable surgery parameters, the set of surgery times is discrete, and in particular, do not accumulate in finite time. Unfortunately, as was known to some mathematicians working in the field of Ricci flow and as pointed out in [**298**], there was an error in this part of Hamilton's argument.

In the recent work of Perelman [**297**], [**298**], building on Hamilton's theory, Ricci flow behavior (especially singularity formation) on 3-manifolds is carefully examined and classified. The overall picture is subtle and technical, with some of the foundations being discussed in this volume. In the following two simplified examples (the first of which continues our discussion above), we try to convey some of its topological flavor, omitting most of the details.

The formation of neckpinch singularities (as described above) in a certain sense reflects the topological connected-sum decomposition of the underlying manifold. Indeed, suppose a neckpinch with two ends occurs on a region identified with $\mathcal{S}^2 \times \mathbb{R}$. Hamilton proposes a surgery process as follows [**189**].[4] One does surgery near the large ends of the long, thin tubes in that part of the manifold identified with $\mathcal{S}^2 \times \mathbb{R}$, capping these off with round 3-balls. Note that Hamilton's theory predicts the existence of such tubes where the curvature is very large at the center and slowly decreases as one moves away from the center along the *relatively* very long length of the tube. In fact his theory predicts that as one approaches the singularity time, the tube becomes arbitrarily close to an exact cylinder and its size slowly increases as one moves away from the center to an arbitrarily much larger but still very small size.[5] Note that Perelman's surgery process in [**298**] is a modification of the surgery process proposed earlier by Hamilton.[6]

---

[4]More precisely, he considers the 4-dimensional version of this.

[5]More precisely, the tube is *conformally* close to a round product cylinder, where the conformal factor changes very slowly as one moves away from the center.

[6]Huisken and Sinestrari have considered an analogue of Hamilton's surgery theory for the mean curvature flow.

After the surgery, one continues the Ricci flow on the resulting (possibly disconnected) manifold, taking the glued (smoothed) metric as initial data. Heuristically, this surgery procedure corresponds to Kneser's sphere decomposition theorem. It is known that a 2-sphere removed from the neck may bound a 3-ball. Thus for an arbitrary initial 3-manifold $\mathcal{M}$, there is no guarantee that this surgery process will stop in finitely many steps.

The finiteness theorem of Kneser is in some sense related to a finiteness conjecture of Hamilton for Ricci flow. The latter states that if one runs the unnormalized Ricci flow on a closed 3-manifold and performs geometric-topological surgeries whenever the Ricci flow develops a finite time singularity, then after finitely many such surgeries, the Ricci flow will have a nonsingular solution for all time. (Since one discards $\mathcal{S}^3$ and $\mathcal{S}^2 \times \mathcal{S}^1$ factors, this solution may well be empty.) Using Kneser's finiteness theorem, one sees that if Hamilton's conjecture does not hold, then all but finitely many geometric-topological surgeries at finite time singularities of the Ricci flow must split off 3-spheres. (In this regard, see the following recent papers: Perelman [**299**] and Colding and Minicozzi [**116**].) In this context, Hamilton's finiteness conjecture may be regarded as a geometric refinement of Kneser's theorem.

Another type of Ricci flow behavior on 3-manifolds reflects the torus decomposition. This was first noticed in the work of Hamilton [**190**]. For simplicity, we recall Hamilton's formulation. He assumes that a solution to the normalized Ricci flow on a closed 3-manifold $\mathcal{M}$ exists for all time $t \in [0, \infty)$ with uniformly bounded curvature. In this scenario, as time approaches infinity, it may happen that a manifold $\mathcal{M}$ can be decomposed into two parts $\mathcal{M} = \mathcal{M}_{\text{thin}} \cup \mathcal{M}_{\text{thick}}$. In the components of $\mathcal{M}_{\text{thin}}$, the metrics are collapsing with bounded curvature. (Recall that a manifold is said to be **collapsible** if it admits a sequence of Riemannian metrics of uniformly bounded curvature and volumes tending to zero.) In the components of $\mathcal{M}_{\text{thick}}$, the metrics converge to complete hyperbolic metrics of finite volume. Furthermore, using minimal surface techniques, Hamilton proves that the fundamental group of $\mathcal{M}_{\text{thin}} \cap \mathcal{M}_{\text{thick}}$ injects into the fundamental group of $\mathcal{M}$. By the work of Cheeger and Gromov on collapsing manifolds, one concludes that $\mathcal{M}_{\text{thin}}$ is a Cheeger–Gromov graph manifold and hence a connected sum of Seifert spaces and sol-geometry manifolds. As a consequence, Hamilton was able to establish the geometrization conjecture under the (restrictive) hypotheses of long-time existence and uniformly bounded curvature. Perelman's recent work [**298**], in conjunction with Shioya and Yamaguchi [**332**], claims to establish a similar picture without the assumption of bounded curvature.

4. Notes and commentary

We would like to thank Ian Agol for carefully reading this chapter and for pointing out mistakes in a draft version. We also note that Milnor's

[**267**] and Morgan's [**272**] are two excellent survey papers introducing Ricci flow on 3-manifolds and giving an overall picture of the current state of the research in the field.

APPENDIX A

Basic Ricci Flow Theory

> Heat, like gravity, penetrates every substance of the universe, its rays occupy all parts of space. The object of our work is to set forth the mathematical laws which this element obeys. The theory of heat will hereafter form one of the most important branches of general physics. – Joseph Fourier

In this appendix we recall some basic Ricci flow notation, formulas, and results, mostly from Volume One. Unless otherwise indicated, all page numbers, theorem references, chapter and section numbers, etc., refer to Volume One. Some of the results below are slight modifications of those stated therein. If an unnumbered formula appears on p. $\heartsuit\spadesuit$ of Volume One, we refer to it as (V1-p. $\heartsuit\spadesuit$); if the equation is numbered $\diamondsuit.\clubsuit$, then we refer to it as (V1-$\diamondsuit.\clubsuit$).

The reader who has read or is familiar with Volume One may essentially skip this chapter, referring to it only when necessary.

1. Riemannian geometry

1.1. Notation. Let $(\mathcal{M}, g)$ be a Riemannian manifold. Throughout this appendix we shall often sum over repeated indices and not bother to raise (or lower) indices. For example, $a_{ij}b_{ij} \doteqdot g^{ik}g^{j\ell}a_{ij}b_{k\ell}$.

- If α is a 1-form, then $\alpha^\sharp$ denotes the dual vector field. Conversely, if X is a vector field, then $X^\flat$ denotes the dual 1-form.
- $T\mathcal{M}$, $T^*\mathcal{M}$, $\Lambda^2 T^*\mathcal{M}$, and $S_2 T^*\mathcal{M}$ denote the tangent, cotangent, 2-form, and symmetric $(2,0)$-tensor bundles, respectively.
- Γ, ∇, and Δ denote the Christoffel symbols, covariant derivative, and Laplacian, respectively.
- R, Rc, and Rm denote the scalar, Ricci, and Riemann curvature tensors, respectively.
- r often denotes the average scalar curvature (assuming $\mathcal{M}$ is compact).
- The upper index on the Riemann $(3,1)$-tensor is lowered into the 4-th position: $R_{ijk\ell} = R_{ijk}^m g_{m\ell}$.
- $\lambda \geq \mu \geq \nu$ denote the eigenvalues of the Riemann curvature operator of a 3-manifold, in decreasing order.
- $d = $ dist, diam, and inj denote the Riemannian distance, diameter, and injectivity radius, respectively.
- L, $A = $ Area, and V denote length, area, and volume, respectively.

- tr_g denotes the trace with respect to g (e.g., of a symmetric $(2,0)$-tensor).
- $\mathcal{S}^n$ usually denotes the unit n-sphere.
- For tensors A and B, $A * B$ denotes a linear combination of contractions of the tensor product of A and B.
- $\stackrel{\mathrm{G}}{=}$ denotes an equality which holds on gradient Ricci solitons.
- $\stackrel{\mathrm{E}}{=}$ denotes an equality which holds on expanding gradient Ricci solitons.

1.2. Basic Riemannian geometry formulas in local coordinates.
In Ricci flow, where the metric is time-dependent, it is convenient to compute in a local coordinate system.

Let $(\mathcal{M}^n, g)$ be an n-dimensional Riemannian manifold. Almost everywhere we shall assume the metric g is complete. Let $\{x^i\}$ be a local coordinate system and let $\partial_i \doteqdot \frac{\partial}{\partial x^i}$. The components of the metric are $g_{ij} \doteqdot g(\partial_i, \partial_j)$. The Christoffel symbols for the Levi-Civita connection, defined by $\nabla_{\partial_i} \partial_j \doteqdot \Gamma_{ij}^k \partial_k$, are

(V1-p. 24) $$\Gamma_{ij}^k = \frac{1}{2} g^{k\ell} \left(\partial_i g_{j\ell} + \partial_j g_{i\ell} - \partial_\ell g_{ij} \right),$$

where (g^{ij}) is the inverse matrix of (g_{ij}). The components of the Riemann curvature $(3,1)$-tensor, defined by

(V1-p. 286) $$R\left(\frac{\partial}{\partial x^i}, \frac{\partial}{\partial x^j}\right) \frac{\partial}{\partial x^k} \doteqdot R_{ijk}^\ell \frac{\partial}{\partial x^\ell},$$

are

(V1-p. 68) $$R_{ijk}^\ell = \partial_i \Gamma_{jk}^\ell - \partial_j \Gamma_{ik}^\ell + \Gamma_{jk}^p \Gamma_{ip}^\ell - \Gamma_{ik}^p \Gamma_{jp}^\ell.$$

The Ricci tensor is given by

(V1-p. 92) $$R_{ij} = R_{pij}^p = \partial_p \Gamma_{ij}^p - \partial_i \Gamma_{pj}^p + \Gamma_{ij}^q \Gamma_{pq}^p - \Gamma_{pj}^q \Gamma_{iq}^p.$$

The scalar curvature is $R = g^{ij} R_{ij}$. If $\mathcal{M}$ is oriented and the local coordinates $\{x^i\}_{i=1}^n$ have positive orientation, then the volume form is

(V1-p. 70) $$d\mu = \sqrt{\det g}\, dx^1 \wedge \cdots \wedge dx^n.$$

The Bianchi identities.

(1) **First Bianchi identity**:

(V1-3.17) $$0 = R_{ijk\ell} + R_{ik\ell j} + R_{i\ell jk}.$$

(2) **Second Bianchi identity**:

(V1-3.18) $$0 = \nabla_q R_{ijk\ell} + \nabla_i R_{jqk\ell} + \nabla_j R_{qik\ell},$$

where we take $\nabla \operatorname{Rm}$, cyclically permute the first three indices of $\nabla_q R_{ijk\ell}$ (components), and sum to get zero.

(3) **Contracted second Bianchi identity**:

(V1-3.13) $$\nabla^j R_{ij} = \frac{1}{2}\nabla_i R,$$

which is obtained from (V1-3.18) by taking two traces (e.g., multiplying by $g^{q\ell}g^{jk}$ and summing over q, ℓ, j, k).

1.3. Cartan structure equations. For metrics with symmetry, such as rotationally symmetric metrics, it is convenient to calculate with respect to a local orthonormal frame, also called a moving frame.

Let $\{e_i\}_{i=1}^n$ be a local orthonormal frame field in an open set $\mathcal{U} \subset \mathcal{M}^n$. Denote the dual orthonormal basis of $T^*\mathcal{M}$ by $\{\omega^i\}_{i=1}^n$ so that $g = \sum_{i=1}^n \omega^i \otimes \omega^i$. The connection 1-forms $\omega_i^j \in \Omega^1(\mathcal{U})$ are defined by

(V1-p. 106a) $$\nabla_X e_i \doteqdot \sum_{j=1}^n \omega_i^j(X) e_j,$$

for all $i = 1, \ldots, n$ and $X \in C^\infty(T\mathcal{M}|_\mathcal{U})$. They are antisymmetric: $\omega_i^j = -\omega_j^i$. The first and second Cartan structure equations are

(V1-p. 106b) $$d\omega^i = \omega^j \wedge \omega_j^i,$$

(V1-p. 106c) $$\operatorname{Rm}_i^j \equiv \Omega_i^j = d\omega_i^j - \omega_i^k \wedge \omega_k^j.$$

The following formula is useful for computing the connection 1-forms:

(A.1) $$\omega_i^k(e_j) = d\omega^i(e_j, e_k) + d\omega^j(e_i, e_k) - d\omega^k(e_j, e_i).$$

1.4. Curvature under conformal change of the metric. Let g and $\tilde{g}$ be two Riemannian metrics on a manifold $\mathcal{M}^n$ conformally related by $\tilde{g} = e^{2u}g$, where $u : \mathcal{M} \to \mathbb{R}$. If $\{e_i\}_{i=1}^n$ is an orthonormal frame field for g, then $\{\tilde{e}_i\}_{i=1}^n$, where $\tilde{e}_i = e^{-u}e_i$, is an orthonormal frame field for $\tilde{g}$. The Ricci tensors of g and $\tilde{g}$ are related by

(A.2) $$\widetilde{\operatorname{Rc}}(\tilde{e}_\ell, \tilde{e}_i) = e^{-2u}\left(\begin{array}{c}\operatorname{Rc}(e_\ell, e_i) + (2-n)\nabla_{e_\ell}\nabla_{e_i}u - \delta_{\ell i}\Delta u \\ + |\nabla u|^2(2-n)\delta_{i\ell} - (2-n)e_\ell(u)e_i(u)\end{array}\right).$$

Tracing this, we see that the scalar curvatures of g and $\tilde{g}$ are related by

(A.3) $$\tilde{R} = e^{-2u}\left(R - 2(n-1)\Delta u - (n-2)(n-1)|\nabla u|^2\right).$$

For derivations of the formulas above, which are standard, see subsection 7.2 of Chapter 1 in [**111**] for example.

1.5. Variations and evolution equations of geometric quantities. The Ricci flow is an evolution equation where the variation of the metric (i.e., time-derivative of the metric) is minus twice the Ricci tensor. More generally, we may consider arbitrary variations of the metric. Given a variation of the metric, we recall the corresponding variations of the Levi-Civita connection and curvatures. (In Volume One, see Section 1 of Chapter 3 for the derivations, or see Lemma 6.5 on p. 174 for a summary.)

448   A. BASIC RICCI FLOW THEORY

LEMMA A.1 (Metric variation formulas). *Suppose that $g(s)$ is a smooth 1-parameter family of metrics on a manifold $\mathcal{M}^n$ such that $\frac{\partial}{\partial s} g = v$.*

(1) *The Levi-Civita connection Γ of g evolves by*

(V1-3.3) $$\frac{\partial}{\partial s}\Gamma^k_{ij} = \frac{1}{2} g^{k\ell} \left(\nabla_i v_{j\ell} + \nabla_j v_{i\ell} - \nabla_\ell v_{ij} \right).$$

(2) *The $(3,1)$-Riemann curvature tensor Rm of g evolves by*

(V1-3.4) $$\frac{\partial}{\partial s} R^\ell_{ijk} = \frac{1}{2} g^{\ell p} \begin{pmatrix} \nabla_i \nabla_j v_{kp} + \nabla_i \nabla_k v_{jp} - \nabla_i \nabla_p v_{jk} \\ -\nabla_j \nabla_i v_{kp} - \nabla_j \nabla_k v_{ip} + \nabla_j \nabla_p v_{ik} \end{pmatrix}$$

(V1-p. 69a) $$= \frac{1}{2} g^{\ell p} \begin{pmatrix} \nabla_i \nabla_k v_{jp} + \nabla_j \nabla_p v_{ik} - \nabla_i \nabla_p v_{jk} - \nabla_j \nabla_k v_{ip} \\ -R^q_{ijk} v_{qp} - R^q_{ijp} v_{kq} \end{pmatrix}.$$

(3) *The Ricci tensor Rc of g evolves by*

(V1-3.5) $$\frac{\partial}{\partial s} R_{jk} = \frac{1}{2} g^{pq} \left(\nabla_q \nabla_j v_{kp} + \nabla_q \nabla_k v_{jp} - \nabla_q \nabla_p v_{jk} - \nabla_j \nabla_k v_{qp} \right)$$

(V1-p. 69b) $$= -\frac{1}{2} \left[\Delta_L v_{jk} + \nabla_j \nabla_k (\mathrm{tr}_g v) + \nabla_j (\delta v)_k + \nabla_k (\delta v)_j \right],$$

where Δ_L denotes the **Lichnerowicz Laplacian** of a $(2,0)$-tensor, which is defined by

(V1-3.6) $$(\Delta_L v)_{jk} \doteq \Delta v_{jk} + 2 g^{qp} R^r_{qjk} v_{rp} - g^{qp} R_{jp} v_{qk} - g^{qp} R_{kp} v_{jq}.$$

(4) *The scalar curvature R of g evolves by*

(V1-p. 69c) $$\frac{\partial}{\partial s} R = g^{ij} g^{k\ell} \left(-\nabla_i \nabla_j v_{k\ell} + \nabla_i \nabla_k v_{j\ell} - v_{ik} R_{j\ell} \right)$$

(V1-p. 69d) $$= -\Delta V + \mathrm{div}\,(\mathrm{div}\, v) - \langle v, \mathrm{Rc} \rangle,$$

where $V \doteq g^{ij} v_{ij}$ is the trace of v.

(5) *The volume element $d\mu$ evolves by*

(V1-p. 70) $$\frac{\partial}{\partial s} d\mu = \frac{V}{2} d\mu.$$

(6) *Let γ_s be a smooth family of curves with fixed endpoints in $\mathcal{M}^n$ and let L_s denote the length with respect to $g(s)$. Then*

(V1-3.8) $$\frac{d}{ds} L_s(\gamma_s) = \frac{1}{2} \int_{\gamma_s} v(T,T)\, d\sigma - \int_{\gamma_s} \langle \nabla_T T, U \rangle\, d\sigma,$$

where σ is arc length, $T \doteq \frac{\partial \gamma_s}{\partial \sigma}$, and $U \doteq \frac{\partial}{\partial s}$.

1.6. Commuting covariant derivatives. In deriving how geometric quantities evolve when the metric evolves by Ricci flow, commutators of covariant derivatives often enter the calculations. (For the following, see p. 286 in Section 6 of Appendix A in Volume One.) If X is a vector field, then

(V1-p. 286a) $$[\nabla_i, \nabla_j] X^\ell \doteq \nabla_i \nabla_j X^\ell - \nabla_j \nabla_i X^\ell = R^\ell_{ijk} X^k.$$

If θ is a 1-form, then

(V1-p. 286b) $$[\nabla_i, \nabla_j]\theta_k \doteqdot \nabla_i\nabla_j\theta_k - \nabla_j\nabla_i\theta_k = -R^\ell_{ijk}\theta_\ell.$$

More generally, if A is any (p,q)-tensor field, one has the commutator

(V1-p. 286)
$$[\nabla_i, \nabla_j] A^{\ell_1\cdots\ell_q}_{k_1\cdots k_p} \doteqdot \nabla_i\nabla_j A^{\ell_1\cdots\ell_q}_{k_1\cdots k_p} - \nabla_j\nabla_i A^{\ell_1\cdots\ell_q}_{k_1\cdots k_p}$$
$$= \sum_{r=1}^{q} R^{\ell_r}_{ijm} A^{\ell_1\cdots\ell_{r-1}\,m\,\ell_{r+1}\cdots\ell_q}_{k_1\cdots k_p} - \sum_{s=1}^{p} R^{m}_{ijk_s} A^{\ell_1\cdots\ell_q}_{k_1\cdots k_{s-1}\,m\,k_{s+1}\cdots k_p}.$$

1.7. Lie derivative. Because of the diffeomorphism invariance of the Ricci flow, the effect of infinitesimal diffeomorphisms on tensors, e.g., the Lie derivative, enters the Ricci flow. (See p. 282 in Section 2 of Appendix A in Volume One.)

The Lie derivative of the metric satisfies

(V1-p. 282a) $$(\mathcal{L}_X g)(Y, Z) = g(\nabla_Y X, Z) + g(Y, \nabla_Z X)$$

for all vector fields X, Y, Z. In local coordinates

(V1-p. 282b) $$(\mathcal{L}_X g)_{ij} = (\mathcal{L}_X g)\left(\frac{\partial}{\partial x^i}, \frac{\partial}{\partial x^j}\right) = \nabla_i X_j + \nabla_j X_i.$$

In particular, if $X = \nabla f$ is a gradient vector field, then

$$(\mathcal{L}_{\nabla f} g)_{ij} = 2\nabla_i\nabla_j f.$$

1.8. Bochner formulas. (See p. 284 in Section 4 of Appendix A in Volume One.) The **rough Laplacian** denotes the operators

$$\Delta : C^\infty\left(T^q_p\mathcal{M}^n\right) \to C^\infty\left(T^q_p\mathcal{M}^n\right),$$

where $T^q_p\mathcal{M} \doteqdot \bigotimes^p T^*\mathcal{M} \otimes \bigotimes^q T\mathcal{M}$, defined by

(V1-p. 284a)
$$(\Delta A)(Y_1, \ldots, Y_p; \theta_1, \ldots, \theta_q) = \sum_{i=1}^{n} (\nabla^2 A)(e_i, e_i, Y_1, \ldots, Y_p; \theta_1, \ldots, \theta_q)$$

for all (p,q)-tensors A, all vector fields $Y_1, \ldots, Y_p$, and all covector fields $\theta_1, \ldots, \theta_q$, where $\{e_i\}_{i=1}^n$ is a (local) orthonormal frame field. The **Hodge–de Rham Laplacian** $-\Delta_d : \Omega^p(\mathcal{M}) \to \Omega^p(\mathcal{M})$ is defined by

(V1-p. 284b) $$-\Delta_d \doteqdot d\delta + \delta d.$$

In particular, if θ is a 1-form, then

(V1-p. 284c) $$\Delta_d \theta = \Delta\theta - \mathrm{Rc}(\theta).$$

For any function $f : \mathcal{M} \to \mathbb{R}$

(A.4) $$\Delta\nabla f = \nabla\Delta f + \mathrm{Rc}(\nabla f)$$

and

(A.5) $$\Delta|\nabla f|^2 = 2|\nabla\nabla f|^2 + 2\mathrm{Rc}(\nabla f, \nabla f) + 2\nabla f \cdot \nabla(\Delta f),$$

where the dot denotes the metric inner product, i.e., $X \cdot Y = \langle X, Y \rangle = g_{ij} X^i Y^j$.

If $\frac{\partial}{\partial t} g_{ij} = -2 R_{ij}$ on $\mathcal{M} \times (\alpha, \omega)$ and $f : \mathcal{M} \times (\alpha, \omega) \to \mathbb{R}$, then

$$\left(\Delta - \frac{\partial}{\partial t} \right) |\nabla f|^2 = 2 |\nabla \nabla f|^2 + 2 \nabla f \cdot \nabla \left(\left(\Delta - \frac{\partial}{\partial t} \right) f \right).$$

1.9. The cylinder-to-ball rule. The following is an obvious modification of Lemma 2.10 on p. 29 of Volume One.

LEMMA A.2 (Cylinder-to-ball rule). *Let $0 < L \leq \infty$ and let g be a warped-product metric on the topological cylinder $(0, L) \times \mathcal{S}^n$ of the form*

$$g = dr^2 + w(r)^2 \, g_{\mathrm{can}},$$

where $w : (0, L) \to \mathbb{R}_+$ and g_{can} is the canonical round metric of radius 1 on $\mathcal{S}^n$. Then g extends to a smooth metric on $B\left(\vec{0}, L\right)$ (as $r \to 0_+$) if and only if

(V1-2.16) $$\lim_{r \to 0_+} w(r) = 0,$$

(V1-2.17) $$\lim_{r \to 0_+} w'(r) = 1,$$

and

(V1-2.18) $$\lim_{r \to 0_+} \frac{d^{2k} w}{dr^{2k}}(r) = 0 \quad \text{for all } k \in \mathbb{N}.$$

1.10. Volume comparison. We recall the Bishop–Gromov volume comparison (BGVC) theorem.

THEOREM A.3 (Bishop–Gromov volume comparison). *Let $(\mathcal{M}^n, g)$ be a complete Riemannian manifold with $\mathrm{Rc} \geq (n-1) K$, where $K \in \mathbb{R}$. Then for any $p \in \mathcal{M}$,*

$$\frac{\mathrm{Vol} \, B(p, r)}{\mathrm{Vol}_K \, B(p_K, r)}$$

is a nonincreasing function of r, where p_K is a point in the n-dimensional simply-connected space form of constant curvature K and Vol_K denotes the volume in the space form. In particular

(A.6) $$\mathrm{Vol} \, B(p, r) \leq \mathrm{Vol}_K \, B(p_K, r)$$

for all $r > 0$. Given p and $r > 0$, equality holds in (A.6) if and only if $B(p, r)$ is isometric to $B(p_K, r)$.

If $\mathrm{Rc} \geq 0$, we then have the following.

COROLLARY A.4 (BGVC for $\mathrm{Rc} \geq 0$). *If $(\mathcal{M}^n, g)$ is a complete Riemannian manifold with $\mathrm{Rc} \geq 0$, then for any $p \in \mathcal{M}$, the volume ratio $\frac{\mathrm{Vol} \, B(p,r)}{r^n}$ is a nonincreasing function of r. We have $\frac{\mathrm{Vol} \, B(p,r)}{r^n} \leq \omega_n$ for all $r > 0$, where ω_n is the volume of the Euclidean unit n-ball. Equality holds if and only if $(\mathcal{M}^n, g)$ is isometric to Euclidean space.*

As a consequence, we have the following characterization of Euclidean space.

COROLLARY A.5 (Volume characterization of $\mathbb{R}^n$). *If $(\mathcal{M}^n, g)$ is a complete noncompact Riemannian manifold with $\mathrm{Rc} \geq 0$ and if for some $p \in \mathcal{M}$,*
$$\lim_{r \to \infty} \frac{\mathrm{Vol}\, B(p, r)}{r^n} = \omega_n,$$
then $(\mathcal{M}, g)$ is isometric to Euclidean space.

The following result about the volume growth of complete manifolds with nonnegative Ricci curvature is due to Yau (compare with the proof of Theorem 2.92).

COROLLARY A.6 ($\mathrm{Rc} \geq 0$ has at least linear volume growth). *There exists a constant $c(n) > 0$ depending only on n such that if $(\mathcal{M}^n, g)$ is a complete Riemannian manifold with nonnegative Ricci curvature and $p \in \mathcal{M}^n$, then*
$$\mathrm{Vol}\, B(p, r) \geq c(n) \mathrm{Vol}\, B(p, 1) \cdot r$$
for any $r \in [1, 2\,\mathrm{diam}(\mathcal{M}))$.[1]

The **asymptotic volume ratio** of a complete Riemannian manifold $(\mathcal{M}^n, g)$ with $\mathrm{Rc} \geq 0$ is defined by

(A.7) $$\mathrm{AVR}(g) \doteqdot \lim_{r \to \infty} \frac{\mathrm{Vol}\, B(p, r)}{\omega_n r^n},$$

where ω_n is the volume of the unit ball in $\mathbb{R}^n$. By the Bishop–Gromov volume comparison theorem, $\mathrm{AVR}(g) \leq 1$. Again assuming $\mathrm{Rc}(g) \geq 0$, we have for $s \geq r$,

(1)
$$A(s) \leq A(r) \frac{s^{n-1}}{r^{n-1}},$$
where $A(s) \doteqdot \mathrm{Vol}\, \partial B(p, s)$,

(2)

(A.8) $$\frac{\mathrm{Vol}\, B(p, r)}{\omega_n r^n} \geq \frac{A(s)}{n \omega_n s^{n-1}} \geq \mathrm{AVR}(g).$$

We have the following relation between volume ratios and the injectivity radius in the presence of a curvature bound (see for example Theorem 5.42 of [111]).

THEOREM A.7 (Cheeger, Gromov, and Taylor). *Given $c > 0$, $r_0 > 0$, and $n \in \mathbb{N}$, there exists $\iota_0 > 0$ such that if $(\mathcal{M}^n, g)$ is a complete Riemannian manifold with $|\mathrm{sect}| \leq 1$ and if $p \in \mathcal{M}$ is such that*
$$\frac{\mathrm{Vol}\, B(p, r_0)}{r_0^n} \geq c,$$

---

[1] We allow the noncompact case where $\mathrm{diam}(\mathcal{M}) = \infty$.

then
$$\operatorname{inj}(p) \geq \iota_0.$$

1.11. Laplacian and Hessian comparison theorems. Given $K \in \mathbb{R}$ and $r > 0$, let

$$H_K(r) \doteqdot \begin{cases} (n-1)\sqrt{K}\cot\left(\sqrt{K}r\right) & \text{if } K > 0, \\ \frac{n-1}{r} & \text{if } K = 0, \\ (n-1)\sqrt{|K|}\coth\left(\sqrt{|K|}r\right) & \text{if } K < 0, \end{cases}$$

where if $K > 0$ we assume $r < \frac{\pi}{2\sqrt{K}}$. The function $H_K(r)$ is equal to the mean curvature of the $(n-1)$-sphere of radius r in the complete simply-connected Riemannian manifold of constant sectional curvature K.

THEOREM A.8 (Laplacian comparison). *Let $(\mathcal{M}^n, g)$ be a complete Riemannian manifold with $\operatorname{Rc} \geq (n-1)K$, where $K \in \mathbb{R}$. For any $p \in \mathcal{M}^n$ and $x \in \mathcal{M}^n$ at which $d_p(x)$ is smooth, we have*

$$\text{(A.9)} \qquad \Delta d_p(x) \leq H_K(d_p(x)).$$

On the whole manifold, the Laplacian comparison theorem (A.9) *holds in the **sense of distributions**. That is, for any nonnegative C^∞ function φ on $\mathcal{M}^n$ with compact support, we have*

$$\int_{\mathcal{M}^n} d_p(x)\,\Delta\varphi(x)\,d\mu(x) \leq \int_{\mathcal{M}^n} c_K(d_p(x))\,\varphi(x)\,d\mu(x).$$

The following is a special case of the Hessian comparison theorem.

THEOREM A.9 (Hessian comparison theorem). *If $(\mathcal{M}^n, g)$ is a complete Riemannian manifold with $\operatorname{sect} \geq K$, then for any point $p \in \mathcal{M}$ the distance function satisfies*

$$\text{(A.10)} \qquad (\nabla_i \nabla_j d_p)(x) \leq \frac{1}{n-1} H_K(d_p(x))\, g_{ij}$$

*at all points where d_p is smooth (i.e., away from p and the cut locus). On all of $\mathcal{M}$ the above inequality holds in the **sense of support functions**. That is, for every point $x \in \mathcal{M}$ and unit tangent vector $V \in T_x\mathcal{M}$, there exists a C^2 function $v : (-\varepsilon, \varepsilon) \to \mathbb{R}$ with $v(0) = d_p(x)$,*

$$d_p(\exp_x(tV)) \leq v(t), \quad \text{for } t \in (-\varepsilon, \varepsilon),$$

and

$$\left.\frac{d^2}{dt^2}\right|_{t=0} v(t) \leq \frac{1}{n-1} H_K(d_p(x)).$$

Note that $\frac{1}{n-1} H_K(r)$ is equal to the principal curvature of the totally umbilic $(n-1)$-sphere of radius r in the complete simply-connected Riemannian manifold of constant sectional curvature K. In fact Δd_p is the mean curvature of the distance sphere in $\mathcal{M}$ whereas $\nabla\nabla d_p$ is the second fundamental form of the distance sphere in $\mathcal{M}$.

1. RIEMANNIAN GEOMETRY

1.12. Li–Yau differential Harnack estimate. Let $u : \mathcal{M}^n \times [0, \infty) \to \mathbb{R}$ be a positive solution to the heat equation $\frac{\partial u}{\partial t} = \Delta u$ on a complete Riemannian manifold $(\mathcal{M}^n, g)$. Define f by

$$u \doteqdot (4\pi t)^{-n/2} e^{-f}, \tag{A.11}$$

so that $\frac{\partial f}{\partial t} = \Delta f - |\nabla f|^2 - \frac{n}{2t}$.

THEOREM A.10 (Li–Yau differential Harnack estimate). *If $(\mathcal{M}^n, g)$ has nonnegative Ricci curvature, then*

$$\Delta f - \frac{n}{2t} = \frac{\partial f}{\partial t} + |\nabla f|^2 \leq 0. \tag{A.12}$$

Integrating (A.12) yields the following sharp version of the classical Harnack estimate:

$$f(x_2, t_2) - f(x_1, t_1) \leq \frac{d(x_1, x_2)^2}{4(t_2 - t_1)} \tag{A.13}$$

for all $x_1, x_2 \in \mathcal{M}$ and $t_2 > t_1$. If $u = H$ is a fundamental solution centered at a point $x \in \mathcal{M}$, then taking $t_1 \to 0$ implies the Cheeger–Yau estimate:

$$f(y, t) \leq \frac{d(x, y)^2}{4t}. \tag{A.14}$$

In terms of u, the positive solution to the heat equation, on a complete Riemannian manifolds with nonnegative Ricci curvature, we have

$$\frac{\partial}{\partial t} \log u - |\nabla \log u|^2 = \Delta \log u \geq -\frac{n}{2t}$$

and

$$\frac{u(x_2, t_2)}{u(x_1, t_1)} \geq \left(\frac{t_2}{t_1}\right)^{-n/2} \exp\left\{-\frac{d(x_1, x_2)^2}{4(t_2 - t_1)}\right\}.$$

For a fundamental solution $u = H$,

$$H(y, t) \geq (4\pi t)^{-n/2} \exp\left\{-\frac{d(x, y)^2}{4t}\right\}.$$

1.13. Calabi's trick. In this subsection we give an example of Calabi's trick which is useful in the study of heat-type equations and analytic aspects of the Ricci flow. In particular, a slight modification of the discussion below applies to the proof of the local first derivative of curvature estimate for the Ricci flow (see Theorem A.30).

First, let us recall some facts about the distance function. Let $(\mathcal{M}^n, g)$ be a Riemannian manifold. Given $p \in \mathcal{M}$, the distance function $r(x) \doteqdot d(x, p)$ is Lipschitz on $\mathcal{M}$ with Lipschitz constant 1. Let $\mathrm{Cut}(p)$ denote the cut locus of p and let

$$C_p \doteqdot \{V \in T_p\mathcal{M} : d(p, \exp_p(V)) = |V|\},$$

so that $\operatorname{Cut}(p) = \exp_p(\partial C_p)$. The cut locus is a closed set with measure zero. We have
$$\exp_p|_{\operatorname{int} C_p} : C_p \backslash \partial C_p \to \mathcal{M} \backslash \operatorname{Cut}(p)$$
is a diffeomorphism. Let $\partial/\partial r = \frac{1}{|x|}\sum_{i=1}^n x^i \frac{\partial}{\partial x^i}$ denote the unit radial vector field on $T_p\mathcal{M} - \{\vec{0}\}$. If $x \notin \operatorname{Cut}(p) \cup \{p\}$, then r is smooth at x, $\nabla r(x) = (\exp_p)_* \partial/\partial r$, and $|\nabla r(x)| = 1$.

Suppose that we have a function $F : \mathcal{M} \times [0, T) \to \mathbb{R}$ which satisfies the differential inequality

(A.15) $$\left(\frac{\partial}{\partial t} - \Delta\right) F \leq C^2 - F^2$$

for some constant C. For an example of such a function, see the proof of the local first derivative of curvature estimate in Part II of this volume.

If $\mathcal{M}$ is a *closed* manifold, then we can apply the maximum principle to F to obtain the estimate
$$F(x, t) \leq C \coth(Ct),$$
where the RHS is the solution to the ODE $\frac{df}{dt} = C^2 - f^2$ with $\lim_{t \searrow 0} f(t) = +\infty$. On the other hand, if $\mathcal{M}$ is *noncompact*, then one way of obtaining an estimate for F is to *localize* the equation by introducing a cut-off function.

In particular, suppose $(\mathcal{M}^n, g)$ satisfies $\operatorname{Rc} \geq -(n-1)K$ for some $K > 0$. Let $\eta : [0, \infty) \to \mathbb{R}$ be a smooth nonincreasing function satisfying $\eta(s) = 1$ for $0 \leq s \leq \frac{1}{2}$ and $\eta(s) = 0$ for $s \geq 1$. We may assume

(A.16) $$0 \geq \eta' \geq -6\sqrt{\eta} \quad \text{and} \quad -C_0\sqrt{\eta} \leq \eta'' \leq C_0,$$

where C_0 is a universal constant.[2] Given $p \in \mathcal{M}$ and $A > 0$, define $\phi(x) \doteqdot \eta\left(\frac{d(x,p)}{A}\right)$. Recall that in $\mathcal{M} \backslash (\{p\} \cup \operatorname{Cut}(p))$
$$|\nabla d(\cdot, p)| = 1, \quad \Delta d(\cdot, p) \leq (n-1)\sqrt{K} \coth\left(\sqrt{K} d(\cdot, p)\right),$$
where the Laplacian estimate follows from (A.9) and the assumption $\operatorname{Rc} \geq -(n-1)K$. Hence at points $x \notin \operatorname{Cut}(p)$, ϕ is smooth and

(A.17) $$|\nabla \phi|^2 \leq C\phi, \quad \Delta \phi = \frac{1}{A}\eta' \cdot \Delta d + \frac{1}{A^2}\eta'' |\nabla d|^2 \geq -C\sqrt{\phi},$$

where C depends on A (and where we used (A.16), $\eta' \leq 0$, and the fact that the support of $\eta'\left(\frac{d(\cdot,p)}{A}\right)$ is contained in $B(p, A) \backslash B\left(p, \frac{A}{2}\right)$).

We calculate that for $x \notin \operatorname{Cut}(p)$,
$$\phi\left(\frac{\partial}{\partial t} - \Delta\right)(\phi F) = \phi^2\left(\frac{\partial}{\partial t} - \Delta\right)F - \phi(\Delta\phi)F - 2\phi\nabla\phi \cdot \nabla F$$
$$\leq -2\nabla\phi \cdot \nabla(\phi F) + \phi^2 C^2 - \phi^2 F^2 - \phi F \Delta\phi + 2F|\nabla\phi|^2$$

---

[2] Let ζ be a cut-off function with $0 \geq \zeta' \geq -3$ and $|\zeta''| \leq C$, where C is a universal constant, and define $\eta = \zeta^2$.

1. RIEMANNIAN GEOMETRY

(the above calculation holds for any C^2 function ϕ). That is, if $x \notin \text{Cut}(p)$, then by (A.17) we have wherever $F \geq 0$,

$$\left[\phi\left(\frac{\partial}{\partial t} - \Delta\right) + 2\nabla\phi \cdot \nabla\right](\phi F) \leq \phi^2 C^2 - (\phi F)^2 + C\phi^{3/2} F + C\phi F$$

$$\leq \frac{1}{2}\left(C_1^2 - (\phi F)^2\right)$$

for some constant $C_1 < \infty$. Hence

$$\left[\phi\left(\frac{\partial}{\partial t} - \Delta\right) + 2\nabla\phi \cdot \nabla\right](t\phi F) \leq \frac{t}{2}\left(C_1^2 - (\phi F)^2\right) + \phi^2 F$$

$$\leq \frac{1}{2t}\left(C_1^2 t^2 - (t\phi F)^2 + 2t\phi F\right).$$

Since $\lim_{t\to 0}(t\phi F) = 0$ and $t\phi F$ has compact support, we may apply the maximum principle to conclude

$$C_1^2 t^2 - (t\phi F)^2 + 2t\phi F \geq 0$$

at a maximum point of $t\phi F$ on $\mathcal{M} \times [0, \bar{t}]$ for any $\bar{t} \in (0, T)$. In particular,

$$t\phi(x) F(x,t) \leq C_2$$

on $\mathcal{M} \times [0, T)$, where C_2 depends only on C in (A.15) and A. In particular,

$$F(x,t) \leq \frac{C_2}{t}$$

on $B(p, A/2) \times [0, T)$.

EXERCISE A.11. What happens to the constant C_2 as $A \to \infty$?

In the above we have assumed that at the choice (x_0, t_0) of maximum point of $t\phi F$, the function ϕ is C^2. However ϕ is only Lipschitz continuous since the distance function $d(\cdot, p)$ is only Lipschitz. To solve this problem, we apply **Calabi's trick** (see [**42**]). In particular, suppose for our choice of maximum point (x_0, t_0) of $t\phi F$, we have that ϕ is not smooth at x_0, i.e., $x_0 \in \text{Cut}(p)$. Let

$$\gamma : [0, d(x_0, p)] \to \mathcal{M}$$

be a unit speed minimal geodesic joining x_0 to p. Consider $q \doteqdot \gamma(d(x_0, p) - \delta)$ for any small $\delta = d(q, p) > 0$. We have $d(x_0, q) + \delta = d(x_0, p)$ and also q is not in the cut locus of x_0 since $\delta > 0$. Consider the function

$$G(x,t) \doteqdot t\eta\left(\frac{d(x,q) + \delta}{A}\right) F(x,t).$$

Since $d(x,q) + \delta \geq d(x,p)$ and η is nonincreasing, we have

$$G(x,t) \leq (t\phi F)(x,t)$$

at any point (x,t) where $F(x,t) \geq 0$. Since $G(x_0, t_0) = t_0\phi(x_0) F(x_0, t_0) = \max_{\mathcal{M} \times [0,\bar{t}]}(t\phi F)$, we conclude that $G(x,t)$ also achieves its maximum at

(x_0, t_0). However, since $d(\cdot, q)$ is C^2 at x_0, we may apply the maximum principle to the equation for G. One checks (Exercise: Prove this) that

$$G(x, t) \leq C_2'$$

on $\mathcal{M} \times [0, T)$, where $C_2' \to C_2$ as $\delta \to 0$. Hence

$$\max_{\mathcal{M} \times [0, \bar{t}]} (t\phi F) \leq C_2.$$

2. Basic Ricci flow

The (unnormalized) Ricci flow equation on a manifold $\mathcal{M}^n$ is

(A.18) $$\frac{\partial}{\partial t} g_{ij} = -2 R_{ij},$$

whereas its cousin, the (volume-preserving) normalized Ricci flow equation on a closed manifold, is

(A.19) $$\frac{\partial}{\partial t} g_{ij} = -2 R_{ij} + \frac{2r}{n} g_{ij},$$

where $r(t) \doteqdot \left(\int_{\mathcal{M}} R d\mu / \int_{\mathcal{M}} d\mu \right)(t)$ is the average scalar curvature.

REMARK A.12. The variation of a Riemannian metric in the direction of the Ricci curvature was considered by Bourguignon (see Proposition VIII.4 of [**32**]). A fundamental work using a nonlinear heat-type equation (the harmonic map heat flow) is by Eells and Sampson [**135**].

Substituting $h = -2 \operatorname{Rc}$ into Lemma A.1 yields the following result, which gives the evolution equations for the Levi-Civita connection and curvatures under the Ricci flow (A.18). (See Corollary 6.6 (1) on p. 175, Lemma 6.15 on p. 179, and Lemmas 6.9 and 6.7 on p. 176 of Volume One.)

COROLLARY A.13 (Ricci flow evolutions). *Suppose $g(t)$ is a solution of the Ricci flow: $\frac{\partial}{\partial t} g = -2 \operatorname{Rc}$.*

(1) *The Levi-Civita connection Γ of g evolves by*

(V1-6.1) $$\frac{\partial}{\partial t} \Gamma_{ij}^k = -g^{k\ell} (\nabla_i R_{j\ell} + \nabla_j R_{i\ell} - \nabla_\ell R_{ij}).$$

(2) *Under the Ricci flow, the $(4, 0)$-Riemann curvature tensor evolves by*

(V1-6.17) $$\frac{\partial}{\partial t} R_{ijk\ell} = \Delta R_{ijk\ell} + 2 (B_{ijk\ell} - B_{ij\ell k} + B_{ikj\ell} - B_{i\ell jk})$$
$$- \left(R_i^p R_{pjk\ell} + R_j^p R_{ipk\ell} + R_k^p R_{ijp\ell} + R_\ell^p R_{ijkp} \right),$$

where

(V1-6.16) $$B_{ijk\ell} \doteqdot -g^{pr} g^{qs} R_{ipjq} R_{kr\ell s} = -R_{pij}^q R_{q\ell k}^p.$$

(3) The Ricci tensor Rc of g evolves by

(V1-6.7) $$\frac{\partial}{\partial t}R_{jk} = \Delta_L R_{jk} = \Delta R_{jk} + 2g^{pq}g^{rs}R_{pjkr}R_{qs} - 2g^{pq}R_{jp}R_{qk}.$$

In dimension 3, this equation becomes

(V1-6.10) $$\frac{\partial}{\partial t}R_{jk} = \Delta R_{jk} + 3RR_{jk} - 6g^{pq}R_{jp}R_{qk} + \left(2\left|\text{Rc}\right|^2 - R^2\right)g_{jk}.$$

(4) The scalar curvature R of g evolves by

(V1-6.6) $$\frac{\partial R}{\partial t} = \Delta R + 2\left|\text{Rc}\right|^2.$$

(5) The volume form $d\mu$ evolves by

(V1-6.5) $$\frac{\partial}{\partial t}d\mu = -R\,d\mu.$$

(6) If γ_t is a smooth 1-parameter family of geodesic loops, then

$$\left.\frac{d}{dt}L_t\left(\gamma_t\right)\right|_{t=\tau} = -\int_{\gamma_\tau}\text{Rc}\left(\frac{\partial\gamma_\tau}{\partial s},\frac{\partial\gamma_\tau}{\partial s}\right)ds,$$

where s is the arc length parameter.

Part (6) is similar to Lemma 5.71 on p. 152 of Volume One.

EXERCISE A.14. Derive the corresponding formulas for the normalized Ricci flow. For example, under (A.19) we have

$$\frac{\partial R}{\partial t} = \Delta R + 2\left|\text{Rc}\right|^2 - \frac{2}{n}rR.$$

2.1. Short- and long-time existence. Any smooth metric on a closed manifold will flow uniquely, at least for a little while (Theorem 3.13 on p. 78 of Volume One).

THEOREM A.15 (Short-time existence for $\mathcal{M}$ closed). *If $(\mathcal{M}^n, g_0)$ is a closed Riemannian manifold, then there exists a unique solution $g(t)$ to the Ricci flow defined on some positive time interval $[0,\varepsilon)$ such that $g(0) = g_0$.*

As long as the curvature stays bounded, the solution exists (Corollary 7.2 on p. 224 of Volume One).

THEOREM A.16 (Long-time existence for $\mathcal{M}$ closed). *If $(\mathcal{M}^n, g(t))$, $t \in (0,T)$, where $T < \infty$, is a solution to the Ricci flow on a closed manifold with $\sup_{\mathcal{M}\times(0,T)}\left|\text{Rm}\right| < \infty$, then the solution $g(t)$ can be uniquely extended past time T.*

In the theorem above the condition $\sup_{\mathcal{M}\times(0,T)}\left|\text{Rm}\right| < \infty$ may be replaced by $\sup_{\mathcal{M}\times(0,T)}\left|\text{Rc}\right| < \infty$; this was proved by Sesum [**321**].

W.-X. Shi generalized Hamilton's short-time existence theorem to complete solutions with bounded curvature on noncompact manifolds.

458   A. BASIC RICCI FLOW THEORY

THEOREM A.17 (Short-time existence on noncompact manifolds). *Let M be a noncompact manifold and let g_0 be a complete metric with bounded sectional curvature. There exists a complete solution $g(t)$, $t \in [0,T)$, of the Ricci flow with $g(0) = g_0$ and curvature bounded on compact time intervals. This solution is unique in the class of complete solutions with curvature bounded on compact time intervals.*

2.2. Maximum principles for scalars, tensors and systems. A form of the scalar maximum principle useful for the Ricci flow is the following (Theorem 4.4 on p. 96 of Volume One).

THEOREM A.18 (Scalar maximum principle: ODE to PDE). *Let $u : \mathcal{M}^n \times [0,T) \to \mathbb{R}$ be a C^2 function on a closed manifold satisfying*

$$\frac{\partial u}{\partial t} \leq \Delta_{g(t)} u + \langle X, \nabla u \rangle + F(u)$$

and $u(x,0) \leq C$ for all $x \in \mathcal{M}$, where $g(t)$ is a 1-parameter family of metrics and F is locally Lipschitz. Let $\varphi(t)$ be the solution to the initial-value problem

$$\frac{d\varphi}{dt} = F(\varphi),$$
$$\varphi(0) = C.$$

Then

$$u(x,t) \leq \varphi(t)$$

for all $x \in \mathcal{M}$ and $t \in [0,T)$ such that $\varphi(t)$ exists.

Since under the Ricci flow, by (V1-6.6) $\frac{\partial R}{\partial t} \geq \Delta R + \frac{2}{n} R^2$, we have the following (see the proof of Lemma 6.53 on pp. 209–210 of Volume One).

COROLLARY A.19 (Scalar curvature lower bound). *Let $(\mathcal{M}^n, g(t))$, where $0 \leq t < T$, be a solution of the Ricci flow for which the maximum principle holds. If $\inf_{x \in \mathcal{M}^n} R(x,t_0) \doteqdot \rho > 0$ for some $t_0 \in [0,T)$, then*

$$R_{\inf}(t) \doteqdot \inf_{x \in \mathcal{M}} R(x,t) \geq \frac{1}{\frac{1}{\rho} - \frac{2}{n}(t-t_0)}.$$

In particular, $g(t)$ becomes singular in finite time.

Moreover, we have the following.

LEMMA A.20 (Minimum scalar curvature monotonicity).
(1) *Under the unnormalized Ricci flow the minimum scalar curvature is a nondecreasing function of time.*
(2) *Under the normalized Ricci flow the minimum scalar curvature is nondecreasing as long as it is nonpositive.*

PROOF. Let $\rho(t) \doteqdot R_{\min}(t)$. Under the unnormalized Ricci flow,
$$\frac{d\rho}{dt} \geq \frac{2}{n}\rho^2 \geq 0.$$
Under the normalized Ricci flow,
$$\frac{d\rho}{dt} \geq \frac{2}{n}\rho(\rho - r) \geq 0$$
as long as $\rho \leq 0$ (note that $\rho - r \leq 0$ always). $\square$

The weak maximum principle as applied to symmetric 2-tensors says the following (see Theorem 4.6 on p. 97 of Volume One).[3]

THEOREM A.21 (Maximum principle for 2-tensors). *Let $g(t)$ be a smooth 1-parameter family of Riemannian metrics on a closed manifold $\mathcal{M}^n$. Let $\alpha(t) \in C^\infty(T^*\mathcal{M} \otimes_S T^*\mathcal{M})$ be a symmetric $(2,0)$-tensor satisfying the semilinear heat equation*
$$\frac{\partial}{\partial t}\alpha \geq \Delta_{g(t)}\alpha + \beta,$$
*where $\beta(\alpha, g, t)$ is a symmetric $(2,0)$-tensor which is locally Lipschitz in all its arguments and satisfies the **null eigenvector assumption** that*
$$\beta(V, V)(x, t) = \left(\beta_{ij} V^i V^j\right)(x, t) \geq 0$$
at any point and time (x, t) where $\left(\alpha_{ij} W^i W^j\right)(x, t) \geq 0$ for all W and
$$\left(\alpha_{ij} V^j\right)(x, t) = 0.$$
If $\alpha(0) \geq 0$ (that is, if $\alpha(0)$ is positive semidefinite), then $\alpha(t) \geq 0$ for all $t \geq 0$ such that the solution exists.

Applying this to the evolution equation (V1-6.10) for the Ricci tensor in dimension 3 yields the following (Corollary 6.11 on p. 177 of Volume One).

COROLLARY A.22 (3d positive Ricci curvature persists). *Let $g(t)$ be a solution of the Ricci flow on a closed 3-manifold with $g(0) = g_0$. If g_0 has positive (nonnegative) Ricci curvature, then $g(t)$ has positive (nonnegative) Ricci curvature for as long as the solution exists.*

2.3. Uhlenbeck's trick. **Uhlenbeck's trick** allows us to put the evolution equation (V1-6.17) satisfied by Rm into a particularly nice form. (See pp. 180–183 in Section 2 of Chapter 6 in Volume One.) Let $(\mathcal{M}^n, g(t))$, $t \in [0, T)$, be a solution to the Ricci flow with $g(0) = g_0$. Let V be a vector bundle over $\mathcal{M}$ isomorphic to $T\mathcal{M}$, and let $\iota_0 : V \to T\mathcal{M}$ be a bundle isomorphism. Then if we define a metric h on V by

(V1-p. 181a) $$h \doteqdot \iota_0^*(g_0),$$

we automatically obtain a bundle isometry

(V1-p. 181b) $$\iota_0 : (V, h) \to (T\mathcal{M}, g_0).$$

---

[3]The statement we give here is slightly stronger, since in the null eigenvector assumption we also assume that $\left(\alpha_{ij} V^i V^j\right)(x, t) \geq 0$ at (x, t).

LEMMA A.23. *If we evolve the isometry $\iota(t)$ by*

(V1-6.19a) $$\frac{\partial}{\partial t}\iota = \operatorname{Rc}\circ\iota,$$

(V1-6.19b) $$\iota(0) = \iota_0,$$

then the bundle maps

$$\iota(t) : (V, h) \to (T\mathcal{M}, g(t))$$

remain isometries.[4]

We define the Laplacian acting on tensor bundles of $T\mathcal{M}$ and V by

$$\Delta_D \doteqdot \operatorname{tr}_g (\nabla_D \circ \nabla_D),$$

where $(\nabla_D)_X(\xi) = \iota^{-1}(\nabla_X(\iota(\xi)))$ acting on sections of V and ∇_D is naturally extended to act on tensor bundles. For $x \in \mathcal{M}$ and $X, Y, Z, W \in V_x$, the tensor

(V1-p. 182) $$\iota^* \operatorname{Rm} \in C^\infty \left(\Lambda^2 V \otimes_S \Lambda^2 V \right)$$

is defined by

(V1-6.20) $$(\iota^* \operatorname{Rm})(X, Y, Z, W) \doteqdot \operatorname{Rm}(\iota(X), \iota(Y), \iota(Z), \iota(W)).$$

Let R_{abcd} denote the components of $\iota^* \operatorname{Rm}$ with respect to an orthonormal basis of sections of V. We have

(V1-6.21) $$\frac{\partial}{\partial t} R_{abcd} = \Delta_D R_{abcd} + 2 (B_{abcd} - B_{abdc} + B_{acbd} - B_{adbc}),$$

where

$$B_{abcd} \doteqdot -h^{eg} h^{fh} R_{aebf} R_{cgdh}.$$

We may rewrite the above equation in a more elegant way. (See pp. 183–187 in Section 3 of Chapter 6 in Volume One.) Let $\mathfrak{g}$ be a Lie algebra endowed with an inner product $\langle \cdot, \cdot \rangle$. Choose a basis $\{\varphi^\alpha\}$ of $\mathfrak{g}$ and let $C^{\alpha\beta}_\gamma$ denote the structure constants defined by $[\varphi^\alpha, \varphi^\beta] \doteqdot \sum_\gamma C^{\alpha\beta}_\gamma \varphi^\gamma$. We define the **Lie algebra square** $L^\# \in \mathfrak{g} \otimes_S \mathfrak{g}$ of L by

(V1-6.24) $$(L^\#)_{\alpha\beta} \doteqdot C^{\gamma\delta}_\alpha C^{\varepsilon\zeta}_\beta L_{\gamma\varepsilon} L_{\delta\zeta}.$$

For each $x \in \mathcal{M}^n$, the vector space $\Lambda^2 T_x^* \mathcal{M}$ can be given the structure of a Lie algebra $\mathfrak{g}$ isomorphic to $\mathfrak{so}(n)$. Given $U, V \in \Lambda^2 T_x^* \mathcal{M}$, we define their Lie bracket by

(V1-6.25) $$[U, V]_{ij} \doteqdot g^{k\ell} (U_{ik} V_{\ell j} - V_{ik} U_{\ell j}).$$

THEOREM A.24 (Rm evolution after Uhlenbeck's trick). *If $g(t)$ is a solution of the Ricci flow, then the curvature $\iota^* \operatorname{Rm}$ defined in (V1-6.20) evolves by*

(V1-6.27) $$\frac{\partial}{\partial t}(\iota^* \operatorname{Rm}) = \Delta_D(\iota^* \operatorname{Rm}) + (\iota^* \operatorname{Rm})^2 + (\iota^* \operatorname{Rm})^\#.$$

---

[4] The statement here is the one we intended in Claim 6.21 of Volume One.

(See pp. 187–189 in Section 4 of Chapter 6 in Volume One.)

The PDE (V1-6.27) governing the behavior of Rm corresponds to the ODE

(V1-6.28) $$\frac{d}{dt}\mathbb{M} = \mathbb{M}^2 + \mathbb{M}^\#.$$

In dimension 3, if $\mathbb{M}_0$ is diagonal, then $\mathbb{M}(t)$ remains diagonal, and its eigenvalues satisfy

(V1-6.32) $$\begin{aligned}\frac{d\lambda}{dt} &= \lambda^2 + \mu\nu, \\ \frac{d\mu}{dt} &= \mu^2 + \lambda\nu, \\ \frac{d\nu}{dt} &= \nu^2 + \lambda\mu.\end{aligned}$$

From now on we shall assume $\lambda \geq \mu \geq \nu$, a condition which is preserved under the ODE.

Theorem 4.8 on p. 101 of Volume One applied to the Riemann curvature operator Rm yields the following.

THEOREM A.25 (Maximum principle for Rm: ODE to PDE). *Let $g(t)$ be a solution to the Ricci flow on a closed manifold $\mathcal{M}^n$ and let $\mathcal{K}(t)$ be a closed subset of $\mathcal{E} \doteqdot \Lambda^2 V \otimes_S \Lambda^2 V$ for all $t \in [0, T)$ satisfying the following properties:*

(1) *the space-time track $\bigcup_{t \in [0,T)} (\mathcal{K}(t) \times \{t\})$ is a closed subset of $\mathcal{E} \times [0, T)$;*
(2) *$\mathcal{K}(t)$ is invariant under parallel translation by $\bar{\nabla}(t)$ for all $t \in [0, T)$;*
(3) *$\mathcal{K}_x(t) \doteqdot \mathcal{K}(t) \cap \pi^{-1}(x)$ is a closed convex subset of $\mathcal{E}_x$ for all $x \in \mathcal{M}$ and $t \in [0, T)$; and*
(4) *every solution $\mathbb{M}$ of the ODE (V1-6.28) with Rm $(t_0) \in \mathcal{K}_x(t_0)$ defined in each fiber $\mathcal{E}_x$ remains in $\mathcal{K}_x(t)$ for all $t \geq t_0$ and $t_0 \in [0, T)$.*

If $(\iota^* \text{Rm})(0) \in \mathcal{K}(0)$, then $(\iota^* \text{Rm})(t) \in \mathcal{K}(t)$ for all $t \in [0, T)$.

2.4. 3-manifolds with positive Ricci curvature. The following famous theorem of Hamilton started the Ricci flow (RF). (See Theorem 6.3 on p. 173 of Volume One.)

THEOREM A.26 (RF on closed 3-manifolds with Rc > 0). *Let $(\mathcal{M}^3, g_0)$ be a closed Riemannian 3-manifold of positive Ricci curvature. Then a unique solution $g(t)$ of the normalized Ricci flow with $g(0) = g_0$ exists for all positive time; and as $t \to \infty$, the metrics $g(t)$ converge exponentially fast in every C^k-norm, $k \in \mathbb{N}$, to a metric g_∞ of constant positive sectional curvature.*

The above result generalizes to orbifolds (see [**191**]).

THEOREM A.27 (RF on closed 3-orbifolds with Rc > 0). *If $(\mathcal{V}^3, g_0)$ is a closed Riemannian 3-orbifold of positive Ricci curvature, then a unique solution $g(t)$ of the normalized Ricci flow with $g(0) = g_0$ exists for all $t > 0$, and as $t \to \infty$, the $g(t)$ converge to a metric g_∞ of constant positive sectional curvature. In particular, $\mathcal{V}^3$ is diffeomorphic to the quotient of $\mathcal{S}^3$ by a finite group of isometries.*

One of the main ideas in the proof of Theorem A.26 is to apply Theorem A.25 to obtain pointwise curvature estimates which lead to the curvature tending to constant as the solution evolves. From now on we shall assume that $\lambda(t) \geq \mu(t) \geq \nu(t)$ are solutions of the ODE system (V1-6.32). The evolutions of various quantities and their applications to the Ricci flow on closed 3-manifolds via the maximum principle for systems are given as follows.

(1)

$$\text{(A.20)} \qquad \frac{d}{dt}(\nu + \mu) = \nu^2 + \mu^2 + (\nu + \mu)\lambda \geq 0,$$

with the inequality holding whenever $\mu + \nu \geq 0$. So Rc ≥ 0 is preserved in dimension 3 under the Ricci flow.

(2)

$$\text{(A.21)} \qquad \frac{d}{dt}\log\left(\frac{\lambda}{\nu + \mu}\right) = \frac{\mu^2(\nu - \lambda) + \nu^2(\mu - \lambda)}{\lambda(\nu + \mu)} \leq 0.$$

If g_0 has positive Ricci curvature, then so does $g(t)$ and there exists a constant $C_1 < \infty$ such that

$$\text{(A.22)} \qquad \lambda(\text{Rm}) \leq C_1[\nu(\text{Rm}) + \mu(\text{Rm})].$$

(3) If $\nu + \mu > 0$, then

$$\frac{d}{dt}\log\left(\frac{\lambda - \nu}{(\nu + \mu + \lambda)^{1-\delta}}\right)$$
$$= \delta(\nu + \lambda - \mu) - (1 - \delta)\frac{(\nu + \mu)\mu + (\mu - \nu)\lambda + \mu^2}{\nu + \mu + \lambda}$$
$$\leq \delta(\nu + \lambda - \mu) - (1 - \delta)\frac{\mu^2}{\nu + \mu + \lambda}.$$

Since $\nu + \lambda - \mu \leq \lambda \leq 2C_1\mu$ and $\frac{\mu}{\nu+\mu+\lambda} \geq \frac{\nu+\mu}{6\lambda} \geq \frac{1}{6C_1}$, choosing $\delta > 0$ small enough so that $\frac{\delta}{1-\delta} \leq \frac{1}{12C_1^2}$, we have

$$\frac{d}{dt}\log\left(\frac{\lambda - \nu}{(\nu + \mu + \lambda)^{1-\delta}}\right) \leq 0.$$

So if g_0 has positive Ricci curvature, then there exist constants $C < \infty$ and $\delta > 0$ such that

(A.23) $$\frac{\lambda(\mathrm{Rm}) - \nu(\mathrm{Rm})}{R^{1-\delta}} \leq C.$$

We shall call (A.23) the **'pinching improves' estimate**.

Next we consider estimates for the derivatives of Rm.

2.5. Global derivative estimates. For solutions to the Ricci flow on a closed 3-manifold with positive Ricci curvature, we have the following estimate for the gradient of the scalar curvature. (See Theorem 6.35 on p. 194 of Volume One.)

THEOREM A.28 (3-dimensional gradient of scalar curvature estimate). *Let $\left(\mathcal{M}^3, g(t)\right)$ be a solution of the Ricci flow on a closed 3-manifold with $g(0) = g_0$. If $\mathrm{Rc}(g_0) > 0$, then there exist $\bar{\beta}, \bar{\delta} > 0$ depending only on g_0 such that for any $\beta \in \left[0, \bar{\beta}\right]$, there exists C depending only on β and g_0 such that*
$$\frac{|\nabla R|^2}{R^3} \leq \beta R^{-\bar{\delta}/2} + C R^{-3}.$$

After a short time, the higher derivatives of the curvature are bounded in terms of the space-time bound for the curvature. (See Theorem 7.1 on pp. 223–224 of Volume One.)

THEOREM A.29 (Bernstein–Bando–Shi estimate). *Let $(\mathcal{M}^n, g(t))$ be a solution of the Ricci flow for which the maximum principle applies to all the quantities that we consider. (This is true in particular if $\mathcal{M}$ is compact.) Then for each $\alpha > 0$ and every $m \in \mathbb{N}$, there exists a constant $C(m, n, \alpha)$ depending only on m, and n, and $\max\{\alpha, 1\}$ such that if*
$$|\mathrm{Rm}(x,t)|_{g(t)} \leq K \quad \text{for all } x \in \mathcal{M} \text{ and } t \in [0, \tfrac{\alpha}{K}],$$
then for all $x \in \mathcal{M}$ and $t \in (0, \tfrac{\alpha}{K}]$,

(V1-p. 224) $$|\nabla^m \mathrm{Rm}(x,t)|_{g(t)} \leq \frac{C(m,n,\alpha) K}{t^{m/2}}.$$

With all of the above estimates and some more work, one obtains Theorem A.26.

Finally we mention that an important local version of Theorem A.29 is the following.

THEOREM A.30 (Shi—local first derivative estimate). *For any $\alpha > 0$ there exists a constant $C(n, K, r, \alpha)$ depending only on K, r, α and n such that if $\mathcal{M}^n$ is a manifold, $p \in \mathcal{M}$, and $g(t)$, $t \in [0, \tau]$, $0 < \tau \leq \alpha/K$, is a solution to the Ricci flow on an open neighborhood U of p containing $\bar{B}_{g(0)}(p, r)$ as a compact subset and if*
$$|\mathrm{Rm}(x,t)| \leq K \text{ for all } x \in U \text{ and } t \in [0, \tau],$$

then

(A.24) $$|\nabla \operatorname{Rm}(y,t)| \leq \frac{C(n,K,r,\alpha)}{\sqrt{t}} = \frac{C(n,\sqrt{K}r,\alpha)K}{\sqrt{t}}$$

for all $(y,t) \in B_{g(0)}(p,r/2) \times (0,\tau]$. Given in addition $\beta > 0$ and $\gamma > 0$, if also $\gamma/\sqrt{K} \leq r \leq \beta/\sqrt{K}$, then there exists $C(n,\alpha,\beta,\gamma)$ such that under the above assumptions

$$|\nabla \operatorname{Rm}| \leq C(n,\alpha,\beta,\gamma) \frac{K}{\sqrt{t}}$$

in $B_{g(0)}(p,r/2) \times (0,\tau]$.

For a proof and applications, see W.-X. Shi [**329**], [**330**], Hamilton [**186**], [**111**], or Part II of this volume.

2.6. The Hamilton–Ivey estimate. The following result reveals the precise sense in which all sectional curvatures of a complete 3-manifold evolving by the Ricci flow are dominated by the positive sectional curvatures. (See [**186**] or Theorem 9.4 on p. 258 of Volume One.)

THEOREM A.31 (3d Hamilton–Ivey curvature estimate). *Let $(\mathcal{M}^3, g(t))$ be any solution of the Ricci flow on a closed 3-manifold for $0 \leq t < T$. Let $\nu(x,t)$ denote the smallest eigenvalue of the curvature operator. If $\inf_{x \in \mathcal{M}} \nu(x,0) \geq -1$, then at any point $(x,t) \in \mathcal{M} \times [0,T)$ where $\nu(x,t) < 0$, the scalar curvature is estimated by*

(A.25) $$R \geq |\nu|(\log|\nu| + \log(1+t) - 3).$$

2.7. Ricci solitons. If g is a Ricci soliton, then

$$\operatorname{Rc} - \frac{\rho}{n}g = \mathcal{L}_{X^\flat}g$$

for some $\rho \in \mathbb{R}$ and 1-form X. Under this equation we have

(V1-5.16) $$\Delta(R-\rho) + \langle \nabla(R-\rho), X \rangle + 2\left|\operatorname{Rc} - \frac{\rho}{n}g\right|^2 + \frac{2\rho}{n}(R-\rho) = 0.$$

Using this formula, in Proposition 5.20 on p. 117 of Volume One, the following classification result for Ricci solitons was proved. (See Chapter 1 of this volume for the relevant definitions.)

PROPOSITION A.32 (Expanders or steadies on closed manifolds are Einstein). *Any expanding or steady Ricci soliton on a closed n-dimensional manifold is Einstein. A shrinking Ricci soliton on a closed n-dimensional manifold has positive scalar curvature.*

In dimension 2, all solitons have constant curvature. (See Proposition 5.21 on p. 118 of Volume One.)

PROPOSITION A.33 (Ricci solitons on closed surfaces are trivial). *If $(\mathcal{M}^2, g(t))$ is a self-similar solution of the normalized Ricci flow on a Riemannian surface, then $g(t) \equiv g(0)$ is a metric of constant curvature.*

3. Basic singularity theory for Ricci flow

The knowledge of which geometry aims is the knowledge of the eternal.
– Plato

Geometry is knowledge of the eternally existent. – Pythagoras

And perhaps, posterity will thank me for having shown it that the ancients did not know everything. – Pierre Fermat

In this section we review some basic singularity theory as developed by Hamilton and discussed in Volume One.

3.1. Long-existing solutions and singularity types.
For the following, see pp. 234–236 in Section 1 of Chapter 8 in Volume One.

DEFINITION A.34.
- An **ancient solution** is a solution that exists on a past time interval $(-\infty, \omega)$.
- An **immortal solution** is a solution that exists on a future time interval (α, ∞).
- An **eternal solution** is a solution that exists for all time $(-\infty, \infty)$.

DEFINITION A.35 (Singularity types). Let $(\mathcal{M}^n, g(t))$ be a solution of the Ricci flow that exists up to a maximal time $T \leq \infty$.

- One says $(\mathcal{M}, g(t))$ forms a **Type I singularity** if $T < \infty$ and
$$\sup_{\mathcal{M} \times [0,T)} (T-t) |\mathrm{Rm}(\cdot, t)| < \infty.$$

- One says $(\mathcal{M}, g(t))$ forms a **Type IIa singularity** if $T < \infty$ and
$$\sup_{\mathcal{M} \times [0,T)} (T-t) |\mathrm{Rm}(\cdot, t)| = \infty.$$

- One says $(\mathcal{M}, g(t))$ forms a **Type IIb singularity** if $T = \infty$ and
$$\sup_{\mathcal{M} \times [0,\infty)} t |\mathrm{Rm}(\cdot, t)| = \infty.$$

- One says $(\mathcal{M}, g(t))$ forms a **Type III singularity** if $T = \infty$ and
$$\sup_{\mathcal{M} \times [0,\infty)} t |\mathrm{Rm}(\cdot, t)| < \infty.$$

To this we add the following.

DEFINITION A.36 (More singularity types). If $g(t)$ is defined on $(0, T]$, where $T < \infty$, then

- one says $(\mathcal{M}, g(t))$ forms a **Type IIc singularity** as $t \to 0$ if
$$\sup_{\mathcal{M} \times (0,T]} t |\mathrm{Rm}(\cdot, t)| = \infty;$$

- one says $(\mathcal{M}, g(t))$ forms a **Type IV singularity** as $t \to 0$ if
$$\sup_{\mathcal{M} \times (0,T]} t\, |\mathrm{Rm}(\cdot, t)| < \infty.$$

We have the following examples of singularities. A neckpinch forms a Type I singularity (Section 5 in Chapter 2 of Volume One). A degenerate neckpinch, if it exists, forms a Type IIa singularity (Section 6 in Chapter 2 of Volume One). Many homogeneous solutions form Type III singularities (Chapter 1 of Volume One).

CONJECTURE A.37 (Degenerate neckpinch existence). *There exist solutions to the Ricci flow on closed manifolds which form degenerate neckpinches.*

The analogue of the above conjecture has been proved for the mean curvature flow [**9**].

CONJECTURE A.38 (Nonexistence of Type IIb on closed 3-manifolds). *If $(\mathcal{M}^3, g(t))$, $t \in [0, \infty)$, is a solution to the Ricci flow on a closed 3-manifold, then $g(t)$ forms a Type III singularity.*

Similar to the division of types for finite time singular solutions, we may divide ancient solutions into types.

DEFINITION A.39 (Ancient solution types). Let $(\mathcal{M}^n, g(t))$ be a solution of the Ricci flow defined on $(-\infty, 0)$.

- We say $(\mathcal{M}, g(t))$ is a **Type I ancient solution** if
$$\sup_{\mathcal{M} \times (-\infty, -1]} |t|\, |\mathrm{Rm}(\cdot, t)| < \infty.$$

- We say $(\mathcal{M}, g(t))$ is a **Type II ancient solution** if
$$\sup_{\mathcal{M} \times (-\infty, -1]} |t|\, |\mathrm{Rm}(\cdot, t)| = \infty.$$

3.2. Ancient solutions have nonnegative curvature. Every ancient solution (of any dimension) has nonnegative scalar curvature. (See Lemma 9.15 on p. 271 of Volume One.)

LEMMA A.40 (Ancient solutions have $R \geq 0$). *Let $(\mathcal{M}^n, g(t))$ be a complete ancient solution of the Ricci flow. Assume that the function $R_{\min}(t) \doteqdot \inf_{x \in \mathcal{M}^n} R(x, t)$ is finite for all $t \leq 0$ and that there is a continuous function $K(t)$ such that $|\mathrm{sect}\,[g(t)]| \leq K(t)$. Then $g(t)$ has nonnegative scalar curvature for as long as it exists.*

A particular consequence of the Hamilton–Ivey estimate is that ancient 3-dimensional solutions of the Ricci flow have nonnegative sectional curvature. (See Corollary 9.8 on p. 261 of Volume One.)

COROLLARY A.41 (Ancient 3-dimensional solutions have $\mathrm{Rm} \geq 0$). *Let $(\mathcal{M}^3, g(t))$ be a complete ancient solution of the Ricci flow. Assume that there exists a continuous function $K(t)$ such that $|\mathrm{sect}\,[g(t)]| \leq K(t)$. Then $g(t)$ has nonnegative sectional curvature for as long as it exists.*

3. BASIC SINGULARITY THEORY FOR RICCI FLOW

3.3. Trace Harnack inequality. Given a surface $(\mathcal{M}^2, g)$ with positive curvature, the trace Harnack quantity is defined by

$$\text{(V1-5.35)} \qquad Q = \Delta \log R + R - r = \frac{\partial}{\partial t} \log R - |\nabla \log R|^2.$$

(Also see Lemma 5.35 on p. 144 of Volume One.)

We have the following differential Harnack estimate of Li–Yau–Hamilton-type. (See Corollary 5.56 on p. 145 of Volume One.)

COROLLARY A.42 (2d trace Harnack evolution and estimate). *On any solution of the normalized Ricci flow* (A.19) *on a complete surface with bounded positive scalar curvature, Q satisfies the evolutionary inequality*

$$\text{(V1-5.38)} \qquad \frac{\partial}{\partial t} Q \geq \Delta Q + 2 \langle \nabla Q, \nabla L \rangle + Q^2 + rQ.$$

For the unnormalized *flow* (A.18), *the analogous quantity*

$$\text{(V1-p. 169a)} \qquad \tilde{Q} = \Delta \log R + R = \frac{\partial}{\partial t} \log R - |\nabla \log R|^2$$

satisfies

$$\text{(V1-5.57)} \qquad \frac{\partial}{\partial t} \tilde{Q} \geq \Delta \tilde{Q} + 2 \left\langle \nabla \tilde{Q}, \nabla L \right\rangle + \tilde{Q}^2.$$

By the maximum principle,

$$\text{(V1-p.169b)} \qquad \tilde{Q}(x, t) + \frac{1}{t} \geq 0$$

for all $x \in \mathcal{M}$ and $t > 0$.

In all dimensions, we have the following. (See Proposition 9.20 on p. 274 of Volume One.)

PROPOSITION A.43 (Trace Harnack estimate). *If $(\mathcal{M}^n, g(t))$ is a solution of the Ricci flow on a complete manifold with bounded positive curvature operator, then for any vector field X on $\mathcal{M}$ and all times $t > 0$ such that the solution exists, one has*

$$\text{(V1-p.274)} \qquad \frac{\partial R}{\partial t} + \frac{R}{t} + 2 \langle \nabla R, X \rangle + 2 \operatorname{Rc}(X, X) \geq 0.$$

The proof of Proposition A.43 will be given in Part II. When $n = 2$, by choosing the minimizing vector field $X = -R^{-1} \nabla R$, it can be seen that (V1-p.274) is equivalent to (V1-p.169b).

One also has Corollary 9.21 on p. 274 of Volume One, namely

COROLLARY A.44 (Trace Harnack consequence, tR monotonicity). *If $(\mathcal{M}^n, g(t))$ is a solution of the Ricci flow on a complete manifold with bounded curvature operator, then the function tR is pointwise nondecreasing for all $t \geq 0$ for which the solution exists. If $(\mathcal{M}, g(t))$ is also ancient, then R itself is pointwise nondecreasing.*

3.4. Surface entropy formulas. The surface entropy N is defined for a metric of strictly positive curvature on a closed surface $\mathcal{M}^2$ by

(V1-p. 133) $$N(g) \doteqdot \int_{\mathcal{M}^2} R \log R \, d\mu.$$

Let f be the potential function, defined up to an additive constant by

(V1-5.8) $$\Delta f = R - r.$$

(See Lemma 5.38 on p. 133 and Proposition 5.39 on p. 134 of Volume One.)

PROPOSITION A.45 (Surface entropy formula). *If $(\mathcal{M}^2, g(t))$ is a solution of the normalized Ricci flow on a compact surface with $R(\cdot, 0) > 0$, then*

(V1-5.25) $$\frac{dN}{dt} = -\int_{\mathcal{M}^2} \frac{|\nabla R|^2}{R} dA + \int_{\mathcal{M}^2} (R-r)^2 dA$$

(V1-p. 134) $$= -\int_{\mathcal{M}^2} \frac{|\nabla R + R \nabla f|^2}{R} dA$$

$$- 2 \int_{\mathcal{M}^2} \left| \nabla \nabla f - \frac{1}{2} \Delta f \cdot g \right|^2 dA$$

$$\leq 0.$$

3.5. Ancient 2-dimensional solutions.

3.5.1. *Examples.* (See pp. 24–28 in Section 2 of Chapter 2 in Volume One.)

Hamilton's **cigar soliton** is the complete Riemannian surface $(\mathbb{R}^2, g_\Sigma)$, where

(V1-2.4) $$g_\Sigma \doteqdot \frac{dx \otimes dx + dy \otimes dy}{1 + x^2 + y^2}.$$

This manifold is also known in the physics literature as **Witten's black hole**. In polar coordinates

(V1-2.5) $$g_\Sigma = \frac{dr^2 + r^2 \, d\theta^2}{1 + r^2}.$$

If we define

(V1-p. 25a) $$s \doteqdot \operatorname{arcsinh} r = \log\left(r + \sqrt{1 + r^2}\right),$$

then we may rewrite g_Σ as

(V1-2.7) $$g_\Sigma = ds^2 + \tanh^2 s \, d\theta^2.$$

The scalar curvature of g_Σ is

(V1-p. 25b) $$R_\Sigma = \frac{4}{1 + r^2} = \frac{4}{\cosh^2 s} = \frac{16}{(e^s + e^{-s})^2}.$$

(See pp. 31–34 in subsection 3.3 of Chapter 2 in Volume One.) Let h be the flat metric on the manifold $\mathcal{M}^2 = \mathbb{R} \times \mathcal{S}_1^1$, where $\mathcal{S}_1^1$ is the circle of radius 1. Give $\mathcal{M}^2$ coordinates $x \in \mathbb{R}$ and $\theta \in \mathcal{S}_1^1 = \mathbb{R}/2\pi\mathbb{Z}$. The **Rosenau**

solution or **sausage model** (see [**311**] or [**141**]) of the Ricci flow is the metric $g = u \cdot h$ defined for $t < 0$ by

$$u(x,t) = \frac{\lambda^{-1} \sinh(-\lambda t)}{\cosh x + \cosh \lambda t}, \tag{V1-2.22}$$

where $\lambda > 0$.

LEMMA A.46 (Rosenau solution and its backward limit). *The metric defined by* (V1-2.22) *for* $t < 0$ *extends to an ancient solution with positive curvature of the Ricci flow on* $\mathcal{S}^2$. *The Rosenau solution is a Type II ancient solution which gives rise to an eternal solution if we take a limit looking infinitely far back in time. In particular, if one takes a limit of the Rosenau solution at either pole* $x = \pm\infty$ *as* $t \to -\infty$, *one gets a copy of the cigar soliton.*

Note that the sausage model is an ancient Type II solution which encounters a Type I singularity.

3.5.2. *Classification results.* The following provides a characterization of the cigar soliton. (See Lemma 5.96 on p. 168 of Volume One.)

LEMMA A.47 (Eternal solutions are steady solitons, 2d case). *The only ancient solution of the Ricci flow on a surface of strictly positive curvature that attains its maximum curvature in space and time is the cigar* $(\mathbb{R}^2, g_\Sigma(t))$.

The following classifies 2-dimensional complete ancient Type I solutions. (See Proposition 9.23 on p. 275 of Volume One.)

PROPOSITION A.48 (Nonflat Type I ancient surface solution is round $\mathcal{S}^2$). *A complete ancient Type I solution* $(\mathcal{N}^2, h(t))$ *of the Ricci flow on a surface is a quotient of either a shrinking round* $\mathcal{S}^2$ *or a flat* $\mathbb{R}^2$.

We have the following result for 2-dimensional Type II solutions. (See Proposition 9.24 on p. 277 of Volume One.)

PROPOSITION A.49 (Type II ancient solution backward limit is a steady, 2d case). *Let* $(\mathcal{M}^2, g(t))$ *be a complete Type II ancient solution of the Ricci flow defined on an interval* $(-\infty, \omega)$, *where* $\omega > 0$. *Assume there exists a function* $K(t)$ *such that* $|R| \leq K(t)$. *Then either* $g(t)$ *is flat or else there exists a backwards limit that is the cigar soliton.*

Combining the above results, we obtain the following. (See Corollary 9.25 on p. 277 of Volume One.)

COROLLARY A.50 (Ancient surface solutions). *Let* $(\mathcal{M}^2, g(t))$ *be a complete ancient solution defined on* $(-\infty, \omega)$, *where* $\omega > 0$. *Assume that its curvature is bounded by some function of time alone. Then either the solution is flat or it is a round shrinking sphere or there exists a backwards limit that is the cigar.*

3.6. Necklike points in Type I solutions.
(See Section 4 in Chapter 9 of Volume One.) We say that (x,t) is a **Type I c-essential point** if

(V1-p. 262a) $$|\operatorname{Rm}(x,t)| \geq \frac{c}{T-t} > 0.$$

We say that (x,t) is a **δ-necklike point** if there exists a unit 2-form θ at (x,t) such that

(V1-p. 262b) $$|\operatorname{Rm} - R(\theta \otimes \theta)| \leq \delta |\operatorname{Rm}|.$$

The following result can be used to show that necks must form in Type I solutions where the underlying manifold is not diffeomorphic to a spherical space form. (See Theorem 9.9 on p. 262 of Volume One.)

THEOREM A.51 (Necklike points in 3d Type I singular solutions). *Let $(\mathcal{M}^3, g(t))$ be a closed solution of the Ricci flow on a maximal time interval $0 \leq t < T < \infty$. If the normalized flow does not converge to a metric of constant positive sectional curvature, then there exists a constant $c > 0$ such that for all $\tau \in [0, T)$ and $\delta > 0$, there are $x \in \mathcal{M}$ and $t \in [\tau, T)$ such that (x,t) is a Type I c-essential point and a δ-necklike point.*

The analogous result for ancient solutions is the following. (See Theorem 9.19 on p. 272 of Volume One.)

THEOREM A.52 (Necklike points in 3d Type I ancient solutions). *Let $(\mathcal{M}^3, g(t))$ be a complete ancient solution of the Ricci flow with positive sectional curvature. Suppose that*

$$\sup_{\mathcal{M} \times (-\infty, 0]} |t|^\gamma R(x,t) < \infty$$

for some $\gamma > 0$. Then either $(\mathcal{M}, g(t))$ is isometric to a spherical space form or else there exists a constant $c > 0$ such that for all $\tau \in (-\infty, 0]$ and $\delta > 0$, there are $x \in \mathcal{M}$ and $t \in (-\infty, \tau)$ such that (x,t) is an ancient Type I c-essential point and a δ-necklike point.

4. More Ricci flow theory and ancient solutions

In this section we summarize some additional basic aspects of Ricci flow. We warn the reader that 'basic' here (and in the previous sections) means neither 'should be obvious' nor 'should be known by every graduate student' nor 'should be easy to prove'.

4.1. Strong maximum principle.
For solutions with nonnegative curvature operator, the strong maximum principle implies a certain type of rigidity in the case where the curvature operator is not strictly positive (see [**179**] or Theorem 6.60 in [**111**]).

THEOREM A.53 (Strong maximum principle for Rm). *Let $(\mathcal{M}^n, g(t))$, $t \in [0, T)$, be a solution of the Ricci flow with nonnegative curvature operator. There exists $\delta > 0$ such that for each $t \in (0, \delta)$, the set*

$$\text{Image}\,(\text{Rm}\,[g\,(t)]) \subset \Lambda^2 T^* \mathcal{M}$$

is a smooth subbundle which is invariant under parallel translation and constant in time. Moreover, $\text{Image}\,(\text{Rm}\,[g\,(x,t)])$ is a Lie subalgebra of $\Lambda^2 T_x^* \mathcal{M} \cong \mathfrak{so}\,(n)$ for all $x \in \mathcal{M}$ and $t \in (0, \delta)$.

As an application of the strong maximum principle we have the following classification result due to W.-X. Shi.

THEOREM A.54 (Complete noncompact 3-manifolds with Rc ≥ 0). *If $(\mathcal{M}^3, g(t))$, $t \in [0, T)$, is a complete solution to the Ricci flow on a 3-manifold with nonnegative sectional (Ricci) curvature, then for $t \in (0, T)$ the universal covering solution $(\tilde{\mathcal{M}}^3, \tilde{g}(t))$ is either*

(1) $\mathbb{R}^3$ *with the standard flat metric,*
(2) *the product $(\mathcal{N}^2, h\,(t)) \times \mathbb{R}$, where $h\,(t)$ is a solution to the Ricci flow with positive curvature and $\mathcal{N}^2$ is diffeomorphic to either $\mathcal{S}^2$ or $\mathbb{R}^2$ or*
(3) $g(t)$ *and $\tilde{g}(t)$ have positive sectional (Ricci) curvature and hence $\tilde{\mathcal{M}}^3$ is diffeomorphic to $\mathcal{S}^3$ or $\mathbb{R}^3$ (in the former case $\mathcal{M}^3$ is diffeomorphic to a spherical space form).*

4.2. Hamilton's matrix Harnack estimate. Motivated by the consideration of expanding gradient Ricci solitons, Hamilton proved the following (see [**181**]).

THEOREM A.55 (Matrix Harnack estimate for RF). *If $(\mathcal{M}^n, g(t))$, $t \in [0, T)$, is a complete solution to the Ricci flow with bounded nonnegative curvature operator, then for any 1-form $W \in C^\infty\,(\Lambda^1 \mathcal{M})$ and 2-form $U \in C^\infty\,(\Lambda^2 \mathcal{M})$, we have*

(A.26) $$Q\,(U \oplus W) = M_{ij} W^i W^j + 2 P_{pij} U^{pi} W^j + R_{pijq} U^{pi} U^{qj} \geq 0,$$

where

$$M_{ij} \doteqdot \triangle R_{ij} - \frac{1}{2} \nabla_i \nabla_j R + 2 R^p_{\ell i j} R^\ell_p - R^p_i R_{pj} + \frac{1}{2t} R_{ij}$$

and

$$P_{pij} \doteqdot \nabla_p R_{ij} - \nabla_i R_{pj}.$$

REMARK A.56. See the discussion in Section 2 of Chapter 1 for a motivation for defining M_{ij} and P_{pij}.

Choosing an orthonormal basis of cotangent vectors $\{\omega^a\}_{a=1}^n$ at any point (x, t), letting $W = \omega^a$ and $U = \omega^a \wedge X$ for any fixed 1-form X, and summing over a, yield the trace Harnack estimate for the Ricci flow (Proposition A.43).

The following is a generalization of the trace Harnack estimate (see [**105**] and [**290**]).

THEOREM A.57 (Linear trace Harnack estimate). *Let $(\mathcal{M}^n, g(t))$ and $h(t)$, $t \in [0, T)$, be a solution to the linearized Ricci flow system:*

$$\frac{\partial}{\partial t} g_{ij} = -2R_{ij},$$
$$\frac{\partial}{\partial t} h_{ij} = (\Delta_L h)_{ij}$$

such that $(\mathcal{M}, g(t))$ is complete with bounded and nonnegative curvature operator, $h(0) \geq 0$, and $|h(t)|_{g(t)} \leq C$ for some constant $C < \infty$. Then $h(t) \geq 0$ for $t \in [0, T)$ and for any vector X we have

(A.27) $$\nabla^i \nabla^j h_{ij} + R_{ij} h_{ij} + 2 \left(\nabla^j h_{ij} \right) X^i + h_{ij} X^i X^j + \frac{H}{2t} \geq 0,$$

where $H = g^{ij} h_{ij}$.

Indeed, (A.27) generalizes Hamilton's trace Harnack estimate since we may take $h_{ij} = R_{ij}$ (under the Ricci flow we have $\frac{\partial}{\partial t} R_{ij} = (\Delta_L \operatorname{Rc})_{ij}$).

4.3. Geometry of gradient Ricci solitons. The asymptotic scalar curvature ratio of a complete noncompact Riemannian manifold $(\mathcal{M}^n, g)$ is defined by

$$\operatorname{ASCR}(g) = \limsup_{d(x, O) \to \infty} R(x) d(x, O)^2,$$

where $O \in \mathcal{M}$ is a choice of origin. This definition is independent of the choice of O.

Theorem 9.44 on p. 354 of [**111**]:

THEOREM A.58 (Asymptotic scalar curvature ratio is infinite on steady solitons, $n \geq 3$). *If $(\mathcal{M}^n, g, f)$, $n \geq 3$, is a complete steady gradient Ricci soliton with $\operatorname{sect}(g) \geq 0$, $\operatorname{Rc}(g) > 0$, and if $R(g)$ attains its maximum at some point, then $\operatorname{ASCR}(g) = \infty$.*

Theorem 8.46 on p. 318 of [**111**]:

THEOREM A.59 (Dimension reduction). *Let $(\mathcal{M}^n, g(t))$, $t \in (-\infty, \omega)$, $\omega > 0$, be a complete noncompact ancient solution of the Ricci flow with bounded nonnegative curvature operator. Suppose there exist sequences $x_i \in \mathcal{M}$, $r_i \to \infty$, and $A_i \to \infty$ such that $\frac{d_0(p, x_i)}{r_i} \geq A_i$ and*

(A.28) $$R(y, 0) \leq r_i^{-2} \quad \text{for all } y \in B_0(x_i, A_i r_i).$$

Assume further that there exists an injectivity radius lower bound at $(x_i, 0)$; namely, $\operatorname{inj}_{g(0)}(x_i) \geq \delta r_i$ for some $\delta > 0$. Then a subsequence of solutions $(\mathcal{M}^n, r_i^{-2} g(r_i^2 t), x_i)$ converges to a complete limit solution $(\mathcal{M}_\infty^n, g_\infty(t), x_\infty)$ which is the product of an $(n-1)$-dimensional solution (with bounded nonnegative curvature operator) with a line.

Theorem A.58 says the following. (See Chapter 6 for a definition of κ-noncollapsed.)

COROLLARY A.60 (Dimension reduction of steady solitons). *If $(\mathcal{M}^n,g,f)$, $n \geq 3$, is a complete steady gradient Ricci soliton which is κ-noncollapsed on all scales for some $\kappa > 0$ and if $\operatorname{sect}(g) \geq 0$, $\operatorname{Rc}(g) > 0$, and if $R(g)$ attains its maximum at some point, then a dilation about a sequence of points tending to spatial infinity at time $t = 0$ converges to a complete solution $(\mathcal{M}_\infty^n, g_\infty(t), x_\infty)$ which is the product of an $(n-1)$-dimensional solution[5] with $\mathbb{R}$.*

Proposition 9.46 on p. 356 of [**111**]:

PROPOSITION A.61 ($\operatorname{Rc} > 0$ expanders have AVR > 0). *If $(\mathcal{M}^n, g(t))$, $t > 0$, is a complete noncompact expanding gradient Ricci soliton with $\operatorname{Rc} > 0$, then $\operatorname{AVR}(g(t)) > 0$.*

Theorem 9.56 on p. 362 of [**111**]:

THEOREM A.62 (Steady or expander with pinched Ricci has R exponential decay). *If (M^n, g) is a gradient Ricci soliton on a noncompact manifold with pinched Ricci curvature in the sense that $R_{ij} \geq \varepsilon R g_{ij}$ for some $\varepsilon > 0$, where $R \geq 0$, then the scalar curvature R has exponential decay.*

Theorem 9.79 on pp. 375–376 of [**111**]:

THEOREM A.63 (Classification of 3-dimensional gradient shrinking solitons with $\operatorname{Rm} \geq 0$). *In dimension 3, any nonflat complete shrinking gradient Ricci soliton with bounded nonnegative sectional curvature is either a quotient of the 3-sphere or a quotient of $S^2 \times \mathbb{R}$.*

4.4. Ancient solutions. Theorem 10.48 on p. 417 of [**111**]:

THEOREM A.64 (Ancient solution with $\operatorname{Rm} \geq 0$ and attaining $\sup R$ is steady gradient soliton). *If $(\mathcal{M}^n, g(t))$, $t \in (-\infty, \omega)$, is a complete solution to the Ricci flow with nonnegative curvature operator, positive Ricci curvature, and such that $\sup_{\mathcal{M} \times (-\infty, \omega)} R$ is attained at some point in space and time, then $(\mathcal{M}^n, g(t))$ is a steady gradient Ricci soliton.*

Analogous to the above result is the following:

THEOREM A.65 (Immortal solution with $\operatorname{Rm} \geq 0$ and attaining $\sup tR$ is gradient expander). *If $(\mathcal{M}^n, g(t))$, $t \in (0, \infty)$, is a complete solution to the Ricci flow with nonnegative curvature operator, positive Ricci curvature, and such that $\sup_{\mathcal{M} \times (0, \infty)} tR$ is attained at some point in space and time, then $(\mathcal{M}^n, g(t))$ is an expanding gradient Ricci soliton.*

Proposition 9.29 on p. 344 of [**111**]:

---

[5]With bounded nonnegative sectional curvature.

PROPOSITION A.66 ($n \geq 2$ backward limit of Type II ancient solution with Rm ≥ 0 and sect > 0). *Let $(\mathcal{M}^n, g(t))$, $t \in (-\infty, \omega)$, $\omega \in (0, \infty]$, be a complete Type II ancient solution of the Ricci flow with bounded nonnegative curvature operator and positive sectional curvature. Assume either*

(1) *$\mathcal{M}$ is noncompact,*
(2) *n is even and $\mathcal{M}$ is orientable, or*
(3) *$g(t)$ is κ-noncollapsed on all scales.*

Then there exists a sequence of points and times (x_i, t_i) with $t_i \to -\infty$ such that $(\mathcal{M}, g_i(t), x_i)$, where $g_i(t) \doteqdot R_i g\left(t_i + R_i^{-1} t\right)$, limits in the C^∞ pointed Cheeger–Gromov sense to a complete nonflat steady gradient Ricci soliton $(\mathcal{M}_\infty^n, g_\infty(t), x_\infty)$ with bounded nonnegative curvature operator.

Theorem 9.30 on p. 344 of [**111**]:

THEOREM A.67 (Ancient has AVR $= 0$). *Let $(\mathcal{M}^n, g(t))$, $t \in (-\infty, 0]$, be a complete noncompact nonflat ancient solution of the Ricci flow. Suppose $g(t)$ has nonnegative curvature operator and*

$$\sup_{(x,t) \in \mathcal{M} \times (-\infty, 0]} |\operatorname{Rm}_{g(t)}(x)|_{g(t)} < \infty.$$

Then the asymptotic volume ratio $\operatorname{AVR}(g(t)) = 0$ for all t.

Theorem 9.32 on p. 345 of [**111**]:

THEOREM A.68 (Type I ancient has ASCR $= \infty$). *If $(\mathcal{M}^n, g(t))$, $-\infty < t < \omega$, is a complete noncompact Type I ancient solution of the Ricci flow with bounded positive curvature operator, then the asymptotic scalar curvature ratio $\operatorname{ASCR}(g(t)) = \infty$ for all t.*

5. Classical singularity theory

In this section we continue the discussion of Hamilton's singularity theory and recall some further results concerning the classifications of singularities, especially in dimension 3. An exposition of some of these results, which were originally proved by Hamilton in [**186**] and [**190**], is given in [**111**]. One of the differences between Hamilton's and Perelman's singularity theories is that in Hamilton's theory, singularities are divided into types, e.g., for finite time singularities, Type I and Type IIa. In Perelman's theory, a more natural space-time approach is taken where singularity analysis is approached via the reduced distance function. Throughout most of this section we shall consider the case of dimension 3.

We first consider the case of Type I singularities, which was essentially treated in Volume One (see Theorems A.51 and A.52 above). Applying the compactness theorem and the classification of Type I ancient surface solutions to Theorem A.51 yields the following.

THEOREM A.69 (3d Type I — existence of necks). *If $(\mathcal{M}^3, g(t))$ is a Type I singular solution of the Ricci flow on a closed 3-manifold on a maximal time interval $0 \leq t < T < \infty$, then there exists a sequence of points and times (x_i, t_i) with $t_i \to T$ such that the corresponding sequence of dilated solutions $(\mathcal{M}^3, g_i(t), x_i)$ converges to the geometric quotient of a round shrinking product cylinder $\mathcal{S}^2 \times \mathbb{R}$.*

For the rest of this section we consider Type IIa singular solutions. In this case we invoke Perelman's no local collapsing theorem (see Chapter 6). This has the following two effects on Hamilton's theory. It enables one to apply the compactness theorem to the dilation of Type IIa singular solutions. It rules out the formation of the cigar soliton as a product factor in a singularity model.

Classical point picking plus the no local collapsing theorem yield the following result (see Proposition 8.17 of [**111**]).

PROPOSITION A.70 (Type IIa singularity models are eternal). *Choose any sequence $T_i \nearrow T$. For a Type IIa singular solution on a closed manifold satisfying*

(A.29) $$|\mathrm{Rm}| \leq CR + C$$

and for any sequence $\{(x_i, t_i)\}$, where $t_i \to T$, satisfying[6]

(A.30) $$(T_i - t_i) R(x_i, t_i) = \max_{\mathcal{M} \times [0, T_i]} (T_i - t) R(x, t),$$

the sequence $(\mathcal{M}, \tilde{g}_i(t), x_i)$, where

(A.31) $$\tilde{g}_i(t) \doteq R_i g(t_i + R_i^{-1} t) \quad \text{with } R_i \doteq R(x_i, t_i),$$

preconverges to a complete eternal solution $(\tilde{\mathcal{M}}_\infty^n, \tilde{g}_\infty(t), x_\infty)$, $t \in (-\infty, \infty)$, with bounded curvature. The singularity model $(\tilde{\mathcal{M}}_\infty, \tilde{g}_\infty(t))$ is nonflat, κ-noncollapsed on all scales for some $\kappa > 0$, and satisfies

$$\sup_{\tilde{\mathcal{M}}_\infty \times (-\infty, \infty)} R(\tilde{g}_\infty(t)) = 1 = R(\tilde{g}_\infty)(x_\infty, 0).$$

If $n = 3$, then the singularity model has nonnegative sectional curvature.

In particular, we have that if $(\mathcal{M}^3, g(t))$ is a Type IIa singular solution of the Ricci flow on a closed 3-manifold, then by Theorem A.64, *the singularity model $(\tilde{\mathcal{M}}_\infty^3, \tilde{g}_\infty(t))$ obtained in Proposition A.70 is a steady gradient soliton.* Since $(\tilde{\mathcal{M}}_\infty, \tilde{g}_\infty(t))$ is nonflat and κ-noncollapsed on all scales for some $\kappa > 0$, the sectional curvatures of $\tilde{g}_\infty(t)$ are positive.[7] Now by Theorem A.58, the asymptotic scalar curvature ratio of $(\tilde{\mathcal{M}}_\infty, \tilde{g}_\infty(t))$ is

---

[6]This is a special case of the point picking method described in subsection 4.2 of Chapter 8 in Volume One.

[7]In the splitting case, we obtain a cigar, contradicting the no local collapsing theorem.

equal to infinity. Thus we can apply dimension reduction, Theorem A.59, to get a second limit which splits as the product of a surface solution and a line. This second limit is a shrinking round product cylinder $\mathcal{S}^2 \times \mathbb{R}$.[8]

Hence, as a consequence of Hamilton's singularity theory and Perelman's no local collapsing theorem, we have the following result, which complements Theorem A.69.

THEOREM A.71 (3d Type IIa — existence of necks). *If $(\mathcal{M}^3, g(t))$ is a Type IIa singular solution of the Ricci flow on a closed 3-manifold, then there exists a sequence of points and times (x_i, t_i) such that the corresponding sequence $(\mathcal{M}, g_i(t), x_i)$ converges to a round shrinking product cylinder $\mathcal{S}^2 \times \mathbb{R}$.*

A precursor to the above result is Theorem 9.9 in Volume One, which basically says that even for a Type IIa singular solution on $\mathcal{M}^3 \times [0, T)$, there exists a sequence of points (x_i, t_i) with $t_i \to T$ whose curvatures satisfy $(T - t_i) |\text{Rm}(x_i, t_i)| \geq c$ for some $c > 0$ independent of i and at the points (x_i, t_i) the curvature operators approach that of $\mathcal{S}^2 \times \mathbb{R}$ after rescaling. However Theorem 9.9 in Volume One does not directly imply the existence of a cylinder limit because, for the sequence $(\mathcal{M}, g_i(t), x_i)$, it not a priori clear that the curvatures are bounded in space at finite distances from x_i independent of i, even at time 0. The reason for this is that globally, the curvature of $g_i(0)$, whose norm is 1 at x_i, may be unbounded since the solution is Type IIa whereas the point (x_i, t_i) may, for example, have curvature $(T - t_i) |\text{Rm}(x_i, t_i)| \leq C$ for some $C < \infty$ independent of i.

Since finite time singularities are either Type I or Type IIa, we obtain the existence of necks for all finite time singular solutions on closed 3-manifolds.

---

[8] Otherwise we again get a cigar limit.

APPENDIX B

Other Aspects of Ricci Flow and Related Flows

1. Convergence to Ricci solitons

Given that convergence to a soliton plays a role in proving the convergence of the Ricci flow on compact surfaces (see Chapter 5 in Volume One), it is reasonable to ask if noncompact steady solitons play a role as limiting geometries for the Ricci flow on complete surfaces (or higher-dimensional manifolds).

Since steady soliton solutions are, by definition, evolving by diffeomorphism, even if we start near a soliton metric, we cannot expect pointwise convergence of the Ricci flow unless we take the diffeomorphisms into account. We use the following notion of convergence [**373**]:

DEFINITION B.1. Let $g(t)$ satisfy the Ricci flow on a noncompact manifold $\mathcal{M}^n$ for $0 \leq t < \infty$ and let $\bar{g}$ be a metric on $\mathcal{M}$. We say that $g(t)$ has **modified subsequence convergence** to $\bar{g}$ if there exist a sequence of times $t_i \to \infty$ and a sequence of diffeomorphisms ϕ_i of $\mathcal{M}$ such that $\phi_i^* g(t_i)$ converges uniformly to $\bar{g}$ on any compact set.

Since the Ricci flow is conformal on surfaces and preserves completeness, it makes sense to begin with metrics in the conformal class of the Euclidean metric, i.e., metrics of the form

(B.1) $$g = e^{u(x,y)} \left(dx^2 + dy^2 \right).$$

We can describe the overall 'shape' of such metrics in terms of the aperture and circumference at infinity. First, the **aperture** is defined by

$$A(g) \doteqdot \lim_{r \to \infty} \frac{L(\partial B(O, r))}{2\pi r},$$

where the $B(O, r)$ are geodesic balls about some chosen point $O \in \mathbb{R}^2$. (The choice does not affect the value of the limit.) For example, the flat metric has aperture one, the cigar metric has aperture zero, and surfaces in $\mathbb{R}^3$ that are asymptotic to cones have aperture $\alpha/(2\pi)$, where α is the cone angle.

The **circumference at infinity** is defined by

$$C_\infty(g) \doteqdot \sup_K \inf_D \{ L(\partial D) \mid K \text{ is compact}, D \text{ is open and } K \subset D \}.$$

For flat and conical metrics, C_∞ is infinite, while C_∞ is finite for the cigar.

Let $R_- = \max\{-R, 0\}$.

THEOREM B.2 (Wu [**373**]). *Let g_0 be a complete metric on $\mathbb{R}^2$, of the form* (B.1), *with R and $|\nabla u|$ (measured using g_0) bounded, and $\int R_- \, d\mu$ finite. Then a solution to the Ricci flow exists for all time, the aperture and circumference at infinity are preserved, and the metric has bounded subsequence convergence as $t \to \infty$. If $R(g_0) > 0$ and $A(g_0) > 0$, then the limit is flat. If $R(g_0) > 0$ and $C_\infty(g_0) < \infty$, then the limit is the cigar.*

Note that C_∞ being finite implies that $A = 0$. However, there are plenty of complete metrics with $A = 0$ and $C_\infty = \infty$, for which the limit of the Ricci flow is not classified. An example of a surface of positive curvature with $A = 0$ and $C_\infty = \infty$ is the paraboloid

$$\left(\mathcal{M}^2, g\right) \doteqdot \left\{(x, y, z) \in \mathbb{R}^3 : z = x^2 + y^2\right\}.$$

OUTLINE OF PROOF. Short-time existence follows from the Bernstein–Bando–Shi estimates. As with the Ricci flow on compact surfaces (see Section 3 of Chapter 5 in Volume One), long-time existence is proved by using a potential f such that $\Delta f = R$ (where Δ denotes the Laplacian with respect to g) and examining the evolution of the quantity

$$h \doteqdot R + |\nabla f|^2.$$

In fact, for metrics of the form (B.1) we can use $f = -u$. Because

$$\frac{\partial u}{\partial t} = \Delta u = -R$$

and

$$\frac{\partial}{\partial t} |\nabla u|^2 = \Delta |\nabla u|^2 - 2|\nabla^2 u|^2,$$

we have

$$\frac{\partial h}{\partial t} = \Delta h - 2|M|^2,$$

where M is the symmetric tensor with components

$$M_{ij} = \nabla_i \nabla_j u + \tfrac{1}{2} R g_{ij}.$$

Long-time bounds for $|\nabla u|$ and R (and higher derivatives) follow.

The bounds on $|R|$ imply that the metric remains complete; in particular, the length of a given curve at time $t > 0$ is bounded above and below by multiples of its length at time zero. By a theorem of Huber [**210**], the hypothesis $\int R_- \, d\mu < \infty$ implies that $\int R \, d\mu \leq 4\pi\chi(\mathcal{M})$ on a complete surface $\mathcal{M}$. In particular, $\int |R| \, d\mu$ is finite, and this is preserved by the Ricci flow. Finite total curvature, together with bounds on $|\nabla R|$ at any positive time, imply that R decays to zero at infinity. One then shows that $C_\infty(g)$ and $A(g)$ are preserved under the flow.

In the special case when $R(g_0) > 0$, $\lim_{t \to \infty} e^{u(x,y,t)}$ exists pointwise and is either identically zero or positive everywhere. In the latter case, $\partial u/\partial t = -R$ implies that $\int_0^\infty R \, dt$ is bounded, and hence the limiting metric is flat. In the general case, we may define diffeomorphisms $\phi_t(x, y) = e^{-u(0,0,t)/2}(x, y)$, so that $\tilde{g}(t) = \phi_t^* g(t)$ is constant at the origin. Then uniform bounds on the derivative of the conformal factor for $\tilde{g}$ give subsequence convergence, on any compact set, to a metric $\bar{g}$.

If $R(g_0) > 0$ and $C_\infty < \infty$, then using the Bernstein–Bando–Shi estimates, one can show that after a short time τ, $\int_{\mathbb{R}^2} |M|^2 d\mu$ is bounded uniformly in time (where $d\mu$ indicates measure with respect to the evolving metric). The evolution equation for $\int |M|^2 d\mu$ then implies that

$$\int_\tau^\infty \left(\int_{\mathbb{R}^2} 2|\nabla M|^2 + 3R|M|^2 d\mu \right) dt < \infty.$$

Thus, either M vanishes or R vanishes for the limiting metric $\bar{g}$. If $M = 0$, then $\bar{g}$ is a gradient soliton; by Proposition 1.25, it is either flat or the cigar metric. However, $C_\infty < \infty$ precludes a flat limit.

If $R(g_0) > 0$ and $A > 0$, then one may use the Harnack inequality to show that $tR_{\max}$ is bounded. It follows that the limit $\bar{g}$ is flat in this case. $\square$

One may also study the Ricci flow for a solution of the form (B.1) in terms of the nonlinear diffusion equation satisfied by the conformal factor $v \doteq e^u$,

(B.2) $$\frac{\partial v}{\partial t} = \bar{\Delta} \log v,$$

where $\bar{\Delta}$ is the standard Laplacian on $\mathbb{R}^2$. (Note that $v = 1/(k + |\mathbf{x}|^2)$ for the cigar, where we write $\mathbf{x} = (x, y)$.)

REMARK B.3. As pointed out by Angenent in an appendix to [**373**], the equation (B.2) is a limiting case of the **porous medium equation**

(B.3) $$\frac{\partial v}{\partial t} = \bar{\Delta} v^m$$

as the positive exponent m tends to zero. For, substituting $t = \tau/m$ in (B.3) gives

$$\frac{\partial v}{\partial \tau} = \bar{\Delta}\left(\frac{v^m - 1}{m}\right)$$

and taking $m \to 0$ gives $\partial v/\partial \tau = \bar{\Delta}(\log v)$.

In connection with Ricci flow on $\mathbb{R}^2$ we also have the following result, which says that metrics that start near the cigar converge to a cigar (in the same conformal class), but under weaker assumptions than in the above theorem.

THEOREM B.4 (Hsu [**206**]). *Suppose that $v_0 \in L^p_{\text{loc}}(\mathbb{R}^2)$ for $p > 1$, $0 \leq v_0 \leq 2/(\beta|\mathbf{x}|^2)$ for $\beta > 0$, and $v_0 - 2/(\beta(|\mathbf{x}|^2 + k_0)) \in L^1(\mathbb{R}^2)$ for $k_0 > 0$. Then there exists a unique positive solution of (B.2), defined for $0 < t < \infty$, such that $\lim_{t \to 0} v = v_0$ in L^1 on any compact set and such that*

$$\lim_{t \to \infty} e^{2\beta t} v(e^{2\beta t} \mathbf{x}, t) = \frac{2}{\beta}(|\mathbf{x}|^2 + k_1)^{-1}$$

in $L^1(\mathbb{R}^2)$, for some $k_1 > 0$.

In higher dimensions, there are not many results. However, consider complete warped product metrics on $\mathbb{R}^3$, of the form

$$g = dr^2 + w(r)^2 g_{\text{can}}, \tag{B.4}$$

where g_{can} is the standard metric on $\mathcal{S}^2$. (Recall that the sectional curvatures ν_1, ν_2 of such metrics are given in terms of w by (1.58).) For such metrics, we can prove convergence to a soliton if the curvature is positive and bounded and the manifold "opens up" like a paraboloid.

THEOREM B.5 (Ivey [**220**]). *Let g be a complete metric on $\mathbb{R}^3$ of the form* (B.4). *Suppose*

$$0 < \nu_1, \nu_2 \leq C \tag{B.5}$$

and suppose

$$\nu_2 \leq Z\nu_1 \tag{B.6}$$

for positive constants C and Z, and

$$\liminf_{r \to \infty} \left(\frac{\partial}{\partial r} w^2 \right) > 0. \tag{B.7}$$

Then the solution of the Ricci flow with $g(0) = g$ exists for all time. If, in addition,

$$\limsup_{r \to \infty} \left(\frac{\partial}{\partial r} w^2 \right) < \infty, \tag{B.8}$$

then the flow converges to a rotationally symmetric steady gradient soliton, in the sense of the C^∞-Cheeger–Gromov topology (see Theorem 3.10).

Intuitively, the condition (B.7) means that the area of a sphere centered at the origin grows at least as fast as for a paraboloid, while (B.8) means that the sphere area grows no faster than a paraboloid. In other words, the metric becomes flat as $r \to \infty$, but not too flat. (By contrast, the result of Shi [**330**], which gives convergence of the Ricci flow to a flat metric, assumes that sectional curvatures fall off like $r^{-(2+\varepsilon)}$.) The condition $\nu_2 \leq Z\nu_1$ means that as the sectional curvature ν_1 along the planes tangent to the spheres becomes flat as $r \to +\infty$, the sectional curvature ν_2 of the perpendicular planes also becomes flat. Of course, for the Bryant soliton of subsection 3.2 of Chapter 1, ν_2 falls off much faster than ν_1. However, it is not difficult to construct other metrics that satisfy the conditions in the theorem; see [**220**] for details.

For metrics on $\mathbb{R}^3$ of the form (B.4), the warping function satisfies a quasilinear heat equation

$$\frac{\partial w}{\partial t} = w'' - \frac{1 - (w')^2}{w} = -(\nu_1 + \nu_2)w.$$

1. CONVERGENCE TO RICCI SOLITONS

However, the time-derivative $\partial/\partial t$ under the Ricci flow and the radial-derivative $\partial/\partial r$ do not commute; in fact,

$$\left[\frac{\partial}{\partial t}, \frac{\partial}{\partial r}\right] = 2\nu_2 \frac{\partial}{\partial r}.$$

We now outline the proof of Theorem B.5; again, details may be found in [**220**].

First, long-time existence is proved by obtaining an estimate, depending only C and Z, for $|ww'''|$. Then, convergence to a soliton is proved by examining the evolution of the quantity

$$Q \doteqdot R + \left((\nu_1 + \nu_2)w/w'\right)^2.$$

Comparing with (1.58) shows that Q coincides with $R + |\nabla f|^2$ when g is the Bryant soliton. Thus, Q is constant for the soliton. In general, it evolves by the equation

$$(B.9) \qquad \left(\frac{\partial}{\partial t} - \Delta\right) Q = -\frac{(ww')^2}{(1 + (w')^2 + w\nu_2)^2}\left(\frac{\partial Q}{\partial r}\right)^2$$
$$- 2\left(\frac{w\nu_2}{w'} + \frac{ww'}{1 + (w')^2 + w\nu_2}(\nu_1 + \nu_2)\right)\frac{\partial Q}{\partial r},$$

where

$$\Delta = \left(\frac{\partial}{\partial r}\right)^2 - 2\frac{w'}{w}\frac{\partial}{\partial r}$$

is the Laplacian with respect to g for functions depending on r and t only.

Given (B.5) and (B.6), the paraboloid condition (B.8) is equivalent to Q having a positive lower bound. In fact, (B.6) implies that

$$(B.10) \qquad (ww')^2 \leq (1+Z)^2/Q.$$

As part of the proof of long-time existence, one shows that conditions (B.5) and (B.6) persist in time. These, together with (B.10), imply that

$$\left(\frac{\partial}{\partial t} - \Delta\right) \geq -\frac{A}{Q}\left(\frac{\partial Q}{\partial r}\right)^2 - B\left|\frac{\partial Q}{\partial r}\right|$$

for some positive constants A, B. Applying the maximum principle to the corresponding inequality for $\phi = 1/Q$ shows that the lower bound on Q persists in time.

Finally, existence of a limiting metric is proven using the compactness theorem, and showing that it is a nontrivial soliton comes by appealing to Theorem A.64 and the positive lower bound for Q.

REMARK B.6. The generalization of Theorem B.5 to rotationally symmetric Ricci flow in higher dimensions should be straightforward. It may even be possible to generalize at least the long-time existence to the non-rotationally-symmetric case by finding a generalization for the quantity Q (open problem). It would also be interesting to find conditions under which the flow converges to the product of the cigar metric with the real line.

2. The mean curvature flow

In this section we give a brief introduction to the mean curvature flow and some monotonicity formulas. It is interesting to compare these monotonicity formulas for the mean curvature flow to those for the Ricci flow. We also refer the reader to the books by Ecker [**132**] and X.-P. Zhu [**386**].

2.1. Mean curvature flow of hypersurfaces in Riemannian manifolds.
Let $\left(\mathcal{P}^{n+1}, g_\mathcal{P}\right)$ be an orientable Riemannian manifold and let $\mathcal{M}^n$ be an orientable differentiable manifold. The first fundamental form of an embedded hypersurface $X : \mathcal{M} \to \mathcal{P}$ is defined by

$$g(V, W) \doteqdot g_\mathcal{P}(V, W)$$

for $V, W \in T_x X(\mathcal{M})$. More generally, for an immersed hypersurface, we define

$$g(V, W) \doteqdot \langle X_* V, X_* W \rangle$$

for $V, W \in T_p \mathcal{M}$. Let ν denote the choice of a smooth unit normal vector field to $\mathcal{M}$. The **second fundamental form** is defined by

$$h(V, W) \doteqdot \langle D_V \nu, W \rangle = -\langle D_V W, \nu \rangle$$

for $V, W \in T_x X(\mathcal{M})$, where D denotes the Riemannian covariant derivative of $(\mathcal{P}, g_\mathcal{P})$ and $\langle \, , \, \rangle \doteqdot g_\mathcal{P}(\, , \,)$. To get the second equality in the line above, we extend W to a tangent vector field in a neighborhood of x and use $\langle \nu, W \rangle \equiv 0$. In particular, $0 = V \langle \nu, W \rangle = \langle D_V \nu, W \rangle + \langle D_V W, \nu \rangle$. The **mean curvature** is the trace of the second fundamental form:

$$H \doteqdot \sum_{i=1}^n h(e_i, e_i),$$

where $\{e_i\}$ is an orthonormal frame on $X(\mathcal{M})$.

A time-dependent immersion $X_t = X(\cdot, t) : \mathcal{M} \to \mathcal{P}$, $t \in [0, T)$, is a solution of the **mean curvature flow** (**MCF**) of a hypersurface in a Riemannian manifold if

$$(\text{B.11}) \quad \frac{\partial X}{\partial t}(p, t) = \vec{H}(x(p, t)) \doteqdot -H(p, t) \cdot \nu(p, t), \quad p \in \mathcal{M}, \ t \in [0, T),$$

where $\vec{H}$ is called the **mean curvature vector**. When the X_t are embeddings, we define $\mathcal{M}_t \doteqdot X_t(\mathcal{M})$. From now on we shall assume that the X_t are embeddings, although for the most part, the following discussion holds for immersed hypersurfaces.

Let $\{x^i\}_{i=1}^n$ denote local coordinates on $\mathcal{M}$ so that $\left\{\frac{\partial X_t}{\partial x^i}\right\}$ are local coordinates on $\mathcal{M}_t$. We have

$$g_{ij} \doteqdot g\left(\frac{\partial X}{\partial x^i}, \frac{\partial X}{\partial x^j}\right) = \left\langle \frac{\partial X}{\partial x^i}, \frac{\partial X}{\partial x^j} \right\rangle,$$

$$h_{ij} \doteqdot h\left(\frac{\partial X}{\partial x^i}, \frac{\partial X}{\partial x^j}\right) = \left\langle \frac{\partial X}{\partial x^i}, D_{\frac{\partial X}{\partial x^j}} \nu \right\rangle = -\left\langle D_{\frac{\partial X}{\partial x^i}} \frac{\partial X}{\partial x^j}, \nu \right\rangle.$$

Note that $H = g^{ij}h_{ij}$ and

$$(B.12) \qquad D_{\frac{\partial X}{\partial x^j}}\nu = h_{jk}g^{k\ell}\frac{\partial X}{\partial x^\ell}.$$

We have the following basic formulas for solutions of the mean curvature flow.

LEMMA B.7 (Huisken). *The evolution of the first fundamental form (induced metric), normal, and second fundamental form are given by the following:*

$$(B.13) \qquad \frac{\partial}{\partial t}g_{ij} = -2Hh_{ij},$$

$$(B.14) \qquad D_{\frac{\partial}{\partial t}}\nu = \nabla H,$$

$$(B.15) \qquad \frac{\partial}{\partial t}h_{ij} = \nabla_i\nabla_j H - Hh_{jk}g^{k\ell}h_{\ell i} + H\,(\operatorname{Rm}\mathcal{P})_{\nu ij\nu}$$

$$\begin{aligned}(B.16)\qquad &= \Delta h_{ij} - 2Hh_{ik}h_{kj} + |h|^2\,h_{ij} \\ &\quad + h_{ij}\,(\operatorname{Rc}\mathcal{P})_{\nu\nu} - h_{ik}\,(\operatorname{Rc}\mathcal{P})_{jk} - h_{kj}\,(\operatorname{Rc}\mathcal{P})_{ik} \\ &\quad + 2\,(\operatorname{Rm}\mathcal{P})_{kij\ell}\,h_{k\ell} + h_{ki}\,(\operatorname{Rm}\mathcal{P})_{\nu kj\nu} + h_{kj}\,(\operatorname{Rm}\mathcal{P})_{\nu ki\nu} \\ &\quad - D_i\,(\operatorname{Rc}\mathcal{P})_{j\nu} - D_j\,(\operatorname{Rc}\mathcal{P})_{i\nu} + D_\nu\,(\operatorname{Rc}\mathcal{P})_{ij}\,,\end{aligned}$$

$$(B.17) \qquad \frac{\partial}{\partial t}H = \Delta H + |h|^2\,H + H\,(\operatorname{Rc}\mathcal{P})_{\nu\nu}\,,$$

$$(B.18) \qquad \frac{\partial}{\partial t}d\mu = -H^2 d\mu,$$

where ∇ is the covariant derivative with respect to the induced metric on the hypersurface and $d\mu$ is the volume form of the evolving hypersurface.

REMARK B.8. Technically, we should consider these tensors (or sections of bundles) as existing on the domain manifold $\mathcal{M}$. However, we shall often view them as tensors on the evolving hypersurface $\mathcal{M}_t = X_t(\mathcal{M})$. The time-derivative of the unit normal is expressed slightly differently since it is actually the covariant derivative of ν in the direction $\frac{\partial X}{\partial t}$ along the path $t \mapsto X_t(p)$ for $p \in \mathcal{M}$ fixed. On the other hand, if we view g and h as on the fixed manifold $\mathcal{M}$, we have the ordinary time-derivatives, whereas if we view g and h as on the evolving $\mathcal{M}_t$, then the time-derivatives are actually $D_{\frac{\partial}{\partial t}}$.

PROOF. While carrying out the computations below, keep in mind that the inner product of a tangential vector with a normal vector is zero. The evolution of the metric is given by

$$\frac{\partial}{\partial t} g_{ij} = \left\langle D_{\frac{\partial X}{\partial x^i}} \left(\frac{\partial X}{\partial t} \right), \frac{\partial X}{\partial x^j} \right\rangle + \left\langle \frac{\partial X}{\partial x^i}, D_{\frac{\partial X}{\partial x^j}} \left(\frac{\partial X}{\partial t} \right) \right\rangle$$

$$= -H \left\langle D_{\frac{\partial X}{\partial x^i}} \nu, \frac{\partial X}{\partial x^j} \right\rangle - H \left\langle \frac{\partial X}{\partial x^i}, D_{\frac{\partial X}{\partial x^j}} \nu \right\rangle$$

$$= -2H h_{ij},$$

where we used $\langle \nu, \frac{\partial X}{\partial x^j} \rangle = \langle \frac{\partial X}{\partial x^i}, \nu \rangle = 0$. This is (B.13).

Since $\langle \frac{\partial \nu}{\partial t}, \nu \rangle = 0$ and

$$0 = \frac{\partial}{\partial t} \left\langle \nu, \frac{\partial X}{\partial x^i} \right\rangle = \left\langle D_{\frac{\partial X}{\partial t}} \nu, \frac{\partial X}{\partial x^i} \right\rangle + \left\langle \nu, D_{\frac{\partial X}{\partial x^i}} (-H\nu) \right\rangle$$

$$= \left\langle D_{\frac{\partial X}{\partial t}} \nu, \frac{\partial X}{\partial x^i} \right\rangle - \frac{\partial H}{\partial x^i},$$

the normal ν evolves by

(B.19) $$D_{\frac{\partial X}{\partial t}} \nu = g^{ij} \frac{\partial H}{\partial x^i} \frac{\partial X}{\partial x^j} = \nabla H,$$

where ∇H is the gradient on the hypersurface of H. This is (B.14).

The evolution of the second fundamental form is (we use $\left[\frac{\partial X}{\partial t}, \frac{\partial X}{\partial x^j} \right] = 0$)

$$\frac{\partial}{\partial t} h_{ij} = -\frac{\partial}{\partial t} \left\langle D_{\frac{\partial X}{\partial x^i}} \frac{\partial X}{\partial x^j}, \nu \right\rangle$$

$$= -\left\langle D_{\frac{\partial X}{\partial t}} \left(D_{\frac{\partial X}{\partial x^i}} \frac{\partial X}{\partial x^j} \right), \nu \right\rangle - \left\langle D_{\frac{\partial X}{\partial x^i}} \frac{\partial X}{\partial x^j}, D_{\frac{\partial X}{\partial t}} \nu \right\rangle$$

$$= -\left\langle D_{\frac{\partial X}{\partial x^i}} \left(D_{\frac{\partial X}{\partial x^j}} \left(\frac{\partial X}{\partial t} \right) \right), \nu \right\rangle - \left\langle \mathrm{Rm}_{\mathcal{P}} \left(\frac{\partial X}{\partial t}, \frac{\partial X}{\partial x^i} \right) \frac{\partial X}{\partial x^j}, \nu \right\rangle$$

$$- \left\langle D_{\frac{\partial X}{\partial x^i}} \frac{\partial X}{\partial x^j}, D_{\frac{\partial X}{\partial t}} \nu \right\rangle$$

$$= \left\langle D_{\frac{\partial X}{\partial x^i}} \left(\frac{\partial H}{\partial x^j} \nu + H D_{\frac{\partial X}{\partial x^j}} \nu \right), \nu \right\rangle + H \left\langle \mathrm{Rm}_{\mathcal{P}} \left(\nu, \frac{\partial X}{\partial x^i} \right) \frac{\partial X}{\partial x^j}, \nu \right\rangle$$

$$- \left\langle D_{\frac{\partial X}{\partial x^i}} \frac{\partial X}{\partial x^j}, \nabla H \right\rangle$$

$$= \frac{\partial^2 H}{\partial x^i \partial x^j} + H h_{jk} g^{k\ell} \left\langle D_{\frac{\partial X}{\partial x^i}} \frac{\partial X}{\partial x^\ell}, \nu \right\rangle + H \left(\mathrm{Rm}_{\mathcal{P}} \right)_{\nu i j \nu} - \Gamma^k_{ij} \frac{\partial H}{\partial x^k}$$

$$= \nabla_i \nabla_j H - H h_{jk} g^{k\ell} h_{\ell i} + H \left(\mathrm{Rm}_{\mathcal{P}} \right)_{\nu i j \nu},$$

where we used (B.11), (B.19), and (B.12), and where

$$\Gamma^k_{ij} = g^{k\ell} \left\langle D_{\frac{\partial X}{\partial x^i}} \frac{\partial X}{\partial x^j}, \frac{\partial X}{\partial x^\ell} \right\rangle$$

are the Christoffel symbols and ∇ is the covariant derivative with respect to the induced metric on the hypersurface. This is (B.15).

Tracing the above formula, we see that the mean curvature evolves by

$$\frac{\partial H}{\partial t} = -\frac{\partial}{\partial t}g_{ij} \cdot h_{ij} + g^{ij}\frac{\partial}{\partial t}h_{ij}$$
$$= 2H\left|h_{ij}\right|^2 + \left(\Delta H - H\left|h_{ij}\right|^2 + H\left(\operatorname{Rc}\mathcal{P}\right)_{\nu\nu}\right)$$
$$= \Delta H + H\left|h_{ij}\right|^2 + H\left(\operatorname{Rc}\mathcal{P}\right)_{\nu\nu}.$$

This is (B.17).

Equation (B.18) follows from (B.13) and

$$\frac{\partial}{\partial t}d\mu = \frac{1}{2}g^{ij}\left(\frac{\partial}{\partial t}g_{ij}\right)d\mu.$$

Finally we rewrite the evolution equation for h_{ij} to see (B.16). Recall that the Gauss equations say that for any tangent vectors X, Y, Z, W on $\mathcal{M}$,

$$(\operatorname{Rm}_{\mathcal{M}})(X, Y, Z, W) = (\operatorname{Rm}_{\mathcal{P}})(X, Y, Z, W)$$
$$+ h(X, W)h(Y, Z) - h(X, Z)h(Y, W).$$

Tracing implies (in index notation)

$$(\operatorname{Rc}_{\mathcal{M}})_{i\ell} = (\operatorname{Rc}_{\mathcal{P}})_{i\ell} - (\operatorname{Rm}_{\mathcal{P}})_{\nu i\ell\nu} + Hh_{i\ell} - h_{i\ell}^2.$$

The Codazzi equations say for any $X, Y, Z \in T\mathcal{M}$,

$$(\nabla_X h)(Y, Z) - (\nabla_Y h)(X, Z) = -\langle \operatorname{Rm}_{\mathcal{P}}(X, Y)Z, \nu \rangle.$$

In index notation, this is

$$\nabla_i h_{kj} = \nabla_k h_{ij} - (\operatorname{Rm}_{\mathcal{P}})_{ikj\nu}.$$

Tracing over the Y and Z components in the hypersurface directions, we have

$$\nabla H - \operatorname{div}(h) = -\operatorname{Rc}_{\mathcal{P}}(\nu).$$

To obtain (B.16) from (B.15) we note that, by using the Codazzi equations and Gauss equations, we obtain

(B.20) $$\nabla_i \nabla_j H = \nabla_i \nabla_k h_{kj} - \nabla_i (\operatorname{Rc}_{\mathcal{P}})_{j\nu},$$

where $\operatorname{Rc}_{\mathcal{P}}(\nu) \doteqdot (\operatorname{Rc}_{\mathcal{P}})_{j\nu}dx^j$ is a 1-form on $\mathcal{M}$. Note that

$$(\Gamma_{\mathcal{P}})_{ij}^{\nu} = -\frac{1}{2}\nu\left\langle \frac{\partial X}{\partial x^i}, \frac{\partial X}{\partial x^j}\right\rangle = -h_{ij}$$

and similarly

$$(\Gamma_{\mathcal{P}})_{i\nu}^{k} = \frac{1}{2}g^{k\ell}\nu\left\langle \frac{\partial X}{\partial x^i}, \frac{\partial X}{\partial x^\ell}\right\rangle = g^{k\ell}h_{i\ell}.$$

Thus, considering $\mathrm{Rc}_{\mathcal{P}}$ as a 2-tensor on $\mathcal{P}$ and changing its covariant derivative to the one with respect to $g_{\mathcal{P}}$ in the formula (B.20), we have

$$\begin{aligned}\nabla_i \nabla_j H &= \nabla_i \nabla_k h_{kj} - D_i \left(\mathrm{Rc}_{\mathcal{P}}\right)_{j\nu} \\&\quad - (\Gamma_{\mathcal{P}})_{ij}^{\nu} \left(\mathrm{Rc}_{\mathcal{P}}\right)_{\nu\nu} - (\Gamma_{\mathcal{P}})_{i\nu}^{k} \left(\mathrm{Rc}_{\mathcal{P}}\right)_{jk} \\&= \nabla_i \nabla_k h_{kj} - D_i \left(\mathrm{Rc}_{\mathcal{P}}\right)_{j\nu} \\&\quad + h_{ij} \left(\mathrm{Rc}_{\mathcal{P}}\right)_{\nu\nu} - h_i^k \left(\mathrm{Rc}_{\mathcal{P}}\right)_{jk}.\end{aligned}$$

Commuting derivatives and applying the Codazzi and Gauss equations, we compute

$$\begin{aligned}\nabla_i \nabla_j H &= \nabla_k \nabla_i h_{kj} - (\mathrm{Rc}_{\mathcal{M}})_{i\ell} h_{\ell j} - (\mathrm{Rm}_{\mathcal{M}})_{ikj\ell} h_{k\ell} - D_i (\mathrm{Rc}_{\mathcal{P}})_{j\nu} \\&\quad + h_{ij} (\mathrm{Rc}_{\mathcal{P}})_{\nu\nu} - h_i^k (\mathrm{Rc}_{\mathcal{P}})_{jk} \\&= \nabla_k \nabla_k h_{ij} - \nabla_k (\mathrm{Rm}_{\mathcal{P}})_{ikj\nu} \\&\quad - \left(H h_{i\ell} - h_{i\ell}^2\right) h_{\ell j} - (h_{i\ell} h_{kj} - h_{ij} h_{k\ell}) h_{k\ell} \\&\quad - (\mathrm{Rc}_{\mathcal{P}})_{i\ell} h_{\ell j} + (\mathrm{Rm}_{\mathcal{P}})_{\nu i \ell \nu} h_{\ell j} - (\mathrm{Rm}_{\mathcal{P}})_{ikj\ell} h_{k\ell} \\&\quad - D_i (\mathrm{Rc}_{\mathcal{P}})_{j\nu} + h_{ij} (\mathrm{Rc}_{\mathcal{P}})_{\nu\nu} - h_i^k (\mathrm{Rc}_{\mathcal{P}})_{jk} \\&= \Delta h_{ij} - H h_{ij}^2 + |h|^2 h_{ij} - D_k (\mathrm{Rm}_{\mathcal{P}})_{ikj\nu} \\&\quad + h_{ki} (\mathrm{Rm}_{\mathcal{P}})_{\nu kj\nu} + h_{kk} (\mathrm{Rm}_{\mathcal{P}})_{i\nu j\nu} - h_k^{\ell} (\mathrm{Rm}_{\mathcal{P}})_{ikj\ell} \\&\quad - (\mathrm{Rc}_{\mathcal{P}})_{i\ell} h_{\ell j} + (\mathrm{Rm}_{\mathcal{P}})_{\nu i \ell \nu} h_{\ell j} - (\mathrm{Rm}_{\mathcal{P}})_{ikj\ell} h_{k\ell} \\&\quad - D_i (\mathrm{Rc}_{\mathcal{P}})_{j\nu} + h_{ij} (\mathrm{Rc}_{\mathcal{P}})_{\nu\nu} - h_i^k (\mathrm{Rc}_{\mathcal{P}})_{jk},\end{aligned}$$

where $h_{ij}^2 \doteqdot h_{ik} h_{kj}$ and in the third line $(\mathrm{Rm}_{\mathcal{P}})_{ikj\nu} \, dx^i \otimes dx^k \otimes dx^j$ is considered as a 3-tensor. This is known as Simons' identity. Hence, under the mean curvature flow

$$\begin{aligned}\frac{\partial}{\partial t} h_{ij} &= \nabla_i \nabla_j H - H h_{jk} g^{k\ell} h_{\ell i} + H (\mathrm{Rm}_{\mathcal{P}})_{\nu i j \nu} \\&= \Delta h_{ij} - 2 H h_{ij}^2 + |h|^2 h_{ij} \\&\quad - D_j (\mathrm{Rc}_{\mathcal{P}})_{i\nu} + D_{\nu} (\mathrm{Rc}_{\mathcal{P}})_{ij} - D_i (\mathrm{Rc}_{\mathcal{P}})_{j\nu} \\&\quad - (\mathrm{Rc}_{\mathcal{P}})_{i\ell} h_{\ell j} + (\mathrm{Rm}_{\mathcal{P}})_{\nu i \ell \nu} h_{\ell j} - (\mathrm{Rm}_{\mathcal{P}})_{ikj\ell} h_{k\ell} \\&\quad + h_{ij} (\mathrm{Rc}_{\mathcal{P}})_{\nu\nu} - h_i^k (\mathrm{Rc}_{\mathcal{P}})_{jk} \\&\quad + h_{ki} (\mathrm{Rm}_{\mathcal{P}})_{\nu kj\nu} - h_k^{\ell} (\mathrm{Rm}_{\mathcal{P}})_{ikj\ell},\end{aligned}$$

where we used the second Bianchi identity:

$$-D_k (\mathrm{Rm}_{\mathcal{P}})_{ikj\nu} = -D_j (\mathrm{Rc}_{\mathcal{P}})_{i\nu} + D_{\nu} (\mathrm{Rc}_{\mathcal{P}})_{ij}.$$

□

EXERCISE B.9. Compute $\frac{\partial}{\partial t}\left(|h|^2 - \frac{1}{n}H^2\right)$. See §5 of Huisken [**212**] for a study of under what conditions on $(\mathcal{P}, g_{\mathcal{P}})$ the pinching estimate

$$|h|^2 - \frac{1}{n}H^2 \leq C \cdot H^{2-\delta}$$

holds for some $\delta > 0$ and $C < \infty$.

EXERCISE B.10. Let $X_t : \mathcal{M}^n \to \mathcal{P}^{n+1}$ be a hypersurface evolving by mean curvature flow where the metric $g_{\mathcal{P}}(t)$ on $\mathcal{P}$ evolves by Ricci flow. Compute $\frac{\partial}{\partial t} h_{ij}$.

HINT. The terms $-D_i (\operatorname{Rc}_{\mathcal{P}})_{j\nu} - D_j (\operatorname{Rc}_{\mathcal{P}})_{i\nu} + D_\nu (\operatorname{Rc}_{\mathcal{P}})_{ij}$ are cancelled by new terms introduced by the Ricci flow. Note that the above terms represent the evolution of the Christoffel symbols under the Ricci flow.

2.2. Huisken's monotonicity formula for mean curvature flow. When the ambient manifold $\mathcal{P}^{n+1}$ is euclidean space $\mathbb{R}^{n+1}$, Lemma B.7 says the following.

LEMMA B.11. *The evolution of the first fundamental form (i.e., induced metric), normal, and second fundamental form are given by the following:*

(B.21) $$\frac{\partial}{\partial t} g_{ij} = -2 H h_{ij},$$

(B.22) $$D_{\frac{\partial X}{\partial t}} \nu = \nabla H,$$

(B.23) $$\frac{\partial}{\partial t} h_{ij} = \Delta h_{ij} - 2 H h_{ik} h_{kj} + |h|^2 h_{ij},$$

(B.24) $$\frac{\partial}{\partial t} H = \Delta H + |h|^2 H,$$

(B.25) $$\frac{\partial}{\partial t} d\mu = -H^2 d\mu.$$

For the mean curvature flow there is a monotonicity formula due to Gerhard Huisken (see [**213**], Theorem 3.1).

THEOREM B.12 (Huisken's monotonicity formula). *Let $X_t : \mathcal{M}^n \to \mathbb{R}^{n+1}$, $t \in [0, T)$, be a smooth solution to the mean curvature flow* (B.11). *Then any closed hypersurface $\mathcal{M}_t \doteqdot X_t(\mathcal{M})$ evolving by MCF satisfies*
(B.26)
$$\frac{d}{dt} \int_{\mathcal{M}_t} (4\pi\tau)^{-\frac{n}{2}} e^{-\frac{|X|^2}{4\tau}} d\mu = -\int_{\mathcal{M}_t} \left(H - \frac{\langle X, \nu \rangle}{2\tau}\right)^2 (4\pi\tau)^{-\frac{n}{2}} e^{-\frac{|X|^2}{4\tau}} d\mu,$$
where $\tau \doteqdot T - t$.

PROOF. Let $u \doteqdot (4\pi\tau)^{-\frac{n}{2}} e^{-\frac{|X|^2}{4\tau}}$. Using (B.11) and (B.25), we compute

(B.27) $$\frac{d}{dt} \int_{\mathcal{M}_t} u \, d\mu = \int_{\mathcal{M}_t} \left(\frac{n}{2\tau} + \frac{H \langle X, \nu \rangle}{2\tau} - \frac{|X|^2}{4\tau^2} - H^2\right) u \, d\mu.$$

Using the facts that

$$(B.28) \quad \Delta X = -H\nu, \quad |\nabla X|^2 = n, \quad \text{and} \quad \left|\nabla |X|^2\right|^2 = 4\left|X^T\right|^2,$$

where $X^T \doteq X - \langle X, \nu \rangle \nu$ (Δ is the Laplacian with respect to the induced metric), we have

$$\int_{\mathcal{M}_t} H \langle X, \nu \rangle u\, d\mu = -\int_{\mathcal{M}_t} \langle X, \Delta X \rangle u\, d\mu$$
$$= \int_{\mathcal{M}_t} \left(|\nabla X|^2 u + \frac{1}{2} \langle \nabla |X|^2, \nabla u \rangle\right) d\mu$$
$$= \int_{\mathcal{M}_t} \left(nu - \frac{1}{2\tau} |X^T|^2 u\right) d\mu.$$

Multiplying this by $\frac{1}{2\tau}$ and adding the difference (which is zero) into (B.27), we have

$$\frac{d}{dt} \int_{\mathcal{M}_t} u\, d\mu = \int_{\mathcal{M}_t} \left(\frac{H \langle X, \nu \rangle}{\tau} - \frac{\langle X, \nu \rangle^2}{4\tau^2} - H^2\right) u\, d\mu$$
$$= -\int_{\mathcal{M}_t} \left(H - \frac{\langle X, \nu \rangle}{2\tau}\right)^2 u\, d\mu.$$

$\square$

EXERCISE B.13. Verify the formulas in (B.28).

SOLUTION TO EXERCISE B.13. We have

$$\nabla_i \nabla_j X = \frac{\partial^2 X}{\partial x^i \partial x^j} - \Gamma_{ij}^k \frac{\partial X}{\partial x^k}$$

so that

$$\left\langle \nabla_i \nabla_j X, \frac{\partial X}{\partial x^\ell} \right\rangle = \left\langle \frac{\partial^2 X}{\partial x^i \partial x^j}, \frac{\partial X}{\partial x^\ell} \right\rangle - \Gamma_{ij}^k \left\langle \frac{\partial X}{\partial x^k}, \frac{\partial X}{\partial x^\ell} \right\rangle$$
$$= 0.$$

On the other hand,

$$\langle \nabla_i \nabla_j X, \nu \rangle = \left\langle \frac{\partial^2 X}{\partial x^i \partial x^j}, \nu \right\rangle = -h_{ij}.$$

Hence

$$\nabla_i \nabla_j X = -h_{ij} \nu.$$

Tracing this yields $\Delta X = -H\nu$.

Next we compute

$$|\nabla X|^2 = g^{ij} \frac{\partial X}{\partial x^i} \frac{\partial X}{\partial x^j} = g^{ij} g_{ij} = n$$

and

$$\left|\nabla |X|^2\right|^2 = 4 g^{ij} \left\langle \frac{\partial X}{\partial x^i}, X \right\rangle \left\langle \frac{\partial X}{\partial x^j}, X \right\rangle = 4 \left|X^T\right|^2.$$

Formula (B.26) is useful for studying **Type I singularities** of the mean curvature flow, where

$$\sup_{\mathcal{M}^n \times [0,T)} (T-t)|h|^2 < \infty.$$

For Type I singularities there is a natural way to rescale the MCF equation:

$$\tilde{X}(p,\tilde{t}) \doteqdot \frac{1}{\sqrt{2(T-t)}} X(p,t), \quad \text{where } \tilde{t}(t) \doteqdot \log \frac{1}{\sqrt{T-t}}.$$

Then $\tilde{\mathcal{M}}_{\tilde{t}} \doteqdot \tilde{X}_{\tilde{t}}(\mathcal{M}^n)$,

$$\frac{d\tilde{X}}{d\tilde{t}} = -\tilde{H}\tilde{\nu} + \tilde{X}$$

and (B.26) becomes the normalized monotonicity formula:

$$\frac{d}{dt} \int_{\tilde{\mathcal{M}}_{\tilde{t}}} \exp\left(-\frac{1}{2}|\tilde{X}|^2\right) d\tilde{\mu} = -\int_{\tilde{\mathcal{M}}_{\tilde{t}}} \left(\tilde{H} - \langle \tilde{X}, \tilde{\nu} \rangle\right)^2 \exp\left(-\frac{1}{2}|\tilde{X}|^2\right) d\tilde{\mu}.$$

From this one can prove the following (see Proposition 3.4 and Theorem 3.5 in [**213**]).

THEOREM B.14 (MCF convergence to self-similar). *Suppose $X_t : \mathcal{M}^n \to \mathbb{R}^{n+1}$, $t \in [0,T)$, is a Type I singular solution to the mean curvature flow. For every sequence of times $\tilde{t}_i \to \infty$, there exists a subsequence such that $\tilde{\mathcal{M}}_{\tilde{t}_i}$ converges smoothly to a smooth immersed limit hypersurface $\tilde{\mathcal{M}}_\infty$ which is self-similar:*

$$\tilde{H}_\infty = \langle \tilde{X}_\infty, \tilde{\nu}_\infty \rangle.$$

Huisken's monotonicity formula was generalized by Hamilton [**184**] as follows. Let $X_t : \mathcal{M}^n \to \mathcal{P}^N$, $t \in [0,T)$, be a submanifold of a Riemannian manifold $(\mathcal{P}^N, g)$ evolving by the mean curvature flow[1] $\frac{\partial}{\partial t} X = \vec{H}$ and suppose $u : \mathcal{P}^N \times [0,T) \to \mathbb{R}$ is a positive solution of the backward heat equation:

$$\frac{\partial u}{\partial t} = -\Delta_\mathcal{P} u.$$

Then

$$\frac{d}{dt}\left[(T-t)^{(N-n)/2} \int_{X_t(\mathcal{M})} u\, d\mu\right]$$
$$= -(T-t)^{(N-n)/2} \int_{X_t(\mathcal{M})} \left|\vec{H} - (\nabla \log u)^\perp\right|^2 u\, d\mu$$
$$- (T-t)^{(N-n)/2} \int_{X_t(\mathcal{M})} \operatorname{tr}^\perp\left(\nabla\nabla \log u + \frac{1}{2(T-t)} g\right) u\, d\mu,$$

where $d\mu$ is the volume form of the submanifold $X_t(\mathcal{M})$ and $\operatorname{tr}^\perp$ denotes the trace restricted to the normal bundle of $X_t(\mathcal{M}) \subset \mathcal{P}^N$. By Hamilton's

---

[1] In all codimensions the mean curvature vector $\vec{H}$ is defined by tracing the second fundamental form, which takes values in the normal bundle.

matrix Harnack inequality for the heat equation (see Part II of this volume), if $(\mathcal{P}^N, g)$ has parallel Ricci tensor and nonnegative sectional curvature, then $\nabla\nabla \log u + \frac{1}{2(T-t)} g \geq 0$ and hence

$$\frac{d}{dt} \left[(T-t)^{(N-n)/2} \int_{X_t(\mathcal{M})} u \, d\mu \right] \leq 0.$$

Equality holds if and only if

$$\vec{H} - (\nabla \log u)^\perp = 0$$

and

$$\left(\nabla\nabla \log u + \frac{1}{2(T-t)} g \right)^\perp = 0.$$

EXERCISE B.15. Prove Hamilton's generalization of Huisken's monotonicity formula.

2.3. Monotonicity for the harmonic map heat flow. Similar inequalities hold for the harmonic map heat flow and the Yang–Mills heat flow. The original monotonicity formula for the harmonic map heat flow was discovered by Struwe [**342**] and extended by Chen and Struwe [**90**] and Hamilton [**161**]. The monotonicity formula for the Yang–Mills heat flow appears in [**184**].

THEOREM B.16 (Struwe and Hamilton). *Let $(\mathcal{M}^n, g)$ and $(\mathcal{N}^m, h)$ be Riemannian manifolds where $\mathcal{M}$ is closed. If $F : (\mathcal{M}, g) \to (\mathcal{N}, h)$ is a solution to the* **harmonic map heat flow**

$$\frac{\partial F}{\partial t} = \Delta_{g,h} F,$$

$u : \mathcal{M} \to \mathbb{R}$ is a positive solution to the backward heat equation $\frac{\partial u}{\partial t} = -\Delta_g u$, and $\tau \doteqdot T - t$, then

$$\frac{d}{dt} \left(\tau \int_{\mathcal{M}} |dF|^2 \, u \, d\mu \right) = -2\tau \int_{\mathcal{M}} \left| \Delta F + \left\langle \frac{du}{u}, dF \right\rangle \right|^2 u \, d\mu$$

$$- 2\tau \int_{\mathcal{M}} \left(\nabla_i \nabla_j u - \frac{\nabla_i u \nabla_j u}{u} + \frac{1}{2\tau} u g_{ij} \right) d_i F \cdot d_j F \, d\mu,$$

where $d_i F \doteqdot dF(\partial/\partial x^i)$.

See the above references for applications of the monotonicity formula to the size of the singular set of a solution.

3. The cross curvature flow

In this section we present the details to results for the cross curvature flow stated in subsection 4.2 of Chapter 3 of Volume One.

3. THE CROSS CURVATURE FLOW

3.1. The cross curvature tensor. Let $(\mathcal{M}^3, g)$ be a Riemannian 3-manifold with either negative sectional curvature everywhere or positive sectional curvature everywhere. Recall the **cross curvature tensor** c is defined on p. 87 of Volume One by

$$c_{ij} \doteqdot \det E \left(E^{-1}\right)_{ij} = \frac{1}{2}\mu^{ipq}\mu^{jrs}E_{pr}E_{qs} \tag{B.29}$$

$$= \frac{1}{8}\mu^{pqk}\mu^{rs\ell}R_{i\ell pq}R_{kjrs}, \tag{B.30}$$

where $E_{ij} \doteqdot R_{ij} - \frac{1}{2}Rg_{ij}$ is the Einstein tensor, $\det E \doteqdot \frac{\det E_{ij}}{\det g_{ij}}$, and μ^{ijk} are the components of $d\mu$ with indices raised.

The equalities in the definition of c_{ij} are a consequence of the following identities.

LEMMA B.17 (Elementary curvature identities).

(a)

$$\mu^{pqk}\mu^{rs\ell}R_{kjrs} = 2E^{m\ell}\left(\delta_j^p\delta_m^q - \delta_m^p\delta_j^q\right), \tag{B.31}$$

(b)

$$\det E \left(E^{-1}\right)_{ij} = \frac{1}{8}\mu^{pqk}\mu^{rs\ell}R_{i\ell pq}R_{kjrs}.$$

PROOF. (a) We compute in an oriented orthonormal basis so that $\mu_{123} = \mu^{123} = 1$ and E_{ij} is diagonal. In such a frame, we have $E_{11} = -R_{2332}$, etc. Given ℓ, define a and b so that $ab\ell$ is a cyclic permutation of 123. Then

$$\mu^{pqk}\mu^{rs\ell}R_{kjrs} = 2\mu^{pqk}R_{kjab} = 2\mu^{pqa}\delta_j^b R_{abab} + 2\mu^{pqb}\delta_j^a R_{baab}$$

$$= 2\left(\mu^{pqa}\delta_j^b - \mu^{pqb}\delta_j^a\right)E_{\ell\ell}$$

$$= 2\left(\delta_b^p\delta_\ell^q\delta_j^b - \delta_\ell^p\delta_b^q\delta_j^b - \delta_a^q\delta_\ell^p\delta_j^a + \delta_\ell^q\delta_a^p\delta_j^a\right)E_{\ell\ell}$$

$$= 2\left(\delta_\ell^q\left(\delta_j^p - \delta_\ell^p\delta_j^\ell\right) - \delta_\ell^p\left(\delta_j^q - \delta_\ell^q\delta_j^\ell\right)\right)E_{\ell\ell}$$

$$= 2E^{q\ell}\delta_j^p - 2E^{p\ell}\delta_j^q.$$

(b) By part (a), we have

$$\frac{1}{8}\mu^{pqk}\mu^{rs\ell}R_{i\ell pq}R_{kjrs} = \frac{1}{4}R_{i\ell pq}E^{m\ell}\left(\delta_j^p\delta_m^q - \delta_m^p\delta_j^q\right)$$

$$= \frac{1}{4}\left(E^{q\ell}R_{i\ell jq} - E^{p\ell}R_{i\ell pj}\right)$$

$$= -\frac{1}{2}E^{pr}R_{irpj}.$$

In a frame as in part (a), $a_{ij} \doteqdot -\frac{1}{2}E^{pr}R_{irpj}$ is diagonal and

$$a_{11} = -\frac{1}{2}E^{22}R_{1221} - \frac{1}{2}E^{33}R_{1331}$$

$$= E_{22}E_{33} = \det E \left(E^{-1}\right)_{11},$$

and similarly for a_{22} and a_{33}. □

A nice property of the cross curvature tensor is the following result due to Hamilton.

LEMMA B.18 (Bianchi-type identity for c_{ij}). *If $(\mathcal{M}^3, g)$ has negative sectional curvature (or positive sectional curvature), then c is a metric and*

(B.32) $$\left(c^{-1}\right)^{ij} \nabla_i c_{jk} = \frac{1}{2} \left(c^{-1}\right)^{ij} \nabla_k c_{ij}.$$

This implies id $: (\mathcal{M}^3, c) \to (\mathcal{M}^3, g)$ *is a harmonic map.*

REMARK B.19. We may think of this result as dual to the fact that if Rc is positive (negative) definite, then the identity map id $: (\mathcal{M}^n, g) \to (\mathcal{M}^n, \pm \text{Rc})$ is a harmonic map (Corollary 3.20 on p. 86 of Volume One).

PROOF. Using $\nabla_i E^{ij} = 0$ (which follows from the contracted second Bianchi identity), we compute

$$\left(c^{-1}\right)^{ij} \nabla_i c_{jk} = (\det E)^{-1} E^{ij} \nabla_i \left(\det E \left(E^{-1}\right)_{jk}\right)$$

$$= (\det E)^{-1} \nabla_k \det E = \frac{1}{2} \nabla_k \log \det c$$

$$= \frac{1}{2} \left(c^{-1}\right)^{ij} \nabla_k c_{ij},$$

where $\det c \doteqdot \det c_{ij} / \det g_{ij}$.

Now given two Riemannian metrics c and g on a manifold $\mathcal{M}$, the Laplacian of the identity map id $: (\mathcal{M}, c) \to (\mathcal{M}, g)$ is given by

$$(\Delta \, \text{id})^k = \left(c^{-1}\right)^{ij} \left(\Gamma^k_{ij} - \Delta^k_{ij}\right)$$

(B.33) $$= - \left(c^{-1}\right)^{k\ell} \left(c^{-1}\right)^{ij} \left(\nabla_i c_{j\ell} + \nabla_j c_{i\ell} - \nabla_\ell c_{ij}\right) = 0,$$

where Γ^k_{ij} and Δ^k_{ij} denote the Christoffel symbols of g and c, respectively. Here we used (B.32). □

REMARK B.20. The above proof is the solution to Exercise 3.23 of Volume One.

3.2. The cross curvature flow and short-time existence. We say that a 1-parameter family of 3-manifolds $(\mathcal{M}^3, g(t))$ with negative sectional curvature is a solution of the **cross curvature flow** (XCF) if

$$\frac{\partial}{\partial t} g = 2c.$$

Likewise, if $g(t)$ has positive sectional curvature, we say that $(\mathcal{M}^3, g(t))$ is a solution if

$$\frac{\partial}{\partial t} g = -2c.$$

We have the following result due to Buckland [34].

3. THE CROSS CURVATURE FLOW

THEOREM B.21 (XCF short-time existence). *If $(\mathcal{M}^3, g_0)$ is a closed 3-manifold with either negative sectional curvature everywhere or positive sectional curvature everywhere, then a solution $g(t)$, $t \in [0, \varepsilon)$, to the cross curvature flow with $g(0) = g_0$ exists for a short time.*

PROOF. We only consider the case of negative sectional curvature and leave the case of positive sectional curvature as an exercise for the reader. By Remark 3.4 on p. 69 of Volume One, if $\frac{\partial}{\partial s} g_{ij} = v_{ij}$ is a variation of the metric g_{ij}, then

$$\frac{\partial}{\partial t} R_{ijk\ell} = \frac{1}{2} \left(\nabla_i \nabla_\ell v_{jk} + \nabla_j \nabla_k v_{i\ell} - \nabla_i \nabla_k v_{j\ell} - \nabla_j \nabla_\ell v_{ik} \right)$$

(B.34)
$$+ \frac{1}{2} g^{pq} \left(R_{ijkp} v_{q\ell} + R_{ijp\ell} v_{qk} \right),$$

so that

$$\frac{\partial}{\partial s} R_{ijk\ell} = \frac{1}{2} \left(\frac{\partial^2 v_{jk}}{\partial x^i \partial x^\ell} + \frac{\partial^2 v_{i\ell}}{\partial x^j \partial x^k} - \frac{\partial^2 v_{j\ell}}{\partial x^i \partial x^k} - \frac{\partial^2 v_{ik}}{\partial x^j \partial x^\ell} \right) + \cdots,$$

where the dots denote terms with 1 or fewer derivatives of v. Applying (B.31) to (B.30), we obtain

$$\frac{\partial}{\partial s} c_{ij} = \frac{1}{8} \left(\frac{\partial}{\partial s} R_{i\ell pq} \right) \mu^{pqk} \mu^{rs\ell} R_{kjrs} + \frac{1}{8} \left(\frac{\partial}{\partial s} R_{kjrs} \right) \mu^{pqk} \mu^{rs\ell} R_{i\ell pq} + \cdots$$

$$= -\frac{1}{8} \left(\frac{\partial^2 v_{\ell p}}{\partial x^i \partial x^q} + \frac{\partial^2 v_{iq}}{\partial x^\ell \partial x^p} - \frac{\partial^2 v_{\ell q}}{\partial x^i \partial x^p} - \frac{\partial^2 v_{ip}}{\partial x^\ell \partial x^q} \right) E^{m\ell} \left(\delta_j^p \delta_m^q - \delta_m^p \delta_j^q \right)$$

$$- \frac{1}{8} \left(\frac{\partial^2 v_{\ell p}}{\partial x^j \partial x^q} + \frac{\partial^2 v_{jq}}{\partial x^\ell \partial x^p} - \frac{\partial^2 v_{\ell q}}{\partial x^j \partial x^p} - \frac{\partial^2 v_{jp}}{\partial x^\ell \partial x^q} \right) E^{m\ell} \left(\delta_i^p \delta_m^q - \delta_m^p \delta_i^q \right)$$

$$+ \cdots.$$

Thus the linearization of the map X which takes g to $2c$ is a second-order partial differential operator. Its symbol is obtained from $2 \frac{\partial}{\partial s} c_{ij}$ by replacing $\frac{\partial}{\partial x^i}$ by a cotangent vector ζ_i in the second-order terms:

$$[\sigma DX(g)(\zeta) v]_{ij} = E^{m\ell} \left(-\zeta_i \zeta_m v_{\ell j} - \zeta_\ell \zeta_j v_{im} + \zeta_i \zeta_j v_{\ell m} + \zeta_\ell \zeta_m v_{ij} \right);$$

here $DX(g)$ denotes the linearization of X at g. Since the sectional curvature is negative, $E^{m\ell}$ is positive, which in turn implies that the eigenvalues of the symbol are nonnegative.

In analyzing the symbol $\sigma DX(g)(\zeta)$, without loss of generality we may assume $\zeta_1 = 0$ and $\zeta_2 = \zeta_3 = 0$. Then

$$[\sigma DX(g)(\zeta) v]_{ij} = -\delta_{i1} E^{1\ell} v_{\ell j} - \delta_{j1} E^{m1} v_{im} + \delta_{i1} \delta_{j1} E^{m\ell} v_{\ell m} + E^{11} v_{ij}.$$

Hence, if we take $e_1 \otimes e_1, e_1 \otimes e_2 + e_2 \otimes e_1, e_1 \otimes e_3 + e_3 \otimes e_1, e_2 \otimes e_2, e_3 \otimes e_3, e_2 \otimes e_3 + e_3 \otimes e_2$ as a basis for $S^2 T^* M$, then

$$\sigma DX(g)(\zeta) \begin{pmatrix} v_{11} \\ v_{12} \\ v_{13} \\ v_{22} \\ v_{33} \\ v_{23} \end{pmatrix} = \begin{pmatrix} -2E^{1m}v_{1m} + E^{m\ell}v_{\ell m} + E^{11}v_{11} \\ -E^{1\ell}v_{\ell 2} + E^{11}v_{12} \\ -E^{1\ell}v_{\ell 3} + E^{11}v_{13} \\ E^{11}v_{22} \\ E^{11}v_{33} \\ E^{11}v_{23} \end{pmatrix}$$

$$= \begin{pmatrix} E^{22}v_{22} + E^{33}v_{33} + 2E^{23}v_{23} \\ -E^{12}v_{22} - E^{13}v_{23} \\ -E^{12}v_{23} - E^{13}v_{33} \\ E^{11}v_{22} \\ E^{11}v_{33} \\ E^{11}v_{23} \end{pmatrix}.$$

The symbol as a matrix is given by

$$\sigma DX(g)(\zeta) = \begin{pmatrix} 0 & 0 & 0 & E^{22} & E^{33} & 2E^{23} \\ 0 & 0 & 0 & -E^{12} & 0 & -E^{13} \\ 0 & 0 & 0 & 0 & -E^{13} & -E^{12} \\ 0 & 0 & 0 & E^{11} & 0 & 0 \\ 0 & 0 & 0 & 0 & E^{11} & 0 \\ 0 & 0 & 0 & 0 & 0 & E^{11} \end{pmatrix}.$$

Since this matrix is upper triangular, its eigenvalues are 0 and E^{11}, which are nonnegative.

To eliminate the degeneracy, we apply DeTurck's trick. As in (3.29) on p. 80 of Volume One, given a fixed torsion-free connection $\tilde{\Gamma}$, define the vector field $W = W\left(g, \tilde{\Gamma}\right)$ by

$$W^k \doteqdot g^{pq} \left(\Gamma^k_{pq} - \tilde{\Gamma}^k_{pq} \right).$$

Consider the second-order operator

$$Y(g) \doteqdot \mathcal{L}_W g.$$

We have

$$\frac{\partial}{\partial s} Y(g)_{ij} = \frac{\partial}{\partial x^i} \left(g^{pq} g_{jk} \frac{\partial}{\partial s} \Gamma^k_{pq} \right) + \frac{\partial}{\partial x^j} \left(g^{pq} g_{ik} \frac{\partial}{\partial s} \Gamma^k_{pq} \right) + \cdots$$

$$= \frac{1}{2} g^{pq} \left(\frac{\partial^2 v_{qj}}{\partial x^i \partial x^p} + \frac{\partial^2 v_{pj}}{\partial x^i \partial x^q} - \frac{\partial^2 v_{pq}}{\partial x^i \partial x^j} \right)$$

$$+ \frac{1}{2} g^{pq} \left(\frac{\partial^2 v_{qi}}{\partial x^j \partial x^p} + \frac{\partial^2 v_{pi}}{\partial x^j \partial x^q} - \frac{\partial^2 v_{pq}}{\partial x^i \partial x^j} \right) + \cdots.$$

3. THE CROSS CURVATURE FLOW

Hence
$$[\sigma DY(g)(\zeta)v]_{ij} = g^{pq}(\zeta_i\zeta_p v_{qj} + \zeta_j\zeta_p v_{qi} - \zeta_i\zeta_j v_{pq})$$
$$= \delta_{i1}v_{1j} + \delta_{j1}v_{1i} - \delta_{i1}\delta_{j1}g^{pq}v_{pq},$$

assuming $\zeta_1 = 0$ and $\zeta_2 = \zeta_3 = 0$. In other words,

$$\sigma DY(g)(\zeta)\begin{pmatrix} v_{11} \\ v_{12} \\ v_{13} \\ v_{22} \\ v_{33} \\ v_{23} \end{pmatrix} = \begin{pmatrix} v_{11} - v_{22} - v_{33} \\ v_{12} \\ v_{13} \\ 0 \\ 0 \\ 0 \end{pmatrix},$$

or in matrix form,

$$\sigma DY(g)(\zeta) = \begin{pmatrix} 1 & 0 & 0 & -1 & -1 & 0 \\ 0 & 1 & 0 & 0 & 0 & 0 \\ 0 & 0 & 1 & 0 & 0 & 0 \\ 0 & 0 & 0 & 0 & 0 & 0 \\ 0 & 0 & 0 & 0 & 0 & 0 \\ 0 & 0 & 0 & 0 & 0 & 0 \end{pmatrix}.$$

Hence

$$\sigma D(X+Y)(g)(\zeta) = \begin{pmatrix} 1 & 0 & 0 & E^{22}-1 & E^{33}-1 & 2E^{23} \\ 0 & 1 & 0 & -E^{12} & 0 & -E^{13} \\ 0 & 0 & 1 & 0 & -E^{13} & -E^{12} \\ 0 & 0 & 0 & E^{11} & 0 & 0 \\ 0 & 0 & 0 & 0 & E^{11} & 0 \\ 0 & 0 & 0 & 0 & 0 & E^{11} \end{pmatrix},$$

which has all positive eigenvalues.

Hence the equation

(B.35) $$\frac{\partial}{\partial t}g = 2c + \mathcal{L}_W g$$

is strictly parabolic. Hence for any metric g_0 with negative sectional curvature, a solution $g(t)$ to (B.35) with $g(0) = g_0$ exists for a short time. Similarly to the case of the Ricci flow, we solve the following ODE at each point in $\mathcal{M}$ (see (3.35) on p. 81 of Volume One):

(B.36) $$\frac{\partial}{\partial t}\varphi_t = -W^*,$$
$$\varphi_0 = \text{id},$$

where $W^*(t)$ is the vector field dual to $W(t)$ with respect to $g(t)$. Pulling back $g(t)$ by the diffeomorphisms φ_t, we obtain a solution

(B.37) $$\bar{g}(t) \doteqdot \varphi_t^* g(t)$$

to the cross curvature flow with $\bar{g}(0) = g_0$. □

REMARK B.22. The above proof follows [**34**], where it is pointed out that the proof of short-time existence in [**106**] is incomplete.

3.3. Monotonicity formulas for the XCF. Given short-time existence of the flow, one would like to prove long-time existence and convergence of the flow on closed 3-manifolds. Although this problem is still open, we have the following result due to Hamilton, from [**106**], which is Proposition 3.24 on p. 88 of Volume One.

PROPOSITION B.23 (Monotonicity formulae). *If $\left(\mathcal{M}^3, g(t)\right)$ is a solution to the cross curvature flow with negative sectional curvature on a closed 3-manifold, then*

(1) (*volume of Einstein tensor increases*)

$$\text{(B.38)} \qquad \frac{\partial}{\partial t} \operatorname{Vol}(E) \geq 0,$$

(2) (*integral difference from hyperbolic decreases*)

$$\text{(B.39)} \qquad \frac{d}{dt} \int_{\mathcal{M}^3} \left(\frac{1}{3} \left(g^{ij} E_{ij} \right) - \left(\frac{\det E}{\det g} \right)^{1/3} \right) d\mu \leq 0.$$

REMARK B.24. Applying the arithmetic-geometric mean inequality to the eigenvalues of E_{ij} with respect to g_{ij}, we see that the integrand in part (2) is nonnegative, and it is identically zero if and only if $E_{ij} \equiv \frac{1}{3} E g_{ij}$, that is, if and only if g_{ij} has constant sectional curvature.

In the remainder of this section, we prove the above two monotonicity formulas. We begin by computing the evolution of the Einstein tensor.

LEMMA B.25.
$$\frac{\partial}{\partial t} E^{ij} = \nabla_k \nabla_\ell \left(E^{k\ell} E^{ij} - E^{ik} E^{j\ell} \right) - \det E \, g^{ij} - C \, E^{ij},$$
where $C \doteqdot g^{ij} c_{ij}$.

PROOF. Taking $q = j$ in (B.31), we may rewrite the Einstein tensor as

$$\text{(B.40)} \qquad E^{mn} = -\frac{1}{4} \mu^{ijm} \mu^{k\ell n} R_{ijk\ell}.$$

From (B.40) and (B.34) with $v_{ij} = 2c_{ij}$, we compute

$$\frac{\partial}{\partial t} E^{mn} = -\frac{1}{4} \mu^{ijm} \mu^{k\ell n} \left(\nabla_i \nabla_\ell c_{jk} + \nabla_j \nabla_k c_{i\ell} - \nabla_i \nabla_k c_{j\ell} - \nabla_j \nabla_\ell c_{ik} \right)$$
$$- \frac{1}{4} \mu^{ijm} \mu^{k\ell n} g^{pq} \left(R_{ijkp} c_{q\ell} + R_{ijp\ell} c_{qk} \right) - 2C E^{mn}$$
$$= \mu^{ijm} \mu^{k\ell n} \nabla_i \nabla_k c_{j\ell} - \frac{1}{2} \mu^{ijm} \mu^{k\ell n} g^{pq} R_{ijp\ell} c_{qk} - 2C E^{mn},$$

where we used $\frac{\partial}{\partial t} \mu^{ijk} = -C\mu^{ijk}$. By (B.29),

$$\begin{aligned}\mu^{ijm}\mu^{k\ell n}c_{j\ell} &= \frac{1}{2}\mu^{ijm}\mu^{jpq}\mu^{k\ell n}\mu^{\ell rs}E_{pr}E_{qs}\\ &= \frac{1}{2}\left(-g^{ip}g^{mq}+g^{iq}g^{mp}\right)\left(-g^{kr}g^{ns}+g^{ks}g^{nr}\right)E_{pr}E_{qs}\\ &= E^{ik}E^{mn}-E^{in}E^{mk}.\end{aligned}$$

Hence

$$\mu^{ijm}\mu^{k\ell n}\nabla_i\nabla_k c_{j\ell} = \nabla_i\nabla_k\left(E^{ik}E^{mn}-E^{in}E^{mk}\right).$$

With this, the lemma follows from the identity

$$\frac{1}{2}\mu^{ijm}\mu^{k\ell n}g^{pq}R_{ijp\ell}c_{qk} + CE^{mn} = \det E\, g^{mn}.$$

To see this, we choose a basis where $g_{ij} = \delta_{ij}$, E^{ij} and c_{ij} are diagonal, $\mu^{123} = 1$, and $R_{ijk\ell} \neq 0$ only if $(i,j) = (k,\ell)$ as unordered pairs. We compute

$$\begin{aligned}\frac{1}{2}\mu^{ij1}\mu^{k\ell 1}g^{pq}R_{ijp\ell}c_{qk} &= \mu^{k\ell 1}g^{pq}R_{23p\ell}c_{qk}\\ &= R_{2323}c_{22} - R_{2332}c_{33}\\ &= -E^{11}\left(c_{22}+c_{33}\right),\end{aligned}$$

so that $\frac{1}{2}\mu^{ij1}\mu^{k\ell 1}g^{pq}R_{ijp\ell}c_{qk} + CE^{11} = E^{11}c_{11} = \det E$, similarly for the other diagonal components. We leave it as an exercise to check that the off-diagonal components are zero. $\square$

As we shall show below, part (1) of Proposition B.23 follows from the following more general computation. Let

$$T^{ijk} \doteq E^{i\ell}\nabla_\ell E^{jk}, \quad T^i \doteq \left(E^{-1}\right)_{jk}T^{ijk} = E^{ij}\nabla_j \log \det E.$$

LEMMA B.26. *For any $\alpha \in \mathbb{R}$*

(B.41)
$$\frac{d}{dt}\int_{\mathcal{M}}(\det E)^\alpha\, d\mu = \alpha \int_{\mathcal{M}}\left(\frac{1}{2}\left|T^{ijk}-T^{jik}\right|^2_{E^{-1}} - \alpha\left|T^i\right|^2_{E^{-1}}\right)(\det E)^\alpha\, d\mu$$
$$+ (1-2\alpha)\int_{\mathcal{M}}(\det E)^\alpha C\, d\mu.$$

Here $\left|T^i\right|^2_{E^{-1}} = \left(E^{-1}\right)_{ij}T^iT^j$ and

$$\left|A^{ijk}\right|^2_{E^{-1}} = \left(E^{-1}\right)_{ip}\left(E^{-1}\right)_{jq}\left(E^{-1}\right)_{kr}A^{ijk}A^{pqr}.$$

PROOF. We compute using the evolution equation for E^{ij} that

$$\frac{d}{dt}\int_{\mathcal{M}}(\det E)^{\alpha}\,d\mu$$

$$=\int_{\mathcal{M}}(\det E)^{\alpha}\left(\alpha\left((E^{-1})_{ij}\frac{\partial}{\partial t}E^{ij}-g_{ij}\left(\frac{\partial}{\partial t}g^{ij}\right)\right)+C\right)d\mu$$

$$=\alpha\int_{\mathcal{M}}(\det E)^{\alpha}\,(E^{-1})_{ij}\nabla_{k}\nabla_{\ell}\left(E^{k\ell}E^{ij}-E^{ik}E^{j\ell}\right)d\mu$$

$$+\int_{\mathcal{M}}(\det E)^{\alpha}\left(\alpha\left((E^{-1})_{ij}\left(-\det E g^{ij}-CE^{ij}\right)+2C\right)+C\right)d\mu$$

$$=-\alpha\int_{\mathcal{M}}\nabla_{k}\left[(\det E)^{\alpha}\,(E^{-1})_{ij}\right]\left(E^{k\ell}\nabla_{\ell}E^{ij}-E^{j\ell}\nabla_{\ell}E^{ik}\right)d\mu$$

$$+(1-2\alpha)\int_{\mathcal{M}}(\det E)^{\alpha}\,C\,d\mu.$$

Since

$$\nabla_{k}\left[(\det E)^{\alpha}\,(E^{-1})_{ij}\right]$$

$$=(\det E)^{\alpha}\left(\alpha\,(E^{-1})_{pq}(\nabla_{k}E^{pq})\,(E^{-1})_{ij}-(E^{-1})_{ip}(E^{-1})_{jq}\nabla_{k}E^{pq}\right),$$

we have

$$\int_{\mathcal{M}}\nabla_{k}\left[(\det E)^{\alpha}\,(E^{-1})_{ij}\right]\left(E^{k\ell}\nabla_{\ell}E^{ij}-E^{j\ell}\nabla_{\ell}E^{ik}\right)d\mu$$

$$=\alpha\int_{\mathcal{M}}(\det E)^{\alpha}\,(E^{-1})_{pq}\nabla_{k}E^{pq}\left(E^{k\ell}\,(E^{-1})_{ij}\nabla_{\ell}E^{ij}-\nabla_{i}E^{ik}\right)d\mu$$

$$-\int_{\mathcal{M}}(\det E)^{\alpha}\,(E^{-1})_{ip}(E^{-1})_{jq}\nabla_{k}E^{pq}\left(E^{k\ell}\nabla_{\ell}E^{ij}-E^{j\ell}\nabla_{\ell}E^{ik}\right)d\mu,$$

and the lemma follows from $\nabla_{i}E^{ik}=0$. $\square$

To see the monotonicity in the $\alpha=1/2$ case, we decompose the 3-tensor T^{ijk} into its irreducible components. In particular, the orthogonal group $O(3)$ with respect to the metric E^{-1} acts on the bundle of 3-tensors which are symmetric in the last two components. The irreducible decomposition is given by

$$T^{ijk}\doteqdot U^{ijk}-\frac{1}{10}\left(E^{ij}T^{k}+E^{ik}T^{j}\right)+\frac{2}{5}E^{jk}T^{i},$$

where the coefficients $-\frac{1}{10}$ and $\frac{2}{5}$ are chosen so that U is trace-free:

$$(E^{-1})_{ij}U^{ijk}=(E^{-1})_{ik}U^{ijk}=(E^{-1})_{jk}U^{ijk}=0.$$

Since this decomposition is orthogonal with respect to E^{-1}, we have

$$\text{(B.42)} \quad \left|T^{ijk} - T^{jik}\right|^2_{E^{-1}} = \left|U^{ijk} - U^{jik}\right|^2_{E^{-1}} + \left|-\frac{1}{2}E^{ik}T^j + \frac{1}{2}E^{jk}T^i\right|^2_{E^{-1}}$$

$$\text{(B.43)} \quad = \left|U^{ijk} - U^{jik}\right|^2_{E^{-1}} + \left|T^i\right|^2_{E^{-1}}.$$

Taking $\alpha = 1/2$ in the lemma, we conclude that

$$\frac{d}{dt}\int_\mathcal{M}(\det E)^{1/2}\,d\mu = \frac{1}{4}\int_\mathcal{M}\left|U^{ijk} - U^{jik}\right|^2_{E^{-1}}(\det E)^{1/2}\,d\mu \geq 0$$

and part (1) of Proposition B.23 follows.

We now give the proof of part (2) of Proposition B.23. Define

$$J \doteq \int_\mathcal{M}\left(\frac{g^{ij}E_{ij}}{3} - (\det E)^{1/3}\right)d\mu,$$

where $g_{ij}E^{ij} = -\frac{1}{2}R$. We compute

$$\frac{d}{dt}\int_\mathcal{M}(g^{ij}E_{ij})\,d\mu = \int_\mathcal{M}\left[\left(\frac{\partial}{\partial t}g_{ij}\right)E^{ij} + g_{ij}\frac{\partial}{\partial t}E^{ij} + (g^{ij}E_{ij})\,C\right]d\mu$$

$$= \int_\mathcal{M}\left(2c_{ij}E^{ij} - g_{ij}\left(\det E\,g^{ij} + C\,E^{ij}\right) + (g^{ij}E_{ij})\,C\right)d\mu$$

$$= 3\int_\mathcal{M}\det E\,d\mu.$$

Combining this equation with (B.41) for $\alpha = 1/3$ and (B.43), we conclude

$$\frac{dJ}{dt} = -\frac{1}{3}\int_\mathcal{M}\left(\frac{1}{2}\left|T^{ijk} - T^{jik}\right|^2_{E^{-1}} - \frac{1}{3}\left|T^i\right|^2_{E^{-1}}\right)(\det E)^{1/3}\,d\mu$$

$$+ \int_\mathcal{M}\det E\,d\mu - \frac{1}{3}\int_\mathcal{M}(\det E)^{1/3}\,C\,d\mu$$

$$= -\frac{1}{6}\int_\mathcal{M}\left(\left|U^{ijk} - U^{jik}\right|^2_{E^{-1}} + \frac{1}{3}\left|T^i\right|^2_{E^{-1}}\right)(\det E)^{1/3}\,d\mu$$

$$- \int_\mathcal{M}\left(\frac{C}{3} - (\det c)^{1/3}\right)(\det E)^{1/3}\,d\mu$$

$$\leq 0,$$

where we used $\det E = (\det c)^{1/3}(\det E)^{1/3}$.

3.4. Cross curvature solitons. A solution to the XCF is a **cross curvature soliton** if there exists a vector field V and $\lambda \in \mathbb{R}$ such that at some time

$$\text{(B.44)} \quad 2c_{ij} + \nabla_i V_j + \nabla_j V_i = \lambda g_{ij}.$$

More generally, a solution to the XCF is a **cross curvature breather** if there exist times $t_1 < t_2$, a diffeomorphism $\varphi : \mathcal{M} \to \mathcal{M}$, and $a > 0$ such that

(B.45) $$g(t_2) = a\varphi^* g(t_1).$$

We have the following nonexistence result.

LEMMA B.27 (XCF breathers are trivial). *If $(\mathcal{M}^3, g(t))$ is a cross curvature breather with negative sectional curvature, then $g(t)$ has constant sectional curvature.*

PROOF. We have
$$\frac{d}{dt} \operatorname{Vol}(g(t)) = \int_{\mathcal{M}} g^{ij} c_{ij} d\mu > 0,$$
so that the breather equation (B.45), i.e., $g(t_2) = a\varphi^* g(t_1)$ for $t_1 < t_2$, implies $a > 1$. On the other hand, $J(g(t_2)) = a^{1/2} J(g(t_1)) \geq 0$, which contradicts the monotonicity formula (B.39) unless $J(g(t_2)) = J(g(t_1)) = 0$, in which case $g(t)$ has constant sectional curvature. □

4. Notes and commentary

Lemma B.7 is Lemma 3.3, Theorem 3.4 and Corollary 3.5(i) in [212].

See Ma and Chen [259] for a study of the cross curvature flow for certain classes of metrics on sphere and torus bundles over the circle. For work on the stability of the cross curvature flow at a hyperbolic metric, see Young and one of the authors [235].

APPENDIX C

Glossary

adjoint heat equation. When associated to the Ricci flow, the equation is
$$\frac{\partial u}{\partial t} + \Delta u - Ru = 0.$$
For a fixed metric, the adjoint heat equation is just the backward heat equation $\frac{\partial u}{\partial t} + \Delta u = 0$.

ancient solution. A solution of the Ricci flow which exists on a time interval of the form $(-\infty, \omega)$, where $\omega \in (-\infty, \infty]$. Limits of dilations about finite time singular solutions on closed manifolds are ancient solutions which are κ-noncollapsed at all scales. For this reason a substantial part of the subject of Ricci flow is devoted to the study of ancient solutions.

asymptotic scalar curvature ratio (ASCR). For a complete noncompact Riemannian manifold,
$$\mathrm{ASCR}(g) = \limsup_{d(x,O) \to \infty} R(x) d(x,O)^2,$$
where $O \in \mathcal{M}^n$ is any choice of origin. This definition is independent of the choice of $O \in \mathcal{M}$. For a complete ancient solution of the Ricci flow $g(t)$ on a noncompact manifold with bounded nonnegative curvature operator, $\mathrm{ASCR}(g(t))$ is independent of time. The ASCR is used to study the geometry at infinity of solutions of the Ricci flow and in particular to perform dimension reduction when $\mathrm{ASCR} = \infty$.

asymptotic volume ratio (AVR). On a complete noncompact Riemannian with nonnegative Ricci curvature, AVR is the limit of the monotone quantity $\frac{\mathrm{Vol}\, B(p,r)}{r^n}$ as $r \to \infty$. This definition is independent of the choice of $p \in \mathcal{M}$. Ancient solutions with bounded nonnegative curvature operator have $\mathrm{AVR} = 0$. Expanding gradient Ricci solitons with positive Ricci curvatures have $\mathrm{AVR} > 0$.

backward Ricci flow. The equation is
$$\frac{\partial}{\partial \tau} g = 2 \operatorname{Rc}.$$
Usually obtained by taking a solution $\tilde{g}(t)$ to the Ricci flow and defining $\tau(t) \doteqdot t_0 - t$ for some t_0.

Bernstein–Bando–Shi estimates (also **BBS estimates**). Short-time estimates for the derivatives of the curvatures of solutions of the Ricci flow assuming global pointwise bounds on the curvatures. Roughly, given a solution of the Ricci flow, if $|\operatorname{Rm}(t)|_{g(t)} \leq K$ on a time interval $[0, T)$ of length on the order of K^{-1}, then

$$|\nabla^m \operatorname{Rm}(t)|_{g(t)} \leq \frac{C_m K}{t^{m/2}}$$

on that interval. The BBS estimates are used to obtain higher derivative of curvature estimates from pointwise bounds on the curvatures. In particular, they are used in the proof of the C^∞ compactness theorem for sequences of solutions assuming only uniform pointwise bounds on the curvatures and injectivity radius estimates.

Bianchi identity. The first and second Bianchi identities are

$$R_{ijk\ell} + R_{jki\ell} + R_{kij\ell} = 0,$$
$$\nabla_i R_{jk\ell m} + \nabla_j R_{ki\ell m} + \nabla_k R_{ij\ell m} = 0,$$

respectively. The Bianchi identities reflect the diffeomorphism invariance of the curvature. In the Ricci flow they are used to derive various evolution equations including the heat-like equation for the Riemann curvature tensor.

Bishop–Gromov volume comparison theorem. An upper bound for the volume of balls given a lower bound for the Ricci curvature. This bound is sharp in the sense that equality holds for complete, simply-connected manifolds with constant sectional curvature.

Bochner formula. A class of formulas where one computes the Laplacian of some quantity (such as a gradient quantity). Such formulas are often used to prove the nonexistence of nontrivial solutions to certain equations. For example, harmonic 1-forms on closed manifolds with negative Ricci curvatures are trivial. In the Ricci flow, Bochner-type formulas (where the Laplacian is replaced by the heat operator) take the form of evolution equations which yield estimates after the application of the maximum principle.

bounded geometry. A sequence or family of Riemannian manifolds has bounded geometry if the curvatures and their derivatives are uniformly bounded (depending on the number of derivatives).

breather solution. A solution of the Ricci flow which, in the space of metrics modulo diffeomorphisms, is a periodic orbit.

Bryant soliton. The complete, rotationally symmetric steady gradient Ricci soliton on Euclidean 3-space. The Bryant soliton has sectional curvatures decaying inverse linearly in the distance to the origin and hence has ASCR $= \infty$ and AVR $= 0$. It is also expected to be the limit of the conjectured degenerate neckpinch.

Buscher duality. A duality transformation of gradient Ricci solitons on warped products with circle or torus fibers.

Calabi's trick. A typical way to localize maximum principle arguments is to multiply the quantity being estimated by a cut-off function depending on the distance to a point. Calabi's trick is a way to deal with the issue of the cut-off function being only Lipschitz continuous (since the distance function is only Lipschitz continuous).

Cartan structure equations. Given an orthonormal frame $\{e_i\}$ and dual coframe $\{\omega^j\}$, they are the identities satisfied by the connection 1-forms $\{\omega^i_j\}$ and the curvature 2-forms $\{\operatorname{Rm}^j_i\}$:

$$d\omega^i = \omega^j \wedge \omega^i_j,$$
$$\operatorname{Rm}^j_i = d\omega^j_i - \omega^k_i \wedge \omega^j_k.$$

The Cartan structure equations are useful for computing curvatures, especially for metrics with some sort of symmetry. They may also be used for general calculations in geometric analysis.

Cheeger–Gromov convergence. (See **compactness theorem**.)

Christoffel symbols. The components of the Levi-Civita connection with respect to a local coordinate system:

$$\nabla_{\partial/\partial x^i} \partial/\partial x^j = \Gamma^k_{ij} \partial/\partial x^k.$$

The variation formula for the Christoffel symbols is the first step in computing the variation formula for the curvatures. The evolution equation for the Christoffel symbols is also used to derive evolution equations for quantities involving covariant derivatives.

cigar soliton. The rotationally symmetric steady gradient Ricci soliton on the plane defined by

$$g_\Sigma(t) = \frac{dx^2 + dy^2}{e^{4t} + x^2 + y^2}.$$

The scalar curvature of the cigar is

$$R_\Sigma = \frac{4}{1 + x^2 + y^2} = 4\operatorname{sech}^2 s,$$

where s is the distance to the origin. Note that g_Σ is asymptotic to a cylinder and R_Σ decays exponentially fast. Perelman's no local collapsing theorem implies the cigar soliton and its product cannot occur as a limit of a finite time singularity on a closed manifold.

classical entropy. An integral of the form $\int_\mathcal{M} f \log f \, d\mu$.

collapsible manifold. A manifold admitting a sequence of metrics with uniformly bounded curvature and maximum injectivity radius tending to zero.

compactness theorem (Cheeger–Gromov-type). If a pointed sequence of complete metrics or solutions of Ricci flow has uniformly bounded curvature and injectivity radius at the origins uniformly bounded from below, then there exists a subsequence which converges to a complete metric or solution. The convergence is after the pull-back by diffeomorphisms, which we call Cheeger–Gromov convergence.

conjugate heat equation. (See **adjoint heat equation**.)

cosmological constant. A constant c introduced into the Ricci flow equation:
$$\frac{\partial}{\partial t} g_{ij} = -2\left(R_{ij} + cg_{ij}\right).$$
The case $c = \frac{1}{2}$ is useful in converting expanding Ricci solitons to steady Ricci solitons.

cross curvature flow. A fully nonlinear flow of metrics on 3-manifolds with either negative sectional curvature everywhere or positive sectional curvature everywhere.

curvature gap estimate. For long existing solutions, a time-dependent lower bound for the spatial supremums of the curvatures. See Lemmas 8.7, 8.9, and 8.11 in [**111**].

curvature operator. The self-adjoint fiberwise-linear map
$$\operatorname{Rm} : \Lambda^2 \mathcal{M}^n \to \Lambda^2 \mathcal{M}^n$$
defined by $\operatorname{Rm}(\alpha)_{ij} \doteq R_{ijk\ell}\alpha_{\ell k}$.

curve shortening flow (CSF). The evolution equation for a plane curve given by
$$\frac{\partial x}{\partial t} = -\kappa \nu,$$
where κ is the curvature and ν is the unit outward normal. It is useful to compare the CSF with the Ricci flow (especially on surfaces).

degenerate neckpinch. A conjectured Type IIa singularity on the n-sphere where a neck pinches at the same time its cap shrinks leading to a cusp-like singularity. Such a singularity has been proven by Angenent and Velazquez to occur for the mean curvature flow.

DeTurck's trick. (See also **Ricci–DeTurck flow**.) A method to prove short-time existence of the Ricci flow using the Ricci–DeTurck flow.

differential Harnack estimate. Any of a class of gradient-like estimates for solutions of parabolic and heat-type equations.

dilaton. (See **Perelman's energy**.)

dimension reduction. For certain classes of complete, noncompact solutions of the Ricci flow, a method of picking points tending to spatial infinity and blowing down the corresponding pointed sequence of solutions to obtain a limit solution which splits off a line.

Einstein–Hilbert functional. The functional of Riemannian metrics: $E(g) = \int_{\mathcal{M}^n} R d\mu$, where R is the scalar curvature.

Einstein metric. A metric with constant Ricci curvature.

Einstein summation convention. The convention in tensor calculus where repeated indices are summed. Strictly speaking the summed indices should be one lower and one upper, but in practice we do not always bother to lower and raise indices.

energy. (See **Perelman's energy**.)

entropy. (See **classical entropy**, **Hamilton's entropy for surfaces**, and **Perelman's entropy**.)

eternal solution. A solution of the Ricci flow existing on the time interval $(-\infty, \infty)$. Note that eternal solutions are ancient solutions which are immortal.

expanding gradient Ricci soliton (a.k.a. **expander**). A gradient Ricci soliton which is evolving by the pull-back by diffeomorphisms and scalings greater than 1.

Gaussian soliton. Euclidean space as either a shrinking or expanding soliton.

geometrization conjecture. (See **Thurston's geometrization conjecture.**)

gradient flow. The evolution of a geometric object in the direction of steepest ascent of a functional.

gradient Ricci soliton. A Ricci soliton which is flowing along diffeomorphisms generated by a gradient vector field.

Gromoll–Meyer theorem. Any complete noncompact Riemannian manifold with positive sectional curvature is diffeomorphic to Euclidean space.

Hamilton's entropy for surfaces. The functional $E(g) = \int_{\mathcal{M}^2} \log R \cdot R d\mu$ defined for metrics on surfaces with positive curvature.

Hamilton–Ivey estimate. A pointwise estimate for the curvatures of solutions of the Ricci flow on closed 3-manifolds (with normalized initial data) which implies that, at a point where there is a sufficiently large (in magnitude) negative sectional curvature, the largest sectional curvature at that point is both positive and much larger in magnitude. In dimension 3 the Hamilton–Ivey estimate implies that the singularity models of finite time singular solutions have nonnegative sectional curvature.

harmonic map. A map between Riemannian manifolds $f : (\mathcal{M}^n, g) \to (\mathcal{N}^m, h)$ satisfying $\Delta_{g,h} f = 0$, where $\Delta_{g,h}$ is the map Laplacian. (See **map Laplacian.**)

harmonic map heat flow. The equation is $\frac{\partial f}{\partial t} = \Delta_{g,h} f$.

Harnack estimate (See **differential Harnack estimate.**)

heat equation. For functions on a Riemannian manifold: $\frac{\partial u}{\partial t} = \Delta u$. This equation is the basic analytic model for geometric evolution equations including the Ricci flow.

heat operator. The operator $\frac{\partial}{\partial t} - \Delta$ appearing in the heat equation.

Hodge Laplacian. Acting on differential forms: $\Delta_d = -(d\delta + \delta d)$.

homogeneous space. A Riemannian manifold $(\mathcal{M}^n, g)$ such that for every $x, y \in \mathcal{M}$ there is an isometry $\iota : \mathcal{M} \to \mathcal{M}$ with $\iota(x) = y$.

Huisken's monotonicity formula. An integral monotonicity formula for hypersurfaces in Euclidean space evolving by the mean curvature flow using the fundamental solution to the adjoint heat equation.

immortal solution. A solution of the Ricci flow which exists on a time interval of the form (α, ∞), where $\alpha \in [-\infty, \infty)$.

isoperimetric estimate. A monotonicity formula for the isoperimetric ratios of solutions of the Ricci flow. Examples are Hamilton's estimates for solutions on closed surfaces and Type I singular solutions on closed 3-manifolds.

Jacobi field. A variation vector field of a 1-parameter family of geodesics.

Kähler–Ricci flow. The Ricci flow of Kähler metrics. Note that on a closed manifold, an initial metric which is Kähler remains Kähler under the Ricci flow.

κ-noncollapsed at all scales. A metric (or solution) which is κ-noncollapsed below scale ρ for all $\rho < \infty$.

κ-noncollapsed below the scale ρ. A Riemannian manifold satisfying $\frac{\text{Vol}\, B(x,r)}{r^n} \geq \kappa$ for any metric ball $B(x,r)$ with $|\operatorname{Rm}| \leq r^{-2}$ in $B(x,r)$ and $r < \rho$.

κ-solution (or **ancient κ-solution**). A complete ancient solution which is κ-noncollapsed on all scales, has bounded nonnegative curvature operator, and is not flat. In dimension 3, a large part of singularity analysis in Ricci flow is to classify ancient κ-solutions.

L-distance. A space-time distance-like function for solutions of the backward Ricci flow obtained by taking the infimum of the $\mathcal{L}$-length. The L-distance between two points may not always be nonnegative.

ℓ-distance function. (See **reduced distance**.)

$\mathcal{L}$-exponential map. The Ricci flow analogue of the Riemannian exponential map.

$\mathcal{L}$-geodesic. A time-parametrized path in a solution of the backward Ricci flow which is a critical point of the $\mathcal{L}$-length functional.

$\mathcal{L}$-Jacobi field. A variation vector field of a 1-parameter family of $\mathcal{L}$-geodesics.

$\mathcal{L}$-length. A length-like functional for time-parametrized paths in solutions of the backward Ricci flow. The $\mathcal{L}$-length of a path may not be positive.

Laplacian (or **rough Laplacian**). On Euclidean space the operator $\Delta = \sum_{i=1}^n \frac{\partial^2}{\partial (x^i)^2}$. On a Riemannian manifold, the second-order linear differential operator $\Delta = g^{ij} \nabla_i \nabla_j$ acting on tensors.

Levi-Civita connection. The unique linear torsion-free connection on the tangent bundle compatible with the metric. (Also called the Riemannian connection.)

Lichnerowicz Laplacian. The second-order differential operator Δ_L acting on symmetric 2-tensors defined by (V1-3.6), i.e.,

$$\Delta_L v_{ij} \doteq \Delta v_{jk} + 2g^{qp} R^r_{qjk} v_{rp} - g^{qp} R_{jp} v_{qk} - g^{qp} R_{kp} v_{jq}.$$

Lichnerowicz Laplacian heat equation. The heat-like equation $\frac{\partial}{\partial t} v_{ij} = (\Delta_L v)_{ij}$ for symmetric 2-tensors.

linear trace Harnack estimate. A differential Harnack estimate for nonnegative solutions of the Lichnerowicz Laplacian heat equation coupled to a solution of the Ricci flow with nonnegative curvature operator, which generalizes the trace Harnack estimate. (See **trace Harnack estimate**.)

little loop conjecture. Hamilton's conjecture which is essentially equivalent to Perelman's no local collapsing theorem (e.g., the conjecture is now a theorem).

Li–Yau–Hamilton (LYH) inequality. (See **differential Harnack estimate**.)

locally homogeneous space. (See also **homogeneous space**.) A Riemannian manifold $(\mathcal{M}^n, g)$ such that for every $x, y \in \mathcal{M}$ there exist open neighborhoods U of x and V of y and an isometry $\iota : U \to V$ with $\iota(x) = y$. A complete simply-connected locally homogeneous space is a homogeneous space.

locally Lipschitz function. A function which locally has a finite Lipschitz constant.

logarithmic Sobolev inequality. A Sobolev-type inequality which essentially bounds the classical entropy of a function by the L^2-norm of the first derivative of the function.

long existing solutions. Solutions which exist on a time interval of infinite duration.

long-time existence. The existence of a solution of the Ricci flow on a closed manifold as long as the Riemann curvature tensor remains bounded. By a result of Sesum, in the above statement the Riemann curvature tensor may be replaced by the Ricci tensor.

map Laplacian. Given a map $f : (\mathcal{M}^n, g) \to (\mathcal{N}^m, h)$ between Riemannian manifolds, $\Delta_{g,h} f$ is the trace with respect to g of the second covariant derivative of f. (See (3.39) in Volume One.)

matrix Harnack estimate. In Ricci flow a certain tensor inequality of Hamilton for solutions with bounded nonnegative curvature operator. A consequence is the trace Harnack estimate. One application of the matrix Harnack estimate is in the proof that an ancient solution with nonnegative curvature operator and which attains the space-time maximum of the scalar curvature is a steady gradient Ricci soliton.

maximum principle. (Also called the **weak maximum principle**.) The first and second derivative tests applied to heat-type equations to obtain bounds for their solutions. The basic idea is to use the inequalities $\Delta u \geq 0$ and $\nabla u = 0$ at a spatial minimum of a function u and the inequality $\frac{\partial u}{\partial t} \leq 0$ at a minimum in space-time up to that time. This principle applies to scalars, tensors, and systems.

maximum volume growth. A noncompact manifold has maximum volume growth if for some point p there exists $c > 0$ such that $\text{Vol } B(p, r) \geq cr^n$ for all $r > 0$. This is the same as $\text{AVR} > 0$ (when $\text{Rc} \geq 0$).

mean curvature flow. The evolution of a submanifold in a Riemannian manifold in the direction of its mean curvature vector.

modified Ricci flow. An equation of the form $\frac{\partial}{\partial t} g = -2 (\text{Rc} + \nabla \nabla f)$, where f is a function on space-time. Often coupled to an equation for f such as Perelman's equation $\frac{\partial f}{\partial t} = -R - \Delta f$.

monotonicity formula. Any formula which implies the monotonicity of a pointwise or integral quantity under a geometric evolution equation. Examples are entropy and Harnack estimates.

μ-invariant. An invariant of a metric and a positive number (scale) obtained from Perelman's entropy functional $\mathcal{W}(g, f, \tau)$ by taking the infimum over all f satisfying the constraint $\int_{\mathcal{M}^n} (4\pi\tau)^{-n/2} e^{-f} d\mu = 1$. There is a monotonicity formula for this invariant under the Ricci flow.

neckpinch. A finite time (Type I) singular solution of the Ricci flow where a region of the manifold asymptotically approaches a shrinking round cylinder $S^{n-1} \times \mathbb{R}$. Sufficient conditions for initial metrics on S^n for a neckpinch have been obtained by Angenent and one of the authors.

no local collapsing. (Also abbreviated **NLC**.) A fundamental theorem of Perelman which applies to all finite time solutions of the Ricci flow on closed manifolds. It says that given such a solution of the Ricci flow and a finite scale $\rho > 0$, there exists a constant $\kappa > 0$ such that for any ball of radius $r < \rho$ with curvature bounded by r^{-2} in the ball, the volume ratio of the ball is at least κ. We say that the solution is κ-**noncollapsed below the scale** ρ.

ν-invariant. A metric invariant obtained from the μ-invariant by taking the infimum over all $\tau > 0$. This invariant may be $-\infty$.

null-eigenvector assumption. A condition, in the statement of the maximum principle for 2-tensors, on the form of a heat-type equation which ensures that the nonnegativity of the 2-tensor is preserved under this heat-type equation.

parabolic equation. In the context of Ricci flow, a heat-type equation (which is second-order). In general, parabolicity of a nonlinear partial differential equation is defined using the symbol of its linearization.

Perelman's energy. The functional

$$\mathcal{F}(g, f) = \int_{\mathcal{M}^n} \left(R + |\nabla f|^2 \right) e^{-f} d\mu.$$

This invariant appeared previously in mathematical physics (e.g., string theory) and f is known as the **dilaton**.

Perelman's entropy. The following functional of the triple of a metric, a function, and a positive constant:

$$\mathcal{W}(g, f, \tau) = \int_{\mathcal{M}^n} \left(\tau \left(R + |\nabla f|^2 \right) + (f - n) \right) (4\pi\tau)^{-n/2} e^{-f} d\mu.$$

Perelman's Harnack (LYH) estimate. A differential Harnack (e.g., gradient) estimate for fundamental solutions of the adjoint heat equation coupled to the Ricci flow.

Perelman solution. The non-explicit 3-dimensional analogue of the Rosenau solution. The Perelman solution is rotationally symmetric and has positive sectional curvature. Its backward limit as $t \to -\infty$ is the Bryant soliton.

Poincaré conjecture. The conjecture that any simply-connected closed 3-manifold is diffeomorphic to the 3-sphere. (In dimension 3, the topological and differentiable categories are the same.) Hamilton's program and Perelman's work aim to complete a proof of the Poincaré conjecture using Ricci flow.

positive (nonnegative) curvature operator. The eigenvalues of the curvature operator are positive (nonnegative).

potentially infinite dimensions. A device which Perelman combined with the space-time approach to the Ricci flow to embed solutions of the Ricci flow into a potentially Ricci flat manifold with potentially infinite dimension.

preconvergent sequence. A sequence for which a subsequence converges.

quasi-Einstein metric. The mathematical physics jargon for a non-Einstein gradient Ricci soliton.

Rademacher's Theorem. The result that a locally Lipschitz function is differentiable almost everywhere.

reaction-diffusion equation. A heat-type equation consisting of the heat equation plus a nonlinear term which is zeroth order in the solution.

reduced distance. The distance-like function for solutions of the backward Ricci flow defined by $\ell(q, \tau) = \frac{1}{2\sqrt{\tau}} L(q, \tau)$. (See ***L*-distance**.) Partly motivated by consideration of the heat kernel and the Li–Yau distance function for positive solutions of the heat equation.

reduced volume. For a solution to the backward Ricci flow, the time-dependent invariant

$$\tilde{V}(\tau) = \int_{\mathcal{M}^n} (4\pi\tau)^{-n/2} e^{-\ell(q,\tau)} d\mu_{g(\tau)}(q).$$

Ricci flow. The equation for metrics is $\frac{\partial}{\partial t} g_{ij} = -2R_{ij}$. This equation was discovered and developed by Richard Hamilton and is the subject of this book.

Ricci soliton. (See also **gradient Ricci soliton**.) A self-similar solution of the Ricci flow. That is, the solution evolves by scaling plus the Lie derivative of the metric with respect to some vector field.

Ricci tensor. The trace of the Riemann curvature operator:

$$\operatorname{Rc}(X, Y) = \operatorname{trace}(X \mapsto \operatorname{Rm}(X, Y) Z) = \sum_{i=1}^{n} \langle \operatorname{Rm}(e_i, X) Y, e_i \rangle.$$

Ricci–DeTurck flow. A modification of the Ricci flow which is a strictly parabolic system. This equation is essentially equivalent to the Ricci flow via the pull-back by diffeomorphisms and is used to prove short-time existence for solutions on closed manifolds.

Riemann curvature operator. (See **curvature operator**.)

Riemann curvature tensor. The curvature $(3, 1)$-tensor obtained by anti-commuting in a tensorial way the covariant derivatives acting on vector

fields. Formally it is defined by
$$\operatorname{Rm}(X,Y)Z \doteq \nabla_X \nabla_Y Z - \nabla_Y \nabla_X Z - \nabla_{[X,Y]} Z.$$

Rosenau solution. An explicit rotationally symmetric ancient solution on the 2-sphere with positive curvature. Its backward limit as $t \to -\infty$ (without rescaling) is the cigar soliton.

scalar curvature. The trace of the Ricci tensor: $R = \sum_{i=1}^n \operatorname{Rc}(e_i, e_i) = g^{ij} R_{ij}$.

sectional curvature. The number
$$K(P) = \langle \operatorname{Rm}(e_1, e_2) e_2, e_1 \rangle$$
associated to a 2-plane in a tangent space $P \subset T_x \mathcal{M}$, where $\{e_1, e_2\}$ is an orthonormal basis of P.

self-similar solution. (For the Ricci flow, see **Ricci soliton**.)

Shi's local derivative estimate. A local estimate for the covariant derivatives of the Riemann curvature tensor.

short-time existence. The existence, when it holds, of a solution to the initial-value problem for the Ricci flow on some nontrivial time interval. For example, for a smooth initial metric on a closed manifold.

shrinking gradient Ricci soliton (a.k.a. **shrinker**). A gradient Ricci soliton which is evolving by the pull-back by diffeomorphisms and scalings less than 1.

singular solution. A solution on a maximal time interval. If the maximal time interval $[0, T)$ is finite, then
$$\sup_{\mathcal{M}^n \times [0,T)} |\operatorname{Rm}| < \infty.$$

singularity. For example, if T is the singular (i.e., maximal) time, we say that the solution forms a singularity at time T.

singularity model. The limit of dilations of a singular solution. For finite time singular solutions, singularity models are ancient solutions. For infinite time singular solutions, singularity models are immortal solutions.

singularity time. For a solution, the time $T \in (0, \infty]$, where $[0, T)$ is the maximal time interval of existence.

Sobolev inequality. A class of inequalities where the L^q-norms of functions are bounded by the L^p-norms of their derivatives, where p and q are related by the dimension and number of derivatives.

space form. A complete Riemannian manifold with constant sectional curvature.

space-time. The manifold $\mathcal{M}^n \times \mathcal{I}$ of a solution $(\mathcal{M}^n, g(t))$ defined on the time interval $\mathcal{I}$.

space-time connection. Any of a class of natural connections on the tangent bundle of space-time

spherical space form. A complete Riemannian manifold with positive constant sectional curvature. Such a manifold is, after scaling, isometric to a quotient of the unit sphere by a group of linear isometries.

steady gradient Ricci soliton. A gradient Ricci soliton which is evolving by the pull-back by diffeomorphisms only. In particular, the metrics at different times are isometric.

strong maximum principle. For the scalar heat equation it says that a nonnegative solution which is zero at some point (x_0, t_0) vanishes for all points (x, t) with $t \leq t_0$. For the curvature operator under the Ricci flow **Hamilton's strong maximum principle** says that given a solution with nonnegative curvature operator, for positive short time $t \in (0, \delta)$ the image of the curvature operator image $(\operatorname{Rm}(t)) \subset \Lambda^2 T^* \mathcal{M}^n$ is invariant under parallel translation and constant in time. At each point $(x, t) \in \mathcal{M}^n \times (0, \delta)$ the image of the curvature operator is a Lie subalgebra of $\Lambda^2 T^* \mathcal{M}^n_x \cong \mathfrak{so}(n)$.

tensor. A section of some tensor product of the tangent and cotangent bundles.

Thurston's geometrization conjecture. The conjecture of William Thurston that any closed 3-manifold admits a decomposition into geometric pieces. This subsumes the Poincaré conjecture.

total scalar curvature. (See **Einstein-Hilbert functional**.)

trace Harnack estimate. Given a solution of the Ricci flow with nonnegative curvature operator, the estimate

$$\frac{\partial R}{\partial t} + \frac{R}{t} + 2 \langle \nabla R, V \rangle + 2 \operatorname{Rc}(V, V) \geq 0$$

holding for any vector V. Taking $V = 0$, we have $\frac{\partial}{\partial t}(tR(x,t)) \geq 0$.

Type I, IIa, IIb, III singularity. The classification of singular solutions according to the growth or decay rates of the curvatures. For $T < \infty$, Type I is when

$$\sup_{\mathcal{M} \times [0,T)} (T-t) |\operatorname{Rm}| < \infty$$

and Type IIa is when

$$\sup_{\mathcal{M} \times [0,T)} (T-t) |\operatorname{Rm}| = \infty.$$

For $T = \infty$, Type III is when

$$\sup_{\mathcal{M} \times [0,T)} t |\operatorname{Rm}| < \infty$$

and Type IIb is when

$$\sup_{\mathcal{M} \times [0,T)} t |\operatorname{Rm}| = \infty.$$

Examples of Type I singularities are shrinking spherical space forms and neckpinches.

Uhlenbeck's trick. In the Ricci flow it is the method of pulling back tensors to a bundle isomorphic to the tangent bundle with a fixed metric. This method simplifies the formulas for various evolution equations involving the Riemann curvature tensor, its derivatives and contractions.

variation formula. Given a variation of a geometric quantity such as a metric or a submanifold, the corresponding variation for an associated geometric quantity. For example, given a variation of a metric, we have variation formulas for the Christoffel symbols and the Riemann, Ricci, and scalar curvature tensors.

volume form. On an oriented n-dimensional Riemannian manifold, the n-form $d\mu = \sqrt{\det(g_{ij})}\, dx^1 \wedge \cdots \wedge dx^n$ in a positively oriented local coordinate system. The integral of the volume form is the volume of the manifold.

volume ratio. For a ball $B(p,r)$, the quantity
$$\frac{\operatorname{Vol} B(p,r)}{r^n}.$$

$\mathcal{W}$-functional. (See **Perelman's entropy**.)

warped product metric. Given Riemannian manifolds $(\mathcal{M}^n, g)$ and $(\mathcal{N}^m, h)$, the metric on the product $\mathcal{M} \times \mathcal{N}$ of the form $g(x) + f(x)h(y)$, where $f : \mathcal{M} \to \mathbb{R}$. The natural space-time metrics are similar to warped products.

Witten's black hole. (See **cigar soliton**.)

Yamabe flow. The geometric evolution equation of metrics: $\frac{\partial}{\partial t} g_{ij} = -R g_{ij}$. When $n = 2$, this is the same as the Ricci flow. A technique applied to the Yamabe flow described in this book is the Aleksandrov reflection method. It is expected that the Yamabe flow evolves metrics on closed manifolds to constant scalar curvature metrics.

Yamabe problem. The problem of showing that in any conformal class there exists a metric with constant scalar curvature. This problem has been solved through the works of Yamabe, Trudinger, Aubin, and Schoen. Schoen's work uses the positive mass theorem of Schoen and Yau.

Bibliography

[1] Alexander, J.W. *On the subdivision of 3-space by a polyhedron.* Proc. Nat. Acad. Seci, USA, **10**, 6-8, 1924.

[2] Alvarez, E.; Kubyshin, Y. *Is the string coupling constant invariant under T-duality?* hep-th/9610032.

[3] Anderson, Greg; Chow, Bennett, *A pinching estimate for solutions of the linearized Ricci flow system on 3-manifolds.* Calculus of Variations **23** (2005), no. 1, 1-12.

[4] Anderson, Michael T. *Degeneration of metrics with bounded curvature and applications to critical metrics of Riemannian functionals.* In Differential geometry: Riemannian geometry (Los Angeles, CA, 1990), 53–79, Proc. Sympos. Pure Math., 54, Part 3, Amer. Math. Soc., Providence, RI, 1993.

[5] Anderson, Michael T. *Geometrization of 3-manifolds via the Ricci flow.* Notices Amer. Math. Soc. **51** (2004), no. 2, 184–193.

[6] Anderson, Michael T.; Cheeger, Jeff. *C^α-compactness for manifolds with Ricci curvature and injectivity radius bounded below.* J. Differential Geom. **35** (1992), no. 2, 265–281.

[7] Angenent, Sigurd B.; Knopf, Dan. *An example of neckpinching for Ricci flow on S^{n+1}.* Math. Res. Lett. **11** (2004), no. 4, 493–518.

[8] Angenent, Sigurd B.; Knopf, Dan. *Precise asymptotics for the Ricci flow neckpinch.* arXiv:math.DG/0511247.

[9] Angenent, Sigurd B.; Velázquez, J. J. L. *Degenerate neckpinches in mean curvature flow.* J. Reine Angew. Math. **482** (1997), 15–66.

[10] Apostol, Tom M. *Calculus. Vol. I: One-variable calculus, with an introduction to linear algebra.* Second edition, Blaisdell Publishing Co., Ginn and Co., Waltham, Mass.-Toronto, Ont.-London, 1967.

[11] Aubin, Thierry. *Équations du type Monge-Ampère sur les variétés kähleriennes compactes.* C. R. Acad. Sci. Paris Sér. A-B **283** (1976), no. 3, Aiii, A119–A121.

[12] Aubin, Thierry. *Équations du type Monge-Ampère sur les variétés kählériennes compactes.* Bull. Sci. Math. (2) **102** (1978), no. 1, 63–95.

[13] Aubin, Thierry. *Nonlinear analysis on manifolds. Monge-Ampère equations.* Grundlehren der Mathematischen Wissenschaften, **252**. Springer-Verlag, New York, 1982.

[14] Aubin, Thierry. *Réduction du cas positif de l'équation de Monge-Ampère sur les variétés Kählériennes compactes à la démonstration d'une inégalité.* J. Funct. Anal. **57** (1984), 143–153.

[15] Baird, Paul; Danielo, Laurent. *Three-dimensional Ricci solitons which project to surfaces.* Preprint.

[16] Bakry, D. *Diffusion on compact Riemannian manifolds and logarithmic Sobolev inequalities,* J. Funct. Anal. **42** (1981) 102–109.

[17] Bakry, D.; Concordet, D.; Ledoux, M. *Optimal heat kernel bounds under logarithmic Sobolev inequalities.* ESAIM Probab. Statist. **1** (1995/97), 391–407 (electronic).

[18] Bakry, D.; Émery, Michel. *Diffusions hypercontractives. (French) [Hypercontractive diffusions]* Séminaire de probabilités, XIX, 1983/84, 177–206, Lecture Notes in Math., **1123**, Springer, Berlin, 1985.

[19] Bando, Shigetoshi. *On the classification of three-dimensional compact Kaehler manifolds of nonnegative bisectional curvature.* J. Differential Geom. **19** (1984), no. 2, 283–297.

[20] Bando, Shigetoshi. *Real analyticity of solutions of Hamilton's equation,* Math. Zeit. **195** (1987), 93–97.

[21] Bando, Shigetoshi; Mabuchi, Toshiki. *Uniqueness of Einstein Kähler metrics modulo connected group actions.* Algebraic geometry, Sendai, 1985, 11–40, Adv. Stud. Pure Math., **10**, North-Holland, Amsterdam, 1987.

[22] Barenblatt, Grigory I. *On self-similar motions of compressible fluid in a porous medium.* Prikladnaya Matematika i Mekhanika (Applied Mathematics and Mechanics (PMM)), **16** (1952), No. 6, 679-698 (in Russian).

[23] Beckner, William; Pearson, Michael. *On sharp Sobolev embedding and the logarithmic Sobolev inequality.* Bull. London Math. Soc. **30** (1998), no. 1, 80–84.

[24] Bérard-Bergery, L. *Sur de nouvelles variétiés riemannienes d'Einstein.* Publ. Inst. E. Cartan # 4, U. de Nancy (1982).

[25] Berndt, Jürgen; Tricerri, Franco; Vanhecke, Lieven. *Generalized Heisenberg groups and Damek-Ricci harmonic spaces.* Lecture Notes in Mathematics, **1598**. Springer-Verlag, Berlin, 1995.

[26] Bernstein, Sergi N. *A limitation on the moduli of a sequence of derivatives of solutions of equations of parabolic type.* (Russian) Dokl. Akad. Nauk SSSR **18** (1938) 385–388.

[27] Besse, Arthur, *Einstein manifolds.* Ergebnisse der Mathematik und ihrer Grenzgebiete, **10**. Springer-Verlag, Berlin, 1987.

[28] Bing, R. H. *An alternative proof that 3-manifolds can be triangulated.* Ann. of Math. (2) **69** (1959), 37–65.

[29] Böhm, Christoph; Kerr, Megan M. *Low-dimensional homogeneous Einstein manifolds.* Trans. Amer. Math. Soc. **358** (2006), no. 4, 1455–1468.

[30] Böhm, Christoph; Wilking, Burkhard. *Manifolds with positive curvature operators are space forms.* arXiv:math.DG/0606187.

[31] Böhm, Christoph; Wilking, Burkhard. *Nonnegatively curved manifolds with finite fundamental groups admit metrics with positive Ricci curvature.* Preprint.

[32] Bourguignon, Jean-Pierre. *Une stratification de l'espace des structures riemanniennes.* (French) Compositio Math. **30** (1975), 1–41.

[33] Bourguignon, Jean-Pierre. *Ricci curvature and Einstein metrics.* Global differential geometry and global analysis (Berlin, 1979), pp. 42–63, Lecture Notes in Math., **838**, Springer, Berlin-New York, 1981.

[34] Buckland, John A. *Short-time existence of solutions to the cross curvature flow on 3-manifolds.* Proc. Amer. Math. Soc. **134** (2006), no. 6, 1803–1807.

[35] Bryant, Robert. *Gradient Kähler Ricci solitons.* arXiv:math.DG/0407453.

[36] Bryant, R. L.; Chern, S. S.; Gardner, R. B.; Goldschmidt, H. L.; Griffiths, P. A. *Exterior differential systems.* Mathematical Sciences Research Institute Publications, **18**. Springer-Verlag, New York, 1991.

[37] Burago, D.; Burago, Y.; Ivanov, S. *A course in metric geometry,* Grad Studies Math. **33**, Amer. Math. Soc., Providence, RI, 2001. Corrections of typos and small errors to the book "A Course in Metric Geometry": http://www.pdmi.ras.ru/staff/burago.html#English

[38] Buscher, T. H. *A symmetry of the string background field equations.* Phys. Lett. **B194** (1987), 59–62.

[39] Buscher, T. H. *Path integral derivation of quantum duality in nonlinear sigma models.* Phys. Lett. **B201** (1988), 466.

[40] Buser, Peter; Karcher, Hermann. *Gromov's almost flat manifolds.* Astérisque, **81**. Société Mathématique de France, Paris, 1981.

[41] Cabré, Xavier. *Nondivergent elliptic equations on manifolds with nonnegative curvature.* Comm. Pure Appl. Math. **50** (1997) no. 7, 623-665.

[42] Calabi, Eugenio. *An extension of E. Hopf's maximum principle with an application to Riemannian geometry.* Duke Math. J. **24** (1957), 45-56.

[43] Calabi, Eugenio. *On Kähler manifolds with vanishing canonical class.* Algebraic geometry and topology. A symposium in honor of S. Lefschetz, pp. 78–89. Princeton University Press, Princeton, N. J., 1957.

[44] Calabi, Eugenio. *Métriques kählériennes et fibrés holomorphes.* Ann. Sci. École Norm. Sup. (4) **12** (1979), no. 2, 269–294.

[45] Cao, Huai-Dong. *Deformation of Kähler metrics to Kähler-Einstein metrics on compact Kähler manifolds.* Invent. Math. **81** (1985), no. 2, 359–372.

[46] Cao, Huai-Dong. *On Harnack's inequalities for the Kähler–Ricci flow.* Invent. Math. **109** (1992), 247–263.

[47] Cao, Huai-Dong. *Existence of gradient Ricci–Kähler solitons.* In Elliptic and parabolic methods in geometry (Minneapolis, MN, 1994), 1–16; A K Peters, Wellesley, MA, 1996.

[48] Cao, Huai-Dong. *Limits of solutions to the Kähler–Ricci flow.* J. Differential Geom. 45 (1997), no. 2, 257–272.

[49] Cao, Huai-Dong; Chen, Bing-Long; Zhu, Xi-Ping. *Ricci flow on compact Kähler manifolds of positive bisectional curvature.* C. R. Math. Acad. Sci. Paris **337** (2003), no. 12, 781–784.

[50] Cao, Huai-Dong; Chow, Bennett. *Compact Kähler manifolds with nonnegative curvature operator.* Invent. Math. **83** (1986), no. 3, 553–556.

[51] Cao, Huai-Dong; Chow, Bennett. *Recent developments on the Ricci flow.* Bull. Amer. Math. Soc. (N.S.) **36** (1999), no. 1, 59–74.

[52] Cao, Huai-Dong; Chow, Bennett; Chu, Sun-Chin; Yau, Shing-Tung, editors. *Collected papers on Ricci flow.* Internat. Press, Somerville, MA, 2003.

[53] Cao, Huai-Dong; Hamilton, Richard S.; Ilmanen, Tom. *Gaussian densities and stability for some Ricci solitons.* arXiv:math.DG/0404165.

[54] Cao, Huai-Dong; Ni, Lei. *Matrix Li–Yau–Hamilton estimates for the heat equation on Kähler manifolds.* Math. Ann. **331** (2005), no. 4, 795–807.

[55] Cao, Huai-Dong; Sesum, Natasa. *The compactness result for Kähler Ricci solitons.* arXiv:math.DG/0504526.

[56] Cao, Huai-Dong; Zhu, Xi-Ping. *A complete proof of the Poincaré and geometrization conjectures — application of the Hamilton-Perelman theory of the Ricci flow.* Asian J. Math. **10** (2006), 165-498.

[57] Carfora; M.; Marzuoli, A. *Model geometries in the space of Riemannian structures and Hamilton's flow.* Classical Quantum Gravity **5** (1988), no. 5, 659–693.

[58] Carlen, E. A.; Kusuoka, S.; Strook, D. W. *Upper bounds for symmetric Markov transition functions.* Ann. Inst. H. Poincaré Probab. Statist. **23** (1987) 245-287.

[59] Carron, G. *Inégalités isopérimétriques de Faber-Krahn et conséquences.* Actes de la Table Ronde de Géométrie Différentielle (Luminy, 1992), Sémin. Congr., 1, Soc. Math. France, Paris 1996, 205-232.

[60] Cascini, Paolo; La Nave, Gabriele. *Kähler–Ricci Flow and the minimal model program for projective varieties.* arXiv:math.AG/0603064.

[61] Casson, Andrew; Jungreis, Douglas. *Convergence groups and Seifert fibered 3-manifolds.* Invent. Math. **118** (1994), no. 3, 441–456.

[62] Chang, Shu-Cheng; Lu, Peng. *Evolution of Yamabe constants under Ricci flow.* Preprint.

[63] Chau, Albert; Tam, Luen-Fai. *Gradient Kähler-Ricci solitons and a uniformization conjecture.* arXiv:math.DG/0310198.

[64] Chau, Albert; Tam, Luen-Fai. *A note on the uniformization of gradient Kähler–Ricci solitons.* Math. Res. Lett. **12** (2005), no. 1, 19–21.

[65] Chau, Albert; Tam, Luen-Fai. *On the complex structure of Kähler manifolds with nonnegative curvature.* arXiv:math.DG/0504422.

[66] Chau, Albert; Tam, Luen-Fai. *Nonnegatively curved Kähler manifolds with average quadratic curvature decay.* arXiv:math.DG/0510252.

[67] Chavel, Isaac. *Eigenvalues in Riemannian geometry.* Including a chapter by Burton Randol. With an appendix by Jozef Dodziuk. Pure and Applied Mathematics, 115. Academic Press, Inc., Orlando, FL, 1984.

[68] Chavel, Isaac. *Riemannian geometry — a modern introduction.* Cambridge Tracts in Mathematics, **108**. Cambridge University Press, Cambridge, 1993.

[69] Cheeger, Jeff. *Some examples of manifolds of nonnegative curvature.* J. Differential Geometry **8** (1973), 623–628.

[70] Cheeger, Jeff. *Finiteness theorems for Riemannian manifolds.* Amer. J. Math. **92** (1970), 61–74.

[71] Cheeger, Jeff; Colding, Tobias H. *Lower bounds on Ricci curvature and the almost rigidity of warped products.* Ann. of Math. (2) **144** (1996), no. 1, 189–237.

[72] Cheeger, Jeff; Ebin, David G. *Comparison theorems in Riemannian geometry.* North-Holland Mathematical Library, Vol. 9. North-Holland Publishing Co., Amsterdam-Oxford; American Elsevier Publishing Co., Inc., New York, 1975.

[73] Cheeger; Jeff; Gromov, Mikhail. *Collapsing Riemannian manifolds while keeping their curvature bounded, I,* J. Differential Geom. **23** (1986) 309-346.

[74] Cheeger; Jeff; Gromov, Mikhail. *Collapsing Riemannian manifolds while keeping their curvature bounded, II,* J. Differential Geom. **32** (1990) 269-298.

[75] Cheeger; Jeff; Gromov, Mikhail; Taylor, Michael. *Finite propagation speed, kernel estimates for functions of the Laplace operator, and the geometry of complete Riemannian manifolds,* J. Differential Geom. **17** (1982) 15–53.

[76] Cheeger; Jeff; Yau, Shing-Tung. *A lower bound for the heat kernel.* Comm. Pure Appl. Math. **34** (1981), no. 4, 465–480.

[77] Chen, Bing-Long; Zhu, Xi-Ping. *Complete Riemannian manifolds with pointwise pinched curvature.* Invent. Math. **140** (2000), no. 2, 423–452.

[78] Chen, Bing-Long; Zhu, Xi-Ping. *A property of Kähler–Ricci solitons on complete complex surfaces.* Geometry and nonlinear partial differential equations (Hangzhou, 2001), 5–12, AMS/IP Stud. Adv. Math., **29**, Amer. Math. Soc., Providence, RI, 2002.

[79] Chen, Bing-Long; Zhu, Xi-Ping. *On complete noncompact Kähler manifolds with positive bisectional curvature.* Math. Ann. **327** (2003) 1-23.

[80] Chen, Bing-Long; Zhu, Xi-Ping. *Volume growth and curvature decay of positively curved Kähler manifolds.* arXiv:math.DG/0211374.

[81] Chen, Bing-Long; Zhu, Xi-Ping. *Ricci flow with surgery on four-manifolds with positive isotropic curvature.* arXiv:math.DG/0504478.

[82] Chen, Bing-Long; Zhu, Xi-Ping. *Uniqueness of the Ricci flow on complete noncompact manifolds.* arXiv:math.DG/0505447.

[83] Chen, Xiuxiong. *The space of Kähler metrics.* J. Differential Geom. **56** (2000), no. 2, 189–234.

[84] Chen, Xiuxiong. *On the lower bound of energy functional $E_1(I)$ – a stability theorem on the Kähler Ricci flow.* arXiv:math.DG/0502196.

[85] Chen, Xiuxiong; Li, H. *The Kähler–Ricci flow on Kähler manifolds with 2-traceless bisectional curvature operator.* arXiv:math.DG/0503645.

[86] Chen, Xiuxiong; Lu, Peng; Tian, Gang. *A note on uniformization of Riemann surfaces by Ricci flow.* Proc. Amer. Math. Soc. **134** (2006), 3391–3393.

[87] Chen, Xiuxiong; Tian, Gang. *Ricci flow on Kähler–Einstein surfaces.* Invent. Math. **147** (2002), no. 3, 487–544.

[88] Chen, Xiuxiong; Tian, Gang. *Ricci flow on Kähler–Einstein manifold.* Duke Math. J. **131** (2006), 17–73.

[89] Chen, Ya-Zhe. *Second-order parabolic partial differential equations,* Beijing University Press, 2003, Beijing.

[90] Chen, Yun Mei; Struwe, Michael. *Existence and partial regularity results for the heat flow for harmonic maps.* Math. Z. **201** (1989), no. 1, 83–103.

[91] Cheng, Hsiao-Bing. *A new Li–Yau–Hamilton estimate for the Ricci flow.* Comm. Anal. Geom. **14** (2006), 551–564.

[92] Cheng, Shiu-Yuen. *Eigenvalue comparison theorems and its geometric applications.* Math. Zeit. **143** (1975), no. 3, 289–297.

[93] Cheng, Shiu-Yuen; Li, Peter; Yau, Shing-Tung. *On the upper estimate of the heat kernel of a complete Riemannian manifold,* Amer. J. Math. **103** (1981), no. 5, 1021–1063.

[94] Cheng, Shiu-Yuen; Yau, Shing-Tung. *On the existence of a complete Kähler metric on noncompact complex manifolds and the regularity of Fefferman's equation.* Comm. Pure Appl. Math. **33** (1980), 507–544.

[95] Chern, Shiing Shen. *Complex manifolds without potential theory. With an appendix on the geometry of characteristic classes.* Second edition. Universitext. Springer-Verlag, New York-Heidelberg, 1979.

[96] Chow, Bennett. *On Harnack's inequality and entropy for the Gaussian curvature flow.* Comm. Pure Appl. Math. **44** (1991), no. 4, 469–483.

[97] Chow, Bennett. *On the entropy estimate for the Ricci flow on compact 2-orbifolds.* J. Differential Geom. **33** (1991) 597-600.

[98] Chow, Bennett. *Interpolating between Li–Yau's and Hamilton's Harnack inequalities on a surface.* J. Partial Diff. Equations (China) **11** (1998) 137–140.

[99] Chow, Bennett. *On an alternate proof of Hamilton's matrix Harnack inequality of Li–Yau type for the Ricci flow.* Indiana Univ. Math. J. **52** (2003), 863-874.

[100] Chow, Bennett; Chu, Sun-Chin. *A geometric interpretation of Hamilton's Harnack inequality for the Ricci flow,* Math. Res. Lett. **2** (1995), 701–718.

[101] Chow, Bennett; Chu, Sun-Chin, *A geometric approach to the linear trace Harnack inequality for the Ricci flow,* Math. Res. Lett. **3** (1996), 549-568.

[102] Chow, Bennett; Chu, Sun-Chin; Lu, Peng; Ni, Lei. *Notes on Perelman's papers on Ricci flow.* Unpublished.

[103] Chow, Bennett; Glickenstein, David; Lu, Peng. *Metric transformations under collapsing of Riemannian manifolds.* Math Res Lett. **10** (2003), No. 5-6, 737-746.

[104] Chow, Bennett; Guenther, Christine. *Monotonicity formulas for parabolic equations in geometry.* In preparation.

[105] Chow, Bennett; Hamilton, Richard S. *Constrained and linear Harnack inequalities for parabolic equations.* Invent. Math. **129** (1997), 213–238.

[106] Chow, Bennett; Hamilton, Richard S. *The cross curvature flow of 3-manifolds with negative sectional curvature.* Turkish Journal of Mathematics **28** (2004), 1–10.

[107] Chow, Bennett; Knopf, Dan. *New Li–Yau–Hamilton inequalities for the Ricci flow via the space-time approach.* J. of Diff. Geom. **60** (2002), 1–54.

[108] Chow, Bennett; Knopf, Dan. *The Ricci flow: An introduction.* Mathematical Surveys and Monographs, AMS, Providence, RI, 2004.

[109] Chow, Bennett; Knopf, Dan; Lu, Peng. *Hamilton's injectivity radius estimate for sequences with almost nonnegative curvature operators.* Comm. Anal. Geom. **10** (2002), no. 5, 1151–1180.

[110] Chow, Bennett; Lu, Peng. *The time-dependent maximum principle for systems of parabolic equations subject to an avoidance set.* Pacific J. Math. **214** (2004), no. 2, 201–222.

[111] Chow, Bennett; Lu, Peng; Ni, Lei. *Hamilton's Ricci flow*. Lectures in Contemporary Mathematics, **3**, Science Press and Graduate Studies in Mathematics, **77**, American Mathematical Society (co-publication), 2006.

[112] Chow, Bennett; Wu, Lang-Fang. *The Ricci flow on compact 2-orbifolds with curvature negative somewhere*. Comm. Pure Appl. Math. **44** (1991), no. 3, 275–286.

[113] Chu, Sun-Chin. *Basic Properties of Gradient Ricci Solitons*. In Geometric Evolution Equations, Contemporary Mathematics **367**, S.-C. Chang, B. Chow, S.-C. Chu. C.-S. Lin, eds., American Mathematical Society, 2005.

[114] Chu, Sun-Chin. *Geometry of 3-dimensional gradient Ricci solitons with positive curvature*. Comm. Anal. Geom. **13** (2005), no. 1, 129–150.

[115] Cohn-Vossen, S. *Kürzete Wege und Totalkrümmung auf Flächen*. Comp. Math. **2** (1935), 69–133.

[116] Colding, Tobias; Minicozzi, William P. II, *Estimates for the extinction time for the Ricci flow on certain 3-manifolds and a question of Perelman*. J. Amer. Math. Soc. **18** (2005), 561-569.

[117] Courant, R.; Hilbert, David. *Methods of mathematical physics*. Vols. I and II. Interscience Publishers, Inc., New York, N.Y., 1953 and 1962 (reprinted in 1989).

[118] Craioveanu, Mircea; Puta, Mircea; Rassias, Themistocles M. *Old and new aspects in spectral geometry*. Mathematics and its Applications, **534**. Kluwer Academic Publishers, Dordrecht, 2001.

[119] Daskalopoulos, Panagiota; del Pino, Manuel A. *On a singular diffusion equation*. Comm. Anal. Geom. **3** (1995), 523–542.

[120] Daskalopoulos, Panagiota; Hamilton, Richard S. *Geometric estimates for the logarithmic fast diffusion equation*. Comm. Anal. Geom. **12** (2004), 143–164.

[121] Daskalopoulos, Panagiota; Sesum, Natasa. *Eternal Solutions to the Ricci Flow on $\mathbb{R}^2$*. arXiv:math.AP/0603525.

[122] Davies, E. B. *Heat kernel and spectral theory,* Cambridge Univ. Press, Cambridge, 1989.

[123] DeTurck, Dennis, *Deforming metrics in the direction of their Ricci tensors*, J. Differential Geom. **18** (1983) 157-162.

[124] DeTurck, Dennis M. *Deforming metrics in the direction of their Ricci tensors, improved version,* Collected Papers on Ricci Flow, H.-D. Cao, B. Chow, S.-C. Chu, and S.-T. Yau, eds. Internat. Press, Somerville, MA, 2003.

[125] Dijkgraaf, Robbert; Verlinde, Herman; Verlinde, Erik. *String propogation in a black hole geometry*. Nucl. Phys. B **371** (1992), 269-314.

[126] Ding, Yu. *Notes on Perelman's second paper*. http://math.uci.edu/~yding/perelman.pdf .

[127] do Carmo, Manfredo Perdigão. *Riemannian geometry*. Translated from the second Portuguese edition by Francis Flaherty. Mathematics: Theory & Applications. Birkhäuser Boston, Inc., Boston, MA, 1992.

[128] Donaldson, Simon K. *Anti self-dual Yang–Mills connections over complex algebraic surfaces and stable vector bundles*. Proc. London Math. Soc. (3) **50** (1985), no. 1, 1–26.

[129] Donaldson, Simon K. *Scalar curvature and stability of toric varieties*. J. Differential Geom. **62** (2002), no. 2, 289–349.

[130] Donaldson, Simon K. *Lower bounds on the Calabi functional*. J. Differential Geom. **70** (2005), no. 3, 453–472.

[131] Drazin, P.; Johnson, N. *Solitons: an introduction*. Cambridge, 1989.

[132] Ecker, Klaus. *Regularity theory for mean curvature flow*. Progress in Nonlinear Differential Equations and their Applications, **57**. Birkhäuser Boston, Inc., Boston, MA, 2004.

[133] Ecker, Klaus. *A local monotonicity formula for mean curvature flow*. Ann. of Math. (2) **154** (2001), no. 2, 503–525.

[134] Ecker, Klaus; Knopf, Dan; Ni, Lei; Topping, Peter. *Local monotonicity and mean value formulas for evolving Riemannian manifolds.* Preprint.

[135] Eells, James, Jr.; Sampson, J. H. *Harmonic mappings of Riemannian manifolds.* Amer. J. Math. **86** (1964), 109–160.

[136] Epstein, D. B. A. *Periodic flows on three-manifolds.* Ann. of Math. (2) **95** 1972 66–82.

[137] Evans, Lawrence. *Partial differential equations.* AMS. Providence, 1998.

[138] Evans, Lawrence C. *Entropy and partial differential equations.* Lecture Notes at UC Berkeley. 212 pp. http://math.berkeley.edu/~evans/entropy.and.PDE.pdf

[139] Evans, Lawrence; Gariepy, R. F. *Measure theory and fine properties of functions.* CRC Press, Boca Raton, 1992.

[140] Fabes E. B.; Garofalo, N. *Mean value properties of solutions to parabolic equations with variable coefficients,* Jour. Math. Anal. Appl. **121** (1987), 305-316.

[141] Fateev, V. A. ; Onofri, E.; Zamolodchikov, A. B. *The sausage model (integrable deformations of O(3) sigma model).* Nucl. Phys. B **406** (1993), 521-565.

[142] Feldman, Mikhail; Ilmanen, Tom; Knopf, Dan. *Rotationally symmetric shrinking and expanding gradient Kähler-Ricci solitons.* J. Differential Geom. **65** (2003), no. 2, 169–209.

[143] Feldman, Mikhail; Ilmanen, Tom; Ni, Lei. *Entropy and reduced distance for Ricci expanders.* arXiv:math.DG/0405036.

[144] Frankel, Theodore. *Manifolds with positive curvature.* Pacific J. Math. **11** (1961), 165–174.

[145] Friedan, Daniel Harry. *Nonlinear models in $2 + \varepsilon$ dimensions.* Ann. Physics **163** (1985), no. 2, 318–419.

[146] Friedman, Avner. *Partial differential equations of parabolic type.* Robert E. Krieger Publishing Company, 1983, Malabar, Florida.

[147] Futaki, Akito. *An obstruction to the existence of Einstein–Kähler metrics.* Invent. Math. **73** (1983), no. 3, 437–443.

[148] Futaki, Akito. *Kähler–Einstein metrics and integral invariants.* Lecture Notes in Math., 1314, Springer-Verlag, Berlin, 1988.

[149] Gabai, David. *Convergence groups are Fuchsian groups.* Ann. of Math. (2) **136** (1992), no. 3, 447–510.

[150] Gabai, David; Meyerhoff, G. Robert; Thurston, Nathaniel. *Homotopy hyperbolic 3-manifolds are hyperbolic.* Ann. of Math. (2) **157** (2003), no. 2, 335–431.

[151] Gage, M.; Hamilton, Richard S. *The heat equation shrinking convex plane curves.* J. Differential Geom. **23** (1986), no. 1, 69–96.

[152] Gao, L. Zhiyong. *Convergence of Riemannian manifolds; Ricci and $L^{n/2}$-curvature pinching.* J. Differential Geom. **32** (1990), no. 2, 349–381.

[153] Garofalo, N.; Lanconelli, E. *Asymptotic behavior of fundamental solutions and potential theory of parabolic operators with variable coefficients.* Math. Ann. **283** (1989) no. 2, 211-239.

[154] Gastel, Andreas; Kronz, Manfred. *A family of expanding Ricci solitons.* Variational problems in Riemannian geometry, 81–93, Progr. Nonlinear Differential Equations Appl., **59**, Birkhhauser, Basel, 2004.

[155] Gilbarg, David; Trudinger, Neil S. *Elliptic partial differential equations of second order.* Reprint of the 1998 edition. Classics in Mathematics. Springer-Verlag, Berlin, 2001.

[156] Glickenstein, David. *Precompactness of solutions to the Ricci flow in the absence of injectivity radius estimates.* Geom. Topol. **7** (2003), 487–510 (electronic).

[157] Goldberg, Samuel I. *Curvature and homology.* Revised reprint of the 1970 edition. Dover Publications, Inc., Mineola, NY, 1998.

[158] Gordon, C. McA.; Heil, Wolfgang. *Cyclic normal subgroups of fundamental groups of 3-manifolds.* Topology **14** (1975), no. 4, 305–309.

[159] Gray, Brayton. *Homotopy theory. An introduction to algebraic topology.* Pure and Applied Mathematics, Vol. 64. Academic Press [Harcourt Brace Jovanovich, Publishers], New York-London, 1975.

[160] Grayson, Matthew A. *Shortening embedded curves.* Ann. of Math. (2) **129** (1989), no. 1, 71–111.

[161] Grayson, Matthew; Hamilton, Richard S. *The formation of singularities in the harmonic map heat flow.* Comm. Anal. Geom. **4** (1996), no. 4, 525–546.

[162] Green, M.B.; Schwarz, J.H.; Witten, E. *Superstring Theory, Volumes 1 & 2.* Cambridge University Press, Cambridge, 1987.

[163] Greene, R. E.; Wu, H. C^∞ *convex functions and manifolds of positive curvature.* Acta Math. **137** (1976), no. 3-4, 209-245.

[164] Greene, R. E.; Wu, H. *Function theory on manifolds which possess a pole.* Lecture Notes in Mathematics, **699**. Springer, Berlin, 1979.

[165] Greene, R. E.; Wu, H. *Lipschitz convergence of Riemannian manifolds.* Pacific J. Math. **131** (1988), no. 1, 119–141.

[166] Griffiths, Phillip; Harris, Joseph. *Principles of algebraic geometry.* Reprint of the 1978 original. Wiley Classics Library. John Wiley & Sons, Inc., New York, 1994.

[167] Grigor'yan, Alexander. *Upper bounds of derivatives of the heat kernel on an arbitrary complete manifold.* J. Funct. Anal. **127** (1995), no. 2, 363–389.

[168] Grigor'yan, Alexander. *Gaussian upper bounds for heat kernel on arbitrary manifolds,* J. Differential Geom. **45** (1997), 33–52.

[169] Gromov, Misha. *Metric structures for Riemannian and non-Riemannian spaces.* Based on the 1981 French original. With appendices by M. Katz, P. Pansu and S. Semmes. Translated from the French by Sean Michael Bates. Progress in Mathematics, **152**. Birkhäuser Boston, Inc., Boston, MA, 1999.

[170] Gross, Leonard. *Logarithmic Sobolev inequalities.* Amer. J. Math. **97** (1975), no. 4, 1061–1083.

[171] Guenther, Christine M. *The fundamental solution on manifolds with time-dependent metrics.* J. Geom. Anal. **12** (2002), no. 3, 425–436.

[172] Guenther, Christine; Isenberg, Jim; Knopf, Dan, *Stability of the Ricci flow at Ricci-flat metrics,* Comm. Anal. Geom. **10** (2002), no. 4, 741–777.

[173] Gursky, Matthew. *The Weyl functional, de Rham cohomology, and Kähler–Einstein metrics,* Ann. of Math. **148** (1998), 315-337.

[174] Gutperle, Michael; Headrick, Matthew; Minwalla, Shiraz; Schomerus, Volker. *Space-time energy decreases under world-sheet RG flow.* arXiv:hep-th/0211063.

[175] Haagensen, Peter E., *Duality Transformations Away From Conformal Points.* Phys. Lett. B **382** (1996), 356–362.

[176] Haken, Wolfgang. *Theorie der Normalflächen.* Acta Math. **105** (1961), 245–375.

[177] Hamilton, Richard S. *Harmonic maps of manifolds with boundary.* Lecture Notes in Mathematics, Vol. 471. Springer-Verlag, Berlin-New York, 1975.

[178] Hamilton, Richard S. *Three-manifolds with positive Ricci curvature.* J. Differential Geom. **17** (1982), no. 2, 255–306.

[179] Hamilton, Richard S. *Four-manifolds with positive curvature operator.* J. Differential Geom. **24** (1986), no. 2, 153–179.

[180] Hamilton, Richard S. *The Ricci flow on surfaces.* Mathematics and general relativity (Santa Cruz, CA, 1986), 237–262, Contemp. Math., **71**, Amer. Math. Soc., Providence, RI, 1988.

[181] Hamilton, Richard S. *The Harnack estimate for the Ricci flow.* J. Differential Geom. **37** (1993), no. 1, 225–243.

[182] Hamilton, Richard S. *Eternal solutions to the Ricci flow,* J. Differential Geom. **38** (1993), 1-11.

[183] Hamilton, Richard. *A matrix Harnack estimate for the heat equation,* Comm. Anal. Geom. **1** (1993), 113–126.

[184] Hamilton, Richard. *Monotonicity formulas for parabolic flows on manifolds*, Comm. Anal. Geom. **1** (1993), 127–137.

[185] Hamilton, Richard S. *Convex hypersurfaces with pinched second fundamental form.* Comm. Anal. Geom. **2** (1994), no. 1, 167–172.

[186] Hamilton, Richard S. *The formation of singularities in the Ricci flow.* Surveys in differential geometry, Vol. II (Cambridge, MA, 1993), 7–136, Internat. Press, Cambridge, MA, 1995.

[187] Hamilton, Richard S. *A compactness property for solutions of the Ricci flow.* Amer. J. Math. **117** (1995), no. 3, 545–572.

[188] Hamilton, Richard S. *Harnack estimate for the mean curvature flow,* J. Diff. Geom. **41** (1995) 215-226.

[189] Hamilton, Richard S. *Four-manifolds with positive isotropic curvature.* Comm. Anal. Geom. **5** (1997), no. 1, 1–92.

[190] Hamilton, Richard S. *Non-singular solutions of the Ricci flow on three-manifolds.* Comm. Anal. Geom. **7** (1999), no. 4, 695–729.

[191] Hamilton, Richard S. *Three-orbifolds with positive Ricci curvature.* In Collected Papers on Ricci Flow, H.-D. Cao, B. Chow, S.-C. Chu, and S.-T. Yau, eds. Internat. Press, Somerville, MA, 2003.

[192] Hamilton, Richard S. *Differential Harnack estimates for parabolic equations,* preprint.

[193] Hamilton, Richard S.; Sesum, Natasa. *Properties of the solutions of the conjugate heat equation.* arXiv:math.DG/0601415.

[194] Hamilton, Richard S.; Yau, Shing-Tung. *The Harnack estimate for the Ricci flow on a surface—revisited.* Asian J. Math. **1** (1997), no. 3, 418–421.

[195] Han, Qing; Lin, Fanghua. *Elliptic Partial Differential Equations.* AMS/New York University Press, 1997.

[196] Hartman, Philip. *Ordinary differential equations.* Corrected reprint of the second (1982) edition. Birkhäuser, Boston, MA.

[197] Hass, Joel; Morgan, Frank. *Geodesics and soap bubbles in surfaces.* Math. Z. **223** (1996), no. 2, 185–196.

[198] Heber, Jens. *Noncompact homogeneous Einstein spaces.* Invent. Math. **133** (1998), no. 2, 279–352.

[199] Hebey, Emmanuel. *Introduction à l'analyse non linéaire sur les variétés.* Diderot Editeur, Arts et Sciences, 1997.

[200] Helgason, Sigurdur. *Differential geometry, Lie groups, and symmetric spaces.* Pure and Applied Mathematics, **80**. Academic Press, Inc. [Harcourt Brace Jovanovich], New York-London, 1978.

[201] Hempel, John. *3-Manifolds.* Ann. of Math. Studies, No. 86. Princeton University Press, Princeton, N. J.; University of Tokyo Press, Tokyo, 1976.

[202] Hiriart-Urruty, Jean-Baptiste; Lemarechal, Claude. *Fundamentals of convex analysis.* Springer, 2001.

[203] Hirsch, Morris W. *Obstruction theories for smoothing manifolds and maps.* Bull. Amer. Math. Soc. **69** (1963), 352–356.

[204] Hirzebruch, F.; Kodaira, K. *On the complex projective spaces.* J. Math. Pures Appl. **36** (1957), 201–216.

[205] Hörmander, Lars. *Notions of convexity.* Birkhauser, 1994.

[206] Hsu, Shu-Yu. *Large time behaviour of solutions of the Ricci flow equation on $\mathbb{R}^2$.* Pacific J. Math. **197** (2001), no. 1, 25–41.

[207] Hsu, Shu-Yu. *Global existence and uniqueness of solutions of the Ricci flow equation.* Differential Integral Equations **14** (2001), no. 3, 305–320.

[208] Hsu, Shu-Yu. *A simple proof on the nonexistence of shrinking breathers for the Ricci flow.* arXiv: math.DG/0509087.

[209] Huang, Hong. *A note on Morse's index theorem for Perelman's $\mathcal{L}$-length.* arXiv:math.DG/0602090.

[210] Huber, Alfred. *On subharmonic functions and differential geometry in the large.* Comment. Math. Helv. **32** (1957), 13–72.

[211] Huisken, Gerhard. *Flow by mean curvature of convex surfaces into spheres,* J. Differential Geom. **20** (1984), no. 1, 237-266.

[212] Huisken, Gerhard. *Ricci deformation of the metric on a Riemannian manifold.* J. Differential Geom. **21** (1985), no. 1, 47–62.

[213] Huisken, Gerhard. *Asymptotic behavior for singularities of the mean curvature flow.* J. Differential Geom. **31** (1990), no. 1, 285–299.

[214] Ishii, Hitoshi; Lions, Pierre-Louis. *Viscosity solutions of fully nonlinear second-order elliptic partial differential equations.* J. Differential Equations **83** (1990), no. 1, 26–78.

[215] Iskovskih, V. A. *Fano threefolds. I.* Izv. Akad. Nauk SSSR Ser. Mat. **41** (1977), no. 3, 516–562, 717.

[216] Iskovskih, V. A. *Fano threefolds. II.* Izv. Akad. Nauk SSSR Ser. Mat. **42** (1978), no. 3, 506–549.

[217] Ivey, Thomas. *On solitons for the Ricci Flow.* PhD thesis, Duke University, 1992.

[218] Ivey, Thomas. *Ricci solitons on compact three-manifolds.* Diff. Geom. Appl. **3** (1993), 301–307.

[219] Ivey, Thomas. *New examples of complete Ricci solitons.* Proc. Amer. Math. Soc. **122** (1994), 241–245.

[220] Ivey, Thomas. *The Ricci flow on radially symmetric $\mathbb{R}^3$.* Comm. Part. Diff. Eq. **19** (1994), 1481-1500.

[221] Ivey, Thomas. *Local existence of Ricci solitons.* Manuscripta Math. **91** (1996), 151–162.

[222] Ivey, Thomas. *Ricci solitons on compact Kähler surfaces.* Proc. Amer. Math. Soc. **125** (1997), no. 4, 1203–1208.

[223] Ivey, Thomas A.; Landsberg, J. M. *Cartan for beginners: differential geometry via moving frames and exterior differential systems.* Graduate Studies in Mathematics, **61**. American Mathematical Society, Providence, RI, 2003.

[224] Jaco, William; Shalen, Peter B. *Seifert fibered spaces in 3-manifolds.* Geometric topology (Proc. Georgia Topology Conf., Athens, Ga., 1977), pp. 91–99, Academic Press, New York-London, 1979.

[225] Johannson, Klaus. *Homotopy equivalences of 3-manifolds with boundaries.* Lecture Notes in Mathematics, **761**. Springer, Berlin, 1979.

[226] Juutinen, Petri; Lindqvist, Peter; Manfredi, Juan J. *On the equivalence of viscosity solutions and weak solutions for a quasi-linear equation.* SIAM J. Math. Anal. **33** (2001), no. 3, 699–717.

[227] H. Karcher, *Riemannian center of mass and mollifier smoothing,* Comm. Pure Appl. Math. **30** (1977), 509-541.

[228] Karp, Leon; Li, Peter. *The heat equation on complete Riemannian manifolds.* Unpublished manuscript.

[229] Kasue, A. *A compactification of a manifold with asymptotically nonnegative curvature.* Ann. Scient. Ec. Norm. Sup. **21** (1988), 593-622.

[230] Kato, Tosio. *Perturbation theory for linear operators.* Reprint of the 1980 edition. Classics in Mathematics. Springer-Verlag, Berlin, 1995.

[231] Kleiner, Bruce; Lott, John. *Notes on Perelman's papers.* May 16, 2006. http://www.math.lsa.umich.edu/~lott/ricciflow/perelman.html

[232] Knapp, Anthony W. *Lie groups, Lie algebras, and cohomology.* Mathematical Notes, **34**. Princeton University Press, Princeton, NJ, 1988.

[233] Kneser, H. *Geschloßene Flächen in dreidimensionalen Mannigfaltikeiten,* Jahresbericht der Deut. Math. Verein. **38** (1929), 248–260.

[234] Knopf, Dan. *Positivity of Ricci curvature under the Kähler–Ricci flow.* Commun. Contemp. Math. **8** (2006), no. 1, 123–133.

[235] Knopf, Dan; Young, Andrea. *Asymptotic stability of the cross curvature flow at a hyperbolic metric.* Preprint.

[236] Kobayashi, Shoshichi; Nomizu, Katsumi. *Foundations of differential geometry.* Vols. I & II. Reprint of the 1963 and 1969 originals. Wiley Classics Library. A Wiley-Interscience Publication. John Wiley & Sons, Inc., New York, 1996.

[237] Kobayashi, Shoshichi; Ochiai, Takushiro. *Characterizations of complex projective spaces and hyperquadrics.* J. Math. Kyoto Univ. **13** (1973), 31–47.

[238] Kodaira, Kunihiko. *Complex manifolds and deformation of complex structures.* Translated from the 1981 Japanese original by Kazuo Akao. Reprint of the 1986 English edition. Classics in Mathematics. Springer-Verlag, Berlin, 2005.

[239] Koiso, Norihito. *On rotationally symmetric Hamilton's equation for Kähler–Einstein metrics.* Recent topics in differential and analytic geometry, 327–337, Adv. Stud. Pure Math., **18-I**, Academic Press, Boston, MA, 1990.

[240] Kotschwar, Brett. *Hamilton's gradient estimate for the heat kernel on complete manifolds.* Proc. A.M.S. To appear.

[241] Kotschwar, Brett. *A note on the uniqueness of complete expanding Ricci solitons in 2-d.* Preprint.

[242] Krylov, N. V.; Safonov, M. V. *A property of the solutions of parabolic equations with measurable coefficients (Russian),* Izv. Akad. Nauk SSSR Ser. Mat. **44**(1980), no. 1, 161-175.

[243] Ladyženskaja, O. A.; Solonnikov, V. A.; Uralćeva, N. N. *Linear and quasilinear equations of parabolic type.* (Russian) Translated from the Russian by S. Smith. Translations of Mathematical Monographs, Vol. **23**, American Mathematical Society, Providence, R.I., 1967.

[244] Lauret, Jorge. *Ricci soliton homogeneous nilmanifolds.* Math. Ann. **319** (2001), no. 4, 715–733.

[245] Li, Peter. *On the Sobolev constant and the p-spectrum of a compact Riemannian manifold.* Ann. Sc. Ec. Norm. Sup. 4e serie, t. **13** (1980), 451–469.

[246] Li, Peter. *Lecture notes on geometric analysis,* RIMGARC Lecture Notes Series **6**, Seoul National University, 1993.

[247] Li, Peter. *Large time behavior of the heat equation on complete manifolds with nonnegative Ricci curvature.* Ann. of Math. (2) **124** (1986), no. 1, 1–21.

[248] Li, Peter. *Lecture notes on geometric analysis.* http://math.uci.edu/ pli/lecture.pdf.

[249] Li, Peter. *Lecture notes on heat kernel.* Lecture Notes at UCI taken by Jiaping Wang.

[250] Li, Peter. *Lecture notes on harmonic functions.* To appear.

[251] Li, Peter. *Curvature and function theory on Riemannian manifolds.* Surveys in Differential Geometry: Papers dedicated to Atiyah, Bott, Hirzebruch, and Singer. Vol **VII**, International Press (2000), 375-432.

[252] Li, Peter; Tam, Luen-Fai; Wang, Jiaping. *Sharp bounds for the Green's function and the heat kernel.* Math. Res. Lett. **4** (1997), no. 4, 589–602.

[253] Li, Peter; Yau, Shing-Tung. *On the parabolic kernel of the Schrödinger operator.* Acta Math. **156** (1986), no. 3-4, 153–201.

[254] Lichnerowicz, André. *Géométrie des groupes de transformations.* (French) Travaux et Recherches Mathématiques, III, Dunod, Paris, 1958.

[255] Lieberman, Gary M. *Second order parabolic differential equations.* World Scientific Publishing Co., River Edge, NJ, 1996.

[256] Lott, John. *On the long-time behavior of type-III Ricci flow solutions.* arXiv:math.DG/0509639.

[257] Lu, Peng. *A compactness property for solutions of the Ricci flow on orbifolds.* Amer. J. Math. **123** (2001), no. 6, 1103–1134.

[258] Lu, Peng. Unpublished.

[259] Ma, Li; Chen, Dezhong. *Examples for cross curvature flow on 3-manifolds.* Calc. Var. Partial Differential Equations **26** (2006), no. 2, 227–243.

[260] Mabuchi, Toshiki. *$\mathbb{C}^3$-actions and algebraic threefolds with ample tangent bundle.* Nagoya Math. J. **69** (1978), 33–64.

[261] Mabuchi, Toshiki. *K-energy maps integrating Futaki invariants.* Tohoku Math. J. **38** (1986), 575-593.

[262] McMullen, C. *Renormalization and 3-manifolds which fiber over the circle.* Annals of Mathematics Studies 12, Princeton U. Press, 1996.

[263] Meeks, William H., III; Yau, Shing Tung. *Topology of three-dimensional manifolds and the embedding problems in minimal surface theory.* Ann. of Math. (2) **112** (1980), no. 3, 441–484.

[264] Milnor, John. *A unique factorization theorem for 3-manifolds.* Amer. J. Math. **84** (1962) 1–7.

[265] Milnor, John W. *Morse theory.* Based on lecture notes by M. Spivak and R. Wells. Annals of Mathematics Studies, No. 51. Princeton University Press, Princeton, N.J., 1963.

[266] Milnor, John. *Curvatures of left invariant metrics on Lie groups.* Advances in Math. **21** (1976), no. 3, 293–329.

[267] Milnor, John. *Towards the Poincaré conjecture and the classification of 3-manifolds.* Notices AMS **50** (2003), 1226-1233.

[268] Moise, E. E., *Affine structures in 3-manifolds.* Ann. of Math. (2) 56 (1952), 96-114.

[269] Mok, Ngaiming. *The uniformization theorem for compact Kähler manifolds of nonnegative holomorphic bisectional curvature.* J. Differential Geom. **27** (1988), no. 2, 179–214.

[270] Mok, Ngaiming. *Metric rigidity theorems on Hermitian locally symmetric manifolds.* Series in Pure Mathematics, **6**. World Scientific Publishing Co., Inc., Teaneck, NJ, 1989.

[271] Mori, Shigefumi. *Projective manifolds with ample tangent bundles.* Ann. of Math. (2) **110** (1979), no. 3, 593–606.

[272] Morgan, John. *Recent progress on the Poincaré conjecture and the classification of 3-manifolds.* Bull. Amer. Math. Soc. **42** (2005), 57-78.

[273] Morgan, John; Tian, Gang. *Ricci flow and the Poincaré conjecture.* arXiv:math.DG/0607607.

[274] Morrey, C.B. *Multiple integrals in the calculus of variations.* Springer-Verlag, Berlin, 1966.

[275] Morrow, James; Kodaira, Kunihiko. *Complex manifolds.* Holt, Rinehart and Winston, Inc., New York-Montreal, Que.-London, 1971.

[276] Moser, Jürgen. *On Harnack's theorem for elliptic differential equations.* Comm. Pure Appl. Math. **14** (1961), 577-591.

[277] Moser, Jürgen. *A Harnack inequality for parabolic differential equations,* Comm. Pure Appl. Math. **17**(1964) 101-134.

[278] Mukai, Shigeru. *Fano 3-folds.* Complex projective geometry (Trieste, 1989/Bergen, 1989), 255–263, London Math. Soc. Lecture Note Ser., **179**, Cambridge Univ. Press, Cambridge, 1992.

[279] Munkres, James. *Obstructions to the smoothing of piecewise-differentiable homeomorphisms.* Ann. of Math. (2) **72** (1960), 521–554.

[280] Myers, Sumner; Steenrod, Norman. *The group of isometries of a Riemannian manifold.* Annals Math. **40** (1939), 400–416.

[281] Nadel, Alan M. *Multiplier ideal sheaves and Kähler–Einstein metrics of positive scalar curvature.* Ann. of Math. (2) **132** (1990), no. 3, 549–596.

[282] Nash, John. *Continuity of solutions of parabolic and elliptic equations,* Amer. J. Math. **80** (1958), 931-954.

[283] Ni, Lei. *The entropy formula for linear heat equation.* Journal of Geometric Analysis, **14** (2004), 85–98. Addenda, **14** (2004), 369–374.

[284] Ni, Lei. *A monotonicity formula on complete Kähler manifolds with nonnegative bisectional curvature.* J. Amer. Math. Soc. **17** (2004), no. 4, 909–946.

[285] Ni, Lei. *Monotonicity and Kähler–Ricci flow.* In Geometric Evolution Equations, Contemporary Mathematics, Vol. 367, S.-C. Chang, B. Chow, S.-C. Chu. C.-S. Lin, eds. American Mathematical Society, 2005.

[286] Ni, Lei. *Ancient solution to Kähler–Ricci flow.* Math. Res. Lett. **12** (2005), 633–654.

[287] Ni, Lei. *A matrix Li–Yau–Hamilton estimate for Kähler–Ricci flow.* J. Diff. Geom. **75** (2007), 303–358.

[288] Ni, Lei. *A note on Perelman's Li–Yau–Hamilton inequality.* Comm. Anal. Geom. **14** (2006).

[289] Ni, Lei. Unpublished.

[290] Ni, Lei; Tam, Luen-Fai. *Plurisubharmonic functions and the Kähler-Ricci flow.* Amer. J. Math. **125** (2003), 623-645.

[291] Ni, Lei; Tam, Luen-Fai. *Plurisubharmonic functions and the structure of complete Kähler manifolds with nonnegative curvature.* J. Differential Geom. **64** (2003), no. 3, 457–524.

[292] Ni, Lei; Tam, Luen-Fai. *Kähler–Ricci flow and the Poincare–Lelong equation.* Comm. Anal. Geom. **12** (2004), 111–141.

[293] Nitta, Muneto. *Conformal sigma models with anomalous dimensions and Ricci solitons.* Modern Phys. Lett. A **20** (2005), no. 8, 577–584.

[294] Otal, J.-P. *Le théoréme d'hyperbolisation pour les variétés fibrées de dimension 3.* Atérisque 235, 1996.

[295] Pali, Nefton. *Characterization of Einstein–Fano manifolds via the Kähler–Ricci flow.* arXiv:math.DG/0607581.

[296] Pattle, R. E. *Diffusion from an instantaneous point source with a concentration-dependent coefficient.* Quart. J. Mech. Appl. Math. **12** (1959), 407–409.

[297] Perelman, Grisha. *The entropy formula for the Ricci flow and its geometric applications.* arXiv:math.DG/0211159.

[298] Perelman, Grisha. *Ricci flow with surgery on three-manifolds.* arXiv:math.DG/0303109.

[299] Perelman, Grisha. *Finite extinction time for the solutions to the Ricci flow on certain three-manifolds.* arXiv:math.DG/0307245.

[300] Peters, Stefan. *Convergence of Riemannian manifolds.* Compositio Math. **62** (1987), no. 1, 3–16.

[301] Petersen, Peter. *Convergence theorems in Riemannian geometry.* Comparison geometry (Berkeley, CA, 1993–94), 167–202, Math. Sci. Res. Inst. Publ., **30**, Cambridge Univ. Press, Cambridge, 1997.

[302] Petersen, Peter. *Riemannian geometry.* Graduate Texts in Mathematics, **171**. Springer-Verlag, New York, 1998.

[303] Petrunin, A.; Tuschmann, W. *Asymptotic flatness and cone structure at infinity.* Math. Ann. **321** (2001), 775-788.

[304] Phong, D. H.; Sturm, Jacob. *Stability, energy functionals, and Kähler–Einstein metrics.* Comm. Anal. Geom. **11** (2003), no. 3, 565–597.

[305] Phong, D.H.; Sturm, Jacob. *On the Kähler–Ricci flow on complex surfaces.* arXiv:math.DG/0407232.

[306] Phong, D.H.; Sturm, Jacob. *On stability and the convergence of the Kähler–Ricci flow.* arXiv:math.DG/0412185.

[307] Polchinksi, Joe, *String theory.* Vols. I and II. Cambridge Monographs on Mathematical Physics. Cambridge University Press, Cambridge, 1998.

[308] Poor, Walter A. *Differential geometric structures.* McGraw-Hill Book Co., New York, 1981.

[309] Protter, Murray H.; Weinberger, Hans F. *Maximum principles in differential equations*. Corrected reprint of the 1967 original. Springer-Verlag, New York, 1984.

[310] Reed, Michael ; Simon, Barry. *Methods of modern mathematical physics, Analysis of operators, Volume IV*. Academic Press, New York, 1978.

[311] Rosenau, Philip. *On fast and super-fast diffusion*. Phys. Rev. Lett. **74** (1995), 1056–1059.

[312] Rothaus, O. S., *Logarithmic Sobolev inequalities and the spectrum of Schrödinger operators*. J. Funct. Anal. **42** (1981), 110–120.

[313] Royden, Halsey L. *Real analysis. Third edition*. Macmillan Publishing Company, New York, 1988.

[314] Ruan, Wei-Dong. *On the convergence and collapsing of Kähler metrics*. J. Differential Geom. **52** (1999), no. 1, 1–40.

[315] Saloff-Coste, Laurent. *Uniformly elliptic operators on Riemannian manifolds*. J. Differential Geom. **36** (1992), no. 2, 417-450.

[316] Schoen, R.; Yau, S.-T. *Lectures on differential geometry*. Lecture notes prepared by Wei Yue Ding, Kung Ching Chang [Gong Qing Zhang], Jia Qing Zhong and Yi Chao Xu. Translated from the Chinese by Ding and S. Y. Cheng. Preface translated from the Chinese by Kaising Tso. Conference Proceedings and Lecture Notes in Geometry and Topology, I. International Press, Cambridge, MA, 1994.

[317] Schueth, Dorothee. *On the 'standard' condition for noncompact homogeneous Einstein spaces*. Geom. Dedicata **105** (2004), 77–83.

[318] Scott, Peter. *A new proof of the annulus and torus theorems*. Amer. J. Math. **102** (1980), no. 2, 241–277.

[319] Scott, Peter. *The geometries of 3-manifolds*. Bull. London Math. Soc. **15** (1983), no. 5, 401–487.

[320] Sesum, Natasa. *Convergence of Kähler–Einstein orbifolds*. J. Geom. Anal. **14** (2004), no. 1, 171–184.

[321] Sesum, Natasa. *Curvature tensor under the Ricci flow*. Amer. J. Math. **127** (2005), no. 6, 1315–1324.

[322] Sesum, Natasa. *Limiting behaviour of the Ricci flow*. arXiv:math.DG/0402194.

[323] Sesum, Natasa. *Convergence of a Kähler–Ricci flow*. Math. Res. Lett. **12** (2005), no. 5-6, 623–632.

[324] Sesum, Natasa. *Convergence of the Ricci flow toward a soliton*. Comm. Anal. Geom. **14** (2006), no. 2, 283–343.

[325] Sesum, Natasa; Tian, Gang. *Bounding scalar curvature and diameter along the Kähler–Ricci flow (after Perelman)*. Preprint.

[326] Sesum, Natasa; Tian, Gang; Wang, Xiaodong, *Notes on Perelman's paper on the entropy formula for the Ricci flow and its geometric applications*. June 23, 2003.

[327] Shalen, Peter B. *A "piecewise-linear" method for triangulating 3-manifolds*. Adv. in Math. **52** (1984), no. 1, 34–80.

[328] Sharafutdinov, V. A. *The Pogorelov–Klingenberg theorem for manifolds that are homeomorphic to $\mathbb{R}^n$*. (Russian) Sibirsk. Mat. Z. **18** (1977), no. 4, 915–925, 958.

[329] Shi, Wan-Xiong. *Deforming the metric on complete Riemannian manifolds*. J. Differential Geom. **30** (1989), no. 1, 223–301.

[330] Shi, Wan-Xiong. *Ricci deformation of the metric on complete noncompact Riemannian manifolds*. J. Differential Geom. **30** (1989), no. 2, 303–394.

[331] Shi, Wan-Xiong. *Ricci flow and the uniformization on complete noncompact Kähler manifolds*. J. Differential Geom. **45** (1997), no. 1, 94-220.

[332] Shioya, Takashi; Yamaguchi, Takao. *Collapsing three-manifolds under a lower curvature bound*. J. Differential Geom. **56** (2000), no. 1, 1–66.

[333] Simon, Leon. *Schauder estimates by scaling*. Calc. Var. Partial Differential Equations **5** (1997), no. 5, 391–407.

[334] Siu, Yum Tong. *Lectures on Hermitian–Einstein metrics for stable bundles and Kähler-Einstein metrics.* DMV Seminar, **8**. Birkhäuser Verlag, Basel, 1987.

[335] Siu, Yum Tong. *The existence of Kähler–Einstein metrics on manifolds with positive anticanonical line bundle and a suitable finite symmetry group.* Annals Math. **127** (1988), 585–627.

[336] Siu, Yum Tong; Yau, Shing-Tung. *Compact Kähler manifolds of positive bisectional curvature.* Invent. Math. **59** (1980), no. 2, 189–204.

[337] Song, Jian; Tian, Gang. *The Kähler–Ricci flow on surfaces of positive Kodaira dimension.* arXiv:math.DG/0602150.

[338] Song, Jian; Weinkove, Ben. *Energy functionals and canonical Kähler metrics.* Preprint.

[339] Souplet, Philippe; Zhang, Qi. *Sharp gradient estimate and Yau's Liouville theorem for the heat equation on noncompact manifolds.* arXiv:math.DG/0502079.

[340] Stein, Elias. *Singular integrals and differentiability properties of functions.* Princeton Univ. Press, 1970.

[341] Strominger, Andrew; Vafa, Cumrun. *Microscopic origin of the Bekenstein–Hawking entropy.* arXiv:hep-th/9601029.

[342] Struwe, Michael. *On the evolution of harmonic maps in higher dimensions.* J. Differential Geom. **28** (1988), no. 3, 485–502.

[343] Thurston, William P. *Three-dimensional geometry and topology.* Vol. 1. Edited by Silvio Levy. Princeton Mathematical Series, **35**. Princeton University Press, Princeton, NJ, 1997.

[344] Tian, Gang. *On Kähler–Einstein metrics on certain Kähler manifolds with $C_1 > 0$.* Invent. Math. **89** (1987), no. 2, 225–246.

[345] Tian, Gang. *Calabi's conjecture for complex surfaces with positive first Chern class.* Invent. Math. **101** (1990), no. 1, 101–172.

[346] Tian, Gang. *Kähler–Einstein metrics with positive scalar curvature.* Invent. Math. **130** (1997), no. 1, 1–37.

[347] Tian, Gang. *Canonical metrics in Kähler geometry.* Notes taken by Meike Akveld. Lectures in Mathematics ETH Zürich. Birkhäuser Verlag, Basel, 2000.

[348] Tian, Gang; Yau, Shing-Tung. *Kähler–Einstein metrics on complex surfaces with $C_1 > 0$.* Comm. Math. Phys. **112** (1987), no. 1, 175–203.

[349] Tian, Gang; Yau, Shing-Tung. *Complete Kähler manifolds with zero Ricci curvature. I.* J. Amer. Math. Soc. **3** (1990), no. 3, 579–609.

[350] Tian, Gang; Yau, Shing-Tung. *Complete Kähler manifolds with zero Ricci curvature. II.* Invent. Math. **106** (1991), no. 1, 27–60.

[351] Tian, Gang; Zhang, Zhou. Preprint.

[352] Tian, Gang; Zhu, Xiaohua. *Uniqueness of Kähler–Ricci solitons on compact Kähler manifolds.* C. R. Acad. Sci. Paris Sér. I Math. **329** (1999), no. 11, 991–995.

[353] Tian, Gang; Zhu, Xiaohua. *Uniqueness of Kähler–Ricci solitons.* Acta Math. **184** (2000), no. 2, 271–305.

[354] Tian, Gang; Zhu, Xiaohua. *A new holomorphic invariant and uniqueness of Kähler–Ricci solitons.* Comment. Math. Helv. **77** (2002), no. 2, 297–325.

[355] Tian, Gang; Zhu, Xiaohua. Preprint.

[356] Topping, Peter. *Lectures on the Ricci flow.* London Mathematical Society Lecture Note Series (No. 325). Cambridge University Press, 2006.

[357] Topping, Peter. *Diameter control under Ricci flow.* Comm. Anal. Geom. **13** (2005) 1039-1055.

[358] Topping, Peter. *Ricci flow compactness via pseudolocality, and flows with incomplete initial metrics.* Preprint.

[359] Tso, Kaising. *On an Aleksandrov–Bakelman type maximum principle for second-order parabolic equations.* Comm. Partial Diff. Equations **10** (1985), no. 5, 543-553.

[360] Tsuji, Hajime. *Existence and degeneration of Kähler–Einstein metrics on minimal algebraic varieties of general type.* Math. Ann. 281 (1988), no. 1, 123–133.

[361] Varadhan, S. R. S. *On the behavior of the fundamental solution of the heat equation with variable coefficients.* Comm. Pure Appl. Math. 20 (1967), 431–455.

[362] Varopoulos, N. *Hardy–Littlewood theory for semigroups.* J. Funct. Anal. 63 (1985), 240-260.

[363] Waldhausen, Friedhelm. *On irreducible 3-manifolds which are sufficiently large.* Ann. of Math. (2) 87 (1968), 56–88.

[364] Waldhausen, Friedholm. *Gruppen mit Zentrum und 3-dimensionale Mannigfaltigkeiten.* Topology 6 (1967), 505–517.

[365] Wang, McKenzie Y.; Ziller, Wolfgang. *Existence and nonexistence of homogeneous Einstein metrics.* Invent. Math. 84 (1986), no. 1, 177–194.

[366] Wang, Xu-Jia; Zhu, Xiaohua. *Kähler–Ricci solitons on toric manifolds with positive first Chern class.* Adv. Math. 188 (2004), no. 1, 87–103.

[367] Watson, N. A. *A theory of temperatures in several variables.* Jour Proc. London Math. Soc. 26 (1973), 385-417.

[368] Wei, Guofang. *Curvature formulas in Section 6.1 of Perelman's paper (math.DG/0211159).* http://www.math.ucsb.edu/~wei/Perelman.html

[369] Weibel, Charles A. *An introduction to homological algebra.* Cambridge Studies in Advanced Mathematics, 38. Cambridge University Press, Cambridge, 1994.

[370] Weil, André. *Introduction à l'étude des variétés kählériennes.* (French) Publications de l'Institut de Mathématique de l'Université de Nancago, VI. Actualités Sci. Ind., no. 1267. Hermann, Paris, 1958.

[371] Wells, R. O., Jr. *Differential analysis on complex manifolds.* Second edition. Graduate Texts in Mathematics, 65. Springer-Verlag, New York-Berlin, 1980.

[372] Wu, Lang-Fang. *The Ricci flow on 2-orbifolds with positive curvature.* J. Differential Geom. 33 (1991), no. 2, 575–596.

[373] Wu, Lang-Fang. *The Ricci flow on complete $\mathbb{R}^2$.* Comm. Anal. Geom. 1 (1993), no. 3-4, 439–472.

[374] Yang, Deane. *Convergence of Riemannian manifolds with integral bounds on curvature, I.* Ann. Sci. École Norm. Sup. (4) 25 (1992), 77–105.

[375] Yang, Deane. *Convergence of Riemannian manifolds with integral bounds on curvature, II.* Ann. Sci. École Norm. Sup. (4) 25 (1992), 179–199.

[376] Yau, Shing-Tung. *On the curvature of compact Hermitian manifolds.* Invent. Math. 25 (1974), 213–239.

[377] Yau, Shing-Tung. *Harmonic functions on complete Riemannian manifolds.* Comm. Pure Appl. Math. 28 (1975), 201–228.

[378] Yau, Shing-Tung. *Calabi's conjecture and some new results in algebraic geometry.* Proc. Nat. Acad. Sci. U.S.A. 74 (1977), no. 5, 1798–1799.

[379] Yau, Shing-Tung. *On the Ricci curvature of a compact Kähler manifold and the complex Monge-Ampère equation. I.* Comm. Pure Appl. Math. 31 (1978), no. 3, 339–411.

[380] Yau, Shing-Tung. *On the Harnack inequalities of partial differential equations,* Comm. Anal. Geom. 2 (1994), no. 3, 431–450.

[381] Yau, Shing-Tung. *Harnack inequality for non-self-adjoint evolution equations,* Math. Res. Lett. 2 (1995), no. 4, 387–399.

[382] Ye, Rugang. *On the ℓ-function and the reduced volume of Perelman.* http://www.math.ucsb.edu/~yer/reduced.pdf

[383] Zheng, Fangyang. *Complex differential geometry.* AMS/IP Studies in Advanced Mathematics, 18. American Mathematical Society, Providence, RI; International Press, Boston, MA, 2000.

[384] Zhu, Shunhui. *The comparison geometry of Ricci curvature.* In Comparison Geometry, Grove and Petersen, eds. MSRI Publ. 30 (1997), 221–262.

[385] Zhu, Xiaohua. *Kähler–Ricci soliton typed equations on compact complex manifolds with $C_1(M) > 0$.* J. Geom. Anal. **10** (2000), no. 4, 759–774.

[386] Zhu, Xi-Ping. *Lectures on mean curvature flows.* AMS/IP Studies in Advanced Mathematics, **32**. American Mathematical Society, Providence, RI; International Press, Somerville, MA, 2002.

Index

adjoint heat equation, 201, 228
Alexander's Theorem, 433
almost complex manifold, 56
almost complex structure, 56
   integrable, 56
ancient solution, 465
anti-holomorphic tangent bundle, 58
aperture, 477
approximate isometry, 150
arc length
   evolution, 457
   variation formula, 448
Arzela–Ascoli Theorem, 137
associated 2-form, 61
asymptotic scalar curvature ratio, 472
   is infinite on steady solitons, 472
asymptotic volume ratio, 253, 386, 451
   of expanding soliton, 473
   of Type I ancient solutions, 474
average
   nonlinear, 182
average scalar curvature, 6

backward heat equation, 198
backward Ricci flow, 288
Bakry–Emery log Sobolev inequality, 218
Bianchi identity
   second, 66
bisectional curvature, 65
   nonnegative, 95, 96, 103, 118
Bishop volume comparison theorem, 450
Bishop–Gromov volume comparison, 450
boundary condition
   Dirichlet, 269
bounded geometry, 130
bounded variation
   locally, 367

breather solution, 41
Bryant soliton, 17, 440
Buscher duality transformation, 46

Calabi conjecture, 71
Calabi's trick, 453
Calabi–Yau metric, 72
canonical case, 73, 76
Cartan–Kähler Theorem, 10
center of mass, 182
Cheeger–Gromov convergence, 129
Cheeger–Gromov–Taylor, 159
Cheeger–Yau estimate, 453
Cheng's eigenvalue comparison, 269
Chevalley complex, 33
cigar soliton, 14, 102, 440, 468, 469
cigar-paraboloid behavior, 102
circumference at infinity, 477
Classic Dimension, xiv
classical entropy, 124, 214
   monotonicity, 215
Cohn–Vossen inequality, 17
collapsible manifold, 442
compactness
   of isometries, 155
compactness theorem
   for Kähler metrics, 139
   for Kähler–Ricci flow, 140
   for metrics, 130
   for solutions, 131, 138
   local version, 138
complex manifold, 56
complex projective space, 98
complex structure, 56
complex submanifold, 56
complexified cotangent bundle, 58
complexified tangent bundle, 58
concatenated path, 292
conjugate
   complex, 58

conjugate heat equation, 201
connected sum decomposition, 433
convergence
    of maps
        C^∞ on compact sets, 155
        C^p, 155
convergence on $\mathbb{R}^2$, 478
convex, 178
convex function, 367
coupled modified Ricci flow, 198
cross curvature breather, 499
cross curvature flow, 492
    monotonicity formula, 496
    short-time existence, 493
cross curvature soliton, 499
cross curvature tensor, 491

$\partial\bar{\partial}$-Lemma, 70
derivation of a Lie algebra, 33
derivative
    first, 366
    second, 366
derivative of metric bounds, 132
diameter control, 268
differentiable function, 364
dilaton, 49, 191
dilaton shift, 49
dimension reduction, 440, 472
direct limit, 156
directed system, 156
Dirichlet boundary condition, 269
doubly-warped product, 26, 50

effective action, 49
Einstein metric, 5, 53
    homogeneous, 32
    standard type, 32
Einstein summation convention, 59
    extended, 67
energy
    of a path, 286
energy functional, 44, 191
entropy, 124
entropy functional, 222
    Euler–Lagrange equation for
        minimizer, 237
    existence of minimizer, 238
entropy monotonicity
    for gradient flow, 226
    for Ricci flow, 226
ε-entropy, 229
essential 2-sphere, 433
eternal solution, 465

Euclidean space
    characterization of, 451
exhaustion, 128
expanding breathers
    nonexistence, 213
expanding gradient soliton, 46
expanding soliton, 2
exploding soliton, 14
exponential map
    derivatives of, 175
exterior differential systems, 10

$\mathcal{F}$-functional, 191
Fano manifold, 72
first Chern class, 64
first derivative, 366
first fundamental form, 482
first variation
    of $\mathcal{F}$, 192
    of $\mathcal{W}$, 223
    of $\mathcal{W}_\varepsilon$, 231
Frankel Conjecture, 73
fundamental solution
    of the heat equation, 453
funny way, 282
Futaki functional, 73, 98
Futaki invariant, 72

Gaussian soliton, 5
geometrically atoroidal, 436
geometrization conjecture, 437
gradient flow of $\mathcal{F}^m$, 197
gradient soliton structure, 4
graph
    space-time, 292
graph manifold
    Cheeger–Gromov, 439
    topological, 438
Gromov–Hausdorff convergence, 38
Gross's logarithmic Sobolev inequality
    Beckner–Pearson's proof, 249

Haken 3-manifold, 436
Hamilton's surface entropy
    formula, 217
    gradient is matrix Harnack, 218
    monotonicity, 217
harmonic map heat flow, 490
    monotonicity formula, 490
Harnack inequality
    classical, 91
heat equation, 453
    fundamental solution, 453
Hermitian metric, 57

Hessian upper bound, 370
holomorphic coordinates, 55
holomorphic sectional curvature, 65
holomorphic tangent bundle, 58
homogeneous space, 32
Huisken's monotonicity formula, 487
   Hamilton's generalization, 489
hyperbolic space, 269

immortal solution, 465
incompressible surface, 436
index form
   Riemannian, 345
Index Lemma
   Riemannian, 346
infinitesimal automorphism, 57, 97
injectivity radius estimate, 142, 159
   local, 257
integral curves
   of $\nabla \ell$, 329
integration by parts
   for Lipschitz functions, 365
involutive system, 10
irreducible 3-manifold, 434

Jacobi field
   Riemannian, 346
Jacobian
   Riemannian, 359
Jensen's inequality, 235

Kähler class, 61
Kähler form, 61
Kähler identities, 61
Kähler manifold, 57
Kähler metric, 57
Kähler–Einstein metric, 72
Kähler–Ricci flow
   normalized, 76
κ-collapsed
   at the scale r, 252
κ-noncollapsed
   below the scale ρ, 252
Killing vector field, 11
Kneser Finiteness Theorem, 434
Koiso soliton, 44, 98
KRF, 81

$\mathcal{L}$-cut locus, 356
L-distance, 293
   and trace Harnack, 323
   for Ricci flat solutions, 294
   gradient of, 311
   Hessian of, 318

is locally Lipschitz, 308
Laplacian Comparison Theorem, 321
time-derivative of, 312
triangle inequality, 328
$\mathcal{L}$-exponential map, 352
   as $\tau \to 0$, 353
   for Ricci flat solution, 353
$\mathcal{L}$-geodesic, 298
   estimate for speed of , 303
   existence for IVP, 305
   on Einstein solution, 299
$\mathcal{L}$-geodesic equation, 298
$\mathcal{L}$-geodesics
   short ones are minimizing, 308
$\mathcal{L}$-Hopf-Rinow Theorem, 355
$\mathcal{L}$-index form, 359
$\mathcal{L}$-index lemma, 357
$\mathcal{L}$-Jacobi equation, 347
$\mathcal{L}$-Jacobi field, 346
   time-derivative of, 351
$\mathcal{L}$-Jacobian, 360
   of Ricci flat solution, 361
   time-derivative of, 362
$\mathcal{L}$-length, 291
   additivity, 292
   first variation formula, 297
   lower bound, 292
   scaling property, 293
   second variation of, 316
λ-invariant, 204
   lower bound, 206
   monotonicity, 209
   second variation of, 280
   upper bound, 206
Laplacian
   for a Kähler metric, 67
   Hodge–de Rham, 449
   rough, 449
Laplacian comparison
   Riemannian, 376
Laplacian Comparison Theorem, 452
   for L-distance, 321
$\bar{L}$-distance
   as $\bar{\tau} \to 0$, 324
   monotonicity of minimum, 324
   supersolution to heat equation, 323
Levi-Civita connection, 446
   evolution, 456
   Kähler evolution, 77
   of a Kähler metric, 62
   variation formula, 448
Li–Yau inequality, 453
Lichnerowicz Laplacian, 448

for a Kähler metric, 70, 106, 118
Lie algebra cohomology, 33
Lie algebra square, 460
Lie derivative, 449
linear trace Harnack
  and $\mathcal{W}_\varepsilon$, 233
linearized Kähler–Ricci flow, 118
Lipschitz graph, 365
Little Loop Conjecture, 256, 257
locally bounded variation, 367
locally collapsing, 255
locally homogeneous geometries, 438
locally Lipschitz, 364
  vector field, 365
logarithmic Sobolev inequality, 246
  of L. Gross, 247

mapping torus, 37, 38
matrix Harnack formula
  for adjoint heat equation, 281
matrix Harnack quadratic, 9
  for Kähler–Ricci flow, 109
maximal function, 265
maximum principle
  strong tensor, 107, 471
  weak scalar, 458
  weak tensor, 459
maximum volume growth, 96, 253, 507
mean curvature, 482
mean curvature flow, 482
  convergence to self-similar, 489
  evolution equations under, 483
measure, 196
minimal $\mathcal{L}$-geodesic, 293
  existence, 305
minimizer of $\mathcal{F}$
  Euler–Lagrange equation for, 206
  existence of, 205
modified Ricci curvature, 194
modified Ricci flow, 197
modified Ricci tensor, 48
modified scalar curvature, 48, 194
  evolution of, 203
  first variation, 273
modified subsequence convergence, 477
mollifier, 367
Monge–Ampère equation
  complex, 75, 76
Monge-Ampère equation
  complex, 71
monotonicity
  of $\mathcal{W}_\varepsilon$, 233
monotonicity formula

for static metric reduced volume , 382
  for the gradient flow, 197
μ-invariant, 236
  as $\tau \to 0$, 244
  as $\tau \to \infty$, 243
  finiteness, 237
  monotonicity, 239
  under Cheeger-Gromov convergence, 240

necklike point, 470
neckpinch, 439
Newlander-Nirenberg Theorem, 56
Nijenhuis tensor, 56
nilpotent Lie group, 34
NKRF, 81
no local collapsing, 256, 440
  improved version, 267
  proof of, 258
normal holomorphic coordinates, 68
normal surfaces, 435
ν-invariant, 236
  monotonicity, 244
null eigenvector assumption, 459

open problem, 282
orbifold, 12, 141
  bad, 12

parabola, 287
parallel 2-spheres, 434
Perelman's energy functional, 44
Perelman's entropy functional, 45, 222
Perelman's equations
  coupled to Ricci flow, 225
Perelman's Harnack quantity, 227
  evolution, 227
pinching improves, 463
PL category, 433
Poincaré conjecture, 437
pointed Riemannian manifold, 129
pointed solution to the Ricci flow, 129
pointwise convergence
  in C^p , 128
  uniformly on compact sets, 129
pointwise monotonicity
  along $\mathcal{L}$-geodesics, 391
porous medium equation, 479
potential function, 76
potentially infinite dimensions
  and conformal geometry, 194

quasi-Einstein metric, 52

Rademacher's Theorem, 364
real (p,p)-form, 59
real part
  of complex vector, 58
reduced distance, 326
  for Einstein solution, 336
  of a static manifold, 384
  on shrinker, 343
  partial differential inequalities, 327
  regularity properties, 377
  under Cheeger–Gromov convergence, 334
reduced volume
  for Ricci flow, 388
  static, 381
reduced volume monotonicity
  heuristic proof, 389
  proof of, 392, 395
Ricci breather, 203
Ricci curvature
  quasi-positive, 103
Ricci flow geometry, 391
Ricci flow on surfaces
  revisited, 145
Ricci form, 64
Ricci soliton, 2, 464
  canonical form, 3
Ricci soliton structure, 4
Ricci tensor, 446
  evolution, 457
  of a Kähler metric, 64
  variation formula, 448
Riemann curvature tensor
  variation formula, 448
Riemann tensor, 446
  evolution, 456, 460
  evolution for Kähler metric, 104
  of a Kähler metric, 63
Riemannian distance function
  convexity of, 178
  derivatives of, 175
Rosenau solution, 469

scalar curvature, 446
  evolution, 457
  Kähler evolution, 77
  of a Kähler metric, 65
  variation formula, 448
Schoenflies problem, 433
second derivative, 366
second fundamental form, 482
Seifert 3-manifold, 436
self-similar solution, 2

semisimple Lie group, 34
separating 2-sphere, 433
short-time existence
  compact manifolds, 457
  noncompact manifolds, 458
shrinking breathers
  are gradient solitons, 242
shrinking gradient soliton, 45
shrinking soliton, 2
Simons' identity, 486
singularity model, 439
  existence of, 143, 263
smooth category, 433
soliton
  Kähler–Ricci, 96, 97
space-time metric, 387
static reduced volume, 381
steady breathers
  nonexistence, 210
steady gradient soliton, 44
steady soliton, 2
Sterling's formula, 250
strongly κ-collapsed, 401
supersolution, 374
support sense, 373
surface entropy, 468
surgery
  Ricci flow with surgery, 441

T-duality, 46
tensor of type (p,q), 59
3-manifolds with positive Ricci curvature
  revisited, 143
topological category, 433
torus decomposition, 436
trace Harnack quadratic, 467
  for Kähler–Ricci flow, 109
Type I essential point, 470
Type I singularity, 489
Type III solution, 38

Uhlenbeck's trick, 459
Uniformization Theorem, 12, 438
unitary frame, 65

vicosity sense, 374
volume doubling property, 266
volume form, 446
  evolution, 457
  Kähler evolution, 77
  of a Kähler metric, 66
  variation formula, 448
volume ratio, 382

$\mathcal{W}$-functional, 222
warped product, 12, 46, 480
weak sense, 373
weakened no local collapsing, 401
Witten's black hole, 468

Yau's unifomization conjecture, 96